GALOIS THEORY

GALOIS THEORY

DAVID A. COX

Amherst College
Department of Mathematics & Computer Science
Amherst, MA

A JOHN WILEY & SONS, INC., PUBLICATION

Library of Congress Cataloging-in-Publication Data:

Cox, David A.
    Galois theory / David A. Cox.
       p. cm. — (Pure and applied mathematics : a Wiley-Interscience series of texts,
    monographs, and tracts)
    Includes bibliographical references and index.
    ISBN 0-471-43419-1 (cloth : acid-free paper)
    1. Galois theory.   I. Title.   II. Pure and applied mathematics (John Wiley & Sons :
    Unnumbered)

QA214.C69 2004
512'.32—dc22   2004045640

10 9 8 7 6 5 4

Preface

Galois theory is a wonderful part of mathematics. Its historical roots date back to the solution of cubic and quartic equations in the sixteenth century. But besides helping us understand the roots of polynomials, Galois theory also gave birth to many of the central concepts of modern algebra, including groups and fields. In addition, there is the human drama of Évariste Galois, whose death at age 20 left us with the brilliant but not fully developed ideas that eventually led to Galois theory.

Besides being great history, Galois theory is also great mathematics. This is due primarily to two factors: first, its surprising link between group theory and the roots of polynomials, and second, the elegance of its presentation. Galois theory is often described as one of the most beautiful parts of mathematics.

This book was written in an attempt to do justice to both the history and the power of Galois theory. My goal is for students to appreciate the elegance of the theory and simultaneously have a strong sense of where it came from.

The book is intended for undergraduates, so that many graduate-level topics are not covered. On the other hand, the book does discuss a broad range of topics, including symmetric polynomials, angle trisections via origami, Galois's criterion for an irreducible polynomial of prime degree to be solvable by radicals, and Abel's theorem about ruler-and-compass constructions on the lemniscate.

A. Structure of the Text. The text is divided into chapters and sections. We use the following numbering conventions:

- Theorems, lemmas, definitions, examples, etc., are numbered according to chapter and section. For example, the third section of Chapter 7 is called Section 7.3. This section begins with Theorem 7.3.1, Corollary 7.3.2, and Example 7.3.3.

- In contrast, equations are numbered according to the chapter. For example, (4.1) means the first numbered equation of Chapter 4.

Sections are sometimes divided informally into subsections labeled A, B, C, etc. In addition, many sections contain of endnotes of two types:

- Mathematical Notes develop the ideas introduced in the section. Each idea is announced with a black square ∎ .
- Historical Notes explain some of the history behind the concepts introduced in the section.

The symbol □ denotes the end of a proof or the absence of a proof, and ◁▷ denotes the end of an example.

References in the text use one of two formats:

- References to the bibliography at the end of the book are given by the author's last name, as in [Abel]. When there is more than one item by a given author, we add numbers, as in [Jordan1] and [Jordan2].
- Some more specialized references are listed at the end of the chapter in which the reference occurs. These references are listed numerically, so that if you are reading Chapter 10, then [1] means the first reference at the end of that chapter.

The text has numerous exercises, many more than can be assigned during an actual course. Some of the exercises can be used as exam questions. Hints to selected exercises can be found in Appendix B.

The algebra needed for the book is covered in Appendix A. Students should read Sections A.1 and A.2 before starting Chapter 1.

B. The Four Parts. The book is organized into four parts. Part I (Chapters 1 to 3) focuses on polynomials. Here, we study cubic polynomials, symmetric polynomials and prove the Fundamental Theorem of Algebra. In Part II (Chapters 4 to 7), the focus shifts to fields, where we develop their basic properties and prove the Fundamental Theorem of Galois Theory. Part III is concerned with the following applications of Galois theory:

- Chapter 8 discusses solvability by radicals.
- Chapter 9 treats cyclotomic equations.
- Chapter 10 explores geometric constructions.
- Chapter 11 studies finite fields.

Finally, Part IV covers the following further topics:

- Chapter 12 discusses the work of Lagrange, Galois, and Kronecker.
- Chapter 13 explains how to compute Galois groups.
- Chapter 14 treats solvability by radicals for polynomials of prime power degree.
- Chapter 15 proves Abel's theorem on the lemniscate.

C. Notes to the Instructor. Many books on Galois theory have been strongly influenced by Artin's thin but elegant presentation [Artin]. This book is different. In particular:

- Symmetric polynomials and the Theorem of the Primitive Element are used to prove some of the main results of Galois theory.
- The historical context of Galois theory is discussed in detail.

These choices reflect my personal preferences and my conviction that students need to know what an idea really means and where it came from before they can fully appreciate its elegance. The result is a book which is definitely not thin, though I hope that the elegance comes through.

The core of the book consists of Parts I and II (Chapters 1 to 7). It should be possible to cover this material in about 9 weeks, assuming three lectures per week. In remainder of the course, the instructor can pick and choose sections from Parts III and IV. These chapters can also be used for reading courses, student projects, or independent study.

Here are some other comments for the instructor:

- Sections labeled "Optional" can be skipped without loss of continuity. I sometimes assign the optional section on Abelian equations (Section 6.5) as part of a take-home exam.
- Students typically will have seen most but not all of the algebra in Appendix A. My suggestion is to survey the class about what parts of Appendix A are new to them. These topics can then be covered when needed in the text.
- For the most part, the Mathematical Notes and Historical Notes are not used in the subsequent text, though I find that they stimulate some interesting classroom discussions. The exception is Chapter 12, which draws on the Historical Notes of earlier chapters.

D. Acknowledgments. The manuscript of this book was completed during a Mellon 8 sabbatical funded by the Mellon Foundation and Amherst College. I am very grateful for their support. I also want to express my indebtedness to the authors of the many fine presentations of Galois theory listed at the end of the book.

I am especially grateful to Joseph Fineman, Walt Parry, Abe Shenitzer, and Jerry Shurman for their careful reading of the manuscript. I would also like to thank Kamran Divaani-Aazar, Harold Edwards, Alexander Hulpke, Teresa Krick, Barry Mazur, John McKay, Norton Starr, and Siman Wong for their help.

The students who took courses at Amherst College based on preliminary versions of the manuscript contributed many useful comments and suggestions. I thank them all and dedicate this book to students (of all ages) who undertake the study of this wonderful subject.

DAVID A. COX

Amherst, Massachusetts

Contents

Notation

We use the following standard notation:

$\mathbb{Z}$ for the set of integers,
$\mathbb{Q}$ for the set of rational numbers,
$\mathbb{R}$ for the set of real numbers,
$\mathbb{C}$ for the set of complex numbers.

For integers $a, b \in \mathbb{Z}$, we define:

$a|b$ if b is an integer multiple of a,
$a \nmid b$ if b is not an integer multiple of a,
$a \equiv b \bmod n$ if $n|a - b$,

and we also set

$\gcd(a, b) =$ the greatest common divisor of a, b,
$\mathrm{lcm}(a, b) =$ the least common multiple of a, b.

We use the usual notation for union $\cup$ and intersection $\cap$, and we also define

$A \setminus B = \{x \in A \mid x \notin B\}$,
$|S| =$ the number of elements in a finite set S.

We write $A \subset B$ to indicate that A is a subset of B. (Some texts write $A \subseteq B$ for an arbitrary subset and reserve $A \subset B$ for the case when A is strictly smaller than B. We do not follow this practice.) Thus $A = B$ if and only if $A \subset B$ and $B \subset A$. Given sets A and B, their *Cartesian product* is

$$A \times B = \{(a,b) \mid a \in A, b \in B\}.$$

A function $f : A \to B$ is sometimes denoted $x \mapsto f(x)$, where the arrow $\mapsto$ means "maps to." Also, we often write

$$f : A \simeq B$$

to indicate that the map f is one to one and onto. In this situation, we say that f is a *one-to-one correspondence*.

Given any set S, the *identity map* $1_S : S \to S$ is defined by $s \mapsto s$ for all $s \in S$. Also, given any function $f : A \to B$ and a subset $A_0 \subset A$, the *restriction* of f to A_0 is the function

$$f|_{A_0} : A_0 \to B$$

defined by $f|_{A_0}(a) = f(a)$ for $a \in A_0$. Furthermore:

$$A_0 \subset A \quad \text{gives} \quad f(A_0) = \{f(a) \mid a \in A_0\} \subset B,$$
$$B_0 \subset B \quad \text{gives} \quad f^{-1}(B_0) = \{a \in A \mid f(a) \in B_0\} \subset A.$$

Other notation is introduced as needed in the course of the text. A review of the abstract algebra needed for the text can be found in Section A.1 of Appendix A.

Part I

Polynomials

The first three chapters focus on polynomials and their roots.

We begin in Chapter 1 with cubic polynomials. The goal is to derive *Cardan's formulas* and to see how the permutations of the roots influence things.

Then, in Chapter 2, we learn how to express the coefficients of a polynomial as certain *symmetric polynomials* in the roots. This leads to questions about describing all symmetric polynomials. We also discuss the *discriminant*.

Finally, in Chapter 3, we show that all polynomials have roots in a possibly larger field. We also prove the *Fundamental Theorem of Algebra*, which asserts that the roots of a polynomial with complex coefficients are complex numbers.

1

Cubic Equations

The quadratic formula states that the solutions of a quadratic equation

$$ax^2 + bx + c = 0, \quad a, b, c \in \mathbb{C}, \ a \neq 0,$$

are given by

(1.1)
$$x = \frac{-b \pm \sqrt{b^2 - 4ac}}{2a}.$$

In this chapter we will see that the solutions of a cubic equation

$$ax^3 + bx^2 + cx + d = 0, \quad a, b, c, d \in \mathbb{C}, \ a \neq 0,$$

are given by a similar though more complicated formula. Finding the formula will not be difficult, but understanding where it comes from and what it means will lead to some interesting questions.

1.1 CARDAN'S FORMULAS

Given a cubic equation $ax^3 + bx^2 + cx + d = 0$ with $a \neq 0$, we first divide by a to rewrite the equation as

$$x^3 + bx^2 + cx + d = 0, \quad b, c, d \in \mathbb{C},$$

where b/a, c/a and d/a have been replaced with b, c, and d, respectively. Observe that $x^3 + bx^2 + cx + d$ is a monic polynomial and that reducing to the monic case has no effect on the roots.

The next step is to remove the coefficient of x^2 by the substitution

$$x = y - \frac{b}{3}.$$

The binomial theorem implies that

$$x^2 = y^2 - 2y\frac{b}{3} + \left(\frac{b}{3}\right)^2 = y^2 - \frac{2b}{3}y + \frac{b^2}{9},$$

$$x^3 = y^3 - 3y^2\frac{b}{3} + 3y\left(\frac{b}{3}\right)^2 - \left(\frac{b}{3}\right)^3 = y^3 - by^2 + \frac{b^2}{3}y - \frac{b^3}{27},$$

so that

$$0 = x^3 + bx^2 + cx + d$$

$$= \left(y^3 - by^2 + \frac{b^2}{3}y - \frac{b^3}{27}\right) + b\left(y^2 - \frac{2b}{3}y + \frac{b^2}{9}\right) + c\left(y - \frac{b}{3}\right) + d.$$

If we collect terms, then we can write the resulting equation in y as

$$y^3 + py + q = 0,$$

where

(1.2)

$$p = -\frac{b^2}{3} + c,$$

$$q = \frac{2b^3}{27} - \frac{bc}{3} + d.$$

You will verify the details of this calculation in Exercise 1.

We call a cubic of the form $y^3 + py + q = 0$ a *reduced cubic*. If we can find the roots y_1, y_2, y_3 of the reduced cubic, then we get the roots of the original cubic $x^3 + bx^2 + cx + d = 0$ by adding $-b/3$ to each y_i.

To solve $y^3 + py + q = 0$, we use the substitution

(1.3)

$$y = z - \frac{p}{3z}.$$

This change of variable has a dramatic effect on the equation. Using the binomial theorem again, we obtain

$$y^3 = z^3 - 3z^2\frac{p}{3z} + 3z\left(\frac{p}{3z}\right)^2 - \left(\frac{p}{3z}\right)^3 = z^3 - pz + \frac{p^2}{3z} - \frac{p^3}{27z^3}.$$

Combining this with (1.3) gives

$$y^3 + py + q = \left(z^3 - pz + \frac{p^2}{3z} - \frac{p^3}{27z^3}\right) + p\left(z - \frac{p}{3z}\right) + q = z^3 - \frac{p^3}{27z^3} + q.$$

Multiplying by z^3, we conclude that $y^3 + py + q = 0$ is equivalent to the equation

(1.4)
$$z^6 + qz^3 - \frac{p^3}{27} = 0.$$

This equation is the *cubic resolvent* of the reduced cubic $y^3 + py + q = 0$.

At first glance, (1.4) might not seem useful, since we have replaced a cubic equation with one of degree 6. However, upon closer inspection, we see that the cubic resolvent can be written as

$$(z^3)^2 + qz^3 - \frac{p^3}{27} = 0.$$

By the quadratic formula (1.1), we obtain

$$z^3 = \frac{1}{2}\left(-q \pm \sqrt{q^2 + \frac{4p^3}{27}}\right),$$

so that

(1.5)
$$z = \sqrt[3]{\frac{1}{2}\left(-q \pm \sqrt{q^2 + \frac{4p^3}{27}}\right)}.$$

Substituting this into (1.3) gives a root of the reduced cubic $y^3 + py + q$, and then $x = y - b/3$ is a root of the cubic $x^3 + bx^2 + cx + d$.

However, before we can claim to have solved the cubic, there are several questions that need to be answered:

- By setting $y^3 + py + q = 0$, we essentially assumed that a solution exists. What justifies this assumption?
- A cubic equation has three roots, yet the cubic resolvent has degree 6. Why?
- The substitution (1.3) assumes that $z \neq 0$. What happens when $z = 0$?
- $y^3 + py + q$ has coefficients in $\mathbb{C}$, since $b, c, d \in \mathbb{C}$. Thus (1.5) involves square roots and cube roots of complex numbers. How are these described?

The first bullet will be answered in Chapter 3 when we discuss the existence of roots. The second bullet will be considered in Section 1.2, though the ultimate answer will involve Galois theory. For the rest of this section, we will concentrate on the last two bullets. Our strategy will be to study the formula (1.5) in more detail.

First assume that $p \neq 0$ in the reduced cubic $y^3 + py + q$. By Section A.2, every nonzero complex number has n distinct nth roots when $n \in \mathbb{Z}$ is positive. In (1.5), the $\pm$ in the formula indicates that a nonzero complex number has two square roots. Similarly, the cube root symbol denotes any of the *three* cube roots of the complex number under the radical. To understand these cube roots, we use the cube roots of unity $1, \zeta_3, \zeta_3^2$ from Section A.2. We will write ζ_3 as ω. Recall that

$$\omega = \zeta_3 = e^{2\pi i/3} = \frac{-1 + i\sqrt{3}}{2}$$

and that given one cube root of a nonzero complex number, we get the other two cube roots by multiplying by ω and ω^2.

We can now make sense of (1.5). Let

$$\sqrt{q^2 + \frac{4p^3}{27}}$$

denote a fixed square root of $q^2 + 4p^3/27 \in \mathbb{C}$. With this choice of square root, let

$$z_1 = \sqrt[3]{\frac{1}{2}\left(-q + \sqrt{q^2 + \frac{4p^3}{27}}\right)}$$

denote a fixed cube root of $\frac{1}{2}\left(-q + \sqrt{q^2 + 4p^3/27}\right)$. Then we get the other two cube roots by multiplying by ω and ω^2. Note also that $p \neq 0$ implies that $z_1 \neq 0$ and that z_1 is a root of the cubic resolvent (1.4). It follows easily that if we set

$$z_2 = -\frac{p}{3z_1},$$

then

(1.6) $$y_1 = z_1 + z_2 = z_1 - \frac{p}{3z_1}$$

is a root of the reduced cubic $y^3 + py + q$.

To understand z_2, observe that

$$z_1^3 z_2^3 = z_1^3 \left(-\frac{p}{3z_1}\right)^3 = -\frac{p^3}{27}.$$

An easy calculation shows that

$$z_1^3 \cdot \frac{1}{2}\left(-q - \sqrt{q^2 + \frac{4p^3}{27}}\right) = \frac{1}{2}\left(-q + \sqrt{q^2 + \frac{4p^3}{27}}\right) \cdot \frac{1}{2}\left(-q - \sqrt{q^2 + \frac{4p^3}{27}}\right) = -\frac{p^3}{27}.$$

Since $z_1 \neq 0$, these formulas imply that

$$z_2^3 = \frac{1}{2}\left(-q - \sqrt{q^2 + \frac{4p^3}{27}}\right).$$

Hence $z_2 = -p/3z_1$ is a cube root of $\frac{1}{2}\left(-q - \sqrt{q^2 + 4p^3/27}\right)$, so that

(1.7) $$z_1 = \sqrt[3]{\frac{1}{2}\left(-q + \sqrt{q^2 + \frac{4p^3}{27}}\right)} \text{ and } z_2 = \sqrt[3]{\frac{1}{2}\left(-q - \sqrt{q^2 + \frac{4p^3}{27}}\right)}$$

are cube roots with the property that their product is $-p/3$.

From (1.6), we see that $y_1 = z_1 + z_2$ is a root of $y^3 + py + q$ when z_1 and z_2 are the above cube roots. To get the other roots, note that (1.6) gives a root of the cubic whenever the cube roots are chosen so that their product is $-p/3$ (be sure you understand this). For example, if we use the cube root ωz_1, then

$$\omega z_1 \cdot \omega^2 z_2 = z_1 z_2 = -\frac{p}{3}$$

shows that $y_2 = \omega z_1 + \omega^2 z_2$ is also a root. Similarly, using the cube root $\omega^2 z_1$ shows that $y_3 = \omega^2 z_1 + \omega z_2$ is a third root of the reduced cubic.

By (1.7), it follows that the three roots of $y^3 + py + q = 0$ are given by

$$y_1 = \sqrt[3]{\frac{1}{2}\left(-q + \sqrt{q^2 + \frac{4p^3}{27}}\right)} + \sqrt[3]{\frac{1}{2}\left(-q - \sqrt{q^2 + \frac{4p^3}{27}}\right)},$$

$$y_2 = \omega \sqrt[3]{\frac{1}{2}\left(-q + \sqrt{q^2 + \frac{4p^3}{27}}\right)} + \omega^2 \sqrt[3]{\frac{1}{2}\left(-q - \sqrt{q^2 + \frac{4p^3}{27}}\right)},$$

$$y_3 = \omega^2 \sqrt[3]{\frac{1}{2}\left(-q + \sqrt{q^2 + \frac{4p^3}{27}}\right)} + \omega \sqrt[3]{\frac{1}{2}\left(-q - \sqrt{q^2 + \frac{4p^3}{27}}\right)},$$

provided the cube roots in (1.7) are chosen so that their product is $-p/3$. These are *Cardan's formulas* for the roots of the reduced cubic $y^3 + py + q$.

Example 1.1.1. For the reduced cubic $y^3 + 3y + 1$, consider the real cube roots

$$\sqrt[3]{\tfrac{1}{2}(-1 + \sqrt{5})} \quad \text{and} \quad \sqrt[3]{\tfrac{1}{2}(-1 - \sqrt{5})}.$$

Their product is $-1 = -p/3$, so by Cardan's formulas, the roots of $y^3 + 3y + 1$ are

$$y_1 = \sqrt[3]{\tfrac{1}{2}(-1 + \sqrt{5})} + \sqrt[3]{\tfrac{1}{2}(-1 - \sqrt{5})},$$

$$y_2 = \omega \sqrt[3]{\tfrac{1}{2}(-1 + \sqrt{5})} + \omega^2 \sqrt[3]{\tfrac{1}{2}(-1 - \sqrt{5})},$$

$$y_3 = \omega^2 \sqrt[3]{\tfrac{1}{2}(-1 + \sqrt{5})} + \omega \sqrt[3]{\tfrac{1}{2}(-1 - \sqrt{5})}.$$

Note that y_1 is real. In Exercise 2 you will show that y_2 and y_3 are complex conjugates of each other.   ◁▷

Although Cardan's formulas only apply to a reduced cubic, we get formulas for the roots of an arbitrary monic cubic polynomial $x^3 + bx^2 + cx + d \in \mathbb{C}[x]$ as follows. The substitution $x = y - b/3$ gives the reduced cubic $y^3 + py + q = 0$, where p and q are as in (1.2). If z_1 and z_2 are the cube roots in Cardan's formulas for $y^3 + py + q = 0$, then the roots of $x^3 + bx^2 + cx + d = 0$ are given by

$$x_1 = -\frac{b}{3} + z_1 + z_2,$$

$$x_2 = -\frac{b}{3} + \omega z_1 + \omega^2 z_2,$$

$$x_3 = -\frac{b}{3} + \omega^2 z_1 + \omega z_2,$$

where z_1 and z_2 from (1.7) satisfy $z_1 z_2 = -p/3$. Our derivation assumed $p \neq 0$, but these formulas give the correct roots even when $p = 0$ (see Exercise 3).

We will eventually see that Cardan's formulas make perfect sense from the point of view of Galois theory. For example, the quantity under the square root is

$$q^2 + \frac{4p^3}{27}.$$

Up to a constant factor, this is the *discriminant* of the polynomial $y^3 + py + q$. We will give a careful definition of discriminant in Section 1.2, and Section 1.3 will show that the discriminant gives useful information about the roots of a real cubic.

Here is an example of a puzzle that arises when using Cardan's formula.

Example 1.1.2. The cubic equation $y^3 - 3y = 0$ has roots $y = 0, \pm\sqrt{3}$, all of which are real. When we apply Cardan's formulas, we begin with

$$z_1 = \sqrt[3]{\frac{1}{2}\left(-0 + \sqrt{0^2 + \frac{4(-3)^3}{27}}\right)} = \sqrt[3]{i}.$$

To pick a specific value for z_1, notice that $(-i)^3 = i$, so that we can take $z_1 = -i$. Thus $z_2 = -p/3z_1 = i$, since $p = -3$. Then Cardan's formulas give the roots

$$y_1 = -i + i = 0,$$
$$y_2 = \omega(-i) + \omega^2(i) = \sqrt{3},$$
$$y_3 = \omega^2(-i) + \omega(i) = -\sqrt{3}.$$

(You will verify the last two formulas in Exercise 4.)   ◁▷

The surprise is that Cardan's formulas express the real roots of $y^3 - 3y$ in terms of complex numbers. In Section 1.3, we will prove that for *any* cubic with distinct real roots, Cardan's formulas *always* involve complex numbers.

Historical Notes

The quadratic formula is very old, dating back to the Babylonians, circa 1700 B.C. Cubic equations were first studied systematically by Islamic mathematicians such as Omar Khayyam, and by the Middle Ages they had become a popular topic.

For example, when Leonardo of Pisa (also known as Fibonacci) was introduced to Emperor Frederick II in 1225, Fibonacci was asked to solve two problems, the second of which was the cubic equation

$$x^3 + 2x^2 + 10x = 20.$$

Fibonacci's solution was

$$x = 1 + \frac{22}{60} + \frac{7}{60^2} + \frac{42}{60^3} + \frac{33}{60^4} + \frac{4}{60^5} + \frac{40}{60^6}.$$

In decimal notation, this gives $x = 1.368808107853\ldots$, which is correct to 10 decimal places. Not bad for 775 years ago!

Challenges and contests involving cubic equations were not uncommon during the Middle Ages, and one such contest played a crucial role in the development of Cardan's formula. Early in the sixteenth century, Scipio del Ferro found a solution for cubics of the form $x^3 + bx = c$, where b and c are positive. His student Florido knew this solution, and in 1535, Florido challenged Niccolò Fontana (also known as Tartaglia) to a contest involving 30 cubic equations. Working feverishly in preparation for the contest, Tartaglia worked out the solution of this and other cases, and went on to defeat Florido. In 1539, Tartaglia told his solution to Girolamo Cardan (or Cardano), who published it in 1545 in his book *Ars Magna* (see [2]).

Rather than present one solution to the cubic, as we have done here, Cardan's treatment in *Ars Magna* requires 13 cases. For example, Chapter XIV considers $x^3 + 64 = 18x^2$, and Chapter XV does $x^3 + 6x^2 = 40$. The reason is that Cardan prefers positive coefficients. However, he makes systematic use of the substitution $x = y - b/3$ to get rid of the coefficient of x^2, and Cardan was also aware that complex numbers can arise in solutions of quadratic equations.

Numerous other people worked to simplify and understand Cardan's solution. In 1550, Rafael Bombelli considered more carefully the role of complex solutions (see Section 1.3), and in two papers published posthumously in 1615, François Viète (or Vieta, in Latin) found the trigonometric solution to be discussed in Section 1.3 and introduced the substitution (1.3) used in our derivation of Cardan's formulas.

In addition to the cubic, *Ars Magna* also contained a solution for the quartic equation due to Lodovico (or Luigi) Ferrari, a student of Cardan's. We will discuss the solution of the quartic in Chapter 12.

Exercises for Section 1.1

Exercise 1. Complete the demonstration (begun in the text) that the substitution $x = y - b/3$ transforms $x^3 + bx^2 + cx + d$ into $y^3 + py + q$, where p and q are given by (1.2).

Exercise 2. In Example 1.1.1, show that y_2 and y_3 are complex conjugates of each other.

Exercise 3. Show that Cardan's formulas give the roots of $y^3 + py + q$ even when $p = 0$.

Exercise 4. Verify the formulas for y_2 and y_3 in Example 1.1.2.

Exercise 5. The substitution $x = y - b/3$ can be adapted to other equations as follows.

(a) Show that $x = y - b/2$ gets rid of the coefficient of x in the quadratic equation $x^2 + bx + c = 0$. Then use this to derive the quadratic formula.

(b) For the quartic equation $x^4 + bx^3 + cx^2 + dx + e = 0$, what substitution should you use to get rid of the coefficient of x^3?

(c) Explain how part (b) generalizes to a monic equation of degree n.

Exercise 6. Consider the equation $x^3 + x - 2 = 0$. Note that $x = 1$ is a root.

(a) Use Cardan's formulas (carefully) to derive the surprising formula

$$1 = \sqrt[3]{1 + \frac{2}{3}\sqrt{\frac{7}{3}}} + \sqrt[3]{1 - \frac{2}{3}\sqrt{\frac{7}{3}}}.$$

(b) Show that $1 + \frac{2}{3}\sqrt{\frac{7}{3}} = (\frac{1}{2} + \frac{1}{2}\sqrt{\frac{7}{3}})^3$, and use this to explain the result of part (a).

Exercise 7. Cardan's formulas, as stated in the text, express the roots as sums of two cube roots. Each cube root has three values, so there are nine different possible values for the sum of the cube roots. Show that these nine values are the roots of the equations $y^3 + py + q = 0$, $y^3 + \omega py + q = 0$, and $y^3 + \omega^2 py + q = 0$, where as usual $\omega = \frac{1}{2}(-1 + i\sqrt{3})$.

Exercise 8. Use Cardan's formulas to solve $y^3 + 3\omega y + 1 = 0$.

1.2   PERMUTATIONS OF THE ROOTS

In Section 1.1 we learned that the roots of $x^3 + bx^2 + cx + d = 0$ are given by

$$x_1 = -\frac{b}{3} + z_1 + z_2,$$

(1.8)
$$x_2 = -\frac{b}{3} + \omega z_1 + \omega^2 z_2,$$

$$x_3 = -\frac{b}{3} + \omega^2 z_1 + \omega z_2,$$

where z_1 and z_2 are the cube roots (1.7) chosen so that $z_1 z_2 = -p/3$. We also know that z_1 is a root of the cubic resolvent

(1.9)
$$z^6 + qz^3 - \frac{p^3}{27} = 0,$$

and in Exercise 1 you will show that z_2 is also a root of (1.9). The goal of this section is to understand more clearly the relation between x_1, x_2, x_3 and z_1, z_2. We will learn that *permutations*, the *discriminant*, and *symmetric polynomials* play an important role in these formulas.

A.   Permutations. We begin by observing that we can use (1.8) to express z_1, z_2 in terms of x_1, x_2, x_3. We do this by multiplying the second equation by ω^2 and the third by ω. When we add the three resulting equations, we obtain

$$x_1 + \omega^2 x_2 + \omega x_3 = -(1 + \omega^2 + \omega)\frac{b}{3} + 3z_1 + (1 + \omega + \omega^2)z_2.$$

However, ω is a root of $x^3 - 1 = (x - 1)(x^2 + x + 1)$, which easily implies $1 + \omega + \omega^2 = 0$. Thus the above equation simplifies to

$$x_1 + \omega^2 x_2 + \omega x_3 = 3z_1,$$

so that

$$z_1 = \frac{1}{3}(x_1 + \omega^2 x_2 + \omega x_3).$$

Similarly, multiplying the second equation of (1.8) by ω and the third by ω^2 leads to the formula

$$z_2 = \frac{1}{3}(x_1 + \omega x_2 + \omega^2 x_3).$$

This shows that the roots z_1 and z_2 of the cubic resolvent can be expressed in terms of the roots of the original cubic. However, z_1 and z_2 are only two of the six roots of (1.9). What about the other four? In Exercise 1 you will show that the roots of the cubic resolvent (1.9) are

$$z_1, \ z_2, \ \omega z_1, \ \omega z_2, \ \omega^2 z_1, \ \omega^2 z_2,$$

and that these roots are given in terms of x_1, x_2, x_3 by

(1.10)
$$z_1 = \tfrac{1}{3}(x_1 + \omega^2 x_2 + \omega x_3),$$
$$z_2 = \tfrac{1}{3}(x_1 + \omega^2 x_3 + \omega x_2),$$
$$\omega z_1 = \tfrac{1}{3}(x_2 + \omega^2 x_3 + \omega x_1),$$
$$\omega z_2 = \tfrac{1}{3}(x_3 + \omega^2 x_2 + \omega x_1),$$
$$\omega^2 z_1 = \tfrac{1}{3}(x_3 + \omega^2 x_1 + \omega x_2),$$
$$\omega^2 z_2 = \tfrac{1}{3}(x_2 + \omega^2 x_1 + \omega x_3).$$

These expressions for the roots of the resolvent all look similar. What lies behind this similarity is the following crucial fact: *the six roots of the cubic resolvent are obtained from z_1 by permuting x_1, x_2, x_3.* Hence the symmetric group S_3 now enters the picture.

From an intuitive point of view, this is reasonable, since labeling the roots x_1, x_2, x_3 simply lists them in one particular order. If we list the roots in a different order, then we should still get a root of the resolvent. This also explains why the cubic resolvent has degree 6, since $|S_3| = 6$.

B. The Discriminant. We can also use (1.10) to get a better understanding of the square root which appears in Cardan's formulas. If we set

(1.11)
$$D = q^2 + \frac{4p^3}{27},$$

then we can write z_1 and z_2 as

(1.12)
$$z_1 = \sqrt[3]{\tfrac{1}{2}(-q + \sqrt{D})},$$
$$z_2 = \sqrt[3]{\tfrac{1}{2}(-q - \sqrt{D})}.$$

We claim that D can be expressed in terms of the roots x_1, x_2, x_3. To see why, note that the above formulas imply that

$$z_1^3 - z_2^3 = \tfrac{1}{2}(-q + \sqrt{D}) - \tfrac{1}{2}(-q - \sqrt{D}) = \sqrt{D}.$$

However, (A.15) gives the factorization

(1.13) $$z_1^3 - z_2^3 = (z_1 - z_2)(z_1 - \omega z_2)(z_1 - \omega^2 z_2).$$

Using (1.10), we obtain

$$\begin{aligned}
z_1 - z_2 &= \tfrac{1}{3}(x_1 + \omega^2 x_2 + \omega x_3) - \tfrac{1}{3}(x_1 + \omega x_2 + \omega^2 x_3) \\
&= \tfrac{1}{3}(\omega^2 - \omega)(x_2 - x_3) \\
&= \tfrac{-i}{\sqrt{3}}(x_2 - x_3),
\end{aligned}$$

where the last line uses $\omega^2 - \omega = -i\sqrt{3}$. Similarly, one can show that

(1.14)
$$\begin{aligned}
z_1 - \omega z_2 &= \tfrac{i\omega^2}{\sqrt{3}}(x_1 - x_3), \\
z_1 - \omega^2 z_2 &= \tfrac{-i\omega}{\sqrt{3}}(x_1 - x_2)
\end{aligned}$$

(see Exercise 2). Combining these formulas with $z_1^3 - z_2^3 = \sqrt{D}$ and (1.13) easily implies that

(1.15) $$\sqrt{D} = -\frac{i}{3\sqrt{3}}(x_1 - x_2)(x_1 - x_3)(x_2 - x_3).$$

If we square this formula for $\sqrt{D}$ and combine it with (1.11), we obtain

(1.16) $$q^2 + \frac{4p^3}{27} = -\frac{1}{27}(x_1 - x_2)^2(x_1 - x_3)^2(x_2 - x_3)^2.$$

It is customary to define the *discriminant* of $x^3 + bx^2 + cx + d$ to be

$$\Delta = (x_1 - x_2)^2(x_1 - x_3)^2(x_2 - x_3)^2.$$

Thus Δ is the product of the squares of the differences of the roots. In this notation we can write (1.16) as

(1.17) $$q^2 + \frac{4p^3}{27} = -\frac{1}{27}\Delta.$$

Then (1.12) becomes

(1.18) $$z_1 = \sqrt[3]{\tfrac{1}{2}\left(-q + \sqrt{\tfrac{-\Delta}{27}}\right)} \quad \text{and} \quad z_2 = \sqrt[3]{\tfrac{1}{2}\left(-q - \sqrt{\tfrac{-\Delta}{27}}\right)}.$$

Substituting this into (1.8), we get a version of Cardan's formulas which uses the square root of the discriminant.

The discriminant is also important in the quadratic case. By the quadratic formula, the roots of $x^2 + bx + c$ are

$$x_1 = \frac{-b + \sqrt{\Delta}}{2} \quad \text{and} \quad x_2 = \frac{-b - \sqrt{\Delta}}{2},$$

where $\Delta = b^2 - 4c$ is the discriminant. This makes it easy to see that

$$\sqrt{\Delta} = x_1 - x_2 \quad \text{and} \quad \Delta = (x_1 - x_2)^2.$$

Thus the discriminant is the square of the difference of the roots. In Chapter 2 we will study the discriminant of a polynomial of degree n.

C. Symmetric Polynomials. We begin by noting two interesting properties of

$$\Delta = (x_1 - x_2)^2 (x_1 - x_3)^2 (x_2 - x_3)^2.$$

First suppose that we permute x_1, x_2, x_3 in this formula. The observation is that no matter how we do this, we will still have the product of the squares of the differences of the roots. This shows that Δ is unchanged by permutations of the roots. In the language of Chapter 2 we say that Δ is *symmetric* in the roots x_1, x_2, x_3.

Second, we can also express Δ in terms of the coefficients of $x^3 + bx^2 + cx + d$. By (1.17), we know that $\Delta = -4p^3 - 27q^2$. However, we also have

(1.19)
$$p = -\frac{b^2}{3} + c,$$

$$q = \frac{2b^3}{27} - \frac{bc}{3} + d$$

by Exercise 1 of Section 1.1. If we substitute these into (1.17), then a straightforward calculation shows that

(1.20)
$$\Delta = b^2 c^2 + 18bcd - 4c^3 - 4b^3 d - 27d^2$$

(see Exercise 3). When $b = 0$, it follows that $x^3 + cx + d$ has discriminant

$$\Delta = -4c^3 - 27d^2.$$

This will be useful in Section 1.3.

The above formula expresses the discriminant in terms of the coefficients of the original equation, just as the discriminant of $x^2 + bx + c = 0$ is $\Delta = b^2 - 4c$. The *Fundamental Theorem of Symmetric Polynomials*, to be proved in Chapter 2, will imply that *any* symmetric polynomial in x_1, x_2, x_3 can be expressed in terms of the coefficients b, c, d. In order to see why b, c, d are so important, note that if x_1, x_2, x_3 are the roots of $x^3 + bx^2 + cx + d$, then

$$x^3 + bx^2 + cx + d = (x - x_1)(x - x_2)(x - x_3).$$

Multiplying out the right-hand side and comparing coefficients leads to the following formulas for b, c, d:

$$\begin{aligned} b &= -(x_1 + x_2 + x_3), \\ c &= x_1 x_2 + x_1 x_3 + x_2 x_3, \\ d &= -x_1 x_2 x_3. \end{aligned}$$

(1.21)

These formulas show that the coefficients of a cubic can be expressed as symmetric functions of its roots. The polynomials b, c, d are (up to sign) the *elementary symmetric polynomials* of x_1, x_2, x_3. These polynomials (and their generalization to an arbitrary number of variables) will play a crucial role in Chapter 2.

Mathematical Notes

One aspect of the text needs further discussion.

■ *Algebra* **versus** *Abstract Algebra.* High school algebra is very different from a course on groups, rings, and fields, yet both are called "algebra." The evolution of algebra can be seen in the difference between Section 1.1, where we used high school algebra, and this section, where questions about the underlying structure (*why does the cubic resolvent have degree 6?*) led us to realize the importance of permutations. Many concepts in abstract algebra came from high school algebra in this way.

Historical Notes

In 1770 and 1771, Lagrange's magnificent treatise *Réflexions sur la résolution algébrique des équations* appeared in the Nouvelles Mémoires de l'Academie royale des Sciences et Belles-Lettres de Berlin. This long paper covers pages 205–421 in Volume 3 of Lagrange's collected works [Lagrange]. It is a leisurely account of the known methods for solving equations of degree 3 and 4, together with an analysis of these methods from the point of view of permutations. Lagrange wanted to determine whether these methods could be adapted to equations of degree ≥ 5.

One of Lagrange's powerful ideas is that one should study the roots of a polynomial *without regard to their possible numerical value.* When dealing with functions of the roots, such as

$$z_1 = \frac{1}{3}(x_1 + \omega^2 x_2 + \omega x_3)$$

from (1.10), Lagrange says that he is concerned "only with the form" of such expressions and not "with their numerical quantity" [Lagrange, Vol. 3, p. 385]. In modern terms, Lagrange is saying that we should regard the roots as variables. We will learn more about this idea when we discuss the *universal polynomial* in Chapter 2.

We will see in Chapter 12 that many basic ideas from group theory and Galois theory are implicit in Lagrange's work. However, Lagrange's approach fails when the roots take on specific numerical values. This is part of why Galois's work is so important: he was able to treat the case when the roots were arbitrary. The ideas of

Galois, of course, are the foundation of what we now call Galois theory. This will be the main topic of Chapters 4 through 7.

Exercises for Section 1.2

Exercise 1. Let z_1, z_2 be the roots of (1.9) chosen at the beginning of the section.
  (a) Show that $z_1, z_2, \omega z_1, \omega z_2, \omega^2 z_1, \omega^2 z_2$ are the six roots of the cubic resolvent.
  (b) Prove (1.10).

Exercise 2. Prove (1.14) and (1.15).

Exercise 3. Prove (1.20).

Exercise 4. We say that a cubic $x^3 + bx^2 + cx + d$ has a *multiple root* if it can be written as $(x - r_1)^2(x - r_2)$. Prove that $x^3 + bx^2 + cx + d$ has a multiple root if and only if its discriminant is zero.

Exercise 5. Since $\Delta = (x_1 - x_2)^2(x_1 - x_3)^2(x_2 - x_3)^2$, we can define the square root of Δ to be $\sqrt{\Delta} = (x_1 - x_2)(x_1 - x_3)(x_2 - x_3)$. Prove that an even permutation of the roots takes $\sqrt{\Delta}$ to $\sqrt{\Delta}$ while an odd permutation takes $\sqrt{\Delta}$ to $-\sqrt{\Delta}$. In Section 2.4 we will see that this generalizes nicely to the case of degree n.

1.3   CUBIC EQUATIONS OVER THE REAL NUMBERS

The final topic of this chapter concerns cubic equations with coefficients in the field $\mathbb{R}$ of real numbers. As in Section 1.1, we can reduce to equations of the form $y^3 + py + q = 0$, where $p, q \in \mathbb{R}$. Then Cardan's formulas show that the roots y_1, y_2, y_3 lie in the field $\mathbb{C}$ of complex numbers. We will show that the *sign* of the discriminant of $y^3 + py + q = 0$ tells us how many of the roots are real. We will also give an unexpected application of trigonometry when the roots are all real.

A. The Number of Real Roots. The discriminant of $y^3 + py + q$ is

$$\Delta = (y_1 - y_2)^2(y_1 - y_3)^2(y_2 - y_3)^2.$$

As we noted in the discussion following (1.20), Δ can be expressed as

(1.22) $$\Delta = -4p^3 - 27q^2.$$

You will give a different proof of this in Exercise 1.

For the rest of the section we will assume that the cubic $y^3 + py + q$ has *distinct* roots y_1, y_2, y_3. It follows that the discriminant Δ is a nonzero real number. We next show that the *sign* of Δ gives interesting information about the roots.

Theorem 1.3.1. *Suppose that the polynomial $y^3 + py + q \in \mathbb{R}[y]$ has distinct roots and discriminant $\Delta \neq 0$. Then:*
  (a) $\Delta > 0$ *if and only if the roots of $y^3 + py + q = 0$ are all real.*
  (b) $\Delta < 0$ *if and only if $y^3 + py + q = 0$ has only one real root and the other two roots are complex conjugates of each other.*

Proof. First recall from Section A.2 that complex conjugation $z \mapsto \bar{z}$ satisfies $\overline{z + w} = \bar{z} + \bar{w}$ and $\overline{zw} = \bar{z}\,\bar{w}$. It follows that if y_1 is a root of $y^3 + py + q = 0$, then

$$0 = \bar{0} = \overline{y_1^3 + py_1 + q} = \overline{y_1}^3 + p\overline{y_1} + q,$$

so that $\overline{y_1}$ is also a root. This proves the standard fact that the roots of a polynomial with real coefficients either are real (if $\overline{y_1} = y_1$) or come in complex conjugate pairs (if $\overline{y_1} \neq y_1$).

If y_1, y_2, y_3 are all real and distinct, then $\Delta = (y_1 - y_2)^2(y_1 - y_3)^2(y_2 - y_3)^2$ shows that $\Delta > 0$. If the roots aren't all real, then the above discussion shows that we must have one real root, say y_1, and a complex conjugate pair, say y_2 and $\overline{y_2}$. Write $y_2 = u + iv$, where $u, v \in \mathbb{R}$ and $v \neq 0$. Then $\overline{y_2} = u - iv$ and

$$
\begin{aligned}
\Delta &= \big(y_1 - (u + iv)\big)^2 \big(y_1 - (u - iv)\big)^2 \big((u + iv) - (u - iv)\big)^2 \\
&= \big((y_1 - u) - iv\big)^2 \big((y_1 - u) + iv\big)^2 (2iv)^2 \\
&= -4v^2 \big((y_1 - u)^2 + v^2\big)^2.
\end{aligned}
$$

It follows that $\Delta < 0$ when there is only one real root. This completes the proof. $\square$

In Exercises 2–5, we will sketch a different proof of Theorem 1.3.1 which uses curve graphing techniques from calculus.

We next apply the theory developed so far to Cardan's formulas

$$
\begin{aligned}
y_1 &= z_1 + z_2, \\
y_2 &= \omega z_1 + \omega^2 z_2, \\
y_3 &= \omega^2 z_1 + \omega z_2,
\end{aligned}
$$

where the cube roots

$$(1.23) \quad z_1 = \sqrt[3]{\frac{1}{2}\left(-q + \sqrt{q^2 + \frac{4p^3}{27}}\right)} \quad \text{and} \quad z_2 = \sqrt[3]{\frac{1}{2}\left(-q - \sqrt{q^2 + \frac{4p^3}{27}}\right)}$$

are chosen so that $z_1 z_2 = -p/3$.

First, suppose that $\Delta < 0$. Then Theorem 1.3.1 implies that $y^3 + py + q = 0$ has precisely *one* real root. Furthermore, by (1.22), we have

$$\Delta = -4p^3 - 27q^2 < 0.$$

Hence the square root $\sqrt{q^2 + 4p^3/27}$ is real, which means that we can take z_1 to be the unique real cube root. Then $z_1 z_2 = -p/3$ implies that z_2 is also the real cube root. It follows that

$$y_1 = \sqrt[3]{\frac{1}{2}\left(-q + \sqrt{q^2 + \frac{4p^3}{27}}\right)} + \sqrt[3]{\frac{1}{2}\left(-q - \sqrt{q^2 + \frac{4p^3}{27}}\right)}$$

expresses the *real* root of $y^3 + py + q = 0$ in terms of *real* radicals. Furthermore, in the above formulas for y_2 and y_3, we see that $y_3 = \overline{y_2}$, since the cube roots are real and $\omega^2 = \overline{\omega}$. Thus we have a complete understanding of how Cardan's formulas work when the discriminant is negative.

However, the case when $\Delta > 0$ is very different. Here, $y^3 + py + q = 0$ has three real roots by Theorem 1.3.1. Since

$$\Delta = -4p^3 - 27q^2 > 0,$$

one value of the square root $\sqrt{q^2 + 4p^3/27}$ is

$$\sqrt{q^2 + \frac{4p^3}{27}} = \sqrt{\frac{-\Delta}{27}} = i\sqrt{\frac{\Delta}{27}}.$$

Using this and (1.23), we can write z_1 and z_2 as the cube roots

$$z_1 = \sqrt[3]{\frac{1}{2}\left(-q + i\sqrt{\frac{\Delta}{27}}\right)} \quad \text{and} \quad z_2 = \sqrt[3]{\frac{1}{2}\left(-q - i\sqrt{\frac{\Delta}{27}}\right)}.$$

This shows that z_1 and z_2 are both nonreal complex numbers when $\Delta > 0$. You will prove in Exercise 6 that

(1.24)
$$z_1 z_2 = -\frac{p}{3} \implies z_2 = \overline{z_1}.$$

Combining (1.24) with Cardan's formulas, we see that when $\Delta > 0$, the roots of $y^3 + py + q$ can be written

$$\begin{aligned} y_1 &= z_1 + \overline{z_1}, \\ y_2 &= \omega z_1 + \omega^2 \overline{z_1}, \\ y_3 &= \omega^2 z_1 + \omega \overline{z_1}. \end{aligned}$$

The root y_1 is real, since it is expressed as the sum of a complex number and its conjugate. Furthermore, using $\omega^2 = \overline{\omega}$, one easily sees that

$$\overline{\omega z_1} = \omega^2 \overline{z_1} \quad \text{and} \quad \overline{\omega^2 z_1} = \omega \overline{z_1},$$

so that y_2 and y_3 are also real, since they too are the sum of a complex number and its conjugate.

Notice that, unlike the case when $\Delta < 0$, we no longer have a canonical choice of z_1—it is just one cube root of the complex number $\frac{1}{2}\left(-q + i\sqrt{\Delta/27}\right)$. Furthermore, we get y_1, y_2, y_3 by taking the three cube roots of this number and adding each to its conjugate. This explains how Cardan's formulas work when $\Delta > 0$.

The puzzle, of course, is that we are using complex numbers to express the real roots of a real polynomial. Historically, this is referred to as the *casus irreducibilis*. We will have more to say about this below.

Example 1.3.2. In 1550, Rafael Bombelli applied Cardan's formulas to the cubic $y^3 - 15y - 4 = 0$. This polynomial has discriminant $\Delta = -4(-15)^3 - 27(-4)^2 = 13068 > 0$, so that all three roots are real. Bombelli noted that one root is $y = 4$ and used Cardan's formulas to show that

$$4 = \sqrt[3]{2 + 11i} + \sqrt[3]{2 - 11i}$$

for appropriate choices of cube roots. To understand this formula, Bombelli noted that $(2 + i)^3 = 2 + 11i$ and $(2 - i)^3 = 2 - 11i$. Hence the cube roots in the above formula are $2 + i$ and $2 - i$, and their sum is clearly 4.

In Exercise 7 below, you will find the other two roots of the equation and explain how Cardan's formulas give these two roots.   ◁▷

From the point of view of Cardan's formulas, complex numbers are unavoidable when $\Delta > 0$. But is it possible that there are other ways of expressing the roots which only involve real radicals? In Chapter 8 we will prove that when an irreducible cubic has real roots, the answer to this question is no—using Galois theory, we will see that complex numbers are in fact unavoidable when trying to express the roots of an irreducible cubic with positive discriminant in terms of radicals.

B. Trigonometric Solution of the Cubic. Although complex numbers are unavoidable when applying Cardan's formulas to a cubic with positive discriminant, there is a purely "real" solution provided we use trigonometric functions rather than radicals. This is the *trigonometric solution of the cubic*, due to Viète.

Our starting point is the trigonometric identity

$$\cos(3\theta) = 4\cos^3\theta - 3\cos\theta,$$

which you will prove in Exercise 8. If we write this as $4\cos^3\theta - 3\cos\theta - \cos(3\theta) = 0$, then $t_1 = \cos\theta$ is a root of the cubic equation $4t^3 - 3t - \cos(3\theta) = 0$. However, replacing θ with $\theta + \frac{2\pi}{3}$ gives the same cubic polynomial, since $\cos(3(\theta + \frac{2\pi}{3})) = \cos(3\theta)$. It follows that $t_2 = \cos(\theta + \frac{2\pi}{3})$ is another root of $4t^3 - 3t - \cos(3\theta) = 0$, and similarly, $t_3 = \cos(\theta + \frac{4\pi}{3})$ is also a root.

In Exercise 9 you will show that the discriminant of $4t^3 - 3t - \cos(3\theta)$ is $\frac{27}{16}\sin^2(3\theta)$. This is zero if and only if $\sin(3\theta) = 0$, which in turn is equivalent to $\cos(3\theta) = \pm 1$. Thus $\cos(3\theta) \neq \pm 1$ implies that $4t^3 - 3t - \cos(3\theta)$ has roots

(1.25) $t_1 = \cos\theta$, $t_2 = \cos\left(\theta + \frac{2\pi}{3}\right)$, $t_3 = \cos\left(\theta + \frac{4\pi}{3}\right)$.

Hence $4t^3 - 3t - \cos(3\theta) = 0$ is a cubic equation with known roots. Viète's insight was that by a simple change of variable, we can use this to solve *any* cubic equation with positive discriminant. Here is his result.

Theorem 1.3.3. *Let $y^3 + py + q = 0$ be a cubic equation with real coefficients and positive discriminant. Then $p < 0$, and the roots of the equation are*

$$y_1 = 2\sqrt{\frac{-p}{3}}\cos\theta, \ \ y_2 = 2\sqrt{\frac{-p}{3}}\cos\left(\theta + \frac{2\pi}{3}\right), \ \text{and} \ y_3 = 2\sqrt{\frac{-p}{3}}\cos\left(\theta + \frac{4\pi}{3}\right),$$

where θ is the real number defined by

$$\theta = \frac{1}{3}\cos^{-1}\left(\frac{3\sqrt{3}q}{2p\sqrt{-p}}\right).$$

Proof. You will prove this in Exercise 10. □

In Exercise 11 you will explore how this relates to Cardan's formulas.

Historical Notes

When Cardan wrote *Ars Magna* in 1545, he and his contemporaries wanted to find real roots of cubic equations. In fact, they worked almost exclusively with positive roots, although they were aware of the existence of negative roots, which Cardan called "false" or "fictitious". However, Cardan does use complex numbers in Chapter XXXVII when he considers the problem of dividing 10 into two parts so that their product is 40. In modern notation this gives the equations $x + y = 10$ and $xy = 40$. Eliminating y, we get the quadratic equation

$$x^2 - 10x + 40 = 0$$

with roots $5 \pm i\sqrt{15}$. After deriving this solution, Cardan says "Putting aside the mental tortures involved, multiply $5 + \sqrt{-15}$ by $5 - \sqrt{-15}$, making $25 - (-15) \ldots$ Hence this product is 40." Cardan's conclusion is that "This truly is sophisticated" [2, pp. 219–220].

Cardan was also aware of Theorem 1.3.1, though he stated it in very different terms. As an example of a cubic with three real roots, he considers $x^3 + 9 = 12x$, for which he gives the "true" (i.e., positive) solutions 3 and $\sqrt{5\frac{1}{4}} - 1\frac{1}{2}$ and the "false" (i.e., negative) solution $-\sqrt{5\frac{1}{4}} - 1\frac{1}{2}$.

However, Cardan never applies his formulas to cubics like $x^3 + 9 = 12x$. He only considers cases where there is one real root, which can be expressed in terms of real radicals. Yet Cardan must have known that complex numbers appear in the radicals when the discriminant is positive. This is the *casus irreducibilis* ("irreducible case") mentioned above. According to [1], Tartaglia was also aware of the *casus irreducibilis*, and in fact delayed publication of his results because he was so troubled by it. This is part of the reason why Cardan's work appeared first.

One of the first people to comment directly on the *casus irreducibilis* was Rafael Bombelli. In his book *L'algebra*, written around 1550 but not published until 1572, he treats this case in detail, including the formula

(1.26) $$4 = \sqrt[3]{2 + 11i} + \sqrt[3]{2 - 11i}$$

from Example 1.3.2. There we saw how Bombelli explained this formula by showing that $2 + 11i = (2 + i)^3$, so that (1.26) reduces to $4 = (2 + i) + (2 - i)$. Bombelli was pleased with this calculation and commented that

> At first, the thing [equation (1.26)] seemed to me to be based more on sophism
> than on truth, but I searched until I found a proof.

In working out this solution, Bombelli was the first to give systematic rules for adding
and multiplying complex numbers. Exercise 12 will discuss another example of
complex cube roots taken from Bombelli's work.

The moral is that cubic equations forced mathematicians to confront complex
numbers. For quadratic equations, one could pretend that complex solutions don't
exist. But for a cubic with real roots, we've seen that Cardan's formula must involve
complex numbers. So it is impossible to ignore complex numbers in this case.
See the books [1] and [3] for more background and discussion on the discovery of
complex numbers.

We should also say a few words about Viète's trigonometric solution of the cubic.
Once we realize that $\cos(3\theta) = 4\cos^3\theta - 3\cos\theta$ gives a cubic equation with $\cos\theta$
as a root, proving Theorem 1.3.3 is not that difficult. Viète was well aware of
such identities. For example, in 1593, Adrianus Romanus (also called Adriaen van
Roomen) posed the problem of finding a root of the equation

$$
\begin{aligned}
A = \ & x^{45} - 45x^{43} + 945x^{41} - 12300x^{39} + 111150x^{37} - 740259x^{35} \\
& + 3764565x^{33} - 14945040x^{31} + 46955700x^{29} - 117679100x^{27} \\
& + 236030652x^{25} - 37865800x^{23} + 483841800x^{21} - 488494125x^{19} \\
& + 384942237x^{17} - 232676280x^{15} + 105306075x^{13} - 3451207x^{11} \\
& + 7811375x^9 - 1138500x^7 + 95634x^5 - 3795x^3 + 45x,
\end{aligned}
$$

(1.27)

where

(1.28)
$$
A = \sqrt{\frac{7}{4} - \sqrt{\frac{5}{16}} - \sqrt{\frac{15}{8}} - \sqrt{\frac{45}{64}}}.
$$

Viète solved this equation by noting that $2\sin(45\alpha)$ can be expressed as a polynomial
of degree 45 in $2\sin\alpha$ whose coefficients match the right-hand side of (1.27). It
follows that if $A = 2\sin(45\alpha)$, then $x = 2\sin\alpha$ is a root.

Viète also realized that (1.28) can be written

$$
A = 2\sin(\pi/15) = 2\sin(45 \cdot \pi/675),
$$

which easily implies that one root of (1.27) is $x = 2\sin(\pi/675)$. Using the trick of
(1.25), we get the 44 additional solutions

$$
x = 2\sin\left(\frac{\pi}{675} + j\frac{2\pi}{45}\right), \quad j = 1, \ldots, 44.
$$

Viète listed only 23 roots, since he (like Cardan) wanted positive solutions. Nevertheless, Viète's insight is impressive, and his solution of (1.27) makes it clear how
he was able to find the trigonometric solution of the cubic.

Exercises for Section 1.3

Exercise 1. Let $f(y) = y^3 + py + q = (y - y_1)(y - y_2)(y - y_3)$, and set

$$\Delta = (y_1 - y_2)^2 (y_1 - y_3)^2 (y_2 - y_3)^2.$$

The goal of this exercise is to give a different proof of (1.22).

(a) Use the product rule to show that $f'(y_1) = (y_1 - y_2)(y_1 - y_3)$, where f' denotes the derivative of f. Also derive similar formulas for $f'(y_2)$ and $f'(y_3)$.

(b) Conclude that $\Delta = -f'(y_1) f'(y_2) f'(y_3)$. Be sure to explain where the minus sign comes from.

(c) The quadratic $f'(y) = 3y^2 + p$ factors as $f'(y) = 3(y - \alpha)(y - \beta)$, where $\alpha = \sqrt{-p/3}$ and $\beta = -\sqrt{-p/3}$ (when $p > 0$, we let $\sqrt{-p/3} = i\sqrt{p/3}$). Prove that $\Delta = -27 f(\alpha) f(\beta)$.

(d) Use $f(y) = y^3 + py + q$ and $\alpha = \sqrt{-p/3}$ to show that

$$f(\alpha) = (\sqrt{-p/3})^3 + p\sqrt{-p/3} + q = (2/3)p\sqrt{-p/3} + q.$$

Similarly, show that $f(\beta) = -(2/3)p\sqrt{-p/3} + q$.

(e) By combining parts (c) and (d), conclude that $\Delta = -4p^3 - 27q^2$.

Exercise 2. Let $f(y) = y^3 + py + q$. The purpose of Exercises 2–5 is to prove Theorem 1.3.1 geometrically using curve graphing techniques. The proof breaks up into three cases corresponding to $p > 0$, $p = 0$, and $p < 0$. This exercise will consider the case $p > 0$.

(a) Explain why $\Delta < 0$.

(b) Analyze the sign of $f'(y)$, and show that $f(y)$ is always increasing.

(c) Explain why $f(y)$ has only one real root.

Exercise 3. Next, consider the case $p = 0$.

(a) Explain why $\Delta < 0$.

(b) Explain why $f(y)$ has only one real root.

Exercise 4. Finally, consider the case $p < 0$. In this case, $f'(y) = 3y^2 + p$ has roots $\alpha = \sqrt{-p/3}$ and $\beta = -\sqrt{-p/3}$, which are real and distinct.

(a) Show that the graph of $f(y)$ has a local minimum at α and a local maximum at β. Thus $f(\alpha)$ is a local minimum value and $f(\beta)$ is a local maximum value. Also show that $f(\alpha) < f(\beta)$.

(b) Explain why $f(y)$ has three real roots if $f(\alpha)$ and $f(\beta)$ have opposite signs and has one real root if they have the same sign. Illustrate your answer with a drawing of the three cases that can occur.

(c) Conclude that $f(y)$ has three real roots if and only if $f(\alpha) f(\beta) < 0$.

(d) Finally, use part (c) of Exercise 1 to show that the roots are all real if and only if $\Delta > 0$.

Exercise 5. Explain how Theorem 1.3.1 follows from Exercises 2, 3, and 4. Notice that · the quantity $f(\alpha) f(\beta)$, which appeared earlier in part (c) of Exercise 1, arises naturally in Exercise 4.

Exercise 6. Prove (1.24).

Exercise 7. Example 1.3.2 expressed the root $y = 4$ of $y^3 - 15y - 4$ in terms of Cardan's formulas. Find the other two roots, and explain how Cardan's formulas give these roots.

Exercise 8. Derive the trigonometric identity $\cos(3\theta) = 4\cos^3\theta - 3\cos\theta$ using $\cos(x+y) = \cos x \cos y - \sin x \sin y$ and $\cos^2\theta + \sin^2\theta = 1$.

Exercise 9. When divided by 4, $4t^3 - 3t - \cos(3\theta)$ gives $t^3 - \frac{3}{4}t - \frac{1}{4}\cos(3\theta)$, which is monic. Show that the discriminant of this polynomial is $\frac{27}{16}\sin^2(3\theta)$.

Exercise 10. The goal of this exercise is to prove Theorem 1.3.3. Let $y^3 + py + q = 0$ be a cubic equation with positive discriminant. Consider the substitution $y = \lambda t$, which transforms the given equation into $\lambda^3 t^3 + \lambda pt + q = 0$.
(a) Show that Exercises 2 and 3 imply that $p < 0$.
(b) The equation $\lambda^3 t^3 + \lambda pt + q = 0$ can be written as

$$4t^3 - \left(\frac{-4p}{\lambda^2}\right)t - \left(\frac{-4q}{\lambda^3}\right) = 0.$$

Show that this coincides with $4t^3 - 3t - \cos(3\theta) = 0$ if and only if

$$\lambda = 2\sqrt{\frac{-p}{3}} \quad \text{and} \quad \cos(3\theta) = \frac{3\sqrt{3}q}{2p\sqrt{-p}}.$$

Note that $\sqrt{-p}$ is real and nonzero by part (a).
(c) Use $\Delta = -(4p^3 + 27q^2) > 0$ to prove that

$$\left|\frac{3\sqrt{3}q}{2p\sqrt{-p}}\right| < 1.$$

(d) Explain how part (c) implies that the second equation of part (b) can be solved for θ. Also show that $\Delta > 0$ implies that $\cos(3\theta) \neq \pm 1$.
(e) By (1.25), $t_1 = \cos\theta$, $t_2 = \cos\left(\theta + \frac{2\pi}{3}\right)$, and $t_3 = \cos\left(\theta + \frac{4\pi}{3}\right)$ are the three roots of $\lambda^3 t^3 + \lambda pt + q = 0$. Then show that the theorem follows by transforming this back to $y = \lambda t$ via part (b).

Exercise 11. Consider the equation $4t^3 - 3t - \cos(3\theta) = 0$, where $\cos(3\theta) \neq \pm 1$. In (1.25), we expressed the roots in terms of trigonometric functions. In this exercise, you will study what happens when we use Cardan's formulas.
(a) Show that Cardan's formulas give the root

$$t_1 = \frac{1}{2}\sqrt[3]{\cos(3\theta) + i\sin(3\theta)} + \frac{1}{2}\sqrt[3]{\cos(3\theta) - i\sin(3\theta)}.$$

(b) Explain why $\frac{1}{2}e^{i\theta} = \frac{1}{2}(\cos\theta + i\sin\theta)$ is a value of $\frac{1}{2}\sqrt[3]{\cos(3\theta) + i\sin(3\theta)}$, and use this to show that t_1 is just $\cos\theta$.
(c) Similarly, show that Cardan's formulas also give the roots t_2 and t_3 as predicted by (1.25).

Exercise 12. Example 1.3.2 discusses Bombelli's discovery that $\sqrt[3]{2 + 11i} = 2 + i$. But not all cube roots can be expressed so simply. This exercise will show that $\sqrt[3]{4 + \sqrt{11}i}$ is not of the form $a + b\sqrt{11}i$ for $a, b \in \mathbb{Z}$.
(a) Suppose that $4 + \sqrt{11}i = (a + b\sqrt{11}i)^3$ for some $a, b \in \mathbb{Z}$. Show that this implies that $4 = a^3 - 33ab^2$ and $1 = 3a^2b - 11b^3$.
(b) Show that the equations of part (a) imply that $b = \pm 1$ and $a|4$. Conclude that the equation $4 + \sqrt{11}i = (a + b\sqrt{11}i)^3$ has no solutions with $a, b \in \mathbb{Z}$.

(c) Find a cubic polynomial of the form $x^3 + px + q$ with $p, q \in \mathbb{Z}$ which has the number $\sqrt[3]{4 + \sqrt{11}i} + \sqrt[3]{4 - \sqrt{11}i}$ as a root.

In contrast to $\sqrt[3]{2 + 11i} = 2 + i$, Bombelli was not certain that $\sqrt[3]{4 + \sqrt{11}i}$ was a complex number. He calls $\sqrt[3]{4 + \sqrt{11}i}$ "another sort of cubic radical." Bombelli never deals with this radical by itself, but rather considers the sum $\sqrt[3]{4 + \sqrt{11}i} + \sqrt[3]{4 - \sqrt{11}i}$, which is a root of the cubic equation found in part (c).

Exercise 13. Suppose that a quartic polynomial $f = x^4 + bx^3 + cx^2 + dx + e$ in $\mathbb{R}[x]$ has distinct roots $x_1, x_2, x_3, x_4 \in \mathbb{C}$. The *discriminant* of f is defined by the equation

$$\Delta = (x_1 - x_2)^2 (x_1 - x_3)^2 (x_1 - x_4)^2 (x_2 - x_3)^2 (x_2 - x_4)^2 (x_3 - x_4)^2.$$

The theory developed in Chapter 2 will imply that $\Delta \in \mathbb{R}$, and $\Delta \neq 0$, since the x_i are distinct. Adapt the proof of Theorem 1.3.1 to show that

$$\Delta < 0 \iff x^4 + bx^3 + cx^2 + dx + e = 0 \text{ has exactly two real roots.}$$

Exercise 14. In Section 1.1, we discussed the equation $x^3 + 2x^2 + 10x = 20$ considered by Fibonacci.
  (a) Show that this equation has precisely one real root. This is the root Fibonacci approximated so well.
  (b) Use Cardan's formulas and a calculator to work out numerically the three roots of this polynomial.

Exercise 15. Use a calculator and Theorem 1.3.3 to compute the roots of the cubic equation $y^3 - 7y + 3 = 0$ to eight decimal places of accuracy.

REFERENCES

1. I. G. Bashmakova and G. S. Smirnova, *The Beginnings and Evolution of Algebra*, English translation by A. Shenitzer, MAA, Washington, DC, 1999.

2. G. Cardan, *Ars Magna*, Johann Petrieus, Nürnberg, 1545. English Translation *The Great Art* by T. R. Witmer, MIT Press, Cambridge, MA, 1968.

3. B. Mazur, *Imagining Numbers (particularly the square root of minus fifteen)*, Farrar Straus Giroux, New York, 2003.

$$\frac{2}{}$$

Symmetric Polynomials

The goal of this chapter is to provide some tools needed for our study of Galois theory. The basic result is that any polynomial unchanged under all possible permutations of the variables can be expressed in terms of certain special polynomials called the *elementary symmetric polynomials*. After proving this, we will show how to compute with symmetric polynomials and discuss the *discriminant* mentioned in Chapter 1.

2.1 POLYNOMIALS OF SEVERAL VARIABLES

Galois theory often deals with polynomials of more than one variable, especially when studying the roots of a polynomial. This section will introduce polynomials of several variables and the elementary symmetric polynomials.

A. The Polynomial Ring in n Variables. Let $x_1, \ldots, x_n$ be distinct formal symbols called variables. A polynomial in $x_1, \ldots, x_n$ with coefficients in a field F is a finite sum of *terms*, which are expressions of the form

$$c \, x_1^{a_1} \cdots x_n^{a_n}, \qquad c \text{ in } F, \quad a_1, \ldots, a_n \geq 0 \text{ in } \mathbb{Z}.$$

We call the product $x_1^{a_1} \cdots x_n^{a_n}$ a *monomial*, so that a term is an element of F times a monomial. A term is *nonzero* if the constant is nonzero. The *total degree* of a nonzero term $c \, x_1^{a_1} \cdots x_n^{a_n}$ is the sum of its exponents $a_1 + \cdots + a_n$.

We define $F[x_1, \ldots, x_n]$ to be the set of all polynomials in $x_1, \ldots, x_n$ with coefficients in F. It is easy to see that $F[x_1, \ldots, x_n]$ is a ring under addition and multiplication of polynomials. The *total degree* of a nonzero $f \in F[x_1, \ldots, x_n]$, denoted $\deg(f)$, is the maximum of the total degrees of the nonzero terms of f. Since

F is an integral domain, one can prove without difficulty that if $f, g \in F[x_1, \ldots, x_n]$ are nonzero, then

(2.1) $$\deg(fg) = \deg(f) + \deg(g).$$

It follows that $F[x_1, \ldots, x_n]$ is an integral domain. Note that $\deg(0)$ is not defined.
  Since $F[x_1, \ldots, x_n]$ is an integral domain, we can define its field of fractions

$$F(x_1, \ldots, x_n) = \left\{ \frac{f}{g} \mid f, g \in F[x_1, \ldots, x_n], \ g \neq 0 \right\}.$$

This is the *field of rational functions* in n variables. Note that:

- Square brackets, as in $F[x_1, \ldots, x_n]$, refer to *polynomials*.
- Parentheses, as in $F(x_1, \ldots, x_n)$, refer to *quotients of polynomials*.

A nonconstant polynomial in $F[x_1, \ldots, x_n]$ is *irreducible over F* if it is not a product of polynomials of strictly smaller total degree. We can factor polynomials in $F[x_1, \ldots, x_n]$ into irreducibles as follows.

Theorem 2.1.1. *Let $f \in F[x_1, \ldots, x_n]$ be nonconstant. Then there are irreducible polynomials $g_1, \ldots, g_r \in F[x_1, \ldots, x_n]$ such that*

$$f = g_1 \cdots g_r.$$

Furthermore, if there is a second factorization of f into irreducibles

$$f = h_1 \cdots h_s,$$

then $r = s$ and there is a permutation $\sigma \in S_r$ such that for each $1 \leq i \leq r$, there is a nonzero $c_i \in F$ such that $g_i = c_i h_{\sigma(i)}$.

Proof. See the references listed in Section A.5 of Appendix A. □

In Section A.5, we define the general notion of a *unique factorization domain*, or *UFD*. In this terminology, Theorem 2.1.1 states that $F[x_1, \ldots, x_n]$ is a UFD.
  A useful property of $F[x_1, \ldots, x_n]$ is that *evaluation is a ring homomorphism*. Suppose that we have a field F, a ring R containing F, and elements $\alpha_1, \ldots, \alpha_n \in R$. Then the *evaluation map*

$$F[x_1, \ldots, x_n] \longrightarrow R$$

is defined by

(2.2) $$f(x_1, \ldots, x_n) \mapsto f(\alpha_1, \ldots, \alpha_n) \in R.$$

We have the following important result.

Theorem 2.1.2. *Given a field F, a ring R containing F, and $\alpha_1, \ldots, \alpha_n \in R$, the evaluation map (2.2) is a ring homomorphism $F[x_1, \ldots, x_n] \to R$.* □

Proof. The proof is a tedious verification that

$$(f + g)(\alpha_1, \ldots, \alpha_n) = f(\alpha_1, \ldots, \alpha_n) + g(\alpha_1, \ldots, \alpha_n),$$
$$(fg)(\alpha_1, \ldots, \alpha_n) = f(\alpha_1, \ldots, \alpha_n)g(\alpha_1, \ldots, \alpha_n),$$

where fg and $f + g$ are the sum and product of polynomials f and g. □

Once we fix the field F, the variables $x_1, \ldots, x_n$ play two roles. At the beginning of the section, they were formal symbols used in the definition of polynomial. But each variable x_i also has the ability to "take any value." In other words, $x_1, \ldots, x_n$ can take arbitrary values in any ring R containing F. Be sure you understand how Theorem 2.1.2 makes this precise.

B. The Elementary Symmetric Polynomials. How do the roots of a monic polynomial in x relate to its coefficients? To answer this question, we begin with cubic and quartic polynomials. Suppose that $f = x^3 + a_1 x^2 + a_2 x + a_3 \in F[x]$ has roots $\alpha_1, \alpha_2, \alpha_3 \in F$. Then

$$f = (x - \alpha_1)(x - \alpha_2)(x - \alpha_3).$$

If we multiply this out and compare coefficients, then the coefficients can be expressed in terms of the roots as

(2.3)
$$\begin{aligned} a_1 &= -(\alpha_1 + \alpha_2 + \alpha_3), \\ a_2 &= \alpha_1\alpha_2 + \alpha_1\alpha_3 + \alpha_2\alpha_3, \\ a_3 &= -\alpha_1\alpha_2\alpha_3. \end{aligned}$$

(See also (1.21) in Section 1.2.) For $n = 4$, a similar computation shows that if $f = x^4 + a_1 x^3 + a_2 x^2 + a_3 x + a_4 \in F[x]$ has roots $\alpha_1, \alpha_2, \alpha_3, \alpha_4 \in F$, then

$$\begin{aligned} a_1 &= -(\alpha_1 + \alpha_2 + \alpha_3 + \alpha_4), \\ a_2 &= \alpha_1\alpha_2 + \alpha_1\alpha_3 + +\alpha_1\alpha_4 + \alpha_2\alpha_3 + \alpha_2\alpha_4 + \alpha_3\alpha_4, \\ a_3 &= -(\alpha_1\alpha_2\alpha_3 + \alpha_1\alpha_2\alpha_4 + \alpha_1\alpha_3\alpha_4 + \alpha_2\alpha_3\alpha_4), \\ a_4 &= \alpha_1\alpha_2\alpha_3\alpha_4. \end{aligned}$$

Up to sign, a_1 uses the sum of the roots, a_2 takes the roots two a time, and a_3 takes them three at a time. We generalize this pattern as follows.

Definition 2.1.3. *Let $x_1, \ldots, x_n$ be variables over a field F. Then*

$$\sigma_1 = x_1 + \cdots + x_n,$$

$$\sigma_2 = \sum_{i<j} x_i x_j,$$

$$\vdots$$

$$\sigma_r = \sum_{i_1 < \cdots < i_r} x_{i_1} x_{i_2} \cdots x_{i_r},$$

$$\vdots$$

$$\sigma_n = x_1 x_2 \cdots x_n$$

*are the **elementary symmetric polynomials**. Thus $\sigma_1, \ldots, \sigma_n \in F[x_1, \ldots, x_n]$.*

We will sometimes write $\sigma_r = \sigma_r(x_1, \ldots, x_n)$. The following identity is one of the key properties of the elementary symmetric polynomials.

Proposition 2.1.4. *Let* $x_1, \ldots, x_n$ *be variables over a field* F. *Then, given another variable* x, *we have*

$$(2.4) \quad (x - x_1) \cdots (x - x_n) = x^n - \sigma_1 x^{n-1} + \cdots + (-1)^r \sigma_r x^{n-r} + \cdots + (-1)^n \sigma_n.$$

Proof. The proof follows by multiplying out the left-hand side of (2.4) and then computing the coefficient of each power of x. For example, the constant term is obviously the product of constant terms, namely $(-x_1) \cdots (-x_n) = (-1)^n \sigma_n$. Similarly, the coefficient of x^{n-1} is easily seen to be $-x_1 - \cdots - x_n = -\sigma_1$.

For readers interested in the details of how this works in general, observe that we multiply out $(x - x_1) \cdots (x - x_n)$ as follows:

- For each of the n factors $x - x_i$, choose either x or $-x_i$.
- Take the product of these n choices.
- Sum these products over all possible ways of making the n choices.

It follows that the terms involving x^{n-r} in $(x - x_1) \cdots (x - x_n)$ are those products where we chose x exactly $n - r$ times in the first bullet. This means choosing $-x_i$ for the i_1st, i_2nd, ..., i_rth factors and choosing x for the remaining $n - r$ factors. As described in the second bullet, the product of these choices is

$$(-x_{i_1})(-x_{i_2}) \cdots (-x_{i_r}) x^{n-r} = (-1)^r x_{i_1} \cdots x_{i_r} x^{n-r}.$$

When we sum over all possible ways of making the n choices (as described in the third bullet), it follows that the coefficient of x^{n-r} in the left-hand side of (2.4) is

$$(-1)^r \sum_{i_1 < \cdots < i_r} x_{i_1} \cdots x_{i_r} = (-1)^r \sigma_r.$$

This completes the proof of the proposition.   □

Proposition 2.1.4 has the following useful application. Suppose that a monic polynomial $f = x^n + a_1 x^{n-1} + a_2 x^{n-2} + \cdots + a_{n-1} x + a_n \in F[x]$ has roots $\alpha_1, \ldots, \alpha_n$ in a larger field L. This means that

$$x^n + a_1 x^{n-1} + \cdots + a_{n-1} x + a_n = (x - \alpha_1) \cdots (x - \alpha_n).$$

However, since evaluation is a ring homomorphism (Theorem 2.1.2), we can evaluate the identity (2.4) at $x_1 = \alpha_1, \ldots, x_n = \alpha_n$ to obtain

$$(x - \alpha_1) \cdots (x - \alpha_n) = x^n - \sigma_1(\alpha_1, \ldots, \alpha_n) x^{n-1} + \cdots$$
$$+ (-1)^{n-1} \sigma_{n-1}(\alpha_1, \ldots, \alpha_n) x + (-1)^n \sigma_n(\alpha_1, \ldots, \alpha_n).$$

These two formulas give the following corollary of Proposition 2.1.4.

Corollary 2.1.5. *Let* $f = x^n + a_1 x^{n-1} + a_2 x^{n-2} + \cdots + a_{n-1} x + a_n$ *be a monic polynomial of degree* $n > 0$ *with coefficients in a field* F. *If* f *has roots* $\alpha_1, \ldots, \alpha_n$ *in a larger field* L, *then the coefficients of* f *are expressed in terms of its roots as*

$$a_r = (-1)^r \sigma_r(\alpha_1, \ldots, \alpha_n)$$

for $r = 1, \ldots, n$.   □

Here is what happens when $n = 3$.

Example 2.1.6. If $x^3 + a_1 x^2 + a_2 x + a_3$ has roots $\alpha_1, \alpha_2, \alpha_3$, then Corollary 2.1.5 implies that

$$a_1 = -\sigma_1(\alpha_1, \alpha_2, \alpha_3) = -(\alpha_1 + \alpha_2 + \alpha_3),$$
$$a_2 = \sigma_2(\alpha_1, \alpha_2, \alpha_3) = \alpha_1\alpha_2 + \alpha_1\alpha_3 + \alpha_2\alpha_3,$$
$$a_3 = -\sigma_3(\alpha_1, \alpha_2, \alpha_3) = -\alpha_1\alpha_2\alpha_3,$$

in agreement with (2.3).   ◁▷

Mathematical Notes

There are two topics for us to discuss.

■ **Ideals in a Polynomial Ring.** The text makes it seem that $F[x_1, \ldots, x_n]$ behaves like the one-variable case studied in Section A.1. However, once we start talking about ideals, some significant differences emerge. For example, Theorem A.1.17 implies that $F[x]$ is a PID. But as soon as the number of variables is two or more, not all ideals are principal. Exercise 1 will give a simple example.

In fact, $F[x_1, \ldots, x_n]$ has a rich supply of ideals when $n \geq 2$. These are related to solutions of simultaneous sets of polynomial equations, which the subject of *algebraic geometry*. See [2] for an introduction to this area of mathematics.

■ **Coefficients as Polynomials.** There are other ways to think about polynomials in several variables. For example, we can regard $f \in F[x_1, \ldots, x_n]$ as a polynomial in x_n with coefficients in $F[x_1, \ldots, x_{n-1}]$, that is,

$$f = \sum_{i=0}^{m} p_i(x_1, \ldots, x_{n-1}) \, x_n^i, \quad p_i \in F[x_1, \ldots, x_{n-1}].$$

This is expressed more formally as $F[x_1, \ldots, x_n] = F[x_1, \ldots, x_{n-1}][x_n]$. For instance, (2.4) takes place in $F[x_1, \ldots, x_n][x]$. See Exercise 2 for more examples.

Exercises for Section 2.1

Exercise 1. Show that $\langle x, y \rangle = \{xg + yh \mid g, h \in F[x, y]\} \subset F[x, y]$ is not a principal ideal in $F[x, y]$.

Exercise 2. Express each of the following polynomials as a polynomial in y with coefficients that are polynomials in the remaining variables.
(a) $x^2 y + 3y^2 - xy^2 + 3x + xy^2 + 7x^3 y^3$.
(b) $\left(y - (x_1 + x_2)\right)\left(y - (x_1 + x_3)\right)\left(y - (x_2 + x_3)\right)$.

Exercise 3. Given positive integers n and r with $1 \leq r \leq n$, let $\binom{n}{r}$ be the number of ways of choosing r elements from a set with n elements. Recall that $\binom{n}{r} = \frac{n!}{r!(n-r)!}$.
(a) Show that the polynomial σ_r is a sum of $\binom{n}{r}$ terms.

(b) Show that $\sigma_r(-\alpha, \ldots, -\alpha) = (-1)^r \binom{n}{r} \alpha^r$.

(c) Let $f = (x + \alpha)^n$. Use part (b) and Corollary 2.1.5 to prove that

$$(x + \alpha)^n = \sum_{r=0}^{n} \binom{n}{r} \alpha^r x^{n-r},$$

where $\binom{n}{0} = 1$. This shows that the binomial theorem follows from Corollary 2.1.5.

2.2   SYMMETRIC POLYNOMIALS

We will consider polynomials in n variables $x_1, \ldots, x_n$ over a field F.

Definition 2.2.1. *A polynomial $f \in F[x_1, \ldots, x_n]$ is **symmetric** if*

$$f(x_{\sigma(1)}, \ldots, x_{\sigma(n)}) = f(x_1, \ldots, x_n)$$

for all permutations $\sigma \in S_n$.

A.   The Fundamental Theorem.   In Section 2.1, we defined the elementary symmetric polynomials $\sigma_1, \ldots, \sigma_n$. To prove that these are symmetric in the above sense, consider the identity

$$(x - x_1) \cdots (x - x_n) = x^n - \sigma_1 x^{n-1} + \cdots + (-1)^r \sigma_r x^r + \cdots + (-1)^n \sigma_n$$

from Proposition 2.1.4. The product on the left-hand side is symmetric because permuting the x_i simply permutes the factors. Comparing this with the right-hand side, it follows that $\sigma_1, \ldots, \sigma_n$ are symmetric.

Since $\sigma_1, \ldots, \sigma_n$ are symmetric, any polynomial in $\sigma_1, \ldots, \sigma_n$ is also symmetric. The remarkable fact is that *all* symmetric polynomials arise in this way. More precisely, we have the following *Fundamental Theorem of Symmetric Polynomials*.

Theorem 2.2.2. *Any symmetric polynomial in $F[x_1, \ldots, x_n]$ can be written as a polynomial in $\sigma_1, \ldots, \sigma_n$ with coefficients in F.*

Proof. We will follow (with a few changes) the argument given by Gauss in 1816 in his second proof of the Fundamental Theorem of Algebra. The proof will involve an inductive process which requires that we order monomials $x_1^{a_1} \cdots x_n^{a_n}$ in $x_1, \ldots, x_n$. We will use *graded lexicographic order*, which is defined by

$$
\begin{aligned}
x_1^{a_1} \cdots x_n^{a_n} < x_1^{b_1} \cdots x_n^{b_n} \iff{} & a_1 + \cdots + a_n < b_1 + \cdots + b_n, \\
& \text{or } a_1 + \cdots + a_n = b_1 + \cdots + b_n \\
& \quad \text{and } a_1 < b_1, \\
& \text{or } a_1 + \cdots + a_n = b_1 + \cdots + b_n \\
& \quad \text{and } a_1 = b_1 \text{ and } a_2 < b_2, \\
& \text{or } \ldots.
\end{aligned}
$$

(2.5)

We also define $x_1^{b_1} \cdots x_n^{b_n} > x_1^{a_1} \cdots x_n^{a_n}$ to mean $x_1^{a_1} \cdots x_n^{a_n} < x_1^{b_1} \cdots x_n^{b_n}$.

To compare one monomial with another, one first computes the total degree of each monomial, and when these are equal, one checks the two monomials one exponent at a time, starting with x_1, to find the first which differs. For example,

$$x_1^4 x_2^2 x_3 < x_1^2 x_2^3 x_3^3 \qquad \text{(smaller total degree),}$$

$$x_1^4 x_2^2 x_3 > x_1^4 x_2 x_3^2 \qquad \text{(same total degree, equal } x_1 \text{ exponent,}$$

$$\text{greater } x_2 \text{ exponent).}$$

An important property of graded lexicographic order is that there are at most finitely many monomials $x_1^{b_1} \cdots x_n^{b_n}$ such that

(2.6) $$x_1^{b_1} \cdots x_n^{b_n} < x_1^{a_1} \cdots x_n^{a_n} \quad \text{for fixed} \quad a_1, \ldots, a_n.$$

This follows because (2.6) and (2.5) imply that $a_1 + \cdots + a_n \geq b_1 + \cdots + b_n$ (be sure you understand this). Since $N = a_1 + \cdots + a_n$ is fixed and $b_i \geq 0$ for all i, we get the inequality

$$N = a_1 + \cdots + a_n \geq b_1 + \cdots + b_n \geq b_i$$

for all i. Hence there are only $N + 1$ possibilities for each b_i, which easily implies that (2.6) can hold for at most finitely many $x_1^{b_1} \cdots x_n^{b_n}$.

We can apply graded lexicographic order to a nonzero polynomial as follows. We saw in Section 2.1 that such a polynomial is a sum of nonzero terms, each of which is a nonzero element of F times a monomial. Then the *leading term* is the greatest of these monomials—relative to (2.5)—times its coefficient. Thus any nonzero polynomial has a leading term.

For example, the leading term of $\sigma_2 = x_1 x_2 + x_1 x_3 + \cdots + x_2 x_3 + \cdots + x_{n-1} x_n$ is $x_1 x_2$. In other words, $x_1 x_2 > x_i x_j$ when $i < j$ and $(i, j) \neq (1, 2)$. This follows by checking the exponent of x_1 (if $i > 1$) or x_2 (if $i = 1$ and $j > 2$) in $x_i x_j$. You will generalize this in Exercise 1 by showing that $x_1 x_2 \cdots x_r$ is the leading term of the rth elementary symmetric polynomial

$$\sigma_r = \sum_{i_1 < \cdots < i_r} x_{i_1} x_{i_2} \cdots x_{i_r}.$$

We are now ready to prove the theorem. Let $f \in F[x_1, \ldots, x_n]$ be symmetric and nonzero with leading term

(2.7) $$c\, x_1^{a_1} \cdots x_n^{a_n}.$$

We claim that

(2.8) $$a_1 \geq a_2 \geq \cdots \geq a_n.$$

To show this, suppose that $a_i < a_{i+1}$ for some $1 \leq i \leq n - 1$. The symmetry of f implies that interchanging x_i and x_{i+1} gives the same polynomial. Since (2.7) is a term of f, it follows that

(2.9) $$c\, x_1^{a_1} \cdots x_{i+1}^{a_i} x_i^{a_{i+1}} \cdots x_n^{a_n}$$

is also a term of f. To compare this with (2.7), note that both monomials have the same total degree and the same exponents of $x_1, \ldots, x_{i-1}$. However, x_i has exponent a_{i+1} in (2.9) and exponent a_i in (2.7). Then $a_{i+1} > a_i$ implies that (2.9) is a term of f *greater* than (2.7) according to the order relation (2.5). Yet (2.7) is the leading term of f. This contradiction proves (2.8).

Now consider

$$(2.10) \qquad g = \sigma_1^{a_1-a_2} \sigma_2^{a_2-a_3} \cdots \sigma_{n-1}^{a_{n-1}-a_n} \sigma_n^{a_n}.$$

This is a polynomial by (2.8). In Exercise 2, you will prove that the leading term of a product is the product of the leading terms. Since the leading term of σ_r is $x_1 \cdots x_r$, it follows that the leading term of g is

$$
\begin{aligned}
(2.11) \qquad & x_1^{a_1-a_2} (x_1 x_2)^{a_2-a_3} (x_1 x_2 x_3)^{a_3-a_4} \cdots (x_1 \cdots x_{n-1})^{a_{n-1}-a_n} (x_1 \cdots x_n)^{a_n} \\
& = x_1^{a_1-a_2+a_2-a_3+\cdots+a_n} x_2^{a_2-a_3+\cdots+a_n} \cdots x_{n-1}^{a_{n-1}-a_n+a_n} x_n^{a_n} \\
& = x_1^{a_1} \cdots x_n^{a_n}.
\end{aligned}
$$

This shows that f and cg have the same leading term. Hence $f_1 = f - cg$ has a strictly smaller leading term according to the ordering defined in (2.5). Note that f_1 is symmetric, since f and g are.

Now repeat this process, starting with f_1 instead of f. Since f_1 is symmetric, it has a leading term with coefficient c_1 and exponents $b_1 \geq \cdots \geq b_n$. As above, this will give an expression g_1 in the elementary symmetric polynomials such that f_1 and $c_1 g_1$ have the same leading term. It follows that

$$f_2 = f_1 - c_1 g_1 = f - cg - c_1 g_1$$

has a strictly smaller leading term. Continuing in this way, we get polynomials

$$f, \quad f_1 = f - cg, \quad f_2 = f - cg - c_1 g_1, \quad f_3 = f - cg - c_1 g_1 - c_2 g_2, \ldots,$$

where at each stage the leading term gets strictly smaller according to the order defined in (2.5). This process will terminate if we find some m with $f_m = 0$, for the zero polynomial has no leading term. If, on the other hand, we never had $f_m = 0$, then the above would give an infinite sequence of nonzero polynomials with strictly decreasing leading terms. But we showed above that there are only finitely many monomials strictly smaller than the leading term of f. Hence the above process must terminate.

However, once we have $f_m = 0$ for some m, we obtain

$$f = cg + c_1 g_1 + \cdots + c_{m-1} g_{m-1}$$

since $f_m = f - cg - c_1 g_1 - \cdots - c_{m-1} g_{m-1}$. Each g_i is a product of the σ_j to various powers, which proves that f is a polynomial in the elementary symmetric polynomials. This completes the proof. □

In Theorem 2.2.7 below, we will prove that the expression of f as a polynomial in $\sigma_1, \ldots, \sigma_n$ is unique.

The proof of Theorem 2.2.2 can be turned into an algorithm for writing a given symmetric polynomial in terms of the σ_i. For this purpose, we will use the notation

$$\sum_n x_1^{a_1} \cdots x_n^{a_n}$$

to denote the sum of all *distinct* monomials obtained from $x_1^{a_1} \cdots x_n^{a_n}$ by permuting $x_1, \ldots, x_n$. Here are some simple examples.

Example 2.2.3. One easily sees that

$$\sum_2 x_1^2 x_2 = x_1^2 x_2 + x_2^2 x_1$$

and

$$\sum_3 x_1^2 x_2 = x_1^2 x_2 + x_2^2 x_1 + x_1^2 x_3 + x_3^2 x_1 + x_2^2 x_3 + x_3^2 x_2.$$

Also, $\sum_4 x_1^2 x_2$ has 12 terms instead of 24. This is because $x_1^2 x_2 = x_1^2 x_2 x_3^0 x_4^0$ for $n = 4$. Switching the last two variables gives the same monomial, yet $\sum_4 x_1^2 x_2$ uses only the *distinct* monomials we get by permuting the variables. ◁▷

If $c\, x_1^{a_1} \cdots x_n^{a_n}$ is a term of a symmetric polynomial $f \in F[x_1, \ldots, x_n]$, then

$$f = c \sum_n x_1^{a_1} \cdots x_n^{a_n} + \text{a sum of terms involving monomials}$$

(2.12)

$$\text{different from those in } \sum_n x_1^{a_1} \cdots x_n^{a_n}.$$

Do you see how we used this fact in the proof of Theorem 2.2.2?

Here is an example of how to write a symmetric polynomial in terms of the σ_i.

Example 2.2.4. The polynomial in x_1, x_2, x_3, x_4 given by

$$f = \sum_4 x_1^3 x_2^2 x_3,$$

has 24 terms and is symmetric. In the first chapter of his 1782 book *Meditationes Algebraicæ* [6], Edward Waring shows how to express f in terms of $\sigma_1, \sigma_2, \sigma_3, \sigma_4$. His method is similar to what we did in the proof of Theorem 2.2.2. In this case, we proceed as follows (you will supply the details in Exercise 3):

Step 1. The leading term of f is $x_1^3 x_2^2 x_3 = x_1^3 x_2^2 x_3^1 x_4^0$, so that (2.10) becomes

$$\sigma_1^{3-2} \sigma_2^{2-1} \sigma_3^{1-0} \sigma_4^0 = \sigma_1 \sigma_2 \sigma_3.$$

Furthermore, one can use a computer to show that

(2.13) $\quad \sigma_1 \sigma_2 \sigma_3 = \sum_4 x_1^3 x_2^2 x_3 + 3 \sum_4 x_1^3 x_2 x_3 x_4 + 3 \sum_4 x_1^2 x_2^2 x_3^2 + 8 \sum_4 x_1^2 x_2^2 x_3 x_4.$

Hence

$$f_1 = f - \sigma_1 \sigma_2 \sigma_3 = -3 \sum_4 x_1^3 x_2 x_3 x_4 - 3 \sum_4 x_1^2 x_2^2 x_3^2 - 8 \sum_4 x_1^2 x_2^2 x_3 x_4.$$

Step 2. The leading term of f_1 is $-3 x_1^3 x_2 x_3 x_4$, which gives

(2.14) $\quad \sigma_1^{3-1} \sigma_2^{1-1} \sigma_3^{1-1} \sigma_4^1 = \sigma_1^2 \sigma_4 = \sum_4 x_1^3 x_2 x_3 x_4 + 2 \sum_4 x_1^2 x_2^2 x_3 x_4.$

Thus

$$f_2 = f - \sigma_1\sigma_2\sigma_3 + 3\sigma_1^2\sigma_4 = -3\sum_4 x_1^2 x_2^2 x_3^2 - 2\sum_4 x_1^2 x_2^2 x_3 x_4.$$

Step 3. For f_2, we have $-3x_1^2 x_2^2 x_3^2$ as leading term. Since

(2.15) $$\sigma_3^2 = \sum_4 x_1^2 x_2^2 x_3^2 + 2\sum_4 x_1^2 x_2^2 x_3 x_4,$$

we obtain

$$f_3 = f - \sigma_1\sigma_2\sigma_3 + 3\sigma_1^2\sigma_4 + 3\sigma_3^2 = 4\sum_4 x_1^2 x_2^2 x_3 x_4.$$

Step 4. The leading term of f_3 is $4x_1^2 x_2^2 x_3 x_4$, and from

(2.16) $$\sigma_2\sigma_4 = \sum_4 x_1^2 x_2^2 x_3 x_4,$$

we see that

$$f_4 = f - \sigma_1\sigma_2\sigma_3 + 3\sigma_1^2\sigma_4 + 3\sigma_3^2 - 4\sigma_2\sigma_4 = 0.$$

Conclusion. Since $f_4 = 0$, the process terminates and we obtain the formula

$$f = \sigma_1\sigma_2\sigma_3 - 3\sigma_1^2\sigma_4 - 3\sigma_3^2 + 4\sigma_2\sigma_4$$

expressing f in terms of the elementary symmetric polynomials.   ◁▷

In the exercises you will apply these methods to a variety of problems dealing with symmetric polynomials. For readers interested in doing more substantial problems, Section 2.3 will explain how to compute with symmetric polynomials using *Maple* and *Mathematica*.

B. The Roots of a Polynomial. In Galois theory, symmetric polynomials are often evaluated at the roots $\alpha_1, \ldots, \alpha_n$ of a polynomial $f \in F[x]$. The following result will be crucial.

Corollary 2.2.5. *Let $f \in F[x]$ be a monic polynomial of degree $n > 0$ with roots $\alpha_1, \ldots, \alpha_n$ in a larger field L. Then, given any symmetric polynomial $p(x_1, \ldots, x_n)$ with coefficients in F, we have*

$$p(\alpha_1, \ldots, \alpha_n) \in F.$$

Proof. The evaluation map $F[x_1, \ldots, x_n] \to L$ defined by $p \mapsto p(\alpha_1, \ldots, \alpha_n)$ is a ring homomorphism by Theorem 2.1.2.

Since p is symmetric in $x_1, \ldots, x_n$, Theorem 2.2.2 implies that p is a polynomial in the σ_r with coefficients in F. Hence, when we evaluate at $\alpha_1, \ldots, \alpha_n$, we see that $p(\alpha_1, \ldots, \alpha_n)$ is a polynomial in the $\sigma_r(\alpha_1, \ldots, \alpha_n)$ with coefficients in F.

Corollary 2.1.5 tells us that $\sigma_r(\alpha_1, \ldots, \alpha_n)$ is, up to sign, a coefficient of f. Since $f \in F[x]$ by hypothesis, we conclude that $\sigma_r(\alpha_1, \ldots, \alpha_n) \in F$. The corollary now follows immediately from the previous paragraph.   □

Here is an example of how Corollary 2.2.5 works.

Example 2.2.6. Suppose that $f = x^3 + 2x^2 + x + 7 \in \mathbb{Q}[x]$ has roots $\alpha_1, \alpha_2, \alpha_3 \in \mathbb{C}$. Let g be the monic polynomial whose roots are $\alpha_1 + \alpha_2$, $\alpha_1 + \alpha_3$, and $\alpha_2 + \alpha_3$. We claim that g has coefficients in $\mathbb{Q}$. To prove this, note that g can be written

$$
\begin{aligned}
(2.17) \quad g(x) &= \left(x - (\alpha_1 + \alpha_2)\right)\left(x - (\alpha_1 + \alpha_3)\right)\left(x - (\alpha_2 + \alpha_3)\right) \\
&= x^3 - (2\alpha_1 + 2\alpha_2 + 2\alpha_3)\, x^2 \\
&\quad + (\alpha_1^2 + \alpha_2^2 + \alpha_3^2 + 3\alpha_1\alpha_2 + 3\alpha_1\alpha_3 + 3\alpha_2\alpha_3)\, x \\
&\quad - (\alpha_1 + \alpha_2)(\alpha_1 + \alpha_3)(\alpha_2 + \alpha_3).
\end{aligned}
$$

The coefficients of (2.17) are symmetric polynomials evaluated at $\alpha_1, \alpha_2, \alpha_3$. Since the α_i are the roots of a polynomial with coefficients in $\mathbb{Q}$, Corollary 2.2.5 implies that the coefficients of g are in $\mathbb{Q}$. Hence $g \in \mathbb{Q}[x]$.

We can also determine g explicitly. In general, if $f = x^3 + bx^2 + cx + d$ has roots $\alpha_1, \alpha_2, \alpha_3$, and g is the polynomial with roots $\alpha_1 + \alpha_2$, $\alpha_1 + \alpha_3$, and $\alpha_2 + \alpha_3$, as defined in (2.17), then the techniques of this section imply that

$$
g(x) = x^3 + 2b\, x^2 + (b^2 + c)\, x + bc - d.
$$

(see Exercise 4 for the details). For $f = x^3 + 2x^2 + x + 7$, it follows that

$$
g(x) = x^3 + 2 \cdot 2\, x^2 + (2^2 + 1)\, x + 2 \cdot 1 - 7 = x^3 + 4x^2 + 5x - 5
$$

is the polynomial whose roots are the sums of distinct pairs of roots of f. ◁▷

C. Uniqueness. Every symmetric polynomial in $x_1, \ldots, x_n$ is can be written in terms of $\sigma_1, \ldots, \sigma_n$ by Theorem 2.2.2. We now prove that this expression is unique.

Theorem 2.2.7. *A given symmetric polynomial can be expressed as a polynomial in the elementary symmetric polynomials in only one way.*

Proof. We will use the polynomial ring $F[u_1, \ldots, u_n]$, where $u_1, \ldots, u_n$ are new variables. By Theorem 2.1.2, the map sending u_i to $\sigma_i \in F[x_1, \ldots, x_n]$ defines a ring homomorphism

$$
\varphi : F[u_1, \ldots, u_n] \longrightarrow F[x_1, \ldots, x_n].
$$

In other words, if $h = h(u_1, \ldots, u_n)$ is a polynomial in $u_1, \ldots, u_n$ with coefficients in F, then $\varphi(h) = h(\sigma_1, \ldots, \sigma_n)$.

The image of φ is the set of all polynomials in the σ_i with coefficients in F. We denote this image by

$$
F[\sigma_1, \ldots, \sigma_n] \subset F[x_1, \ldots, x_n].
$$

Note that $F[\sigma_1, \ldots, \sigma_n]$ is a subring of $F[x_1, \ldots, x_n]$. In this notation, we can write φ as a map

$$
(2.18) \qquad \varphi : F[u_1, \ldots, u_n] \longrightarrow F[\sigma_1, \ldots, \sigma_n].
$$

This map is onto by the definition of $F[\sigma_1, \ldots, \sigma_n]$, and uniqueness will be proved by showing that φ is one to one. Be sure you understand this.

To prove that φ is one to one, it suffices to show that its kernel is $\{0\}$. Thus we must show that if h is a nonzero polynomial in the u_i, then $h(\sigma_1, \ldots, \sigma_n)$ gives a nonzero polynomial in the x_i. We will sketch the main idea of the argument and leave the details for Exercise 5.

Let $c \, u_1^{b_1} \cdots u_n^{b_n}$ be a nonzero term of h. Applying φ gives $c \, \sigma_1^{b_1} \cdots \sigma_n^{b_n}$, and the argument of (2.11) shows that the leading term of this polynomial is

$$c \, x_1^{b_1 + \cdots + b_n} x_2^{b_2 + \cdots + b_n} \cdots x_n^{b_n}.$$

Since h is the sum of its terms, $\varphi(h)$ is the sum of the corresponding polynomials $c \, \sigma_1^{b_1} \cdots \sigma_n^{b_n}$, each of which has a leading term as displayed above. The crucial fact is that the map

$$(b_1, b_2, \ldots, b_n) \longmapsto (b_1 + \cdots + b_n, b_2 + \cdots + b_n, \ldots, b_n)$$

is one to one, so that the leading terms can't all cancel. Hence $\varphi(h)$ can't be the zero polynomial, and uniqueness follows. See Exercise 5 for the details. $\square$

The proof of Theorem 2.2.7 constructs a ring isomorphism

$$(2.19) \qquad \varphi : F[u_1, \ldots, u_n] \simeq F[\sigma_1, \ldots, \sigma_n],$$

where $F[u_1, \ldots, u_n]$ is the polynomial ring in variables $u_1, \ldots, u_n$ and $u_i \mapsto \sigma_i$. Hence we can regard $\sigma_1, \ldots, \sigma_n$ as independent variables. This leads to the following interesting application.

Using the above variables $u_1, \ldots, u_n$, we call

$$(2.20) \qquad \tilde{f} = x^n - u_1 x^{n-1} + \cdots + (-1)^{n-1} u_{n-1} x + (-1)^n u_n$$

the *universal polynomial of degree n* (the reason for the signs will soon become clear). This name is justified because if $f = x^n + a_1 x^{n-1} + \cdots + a_1 x + a_n \in F[x]$ is *any* monic polynomial of degree n, then the evaluation map sending u_i to $(-1)^i a_i$ takes $\tilde{f}$ to f. Thus the universal polynomial of degree n can be mapped to any monic polynomial of degree n with coefficients in F.

We can construct the roots of $\tilde{f}$ as follows. Under the isomorphism (2.19), the polynomial (2.20) maps to $x^n - \sigma_1 x^{n-1} + \cdots + (-1)^n \sigma_n$. But $F[\sigma_1, \ldots, \sigma_n]$ lies in the larger ring $F[x_1, \ldots, x_n]$, and in this ring, (2.4) gives the factorization

$$x^n - \sigma_1 x^{n-1} + \cdots + (-1)^{n-1} \sigma_{n-1} x + (-1)^n \sigma_n = (x - x_1) \cdots (x - x_n).$$

In other words, $x^n - \sigma_1 x^{n-1} + \cdots + (-1)^n \sigma_n$ has roots $x_1, \ldots, x_n$.

Because of this, we identify (2.20) with its image under (2.19) and call

$$(2.21) \qquad \tilde{f} = x^n - \sigma_1 x^{n-1} + \cdots + (-1)^{n-1} \sigma_{n-1} x + (-1)^n \sigma_n$$

the universal polynomial of degree n. Then $\tilde{f}$ is not only universal in the above sense but also has known roots, namely $x_1, \ldots, x_n$.

As mentioned in the Historical Notes to Section 1.2, Lagrange studied the roots of a polynomial without regard to their numerical value. For a monic polynomial of degree $n > 0$, this means considering its roots as variables $x_1, \ldots, x_n$. The above discussion shows that in modern terms, Lagrange was studying the universal polynomial $\tilde{f}$.

Mathematical Notes

Let us discuss further two ideas that appeared in this section.

■ **Algebraic Independence.** The uniqueness proved in Theorem 2.2.7 implies in particular that the map (2.19) is one to one. Hence there are no nontrivial polynomial relations among the σ_i (since any such relation would give a nonzero element in the kernel). When this happens, we say that $\sigma_1, \ldots, \sigma_n$ are *algebraically independent*. Not all collections of polynomials in $F[x_1, \ldots, x_n]$ are algebraically independent. See Exercise 6 for an example.

■ **Symmetric Rational Functions.** The polynomial ring $F[x_1, \ldots, x_n]$ sits inside $F(x_1, \ldots, x_n)$, the field of rational functions in $x_1, \ldots, x_n$ with coefficients in F. In this situation, one can ask which elements of $F(x_1, \ldots, x_n)$ are *symmetric*, that is, are unchanged under all permutations of the variables. An example is

$$\frac{1}{x_1} + \frac{1}{x_2} + \cdots + \frac{1}{x_n}.$$

Using a common denominator, one can express this as

$$\frac{x_2 \cdots x_n + x_1 x_3 \cdots x_n + \cdots + x_1 \cdots x_{n-1}}{x_1 \cdots x_n} = \frac{\sigma_{n-1}}{\sigma_n}.$$

More generally, one can show that any symmetric rational function in $x_1, \ldots, x_n$ is a rational function in the elementary symmetric polynomials. In other words, all symmetric elements of $F(x_1, \ldots, x_n)$ lie in the subfield $F(\sigma_1, \ldots, \sigma_n)$ of rational functions in $\sigma_1, \ldots, \sigma_n$. This will be proved in Exercises 7 and 8.

In Chapters 6, 7, and 8, we will study

$$F(\sigma_1, \ldots, \sigma_n) \subset F(x_1, \ldots, x_n)$$

from the point of view of Galois theory. We will see that the Galois group of this field extension is the symmetric group S_n. This in turn will enable us to determine when one can solve polynomials of degree n by radicals.

Historical Notes

Symmetric polynomials have been around for a long time. In 1629 Albert Girard published *Invention nouvelle en l'algebre*, which contains a clear description of the elementary symmetric polynomials. Girard also considers the *power sums*

$$s_r = x_1^r + \cdots + x_n^r.$$

In the notation used above, note that $s_r = \sum_n x_1^r$. Girard gives formulas for s_1, s_2, s_3, s_4 in terms of the σ_i (see Exercise 17).

In 1665–1666 Isaac Newton worked out many examples of symmetric polynomials, expressing them in terms of the σ_i. His 1707 book *Arithmetica universalis* shows how power sums relate to elementary symmetric polynomials. For $r = 1$, the relation is trivial, namely $s_1 = \sigma_1$, and for $r > 1$, we have the *Newton identities*, which state that

$$
\begin{aligned}
&s_r = \sigma_1 s_{r-1} - \sigma_2 s_{r-2} + \cdots + (-1)^{r-1} r \sigma_r && \text{if } 1 < r \leq n, \\
&s_r = \sigma_1 s_{r-1} - \sigma_2 s_{r-2} + \cdots + (-1)^{r-1} \sigma_n s_{r-n} && \text{if } r > n.
\end{aligned}
$$

(2.22)

Proofs of these identities can be found in [2, Ch. 7, §1], [3, pp. 62–63, 72–73], and [7, pp. 114–115].

As already noted, Waring's *Meditationes algebraicæ* from 1782 contains an implicit version of the algorithm used in the proof of Theorem 2.2.2, though in examples he often used clever shortcuts. The Fundamental Theorem of Symmetric Polynomials was widely known and used in the eighteenth century, though the first complete proof is due to Gauss. He was also the first to raise the issue of uniqueness, and his proof is the one we used.

One difference between Gauss's proof of Theorem 2.2.2 and ours is that he ordered his polynomials differently. In [Gauss, Vol. III, p. 36], he says:

Dein e duobus terminis

$$
M a^\alpha b^\beta c^\gamma \cdots \text{ et } M a^{\alpha'} b^{\beta'} c^{\gamma'} \cdots
$$

priori ordinem altiorem tribuemus quam posteriori, si fit

$$
\text{vel } \alpha > \alpha', \text{ vel } \alpha = \alpha', \ \beta > \beta', \text{ vel } \alpha = \alpha', \ \beta = \beta', \gamma > \gamma', \text{ vel etc.}
$$

i.e. si e differentiis $\alpha - \alpha', \beta - \beta', \gamma - \gamma'$ etc. prima, quae non evanescit, positiva evadit.

Even though this is in Latin, the meaning is quite clear once one realizes that "vel … vel … vel" means "either … or … or." This is now called *lexicographic order*. In Exercise 9 you will use this order to prove Theorem 2.2.2.

Although our interest in symmetric polynomials is due to their importance in Galois theory, these polynomials also arise naturally in invariant theory, algebraic combinatorics, and representation theory. A basic reference for symmetric polynomials is [4]. See also [Tignol, Chs. 4, 8] for the history of symmetric polynomials.

Exercises for Section 2.2

Exercise 1. Show that the leading term of σ_r is $x_1 x_2 \cdots x_r$.

Exercise 2. This exercise will study the order relation defined in (2.5). Given an *exponent vector* $\alpha = (a_1, \ldots, a_n)$, where each $a_i \geq 0$ is an integer, let x^α denote the monomial

$$
x^\alpha = x_1^{a_1} \cdots x_n^{a_n}.
$$

If α and β are exponent vectors, note that $x^\alpha x^\beta = x^{\alpha+\beta}$. Also, the leading term of a nonzero polynomial $f \in F[x_1, \ldots, x_n]$ will be denoted $\mathrm{LT}(f)$.
   (a) Suppose that $x^\alpha > x^\beta$, and let x^γ be any monomial. Prove that $x^{\alpha+\gamma} > x^{\beta+\gamma}$.
   (b) Suppose that $x^\alpha > x^\beta$ and $x^\gamma > x^\delta$. Prove that $x^{\alpha+\gamma} > x^{\beta+\delta}$.
   (c) Let $f, g \in F[x_1, \ldots, x_n]$ be nonzero. Prove that $\mathrm{LT}(fg) = \mathrm{LT}(f)\mathrm{LT}(g)$.

Exercise 3. Prove (2.13)–(2.16). For (2.13), a computer will be helpful; the others can be proved by hand using the identity

$$(y_1 + \cdots + y_m)^2 = y_1^2 + \cdots + y_m^2 + 2\sum_{i<j} y_i y_j.$$

Exercise 4. Let $f = x^3 + bx^2 + cx + d \in F[x]$ have roots $\alpha_1, \alpha_2, \alpha_3$ in a field L containing F, and let g be the polynomial defined in (2.17). Show carefully that

$$g(x) = x^3 + 2b\,x^2 + (b^2 + c)\,x + bc - d.$$

Exercise 5. This exercise will complete the proof of Theorem 2.2.7. Let $h \in F[u_1, \ldots, u_n]$ be a nonzero polynomial. The goal is to prove that $h(\sigma_1, \ldots, \sigma_n)$ is not the zero polynomial in $x_1, \ldots, x_n$.
   (a) If $c\,u_1^{b_1} \cdots u_n^{b_n}$ is a term of h, then use Exercise 2 to show that the leading term of $c\,\sigma_1^{b_1} \cdots \sigma_n^{b_n}$ is $c\,x_1^{b_1+\cdots+b_n} x_2^{b_2+\cdots+b_n} \cdots x_n^{b_n}$.
   (b) Show that $(b_1, \ldots, b_n) \mapsto (b_1 + \cdots + b_n, b_2 + \cdots + b_n, \ldots, b_n)$ is one to one.
   (c) To see why $h(\sigma_1, \ldots, \sigma_n)$ is nonzero, consider the term of $h(u_1, \ldots, u_n)$ for which the leading term of $c\,\sigma_1^{b_1} \cdots \sigma_n^{b_n}$ is maximal. Prove that this leading term is in fact the leading term of $h(\sigma_1, \ldots, \sigma_n)$, and explain how this proves what we want.

Exercise 6. Here is an example of polynomials which are not algebraically independent. Consider $x_1^2, x_1 x_2, x_2^2 \in F[x_1, x_2]$, and let $\phi : F[u_1, u_2, u_3] \to F[x_1, x_2]$ be defined by

$$\phi(u_1) = x_1^2, \quad \phi(u_2) = x_1 x_2, \quad \phi(u_3) = x_2^2.$$

Show that ϕ is *not* one to one by finding a nonzero polynomial $h \in F[u_1, u_2, u_3]$ such that $\phi(h) = 0$. (Using the notion of *transcendence degree*, one can show that *any* collection of three or more elements in $F[x_1, x_2]$ is algebraically dependent. See, for example, [Jacobson, Vol. II, Sec. 8.12].)

Exercise 7. Given a polynomial $f \in F[x_1, \ldots, x_n]$ and a permutation $\sigma \in S_n$, let $\sigma \cdot f$ denote the polynomial obtained from f by permuting the variables according to σ. Show that $\prod_{\sigma \in S_n} \sigma \cdot f$ and $\sum_{\sigma \in S_n} \sigma \cdot f$ are symmetric polynomials.

Exercise 8. In this exercise, you will prove that if $f \in F(x_1, \ldots, x_n)$ is symmetric, then f is a rational function in $\sigma_1, \ldots, \sigma_n$ with coefficients in F. To begin the proof, we know that $f = A/B$, where A and B are in $F[x_1, \ldots, x_n]$. Note that A and B need not be symmetric; only their quotient $f = A/B$ is. Let

$$C = \prod_{\sigma \in S_n \setminus \{e\}} \sigma \cdot B,$$

where we are using the notation of Exercise 7.
   (a) Use Exercise 7 to show that BC is a symmetric polynomial.
   (b) Then use the symmetry of $f = A/B$ to show that AC is a symmetric polynomial.

(c) Use $f = (AC)/(BC)$ and Theorem 2.2.2 to conclude that f is a rational function in the elementary symmetric polynomials with coefficients in F.

Exercise 9. In the Historical Notes, we gave Gauss's definition of lexicographic order.

(a) Give a definition (in English) of lexicographic order.

(b) In the proof of Theorem 2.2.2, we showed that graded lexicographic order has the property that there are only finitely many monomials less than a given monomial. In contrast, this property fails for lexicographic order. Give an explicit example to illustrate this.

(c) In spite of part (b), lexicographic order does have an interesting finiteness property. Namely, prove that there is no infinite sequence of polynomials $f_1, f_2, f_3, \ldots$ that have strictly decreasing leading terms according to lexicographic order.

(d) Explain how part (c) allows one to prove Theorem 2.2.2 using lexicographic order.

Besides graded lexicographic order and lexicographic order, there are many other ways to order monomials. See [2, Ch. 2, §2].

Exercise 10. Apply the proof of Theorem 2.2.2 to express $\sum_3 x_1^2 x_2$ in terms of $\sigma_1, \sigma_2, \sigma_3$.

Exercise 11. Let the roots of $y^3 + 2y^2 - 3y + 5$ be $\alpha, \beta, \gamma \in \mathbb{C}$. Find polynomials with integer coefficients that have the following roots:

(a) $\alpha\beta$, $\alpha\gamma$, and $\beta\gamma$.

(b) $\alpha + 1$, $\beta + 1$, and $\gamma + 1$.

(c) α^2, β^2, and γ^2.

Exercise 12. Consider the symmetric polynomial $f = \sum_n x_1^{a_1} \cdots x_n^{a_n}$.

(a) Prove that f has $n!$ terms when $a_1, \ldots, a_n$ are distinct.

(b) (More challenging) Suppose that the exponents $a_1, \ldots, a_n$ break up into r disjoint groups so that exponents within the same group are equal, but exponents from different groups are unequal. Let ℓ_i denote the number of elements in the ith group, so that $\ell_1 + \cdots + \ell_r = n$. Prove that the number of terms in f is

$$\frac{n!}{\ell_1! \cdots \ell_r!}.$$

For example, $f = \sum_5 x_1^3 x_2^3 x_3^2 x_4^2 x_5$ has $\ell_1 = \ell_2 = 2$ and $\ell_3 = 1$. It follows that f has $5!/(2!2!1!) = 30$ terms.

Exercises 13–16 will discuss some classic tricks for dealing with symmetric polynomials. A polynomial $g \in F[x_1, \ldots, x_n]$ is *homogeneous of total degree* d if every nonzero term of g has total degree d.

Exercise 13. Let $g_1, g_2 \in F[x_1, \ldots, x_n]$ be homogeneous of total degrees d_1, d_2.

(a) Show that $g_1 g_2$ is homogeneous of total degree $d_1 + d_2$.

(b) When is $g_1 + g_2$ homogeneous?

Exercise 14. We define the *weight* of $\sigma_1^{a_1} \cdots \sigma_n^{a_n}$ to be $a_1 + 2a_2 + 3a_3 + \cdots + na_n$.

(a) Prove that $\sigma_1^{a_1} \cdots \sigma_n^{a_n}$ is homogeneous and that its weight is the same as its total degree when considered as a polynomial in $x_1, \ldots, x_n$.

(b) Let $f \in F[x_1, \ldots, x_n]$ be symmetric and homogeneous of total degree d. Show that f is a linear combination of products $\sigma_1^{a_1} \cdots \sigma_n^{a_n}$ of weight d.

Exercise 15. Given a polynomial $f \in F[x_1, \ldots, x_n]$, let $\deg_i(f)$ be the maximal exponent of x_i which appears in f. Thus $f = x_1^3 x_2 + x_1 x_2^4$ has $\deg_1(f) = 3$ and $\deg_2(f) = 4$.

(a) If f is symmetric, explain why the $\deg_i(f)$ are the same for $i = 1, \ldots, n$.
(b) Show that $\deg_i(\sigma_1^{a_1} \cdots \sigma_n^{a_n}) = a_1 + a_2 + \cdots + a_n$ for $i = 1, \ldots, n$.

Exercise 16. This exercise is based on [7, pp. 110–112] and will express the discriminant $\Delta = (x_1 - x_2)^2 (x_1 - x_3)^2 (x_2 - x_3)^2$ in terms of the elementary symmetric functions without using a computer. We will use the terminology of Exercises 14 and 15. Note that Δ is homogeneous of total degree 6 and $\deg_i(\Delta) = 4$ for $i = 1, 2, 3$.
(a) Find all products $\sigma_1^{a_1} \sigma_2^{a_2} \sigma_3^{a_3}$ of weight 6 and $\deg_i(\sigma_1^{a_1} \sigma_2^{a_2} \sigma_3^{a_3}) \le 4$.
(b) Explain how part (a) implies that there are constants $\ell_1, \ldots, \ell_5$ such that

$$\Delta = \ell_1 \sigma_3^2 + \ell_2 \sigma_1 \sigma_2 \sigma_3 + \ell_3 \sigma_1^3 \sigma_3 + \ell_4 \sigma_2^3 + \ell_5 \sigma_1^2 \sigma_2^2.$$

(c) We will compute the ℓ_i by using the universal property of the elementary symmetric polynomials. For example, to determine ℓ_1, use the roots $1, \omega, \omega^2$ to show that $x^3 - 1$ has discriminant -27. By applying the ring homomorphism defined by $x_1 \mapsto 1, x_2 \mapsto \omega, x_3 \mapsto \omega^2$ to part (b), conclude that $\ell_1 = -27$.
(d) Show that $x^3 - x$ has roots $0, \pm 1$ and discriminant 4. By adapting the argument of part (c), conclude that $\ell_4 = -4$.
(e) Similarly, use $x^3 - 2x^2 + x$ to show that $\ell_5 = 1$.
(f) Next, note that $x^3 - 2x^2 - x + 2$ has roots $\pm 1, 2$, and use this (together with the known values of ℓ_1, ℓ_4, ℓ_5) to conclude that $\ell_2 - 4\ell_3 = 34$.
(g) Finally, use $x^3 - 3x^2 + 3x - 1$ to show $\ell_2 + 3\ell_3 = 6$. Using part (f), this implies $\ell_2 = 18, \ell_3 = -4$ and gives the usual formula for Δ.
Other examples illustrating this method can be found in [1, pp. 442–444].

Exercises 17–20 will study power sums $s_r = x_1^r + \cdots + x_n^r$ and the Newton identities (2.22) discussed in the Historical Notes.

Exercise 17. Use the Newton identities (2.22) to express the power sums s_2, s_3, s_4 in terms of the elementary symmetric polynomials $\sigma_1, \sigma_2, \sigma_3, \sigma_4$.

Exercise 18. Suppose that complex numbers α, β, γ satisfy the equations

$$\alpha + \beta + \gamma = 3,$$
$$\alpha^2 + \beta^2 + \gamma^2 = 5,$$
$$\alpha^3 + \beta^3 + \gamma^3 = 12.$$

Show that $\alpha^n + \beta^n + \gamma^n \in \mathbb{Z}$ for all $n \ge 4$. Also compute $\alpha^4 + \beta^4 + \gamma^4$.

Exercise 19. Suppose that F is a field of characteristic 0.
(a) Use the Newton identities (2.22) and Theorem 2.2.2 to prove that every symmetric polynomial in $F[x_1, \ldots, x_n]$ can be expressed as a polynomial in $s_1, \ldots, s_n$.
(b) Show how to express $\sigma_4 \in F[x_1, x_2, x_3, x_4]$ as a polynomial in s_1, s_2, s_3, s_4.

Exercise 20. Let $\mathbb{F}_2$ be the field with two elements. Show that in $\mathbb{F}_2[x_1, \ldots, x_n]$, it is impossible to express σ_2 as a polynomial in $s_1, \ldots, s_n$ when $n \ge 2$.

2.3   COMPUTING WITH SYMMETRIC POLYNOMIALS (OPTIONAL)

The method described in Section 2.2 for expressing a given symmetric polynomial in terms of $\sigma_1, \ldots, \sigma_n$ is useful for simple problems, but can be cumbersome in more complicated situations. Fortunately, computer algebra programs such as *Maple* or *Mathematica* make it relatively easy to represent symmetric functions in terms of the elementary symmetric polynomials. We will discuss briefly how these powerful programs can be used to manipulate symmetric polynomials.

Although few readers will have access to both *Maple* and *Mathematica*, we suggest reading both discussions in order to better appreciate the underlying ideas.

A. Using *Mathematica*. We begin by using *Mathematica* to write the discriminant

$$\Delta = (x_1 - x_2)^2 (x_1 - x_3)^2 (x_2 - x_3)^2$$

from Section 1.2 in terms of the elementary symmetric polynomials. We can think of this in terms of the system of equations

(2.23)
$$\begin{aligned}
\Delta &= (x_1 - x_2)^2 (x_1 - x_3)^2 (x_2 - x_3)^2, \\
\sigma_1 &= x_1 + x_2 + x_3, \\
\sigma_2 &= x_1 x_2 + x_1 x_3 + x_2 x_3, \\
\sigma_3 &= x_1 x_2 x_3.
\end{aligned}$$

The idea is to *eliminate* x_1, x_2, x_3 from these equations. This will give the desired expression for Δ in terms of $\sigma_1, \sigma_2, \sigma_3$.

We tell *Mathematica* to do this elimination using the command

```
Eliminate[ {Delta == (x1-x2)^2(x1-x3)^2(x2-x3)^2, e1 ==
x1+x2+x3, e2 == x1 x2+x1 x3+x2 x3, e3 == x1 x2 x3}, {x1, x2, x3}]
```

The output is

$$-e1^2 e2^2 + 4\,e1^3 e3 - 18\,e1\,e2\,e3 == -Delta - 4\,e2^3 - 27\,e3^2,$$

which tells us that

$$\Delta = -4\sigma_2^3 - 27\sigma_3^2 + \sigma_1^2 \sigma_2^2 - 4\sigma_1^3 \sigma_3 + 18\sigma_1 \sigma_2 \sigma_3.$$

This agrees with (1.20) from Section 1.2 after the substitution $b = -\sigma_1, c = \sigma_2$, and $d = -\sigma_3$.

Using the Eliminate command is straightforward, though having to enter the elementary symmetric polynomials by hand can be time-consuming, especially when the number of variables is large. This can be avoided by using the *Mathematica* package SymmetricPolynomials that comes with the program. This package is loaded by

<< Algebra`SymmetricPolynomials`

Then the above computation can be done using the command

```
SymmetricReduction[ (x1-x2)^2(x1-x3)^2(x2-x3)^2,
{x1, x2, x3}, {e1, e2, e3}]
```

The output is the two-element list

$$\{\texttt{e1}^2\texttt{e2}^2 - 4\,\texttt{e2}^3 - 4\,\texttt{e1}^3\texttt{e3} + 18\,\texttt{e1}\,\texttt{e2}\,\texttt{e3} - 27\,\texttt{e3}^2, 0\}$$

where the second element, 0, tells us that Δ is in fact symmetric, and the first element is the polynomial expressing Δ in terms of the σ_i. To go directly to the first element of this list, one could give the command

```
SymmetricReduction[ (x1-x2)^2(x1-x3)^2(x2-x3)^2,
{x1,x2,x3},{e1,e2,e3}][[1]]
```

since in general *Mathematica* uses "[[i]]" to extract the ith element of a list.

Here is an example which illustrates one of the interesting things which can be done with symmetric polynomials.

Example 2.3.1. Let $\alpha_1, \alpha_2, \alpha_3 \in \mathbb{C}$ be the roots of $y^3 + 2y^2 - 3y + 5$. Our goal is to use *Mathematica* to find the cubic polynomial whose roots are $\alpha_1^3, \alpha_2^3, \alpha_3^3$.

Let y_1, y_2, y_3 be variables and define the polynomial f in *Mathematica* to be

$$f = (y - y1^3)(y - y2^3)(y - y3^3)$$

Note that the evaluation $y_i \mapsto \alpha_i$ takes f to the polynomial we want. If we multiply out f, we get a polynomial whose coefficients are symmetric in y_1, y_2, y_3. We express these in terms of the elementary symmetric polynomials using the *Mathematica* command

```
Do[ Print[ SymmetricReduction[ Coefficient[ f, y, i ],
{y1,y2,y3},{e1,e2,e3}][[1]]], {i, 0, 2}]
```

This instructs *Mathematica* to print out the coefficients of f expressed in terms of the elementary symmetric polynomials, here denoted e1, e2, e3. The output is

(2.24)

constant term	:	$-\texttt{e3}^3,$
coefficient of y	:	$\texttt{e2}^3 - 3\,\texttt{e1}\,\texttt{e2}\,\texttt{e3} + 3\,\texttt{e3}^2,$
coefficient of y^2	:	$-\texttt{e1}^3 + 3\,\texttt{e1}\,\texttt{e2} - 3\,\texttt{e3}.$

The evaluation $y_i \mapsto \alpha_i$ sends e1 $\mapsto -2$, e2 $\mapsto -3$, e3 $\mapsto -5$. Using (2.24), we see that $y^3 + 41y^2 + 138y + 125$ is the polynomial with roots $\alpha_1^3, \alpha_2^3, \alpha_3^3$. ◁▷

The formulas (2.24) imply that for *any* cubic polynomial, we can find a cubic polynomial whose roots are the cubes of the given one. This is part of the universal aspect of the elementary symmetric polynomials.

B. Using *Maple*. As above, our first *Maple* computation will be to express the discriminant $\Delta = (x_1 - x_2)^2(x_1 - x_3)^2(x_2 - x_3)^2$ in terms of $\sigma_1, \sigma_2, \sigma_3$. We will again use the equations (2.23) to eliminate x_1, x_2, x_3, which will give the desired expression for Δ.

To do this in *Maple*, we proceed as follows. The last three lines of (2.23) give the polynomials

(2.25) $e_1 - x_1 - x_2 - x_3, \quad e_2 - x_1x_2 - x_1x_3 - x_2x_3, \quad e_3 - x_1x_2x_3$

in $\mathbb{C}[x_1, x_2, x_3, e_1, e_2, e_3]$. These generate an ideal in this ring, and we eliminate x_1, x_2, x_3 from Δ by replacing all instances of $x_1 + x_2 + x_3$, $x_1x_2 + x_1x_3 + x_2x_3$, $x_1x_2x_3$ with e_1, e_2, e_3 respectively. This operation can be thought of as the *normal form* of Δ with respect to the ideal generated by (2.25).

The first step is to load the *Maple* package Groebner, which contains the commands we need. This is done by

with(Groebner);

We next tell *Maple* to order the monomials in $\mathbb{C}[x_1, x_2, x_3, e_1, e_2, e_3]$ via

T := lexdeg([x1, x2, x3], [e1, e2, e3]);

The monomial order T is specially designed for elimination and is more efficient than the graded lexicographic order used in the proof of Theorem 2.2.2. We need to specify a monomial order because the precise definition of "normal form" depends on how the monomials are ordered.

Once we have the monomial order T, we compute an intermediate object called a *Gröbner basis* using the command

GB:= gbasis([e1-x1-x2-x3, e2-x1*x2-x1*x3-x2*x3,
e3-x1*x2*x3], T):

Note that *Maple* uses ∗ for multiplication. We also used : to suppress output, since we don't need to see the Gröbner basis. Roughly speaking, the Gröbner basis consists of polynomials that generate the same ideal as (2.25) and are optimized for the monomial order T.

The final step is to compute the normal form. This is precisely what the *Maple* command normalf does:

normalf((x1-x2)^2 * (x1-x3)^2 * (x2-x3)^2, GB, T);

This gives the output

$$-4 * e1^3 * e3 + 18\,e1 * e2 * e3 - 27 * e3^2 + e1^2 * e2^2 - 4 * e2^3$$

which agrees with our earlier computation.

The *Mathematica* command Eliminate described earlier uses a Gröbner basis computation similar to what we did here. Gröbner basis methods can be applied to a variety of elimination problems. The full details can be found in [2].

We should also mention that the Maple package symmpoly can be used to generate the elementary symmetric polynomials automatically. This package can be obtained from the web site

http://www.mapleapps.com/maplelinks/share/symmpoly.shtml

To use `symmpoly`, first load it into *Maple* with the command

$$\text{with(symmpoly);}$$

Then

$$\text{ES := symmpoly([x1, x2, x3]):}$$

generates a list with four elements: the first is the constant polynomial 1, and the second, third, and fourth are σ_1, σ_2, and σ_3 respectively. With this setup, the above Gröbner basis can be computed more efficiently via

$$\text{GB := gbasis([e1-ES[2], e2-ES[3], e3-ES[4]], T):}$$

since in general *Maple* uses "[i]" to extract the ith element of a list.

Notice that once the monomial order T is defined and the Gröbner basis GB is computed, `normalf` can be used repeatedly to write a series of symmetric polynomials in terms of the elementary symmetric polynomials.

Example 2.3.2. In Example 2.3.1, we used *Mathematica* to find a polynomial whose roots were the cubes of the roots $\alpha_1, \alpha_2, \alpha_3$ of $y^3 + 2y^2 - 3y + 5$. Let's redo this example using *Maple*. We first enter the polynomial

$$\text{f := (y - y1^3) * (y - y2^3) * (y - y3^3);}$$

Then the *Maple* command

$$\text{for i from 0 to 2 do print(normalf(coeff(f, y, i), GB, T)) od;}$$

prints out the coefficients of f expressed in terms of the elementary symmetric polynomials. As in (2.24), the result is

$$\text{constant term} \quad : -e3^3,$$
$$\text{coefficient of } y \quad : -3 \text{ e1 e2 e3} + 3 \text{ e3}^2 + e2^3,$$
$$\text{coefficient of } y^2 : -3 \text{ e3} + 3 \text{ e1 e2} - e1^3,$$

and the evaluation $e1 \mapsto -2$, $e2 \mapsto -3$, $e3 \mapsto -5$ shows that the roots of $y^3 + 41y^2 + 138y + 125$ are $\alpha_1^3, \alpha_2^3, \alpha_3^3$. ◁▷

Similar examples are given in the exercises.

Exercises for Section 2.3

Exercise 1. Examples 2.3.1 and 2.3.2 showed that the roots of $y^3 + 41y^2 + 138y + 125$ are the cubes of the roots of $y^3 + 2y^2 - 3y + 5$. Verify this numerically.

Exercise 2. Use the method of Example 2.3.1 or 2.3.2 to find the cubic polynomial whose roots are the fourth powers of the roots of the polynomial $y^3 + 2y^2 - 3y + 5$.

Exercise 3. Express $\sum_3 x_1^3 x_2^2$ in terms of the elementary symmetric polynomials. This example was first done by Newton around 1665.

Exercise 4. Given a cubic $x^3 + bx^2 + cx + d$, what condition must b, c, d satisfy in order that one root be the average of the other two?

Exercise 5. Given a quartic $x^4 + bx^3 + cx^2 + dx + e$, what condition must b, c, d, e satisfy in order that one root be the negative of another?

Exercise 6. Find the quartic polynomial whose roots are obtained by adding 1 to each of the roots of $x^4 + 3x^2 + 4x + 7$.

2.4   THE DISCRIMINANT

Given $n \geq 2$ variables $x_1, \ldots, x_n$ over a field F, the *discriminant* is

$$\Delta = \prod_{i<j} (x_i - x_j)^2 \in F[x_1, \ldots, x_n].$$

There are $\binom{n}{2} = \frac{1}{2}n(n-1)$ factors in this product. Furthermore, since $(x_i - x_j)^2 = -(x_i - x_j)(x_j - x_i)$, we can rewrite the above formula as

$$\Delta = (-1)^{\frac{1}{2}n(n-1)} \prod_{i \neq j} (x_i - x_j).$$

This shows that if we permute the variables, then we still have the product of the differences of all distinct pairs of variables. Thus Δ is symmetric in $x_1, \ldots, x_n$.

Theorem 2.2.2 implies that Δ can be written as a polynomial in the elementary symmetric polynomials $\sigma_1, \ldots, \sigma_n$. In other words,

$$\Delta \in F[\sigma_1, \ldots, \sigma_n].$$

When $n = 3$, the formulas of Section 1.2 (or the methods of Section 2.3) imply that

(2.26) $$\Delta = -4\sigma_2^3 - 27\sigma_3^2 + \sigma_1^2\sigma_2^2 - 4\sigma_1^3\sigma_3 + 18\sigma_1\sigma_2\sigma_3.$$

For general n an explicit formula for Δ in terms of $\sigma_1, \ldots, \sigma_n$ will be given in the Mathematical Notes.

The definition of Δ shows that it has a square root in $F[x_1, \ldots, x_n]$. We define

$$\sqrt{\Delta} = \prod_{i<j} (x_i - x_j).$$

We next describe how $\sqrt{\Delta}$ transforms under permutations.

Proposition 2.4.1. *If $\sigma \in S_n$, then*

$$\sigma \cdot \sqrt{\Delta} = \mathrm{sgn}(\sigma) \sqrt{\Delta},$$

where sgn(σ) *is from (A.3) and* $\sigma \cdot \sqrt{\Delta}$ *is the polynomial obtained from* $\sqrt{\Delta}$ *by permuting the variables* $x_1, \ldots, x_n$ *according to* σ.

Proof. In 1841, Jacobi studied how $\sqrt{\Delta}$ transforms under a transposition $(i \ j)$. His argument, adapted to our notation, goes as follows. We can assume $i < j$. Then observe that there is $\epsilon \in \{+1, -1\}$ such that

$$(2.27) \qquad \sqrt{\Delta} = \epsilon \, (x_i - x_j) \prod_{k \neq i, j} (x_i - x_k)(x_j - x_k) \prod_{\substack{l, m \neq i, j \\ l < m}} (x_l - x_m).$$

This follows because the factors appearing in the right-hand side are, up to sign, the factors of $\sqrt{\Delta}$. For example, when $k \neq i, j$, then

$$x_i - x_k = \begin{cases} x_i - x_k, & i < k, \\ -(x_k - x_i), & k < i. \end{cases}$$

Combining all of these signs gives $\epsilon = \pm 1$ in (2.27). Since the transposition $(i \ j)$ takes $(x_i - x_k)(x_j - x_k)$ to $(x_j - x_k)(x_i - x_k)$ and doesn't affect $x_l - x_m$ for $l, m \neq i, j$, we see that (2.27) implies that $(i \ j) \cdot \sqrt{\Delta} = -\sqrt{\Delta}$.

Now let $\sigma \in S_n$, and write σ as a product of transpositions, say $\sigma = \tau_1 \cdots \tau_\ell$. Then $\tau_i \cdot \sqrt{\Delta} = -\sqrt{\Delta}$ implies that

$$(2.28) \qquad \sigma \cdot \sqrt{\Delta} = (\tau_1 \cdots \tau_\ell) \cdot \sqrt{\Delta} = (-1)^\ell \sqrt{\Delta}.$$

Since sgn(σ) $= (-1)^\ell$ (be sure you understand why), the proposition follows. $\square$

We next define the discriminant of a monic polynomial

$$f = x^n + a_1 x^{n-1} + \cdots + a_i x^{n-i} + \cdots + a_n \in F[x]$$

of degree $n \geq 2$. As in Section 2.2, the universal polynomial

$$\tilde{f} = x^n - \sigma_1 x^{n-1} + \cdots + (-1)^i \sigma_i x^{n-i} + \cdots + (-1)^n \sigma_n$$

maps to f via $\sigma_i \mapsto (-1)^i a_i$. Since Δ is a symmetric polynomial, we can write

$$(2.29) \qquad \Delta = \Delta(\sigma_1, \ldots, \sigma_i, \ldots, \sigma_n) \in F[\sigma_1, \ldots, \sigma_n].$$

Then we define the *discriminant of* f, denoted $\Delta(f)$, to be

$$(2.30) \qquad \Delta(f) = \Delta(-a_1, \ldots, (-1)^i a_i, \ldots, (-1)^n a_n) \in F.$$

Thus the evaluation $\sigma_i \mapsto (-1)^i a_i$ takes $\tilde{f}$ to f and Δ to $\Delta(f)$.

We also define $\Delta(f) = 1$ when f has degree 1. This will be useful later.

Example 2.4.2. Consider $f = x^3 + bx^2 + cx + d$. We saw in (2.26) that

$$\Delta = \Delta(\sigma_1, \sigma_2, \sigma_3) = -4\sigma_2^3 - 27\sigma_3^2 + \sigma_1^2 \sigma_2^2 - 4\sigma_1^3 \sigma_3 + 18\sigma_1 \sigma_2 \sigma_3.$$

Since the evaluation is given by $\sigma_1 \mapsto -b, \sigma_2 \mapsto c$ and $\sigma_3 \mapsto -d$, we obtain

$$
\begin{aligned}
\Delta(f) &= \Delta(-b, c, -d) \\
&= -4c^3 - 27(-d)^2 + (-b)^2c^2 - 4(-b)^3(-d) + 18(-b)c(-d) \\
&= -4c^3 - 27d^3 + b^2c^2 - 4b^3d + 18bcd.
\end{aligned}
$$

This agrees with the formula (1.20) found in Section 1.2.   ◁▷

In the case when we know the roots of a polynomial, we get the following formula for its discriminant.

Proposition 2.4.3. *Suppose that a monic polynomial $f \in F[x]$ of degree $n \geq 2$ has roots $\alpha_1, \ldots, \alpha_n$ in a field L containing F. Then*

$$
\Delta(f) = \prod_{1 \leq i < j \leq n} (\alpha_i - \alpha_j)^2.
$$

Proof. In $F[x_1, \ldots, x_n]$, we know that $\Delta = \prod_{1 \leq i < j \leq n}(x_i - x_j)^2$. Consider the evaluation map that takes x_i to α_i. Since evaluation is a ring homomorphism, this takes Δ to

$$
\prod_{1 \leq i < j \leq n} (\alpha_i - \alpha_j)^2.
$$

If we write $\Delta = \Delta(\sigma_1, \ldots, \sigma_i, \ldots, \sigma_n)$ as in (2.29), then $x_i \to \alpha_i$ takes Δ to

$$
\Delta(\sigma_1(\alpha_1, \ldots \alpha_n), \ldots, \sigma_i(\alpha_1, \ldots \alpha_n), \ldots, \sigma_n(\alpha_1, \ldots \alpha_n)).
$$

By Corollary 2.1.5, this equals

$$
\Delta(-a_1, \ldots, (-1)^i a_i, \ldots, (-1)^n a_n) = \Delta(f)
$$

by the definition of $\Delta(f)$.   □

Mathematical Notes

There are several ideas in this section in need of further discussion.

■ **The Action of the Symmetric Group.** This chapter used the action of S_n on the polynomial ring $F[x_1, \ldots, x_n]$. For $\sigma \in S_n$ and $f \in F[x_1, \ldots, x_n]$, $\sigma \cdot f$ is the polynomial obtained by permuting the variables according to σ. This operation has the following properties:

$$
\begin{aligned}
\sigma \cdot (f + g) &= \sigma \cdot f + \sigma \cdot g, \\
\sigma \cdot (fg) &= (\sigma \cdot f)(\sigma \cdot g), \\
\tau \cdot (\sigma \cdot f) &= (\tau\sigma) \cdot f
\end{aligned}
$$

(2.31)

for $\sigma, \tau \in S_n$ and $f, g \in F[x_1, \ldots, x_n]$. We have used these properties implicitly throughout the chapter, and we will give a formal proof of (2.31) in Chapter 6. The first two imply that $f \mapsto \sigma \cdot f$ is a ring homomorphism from $F[x_1, \ldots, x_n]$ to itself, and the last implies that $(\sigma, f) \mapsto \sigma \cdot f$ is a *group action*, as defined in Section A.4.

▪ **The Alternating Group.** Let F be a field of characteristic different from 2. Proposition 2.4.1 implies that $\sigma \cdot \sqrt{\Delta} = \text{sgn}(\sigma) \sqrt{\Delta}$. Since $-1 \neq +1$ in F,

$$\sigma \cdot \sqrt{\Delta} = \sqrt{\Delta} \iff \text{sgn}(\sigma) = 1 \iff \sigma \in A_n.$$

Thus the alternating group A_n is the subgroup of permutations that fix Δ.

This leads to the question of which other polynomials or rational functions are fixed by A_n. The answer is as follows.

Theorem 2.4.4. *Let F be a field of characteristic $\neq 2$. If $f \in F(x_1, \ldots, x_n)$ is invariant under A_n, then there are $A, B \in F(\sigma_1, \ldots, \sigma_n)$ such that*

$$f = A + B\sqrt{\Delta}.$$

Furthermore, $f \in F[x_1, \ldots, x_n]$ implies that $A, B \in F[\sigma_1, \ldots, \sigma_n]$. ☐

We will prove this theorem in Chapter 7. We will also explain how it relates to the Galois correspondence between subgroups of S_n and subfields of $F(x_1, \ldots, x_n)$ which contain $F(\sigma_1, \ldots, \sigma_n)$.

▪ **The Existence of Roots.** Our discussion of discriminants raises an interesting question about roots. Given a monic polynomial f, our definition of $\Delta(f)$ involves only the coefficients of f. However, if we know the roots of f, then we get the simpler formula given by Proposition 2.4.3. This brings up the fundamental question: *does every polynomial in $F[x]$ have roots in a possibly larger field?* We will answer this question in Chapter 3.

▪ **Discriminant Formulas.** The polynomial expressing Δ in terms of $\sigma_1, \ldots, \sigma_n$ gets more and more complicated as n increases. But if we use determinants, then we get several compact ways to represent both Δ and $\sqrt{\Delta}$. We begin with *Vandermonde's formula* for $\sqrt{\Delta}$.

Proposition 2.4.5.

$$\sqrt{\Delta} = \det \begin{pmatrix} x_1^{n-1} & x_2^{n-1} & \cdots & x_n^{n-1} \\ x_1^{n-2} & x_2^{n-2} & \cdots & x_n^{n-2} \\ \vdots & \vdots & \ddots & \vdots \\ x_1 & x_2 & \cdots & x_n \\ 1 & 1 & \cdots & 1 \end{pmatrix}.$$

*The determinant on the right is called a **Vandermonde determinant**.* ☐

Another nice formula for $\sqrt{\Delta}$ is

$$(2.32) \qquad \sqrt{\Delta} = \det \begin{pmatrix} \frac{\partial \sigma_1}{\partial x_1} & \cdots & \frac{\partial \sigma_n}{\partial x_1} \\ \vdots & & \vdots \\ \frac{\partial \sigma_1}{\partial x_n} & \cdots & \frac{\partial \sigma_n}{\partial x_n} \end{pmatrix}.$$

In Exercise 1, you will use Proposition 2.4.5 to derive the following formula for Δ in terms of the power sums $s_r = x_1^r + \cdots + x_n^r$:

$$(2.33) \qquad \Delta = \det \begin{pmatrix} s_{2n-2} & s_{2n-1} & \cdots & s_{n-1} \\ s_{2n-1} & s_{2n-2} & \cdots & s_{n-2} \\ \vdots & \vdots & \ddots & \vdots \\ s_{n-1} & s_{n-2} & \cdots & s_0 \end{pmatrix}.$$

Finally, in Section 5.3 we will discuss the relation between discriminants and resultants. This leads to a formula that uses the $(2n - 1) \times (2n - 1)$ matrix M defined by

$$\begin{pmatrix} 1 & & & n & & \\ -\sigma_1 & \ddots & & -(n-1)\sigma_1 & n & \\ \sigma_2 & \ddots & 1 & (n-2)\sigma_2 & -(n-1)\sigma_1 & \ddots \\ \vdots & \ddots & -\sigma_1 & \vdots & (n-2)\sigma_2 & \ddots & n \\ \vdots & & \sigma_2 & \vdots & \vdots & \ddots & -(n-1)\sigma_1 \\ (-1)^{n-1}\sigma_{n-1} & & \vdots & (-1)^{n-1}\sigma_{n-1} & \vdots & & (n-2)\sigma_2 \\ (-1)^n\sigma_n & \ddots & \vdots & & (-1)^{n-1}\sigma_{n-1} & & \vdots \\ & \ddots & (-1)^{n-1}\sigma_{n-1} & & & \ddots & \vdots \\ & & (-1)^n\sigma_n & & & & (-1)^{n-1}\sigma_{n-1} \end{pmatrix}$$

$$\underbrace{\qquad\qquad}_{n-1 \text{ columns}} \qquad \underbrace{\qquad\qquad}_{n \text{ columns}}$$

where the empty spaces are filled in with zeros. Then one can prove that

$$\Delta = (-1)^{n(n-1)/2} \det(M).$$

This gives an explicit representation of Δ in terms of $\sigma_1, \ldots, \sigma_n$. See [5] for further comments on computing discriminants.

Historical Notes

The discriminant Δ can be represented as a polynomial in $x_1, \ldots, x_n$ and also as a polynomial in the elementary symmetric polynomials $\sigma_1, \ldots, \sigma_n$. The second of these came first, since the discriminant for $n = 3$ is implicit in Cardan's formulas from Chapter 1. By the 1770s, Lagrange and Vandermonde knew the properties of Δ and $\sqrt{\Delta}$ for small n. For example, when $n = 4$, Lagrange explicitly stated that a transposition changes the sign of $\sqrt{\Delta}$.

The general form of the discriminant was defined independently by Cauchy in 1815 and Gauss in 1816. Cauchy did this in his pioneering studies of the symmetric group S_n. For him, a polynomial was symmetric if it was unchanged by transpositions, so that the next class of functions to study were those which changed sign under a transposition. In essence, Cauchy proved that if F has characteristic different from 2 and $f \in F[x_1, \ldots, x_n]$ satisfies $\tau \cdot f = -f$ for all transpositions τ, then $f = B\sqrt{\Delta}$ for some $B \in F[\sigma_1, \ldots, \sigma_n]$. In Exercise 2 you will show that this follows from Theorem 2.4.4.

Cauchy also considered determinants, drawing on earlier work of Vandermonde and Laplace. He proved Proposition 2.4.5, though he mistakenly attributed it to Vandermonde. In 1841 Jacobi gave the argument used to prove Proposition 2.4.1.

Gauss studied the discriminant in his second proof of the Fundamental Theorem of Algebra. His discussion of Δ is surprisingly modern. Like us, he initially defines Δ as a polynomial in $F[x_1, \ldots, x_n]$ and then shows that it lies in $F[\sigma_1, \ldots, \sigma_n]$. Using the isomorphism

$$\varphi : F[u_1, \ldots, u_n] \simeq F[\sigma_1, \ldots, \sigma_n]$$

from (2.19) in Section 2.2, Gauss defines $\Delta(u_1, \ldots, u_n) \in F[u_1, \ldots, u_n]$ to be the polynomial such that $\Delta(\sigma_1, \ldots, \sigma_n) = \Delta$ in $F[x_1, \ldots, x_n]$. Finally, given a monic polynomial $f \in F[x]$, Gauss defines $\Delta(f)$ just as we did in (2.30). His notation and terminology are different, but his treatment is virtually identical to ours.

Exercises for Section 2.4

Exercise 1. Let M be the $n \times n$ matrix appearing on the right-hand side of the Vandermonde formula given in Proposition 2.4.5. Prove that (2.33) follows from the fact that M and its transpose both have determinant $\sqrt{\Delta}$.

Exercise 2. Let F have characteristic $\neq 2$, and let $f \in F[x_1, \ldots, x_n]$ satisfy $\tau \cdot f = -f$ for all transpositions $\tau \in S_n$. Prove that $f = B\sqrt{\Delta}$ for some $B \in F[\sigma_1, \ldots, \sigma_n]$.

Exercise 3. Let $f = x^2 + bx + c \in F[x]$. Use the definition of discriminant given in the text to show that $\Delta(f) = b^2 - 4c$.

Exercise 4. Let $f \in F[x]$ be monic, and assume that $f = (x - \alpha_1) \cdots (x - \alpha_n)$ in some field L containing F. Prove that $\Delta(f) \neq 0$ if and only if $\alpha_1, \ldots, \alpha_n$ are distinct. This shows that f has distinct roots if and only if its discriminant is nonvanishing.

Exercise 5. Show that $\sqrt{\Delta} \in F[x_1, \ldots, x_n]$ is symmetric if and only if F is a field of characteristic 2.

Exercise 6. Exercise 5 showed how things can differ in a field of characteristic 2. Another example comes from the quadratic formula, which doesn't apply over such fields because of the 2 in the denominator. This exercise will describe how to solve quadratic equations over a field F of characteristic 2.
  (a) Given $b \in F$, we will assume there is a larger field $F \subset L$ such that $b = \beta^2$ for some $\beta \in L$. Show that β is unique and that β is the unique root of $x^2 + b$. Because of this, we denote β by $\sqrt{b}$.
  (b) Now suppose that $f = x^2 + ax + b$ is a quadratic polynomial in $F[x]$ with $a \neq 0$. Suppose also that f is irreducible over F, so that it has no roots in F. We will see in Chapter 3 that f has a root α in a field L containing F. Prove that α cannot be written in the form $\alpha = u + v\sqrt{w}$ where $u, v, w \in F$.
  (c) Part (b) shows that solving a quadratic equation with nonzero x-coefficient requires more than square roots. We do this as follows. If $b \in F$, let $R(b)$ denote a root of $x^2 + x + b$

(possibly lying in some larger field). We call $R(b)$ and $R(b) + 1$ the 2-*roots* of b. Prove that the roots of $x^2 + x + b$ are $R(b)$ and $R(b) + 1$, and explain why adding 1 to the second 2-root gives the first. 2-roots behave like square roots (in characteristic $\neq 2$) in the following sense: if one square root is $\sqrt{c}$, then we multiply by -1 to get the other square root $-\sqrt{c}$. 2-roots work the same way, provided we replace "multiply by -1" with "add 1".

(d) Show that the roots of $f = x^2 + ax + b, a \neq 0$, are $aR(b/a^2)$ and $a(R(b/a^2) + 1)$.

It follows that when F has characteristic 2, then the roots of $x^2 + ax + b \in F[x]$ are

$$
x = \begin{cases} \sqrt{b}, & a = 0, \\ aR(b/a^2), \ a(R(b/a^2) + 1), & a \neq 0. \end{cases}
$$

This is the "quadratic formula" over a field of characteristic 2.

Exercise 7. Explain how the third property of (2.31) was used (implicitly) in (2.28) in the proof of Proposition 2.4.1.

Exercise 8. As explained on page 36, we can regard $F[\sigma_1, \ldots, \sigma_n]$ as a polynomial ring in the variables $\sigma_1, \ldots, \sigma_n$. In this exercise, you will prove that although Δ factors in $F[x_1, \ldots, x_n]$, it is irreducible in $F[\sigma_1, \ldots, \sigma_n]$ when F has characteristic different from 2. To begin the proof, assume that $\Delta = AB$, where $A, B \in F[\sigma_1, \ldots, \sigma_n]$ are nonconstant.

(a) Using the definition of Δ and unique factorization in $F[x_1, \ldots, x_n]$, show that A is divisible in $F[x_1, \ldots, x_n]$ by $x_i - x_j$ for some $1 \leq i < j \leq n$.

(b) Given $1 \leq i < j \leq n$ and $1 \leq l < m \leq n$, show that there is a permutation $\sigma \in S_n$ such that $\sigma(i) = l$ and $\sigma(j) = m$.

(c) Use parts (a) and (b) to show that A is divisible by $x_l - x_m$ for all $1 \leq l < m \leq n$.

(d) Conclude that A is a multiple of $\sqrt{\Delta}$ and that the same is true for B.

(e) Show that part (d) implies that A and B are constant multiples of $\sqrt{\Delta}$ and explain why this contradicts $A, B \in F[\sigma_1, \ldots, \sigma_n]$.

(f) Finally, suppose that F has characteristic 2. Prove that Δ is not irreducible.

Exercise 9. For $n = 4$, the variables x_1, x_2, x_3, x_4 have discriminant

$$
\Delta = (x_1 - x_2)^2 (x_1 - x_3)^2 (x_1 - x_4)^2 (x_2 - x_3)^2 (x_2 - x_4)^2 (x_3 - x_4)^2.
$$

Let $y_1 = x_1 x_2 + x_3 x_4, y_2 = x_1 x_3 + x_2 x_4, y_3 = x_1 x_4 + x_2 x_3$, and consider

$$
\theta(y) = (y - y_1)(y - y_2)(y - y_3).
$$

This is a cubic polynomial in y. As in the text, the discriminant of θ will be denoted $\Delta(\theta)$. Show that $\Delta(\theta) = \Delta$. When we discuss Lagrange's work in Chapter 12, we will see that θ is the *Ferrari resolvent*, which plays an important role in the solution of the quartic equation.

Exercise 10. Let $C, D \in F[\sigma_1, \ldots, \sigma_n]$ be nonzero and relatively prime. This exercise will show that C and D remain relatively prime when regarded as elements of $F[x_1, \ldots, x_n]$.

(a) Show that C^m, D^m are relatively prime in $F[\sigma_1, \ldots, \sigma_n]$ for any positive integer m.

(b) Suppose that $p \in F[x_1, \ldots, x_n]$ is a nonconstant polynomial dividing C and D. Prove that $\sigma \cdot p$ divides C and D for all $\sigma \in S_n$.

(c) As in Exercise 7 of Section 2.2, let $P = \prod_{\sigma \in S_n} \sigma \cdot p$. Show that P divides $C^{n!}$ and $D^{n!}$, and then use part (a) and Exercise 7 of Section 2.2 to obtain a contradiction.

Exercise 11. Exercise 8 of Section 2.2 showed that if $f \in F(x_1, \ldots, x_n)$ is symmetric, then $f \in F(\sigma_1, \ldots, \sigma_n)$. In this exercise, you will refine this result as follows. Suppose that $f \in F(x_1, \ldots, x_n)$ is symmetric, and write $f = A/B$, where $A, B \in F[x_1, \ldots, x_n]$ are relatively prime. The claim is that A, B are themselves symmetric and hence lie in $F[\sigma_1, \ldots, \sigma_n]$. We can assume that A and B are nonzero.

(a) Use the previous exercise and Exercise 8 of Section 2.2 to show that $f = C/D$ where $C, D \in F[\sigma_1, \ldots, \sigma_n]$ are relatively prime in $F[x_1, \ldots, x_n]$.

(b) Show that $AD = BC$ and then use unique factorization in $F[x_1, \ldots, x_n]$ to show that A and B are constant multiples of C and D, respectively.

(c) Conclude that $A, B \in F[\sigma_1, \ldots, \sigma_n]$ as claimed.

Exercise 12. Prove (2.32) when $n = 2$ and 3.

REFERENCES

1. G. Chrystal, *Textbook of Algebra*, Seventh Edition, Chelsea, New York, 1964.

2. D. Cox, J. Little, and D. O'Shea, *Ideals, Varieties and Algorithms*, Second Edition, Springer-Verlag, New York, Berlin, Heidelberg, 1997.

3. D. E. Dobbs and R. Hanks, *A Modern Course on the Theory of Equations*, Polygonal Publishing House, Passaic, NJ, 1980.

4. I. G. MacDonald, *Symmetric Functions and Hall Polynomials*, Oxford U. P., Oxford, 1979.

5. J. McKay, *On computing discriminants*, Amer. Math. Monthly **94** (1987), 523–527.

6. E. Waring, *Meditationes algebraicæ*, English translation by D. Weeks, AMS, Providence, RI, 1991.

7. L. Weisner, *Introduction to the Theory of Equations*, Macmillan, New York, 1938.

3

Roots of Polynomials

This chapter will study the roots of a polynomial in one variable. We will first show that every nonconstant polynomial with coefficients in a field has roots in some possibly larger field. Then, in the special case of a polynomial with coefficients in the field $\mathbb{C}$ of complex numbers, we will show that the roots also lie in $\mathbb{C}$.

3.1 THE EXISTENCE OF ROOTS

In this section, we will show that given a field F and a nonconstant polynomial $f \in F[x]$, there is a field L containing F which also contains all roots of f. We will motivate our construction by considering the complex numbers $\mathbb{C}$.

So far we have assumed the existence of the real and complex numbers. But if we're given just the real numbers $\mathbb{R}$, how do we get $\mathbb{C}$? There are several ways of doing this. For example, in 1835, Hamilton defined

$$\mathbb{C} = \{(a, b) \mid a, b \in \mathbb{R}\},$$

where addition and multiplication are given by

$$(a, b) + (c, d) = (a + b, c + d) \quad \text{and} \quad (a, b) \cdot (c, d) = (ac - bd, ad + bc).$$

It is straightforward (though somewhat tedious) to verify that these operations make the above set into a field with $(1, 0)$ as the multiplicative identity. Furthermore, the formula for multiplication implies that

$$(0, 1) \cdot (0, 1) = (-1, 0) = -(1, 0).$$

If we let 1 denote $(1, 0)$ and i denote $(0, 1)$, then this equation becomes $i^2 = -1$, and we also have

$$(a, b) = a(1, 0) + b(0, 1) = a \cdot 1 + b \cdot i.$$

In this way, we recover the usual description of $\mathbb{C}$ as the set of numbers of the form $a + bi$, where $a, b \in \mathbb{R}$.

A very different definition of $\mathbb{C}$ was given by Cauchy in 1847. He worked in the polynomial ring $\mathbb{R}[x]$ and defined

$$(3.1) \qquad\qquad \phi(x) \equiv \chi(x) \ [\text{mod. } x^2 + 1]$$

to mean that $\phi(x), \chi(x) \in \mathbb{R}[x]$ have the same remainder on division by $x^2 + 1$. Then, to simplify these congruences, he introduced the symbol i as follows:

> ... the *symbolic* letter i, when substituted for the letter x in a polynomial $f(x)$, indicates the value obtained, not by the polynomial $f(x)$, but rather by the remainder of the algebraic division of $f(x)$ by $x^2 + 1$, when one attributes to x the particular value i.

(See [Cauchy, p. 317].) This allowed Cauchy to replace (3.1) with the equivalent statement

$$\phi(i) = \chi(i).$$

To illustrate how this works, Cauchy considers the polynomial

$$f(x) = (a + bx)(c + dx) = ac + bdx^2 + (ad + bc)x.$$

The remainder of $f(x)$ on division by $x^2 + 1$ is easily seen to be $ac - bd + (ad + bc)x$ (be sure you see why), so that by the above quotation, $f(i)$ is defined to be the symbolic expression $ac - bd + (ad + bc)i$. The same process, when applied to $a + bx$ and $c + dx$, yields $a + bi$ and $c + di$ respectively. From this Cauchy concluded that

$$(3.2) \qquad\qquad (a + bi)(c + di) = ac - bd + (ad + bc)i.$$

Thus we have a symbolic construction of the complex numbers using remainders of polynomials in $\mathbb{R}[x]$ on division by $x^2 + 1$.

From a modern point of view, we can explain Cauchy's construction as follows. In the notation of Section A.1, $x^2 + 1$ generates the ideal

$$\langle x^2 + 1 \rangle = \{(x^2 + 1)g \mid g \in \mathbb{R}[x]\}$$

in the ring $\mathbb{R}[x]$. This gives the quotient ring

$$\mathbb{R}[x]/\langle x^2 + 1 \rangle = \{g + \langle x^2 + 1 \rangle \mid g \in \mathbb{R}[x]\},$$

where we are using the coset notation of Section A.1.

Now, following Cauchy, we take $\phi \in \mathbb{R}[x]$ and divide it by $x^2 + 1$ to obtain

$$\phi = q \cdot (x^2 + 1) + a + bx$$

for a unique $q \in \mathbb{R}[x]$ and $a, b \in \mathbb{R}$. Since cosets $g + \langle x^2 + 1 \rangle, h + \langle x^2 + 1 \rangle$ in $\mathbb{R}[x]/\langle x^2 + 1 \rangle$ are equal if and only if $g - h \in \langle x^2 + 1 \rangle$, we see that

$$\phi + \langle x^2 + 1 \rangle = a + bx + \langle x^2 + 1 \rangle.$$

It follows that (3.1) is true if and only if ϕ and χ give the same coset in $\mathbb{R}[x]/\langle x^2+1 \rangle$. Furthermore, since the remainder of $(a + bx)(c + dx)$ on division by $x^2 + 1$ is $ac - bd + (ad + bc)x$, we have

$$\big(a + bx + \langle x^2 + 1 \rangle\big)\big(c + dx + \langle x^2 + 1 \rangle\big) = (a + bx)(c + dx) + \langle x^2 + 1 \rangle$$
$$= ac - bd + (ad + bc)x + \langle x^2 + 1 \rangle.$$

Hence Cauchy's construction of $\mathbb{C}$ is equivalent to the quotient ring $\mathbb{R}[x]/\langle x^2 + 1 \rangle$.

But we can do even better, for we can also interpret Cauchy's symbolic letter i as the coset $x + \langle x^2 + 1 \rangle$. More precisely, if we identify 1 with $1 + \langle x^2 + 1 \rangle$ and i with $x + \langle x^2 + 1 \rangle$, then

$$a + bx + \langle x^2 + 1 \rangle = a \cdot 1 + b \cdot i,$$

and the symbolic multiplication (3.2) becomes the above multiplication of cosets.

However, interpreting Cauchy's construction as $\mathbb{R}[x]/\langle x^2 + 1 \rangle$ gives only a ring structure. In order for this quotient ring to be a field, we need $\langle x^2 + 1 \rangle$ to be a maximal ideal. The following proposition will be useful.

Proposition 3.1.1. *If F is a field and $f \in F[x]$ is nonconstant, then the following are equivalent:*
(a) *The polynomial f is irreducible over F.*
(b) *The ideal $\langle f \rangle = \{ fg \mid g \in F[x] \}$ is a maximal ideal.*
(c) *The quotient ring $F[x]/\langle f \rangle$ is a field.*

Proof. The equivalence (b) $\Leftrightarrow$ (c) is Theorem A.1.12 from Section A.1. It remains to prove (a) $\Leftrightarrow$ (b).

Suppose f is irreducible and I is an ideal of $F[x]$ such that $\langle f \rangle \subset I \subset F[x]$. By Theorem A.1.17 from Section A.1, $I = \langle g \rangle$ for some $g \in F[x]$. Then $f \in \langle f \rangle \subset I = \langle g \rangle$ implies that $f = gh$ for some $h \in F[x]$. Since f is irreducible, g or h must be constant. We will leave it as Exercise 1 for the reader to show that g constant implies $I = F[x]$ and h constant implies $I = \langle f \rangle$. It follows that $\langle f \rangle$ is maximal.

Conversely, suppose that $\langle f \rangle$ is maximal and let $f = gh$ where $g, h \in F[x]$. This gives $\langle f \rangle \subset \langle g \rangle \subset F[x]$. Since $\langle f \rangle$ is maximal, $\langle g \rangle$ must equal either $\langle f \rangle$ or $F[x]$. In Exercise 1 you will show that the former implies that h is constant and the latter implies that g is constant. Thus f is irreducible. $\square$

Since $x^2 + 1$ is irreducible over $\mathbb{R}$, Proposition 3.1.1 implies that $\mathbb{R}[x]/\langle x^2 + 1 \rangle$ is a field. This completes our second construction of $\mathbb{C}$.

One interesting feature of the above two constructions of $\mathbb{C}$ just given is that *neither* contains $\mathbb{R}$. This might seem contradictory, but consider the following:

- In Hamilton's construction, a complex number is an ordered pair (a, b) of real numbers. In order for this definition of $\mathbb{C}$ to contain $\mathbb{R}$, we must identify the number $a \in \mathbb{R}$ with the ordered pair $(a, 0) \in \mathbb{C}$.

- In the modern interpretation of Cauchy's construction, a complex number is a coset $g + \langle x^2 + 1 \rangle$. In order for this to contain $\mathbb{R}$, we must identify the number $a \in \mathbb{R}$ with the coset $a + \langle x^2 + 1 \rangle \in \mathbb{C}$.

Thus both constructions of $\mathbb{C}$ give one-to-one homomorphisms $\mathbb{R} \to \mathbb{C}$ that become inclusions only after we identify $\mathbb{R}$ with its image in $\mathbb{C}$. This motivates the following definition.

Definition 3.1.2. *Given a ring homomorphism of fields $\varphi : F \to L$, we say that L is a **field extension of** F **via** φ. We will usually identify F with its image*

$$\varphi(F) = \{\varphi(a) \mid a \in F\} \subset L$$

and write $F \subset L$.

In Exercise 2 you will show that a ring homomorphism of fields $\varphi : F \to L$ is automatically one to one and induces an isomorphism $\varphi : F \simeq \varphi(F)$. Hence once we identify F with $\varphi(F) \subset L$ via φ, we may regard F as a subfield of L. For the two constructions of $\mathbb{C}$ given above, this gives $\mathbb{R} \subset \mathbb{C}$, as desired.

Armed with this notion of a field extension, we can prove that every irreducible polynomial has a root in an extension field.

Proposition 3.1.3. *If $f \in F[x]$ is irreducible, then there is an extension field $F \subset L$ and $\alpha \in L$ such that $f(\alpha) = 0$.*

Proof. Let $I = \langle f \rangle$, so that $L = F[x]/I$ is a field by Proposition 3.1.1. Furthermore, $a \in F$ gives the constant polynomial $a \in F[x]$, which in turn gives the coset $a + I \in L$. Thus we get a natural map $\varphi : F \to L$. In Exercise 3 you will check that φ is a ring homomorphism, so that using the convention of Definition 3.1.2, we get a field extension $F \subset L$.

It remains to show that there is $\alpha \in L$ such that $f(\alpha) = 0$. This is surprisingly easy. Motivated by Cauchy's symbolic construction, we set $\alpha = x + I$. To prove that $f(\alpha) = 0$, suppose that $f = a_0 x^n + \cdots + a_n$, where $a_i \in F$. Then, recalling our identification of $a \in F$ with the coset $a + I \in L$, we have

$$\begin{aligned}
f(\alpha) &= (a_0 + I)\alpha^n + \cdots + (a_n + I) \\
&= (a_0 + I)(x + I)^n + \cdots + (a_n + I) \\
&= (a_0 x^n + \cdots + a_n) + I = f + I \\
&= 0 + I,
\end{aligned}$$

where the third equality uses the definition of addition and multiplication of cosets, and the last uses $f + I = 0 + I \Leftrightarrow f - 0 \in I$. Since $0 + I$ is the additive identity of L, we have $f(\alpha) = 0$, as claimed. $\qquad \square$

Recall the elementary fact that $\alpha \in L$ is a root of a polynomial $f \in L[x]$ if and only if $x - \alpha$ is a factor of f in $L[x]$ (this is Corollary A.1.15). Thus, to say that a field L contains *all* roots of f means that f factors as

$$f = a_0(x - \alpha_1) \cdots (x - \alpha_n),$$

where $\alpha_1, \ldots, \alpha_n \in L$. When this happens, we say that f *splits completely over* L.

Theorem 3.1.4. *Let $f \in F[x]$ be a polynomial of degree $n > 0$. Then there is an extension field $F \subset L$ such that f splits completely over L.*

Proof. We will prove this using induction on $n = \deg(f)$. If $n = 1$, then $f = a_0 x + a_1$, where $a_0 \neq 0$ and $a_0, a_1 \in F$. Setting $L = F$ and $\alpha_1 = -a_1/a_0$ implies that $f = a_0(x - \alpha_1)$ and proves the theorem in this case.

Now suppose that $\deg(f) = n > 1$ and that the theorem is true for $n - 1$. Since F is a field, $F[x]$ is a UFD by Theorem 2.1.1. In particular, f has an irreducible divisor f_1. If we apply Proposition 3.1.3 to $f_1 \in F[x]$, then we get an extension field $F \subset F_1$ and an element $\alpha_1 \in F_1$ such that $f_1(\alpha_1) = 0$ in F_1.

Since f_1 is a factor of f, we also have $f(\alpha_1) = 0$ in F_1. As noted above, this implies that $x - \alpha_1$ is a factor of f in $F_1[x]$. In other words,

$$f = (x - \alpha_1)g$$

for some $g \in F_1[x]$ of degree $n - 1$. Applying our inductive hypothesis to g, we get a field extension $F_1 \subset L$ and elements $\alpha_2, \ldots, \alpha_n \in L$ such that

$$g = a_0(x - \alpha_2) \cdots (x - \alpha_n).$$

The displayed formulas for f and g show that f splits completely over L. $\square$

Mathematical Notes

This section includes several ideas which are worthy of comment.

■ **Identifications.** In Definition 3.1.2, we wrote a field extension $\varphi : F \to L$ as $F \subset L$ by identifying F with $\varphi(F)$. This might seem like cheating, but it happens all the time in mathematics. For example, consider $\mathbb{Z} \subset \mathbb{Q}$. Since $\mathbb{Q}$ is the field of fractions of the integral domain $\mathbb{Z}$, an element $a/b \in \mathbb{Q}$ is the equivalence class

(3.3) $$\frac{a}{b} = \{(c, d) \mid c, d \in \mathbb{Z}, d \neq 0, ad = bc\}.$$

(See Exercise 4 for the details.) In particular, according to (3.3), an integer $n \in \mathbb{Z}$ doesn't equal the fraction $n/1 \in \mathbb{Q}$, since n is an integer and $n/1$ is an infinite set of ordered pairs of integers. Rather, we have the ring homomorphism

$$\phi : \mathbb{Z} \to \mathbb{Q}$$

which sends n to $n/1$, and we write $\mathbb{Z} \subset \mathbb{Q}$ by identifying $\mathbb{Z}$ with $\phi(\mathbb{Z})$. This is similar to what we did in the discussion preceding Definition 3.1.2.

■ **Construction of Extension Fields.** Beginning students in algebra often have difficulty with cosets and quotient rings. The key insight is that in a quotient ring R/I, elements of the ideal I become zero. This is because $r \in I$ gives the coset $r + I$, which equals the zero coset $0 + I$, since $r - 0 \in I$. Applying this to the situation of Proposition 3.1.3, $f \in \langle f \rangle$ means that $f + \langle f \rangle$ is zero in $F[x]/\langle f \rangle$. But $f + \langle f \rangle$ is f applied to $\alpha = x + \langle f \rangle$, so that α is a root of f in $F[x]/\langle f \rangle$.

When f is irreducible, Proposition 3.1.3 also showed that $L = F[x]/\langle f \rangle$ is a field. But in practice, if we are given a nonzero coset $g + \langle f \rangle$, how do we find its multiplicative inverse in L? In Exercise 5 you will show the following:

- f and g are relatively prime, so that $Af + Bg = 1$ for some $A, B \in F[x]$.
- The multiplicative inverse of $g + \langle f \rangle$ in $L = F[x]/\langle f \rangle$ is the coset $B + \langle f \rangle$.

While it is important to be able to manipulate cosets at an abstract level, it is also often useful to represent them concretely. This means coming up with a method for picking a unique element—a *coset representative*—from each coset. In the case of $L = F[x]/\langle f \rangle$, we will show in Chapter 4 that if f has degree n, then every coset in $F[x]/\langle f \rangle$ can be written uniquely in the form

$$c_0 + c_1 x + \cdots + c_{n-1} x^{n-1} + \langle f \rangle,$$

where $c_0, \ldots, c_{n-1} \in F$. The rough idea is that given a coset $g + \langle f \rangle$, we replace g with its remainder on division by f, which is a polynomial of degree at most $n - 1$. Furthermore, setting $\alpha = x + \langle f \rangle$ as in the proof of Proposition 3.1.3, we can rewrite the above expression as

$$(3.4) \qquad\qquad c_0 + c_1 \alpha + \cdots + c_{n-1} \alpha^{n-1}.$$

When $F = \mathbb{R}$ and $f = x^2 + 1$, this is what Cauchy did in his construction of $\mathbb{C}$.

The idea of representing cosets by remainders can be applied to other quotient rings as well. For example, the theory of Gröbner bases enables one to represent elements of the quotient ring

$$F[x_1, \ldots, x_n]/\langle f_1, \ldots, f_s \rangle, \quad f_1, \ldots, f_s \in F[x_1, \ldots, x_n]$$

uniquely by remainders (see [3, Ch. 5, §2]).

■ **Construction of Splitting Fields.** The proof of Theorem 3.1.4 constructs a field over which $f \in F[x]$ splits completely by iterating the quotient ring construction of Proposition 3.1.3. Hence elements of this field are cosets of cosets of cosets, etc. This seems very abstract until one remembers that in modern algebra, we don't care what the objects are: it's their behavior that counts. Since the field has the desired behavior, we are content.

In Chapter 5, we will give a refined version of Theorem 3.1.4 where L is chosen to be the smallest extension of F over which $f \in F[x]$ splits completely. We will call this a *splitting field*. We will show that splitting fields are unique up to isomorphism.

■ **The Complex Numbers.** This section began with two constructions of the complex numbers $\mathbb{C}$. The one of greater interest to us was Cauchy's, which eventually led to Proposition 3.1.3. The other construction of $\mathbb{C}$, due to Hamilton, used ordered pairs of real numbers. This suggests using triples of real numbers, and Hamilton tried hard to define addition and multiplication so that such triples would form a field. He didn't succeed, but on October 16, 1843 he realized that this idea would work with quadruples (a, b, c, d) provided multiplication wasn't required to be commutative. If the standard basis of $\mathbb{R}^4$ is denoted $1, i, j, k$, then

$$(a, b, c, d) = a\,1 + b\,i + c\,j + d\,k,$$

and Hamilton defined multiplication so that 1 is the multiplicative identity and

$$i^2 = j^2 = k^2 = -1, \quad ij = -ji = k, \quad jk = -kj = i, \quad ki = -ik = j.$$

These are the famous *quaternions*. They form a *division ring*, which is a noncommutative ring where every nonzero element has a multiplicative inverse.

Historical Notes

In solving the cubic and quartic equations, Cardan and Ferrari implicitly assumed the existence of roots, just as we did in Chapter 1. Girard, in the early seventeenth century, was one of the first to assert the existence of roots, real or imaginary, though "imaginary root" did not have a clear meaning in his work. As people became more comfortable with complex numbers, the existence of roots evolved into the existence of complex roots, which come in complex conjugate pairs when the coefficients are real. Thus the eighteenth century version of the Fundamental Theorem of Algebra asserts that *every nonconstant polynomial in $\mathbb{R}[x]$ factors into linear and quadratic factors with coefficients in $\mathbb{R}$.* In Section 3.2, we will prove the equivalent statement that every nonconstant polynomial in $\mathbb{C}[x]$ splits completely over $\mathbb{C}$.

The first attempt to prove the Fundamental Theorem of Algebra was due to D'Alembert in 1746, and at roughly the same time Euler discovered an algebraic proof (still somewhat incomplete), to be discussed in the next section. Like Cardan, Euler implicitly assumed that the roots exist. In 1799, Gauss noted that Euler's proof in effect made the assumption that

> every equation can be satisfied by a real value of the unknown, or by an imaginary value of the form $a + b\sqrt{-1}$, or by a value that is not subsumed under any form.

(See [Gauss, Vol. III, p. 14].) Gauss criticized this assumption as follows:

> How these magnitudes of which we can form no idea whatsoever—these shadows of shadows—are to be added or multiplied cannot be understood with the kind of clarity required by mathematics.

The main result of this section, Theorem 3.1.4, answers Gauss's criticism very nicely. Given $f \in \mathbb{R}[x]$ of positive degree, we can regard f as lying in $\mathbb{C}[x]$. Applying Theorem 3.1.4, we get an extension $\mathbb{C} \subset L$ where f splits completely over L. Then, as Gauss observes in the first part of the quote, each root of f either lies in $\mathbb{R}$, in $\mathbb{C}$, or in L. However, the roots in L are no longer "shadows of shadows" but rather elements of a field which can be manipulated by the usual operations of algebra, just as Euler assumed they could.

Gauss's 1815 proof of the Fundamental Theorem of Algebra uses symbolic methods to avoid assuming the existence of roots, though his actual construction was quite different from what we did in Theorem 3.1.4. We will say more about Gauss's argument in the next section.

In his 1847 construction of the complex numbers, Cauchy defined congruences modulo an arbitrary polynomial f. He also introduced formal sums similar to (3.4). However, Cauchy did not recognize the importance of f being irreducible, which by Proposition 3.1.1 is necessary if we want the quotient ring to be a field.

The general case of this construction is due to Kronecker. He developed an elaborate theory of algebraic quantities in his 1881–1882 treatise *Grundzüge einer arithmetischen Theorie der algebraischen Grössen* [Kronecker, Vol. II, pp. 237–387] and applied these ideas to the existence of roots in his 1887 paper *Ein Fundamentalsatz der allgemeinen Arithmetik* [Kronecker, Vol. III, pp. 209–240]. His version of Theorem 3.1.4 uses the language of congruences (rather than cosets) to construct an extension $F \subset L$ in which $f \in F[x]$ splits completely. In Chapter 12, we will see how Kronecker drew on ideas of Lagrange and Galois to create L using a *single* quotient, rather than the sequence of quotients used in the proof of Theorem 3.1.4.

Exercises for Section 3.1

Exercise 1. This exercise is concerned with the proof of Proposition 3.1.1. Suppose that $f, g, h \in F[x]$ are polynomials such that f is nonzero and $f = gh$. Also let $I = \langle g \rangle$.
(a) Prove that g constant if and only if $I = F[x]$.
(b) Prove that h constant if and only if $I = \langle f \rangle$.

Exercise 2. Let F and L be fields, and let $\varphi : F \to L$ be a ring homomorphism as defined in Section A.1. Prove that φ is one to one and that we get an isomorphism $\varphi : F \simeq \varphi(F)$.

Exercise 3. Let $I \subset F[x]$ be an ideal, and define $\varphi : F \to F[x]/I$ by $\varphi(a) = a + I$. Prove carefully that φ is a ring homomorphism.

Exercise 4. In your abstract algebra text, review the definition of field of fractions and verify that (3.3) is the correct definition of a/b for $a, b \in \mathbb{Z}, b \neq 0$.

Exercise 5. Let $f \in F[x]$ be irreducible, and let $g + \langle f \rangle$ be a nonzero coset in the quotient ring $L = F[x]/\langle f \rangle$.
(a) Show that f and g are relatively prime and conclude that $Af + Bg = 1$, where A, B are polynomials in $F[x]$.
(b) Show that $B + \langle f \rangle$ is the multiplicative inverse of $g + \langle f \rangle$ in L.

Exercise 6. Apply the method of Exercise 5 to find the multiplicative inverse of the coset $1 + x + \langle x^2 + x + 1 \rangle$ in the field $\mathbb{Q}[x]/\langle x^2 + x + 1 \rangle$.

3.2   THE FUNDAMENTAL THEOREM OF ALGEBRA

The *Fundamental Theorem of Algebra* asserts that every nonconstant $f \in \mathbb{C}[x]$ splits completely over $\mathbb{C}$. In other words,

$$f = a_0(x - \alpha_1) \cdots (x - \alpha_n)$$

for some $a_0, \alpha_1, \ldots, \alpha_n \in \mathbb{C}$.

The following proposition shows that there are several different ways of stating the Fundamental Theorem.

Proposition 3.2.1. *The following are equivalent:*
(a) *Every nonconstant $f \in \mathbb{C}[x]$ has at least one root in $\mathbb{C}$.*
(b) *Every nonconstant $f \in \mathbb{C}[x]$ splits completely over $\mathbb{C}$.*
(c) *Every nonconstant $f \in \mathbb{R}[x]$ has at least one root in $\mathbb{C}$.*

Proof. For (a) $\Rightarrow$ (b), we use induction on $n = \deg(f)$. When $n = 1$, writing $f = ax + b = a(x - (-b/a))$ shows that f splits completely over $\mathbb{C}$.

Now suppose that $n > 1$ and that our assertion is true for $n - 1$. If $f \in \mathbb{C}[x]$ has degree n, then assumption (a) implies that $f(\alpha) = 0$ for some $\alpha \in \mathbb{C}$. By Corollary A.1.15, this implies that f is divisible by $x - \alpha$. Thus

$$f = (x - \alpha)\, g$$

for some $g \in \mathbb{C}[x]$ of degree $n - 1$. By our inductive assumption, g splits completely over $\mathbb{C}$, and then the above equation shows that the same is true for f.

The implication (b) $\Rightarrow$ (c) is clear since $\mathbb{R} \subset \mathbb{C}$. To prove (c) $\Rightarrow$ (a), we must show that $f = a_0 x^n + \cdots + a_n \in \mathbb{C}[x]$ has a root in $\mathbb{C}$ when $n > 0$ and $a_0 \neq 0$. Let

(3.5) $$\bar{f} = \overline{a_0} x^n + \cdots + \overline{a_n}$$

denote the polynomial obtained by taking the complex conjugates of the coefficients of f. In Exercise 1 you will prove that if $f, g \in \mathbb{C}[x]$, then

$$\overline{fg} = \bar{f}\bar{g}.$$

Now let $h = f\bar{f} \in \mathbb{C}[x]$. Then

$$\bar{h} = \overline{f\bar{f}} = \bar{f}\bar{\bar{f}} = \bar{f}f = h$$

implies that $h \in \mathbb{R}[x]$. By (c), we can find $\alpha \in \mathbb{C}$ such that $h(\alpha) = 0$. Then $f(\alpha)\bar{f}(\alpha) = 0$, so that $f(\alpha) = 0$ or $\bar{f}(\alpha) = 0$. In the former case, $\alpha \in \mathbb{C}$ is a root of f, and in the latter, Exercise 1 will show that $\bar{\alpha} \in \mathbb{C}$ is a root of f. This completes the proof of the proposition. $\qquad \square$

We next study polynomials of odd degree with real coefficients.

Proposition 3.2.2. *Every $f \in \mathbb{R}[x]$ of odd degree has at least one root in $\mathbb{R}$.*

Proof. We will use the Intermediate Value Theorem (IVT) from calculus. We know that $f \in \mathbb{R}[x]$ is continuous. Thus, if we can find $M > 0$ such that

(3.6) $$f(-M) < 0 < f(M),$$

then the IVT, applied to f on the interval $[-M, M]$, will guarantee that $f(c) = 0$ for some $c \in (-M, M)$.

Given $f \in \mathbb{R}[x]$ of odd degree, we can assume that f is monic by multiplying f by a suitable nonzero constant. Then

$$f = x^n + a_1 x^{n-1} + \cdots + a_n,$$

where n is odd and $a_1, \ldots, a_n \in \mathbb{R}$. If we set

$$M = |a_1| + \cdots + |a_n| + 1,$$

then

(3.7)
$$
\begin{aligned}
|a_1 M^{n-1} + \cdots + a_n| &\le |a_1| M^{n-1} + |a_2| M^{n-2} + \cdots + |a_n| \\
&\le (|a_1| + |a_2| + \cdots + |a_n|) M^{n-1} \\
&< M^n,
\end{aligned}
$$

where the first inequality uses the triangle inequality, the second uses $M \ge 1$, and the third uses $M > |a_1| + |a_2| + \cdots + |a_n|$. It follows that

$$f(M) = M^n + (a_1 M^{n-1} + \cdots + a_n) > 0$$

since the expression in parentheses has absolute value $< M^n$ by (3.7). Similarly,

$$|a_1(-M)^{n-1} + a_2(-M)^{n-2} + \cdots + a_n| < M^n$$

implies that

$$f(-M) = -M^n + (a_1(-M)^{n-1} + a_2(-M)^{n-2} + \cdots + a_n) < 0,$$

since n is odd and the expression in parentheses has absolute value $< M^n$.

Thus M satisfies (3.6). As noted above, the proposition follows. $\qquad\square$

Finally, we note the following simple consequence of the quadratic formula.

Lemma 3.2.3. *Every quadratic polynomial in $\mathbb{C}[x]$ splits completely over $\mathbb{C}$.*

Proof. Given $f = ax^2 + bx + c \in \mathbb{C}[x]$ with $a \ne 0$, the quadratic formula implies that the roots of f are given by

$$\frac{-b \pm \sqrt{b^2 - 4ac}}{2a}.$$

By Section A.2, every complex number has a square root in $\mathbb{C}$. Hence the above roots are complex numbers, which shows that f splits completely over $\mathbb{C}$. $\qquad\square$

We can now prove the Fundamental Theorem of Algebra.

Theorem 3.2.4. *Every nonconstant $f \in \mathbb{C}[x]$ splits completely over $\mathbb{C}$.*

Proof. Our proof will follow a strategy due to Euler, together with a clever idea first used by Laplace. By Proposition 3.2.1, it suffices to prove that

(3.8)   Every $f \in \mathbb{R}[x]$ of degree $n > 0$ has at least one root in $\mathbb{C}$.

We can write n uniquely in the form

$$n = 2^m k, \quad k \text{ odd}.$$

Euler's strategy is to prove (3.8) by induction on m. By Proposition 3.2.2, a polynomial of odd degree in $\mathbb{R}[x]$ has a root in $\mathbb{R} \subset \mathbb{C}$. Hence (3.8) is true when $m = 0$.

Now suppose that $m > 0$ and that (3.8) is true for $m - 1$. Take $f \in \mathbb{R}[x]$ of degree $n = 2^m k$, k odd. We can regard f as a polynomial in $\mathbb{C}[x]$, so that Theorem 3.1.4 implies that there is an extension $\mathbb{C} \subset L$ such that f splits completely over L. We will denote the roots of f by $\alpha_1, \ldots, \alpha_n \in L$.

Laplace's clever idea is to consider the following auxiliary polynomial. Pick a real number λ, and set

$$g_\lambda(x) = \prod_{1 \le i < j \le n} \left(x - (\alpha_i + \alpha_j) + \lambda \alpha_i \alpha_j \right).$$

This has degree $\frac{1}{2} n(n - 1) =$ the number of distinct pairs of variables.

We first claim that g_λ has coefficients in $\mathbb{R}$. To prove this, consider

(3.9) $$G_\lambda(x) = \prod_{1 \le i < j \le n} \left(x - (x_i + x_j) + \lambda x_i x_j \right).$$

The identity

$$x - (x_i + x_j) + \lambda x_i x_j = x - (x_j + x_i) + \lambda x_j x_i$$

shows that G_λ is a product indexed by pairs of distinct variables. It follows easily that G_λ is unaffected by permutations of the x_i. Then multiplying out G_λ gives

$$G_\lambda(x) = \sum_{i=0}^{\frac{1}{2} n(n-1)} p_i(x_1, \ldots, x_n)\, x^i,$$

where the $p_i(x_1, \ldots, x_n)$ are symmetric in $x_1, \ldots, x_n$ since G_λ is. Also note that $p_i(x_1, \ldots, x_n) \in \mathbb{R}[x_1, \ldots, x_n]$ since $\lambda \in \mathbb{R}$. Then Corollary 2.2.5 implies that $p_i(\alpha_1, \ldots, \alpha_n) \in \mathbb{R}$ since $\alpha_1, \ldots, \alpha_n$ are the roots of $f \in \mathbb{R}[x]$. We conclude that

$$g_\lambda(x) = \sum_{i=0}^{\frac{1}{2} n(n-1)} p_i(\alpha_1, \ldots, \alpha_n)\, x^i \in \mathbb{R}[x].$$

We next compute the exponent of 2 in the degree of g_λ. Using $n = 2^m k$, the degree of g_λ is given by

$$\tfrac{1}{2} n(n - 1) = \tfrac{1}{2} 2^m k (2^m k - 1).$$

Since $m > 0$, we can write this as

(3.10) $$\deg(g_\lambda) = 2^{m-1} k (2^m k - 1).$$

Furthermore, k odd and $m > 0$ imply that $k(2^m k - 1)$ is odd. Thus, even though g_λ has larger degree than f, the exponent of 2 has been reduced by one.

It follows that for any real number λ, our inductive assumption (3.8) applies to g_λ. This means that g_λ has a root in $\mathbb{C}$. By definition, the roots of g_λ are $\alpha_i + \alpha_j - \lambda\alpha_i\alpha_j$. Thus one of these lies in $\mathbb{C}$. In other words, for each $\lambda \in \mathbb{R}$, we can find a pair i, j with $1 \le i < j \le n$ such that

$$\alpha_i + \alpha_j - \lambda\alpha_i\alpha_j \in \mathbb{C}.$$

Note that the pair i, j depends on λ—if we switch to a different value of λ, we might get a different pair. But as we vary over the infinitely many possible values of λ, there are only finitely many possibilities for the corresponding pair i, j. This implies that there must exist $\lambda \ne \mu$ that use the same pair i, j. Thus

(3.11) $\alpha_i + \alpha_j - \lambda\alpha_i\alpha_j \in \mathbb{C}$ and $\alpha_i + \alpha_j - \mu\alpha_i\alpha_j \in \mathbb{C}.$

Subtracting, we obtain

$$(\alpha_i + \alpha_j - \lambda\alpha_i\alpha_j) - (\alpha_i + \alpha_j - \mu\alpha_i\alpha_j) = (\mu - \lambda)\alpha_i\alpha_j \in \mathbb{C},$$

and since $\lambda \ne \mu$ are real, it follows that $\alpha_i\alpha_j \in \mathbb{C}$. Then $\alpha_i + \alpha_j - \lambda\alpha_i\alpha_j \in \mathbb{C}$ implies that $\alpha_i + \alpha_j \in \mathbb{C}$. Thus the sum and product of α_i, α_j are complex numbers.

Now consider the quadratic polynomial

$$(x - \alpha_i)(x - \alpha_j) = x^2 - (\alpha_i + \alpha_j)x + \alpha_i\alpha_j.$$

By what we just proved, it has coefficients in $\mathbb{C}$, so that its roots also lie in $\mathbb{C}$ by Lemma 3.2.3. But the roots are clearly α_i and α_j. This proves that $\alpha_i, \alpha_j \in \mathbb{C}$. Hence f has a complex root, which completes the proof of the theorem. $\square$

Mathematical Notes

As usual, this section has some interesting ideas to discuss.

■ **Proofs of the Fundamental Theorem.** There are many proofs of the Fundamental Theorem of Algebra. Students often see a proof in a course on complex analysis, but there are also some lovely proofs which use topology. The book [4] discusses a variety of proofs of the theorem, including a version of the proof given here. See also [6] for another proof and references to some of the many other proofs in the literature.

The proof of the Fundamental Theorem of Algebra given above is one of the more "algebraic" proofs. However, a closer inspection shows that our proof has three main ingredients:

- Every polynomial of odd degree in $\mathbb{R}[x]$ has a root in $\mathbb{R}$ (Proposition 3.2.2).
- Every complex number has a square root in $\mathbb{C}$ (this gives Lemma 3.2.3).
- Every polynomial splits completely over some extension field.

Of these three, only the last is purely algebraic. The proof of Proposition 3.2.2 uses the IVT, and as shown in Exercise 2, square roots of complex numbers reduce to square roots of positive real numbers, which exist by the IVT (if you're unfamiliar with this argument, do Exercise 3). Since the IVT depends on the completeness of $\mathbb{R}$, one could argue that the Fundamental Theorem of Algebra is really a theorem in analysis. See [1] for a discussion of these issues.

Once we have proved the main theorems of Galois theory, we will give an elegant proof of the Fundamental Theorem due to Artin in Theorem 8.5.8.

■ **Algebraically Closed Fields.** The Fundamental Theorem of Algebra leads to the following definition.

Definition 3.2.5. *A field F is called **algebraically closed** if every nonconstant polynomial in $F[x]$ splits completely over F.*

Theorem 3.2.4 shows that $\mathbb{C}$ is algebraically closed. We will see that there are other algebraically closed fields. In general, one can prove that every field has an algebraically closed extension.

■ **Real Closed Fields.** Another approach to the question of algebraic versus analytic is given by the theory of *real closed fields*. The basic idea is to make the above proof as algebraic as possible. We know from Exercise 2 that the existence of square roots of complex numbers follows directly from the existence of real square roots of positive real numbers. Then one defines a *real closed field* to be a field F that has the following properties:

- F has an order relation $>$ compatible with addition and with multiplication by positive elements (an element $\alpha \in F$ is *positive* if it satisfies $\alpha > 0$).
- Every positive element of F has a square root in F.
- Every polynomial of odd degree in F has a root in F.

The field of real numbers is the prototypical example of a real closed field, but it is not the only one.

Given a real closed field F, we can adjoin $i = \sqrt{-1}$ to F using the methods of Section 3.1 (e.g., Cauchy's method). This gives a field $F(i)$, and one can easily adapt the proof of Theorem 3.2.4 to show that $F(i)$ is algebraically closed. Details can be found in Exercises 4 and 5 (see [Jacobson, Vol. I, Sec. 5.1] for a complete treatment). There is also the related idea of a *formally real field*, due to Artin and Schreier. These fields have an interesting relation to Hilbert's Seventeenth Problem and are discussed in [Jacobson, Vol. II, Ch. 11].

Historical Notes

In 1749 Euler attempted to prove the Fundamental Theorem of Algebra for $f \in \mathbb{R}[x]$ using induction on the exponent of 2 in $\deg(f)$. To give the flavor of his proof, consider the case when $\deg(f) = 2^m$. The idea is to write f as a product

$$(3.12) \qquad\qquad f = gh$$

where g, h have degree 2^{m-1}. Euler did this by finding the equations satisfied by the coefficients of g, h and then showing that they have real solutions. It follows that coefficients of g and h can be chosen to be real. Once this is done, our inductive assumption implies that g and h have roots in $\mathbb{C}$.

Euler's proof has some major gaps, and in 1772 Lagrange wrote *Sur la forme des racines imaginaires des équations* [Lagrange, pp. 479–516] to make Euler's argument more rigorous. Lagrange's proof is almost complete—the difficulty comes when some of the polynomials in the proof have multiple roots, which might cause certain denominators to vanish. Lagrange was well aware of this problem and gave some very interesting arguments to deal with multiple roots. Many authors (including [Tignol] and [2]) accept Lagrange's argument as complete, though I think that some subtle gaps still remain. See [1] and [2] for more on the history of all this.

We next turn to Gauss's 1815 proof of the Fundamental Theorem of Algebra, which appears in [Gauss, Vol. III, pp. 31–56] (see [5, pp. 292–306] for an English translation). The overall strategy of Gauss's argument is similar to what we did in Theorem 3.2.4, with one major exception: he never uses the roots α_i of f. He begins instead with the universal situation and defines

$$z = \prod_{1 \le i < j \le n} \left(x - (x_i + x_j)u + x_i x_j \right) \in \mathbb{R}[u, x, x_1, \ldots, x_n].$$

This is similar to the polynomial G_λ defined in (3.9), except that u is now a variable. Gauss observes that z is a polynomial in x and u whose coefficients are symmetric in the x_i. Hence the coefficients are polynomials in the σ_i. He then replaces each σ_i with a new variable u_i. This gives a new polynomial

$$\zeta = \zeta(x, u, u_1, \ldots, u_n) \in \mathbb{R}[x, u, u_1, \ldots, u_n].$$

Thus Gauss is using the isomorphism $\mathbb{R}[u_1, \ldots, u_n] \simeq \mathbb{R}[\sigma_1, \ldots, \sigma_n]$, which we proved in Section 2.2 using arguments taken from this paper of Gauss. Then, given

$$f = x^n + a_1 x^{n-1} + \cdots + a_n \in \mathbb{R}[x],$$

he uses the substitution $u_i \mapsto (-1)^i a_i$ to send ζ to a polynomial

$$(3.13) \qquad\qquad Z = Z(x, u) \in \mathbb{R}[x, u].$$

In this way, Gauss gets an analog of $g_\lambda(x)$ without knowing the roots of f.

From here, Gauss's argument departs from what we did in Theorem 3.2.4. One difference is that he considers only monic polynomials with nonvanishing discriminant (to be called *separable* in Section 5.3). Other aspects of Gauss's proof are discussed in [1].

In his proof, Gauss uses the methods of Lagrange, which apply to the universal polynomial $\tilde{f}$ studied in Chapter 2. Although these methods are powerful, they can be cumbersome to use in practice. What we really need are methods which apply directly to any field. This leads to the language of field extensions, which is the main topic of the next chapter.

Exercises for Section 3.2

Exercise 1. For $f \in \mathbb{C}[x]$, define $\bar{f}$ as in (3.5).
(a) Show carefully that $\overline{fg} = \bar{f}\bar{g}$ for $f, g \in \mathbb{C}[x]$.
(b) Let $\alpha \in \mathbb{C}$. Show that $\bar{f}(\alpha) = 0$ implies that $f(\bar{\alpha}) = 0$.

Exercise 2. In Section A.2, we used polar coordinates to construct square (and higher) roots of complex numbers. In this exercise, you will give an elementary argument that every complex number has a square root. The only fact you will use (besides standard algebra) is that every positive real number has a real square root.
(a) First explain why every real number has a square root in $\mathbb{C}$.
(b) Now fix $a + bi \in \mathbb{C}$ with $b \neq 0$. For $x, y \in \mathbb{R}$, show that the equation $(x + iy)^2 = a + bi$ is equivalent to the equations

$$x^2 - y^2 = a, \quad 2xy = b.$$

(c) Show that the equations of part (b) are equivalent to

$$x^2 = \frac{a \pm \sqrt{a^2 + b^2}}{2}, \quad y = \frac{b}{2x}.$$

Also show that $x \neq 0$ and that $a \pm \sqrt{a^2 + b^2}$ is positive when we choose the $+$ sign in the formula for x^2.
(d) Conclude that $a + bi$ has a square root in $\mathbb{C}$.

Exercise 3. Use the IVT to prove that every positive real number a has a real square root.

Exercise 4. A field F is an *ordered field* if there is a subset $P \subset F$ such that:
(a) P is closed under addition and multiplication.
(b) For any $a \in F$, exactly one of the following is true: $a \in P$, $a = 0$, or $-a \in P$.
One then defines $a > b$ to mean $a - b \in P$ (so that P becomes the set of positive elements). From this, one can prove all of the typical properties of $>$. Now let F be an ordered field. Prove that -1 is not a square in F.

Exercise 5. Let F be a *real closed field*. As in the text, this means that F is an ordered field (see Exercise 4) such that every positive element of F has a square root in F and every $f \in F[x]$ of odd degree has a root in F.
(a) Use Exercise 4 to show that $x^2 + 1$ is irreducible over F. Then define $F(i)$ to be the field $F[x]/\langle x^2 + 1 \rangle$. By the Cauchy construction described in Section 3.1, elements of $F(i)$ can be written $a + bi$ for $a, b \in F$.
(b) Show that every quadratic polynomial in $F(i)$ splits completely over $F(i)$.
(c) Prove that $F(i)$ is algebraically closed.

Exercise 6. Here is yet another way to state the Fundamental Theorem of Algebra.
(a) Suppose that $f(\alpha) = 0$, where $f \in \mathbb{R}[x]$ and $\alpha \in \mathbb{C}$. Prove that $f(\bar{\alpha}) = 0$.
(b) Prove that the Fundamental Theorem of Algebra is equivalent to the assertion that every nonconstant polynomial in $\mathbb{R}[x]$ is a product of linear and quadratic factors with real coefficients.

Exercise 7. Prove that a field F is algebraically closed if and only if every nonconstant polynomial in $F[x]$ has a root in F.

REFERENCES

1. I. G. Bashmakova and A. N. Rudakov (with the assistance of A. N. Parshin and E. I. Slavutin), *Algebra and Algebraic Number Theory*, in *Mathematics of the 19th Century: Mathematical Logic, Algebra, Number Theory, Probability Theory*, edited by A. N. Kolmogorov and A. P. Yushkevich, English translation by A. Shenitzer, Birkhäuser, Boston, Basel, Berlin, 1992. Second Revised Edition, 2001.

2. I. G. Bashmakova and G. S. Smirnova, *The Beginnings and Evolution of Algebra*, English translation by A. Shenitzer, MAA, Washington, DC, 1999.

3. D. Cox, J. Little, and D. O'Shea, *Ideals, Varieties and Algorithms*, Second Edition, Springer-Verlag, New York, Berlin, Heidelberg, 1997.

4. B. Fine and G. Rosenberger, *The Fundamental Theorem of Algebra*, Springer-Verlag, New York, Berlin, Heidelberg, 1997.

5. D. E. Smith, *A Source Book in Mathematics*, Volume One, Ginn, Boston, New York, 1925. Reprint by Dover, New York, 1959.

6. D. Velleman, *Another proof of the Fundamental Theorem of Algebra*, Math. Mag. **70** (1997), 216–217.

Part II

Fields

In the next four chapters, we shift our attention from polynomials to fields.

We begin by developing the basic language of *field extensions* in Chapter 4. One of the key concepts is the *degree* of an extension. We also consider the special role played by *irreducible* polynomials.

Chapter 5 continues our study of fields by considering *splitting fields*, which are fields obtained by adjoining the roots of a given polynomial. This leads naturally to the notion of a *normal extension*. Finally, we introduce the idea of *separability*, which for a polynomial means not having multiple roots.

We introduce the *Galois group* in Chapter 6, and we explain how it relates to permutations of roots in the case of a splitting field. We also give some nontrivial examples and, in a optional section, discuss Abel's notion of an *Abelian equation*.

Finally, Chapter 7 defines the key ideas of a *Galois extension* and the *Galois correspondence*. After stating and proving the *Fundamental Theorem of Galois Theory*, we give some simple applications.

4

Extension Fields

This chapter will develop the language of *algebraic extensions*, which is needed to prove the main theorems of Galois theory. Recall from Chapter 3 that an *extension* of a field F consists of a field L and a ring homomorphism

$$\varphi : F \longrightarrow L.$$

As before, we will identify F with its image $\varphi(F)$ in L. In this way, we will write a field extension as $F \subset L$.

4.1 ELEMENTS OF EXTENSION FIELDS

Given a field extension $F \subset L$, elements of the larger field can relate to the smaller field in two different ways.

Definition 4.1.1. *Let L be an extension of F, and let $\alpha \in L$. Then α is **algebraic** over F if there is a nonconstant polynomial $f \in F[x]$ such that $f(\alpha) = 0$. If α is not algebraic over F, then α is **transcendental** over F.*

For example, $\sqrt{2} \in \mathbb{R}$ is algebraic over $\mathbb{Q}$, since $\sqrt{2}$ is a root of $x^2 - 2 \in \mathbb{Q}[x]$, and $\zeta_n = e^{2\pi i/n} \in \mathbb{C}$ is algebraic over $\mathbb{Q}$, since it is a root of $x^n - 1 \in \mathbb{Q}[x]$. The numbers π and e are transcendental over $\mathbb{Q}$, though this not easy to prove.

Example 4.1.2. To show that $\sqrt{2} + \sqrt{3}$ is algebraic over $\mathbb{Q}$, consider the polynomial

$$(x - \sqrt{2} - \sqrt{3})(x - \sqrt{2} + \sqrt{3})(x + \sqrt{2} - \sqrt{3})(x + \sqrt{2} + \sqrt{3}).$$

Multiplying this out gives $x^4 - 10x^2 + 1$. Thus $\sqrt{2} + \sqrt{3}$ is the root of a nonconstant polynomial in $\mathbb{Q}[x]$. We will return to this example many times. ◁▷

In Section 4.4 we will generalize Example 4.1.2 by showing that if $\alpha, \beta \in L$ are algebraic over F, then so are $\alpha + \beta$ and $\alpha\beta$. Furthermore, in Exercise 1 you will show that if $\alpha \neq 0$ is algebraic over F, then so is $1/\alpha$. This will imply that the set $\{\alpha \in L \mid \alpha$ is algebraic over $F\}$ is a subfield of L.

A. Minimal Polynomials. When $\alpha \in L$ is algebraic over F, there may be many nonconstant polynomials in $F[x]$ with α as a root. One of these polynomials is especially nice.

Lemma 4.1.3. *If $\alpha \in L$ is algebraic over F, then there is a unique nonconstant monic polynomial $p \in F[x]$ with the following two properties:*
(a) *α is a root of p, i.e., $p(\alpha) = 0$.*
(b) *If $f \in F[x]$ is any polynomial with α as a root, then f is a multiple of p.*

Proof. Among all nonconstant polynomials in $F[x]$ with α as a root, there must be one of smallest degree. Pick one such polynomial and call it p. Multiplying by a constant if necessary, we may assume that p is monic.

This polynomial certainly satisfies (a). As for (b), suppose that $f(\alpha) = 0$ for some $f \in F[x]$. The division algorithm from Section A.1 gives us polynomials $q, r \in F[x]$ such that

$$f = qp + r, \quad r = 0 \text{ or } \deg(r) < \deg(p).$$

Evaluating this equation at α gives

$$0 = f(\alpha) = q(\alpha)p(\alpha) + r(\alpha) = r(\alpha),$$

where the last equality uses $p(\alpha) = 0$. If r had strictly smaller degree than p, this would contradict the definition of p, and $r = 0$ follows. Thus p satisfies (b).

Finally, to prove uniqueness, suppose that another monic polynomial $\tilde{p}$ satisfied properties (a) and (b). Then applying (b) for p to $f = \tilde{p}$ implies that p divides $\tilde{p}$, and reversing the roles of p and $\tilde{p}$, we see that $\tilde{p}$ divides p. Since these are monic polynomials, it follows easily that $\tilde{p} = p$ (see Exercise 2 for details). □

It is customary to name the polynomial of Lemma 4.1.3 as follows.

Definition 4.1.4. *Let $\alpha \in L$. If α is algebraic over F, then the polynomial p of Lemma 4.1.3 is called the **minimal polynomial** of α over F.*

Besides the characterization given in Lemma 4.1.3, there are other ways to think about the minimal polynomial.

Proposition 4.1.5. *Let $\alpha \in L$ be algebraic over F, and let $p \in F[x]$ be its minimal polynomial. If $f \in F[x]$ is a nonconstant monic polynomial, then*

$$f = p \iff f \text{ is a polynomial of minimal degree satisfying } f(\alpha) = 0$$
$$\iff f \text{ is irreducible over } F \text{ and } f(\alpha) = 0.$$

Proof. The first equivalence follows from the proof of Lemma 4.1.3. For the second, we prove that the minimal polynomial p is irreducible over F as follows. If $p = gh$, where $g, h \in F[x]$ have strictly smaller degree than p, then $0 = p(\alpha) = g(\alpha)h(\alpha)$ would imply $g(\alpha) = 0$ or $h(\alpha) = 0$, which would contradict the first equivalence.

Conversely, suppose that $f(\alpha) = 0$ and f is irreducible. Hence p divides f by Lemma 4.1.3, so that $f = ph$ with $h \in F[x]$. Since f is irreducible and p is nonconstant, h must be constant. Then $f = p$ follows, since f and p are monic. $\square$

Here are some examples of minimal polynomials.

Example 4.1.6. The minimal polynomial of $\sqrt{2}$ over $\mathbb{Q}$ is $x^2 - 2$. This follows from the irrationality of $\sqrt{2}$, which implies that $\sqrt{2}$ can't be the root of a polynomial of degree 1 in $\mathbb{Q}[x]$. ◁▷

Example 4.1.7. For $\sqrt{2} + \sqrt{3}$, we showed in Example 4.1.2 that $\sqrt{2} + \sqrt{3}$ is a root of $x^4 - 10x^2 + 1$. But is this the minimal polynomial? By Proposition 4.1.5, this is equivalent to $x^4 - 10x^2 + 1$ being irreducible over $\mathbb{Q}$. For an explicit polynomial, the easiest way to check for irreducibility is by computer. For example, the *Mathematica* command

$$\text{Factor[x^4-10x^2+1]}$$

will produce the output $x^4 - 10x^2 + 1$, which means that the polynomial is irreducible over $\mathbb{Q}$. In *Maple*, the command would be

$$\text{factor(x^4-10*x^2+1);}$$

and again the output $x^4 - 10x^2 + 1$ proves irreducibility over $\mathbb{Q}$. Thus $x^4 - 10x^2 + 1$ is the minimal polynomial of $\sqrt{2} + \sqrt{3}$. In Section 4.2 we will say more about using *Mathematica* and *Maple* to check irreducibility. ◁▷

Example 4.1.8. The minimal polynomial of $\zeta_n = e^{2\pi i/n}$ over $\mathbb{Q}$ is called the nth *cyclotomic polynomial* and is denoted $\Phi_n(x)$. In Chapter 9 we will show that $\Phi_n(x)$ has degree $\phi(n)$, where ϕ is the Euler ϕ-function from number theory. ◁▷

B. Adjoining Elements. We next show how to describe some interesting subrings and subfields of a given extension $F \subset L$. Given $\alpha_1, \dots, \alpha_n \in L$, we define

$$F[\alpha_1, \dots, \alpha_n] = \{h(\alpha_1, \dots, \alpha_n) \mid h \in F[x_1, \dots, x_n]\}.$$

Hence $F[\alpha_1, \dots, \alpha_n]$ consists of all polynomial expressions in L that can be formed using $\alpha_1, \dots, \alpha_n$ with coefficients in F. Then let

$$F(\alpha_1, \dots, \alpha_n) = \left\{ \frac{\alpha}{\beta} \,\middle|\, \alpha, \beta \in F[\alpha_1, \dots, \alpha_n], \ \beta \neq 0 \right\}.$$

Thus $F(\alpha_1, \dots, \alpha_n)$ is the set of all rational expressions in the α_i with coefficients in F. We can characterize $F(\alpha_1, \dots, \alpha_n)$ as follows.

Lemma 4.1.9. $F(\alpha_1, \ldots, \alpha_n)$ *is the smallest subfield of the field L containing F and $\alpha_1, \ldots, \alpha_n$.*

Proof. We leave it as Exercise 3 to show that $F(\alpha_1, \ldots, \alpha_n)$ is a subfield of L. Thus, to prove the lemma, we must show that if K is a subfield of L containing F and $\alpha_1, \ldots, \alpha_n$, then $F(\alpha_1, \ldots, \alpha_n) \subset K$. This is what "smallest" means in the statement of the lemma.

Suppose that $K \subset L$ contains F and $\alpha_1, \ldots, \alpha_n$. Since K is closed under multiplication and addition, it follows that $p(\alpha_1, \ldots, \alpha_n) \in K$ for any polynomial $p \in F[x_1, \ldots, x_n]$. This shows that $F[\alpha_1, \ldots, \alpha_n] \subset K$. Then $F(\alpha_1, \ldots, \alpha_n) \subset K$ follows immediately, since K is a field. $\qquad\qquad\square$

Since $F(\alpha_1, \ldots, \alpha_n)$ is a subfield of L containing F, we get extensions

$$F \subset F(\alpha_1, \ldots, \alpha_n) \subset L.$$

We say that $F(\alpha_1, \ldots, \alpha_n)$ is obtained from F by *adjoining* $\alpha_1, \ldots, \alpha_n \in L$. We can use this to construct fields as follows.

Example 4.1.10. Consider the polynomial $x^4 - 2 \in \mathbb{Q}[x]$. Over the complex numbers, this factors as

$$x^4 - 2 = (x - \sqrt[4]{2})(x + \sqrt[4]{2})(x - i\sqrt[4]{2})(x + i\sqrt[4]{2}),$$

since the roots of $x^4 - 2$ are $\pm\sqrt[4]{2}, \pm i\sqrt[4]{2}$. It follows that

$$\mathbb{Q}(\sqrt[4]{2}, -\sqrt[4]{2}, i\sqrt[4]{2}, -i\sqrt[4]{2})$$

is the smallest field over which $x^4 - 2$ splits completely. We will see in Section 5.1 that this is an example of a *splitting field*.

This field can be described more compactly as

$$(4.1) \qquad\qquad \mathbb{Q}(\sqrt[4]{2}, -\sqrt[4]{2}, i\sqrt[4]{2}, -i\sqrt[4]{2}) = \mathbb{Q}(i, \sqrt[4]{2}).$$

To see why, let $K = \mathbb{Q}(\sqrt[4]{2}, -\sqrt[4]{2}, i\sqrt[4]{2}, -i\sqrt[4]{2})$ and $L = \mathbb{Q}(i, \sqrt[4]{2})$. Then $K \subset L$ follows from Lemma 4.1.9, since L contains $\mathbb{Q}$ and $\pm\sqrt[4]{2}, \pm i\sqrt[4]{2}$. For the opposite inclusion, note that

$$i = \frac{i\sqrt[4]{2}}{\sqrt[4]{2}} \in K.$$

Since K obviously contains $\mathbb{Q}$ and $\sqrt[4]{2}$, we have $L \subset K$, and (4.1) follows. $\qquad\triangleleft\triangleright$

Lemma 4.1.9 also implies that we can adjoin elements to a field in stages. More precisely, we have the following corollary.

Corollary 4.1.11. *If $F \subset L$ and $\alpha_1, \ldots, \alpha_n \in L$, then*

$$F(\alpha_1, \ldots, \alpha_n) = F(\alpha_1, \ldots, \alpha_r)(\alpha_{r+1}, \ldots, \alpha_n)$$

for any $1 \leq r \leq n - 1$.

Proof. The field on the right is obtained by first adjoining $\alpha_1, \ldots, \alpha_r$ to F to get the field $F(\alpha_1, \ldots, \alpha_r)$ and then adjoining $\alpha_{r+1}, \ldots, \alpha_n$ to $F(\alpha_1, \ldots, \alpha_r)$ to get the field $F(\alpha_1, \ldots, \alpha_r)(\alpha_{r+1}, \ldots, \alpha_n)$. This field obviously contains F and $\alpha_1, \ldots, \alpha_r, \alpha_{r+1}, \ldots, \alpha_n$. Then Lemma 4.1.9 implies that

$$F(\alpha_1, \ldots, \alpha_n) \subset F(\alpha_1, \ldots, \alpha_r)(\alpha_{r+1}, \ldots, \alpha_n).$$

The opposite inclusion is similar and is left as Exercise 4.   □

Here is a simple example of why this corollary is useful.

Example 4.1.12. Corollary 4.1.11 implies that $\mathbb{Q}(\sqrt{2}, \sqrt{3}) = \mathbb{Q}(\sqrt{2})(\sqrt{3})$. Then

$$\mathbb{Q} \subset \mathbb{Q}(\sqrt{2}) \subset \mathbb{Q}(\sqrt{2})(\sqrt{3}) = \mathbb{Q}(\sqrt{2}, \sqrt{3})$$

shows that we get $\mathbb{Q}(\sqrt{2}, \sqrt{3})$ by first adjoining $\sqrt{2}$ to $\mathbb{Q}$ and then adjoining $\sqrt{3}$ to $\mathbb{Q}(\sqrt{2})$. Representing an extension this way will be very useful.   ◁▷

We next study $F(\alpha_1, \ldots, \alpha_n)$ and $F[\alpha_1, \ldots, \alpha_n]$ when $\alpha_1, \ldots, \alpha_n$ are algebraic over F. We begin with the case of adjoining a single element.

Lemma 4.1.13. *Assume that $F \subset L$ is a field extension, and let $\alpha \in L$ be algebraic over F with minimal polynomial $p \in F[x]$. Then there is a unique ring isomorphism*

$$F[\alpha] \simeq F[x]/\langle p \rangle$$

that is the identity on F and maps α to the coset $x + \langle p \rangle$.

Proof. Consider the ring homomorphism $\varphi : F[x] \to L$ that sends $h(x) \in F[x]$ to $h(\alpha) \in L$. By definition, the image of φ is $F[\alpha]$. As for the kernel, we claim that $\text{Ker}(\varphi) = \langle p \rangle$. To prove this, first note that $g \in F[x]$ implies that

$$\varphi(gp) = \varphi(g)\,\varphi(p) = g(\alpha)p(\alpha) = g(\alpha)\,0 = 0.$$

This shows that $\langle p \rangle \subset \text{Ker}(\varphi)$. For the other inclusion, suppose that $f \in \text{Ker}(\varphi)$. Then $f(\alpha) = 0$, which by part (b) of Lemma 4.1.3 implies that f is a multiple of p. Thus $\text{Ker}(\varphi) \subset \langle p \rangle$, and $\text{Ker}(\varphi) = \langle p \rangle$ follows.

Since we know the image and kernel of φ, the Fundamental Theorem of Ring Homomorphisms (Theorem A.1.9) gives a ring isomorphism

$$F[x]/\langle p \rangle \simeq F[\alpha].$$

This isomorphism is the identity on F and maps the coset $x + \langle p \rangle$ to α. Its inverse is the isomorphism described in the statement of the lemma.

Finally, uniqueness follows since a ring homomorphism defined on $F[\alpha]$ is uniquely determined by its values on F and α.   □

This lemma shows how to represent $F[\alpha]$ as a quotient ring. However, we also know that the minimal polynomial p of α is irreducible (Proposition 4.1.5). As we saw in Proposition 3.1.1, this implies that $F[x]/\langle p \rangle$ is a field. By Lemma 4.1.13, it follows that $F[\alpha]$ is a field when α is algebraic over F. Hence we have proved part of the following proposition.

Proposition 4.1.14. *Assume that $F \subset L$ is a field extension, and let $\alpha \in L$. Then α is algebraic over F if and only if $F[\alpha] = F(\alpha)$.*

Proof. When α is algebraic over F, the above paragraph shows that $F[\alpha]$ is a field containing F and α. Since $F(\alpha)$ is the smallest subfield of L containing F and α (Lemma 4.1.9), it follows that $F(\alpha) \subset F[\alpha]$. The opposite inclusion always holds, so that $F(\alpha) = F[\alpha]$ when α is algebraic over F.

For the other implication, suppose that $F[\alpha] = F(\alpha)$. We may assume that $\alpha \neq 0$ since 0 is obviously algebraic over F. Then $1/\alpha \in F(\alpha) = F[\alpha]$ implies that

$$1/\alpha = a_0 + a_1\alpha + \cdots + a_m\alpha^m$$

for some $a_0, \ldots, a_m \in F$. Thus

$$0 = -1 + a_0\alpha + a_1\alpha^2 + \cdots + a_m\alpha^{m+1},$$

proving that α is algebraic over F. $\qquad\qquad\square$

We next study what happens when we adjoin several algebraic elements to a field.

Proposition 4.1.15. *Let $F \subset L$ be a field extension, and let $\alpha_1, \ldots, \alpha_n \in L$ be algebraic over F. Then*

$$F[\alpha_1, \ldots, \alpha_n] = F(\alpha_1, \ldots, \alpha_n).$$

Proof. By the argument used in the proof of Proposition 4.1.14, it suffices to prove that $F[\alpha_1, \ldots, \alpha_n]$ is a field. We will do this by induction on n. The case $n = 1$ is covered by Proposition 4.1.14. Now suppose that $n > 1$ and that $F[\alpha_1, \ldots, \alpha_{n-1}]$ is a field. We know that $f(\alpha_n) = 0$ for some nonconstant $f \in F[x]$. We can regard f as having coefficients in the larger field $F[\alpha_1, \ldots, \alpha_{n-1}]$, so that α_n is algebraic over $F[\alpha_1, \ldots, \alpha_{n-1}]$. Then Proposition 4.1.14 implies that

$$F[\alpha_1, \ldots, \alpha_{n-1}][\alpha_n]$$

is a field. We leave it as Exercise 5 to show that this equals $F[\alpha_1, \ldots, \alpha_n]$. $\quad\square$

Here is an example of Proposition 4.1.15.

Example 4.1.16. Consider $\mathbb{Q}(\sqrt{2}, \sqrt{3})$. The above proposition shows that this equals $\mathbb{Q}[\sqrt{2}, \sqrt{3}]$, so that every element of $\mathbb{Q}(\sqrt{2}, \sqrt{3})$ is a polynomial in $\sqrt{2}, \sqrt{3}$ with rational coefficients. Furthermore, since $\sqrt{2}^{2n} = 2^n$ and $\sqrt{2}^{2n+1} = 2^n\sqrt{2}$, and similarly for powers of $\sqrt{3}$, it follows easily that

$$(4.2) \qquad \mathbb{Q}(\sqrt{2}, \sqrt{3}) = \{a + b\sqrt{2} + c\sqrt{3} + d\sqrt{6} \mid a, b, c, d \in \mathbb{Q}\}.$$

In Section 4.3, we will show that the representation of elements of $\mathbb{Q}(\sqrt{2}, \sqrt{3})$ given by (4.2) is unique. $\qquad\qquad\qquad\qquad\qquad\qquad\qquad\qquad\qquad\qquad\qquad\triangleleft\triangleright$

Mathematical Notes

Let us discuss two of the ideas that have appeared in this section.

▪ **The Structure of Fields.** Consider a field of the form $F(\alpha_1, \ldots, \alpha_n)$. In Proposition 4.1.15, we studied the case when the α_i are all algebraic over F. The other extreme is when the α_i are not only transcendental over F but also *algebraically independent*, which (as defined in Mathematical Notes to Section 2.2) means that the α_i satisfy no nontrivial polynomial relation with coefficients in F. In Exercise 6 you will show that this implies that $F(\alpha_1, \ldots, \alpha_n)$ is isomorphic to the field of rational functions $F(x_1, \ldots, x_n)$. We call $F(\alpha_1, \ldots, \alpha_n)$ a *purely transcendental* extension of F in this case.

For the general case, a result of Steinitz says that a field $L = F(\alpha_1, \ldots, \alpha_n)$ can always be written in the form

$$F \subset K = F(\beta_1, \ldots, \beta_m) \subset K(\gamma_1, \ldots, \gamma_l) = L$$

where $m \le n$, $\beta_1, \ldots, \beta_m$ are algebraically independent over F (so that $F \subset K$ is purely transcendental), and $\gamma_1, \ldots, \gamma_l$ are algebraic over K. A proof of this theorem can be found in [Jacobson, Vol. II, Sec. 8.12].

▪ **Number Fields.** A field of the form $\mathbb{Q}(\alpha_1, \ldots, \alpha_n)$, where $\alpha_1, \ldots, \alpha_n$ are algebraic over $\mathbb{Q}$, is called a *number field*. The fields appearing in Examples 4.1.10 and 4.1.16 are number fields. These fields and their Galois theory occupy a central role in *algebraic number theory*.

Historical Notes

Fields have been used implicitly ever since the discovery of addition, subtraction, multiplication, and division. Cardan's formulas, dating from the sixteenth century, use $\mathbb{Q}$, $\mathbb{R}$, and $\mathbb{C}$. The field of rational functions in n variables arises naturally when considering symmetric functions, and Lagrange used such fields (implicitly) in his 1770 study of the roots of polynomials. Number fields also appeared around this time. For example, Euler used the fields $\mathbb{Q}(\sqrt{-2})$ and $\mathbb{Q}(\sqrt{-3})$ to study problems in number theory raised by Fermat.

The first reasonably general definition of $F(\alpha_1, \ldots, \alpha_n)$ was given by Galois in 1831, where he says the following:

> One can agree to regard as rational all rational functions in a certain number of determinate quantities, which are supposed to be known a priori. For example, one can chose a certain root of an integer, and regard as rational all rational functions of the radical.
>
> When we agree in this way to regard certain quantities as known, we say that we ADJOIN them to the equation that we are trying to solve.

(See [Galois, p. 45].) This is why we say that $F(\alpha_1, \ldots, \alpha_n)$ is obtained from F by *adjoining* $\alpha_1, \ldots, \alpha_n$. On the other hand, Abel was the first person to understand that $F[\alpha] = F(\alpha)$ when α is algebraic over F (see [Abel, Vol. I, pp. 66–72]).

Abel, Galois, and their predecessors tended to work with explicitly constructed fields. The first truly "abstract" notion of field is due to Dedekind. In 1877, he gave the following definition:

> I call a system A of numbers a (not all zero) a *field* when the sum, difference, product and quotient of any two of numbers in A also belongs to A.

(See [2, p. 107].) This is not completely general, for the "numbers" in this definition are all complex. From our point of view, Dedekind is really defining a subfield of $\mathbb{C}$. But his definition is modern in spirit, in that he allows any set (he says "system" because set theory was not fully established in 1877) that behaves nicely under addition, subtraction, multiplication, and division. This is very different from his great contemporary Kronecker, who took a more conservative view and only dealt with fields that could be constructed explicitly in finitely many steps. It wasn't until 1893 that Weber gave the first fully abstract definition of field. Weber's definition is similar to the one in use today. A discussion of the evolution of the field concept can be found in [6]. See also [7] for the evolution of the ring concept.

Exercises for Section 4.1

Exercise 1. Let $\alpha \in L \setminus \{0\}$ be algebraic over a subfield F. Prove that $1/\alpha$ is also algebraic over F.

Exercise 2. Complete the proof of Lemma 4.1.3 by showing that if f and g are monic polynomials in $F[x]$ each of which divides the other, then $f = g$.

Exercise 3. Suppose that $F \subset L$ is a field extension and that $\alpha_1, \ldots, \alpha_n \in L$. Show that $F[\alpha_1, \ldots, \alpha_n]$ is a subring of L and that $F(\alpha_1, \ldots, \alpha_n)$ is a subfield of L.

Exercise 4. Complete the proof of Corollary 4.1.11 by showing that

$$F(\alpha_1, \ldots, \alpha_r)(\alpha_{r+1}, \ldots, \alpha_n) \subset F(\alpha_1, \ldots, \alpha_n).$$

Exercise 5. Prove carefully that $F[\alpha_1, \ldots, \alpha_{n-1}][\alpha_n] = F[\alpha_1, \ldots, \alpha_n]$.

Exercise 6. Suppose that $F \subset L$ and that $\alpha_1, \ldots, \alpha_n \in L$ are algebraically independent over F (as defined in the Mathematical Notes to Section 2.2). Prove that there is an isomorphism of fields

$$F(\alpha_1, \ldots, \alpha_n) \simeq F(x_1, \ldots, x_n),$$

where $F(x_1, \ldots, x_n)$ is the field of rational functions in variables $x_1, \ldots, x_n$.

Exercise 7. In the proof of Proposition 4.1.14, we used the quotient ring $F[x]/\langle p \rangle$ to show that $F[\alpha]$ is a field when α is algebraic over F with minimal polynomial $p \in F[x]$. Here, you will prove that $F[\alpha]$ is a field without using quotient rings. Since we know that $F[\alpha]$ is a ring, it suffices to show that every nonzero element $\beta \in F[\alpha]$ has a multiplicative inverse in $F[\alpha]$. So pick $\beta \neq 0$ in $F[\alpha]$. Then $\beta = g(\alpha)$ for some $g \in F[x]$.
 (a) Show that g and p are relatively prime in $F[x]$.
 (b) By part (a) and the Euclidean algorithm, we have $Ap + Bg = 1$ for some $A, B \in F[x]$. Prove that $B(\alpha) \in F[\alpha]$ is the multiplicative inverse of $g(\alpha)$.
Do you see how this exercise relates to Exercise 5 of Section 3.1?

Exercise 8. If a polynomial is irreducible over a field F, it may or may not remain irreducible over a larger field. Here are examples of both types of behavior.

(a) Prove that $x^2 - 3$ is irreducible over $\mathbb{Q}(\sqrt{2})$.

(b) In Example 4.1.7, we showed that $x^4 - 10x^2 + 1$ is irreducible over $\mathbb{Q}$ (it is the minimal polynomial of $\alpha = \sqrt{2} + \sqrt{3}$). Show that $x^4 - 10x^2 + 1$ is not irreducible over $\mathbb{Q}(\sqrt{3})$.

4.2 IRREDUCIBLE POLYNOMIALS

Since minimal polynomials are polynomial,irreducible, it should be clear that the notion of irreducibility plays an important role in field theory. However, given an arbitrary polynomial $f \in F[x]$, it may not be obvious that f is irreducible. How do we tell? In this section, we will discuss some ways of answering this question.

A. Using *Maple* and *Mathematica*. In the previous section we saw examples of how *Maple* and *Mathematica* factor polynomials over $\mathbb{Q}$ into irreducibles. These programs can also factor over *number fields*, which as in Section 4.1 are fields of the form $\mathbb{Q}(\alpha_1, \ldots, \alpha_n)$ with $\alpha_1, \ldots, \alpha_n$ algebraic over $\mathbb{Q}$.

We first describe how *Maple* factors polynomials over a number field. In Section 4.1 we used the `factor` command to show that $x^4 - 10x^2 + 1$ is the minimal polynomial of $\sqrt{2} + \sqrt{3}$ over $\mathbb{Q}$. To study this polynomial over $\mathbb{Q}(\sqrt{2})$, we use

```
factor(x^4-10*x^2+1,sqrt(2));
```

which gives the result

$$(x^2 - 2\sqrt{2}x - 1)(x^2 + 2\sqrt{2}x - 1).$$

This implies in particular that $x^2 - 2\sqrt{2}x - 1$ and $x^2 + 2\sqrt{2}x - 1$ are irreducible over $\mathbb{Q}(\sqrt{2})$. Similarly, the command

```
factor(x^4-10*x^2+1,[sqrt(2),sqrt(3)]);
```

gives

(4.3) $\qquad (x - \sqrt{2} - \sqrt{3})(x + \sqrt{2} + \sqrt{3})(x - \sqrt{2} + \sqrt{3})(x + \sqrt{2} - \sqrt{3}).$

This is the factorization of $x^4 - 10x^2 + 1$ over $\mathbb{Q}(\sqrt{2}, \sqrt{3})$.

Not all number fields have such simple descriptions. For example, consider the field $\mathbb{Q}(\sqrt{2} + \sqrt{3})$. Since the minimal polynomial of $\sqrt{2} + \sqrt{3}$ is $x^4 - 10x^2 + 1$, *Maple* would represent this algebraic number using the `RootOf` command. This is done most conveniently via

```
alias(alpha = RootOf(x^4-10*x^2+1)):
```

which makes α a root of $x^4 - 10x^2 + 1$. Then we can factor a polynomial `poly` in $\mathbb{Q}[x]$ or $\mathbb{Q}(\alpha)[x]$ using the command `factor(poly,alpha)`. For example, if we let $f = x^4 - 10x^2 + 1$, then we know that $x - \alpha$ is a factor of f over $\mathbb{Q}(\alpha)$. But what are the other factors? Using the command

```
factor(x^4-10*x^2+1,alpha);
```

we get the result

(4.4) $\qquad (x - \alpha)(x + \alpha)(x - 10\alpha + \alpha^3)(x + 10\alpha - \alpha^3).$

The surprise is that the polynomial factors completely. This has an interesting consequence concerning the fields $\mathbb{Q}(\sqrt{2} + \sqrt{3})$ and $\mathbb{Q}(\sqrt{2}, \sqrt{3})$.

In *Maple*, the factorization (4.4) takes place in $\mathbb{Q}[x]/\langle x^4 - 10x^2 + 1\rangle$. To get something involving numbers, consider the map $x \mapsto \sqrt{2} + \sqrt{3}$. This induces an isomorphism

$$\mathbb{Q}[x]/\langle x^4 - 10x^2 + 1\rangle \simeq \mathbb{Q}(\sqrt{2} + \sqrt{3})$$

and allows us to assume that $\alpha = \sqrt{2} + \sqrt{3}$ in (4.4).

By comparing (4.3) and (4.4), we conclude that $\alpha^3 - 10\alpha = \pm(\sqrt{2} - \sqrt{3})$, and then an easy numerical calculation shows that $\alpha^3 - 10\alpha = \sqrt{2} - \sqrt{3}$. Since $\alpha = \sqrt{2} + \sqrt{3}$, adding these two equations gives $\alpha^3 - 9\alpha = 2\sqrt{2}$, and it follows that $\sqrt{2} \in \mathbb{Q}(\alpha) = \mathbb{Q}(\sqrt{2} + \sqrt{3})$. Then we also have $\sqrt{3} = \alpha - \sqrt{2} \in \mathbb{Q}(\sqrt{2} + \sqrt{3})$. Then Lemma 4.1.9 implies that

$$\mathbb{Q}(\sqrt{2}, \sqrt{3}) \subset \mathbb{Q}(\sqrt{2} + \sqrt{3}).$$

Since the opposite inclusion clearly holds (be sure you can explain why), we get

(4.5) $\qquad \mathbb{Q}(\sqrt{2}, \sqrt{3}) = \mathbb{Q}(\sqrt{2} + \sqrt{3}).$

We can also do these computations in *Mathematica*. For example, factoring $x^4 - 10x^2 + 1$ over $\mathbb{Q}(\sqrt{2})$ is done by the command

```
Factor[x^4-10x^2+1, Extension -> {Sqrt[2]}]
```

and factoring over $\mathbb{Q}(\sqrt{2}, \sqrt{3})$ is done via

```
Factor[x^4-10x^2+1, Extension -> {Sqrt[2], Sqrt[3]}]
```

Finally, to work over the field generated by a root of an irreducible polynomial such as $x^4 - 10x^2 + 1$, one sets

```
a = Root[x^4-10x^2+1, 1]
```

Then the command

```
Factor[x^4-10x^2+1, Extension -> {a}]
```

produces a result similar to (4.4), except that a is replaced with the ungainly expression it represents. To get a nicer result, one should use the command

```
Factor[x^4-10x^2+1, Extension -> {a}] /. a -> b
```

which gives the result

$$(x - b)(x + b)(x - 10b + b^3)(x + 10b - b^3).$$

In general, *Maple* and *Mathematica* have roughly equivalent capabilities for computing with algebraic numbers.

B. Algorithms for Factoring. The use of *Maple* and *Mathematica* to factor polynomials over number fields implies the existence of an algorithm for doing so. To give the reader an idea of how factoring is done, we will describe an algorithm for deciding whether $f \in \mathbb{Z}[x]$ is irreducible over $\mathbb{Q}$. The key tool is Gauss's Lemma, which is Theorem A.3.2. We will use the following corollary of this result.

Corollary 4.2.1. *If $f \in \mathbb{Z}[x]$ has degree > 0 and is reducible (i.e., not irreducible) over $\mathbb{Q}$, then $f = gh$ where $g, h \in \mathbb{Z}[x]$ have degrees strictly smaller than $\deg(f)$.*

Proof. If f is reducible in $\mathbb{Q}[x]$, then $f = g_1 h_1$, where $g_1, h_1 \in \mathbb{Q}[x]$ have degrees $< \deg(f)$. By Gauss's Lemma, there is $\delta \in \mathbb{Q}$ such that $g = \delta g_1$ and $h = \delta^{-1} h_1$ have integer coefficients. Then $f = gh$ is the desired factorization. $\square$

We now describe an algorithm to test the irreducibility of $f \in \mathbb{Z}[x]$. Let $n = \deg(f) > 0$. First note that if $f(i) = 0$ for some $0 \leq i \leq n - 1$, then $x - i$ is a factor of f and we can quit. Hence, when performing the algorithm, we may assume that $f(0), \ldots, f(n - 1)$ are nonzero. Then create a set of polynomials as follows:

- Fix an integer $0 < d < n$.
- Fix divisors $a_0, \ldots, a_d \in \mathbb{Z}$ of $f(0), \ldots, f(d) \in \mathbb{Z}$.
- Use the Lagrange interpolation formula from Exercise 1 to construct a polynomial $g \in \mathbb{Q}[x]$ of degree $\leq d$ such that $g(i) = a_i$ for $i = 0, \ldots, d$.
- Accept g if it has degree d and integer coefficients; reject it otherwise.

Doing this for all $0 < d < n$ and all divisors $a_0 | f(0), \ldots, a_d | f(d)$ gives a set of polynomials $g \in \mathbb{Z}[x]$.

Proposition 4.2.2. *This set of polynomials $g \in \mathbb{Z}[x]$ is finite, and f is irreducible over $\mathbb{Q}$ if and only if it is not divisible by any of the polynomials in this set.*

Proof. We are assuming that $f(0), \ldots, f(d)$ are nonzero, so that each $f(i)$ has only finitely many divisors. Hence there are only finitely many choices for $a_0, \ldots, a_d \in \mathbb{Z}$. Since g is uniquely determined by the a_i, it follows that there are only finitely many such g's.

To finish the proof, we will show that f is reducible if and only if it is divisible by one of these polynomials. One direction is obvious. For the other direction, suppose that f is reducible. By Corollary 4.2.1, $f = gh$, where $g, h \in \mathbb{Z}[x]$ and g has degree d, $0 < d < n$.

Then, for $0 \leq i \leq d$, let $a_i = g(i)$, and note that $a_i | f(i)$, since $f(i) = g(i)h(i)$. The Lagrange interpolation formula gives $\tilde{g} \in \mathbb{Q}[x]$ of degree $\leq d$ with $\tilde{g}(i) = a_i$ for $0 \leq i \leq d$. Since $g - \tilde{g}$ has degree at most d and vanishes at the $d + 1$ numbers $0, \ldots, d$, it must be the zero polynomial. Hence $g = \tilde{g}$ is on our list. $\square$

Since there are known algorithms for factoring integers, there is an algorithm for computing the set of polynomials in $g \in \mathbb{Z}[x]$ used in Proposition 4.2.2. Then dividing these into our given polynomial f via the division algorithm gives an algorithm for deciding whether f is irreducible over $\mathbb{Q}$.

From a computational point of view, this algorithm is dreadful. The methods used by *Maple* and *Mathematica* are much more efficient. The book [1] describes some good algorithms for factoring polynomials over a number field.

C. The Schönemann–Eisenstein Criterion. While algorithms and computers can be extremely helpful in computing examples of irreducible polynomials, there are certain classes of polynomials that can be proved to be irreducible by traditional means. Here, we will prove the *Schönemann–Eisenstein irreducibility criterion*.

Theorem 4.2.3. *Let* $f = a_n x^n + \cdots + a_0 \in \mathbb{Z}[x]$ *have degree* $n > 0$. *If there is a prime* p *such that* $p \nmid a_n$, $p|a_{n-1}, \ldots, p|a_0$, *and* $p^2 \nmid a_0$, *then* f *is irreducible over* $\mathbb{Q}$.

Proof. By Corollary 4.2.1, if f is reducible over $\mathbb{Q}$, then there are $g, h \in \mathbb{Z}[x]$ of degree $< n$ such that $f = g h$. Now consider the ring homomorphism $\mathbb{Z}[x] \to \mathbb{F}_p[x]$ defined by sending $q = b_m x^m + \cdots + b_0 \in \mathbb{Z}[x]$ to $\bar{q} = [b_m]x^m + \cdots + [b_0] \in \mathbb{F}_p[x]$, where $[b] \in \mathbb{F}_p$ is the congruence class modulo p of $b \in \mathbb{Z}$.

Then $f = g h$ implies that $[a_n]x^n = \bar{g}\,\bar{h}$, since $p|a_{n-1}, \ldots, p|a_0$. However, $\mathbb{F}_p$ is a field, which means that unique factorization holds in $\mathbb{F}_p[x]$. Since $p \nmid a_n$, it follows that $\bar{g} = [a]x^r$ and $\bar{h} = [b]x^s$, where $[a][b] = [a_n]$ and $r + s = n$.

If $r = 0$, then $\bar{g} = [a]$ and $\deg(g) > 0$ would imply that the leading term of g is divisible by p. Then $f = g h$ would imply that the same is true for the leading term a_n of f. Thus $p \nmid a_n$ implies that $r > 0$, and $s > 0$ follows similarly.

But then $\bar{g} = [a]x^r$ for $r > 0$ implies that p divides the constant term of g, and the same is true for the constant term of h, since $s > 0$. Since the constant term a_0 of f is the product of the constant terms of g and h, it follows that $p^2|a_0$. This contradicts $p^2 \nmid a_0$ and completes the proof. $\qquad\square$

Here is a simple example to illustrate the Schönemann–Eisenstein criterion.

Example 4.2.4. Consider the polynomial

$$f = x^n + px + p, \quad n \geq 2, \ p \text{ prime}.$$

The Schönemann–Eisenstein criterion for the prime p implies immediately that f is irreducible over $\mathbb{Q}$, no matter what $n \geq 2$ we choose. $\qquad\triangleleft\triangleright$

The interesting feature of this example is that it cannot be done by *Maple* or *Mathematica*. For a specific n and p, we could check irreducibility by computer (assuming n and p aren't too big), but standard computer algebra systems can't factor polynomials with symbolic exponents. On the other hand, only very special polynomials satisfy the Schönemann–Eisenstein criterion. (If α is a root of a polynomial satisfying this criterion, then from the point of view of algebraic number theory, the extension $\mathbb{Q} \subset \mathbb{Q}(\alpha)$ is *totally ramified* at p, which is a rather rare phenomenon.)

We can use the Schönemann–Eisenstein criterion to determine the minimal polynomial of the pth root of unity $\zeta_p = e^{2\pi i/p}$, where p is prime. Using

$$x^p - 1 = (x - 1)(x^{p-1} + \cdots + x + 1),$$

we see that ζ_p is a root of $\Phi_p = x^{p-1} + \cdots + x + 1$. This is called the pth *cyclotomic polynomial*.

Proposition 4.2.5. $\Phi_p = x^{p-1} + \cdots + x + 1$ *is irreducible over* $\mathbb{Q}$ *when* p *is prime.*

Proof. First observe that $\Phi_p(x) = (x^p - 1)/(x - 1)$, so that

$$\Phi_p(x + 1) = \frac{(x + 1)^p - 1}{x}.$$

The binomial theorem tells us that

$$(x + 1)^p = x^p + \binom{p}{1}x^{p-1} + \cdots + \binom{p}{r}x^{p-r} + \cdots + \binom{p}{p-1}x + 1,$$

and then substituting this into the above formula for $\Phi_p(x + 1)$ gives

$$(4.6) \quad \Phi_p(x + 1) = x^{p-1} + \binom{p}{1}x^{p-2} + \cdots + \binom{p}{r}x^{p-r-1} + \cdots + \binom{p}{p-1}.$$

However, for $1 \leq r \leq p - 1$, the integer

$$\binom{p}{r} = \frac{p!}{r!(p-r)!} = \frac{p(p-1)\cdots(p-r+1)}{r!}$$

is divisible by p, since p divides the numerator but not the denominator (remember that p is prime). Furthermore, note that p^2 does not divide $\binom{p}{p-1} = p$. Then $\Phi_p(x + 1)$ is irreducible, since (4.6) satisfies the Schönemann–Eisenstein criterion. From here it is easy to see that Φ_p is irreducible, for a factorization $\Phi_p(x) = g(x)h(x)$ in $\mathbb{Q}[x]$ would imply $\Phi_p(x + 1) = g(x + 1)h(x + 1)$. If g and h have degree $< p - 1$, then the same would be true for $g(x + 1)$ and $h(x + 1)$, which would contradict the irreducibility of $\Phi_p(x + 1)$. This completes the proof. $\square$

It follows that the minimal polynomial of ζ_p over $\mathbb{Q}$ is $x^{p-1} + \cdots + x + 1$. In Chapter 9 we will describe the minimal polynomial of ζ_n for arbitrary n.

D. Prime Radicals. Given a prime p, our final task is to investigate when the polynomial $x^p - a \in F[x]$ is irreducible over F. Note that if α is a root of $x^p - a$, then $\alpha^p = a$, so that the roots of $x^p - a$ are the pth roots of a. Here is our result.

Proposition 4.2.6. *Let p be prime. Then $f = x^p - a \in F[x]$ is irreducible over F if and only if f has no roots in F.*

Proof. One direction is obvious, for if f has a root $\alpha \in F$, then $x - \alpha \in F[x]$ is a factor of f by Corollary A.1.15. Going the other way, we will assume that f is reducible and prove that f has a root in F.

We first study the roots of f. By Theorem 3.1.4, there is a field $F \subset L$ over which f splits completely, say

(4.7) $$f = (x - \alpha_1)(x - \alpha_2) \cdots (x - \alpha_p), \qquad \alpha_1, \ldots, \alpha_p \in L.$$

If $\alpha_1 = 0$, then f has a root in F. Thus we may assume that $\alpha_1 \neq 0$. If we set

$$\zeta_i = \frac{\alpha_i}{\alpha_1}$$

for $1 \leq i \leq p$, then $\alpha_i^p = a$ implies that

$$\zeta_i^p = \frac{\alpha_i^p}{\alpha_1^p} = \frac{a}{a} = 1.$$

It follows that $\alpha_i = \zeta_i \alpha_1$, where ζ_i is a pth root of unity. Hence (4.7) can be written

(4.8) $$f = (x - \zeta_1 \alpha_1)(x - \zeta_2 \alpha_1) \cdots (x - \zeta_p \alpha_1).$$

Now suppose that $f = gh$, where $g, h \in F[x]$ have degree $r, s < p$. We may assume that g, h are monic by multiplying them by suitable constants if necessary. By $f = gh$ and unique factorization, g must be a product of r of the factors of (4.8). After relabeling if necessary, we may assume that

$$g = (x - \zeta_1 \alpha_1)(x - \zeta_2 \alpha_1) \cdots (x - \zeta_r \alpha_1).$$

Since the constant term of g lies in F, this implies that

$$\zeta \alpha_1^r \in F, \quad \text{where } \zeta = \zeta_1 \cdots \zeta_r.$$

Note also that $\zeta^p = 1$.

Since $0 < r < p$ and p is prime, $mr + np = 1$ for some $m, n \in \mathbb{Z}$. Then

$$\zeta^m \alpha_1 = \zeta^m \alpha_1^{mr+np} = \left(\zeta \alpha_1^r\right)^m \left(\alpha_1^p\right)^n \in F$$

since $\zeta \alpha_1^r \in F$ and $\alpha_1^p = a \in F$. It follows that $\zeta^m \alpha_1 \in F$. Thus

$$\left(\zeta^m \alpha_1\right)^p = \left(\zeta^p\right)^m \alpha_1^p = a$$

shows that $\zeta^m \alpha_1$ is a root of $f = x^p - a$ lying in F.   □

The pth roots of unity used in the above proof are more abstract than the roots of unity constructed in Section A.2.

Here is an easy application of Proposition 4.2.6 that will be useful when we study the *casus irreducibilis* in Chapter 8.

Example 4.2.7. Let F be a subfield of $\mathbb{R}$ and p be an odd prime. Given $a \in F$, we define $\sqrt[p]{a}$ to be the real pth root of a. Furthermore, since p is odd, $\sqrt[p]{a}$ is the *only* real pth root of a (be sure you understand why). Then Proposition 4.2.6 implies that $x^p - a$ is irreducible over F if and only if $\sqrt[p]{a} \notin F$.   ◁▷

Historical Notes

The factorization algorithm for polynomials in $\mathbb{Q}[x]$ is due to Kronecker and was part of his constructive approach to algebra. Precise references can be found in [3].

The Schönemann–Eisenstein criterion was published by Schönemann in 1846 and independently by Eisenstein in 1850. Although it is often called the "Eisenstein criterion," Schönemann's name should be included, since he proved it first. See [3] and [8, p. 254] for references to the original papers.

The slick proof of the irreducibility of $\Phi_p(x)$ given in Proposition 4.2.5 is due to Eisenstein. In Chapter 15 we will explore the fascinating mathematics that led Eisenstein to the irreducibility criterion.

Schönemann discovered the criterion in a very different context. He asked whether a polynomial that is reducible modulo p remains reducible modulo p^2. His version of the criterion states that polynomials of the form $(x - a)^n + pF(x)$, where $a \in \mathbb{Z}$, $F[x] \in \mathbb{Z}[x]$, and $p \nmid F(a)$, are always irreducible. You will prove this in Exercise 2, and in Exercise 3 you will use this to give another proof of Proposition 4.2.5.

The first proof of Proposition 4.2.5 is due to Gauss in 1799 as part of his study of regular polygons in *Disquisitiones* [4]. We will say more about this in Chapters 9 and 10. In 1818, Gauss gave an interesting application of Proposition 4.2.5. His sixth proof of quadratic reciprocity used congruences modulo $x^{p-1} + \cdots + x + 1$, which in modern terms means that he was working in the quotient ring

$$\mathbb{Q}[x]/\langle x^{p-1} + \cdots + x + 1 \rangle.$$

Earlier, Gauss had given a proof of quadratic reciprocity (his fourth, in 1811) using the pth root of unity $\zeta_p = e^{2\pi i/p}$. Since $x^{p-1} + \cdots + x + 1$ is irreducible, it is the minimal polynomial of ζ_p. Combining this with Lemma 4.1.13 and Proposition 4.1.14 gives an isomorphism

$$\mathbb{Q}(\zeta_p) = \mathbb{Q}[\zeta_p] \simeq \mathbb{Q}[x]/\langle x^{p-1} + \cdots + x + 1 \rangle.$$

It follows that Gauss's sixth proof of quadratic reciprocity is a version of the fourth, with complex numbers replaced by the above quotient ring.

Abel used many properties of radicals in his proof of the unsolvability of the general quintic. In a manuscript written shortly before his death in 1829, Abel proved Proposition 4.2.6 in the special case when $\zeta_p \in F$ [Abel, Vol. II, p. 229], and the general case is due to Kronecker in 1879 [Kronecker, Vol. IV, pp. 75–76]. An even more general version of Proposition 4.2.6 is the following 1901 theorem of Capelli (see [Chebotarev, p. 294]).

Theorem 4.2.8. *Let* $f = x^m - a \in F[x]$. *Then* f *is reducible over* F *if and only if* m *has a divisor* $d > 1$ *such that*

$$a = \begin{cases} b^d, & b \in F, \text{ or} \\ -4c^4, & d = 4, c \in F. \end{cases}$$

Exercises for Section 4.2

Exercise 1. This exercise will study the *Lagrange interpolation formula*. Suppose that F is a field and that $b_0, \ldots, b_d, c_0, \ldots, c_d \in F$, where $b_0, \ldots, b_d$ are distinct and $d \geq 1$. Then consider the polynomial

$$g(x) = \sum_{i=0}^{d} c_i \prod_{j \neq i} \frac{x - b_j}{b_i - b_j} \in F[x].$$

(a) Explain why $\deg(g) \leq d$, and give an example for $F = \mathbb{R}$ and $d = 2$ where $\deg(g) < 2$.
(b) Show that $g(b_i) = c_i$ for $i = 0, \ldots, d$.
(c) Let h be a polynomial in $F[x]$ with $\deg(h) \leq d$ such that $h(b_i) = c_i$ for $i = 0, \ldots, d$. Prove that $h = g$.

Exercise 2. This exercise deals with Schönemann's version of the irreducibility criterion.
(a) Let $f(x) = (x - a)^n + pF(x)$, where $a \in \mathbb{Z}$ and $F(x) \in \mathbb{Z}[x]$ satisfy $\deg(F) \leq n$ and $p \nmid F(a)$. Prove that f is irreducible over $\mathbb{Q}$.
(b) More generally, let $g(x) \in \mathbb{Z}[x]$ be irreducible modulo p (i.e., reducing its coefficients modulo p gives an irreducible polynomial in $\mathbb{F}_p[x]$). Then let $f(x) = g(x)^n + pF(x)$, where $F[x] \in \mathbb{Z}[x]$ and $g(x)$ and $F(x)$ are relatively prime modulo p. Prove that f is irreducible over $\mathbb{Q}$.

Exercise 3. Use part (a) of Exercise 2 with $a = 1$ to give another proof of Proposition 4.2.5.

Exercise 4. For each of the following polynomials, use a computer to determine whether it is irreducible over the given field.
(a) $x^4 + x^3 + x^2 + x + 2$ over $\mathbb{Q}$.
(b) $3x^6 + 6x^5 + 9x^4 + 2x^3 + 3x^2 + 1$ over $\mathbb{Q}$ and $\mathbb{Q}(\sqrt[3]{2})$.

Exercise 5. Find the minimal polynomial of the 24th root of unity ζ_{24} as follows.
(a) Factor $x^{24} - 1$ over $\mathbb{Q}$.
(b) Decide which of the factors is the minimal polynomial of ζ_{24}.

Exercise 6. Let F be a finite field. Explain why there is an algorithm for deciding whether $f \in F[x]$ is irreducible.

Exercise 7. For each of the following polynomials, determine, without using a computer, whether it is irreducible over the given field.
(a) $x^3 + x + 1$ over $\mathbb{F}_5$.
(b) $x^4 + x + 1$ over $\mathbb{F}_2$.

Exercise 8. Let $a \in \mathbb{Z}$ be a product of distinct prime numbers. Prove that $x^n - a$ is irreducible over $\mathbb{Q}$ for any $n \geq 1$. What does this imply about $\sqrt[n]{a}$?

Exercise 9. Let k be a field, and let $F = k(t)$ be the field of rational functions in t with coefficients in k. Then consider $f = x^p - t \in F[x]$, where p is prime. By Proposition 4.2.6, f is irreducible provided we can show that f has no roots in F. Prove this.

4.3   THE DEGREE OF AN EXTENSION

When F is a subfield of a field L, there is one bit of structure that hasn't been used yet. We know that L is an Abelian group under addition. Furthermore, since $F \subset L$, the ability to multiply elements of L implies that we can multiply elements of F times elements of L. This gives a scalar multiplication, and one can easily check that L becomes a vector space over F.

A. Finite Extensions. The above paragraph leads to the following definition.

Definition 4.3.1. *Let $F \subset L$ be a field extension.*
(a) *L is a **finite extension** of F if L is a finite-dimensional vector space over F.*
(b) *The **degree** of L over F, denoted $[L:F]$, is defined as follows:*

$$[L:F] = \begin{cases} \dim_F L, & \text{if } L \text{ is a finite extension of } F, \\ \infty, & \text{otherwise,} \end{cases}$$

where $\dim_F L$ is the dimension of L as a vector space over F.

Here is a simple example.

Example 4.3.2. For $\mathbb{R} \subset \mathbb{C}$, the usual way of writing complex numbers as $a + bi$ shows that 1 and i form a basis of $\mathbb{C}$ as a vector space over $\mathbb{R}$. Thus $[\mathbb{C}:\mathbb{R}] = 2$. ◁▷

We can also characterize extensions of degree 1.

Lemma 4.3.3. *An extension $F \subset L$ has degree $[L:F] = 1$ if and only if $F = L$.*

Proof. If $[L:F] = 1$, then any nonzero element of L, say $1 \in L$, is a basis. Thus $L = \{a \cdot 1 \mid a \in F\} = F$. The opposite implication is even easier and is omitted. □

In general, we compute the degree of an extension $F \subset F(\alpha)$ as follows.

Proposition 4.3.4. *Suppose that $F \subset L$ is an extension and $\alpha \in L$.*
(a) *α is algebraic over F if and only if $[F(\alpha):F] < \infty$.*
(b) *Let α be algebraic over F. If n is the degree of the minimal polynomial of α over F, then $1, \alpha, \ldots, \alpha^{n-1}$ form a basis of $F(\alpha)$ over F. Thus $[F(\alpha):F] = n$.*

Proof. First suppose that α is algebraic over F with minimal polynomial p, where $n = \deg(p)$. We need to show that $1, \alpha, \ldots, \alpha^{n-1}$ form a basis of $F(\alpha)$ over F. Since $F(\alpha) = F[\alpha]$, every element of $F(\alpha)$ is of the form $g(\alpha)$ for some $g \in F[x]$. Dividing g by p gives

$$g = qp + a_0 + a_1 x + \cdots + a_{n-1} x^{n-1},$$

where $q \in F[x]$ and $a_0, \ldots, a_{n-1} \in F$, and evaluating this at $x = \alpha$ yields

$$(4.9) \qquad g(\alpha) = a_0 + a_1 \alpha + \cdots + a_{n-1} \alpha^{n-1},$$

since $p(\alpha) = 0$. Thus $1, \alpha, \ldots, \alpha^{n-1}$ span $F(\alpha)$ over F. To show linear independence, suppose that

$$0 = a_0 + a_1 \alpha + \cdots + a_{n-1} \alpha^{n-1},$$

where $a_0, \ldots, a_{n-1} \in F$. Then α is a root of $a_0 + a_1 x + \cdots + a_{n-1} x^{n-1} \in F[x]$. Since the minimal polynomial p has degree n, this must be the zero polynomial. Hence $a_i = 0$ for all i, and linear independence is proved. Then $[F(\alpha) : F] = n$ follows from Definition 4.3.1.

This proves part (b) of the proposition and also one implication of part (a). It remains to consider the case when $[F(\alpha) : F] < \infty$. If we let $n = [F(\alpha) : F]$, then $F(\alpha)$ is an n-dimensional vector space over F. This implies that any collection of $n + 1$ elements of $F(\alpha)$ is linearly dependent. In particular, $1, \alpha, \alpha^2, \ldots, \alpha^n$ are linearly dependent over F. Hence there are $a_0, \ldots, a_n \in F$, not all zero, such that

$$(4.10) \qquad a_0 + a_1 \alpha + a_2 \alpha^2 + \cdots + a_n \alpha^n = 0.$$

As in the previous paragraph, it follows that α is a root of

$$(4.11) \qquad a_0 + a_1 x + a_2 x^2 + \cdots + a_n x^n \in F[x],$$

which is nonzero, since the a_i's are not all zero. Hence α is algebraic over F, and the proposition is proved. $\qquad \qquad \square$

This proposition implies that when the minimal polynomial of α has degree n, every $\beta \in F(\alpha)$ can be written uniquely in the form

$$\beta = a_0 + a_1 \alpha + \cdots + a_{n-1} \alpha^{n-1}, \qquad a_0, \ldots, a_{n-1} \in F.$$

In Exercise 1 you will use an argument similar to the proof just given to describe unique coset representatives for elements of $F[x]/\langle f \rangle$.

Looking back at Example 4.3.2, we see that $[\mathbb{C} : \mathbb{R}] = 2$ follows from Proposition 4.3.4, since $\mathbb{C} = \mathbb{R}(i)$ and the minimal polynomial of i over $\mathbb{R}$ is $x^2 + 1$. Here are some other examples of Proposition 4.3.4.

Example 4.3.5. Consider the extension $\mathbb{Q} \subset \mathbb{Q}(\sqrt{2})$. Since the minimal polynomial of $\sqrt{2}$ is $x^2 - 2$, the proposition implies that $[\mathbb{Q}(\sqrt{2}) : \mathbb{Q}] = 2$ and that

$$(4.12) \qquad \mathbb{Q}(\sqrt{2}) = \{ a + b\sqrt{2} \mid a, b \in \mathbb{Q} \}.$$

Note also that this representation is unique. $\qquad \qquad \triangleleft \triangleright$

Example 4.3.6. By Example 4.1.7, the minimal polynomial of $\sqrt{2} + \sqrt{3}$ over $\mathbb{Q}$ is $x^4 - 10x^2 + 1$. Thus $[\mathbb{Q}(\sqrt{2} + \sqrt{3}) : \mathbb{Q}] = 4$, and every $\beta \in \mathbb{Q}(\sqrt{2} + \sqrt{3})$ can be written

$$\beta = a + b(\sqrt{2} + \sqrt{3}) + c(\sqrt{2} + \sqrt{3})^2 + d(\sqrt{2} + \sqrt{3})^3$$

for unique $a, b, c, d \in \mathbb{Q}$. $\qquad \qquad \triangleleft \triangleright$

Example 4.3.7. Let $F(x)$ be the field of rational functions in the variable x with coefficients in F. Then Proposition 4.3.4 implies that $[F(x) : F] = \infty$, since x is not algebraic over F. $\qquad \qquad \triangleleft \triangleright$

B. The Tower Theorem. We can also determine how the degree behaves when we have successive extensions $F \subset K \subset L$. The following result is sometimes called the *Tower Theorem.*

Theorem 4.3.8. *Suppose that we have fields $F \subset K \subset L$.*
(a) *If $[K : F] = \infty$ or $[L : K] = \infty$, then $[L : F] = \infty$.*
(b) *If $[K : F] < \infty$ and $[L : K] < \infty$, then $[L : F] = [L : K][K : F]$.*

Proof. We will prove the contrapositive of part (a): if $[L : F] < \infty$, then $[K : F] < \infty$ and $[L : K] < \infty$. Thus we may assume that L has finite dimension as a vector space over F. Let $\gamma_1, \ldots, \gamma_N$ be a basis. Then:

- One easily sees that $K \subset L$ is a subspace of L over the field F. Since L has finite dimension over F, so does any subspace. Hence $[K : F] = \dim_F K < \infty$.
- Take $\alpha \in L$. Since $\gamma_1, \ldots, \gamma_N$ span L over F, $\alpha = \sum_{i=1}^{N} a_i \gamma_i$, where $a_i \in F$. Since $F \subset K$, we can consider this as a linear combination with coefficients in K. Thus L is spanned over K by a finite set, so that $[L : K] = \dim_K L < \infty$.

To prove part (b), let $m = [K : F]$ and $n = [L : K]$, and pick bases $\alpha_1, \ldots, \alpha_m$ of K over F and $\beta_1, \ldots, \beta_n$ of L over K. We will prove that the mn products

$$\alpha_i \beta_j, \quad 1 \le i \le m, \ 1 \le j \le n,$$

form a basis of L over F. This will prove the theorem.

We first show that the $\alpha_i \beta_j$ span L over F. Take $\gamma \in L$. Since $\beta_1, \ldots, \beta_n$ span L over K, we can write $\gamma = \sum_{j=1}^{n} b_j \beta_j$, where $b_1, \ldots, b_n \in K$. Then, since $\alpha_1, \ldots, \alpha_m$ span K over F, we have $b_j = \sum_{i=1}^{m} a_{ij} \alpha_i$, where $a_{ij} \in F$. Combining these equations, we obtain

$$\gamma = \sum_{j=1}^{n} \left(\sum_{i=1}^{m} a_{ij} \alpha_i \right) \beta_j = \sum_{i=1}^{m} \sum_{j=1}^{n} a_{ij} \, \alpha_i \beta_j.$$

Since $a_{ij} \in F$, this shows that the $\alpha_i \beta_j$ span L over F.

To prove linear independence, suppose that we have a linear relation

$$\sum_{i=1}^{m} \sum_{j=1}^{n} a_{ij} \, \alpha_i \beta_j = 0$$

where $a_{ij} \in F$. As above, we can write this as

$$\sum_{j=1}^{n} \left(\sum_{i=1}^{m} a_{ij} \alpha_i \right) \beta_j = 0.$$

The expressions in the large parentheses all lie in K, and since the β_j are linearly independent over K, we conclude that

$$\sum_{i=1}^{m} a_{ij} \alpha_i = 0 \quad \text{for} \quad 1 \le j \le n.$$

Since the α_i are linearly independent over F and $a_{ij} \in F$, we must have $a_{ij} = 0$ for all i and j. This proves the desired linear independence. $\square$

Here are two examples of the Tower Theorem.

Example 4.3.9. We will analyze $\mathbb{Q} \subset \mathbb{Q}(\sqrt{2}, \sqrt{3})$ using

$$\mathbb{Q} \subset \mathbb{Q}(\sqrt{2}) \subset \mathbb{Q}(\sqrt{2}, \sqrt{3}).$$

Proposition 4.3.4 shows that $1, \sqrt{2}$ form a basis of $\mathbb{Q}(\sqrt{2})$ over $\mathbb{Q}$, since $x^2 - 2$ is the minimal polynomial of $\sqrt{2}$ over $\mathbb{Q}$. Furthermore, part (a) of Exercise 7 of Section 4.1 shows that $x^2 - 3$ is the minimal polynomial of $\sqrt{3}$ over $\mathbb{Q}(\sqrt{2})$, so that $1, \sqrt{3}$ form a basis of $\mathbb{Q}(\sqrt{2}, \sqrt{3})$ over $\mathbb{Q}(\sqrt{2})$. Thus:

- $[\mathbb{Q}(\sqrt{2}, \sqrt{3}) : \mathbb{Q}] = [\mathbb{Q}(\sqrt{2}, \sqrt{3}) : \mathbb{Q}(\sqrt{2})][\mathbb{Q}(\sqrt{2}) : \mathbb{Q}] = 2 \cdot 2 = 4$.
- The proof of Theorem 4.3.8 shows that the products of the bases $1, \sqrt{2}$ and $1, \sqrt{3}$, namely $1, \sqrt{2}, \sqrt{3}, \sqrt{2}\sqrt{3} = \sqrt{6}$, give a basis of $\mathbb{Q}(\sqrt{2}, \sqrt{3})$ over $\mathbb{Q}$.

Example 4.1.16 showed that $1, \sqrt{2}, \sqrt{3}, \sqrt{6}$ span $\mathbb{Q}(\sqrt{2}, \sqrt{3})$ over $\mathbb{Q}$. We now see that $1, \sqrt{2}, \sqrt{3}, \sqrt{6}$ form a basis that arises naturally from Theorem 4.3.8.

In Section 4.2, we used *Maple* and *Mathematica* to show that $\mathbb{Q}(\sqrt{2}, \sqrt{3}) = \mathbb{Q}(\sqrt{2} + \sqrt{3})$ (see (4.5)). We now give a different proof using

$$\mathbb{Q} \subset \mathbb{Q}(\sqrt{2} + \sqrt{3}) \subset \mathbb{Q}(\sqrt{2}, \sqrt{3}).$$

We just showed that $[\mathbb{Q}(\sqrt{2}, \sqrt{3}) : \mathbb{Q}] = 4$, and Example 4.3.6 tells us that the same is true for $\mathbb{Q} \subset \mathbb{Q}(\sqrt{2} + \sqrt{3})$. Then

$$[\mathbb{Q}(\sqrt{2}, \sqrt{3}) : \mathbb{Q}] = [\mathbb{Q}(\sqrt{2}, \sqrt{3}) : \mathbb{Q}(\sqrt{2} + \sqrt{3})][\mathbb{Q}(\sqrt{2} + \sqrt{3}) : \mathbb{Q}]$$

gives $[\mathbb{Q}(\sqrt{2}, \sqrt{3}) : \mathbb{Q}(\sqrt{2} + \sqrt{3})] = 1$. Thus $\mathbb{Q}(\sqrt{2} + \sqrt{3}) = \mathbb{Q}(\sqrt{2}, \sqrt{3})$.   ◁▷

Example 4.3.10. Let $\omega = e^{2\pi i/3}$ and $L = \mathbb{Q}(\omega, \sqrt[3]{2})$. We will compute $[L : \mathbb{Q}]$ using the extension fields

$$\mathbb{Q} \subset \mathbb{Q}(\sqrt[3]{2}) \subset \mathbb{Q}(\omega, \sqrt[3]{2}) = L.$$

To determine $[\mathbb{Q}(\sqrt[3]{2}) : \mathbb{Q}]$, first observe that $x^3 - 2$ is irreducible over $\mathbb{Q}$. Since $x^3 - 2$ has degree 3, one can prove this using Lemma A.1.19 and Proposition A.3.1, though it is quicker to use the Schönemann–Eisenstein criterion (Theorem 4.2.3) with $p = 2$. By Proposition 4.3.4 we conclude that

$$[\mathbb{Q}(\sqrt[3]{2}) : \mathbb{Q}] = 3.$$

We next compute $[L : \mathbb{Q}(\sqrt[3]{2})]$. Recall that $x^2 + x + 1$ has roots ω and ω^2, neither of which is real. Since $\mathbb{Q}(\sqrt[3]{2}) \subset \mathbb{R}$, $x^2 + x + 1$ has no root in this field, so that $x^2 + x + 1$ is the minimal polynomial of ω over $\mathbb{Q}(\sqrt[3]{2})$. Hence

$$[L : \mathbb{Q}(\sqrt[3]{2})] = 2,$$

since $L = \mathbb{Q}(\sqrt[3]{2})(\omega)$. Then Theorem 4.3.8 implies that

$$[L:\mathbb{Q}] = [L:\mathbb{Q}(\sqrt[3]{2})][\mathbb{Q}(\sqrt[3]{2}):\mathbb{Q}] = 2 \cdot 3 = 6.$$

We will return to this example often.

Mathematical Notes

Here is one of the ideas used in this section.

■ **Algebras over a Field.** The key idea of this section is that a field extension $F \subset L$ gives L the structure of a vector space over F, so that L is simultaneously a field and a vector space. In general, there are many examples of rings that are also vector spaces over a field. Here is the general definition.

Definition 4.3.11. *A (possibly noncommutative) ring R is an **algebra** over the field F if R is a vector space over F such that:*
(a) *The vector space addition on R is the same as the ring addition on R.*
(b) *The scalar multiplication on R is compatible with the ring multiplication:*

$$(ab) \cdot r = a \cdot (b \cdot r) \quad \text{for all } a, b \in F \text{ and } r \in R,$$
$$a \cdot (rs) = (a \cdot r)s = r(a \cdot s) \quad \text{for all } a \in F \text{ and } r, s \in R.$$

A field extension $F \subset L$ makes L into an F-algebra. Other examples include the polynomial ring $F[x_1, \dots, x_n]$ and the algebra of $n \times n$ matrices $M_{n \times n}(F)$.

Historical Notes

The idea of representing elements of a field as linear combinations has a long history. For example, in 1847 Cauchy took a polynomial $f \in F[x]$ of degree n and represented elements of $F[x]/\langle f \rangle$ as linear combinations of the cosets of $1, x, \dots, x^{n-1}$. Kronecker also represented elements of extension fields using linear combinations, and he was aware of the importance of linear independence. But in all of this work, the term "degree" applied only to degrees of polynomials.

In 1894 Dedekind developed a theory of field extensions that included the concept of degree. He writes an extension as $A \subset \Omega$ and gives the modern criterion for $w_1, \dots, w_n \in \Omega$ to be linearly independent over A. Furthermore, if the w_i span Ω, then he sets $(\Omega, A) = n$. He also knows Proposition 4.3.4, but only gives special cases of Theorem 4.3.8.

The modern formulation of the results of this section is due to Emil Artin. He developed his approach to Galois theory in the 1920s. He turned Dedekind's (Ω, A) into the degree $[L : F]$ and made it the centerpiece of his theory of finite extensions. Artin profoundly transformed the way people think about Galois theory. We will say more about this in Section 6.1.

For more details on the history of how these concepts developed, we refer the reader to [5] and [6].

Exercises for Section 4.3

Exercise 1. In (4.9) we represented elements of $F(\alpha)$ uniquely using remainders on division by the minimal polynomial of α. In this exercise you will adapt the proof of Proposition 4.3.4 to the case of quotient rings. Suppose that $f \in F[x]$ has degree $n > 0$. Prove that every coset of $F[x]/\langle f \rangle$ can be written as

$$a_0 + a_1 x + \cdots + a_{n-1} x^{n-1} + \langle f \rangle,$$

where $a_0, a_1, \ldots, a_{n-1} \in F$ are unique.

Exercise 2. Compute the degrees of the following extensions:
(a) $\mathbb{Q} \subset \mathbb{Q}(i, \sqrt[4]{2})$.
(b) $\mathbb{Q} \subset \mathbb{Q}(\sqrt{3}, \sqrt[3]{2})$.
(c) $\mathbb{Q} \subset \mathbb{Q}(\sqrt{2 + \sqrt{2}})$.
(d) $\mathbb{Q} \subset \mathbb{Q}(i, \sqrt{2 + \sqrt{2}})$.

Exercise 3. For each of the extensions in Exercise 2, find a basis over $\mathbb{Q}$ using the method of Example 4.3.9.

Exercise 4. Suppose that $F \subset L$ is a finite extension with $[L : F]$ prime.
(a) Show that the only subfields of L containing F are F and L.
(b) Show that $L = F(\alpha)$ for any $\alpha \in L \setminus F$.

Exercise 5. Consider the extension $\mathbb{Q} \subset L = \mathbb{Q}(\sqrt[4]{2}, \sqrt[3]{3})$. We will compute $[L : \mathbb{Q}]$.
(a) Show that $x^4 - 2$ and $x^3 - 3$ are irreducible over $\mathbb{Q}$.
(b) Use $\mathbb{Q} \subset \mathbb{Q}(\sqrt[4]{2}) \subset L$ to show that $4 | [L : \mathbb{Q}]$ and $[L : \mathbb{Q}] \leq 12$.
(c) Use $\mathbb{Q} \subset \mathbb{Q}(\sqrt[3]{3}) \subset L$ to show that $[L : \mathbb{Q}]$ is also divisible by 3.
(d) Explain why parts (b) and (c) imply that $[L : \mathbb{Q}] = 12$. This works because 3 and 4 are relatively prime. Do you see why?

Exercise 6. Suppose that α and β are algebraic over F with minimal polynomials f and g respectively. Prove the *Reciprocity Theorem*: f is irreducible over $F(\beta)$ if and only if g is irreducible over $F(\alpha)$.

Exercise 7. Suppose we have extensions $L_0 \subset L_1 \subset \cdots \subset L_m$. Use induction to prove the following generalization of Theorem 4.3.8:
(a) If $[L_i : L_{i-1}] = \infty$ for some $1 \leq i \leq m$, then $[L_m : L_0] = \infty$.
(b) If $[L_i : L_{i-1}] < \infty$ for all $1 \leq i \leq m$, then

$$[L_m : L_0] = [L_m : L_{m-1}][L_{m-1} : L_{m-2}] \cdots [L_2 : L_1][L_1 : L_0].$$

4.4 ALGEBRAIC EXTENSIONS

Now that we know the basic properties of the degree of an extension, we can continue our study of algebraic elements. We begin with a definition.

Definition 4.4.1. *A field extension $F \subset L$ is **algebraic** if every element of L is algebraic over F.*

It turns out that finite extensions are always algebraic.

Lemma 4.4.2. *Let $F \subset L$ be a finite extension. Then:*
(a) *$F \subset L$ is algebraic.*
(b) *If $\alpha \in L$, then the degree of the minimal polynomial of α over F divides $[L:F]$.*

Proof. An element $\alpha \in L$ gives $F \subset F(\alpha) \subset L$, and then the Tower Theorem implies that $[F(\alpha):F]$ is finite and divides $[L:F]$. Then (a) and (b) follow immediately from Proposition 4.3.4. $\qquad\square$

Exercise 1 will show that the converse of this lemma is false—there are algebraic extensions that are not finite. So a finite extension is an especially nice algebraic extension.

We next explore the structure of finite extensions.

Theorem 4.4.3. *Let $F \subset L$ be a field extension. Then $[L:F] < \infty$ if and only if there are $\alpha_1, \ldots, \alpha_m \in L$ such that each α_i is algebraic over F and $L = F(\alpha_1, \ldots, \alpha_m)$.*

Proof. First suppose that $[L:F] < \infty$. Let $\alpha_1, \ldots, \alpha_m \in L$ be a basis of L as a vector space over F (so that $m = \dim_F L$). Then

$$L = \{a_1\alpha_1 + \cdots + a_m\alpha_m \mid a_1, \ldots, a_m \in F\} \subset F(\alpha_1, \ldots, \alpha_m) \subset L$$

proves that $L = F(\alpha_1, \ldots, \alpha_m)$. Each α_i is algebraic over F by Lemma 4.4.2.

Going the other way, suppose that $L = F(\alpha_1, \ldots, \alpha_m)$ where each α_i is algebraic over F. Let $L_0 = F$ and $L_i = F(\alpha_1, \ldots, \alpha_i)$ for $1 \le i \le m$. Then we get field extensions

(4.13) $$F = L_0 \subset L_1 \subset \cdots \subset L_m = L,$$

and Corollary 4.1.11 shows that

$$L_i = F(\alpha_1, \ldots, \alpha_{i-1}, \alpha_i) = F(\alpha_1, \ldots, \alpha_{i-1})(\alpha_i) = L_{i-1}(\alpha_i)$$

for $1 \le i \le m$. Since α_i is algebraic over F, it is also algebraic over the larger field $L_{i-1} \supset F$. Then Proposition 4.3.4 implies that

$$[L_i : L_{i-1}] = [L_{i-1}(\alpha_i) : L_{i-1}] < \infty,$$

so that every successive extension in (4.13) has finite degree. Then the generalization of the Tower Theorem given in Exercise 7 of Section 4.3 implies that

$$[L:F] = [L_m : L_0] = [L_m : L_{m-1}] \cdots [L_1 : L_0] < \infty.$$

This completes the proof of the theorem. $\qquad\square$

As an application of the theorem just proved, let's show that the sum and product of algebraic elements are algebraic.

Proposition 4.4.4. *Let $F \subset L$ be a field extension. If $\alpha, \beta \in L$ are algebraic over F, then so are $\alpha + \beta$ and $\alpha\beta$.*

Proof. Theorem 4.4.3 implies that $F \subset F(\alpha, \beta)$ is a finite extension and hence is algebraic by Lemma 4.4.2. Thus every element of $F(\alpha, \beta)$ is algebraic over F. Since $\alpha + \beta, \alpha\beta \in F(\alpha, \beta)$, the proposition is proved. $\qquad\square$

Corollary 4.4.5. *Given any field extension $F \subset L$, the subset*

$$M = \{\alpha \in L \mid \alpha \text{ is algebraic over } F\}$$

is a subfield of L containing F.

Proof. We have $F \subset M$ since $a \in F$ is a root of $x - a \in F[x]$, and M is closed under addition and multiplication by Proposition 4.4.4. Since $-1 \in F \subset M$, we see that $\alpha \in M$ implies $-\alpha = -1 \cdot \alpha \in M$. Finally, if $\alpha \neq 0 \in M$, then Exercise 1 of Section 4.1 shows that $1/\alpha \in M$. It follows that M is a subfield of L. $\qquad\square$

Here is a classic example of this corollary.

Example 4.4.6. A complex number $z \in \mathbb{C}$ is called an *algebraic number* if it is algebraic over $\mathbb{Q}$. By Corollary 4.4.5, we have the *field of algebraic numbers*

$$\overline{\mathbb{Q}} = \{z \in \mathbb{C} \mid z \text{ is an algebraic number}\}.$$

Later in the section we will prove that $\overline{\mathbb{Q}}$ is algebraically closed. $\qquad\triangleleft\triangleright$

We next show that being algebraic is transitive in the following sense.

Theorem 4.4.7. *Let $F \subset K \subset L$. If $\alpha \in L$ is algebraic over K and K is algebraic over F, then α is algebraic over F.*

Proof. Let α be a root of $f = \beta_n x^n + \cdots + \beta_0 \in K[x]$, where $\beta_n, \ldots, \beta_0 \in K$ are not all 0. By hypothesis, each β_i is algebraic over F. Then $M = F(\beta_n, \ldots, \beta_0)$ is a finite extension of F by Theorem 4.4.3. Furthermore, M is constructed so that $f \in M[x]$. It follows that α is algebraic over M, which implies that $M \subset M(\alpha)$ is a finite extension. By Theorem 4.3.8,

$$[M(\alpha) : F] = [M(\alpha) : M][M : F] < \infty.$$

Thus $F \subset M(\alpha)$ is finite and hence algebraic. This means that every element of $M(\alpha)$, including α, is algebraic over F. $\qquad\square$

Here is an example of this theorem.

Example 4.4.8. Theorem 4.4.7 implies that every complex solution of the equation

$$(4.14) \qquad x^{11} - (\sqrt{2} + \sqrt{5})x^5 + 3\sqrt[4]{12}\,x^3 + (1 + 3i)x + \sqrt[5]{17} = 0$$

is an algebraic number. This follows because the coefficients are obviously algebraic over $\mathbb{Q}$. (Do you see why there are no real solutions?) In Exercise 2 you will show that the minimal polynomial of a solution of (4.14) has degree at most 1760. $\qquad\triangleleft\triangleright$

Theorem 4.4.7 also has the following immediate corollary.

Corollary 4.4.9. *If we have field extensions $F \subset K \subset L$ where L is algebraic over K and K is algebraic over F, then L is algebraic over F.* □

Mathematical Notes

Here are some of the ideas encountered in this section.

▪ **The Field of Algebraic Numbers.** In Example 4.4.6, we defined $\overline{\mathbb{Q}}$ to be the set of all algebraic numbers in $\mathbb{C}$. This field has the following nice property.

Theorem 4.4.10. *The field $\overline{\mathbb{Q}}$ of algebraic numbers is algebraically closed.*

Proof. By Exercise 3 of Section 3.2 it suffices to show that every nonconstant polynomial in $\overline{\mathbb{Q}}[x]$ has a root in $\overline{\mathbb{Q}}$. Given such a polynomial f, we can regard f as an element of $\mathbb{C}[x]$, since $\overline{\mathbb{Q}} \subset \mathbb{C}$. Then f has a root $\alpha \in \mathbb{C}$ by the Fundamental Theorem of Algebra, and α is algebraic over $\overline{\mathbb{Q}}$ because $f \in \overline{\mathbb{Q}}[x]$. But $\overline{\mathbb{Q}}$ is algebraic over $\mathbb{Q}$ by definition, so that α is algebraic over $\mathbb{Q}$ by Theorem 4.4.7. Thus f has the root $\alpha \in \overline{\mathbb{Q}}$, and we are done. □

One can also show that if $\mathbb{Q} \subset L$ is an extension such that L is algebraic over $\mathbb{Q}$ and L is algebraically closed, then $L \simeq \overline{\mathbb{Q}}$. More generally, if F is any field, then there is a field $\overline{F}$, unique up to isomorphism, such that $\overline{F}$ is algebraic over F and algebraically closed. We call $\overline{F}$ the *algebraic closure* of F. (Strictly speaking, $\overline{F}$ is only unique up to a nonunique isomorphism. Hence we should say "an algebraic closure" rather than "the algebraic closure.") A discussion of algebraic closures can be found in [Jacobson, Vol. II, Sec. 8.1].

▪ **Algebraic Integers.** Finally, in addition to the notion of an algebraic number in $\mathbb{C}$, one can also define an *algebraic integer* to be a complex number that is a root of a monic polynomial with integer coefficients. For example, $\sqrt{2}$ and $\omega = (-1+i\sqrt{3})/2$ are algebraic integers, since they are roots of $x^2 - 2$ and $x^2 + x + 1$ respectively; but one can show that $\omega/2$ is not an algebraic integer (see Exercise 3). Algebraic integers play an important role in number theory. For example, Euler proved Fermat's Last Theorem for $n = 3$ by writing $x^3 + y^3 = z^3$ as

$$x^3 = z^3 - y^3 = (z - y)(z - \omega y)(z - \omega^2 y)$$

and using unique factorization in the ring of algebraic integers $\mathbb{Z}[\omega]$. This subject is called *algebraic number theory*. For an introduction to algebraic number theory, including the details of Euler's argument, see [9, Ch. 9].

Exercises for Section 4.4

Exercise 1. Lemma 4.4.2 shows that a finite extension is algebraic. Here we will give an example to show that the converse is false. The field of algebraic numbers $\overline{\mathbb{Q}}$ is by definition algebraic over $\mathbb{Q}$. You will show that $[\overline{\mathbb{Q}} : \mathbb{Q}] = \infty$ as follows.

(a) Given $n \geq 2$ in $\mathbb{Z}$, use Example 4.2.4 from Section 4.2 to show that $\overline{\mathbb{Q}}$ has a subfield L such that $[L : \mathbb{Q}] = n$.
(b) Explain why part (a) implies that $[\overline{\mathbb{Q}} : \mathbb{Q}] = \infty$.

Exercise 2. Let $\alpha \in \mathbb{C}$ be a solution of (4.14). We will show that the minimal polynomial of α over $\mathbb{Q}$ has degree at most 1760. Let $L = \mathbb{Q}(\sqrt{2}, \sqrt{5}, \sqrt[4]{12}, i, \sqrt[5]{17}, \alpha)$.
(a) Show that $[L : \mathbb{Q}] \leq 1760$.
(b) Use Lemma 4.4.2 to show that the minimal polynomial of α has degree at most 1760.

Exercise 3. In the Mathematical Notes, we defined an algebraic integer to be a complex number $\alpha \in \mathbb{C}$ that is a root of a monic polynomial in $\mathbb{Z}[x]$.
(a) Prove that $\alpha \in \mathbb{C}$ is an algebraic integer if and only if α is an algebraic number whose minimal polynomial over $\mathbb{Q}$ has integer coefficients.
(b) Show that $\omega/2$ is not an algebraic integer, where $\omega = (-1 + i\sqrt{3})/2$.

Exercise 4. Use (4.10) and (4.11) to prove the following weak form of Lemma 4.4.2: if $n = [L : F] < \infty$, then every $\alpha \in L$ is a root of a nonzero polynomial in $F[x]$ of degree $\leq n$.

Exercise 5. In 1873 Hermite proved that the number e is transcendental over $\mathbb{Q}$, and in 1882 Lindemann showed that π is transcendental over $\mathbb{Q}$. It is unknown whether $\pi + e$ and $\pi - e$ are transcendental. Prove that *at least* one of these numbers is transcendental over $\mathbb{Q}$.

Exercise 6. Let F be a field. Show that other than the elements of F itself, no elements of $F(x)$ are algebraic over F. Thus, even though $[F(x) : F] = \infty$ by Example 4.3.7, the field $M = \{\alpha \in F(x) \mid \alpha \text{ is algebraic over } F\}$ of Corollary 4.4.5 is as small as possible, namely F.

Exercise 7. Suppose that F is an algebraically closed field, and let $F \subset L$ be an algebraic extension. Prove that $F = L$.

Exercise 8. In this exercise you will show that every algebraic extension of $\mathbb{R}$ is finite of degree at most 2. To prove this, consider an algebraic extension $\mathbb{R} \subset L$.
(a) Explain why we can find an extension $L \subset K$ such that $x^2 + 1$ has a root $\alpha \in K$.
(b) Prove that $L(\alpha)$ is algebraic over $\mathbb{R}(\alpha)$ and that $\mathbb{R}(\alpha) \simeq \mathbb{C}$.
(c) Now use the previous exercise to conclude that $[L : \mathbb{R}] \leq 2$ and that equality occurs if and only if $L \simeq \mathbb{C}$.

Exercise 9. Prove that $\alpha \in \mathbb{Q}$ is an algebraic integer if and only if $\alpha \in \mathbb{Z}$.

REFERENCES

1. H. Cohen, *A Course in Computational Algebraic Number Theory*, Springer-Verlag, New York, Berlin, Heidelberg, 1993.

2. R. Dedekind, *Theory of Algebraic Integers*, English translation by J. Stillwell, Cambridge U. P., Cambridge, 1996. (Translation of 1877 French edition.)

3. H. L. Dorwart, *Irreducibility of polynomials*, Amer. Math. Monthly **42** (1935), 369–381.

4. C. F. Gauss, *Disquisitiones Arithmeticae*, Leipzig, 1801. Republished in 1863 as Volume I of [Gauss]. French translation, *Recherches Arithmétiques*, Paris,

1807. Reprint by Hermann, Paris, 1910. German translation, *Untersuchungen über Höhere Arithmetik*, Berlin, 1889. Reprint by Chelsea, New York, 1965. English translation, Yale U. P., New Haven, 1966. Reprint by Springer-Verlag, New York, Berlin, Heidelberg, 1986.

5. B. M. Kiernan, *The development of Galois theory from Lagrange to Artin*, Arch. Hist. Exact Sci. **8** (1971), 40–154.

6. I. Kleiner, *Field theory: From equations to axiomatization*, Parts I and II, Amer. Math. Monthly **106** (1999), 677–684, 859–863.

7. I. Kleiner, *The genesis of the abstract ring concept*, Amer. Math. Monthly **103** (1996), 417–424.

8. F. Lemmermeyer, *Reciprocity Laws*, Springer-Verlag, New York, Berlin, Heidelberg, 2000.

9. I. Niven and H. S. Zuckerman, *An Introduction to the Theory of Numbers*, Third Edition, Wiley, New York, 1972.

5

Normal and Separable Extensions

This chapter will study some important properties of field extensions. We will begin with extensions obtained by adjoining all roots of a polynomial. These *splitting fields* will lead to the idea of *normality*. We will also consider the idea of *separability* for both polynomials and field extensions. The chapter will end with the *Theorem of the Primitive Element*

5.1 SPLITTING FIELDS

Given a nonconstant polynomial $f \in F[x]$, Theorem 3.1.4 shows that there is an extension $F \subset L$ over which f splits completely. In this section we will consider the *smallest* such extension.

A. Definition and Examples. We begin with a definition.

Definition 5.1.1. *Let $f \in F[x]$ have degree $n > 0$. Then an extension $F \subset L$ is a splitting field of f over F if*

(a) $f = c(x - \alpha_1) \cdots (x - \alpha_n)$, *where $c \in F$ and $\alpha_i \in L$, and*

(b) $L = F(\alpha_1, \ldots, \alpha_n)$.

Be sure you understand how this captures the idea of the smallest extension over which a polynomial splits completely. The existence of splitting fields follows from Theorem 3.1.4, for if $f \in F[x]$ splits completely as $f = c(x - \alpha_1) \cdots (x - \alpha_n)$ in $L[x]$, then $F(\alpha_1, \ldots, \alpha_n)$ is clearly a splitting field of f over F. We will prove below that all splitting fields of $f \in F[x]$ are isomorphic.

In the subsequent text, whenever we say "L is a splitting field of $f \in F[x]$," we will tacitly assume that f is nonconstant.

Here are some examples of splitting fields.

Example 5.1.2. $\mathbb{Q}(\sqrt{2}, \sqrt{3}) = \mathbb{Q}(\pm\sqrt{2}, \pm\sqrt{3})$ is a splitting field of $(x^2-2)(x^2-3)$ over $\mathbb{Q}$. ◁▷

Example 5.1.3. In Example 4.1.10 we showed that

$$\mathbb{Q}(\sqrt[4]{2}, -\sqrt[4]{2}, i\sqrt[4]{2}, -i\sqrt[4]{2}) = \mathbb{Q}(i, \sqrt[4]{2}).$$

Thus $\mathbb{Q}(i, \sqrt[4]{2})$ is a splitting field of $x^4 - 2$ over $\mathbb{Q}$. ◁▷

Example 5.1.4. In Exercise 1 you will prove that the field $\mathbb{Q}(\omega, \sqrt[3]{2})$ considered in Example 4.3.10 is a splitting field of $x^3 - 2$ over $\mathbb{Q}$. ◁▷

Note that a splitting field of $f \in F[x]$ depends on both the polynomial f and the field F. For instance:

$$\text{a splitting field of } x^2 + 1 \text{ over } \mathbb{Q} \text{ is } \mathbb{Q}(i);$$

$$\text{a splitting field of } x^2 + 1 \text{ over } \mathbb{R} \text{ is } \mathbb{C};$$

$$\text{a splitting field of } x^2 + 1 \text{ over } \mathbb{C} \text{ is } \mathbb{C}.$$

Since the roots of a nonconstant polynomial $f \in F[x]$ are algebraic over F, it follows from Theorem 4.4.3 that a splitting field of f over F is always a finite extension of F. We can bound the degree of this extension as follows.

Theorem 5.1.5. *Let $f \in F[x]$ be a polynomial of degree $n > 0$, and let L be a splitting field of f over F. Then $[L:F] \leq n!$.*

Proof. We will prove this by induction on n. When $n = 1$, $f = ax + b$ has the root $-b/a \in F$, since $a \neq 0$. Thus $L = F$ in this case, and $[L:F] \leq 1!$ is clear.

Now suppose that f has degree $n > 1$, and let $L = F(\alpha_1, \ldots, \alpha_n)$ be a splitting field of f over F. If we write $f = (x - \alpha_1)g$, then the division algorithm implies that $g \in F(\alpha_1)[x]$. Furthermore, the roots of g are obviously $\alpha_2, \ldots, \alpha_n$, so that a splitting field of g over $F(\alpha_1)$ is given by

$$F(\alpha_1)(\alpha_2, \ldots, \alpha_n) = F(\alpha_1, \alpha_2, \ldots, \alpha_n) = L,$$

where the first equality follows from Corollary 4.1.11. Since $g \in F(\alpha_1)[x]$ has degree $n - 1$, our inductive hypothesis implies that

$$[L:F(\alpha_1)] \leq (n - 1)!.$$

To bound the degree of $F \subset L$, we use the extensions $F \subset F(\alpha_1) \subset L$. By the Tower Theorem (Theorem 4.3.8), we have

$$[L:F] = [L:F(\alpha_1)][F(\alpha_1):F] \leq (n - 1)! \, [F(\alpha_1):F].$$

However, we also know that $[F(\alpha_1):F]$ is the degree of the minimal polynomial of α_1 over F, by Proposition 4.3.4. Since $f(\alpha_1) = 0$, we obtain $[F(\alpha_1):F] \leq n$, and then $[L:F] \leq n!$ follows. $\square$

Sometimes the bound in Theorem 5.1.5 is *sharp*, meaning that there are cases where equality occurs, though the inequality can also be strict. For instance:

- By Example 5.1.2, $\mathbb{Q}(\sqrt{2}, \sqrt{3})$ is a splitting field of $(x^2 - 2)(x^2 - 3)$ over $\mathbb{Q}$ and has degree $4 < 4!$ over $\mathbb{Q}$.
- By Example 5.1.4, $\mathbb{Q}(\omega, \sqrt[3]{2})$ is a splitting field of $x^3 - 2$ over $\mathbb{Q}$ and has degree $6 = 3!$ over $\mathbb{Q}$.

We will see in the next chapter that the size of the splitting field is closely related to the size of the Galois group of the extension.

B. Uniqueness. We next study the uniqueness of splitting fields. A given polynomial $f \in F[x]$ will have many distinct splitting fields. For example, $\mathbb{Q}(\sqrt{2})$ and $\mathbb{Q}[t]/\langle t^2 - 2 \rangle$ are splitting fields of $x^2 - 2$ over $\mathbb{Q}$. The key point is that while they are not the same, they are isomorphic.

In order to prove this result for all polynomials, we need to prove something more general. Suppose that we have an isomorphism of fields $\varphi : F_1 \simeq F_2$, and let $f_1 \in F_1[x]$ be a polynomial of degree $n > 0$. Applying φ to the coefficients of f_1 gives a polynomial $f_2 \in F_2[x]$.

Now let L_i be a splitting field of f_i over F_i for $i = 1, 2$. This gives the picture

$$
\begin{array}{ccc}
L_1 & & L_2 \\
\cup & & \cup \\
F_1 & \xrightarrow{\varphi} & F_2 \, .
\end{array}
$$

Although the splitting fields L_1 and L_2 may be constructed in quite different ways, the following theorem tells us that they are always isomorphic.

Theorem 5.1.6. *Given $f_1 \in F_1[x]$ and $\varphi : F_1 \simeq F_2$ as above, there is an isomorphism $\overline{\varphi} : L_1 \simeq L_2$ such that $\varphi = \overline{\varphi}\big|_{F_1}$.*

Proof. We will prove this by induction on $n = \deg(f_1) = \deg(f_2)$. When $n = 1$, we saw in the proof of Theorem 5.1.5 that $L_1 = F_1$ and $L_2 = F_2$. The theorem follows in this case by taking $\overline{\varphi} = \varphi$.

Now suppose that $n > 1$. We know that $L_1 = F_1(\alpha_1, \ldots, \alpha_n)$, where $\alpha_1, \ldots, \alpha_n$ are the roots of f_1. As in the proof of Theorem 5.1.5, we will use the extensions

$$(5.1) \qquad\qquad F_1 \subset F_1(\alpha_1) \subset L_1,$$

where $F_1(\alpha_1) \subset L_1$ is a splitting field of $g_1 = f_1/(x - \alpha_1)$. We now proceed in the following five steps.

Step 1. We first create an abstract model for $F_1(\alpha_1)$. Let $h_1 \in F_1[x]$ be the minimal polynomial of α_1. We know that h_1 is an irreducible factor of $f_1 \in F_1[x]$, since α_1 is a root of f_1. Thus

$$F_1(\alpha_1) = F_1[\alpha_1] \simeq F_1[x]/\langle h_1 \rangle,$$

where we have used Proposition 4.1.14 (for the equality) and Lemma 4.1.13 (for the isomorphism). The resulting isomorphism takes α_1 to $x + \langle h_1 \rangle$.

Step 2. We next find a root of f_2 corresponding to α_1. The key point is that the field isomorphism $\varphi : F_1 \simeq F_2$ induces a ring isomorphism $\widetilde{\varphi} : F_1[x] \simeq F_2[x]$ that takes f_1 to f_2. This isomorphism takes factors to factors and irreducibles to irreducibles. In particular, h_1 will map to an irreducible factor h_2 of f_2. Since f_2 splits completely over L_2, so does h_2 (do you see why?). Hence we can label the roots of f_2 as $\beta_1, \ldots, \beta_n \in L_2$, where β_1 is a root of h_2.

Step 3. The root β_1 of f_2 gives the extensions

(5.2) $$F_2 \subset F_2(\beta_1) \subset L_2,$$

where $F_2(\beta_1) \subset L_2$ is a splitting field of $g_2 = f_2/(x - \beta_1)$. As in Step 1, we also have

$$F_2(\beta_1) = F_2[\beta_1] \simeq F_2[x]/\langle h_2 \rangle$$

since h_2 is the minimal polynomial of β_1. This isomorphism takes β_1 to $x + \langle h_2 \rangle$.

Step 4. Since $\widetilde{\varphi} : F_1[x] \simeq F_2[x]$ takes h_1 to h_2, it must take $\langle h_1 \rangle$ to $\langle h_2 \rangle$. This means that we get an isomorphism of quotient rings

$$F_1[x]/\langle h_1 \rangle \simeq F_2[x]/\langle h_2 \rangle$$

that takes $x + \langle h_1 \rangle$ to $x + \langle h_2 \rangle$ and is φ on the coefficients. Combining this with Steps 1 and 3, we get an isomorphism

$$\varphi_1 : F_1(\alpha_1) \simeq F_1[x]/\langle h_1 \rangle \simeq F_2[x]/\langle h_2 \rangle \simeq F_2(\beta_1)$$

that takes α_1 to β_1 and satisfies $\varphi_1\big|_{F_1} = \varphi$.

Step 5. Finally, since $\varphi_1 : F_1(\alpha_1) \simeq F_2(\beta_1)$ takes α_1 to β_1 and f_1 to f_2, it also takes $g_1 = f_1/(x - \alpha_1)$ to $g_2 = f_2/(x - \beta_1)$. As noted above, L_1 is a splitting field of g_1 over $F_1(\alpha_1)$, and in the same way, L_2 is a splitting field of g_2 over $F_2(\beta_1)$.

We can now prove the existence of the desired isomorphism between L_1 and L_2. If we combine the extensions (5.1) and (5.2) together with the isomorphisms φ and φ_1, then we get the diagram

(5.3)
$$
\begin{array}{ccc}
L_1 & & L_2 \\
\cup & & \cup \\
F_1(\alpha_1) & \xrightarrow{\ \varphi_1\ } & F_2(\beta_1) \\
\cup & & \cup \\
F_1 & \xrightarrow{\ \varphi\ } & F_2
\end{array}
$$

Since $g_1 = f_1/(x - \alpha_1)$ has degree $n - 1$, Step 5 implies that we can apply the inductive hypothesis to $g_1 \in F_1(\alpha_1)[x]$ and $\varphi_1 : F_1(\alpha_1) \simeq F_2(\beta_1)$. This gives $\overline{\varphi_1} : L_1 \simeq L_2$, whose restriction to $F_1(\alpha_1)$ is φ_1. But since $\varphi_1\big|_{F_1} = \varphi$, it follows that the restriction of $\overline{\varphi_1}$ to F_1 is φ. Thus $\overline{\varphi_1}$ is the desired isomorphism. $\square$

When applied to the identity map $1_F : F \to F$ and $f \in F[x]$, Theorem 5.1.6 implies the following uniqueness result for splitting fields.

Corollary 5.1.7. *If L_1 and L_2 are splitting fields of $f \in F[x]$, then there is an isomorphism $L_1 \simeq L_2$ that is the identity on F.*   □

Because of this corollary, we can now speak of *the* splitting field of $f \in F[x]$, provided that we remember that splitting fields are unique up to isomorphism.

One might wonder why we proved Theorem 5.1.6 if all we wanted was Corollary 5.1.7. The answer lies in the inductive nature of the proof: if we begin with the identity map $\varphi = 1_F : F \to F$, then the inductive step (5.3) uses the isomorphism $\varphi_1 : F(\alpha_1) \simeq F(\beta_1)$. So if we had stated Theorem 5.1.6 only for the identity, then our inductive hypothesis would not apply, since φ_1 need not be the identity.

We conclude this section with a further application of Theorem 5.1.6. The idea is that this theorem gives some interesting isomorphisms of a splitting field. More precisely, the following result will play an important role in Chapter 6.

Proposition 5.1.8. *Let L be a splitting field of a polynomial in $F[x]$, and suppose that $h \in F[x]$ is irreducible and has roots $\alpha, \beta \in L$. Then there is a field isomorphism $\sigma : L \to L$ that is the identity on F and takes α to β.*

Proof. Since h is the minimal polynomial of α, we have an isomorphism

$$F(\alpha) = F[\alpha] \simeq F[x]/\langle h \rangle$$

that is the identity on F and sends α to $x + \langle h \rangle$. Similarly, using β, we have

$$F(\beta) = F[\beta] \simeq F[x]/\langle h \rangle$$

that is the identity on F and sends β to $x + \langle h \rangle$. As in Step 4 of the proof of Theorem 5.1.6, we can put these together to get a field isomorphism

$$\varphi : F(\alpha) \simeq F(\beta)$$

such that $\varphi(\alpha) = \beta$ and φ is the identity on F.

Now suppose that L is a splitting field of $f \in F[x]$. Then $f \in F(\alpha)[x]$ and $f \in F(\beta)[x]$, which means that L is a splitting field of f over both $F(\alpha)$ and $F(\beta)$. Thus we have the following diagram of splitting fields

$$
\begin{array}{ccc}
L & & L \\
\cup & & \cup \\
F(\alpha) & \xrightarrow{\varphi} & F(\beta)
\end{array}
$$

where φ takes f to f. Then Theorem 5.1.6 gives $\overline{\varphi} : L \simeq L$ such that $\overline{\varphi}\big|_{F(\alpha)} = \varphi$. Since φ is the identity on F and maps α to β, $\sigma = \overline{\varphi}$ is what we want.   □

Here is an example of this proposition.

Example 5.1.9. $L = \mathbb{Q}(\sqrt{2})$ is the splitting field of $x^2 - 2$ over $\mathbb{Q}$. This polynomial is irreducible over $\mathbb{Q}$ and has roots $\pm\sqrt{2} \in L$. Then Proposition 5.1.8 implies that there is an isomorphism $\sigma : L \to L$ such that $\sigma(\sqrt{2}) = -\sqrt{2}$.   ◁▷

In the terminology of Chapter 6, an isomorphism $\sigma : L \simeq L$ that is the identity on $F \subset L$ is an element of the Galois group $\mathrm{Gal}(L/F)$. We will use Proposition 5.1.8 to construct elements of $\mathrm{Gal}(L/F)$ when L is a splitting field over F.

Exercises for Section 5.1

Exercise 1. Show that a splitting field of $x^3 - 2$ over $\mathbb{Q}$ is $\mathbb{Q}(\omega, \sqrt[3]{2})$, $\omega = e^{2\pi i/3}$.

Exercise 2. Prove that $f \in F[x]$ splits completely over F if and only if F is the splitting field of f over F.

Exercise 3. Prove that an extension $F \subset L$ of degree 2 is a splitting field.

Exercise 4. Find the splitting field of $x^6 - 1 \in \mathbb{Q}[x]$.

Exercise 5. We showed in Section 4.1 that $f = x^4 - 10x^2 + 1$ is irreducible over $\mathbb{Q}$. Show that $L = \mathbb{Q}(\sqrt{2} + \sqrt{3})$ is the splitting field of f over $\mathbb{Q}$.

Exercise 6. Let $f \in \mathbb{Q}[x]$ be the minimal polynomial of $\alpha = \sqrt{2 + \sqrt{2}}$.
(a) Show that $f = x^4 - 4x^2 + 2$. Thus $[\mathbb{Q}(\alpha) : \mathbb{Q}] = 4$.
(b) Show that $\mathbb{Q}(\alpha)$ is the splitting field of f over $\mathbb{Q}$.

Exercise 7. Let $f = x^3 - x + 1 \in \mathbb{F}_3[x]$.
(a) Show that f is irreducible over $\mathbb{F}_3$.
(b) Let L be the splitting field of f over $\mathbb{F}_3$. Prove that $[L : \mathbb{F}_3] = 3$.
(c) Explain why L is a field with 27 elements.

Exercise 8. Let n be a positive integer. Then $f = x^n - 2$ is irreducible over $\mathbb{Q}$ by the Schönemann–Eisenstein criterion for the prime 2.
(a) Determine the splitting field L of f over $\mathbb{Q}$.
(b) Show that $[L : \mathbb{Q}] = n(n - 1)$ when n is prime.

Exercise 9. Let $f \in F[x]$ have degree $n > 0$, and let L be the splitting field of f over F.
(a) Suppose that $[L : F] = n!$. Prove that f is irreducible over F.
(b) Show that the converse of part (a) is false.

Exercise 10. Let $F \subset L$ be the splitting field of $f \in F[x]$, and let K be a field such that $F \subset K \subset L$. Prove that $K \subset L$ is the splitting field of some polynomial in $K[x]$.

Exercise 11. Suppose that $f \in F[x]$ is irreducible of degree $n > 0$, and let L be the splitting field of f over F.
(a) Prove that $n | [L : F]$.
(b) Give an example to show that $n = [L : F]$ can occur in part (a).

Exercise 12. In the situation of Theorem 5.1.6, explain why $[L_1 : F_1] = [L_2 : F_2]$.

Exercise 13. Let $L = \mathbb{Q}(\sqrt{2}, \sqrt{3})$. Use Proposition 5.1.8 to prove that there is an isomorphism $\sigma : L \simeq L$ such that $\sigma(\sqrt{2}) = \sqrt{2}$ and $\sigma(\sqrt{3}) = -\sqrt{3}$.

5.2 NORMAL EXTENSIONS

In this section, we will discover an important further property of splitting fields. This will lead to the concept of a *normal extension*.

Being a splitting field is a very special property of a field extension. For example, we will see below that $\mathbb{Q}(\sqrt[3]{2})$ is not the splitting field of any $f \in \mathbb{Q}[x]$. The basic reason for this lies in the following proposition.

Proposition 5.2.1. *Let L be the splitting field of $f \in F[x]$, and let $g \in F[x]$ be irreducible. If g has one root in L, then g splits completely over L.*

Proof. We can assume that f and g are monic. Then $L = F(\alpha_1, \ldots, \alpha_n)$, where $f = (x - \alpha_1) \cdots (x - \alpha_n)$. If $\beta \in L$ is a root of g, then g is the minimal polynomial of β since g is irreducible and monic. We need to prove that all roots of g lie in L.

Proposition 4.1.15 implies that $L = F[\alpha_1, \ldots, \alpha_n]$, so that β is a polynomial in the α_i, that is, $\beta = h(\alpha_1, \ldots, \alpha_n)$ for some $h \in F[x_1, \ldots, x_n]$. Now consider the polynomial

$$(5.4) \qquad s(x) = \prod_{\sigma \in S_n} \left(x - h(\alpha_{\sigma(1)}, \ldots, \alpha_{\sigma(n)}) \right) \in L[x].$$

This clearly has all of its roots in L. Furthermore, the factor corresponding to $\sigma = e$ is $x - h(\alpha_1, \ldots, \alpha_n) = x - \beta$, so that β is a root of s.

If we could show that $s \in F[x]$, then $g | s$ would follow immediately, since g is the minimal polynomial of β. Since s splits completely over L, this would imply that g also splits completely over L.

Hence it suffices to prove that $s \in F[x]$. We do this by going to the universal situation, as we did for the polynomial g_λ in the proof of Theorem 3.2.4. The polynomial

$$S(x) = \prod_{\sigma \in S_n} \left(x - h(x_{\sigma(1)}, \ldots, x_{\sigma(n)}) \right),$$

has coefficients in $F[x_1, \ldots, x_n]$. Furthermore, permuting $x_1, \ldots, x_n$ permutes the factors of S. It follows that if we multiply out S, then we get an expression

$$S(x) = \sum_{i=0}^{n!} p_i(x_1, \ldots, x_n) \, x^i,$$

where each $p_i(x_1, \ldots, x_n) \in F[x_1, \ldots, x_n]$ is symmetric. Since the α_i are the roots of $f \in F[x]$, Corollary 2.2.5 implies that $p_i(\alpha_1, \ldots, \alpha_n) \in F$. We conclude that

$$s(x) = \sum_{i=0}^{n!} p_i(\alpha_1, \ldots, \alpha_n) \, x^i \in F[x].$$

As explained above, the proposition now follows. $\square$

This proof of Theorem 5.2.1 uses the theory of symmetric polynomials from Chapter 2. See [Stewart, Ch. 10] for a proof that doesn't use symmetric polynomials.

Here is the example promised above.

Example 5.2.2. It is now easy to see why $\mathbb{Q}(\sqrt[3]{2})$ is not the splitting field of any polynomial in $\mathbb{Q}[x]$, since $x^3 - 2$ is irreducible over $\mathbb{Q}$ and obviously has a root in $\mathbb{Q}(\sqrt[3]{2})$. If this field were a splitting field, then Proposition 5.2.1 would force $x^3 - 2$ to split completely over $\mathbb{Q}(\sqrt[3]{2})$. But this is impossible, since $\mathbb{Q}(\sqrt[3]{2}) \subset \mathbb{R}$ doesn't contain the complex roots $\omega\sqrt[3]{2}$, $\omega^2\sqrt[3]{2}$ of $x^3 - 2$.   ◁▷

In Exercise 1 you will prove similarly that $\mathbb{Q}(\sqrt[4]{2})$ is not the splitting field of any polynomial in $\mathbb{Q}[x]$.

The property of Proposition 5.2.1 leads to the following definition.

Definition 5.2.3. *An algebraic extension $F \subset L$ is **normal** if every irreducible polynomial in $F[x]$ that has a root in L splits completely over L.*

In Exercise 2 you will show that $F \subset L$ is normal if and only if the minimal polynomial (relative to F) of every $\alpha \in L$ splits completely over L.

The following result reveals the strong link between normal extensions and splitting fields.

Theorem 5.2.4. *Suppose that $F \subset L$. Then L is the splitting field of some $f \in F[x]$ if and only if the extension $F \subset L$ is normal and finite.*

Proof. First suppose that L is the splitting field of $f \in F[x]$. Then $F \subset L$ is finite by Theorem 5.1.5 and is normal by Proposition 5.2.1.

For the converse, suppose that $F \subset L$ is normal and finite. By Theorem 4.4.3, the finiteness of this extension implies $L = F(\alpha_1, \ldots, \alpha_m)$, where each α_i is algebraic over F. Let $p_i \in F[x]$ be the minimal polynomial of α_i, and set $f = p_1 \cdots p_m$. We will show that L is the splitting field of f over F.

To prove this, first observe that every p_i splits completely over L, since $F \subset L$ is normal and $p_i \in F[x]$ is irreducible with a root $\alpha_i \in L$. It follows that f splits completely over L. Now let $L' \subset L$ be the subfield of L generated by F and the roots of f. Since the roots of f include $\alpha_1, \ldots, \alpha_m$, we have

$$L = F(\alpha_1, \ldots, \alpha_m) \subset L' \subset L.$$

This shows that $L' = L$, so that L is the splitting field of f over F.   □

We will see that normal extensions play an important role in Galois theory.

Historical Notes

Polynomials similar to $s(x)$ in (5.4) appear in the work of Galois. For example, in his first memoir on Galois theory, Galois says the following:

> In fact, by multiplying together all of the factors of the form $V - \varphi(a, b, c, \ldots, d)$, where one operates on the letters by all possible permutations, one will get an equation rational in V that is necessarily divisible by the equation in question.

(See [Galois, p. 51].) Here, $a, b, c, \ldots, d$ are roots of a polynomial $f \in F[x]$, and $\varphi(a, b, c, \ldots, d)$ is an element of the splitting field $F(a, b, c, \ldots, d)$. Then we can interpret Galois's statement as follows:

- By saying "an equation rational in V," Galois is asserting that the resulting polynomial in V has coefficients in F. This is exactly what we proved about the polynomial $s(x)$ in (5.4).
- When Galois says that this is "necessarily divisible by the equation in question," he is referring to the minimal polynomial of $\varphi(a, b, c, \ldots, d)$. This is what we called g in Proposition 5.2.1.

Thus, although normality does not appear explicitly in Galois's work, the above quotation should make it clear that it is implicit in what he does. We will say more about Galois's results in Chapter 12.

Exercises for Section 5.2

Exercise 1. Prove that $\mathbb{Q}(\sqrt[4]{2})$ is not the splitting field of any polynomial in $\mathbb{Q}[x]$.

Exercise 2. Prove that an algebraic extension $F \subset L$ is normal if and only if for every $\alpha \in L$, the minimal polynomial of α over F splits completely over L.

Exercise 3. Determine whether the following extensions are normal. Justify your answers.
(a) $\mathbb{Q} \subset \mathbb{Q}(\zeta_n)$, where $\zeta_n = e^{2\pi i / n}$.
(b) $\mathbb{Q} \subset \mathbb{Q}(\sqrt{2}, \sqrt[3]{2})$.
(c) $F = \mathbb{F}_3(t) \subset F(\alpha)$, where t is a variable and α is a root of $x^3 - t$ in a splitting field.

Exercise 4. Give an example of a normal extension of $\mathbb{Q}$ that is not finite.

5.3   SEPARABLE EXTENSIONS

Given a nonconstant polynomial $f \in F[x]$ with splitting field $F \subset L$, we can write

$$(5.5) \qquad f = a_0(x - \alpha_1) \cdots (x - \alpha_n), \qquad a_0 \in F, \ \alpha_1, \ldots, \alpha_n \in L.$$

It is important to realize that $\alpha_1, \ldots, \alpha_n$ are not always distinct. For example, $f = x^2 - 2x + 1 \in \mathbb{Q}[x]$ has $\alpha_1 = \alpha_2 = 1$. In this section, we will study those special polynomials for which the roots are all different.

We begin with some terminology. Given f as in (5.5), let $\beta_1, \ldots, \beta_r$ be the *distinct* elements of L that appear among $\alpha_1, \ldots, \alpha_n$, and let m_i be the number of times $x - \beta_i$ appears in (5.5). Then we can write (5.5) as

$$f = a_0(x - \beta_1)^{m_1} \cdots (x - \beta_r)^{m_r}, \qquad a_0 \in F, \ \beta_1, \ldots, \beta_r \in L \text{ distinct}.$$

We call m_i the *muliplicity* of β_i and say that β_i is a *simple root* if $m_i = 1$ and a *multiple root* otherwise.

Definition 5.3.1. *A polynomial $f \in F[x]$ is **separable** if it is nonconstant and its roots in a splitting field are all simple.*

In other words, f is separable if it has distinct roots. These definitions are independent of splitting field used, since all splitting fields of f over F are isomorphic.

One tool used to study separability is the *discriminant* $\Delta(f) \in F$ of a monic polynomial $f \in F[x]$. We defined $\Delta(f)$ in Section 2.4 and showed in Proposition 2.4.3 that if $\deg(f) > 1$, then

$$\Delta(f) = \prod_{1 \le i < j \le n} (\alpha_i - \alpha_j)^2 \quad \text{when} \quad f = (x - \alpha_1) \cdots (x - \alpha_n).$$

Another tool we will need is the *formal derivative*, which for a polynomial $g = a_0 x^n + a_1 x^{n-1} + \cdots + a_{n-1} x + a_n \in F[x]$ is defined to be

$$g' = n a_0 x^{n-1} + (n-1) a_1 x^{n-2} + \cdots + a_{n-1}.$$

The operation $g \mapsto g'$ enjoys the usual properties from calculus, including

(5.6)
$$\begin{aligned}
(ag + bh)' &= ag' + bh', \\
(gh)' &= g'h + gh'
\end{aligned}$$

for $g, h \in F[x]$ and $a, b \in F$. See Exercise 1 for a proof of (5.6).

Separability, the discriminant, and the formal derivative are related as follows.

Proposition 5.3.2. *If $f \in F[x]$ is monic and nonconstant, then the following are equivalent:*

(a) *f is separable.*
(b) *$\Delta(f) \ne 0$.*
(c) *f and f' are relatively prime in $F[x]$, that is, $\gcd(f, f') = 1$.*

Proof. If $\deg(f) = 1$, then $\Delta(f) = 1$ by the definition of $\Delta(f)$ given in Section 2.4. It follows easily that (a), (b), and (c) are all true in this case. Hence we may assume that $\deg(f) = n > 1$.

For (a) $\Leftrightarrow$ (b), let $\alpha_1, \ldots, \alpha_n$ be the roots of f in some splitting field. The above formula for $\Delta(f)$ shows that $\Delta(f) \ne 0$ is equivalent to $\alpha_i \ne \alpha_j$ for all $i < j$.

It remains to show (a) $\Leftrightarrow$ (c). Let L be a splitting field of f over F, so that $f = (x - \alpha_1) \cdots (x - \alpha_n)$ in $L[x]$. For a given i, write

$$f(x) = (x - \alpha_i) h_i(x), \quad h_i(x) = \prod_{j \ne i} (x - \alpha_j).$$

Differentiating, we obtain $f'(x) = (x - \alpha_i) h_i'(x) + h_i(x)$ by the product rule, and then evaluating at α_i gives

(5.7)
$$f'(\alpha_i) = h_i(\alpha_i) = \prod_{j \ne i} (\alpha_i - \alpha_j).$$

If (c) is false, then f and f' have a common factor g of positive degree. Since $g \mid f$, we must have $g(\alpha_i) = 0$ for some i, and then $g \mid f'$ implies that $f'(\alpha_i) = 0$. Hence $0 = f'(\alpha_i) = \prod_{j \ne i} (\alpha_i - \alpha_j)$, so that $\alpha_i = \alpha_j$ for some $j \ne i$.

Conversely, if (c) is true, then $1 = Af + Bf'$ for some $A, B \in F[x]$. Evaluating this at α_i gives $1 = B(\alpha_i) f'(\alpha_i)$, so that $f'(\alpha_i) \ne 0$. By (5.7), this implies that $\prod_{j \ne i} (\alpha_i - \alpha_j)$ is nonzero for all i. Hence $\alpha_1, \ldots, \alpha_n$ are distinct. $\qquad\square$

The definition of separable polynomial given in Definition 5.3.1 is nonstandard in that it applies to arbitrary nonconstant polynomials with distinct roots, while most books focus on *irreducible* polynomials with distinct roots. Fortunately, as long as we restrict to irreducible polynomials, Definition 5.3.1 is consistent with the literature.

We can also extend the concept of separability to algebraic extensions.

Definition 5.3.3. *Let $F \subset L$ be an algebraic extension.*
(a) *$\alpha \in L$ is **separable** over F if its minimal polynomial over F is separable.*
(b) *$F \subset L$ is a **separable extension** if every $\alpha \in L$ is separable over F.*

Since minimal polynomials are irreducible, this agrees with the definition of separable extension given in other texts.

We can interpret the separability of a polynomial in terms of its irreducible factors as follows.

Lemma 5.3.4. *A nonconstant polynomial $f \in F[x]$ is separable if and only if f is a product of irreducible polynomials, each of which is separable and no two of which are multiples of each other.*

Proof. First assume that f is separable. If a factor of f fails to have distinct roots in a splitting field, then the same is true for f. Hence any irreducible factor of f must be separable. Also, if the factorization of f into irreducibles includes two factors that are multiples of each other, then the product of these factors would be a nonseparable divisor of f. Hence the factorization of f must consist of separable, irreducible polynomials no two of which are multiples of each other.

Conversely, let $f = g_1 \cdots g_s$, where $g_1, \ldots, g_s$ are separable and irreducible, and no two are multiples of each other. Then, in the splitting field of f, each g_i has distinct roots. Furthermore, suppose that g_i and g_j share a root α for some $i \neq j$. Since g_i and g_j are irreducible, this would imply that each was a constant times the minimal polynomial of α, which is a contradiction. Hence f is separable.   □

In order to make good use of Lemma 5.3.4, we need to understand when an irreducible polynomial is separable. Fortunately, many irreducible polynomials are automatically separable.

Lemma 5.3.5. *Let $f \in F[x]$ be an irreducible polynomial of degree n. Then f is separable if either of the following conditions is satisfied:*
(a) *F has characteristic 0, or*
(b) *F has characteristic $p > 0$, where $p \nmid n$.*

Proof. Let $f = a_0 x^n + \cdots + a_{n-1} x + a_n$, where $n > 0$ and $a_0 \neq 0$. Then $f' = na_0 x^{n-1} + \cdots + a_{n-1}$. Condition (a) or (b) implies that $n \neq 0$ in F, so that $a_0 \neq 0$ implies $na_0 \neq 0$. Hence f' is nonzero and has degree $n - 1$.

Since f is irreducible, its only divisors (up to constant multiples) are 1 and f. In particular, $g = \gcd(f, f')$ must be 1 or f. But $g \mid f'$ and $f' \neq 0$ imply $\deg(g) \leq \deg(f') = n-1$. Hence g cannot be a multiple of f, so that $\gcd(f, f') = g = 1$.   □

One surprise of Lemma 5.3.5 is that separability is related to the characteristic. Here is another example of this phenomenon.

Example 5.3.6. Consider $f = x^n - 1 \in F[x]$, where $n > 0$. By Proposition 5.3.2, f is separable if and only if f is relatively prime to $f' = nx^{n-1}$. However:

- If $n \neq 0$ in F, then the only irreducible factor of f' is x, which clearly doesn't divide f. Thus f is relatively prime to f' in this case.
- If $n = 0$ in F, then f' is identically zero, in which case f divides f'. Hence f is not relatively prime to f' in this case.

It follows that $x^n - 1 \in F[x]$ fails to be separable if and only if F has characteristic p and p divides n.   ◁▷

For the remainder of the section, we will consider fields of characteristic 0 and characteristic p separately. Since we encounter fields of characteristic 0 most often, we will begin with them.

A. Fields of Characteristic 0. Here is an application of Lemmas 5.3.4 and 5.3.5.

Proposition 5.3.7. *If F has characteristic 0, then:*
(a) *Every irreducible polynomial in $F[x]$ is separable.*
(b) *Every algebraic extension of F is separable.*
(c) *A nonconstant polynomial $f \in F[x]$ is separable if and only if f is a product of irreducible polynomials, no two of which are multiples of each other.*

Proof. Part (a) follows immediately from Lemma 5.3.5, and this implies part (b) by Definition 5.3.3. Finally, part (c) follows from part (a) and Lemma 5.3.4.   □

In characteristic 0, we can get rid of multiple roots as follows.

Proposition 5.3.8. *Let F have characteristic 0, and suppose that $f \in F[x]$ has the factorization $f = c\, g_1^{m_1} \cdots g_l^{m_l}$, where $c \in F$, $g_i \in F[x]$ is monic and irreducible for $1 \le i \le l$, and $g_1, \ldots, g_l$ are distinct. Then*

(5.8)
$$\frac{f}{\gcd(f, f')} = c\, g_1 \cdots g_l.$$

Furthermore, $g_1 \cdots g_l$ is separable and has the same roots as f in a splitting field.

Proof. Proposition 5.3.7 implies that $g_1 \cdots g_l$ is separable, and this polynomial and f clearly have the same roots in a splitting field. Hence it suffices to prove (5.8).

The factorization $f = c\, g_1^{m_1} \cdots g_l^{m_l}$ implies that we can compute $\gcd(f, f')$ by finding the highest power of g_i that divides f' (do you see why?). If we write

$$f = g_i^{m_i} h_i, \quad h_i = c \prod_{j \neq i} g_j^{m_j},$$

then differentiating gives

$$f' = m_i g_i^{m_i - 1} g_i' h_i + g_i^{m_i} h_i' = g_i^{m_i - 1}(m_i g_i' h_i + g_i h_i').$$

This shows that $g_i^{m_i-1} | f'$. If we had $g_i^{m_i} | f'$, then $g_i | (m_i g_i' h_i + g_i h_i')$, and thus $g_i | m_i g_i' h_i$. Since g_i is irreducible, this would force $g_i | m_i g_i'$ or $g_i | h_i$. The latter is impossible by the definition of h_i, and the former is impossible because $m_i g_i'$ is nonzero of degree $\deg(g) - 1$ (this is where we use characteristic 0). Hence $g_i^{m_i-1}$ is the highest power of g_i dividing f', which implies that

$$\gcd(f, f') = g_1^{m_1-1} \cdots g_l^{m_l-1}.$$

The desired formula (5.8) follows immediately.   □

This proposition is more powerful than it seems. For example, suppose that we have a polynomial $f \in F[x]$ that has multiple roots in a splitting field, say

$$(5.9) \qquad f = c(x - \alpha_1)^{m_1} \cdots (x - \alpha_s)^{m_s}, \qquad \alpha_1, \dots, \alpha_s \text{ distinct}, \qquad m_i \geq 1.$$

If we ignore the multiplicities, then we get the separable polynomial

$$g = c(x - \alpha_1) \cdots (x - \alpha_s),$$

which has the same roots as f. There are three methods to find g:

- If we know the roots of f, then we get g from the factorization (5.9). This requires knowing the roots, which rarely happens.
- If we know the irreducible factorization $f = c g_1^{m_1} \cdots g_l^{m_l}$ over F, then we get $g = c g_1 \cdots g_l$ by Proposition 5.3.8. This requires knowing the factorization, which can be time-consuming to compute.
- We get g from the gcd computation given in (5.8) of Proposition 5.3.8.

In practice, the third method is the most efficient. Here is an example.

Example 5.3.9. Let $f = x^{11} - x^{10} + 2x^8 - 4x^7 + 3x^5 - 3x^4 + x^3 + 3x^2 - x - 1 \in \mathbb{Q}[x]$. Using the gcd command in *Maple* or the PolynomialGCD command in *Mathematica*, one finds that

$$\gcd(f, f') = x^6 - x^5 + x^3 - 2x^2 + 1.$$

It follows that

$$(5.10) \quad \frac{f}{\gcd(f, f')} = \frac{x^{11} - x^{10} + 2x^8 - 4x^7 + 3x^5 - 3x^4 + x^3 + 3x^2 - x - 1}{x^6 - x^5 + x^3 - 2x^2 + 1}$$
$$= x^5 + x^2 - x - 1$$

is a separable polynomial with the same roots as f.   ◁▷

B. Fields of Characteristic p. We begin with an important property of such fields.

Lemma 5.3.10. *Let F be a field of characteristic p, and assume that $\alpha, \beta \in F$. Then $(\alpha + \beta)^p = \alpha^p + \beta^p$ and $(\alpha - \beta)^p = \alpha^p - \beta^p$.*

Proof. The binomial theorem implies that

$$(\alpha + \beta)^p = \alpha^p + \binom{p}{1}\alpha^{p-1}\beta + \cdots + \binom{p}{r}\alpha^{p-r}\beta^r + \cdots + \binom{p}{p-1}\alpha\beta^{p-1} + \beta^p.$$

In the proof of Proposition 4.2.5, we showed that $p|\binom{p}{r}$ for $1 \le r \le p - 1$. Since F has characteristic p, the above identity reduces to $(\alpha + \beta)^p = \alpha^p + \beta^p$. In Exercise 2 you will use this to prove that $(\alpha - \beta)^p = \alpha^p - \beta^p$. $\square$

In Exercise 3 you will use Lemma 5.3.10 to show that 1 is the only pth root of unity in a field of characteristic p.

Since $(\alpha\beta)^p = \alpha^p\beta^p$, Lemma 5.3.10 implies that the map $\alpha \to \alpha^p$ is a ring homomorphism over any field F of characteristic p. This is the *Frobenius homomorphism* of F. We will use Frobenius when we discuss finite fields in Chapter 11.

Here is our first example of a nonseparable irreducible polynomial.

Example 5.3.11. Let $F = k(t)$, where k has characteristic p and t is a variable. We claim that $f = x^p - t \in F[x]$ is nonseparable and irreducible over F.

To prove this, note that f has no roots in F, by Exercise 9 in Section 4.2. Since p is prime, Proposition 4.2.6 implies that f is irreducible over F. Furthermore, if $\alpha \in L$ is a root of f in its splitting field L, then $\alpha^p = t$. Using Lemma 5.3.10, it follows that

$$(5.11) \qquad\qquad (x - \alpha)^p = x^p - \alpha^p = x^p - t.$$

Thus f does not have distinct roots in its splitting field L and hence is not separable.

The polynomial f also gives an example of a nonseparable finite extension. Namely, $\alpha \in L$ is a root of f, so that f is the minimal polynomial of α over F, since f is irreducible and monic. It follows that $F \subset L$ is not separable.

Note also that by (5.11), α is the *only* root of f. Hence the splitting field is $L = F(\alpha)$. This implies $[L : F] = p$, since f is the minimal polynomial of α. ◁▷

One caution is that over a field of characteristic p, not all irreducible polynomials of degree p fail to be separable. Here is a simple example.

Example 5.3.12. For the field $\mathbb{F}_2$ of two elements, $f = x^2 + x + 1 \in \mathbb{F}_2[x]$ is irreducible, since it has no roots in $\mathbb{F}_2$. It is separable, since $f' = 2x + 1 = 1$ is relatively prime to f. ◁▷

We will say more about characteristic p in the Mathematical Notes.

C. Computations.

To determine whether a monic polynomial $f \in F[x]$ is separable, one can use either $\Delta(f)$ or $\gcd(f, f')$ by Proposition 5.3.2. We will briefly discuss how to compute both using *Maple* and *Mathematica*.

We begin with a gcd computation.

Example 5.3.13. Example 5.3.9 explained how *Maple* and *Mathematica* do this over $\mathbb{Q}$. For example, if

$$(5.12) \qquad\qquad f = x^6 + 10x^3 + 3x^2 + 1 \in \mathbb{Q}[x],$$

one computes $\gcd(f, f') = 1$, so that f is separable. However, since f has integer coefficients, we can reduce modulo p and obtain a polynomial $f_p \in \mathbb{F}_p[x]$. Then we can ask whether f_p is separable over $\mathbb{F}_p$.

For $p = 2$ or 3, we have $f'_p = 0$, since $f' = 6x^5 + 30x^2 + 6x$. Thus $\gcd(f_p, f'_p) = \gcd(f_p, 0) = f_p \neq 1$, so that f_p is not separable for these primes. For a larger prime such as $p = 557$ (the reason for this choice will soon become clear), we compute the gcd over $\mathbb{F}_{557}$ using the *Maple* command

$$\text{Gcd}(x^6 + 10*x^3 + 3*x^2 + 1, 6*x^5 + 30*x^2 + 6*x) \bmod 557;$$

which gives the result $x + 257$. Thus f_{557} is not separable. In *Mathematica*, this computation is done using the command

$$\text{PolynomialGCD}[x^6 + 10x^3 + 3x^2 + 1, 6x^5 + 30x^2 + 6x,$$
$$\text{Modulus} \rightarrow 557]$$

which gives the same answer $x + 257$.

The second approach to studying whether f is separable would be to compute the discriminant $\Delta(f)$. Chapter 2 gave a cumbersome method for computing $\Delta(f)$ that expresses $\Delta = \prod_{i<j}(x_i - x_j)^2$ in terms of the elementary symmetric polynomials and then evaluates the σ_i at the coefficients of f (up to the usual sign). A more efficient approach uses the *resultant*.

We will not discuss resultants in detail, for this would take us too far afield. The idea is that for $f, g \in F[x]$, their resultant

$$\text{Res}(f, g, x) \in F$$

is a polynomial in the coefficients of f and g with the property that

$$\text{Res}(f, g, x) = 0 \iff f \text{ and } g \text{ have a common root in an extension of } F.$$

An introduction to resultants can be found in [1, Ch. 3, §5] and [3, pp. 97–104].

For us, the most important property of resultants is that if $f \in F[x]$ is monic of degree $n > 1$, then

$$(5.13) \qquad \Delta(f) = (-1)^{\frac{1}{2}n(n-1)}\text{Res}(f, f', x)$$

(see [3, pp. 103–104]). In *Maple* and *Mathematica*, the resultant of f, g is computed using the commands $\text{resultant}(f,g,x)$ and $\text{Resultant}[f,g,x]$.

Example 5.3.14. As an example, consider the polynomial $f = x^6 + 10x^3 + 3x^2 + 1$ given by (5.12). This leads to

$$\Delta(f) = (-1)^{\frac{1}{2}6(6-1)}\text{Res}(f, f', x) = -(-649684800) = 2^6 \cdot 3^6 \cdot 5^2 \cdot 557.$$

As before, reducing f modulo p gives $f_p \in \mathbb{F}_p[x]$. In Exercise 4 you will show that $\Delta(f_p) \in \mathbb{F}_p$ is the congruence class of $\Delta(f)$ modulo p. Thus

$$f_p \text{ is separable over } \mathbb{F}_p \iff \Delta(f) \not\equiv 0 \bmod p.$$

It follows that f_p is separable over $\mathbb{F}_p$ if and only if $p \notin \{2, 3, 5, 557\}$. From here, one easily finds that

$$(5.14) \qquad \gcd(f_p, f_p') = \begin{cases} f_p, & p = 2, 3, \\ x^2 + 3, & p = 5, \\ x + 257, & p = 557, \\ 1, & \text{otherwise.} \end{cases}$$

You will compute a similar example in Exercise 5.   ◁▷

Mathematical Notes

Our treatment omits many interesting results about separability.

▪ **Separable Extensions.** Here are some conditions that imply separability.

Theorem 5.3.15.
(a) *If $L = F(\alpha_1, \ldots, \alpha_n)$, where each α_i is separable over F, then $F \subset L$ is separable.*
(b) *If $F \subset L$ is the splitting field of a separable polynomial, then $F \subset L$ separable.*
(c) *If $F \subset K$ and $K \subset L$ are separable extensions, then $F \subset L$ is separable.*   □

We will defer our proof of part (a) until Chapter 7, since it uses some ideas from Galois theory. (For a proof that doesn't use Galois theory, see Corollaries 1 and 3 of [Garling, Sec. 10.2].) In Exercise 6 you will show that part (b) follows from part (a). The proof of part (c) requires the concept of *separable degree*, which is discussed in [Grillet, Sec. 7.2].

▪ **The Structure of Irreducible Polynomials.** Although irreducible polynomials are separable in characteristic 0, things are more complicated in characteristic p. In this case, irreducible polynomials are built from separable ones as follows.

Proposition 5.3.16. *Let F have characteristic p, and let $f \in F[x]$ be irreducible. Then there is an integer $e \geq 0$ and a separable, irreducible polynomial $g \in F[x]$ such that $f(x) = g(x^{p^e})$.*   □

You will prove this in Exercise 7.

▪ **Purely Inseparable Extensions.** If an algebraic extension $F \subset L$ is not separable, then some (but not necessarily all) elements of L have nonseparable minimal polynomials. Here is a simple example.

Example 5.3.17. Suppose that k has characteristic 3, and let t, u be variables. Consider $F = k(t, u)$, and let $F \subset L$ be the splitting field of $f = (x^2 - t)(x^3 - u)$. Thus L contains elements α, β such that $\alpha^2 = t$ and $\beta^3 = u$. In Exercise 8 you will prove the following:

• The minimal polynomial of α over F is $x^2 - t$, which is separable. Thus α is separable over F.

- The minimal polynomial of β over F is $x^3 - u$, which is not separable (remember that F has characteristic 3). Hence β is not separable over F.

Thus some elements of L are separable over F while others are not.   ◁▷

However, some extensions have very few separable elements. Given $F \subset L$, it is clear that every $a \in F$ is separable over F. Thus we say that an algebraic extension $F \subset L$ is *purely inseparable* if no element of $L \setminus F$ is separable over F. For example, you will prove in Exercise 9 that the extension of Example 5.3.11 is purely inseparable.

In general, if $F \subset L$ is purely inseparable, then the minimal polynomial of $\alpha \in L$ is of the form $x^{p^e} - a$ for some $e \geq 0$ and $a \in F$. This implies that the degree of a finite purely inseparable extension is a power of p (see Exercise 10).

Returning to the case of an arbitrary algebraic extension $F \subset L$ in characteristic p, one can "separate" the separable elements from the inseparable ones. More precisely, one can prove the existence of a unique intermediate field $F \subset K \subset L$ such that K is separable over F and L is purely inseparable over K. A proof can be found in Section 8.7 of [Jacobson, Vol. II].

■ **The Squarefree Decomposition of a Polynomial.** Proposition 5.3.8 shows that if F has characteristic 0 and $f \in F[x]$, then $g = f/\gcd(f, f') \in F[x]$ is separable and has the same roots as f. This means that

$$(5.15) \qquad\qquad f = gh,$$

where g is separable and every root of h has multiplicity at least 2. In this situation, we call g the *squarefree part* of f, and (5.15) its *squarefree decomposition*. Squarefree decompositions also exist when F has characteristic p (this is proved in [2, Tutorial 5, pp. 37–38]). The difference is that in characteristic p, the squarefree part g need not have the same roots as f (can you give an example?).

Exercises for Section 5.3

Exercise 1. Prove (5.6).

Exercise 2. Let F have characteristic p, and suppose that $\alpha, \beta \in F$. Lemma 5.3.10 shows that $(\alpha + \beta)^p = \alpha^p + \beta^p$.
  (a) Prove that $(\alpha - \beta)^p = \alpha^p - \beta^p$ if $\alpha, \beta \in F$.
  (b) Prove that $(\alpha + \beta)^{p^e} = \alpha^{p^e} + \beta^{p^e}$ for all $e \geq 0$.

Exercise 3. Let F be a field of characteristic p. The nth roots of unity are defined to be the roots of $x^n - 1$ in the splitting field $F \subset L$ of $x^n - 1$.
  (a) If $p \nmid n$, show that there are n distinct nth roots of unity in L.
  (b) Show that there is only one pth root of unity, namely $1 \in F$.

Exercise 4. Let $f \in \mathbb{Z}[x]$ be monic and nonconstant and have discriminant $\Delta(f)$. Then let $f_p \in \mathbb{F}_p[x]$ be obtained from f by reducing modulo p. Prove that $\Delta(f_p) \in \mathbb{F}_p$ is the congruence class of $\Delta(f)$.

Exercise 5. For $f = x^7 + x + 1$, find all primes for which f_p is not separable, and compute $\gcd(f_p, f_p')$ as in (5.14).

Exercise 6. Use part (a) of Theorem 5.3.15 to show that the splitting field of a separable polynomial gives a separable extension.

Exercise 7. Suppose that F is a field of characteristic p. The goal of this exercise is to prove Proposition 5.3.16. To begin the proof, let $f \in F[x]$ be irreducible.
  (a) Assume that f' is not identically zero. Then use the argument of Lemma 5.3.5 to show that f is separable.
  (b) Now assume that f' is identically zero. Show that there is a polynomial $g_1 \in F[x]$ such that $f(x) = g_1(x^p)$.
  (c) Show that the polynomial g_1 of part (b) is irreducible.
  (d) Now apply parts (a)–(c) to g_1 repeatedly until you get a separable polynomial g, and conclude that $f(x) = g(x^{p^e})$ where $e \geq 0$ and $g \in F[x]$ is irreducible and separable.

Exercise 8. Let $F = k(t, u)$ and $f = (x^2 - t)(x^3 - u)$ be as in Example 5.3.17. Then the splitting field of f contains elements α, β such that $\alpha^2 = t$ and $\beta^3 = u$.
  (a) Prove that $x^2 - t$ is the minimal polynomial of α over F. Also show that $x^2 - t$ is separable.
  (b) Similarly, prove that $x^3 - u$ is the minimal polynomial of β over F, and show that $x^3 - u$ is not separable.

Exercise 9. Let F be a field of characteristic p, and consider $f = x^p - a \in F[x]$. We will assume that f has no roots in F, so that f is irreducible by Proposition 4.2.6. Let α be a root of f in some extension of F.
  (a) Argue as in Example 5.3.11 that $F(\alpha)$ is the splitting field of f and that $[F(\alpha) : F] = p$.
  (b) Let $\beta \in F(\alpha) \setminus F$. Use Lemma 5.3.10 to show that $\beta^p \in F$.
  (c) Use parts (a) and (b) to show that the minimal polynomial of β over F is $x^p - \beta^p$.
  (d) Conclude that $F \subset F(\alpha)$ is purely inseparable.

Exercise 10. Suppose that F has characteristic p and $F \subset L$ is a finite extension.
  (a) Use Proposition 5.3.16 to prove that $F \subset L$ is purely inseparable if and only if the minimal polynomial of every $\alpha \in L$ is of the form $x^{p^e} - a$ for some $e \geq 0$ and $a \in F$.
  (b) Now suppose that $F \subset L$ is purely inseparable. Prove that $[L : F]$ is a power of p.

Exercise 11. Prove that $f \in F[x]$ is separable if and only if f is nonconstant and f and f' have no common roots in any extension of F.

Exercise 12. Let F have characteristic p, and let $F \subset L$ be a finite extension with $p \nmid [L : F]$. Prove that $F \subset L$ is separable.

Exercise 13. Let $F \subset K \subset L$ be field extensions, and assume that L is separable over F. Prove that $F \subset K$ and $K \subset L$ are separable extensions. Note that this is the converse of part (b) of Theorem 5.3.15

Exercise 14. Let f be the polynomial considered in Example 5.3.9. Use *Maple* or *Mathematica* to factor f and to verify that the product of the distinct irreducible factors of f is the polynomial given in (5.10).

Exercise 15. Let F have characteristic p and consider $f = x^p - x + a \in F[x]$.
  (a) Show that f is separable.
  (b) Let α be a root of f in some extension of F. Show that $\alpha + 1$ is also a root.
  (c) Use part (b) to show that f splits completely over $F(\alpha)$.
  (d) Use part (a) of Theorem 5.3.15 to show that $F \subset F(\alpha)$ is a separable and normal.

In Exercise 5 of Section 6.2 you will show that if f is irreducible over F, then $\text{Gal}(F(\alpha)/F) \simeq \mathbb{Z}/p\mathbb{Z}$, generated by the automorphism sending α to $\alpha + 1$. This is related to a theorem of Artin and Schreier, which states that in characteristic p, *every* separable, normal extension of degree p is the splitting field of an irreducible polynomial of the form $x^p - x + a$.

Exercise 16. Let β be a root of a polynomial f.
(a) Assume that $f(x) = (x - \beta)^m h(x)$ for some polynomial $h(x)$, and let $f^{(m)}$ denote the mth derivative of f. Prove that $f^{(m)}(\beta) = m!h(\beta)$.
(b) Assume that we are in characteristic 0. Prove that β has multiplicity m as a root of f if and only if $f(\beta) = f'(\beta) = \cdots = f^{(m-1)}(\beta) = 0$ and $f^{(m)}(\beta) \neq 0$.
(c) Assume that we are in characteristic p. How big does p need to be relative to m in order for the equivalence of part (b) to be still valid?

5.4 THEOREM OF THE PRIMITIVE ELEMENT

Of the extension fields $F \subset L$ studied so far, the nicest case is when $L = F(\alpha)$ for some $\alpha \in L$. When this happens, we say that α is a *primitive element* of $F \subset L$. In this section, we will show that many but not all finite extensions have primitive elements.

Here is the *Theorem of the Primitive Element*.

Theorem 5.4.1. *Let $F \subset L = F(\alpha_1, \ldots, \alpha_n)$ be a finite extension, where each α_i is separable over F. Then there is $\alpha \in L$ separable over F such that $L = F(\alpha)$. Furthermore, if F is infinite, then α can be chosen to be of the form*

$$\alpha = t_1\alpha_1 + \cdots + t_n\alpha_n$$

where $t_1, \ldots, t_n \in F$.

Proof. First assume that F is infinite and that $L = F(\alpha_1, \ldots, \alpha_n)$, where each α_i is separable over F. We will use induction on n to show that there are $t_1, \ldots, t_n \in F$ such that $L = F(t_1\alpha_1 + \cdots + t_n\alpha_n)$ and $t_1\alpha_1 + \cdots + t_n\alpha_n$ is separable over F.

We begin with the case $n = 2$. Given $L = F(\beta, \gamma)$, let $f, g \in F[x]$ be the minimal polynomials of β, γ respectively, and set $\ell = \deg(f)$, $m = \deg(g)$. In a splitting field of fg, the separability of β, γ implies that

$$f \text{ has distinct roots } \beta = \beta_1, \beta_2, \ldots, \beta_\ell,$$
$$g \text{ has distinct roots } \gamma = \gamma_1, \gamma_2, \ldots, \gamma_m.$$

Since F is infinite, we can find $\lambda \in F$ such that

$$\lambda \neq \frac{\beta_i - \beta_r}{\gamma_s - \gamma_j} \quad \text{for } 1 \leq r, i \leq \ell, \ 1 \leq s, j \leq m, \ s \neq j.$$

This easily implies that

(5.16) $\beta_r + \lambda\gamma_s \neq \beta_i + \lambda\gamma_j \quad \text{for } (r, s) \neq (i, j).$

In particular, since $\beta = \beta_1$ and $\gamma = \gamma_1$, we have

(5.17) $$\beta + \lambda\gamma \neq \beta_i + \lambda\gamma_j \quad \text{for } 1 \leq i \leq \ell, \ 2 \leq j \leq m.$$

We first prove that $F(\beta + \lambda\gamma) = F(\beta, \gamma)$. Since $F(\beta + \lambda\gamma) \subset F(\beta, \gamma)$ is obvious, it suffices to show that $\beta, \gamma \in F(\beta + \lambda\gamma)$. We begin with γ. Observe that

- $g(x)$ vanishes at γ and lies in $F[x] \subset F(\beta + \lambda\gamma)[x]$;
- $f(\beta + \lambda\gamma - \lambda x)$ vanishes at γ (check this!) and also lies in $F(\beta + \lambda\gamma)[x]$.

Our strategy will be to study the greatest common divisor of the polynomials $g(x)$ and $f(\beta + \lambda\gamma - \lambda x)$. We first note that if the gcd were 1, then

$$A(x)g(x) + B(x)f(\beta + \lambda\gamma - \lambda x) = 1$$

for some $A, B \in F(\beta + \lambda\gamma)[x]$. By the above bullets, evaluating this at $x = \gamma$ would give $0 = 1$. Hence

$$h(x) = \gcd\left(g(x), f(\beta + \lambda\gamma - \lambda x)\right) \in F(\beta + \lambda\gamma)[x]$$

has degree at least 1. If the degree were > 1, then $h(x)|g(x)$ implies that for some $2 \leq j \leq m$, γ_j would be a root of $h(x)$ (do you see how this uses the separability of g?). But since $h(x)|f(\beta + \lambda\gamma - \lambda x)$, γ_j must also be a root of $f(\beta + \lambda\gamma - \lambda x)$, that is, $f(\beta + \lambda\gamma - \lambda\gamma_j) = 0$. Since the roots of f are $\beta = \beta_1, \ldots, \beta_\ell$, this implies

$$\beta + \lambda\gamma - \lambda\gamma_j = \beta_i \quad \text{for some } 1 \leq i \leq \ell,$$

which contradicts (5.17). Hence h has degree 1, and then $h = x - \gamma$ follows, since γ is a root. But we also know $h \in F(\beta + \lambda\gamma)[x]$, so that $\gamma \in F(\beta + \lambda\gamma)$. Then $\beta = (\beta + \lambda\gamma) - \lambda \cdot \gamma \in F(\beta + \lambda\gamma)$ follows immediately, since $\lambda \in F$. This completes the proof that $F(\beta, \gamma) = F(\beta + \lambda\gamma)$.

Next let $p \in F[x]$ be the minimal polynomial of $\beta + \lambda\gamma$ over F. We need to show that p is separable. For this purpose, consider

(5.18) $$s(x) = \prod_{j=1}^{m} f(x - \lambda\gamma_j).$$

Note that $\beta + \lambda\gamma$ is a root of s, since $\beta = \beta_1$. Furthermore, since $f \in F[x], \lambda \in F$, and $\gamma_1, \ldots, \gamma_m$ are the roots of $g \in F[x]$, one can easily show that $s \in F[x]$ using the techniques used in the proofs of Theorem 3.2.4 and Proposition 5.2.1. We leave the details as Exercise 1. It follows that p divides s in $F[x]$. However, we also have $f = (x - \beta_1) \cdots (x - \beta_\ell)$, which when combined with (5.18) gives the formula

(5.19) $$s(x) = \prod_{i=1}^{\ell} \prod_{j=1}^{m} \left(x - (\beta_i + \lambda\gamma_j)\right).$$

Then (5.16) implies that s has distinct roots. Hence p is also separable (it divides s), which proves that $\beta + \lambda\gamma$ is separable over F. Letting $t_1 = 1$ and $t_2 = \lambda$, we see that the theorem is true for $n = 2$.

Now suppose that $n > 2$ and that $L = F(\alpha_1, \ldots, \alpha_n)$, where each α_i is separable over F. By our inductive hypothesis, we can find $t_1, \ldots, t_{n-1} \in F$ such that $F(\alpha_1, \ldots, \alpha_{n-1}) = F(\alpha_0)$, where $\alpha_0 = t_1\alpha_1 + \cdots + t_{n-1}\alpha_{n-1}$ is separable over F. Then

$$L = F(\alpha_1, \ldots, \alpha_n) = F(\alpha_1, \ldots, \alpha_{n-1})(\alpha_n) = F(\alpha_0)(\alpha_n) = F(\alpha_0, \alpha_n).$$

By the proof for $n = 2$, we have $F(\alpha_0, \alpha_n) = F(\alpha_0 + \lambda\alpha_n)$ for some $\lambda \in F$, where $\alpha_0 + \lambda\alpha_n$ is separable over F. If we set $t_n = \lambda$, then $\alpha_0 + \lambda\alpha_n = t_1\alpha_1 + \cdots + t_n\alpha_n$ is the desired separable primitive element. This completes the proof when F is infinite.

The proof of the theorem is very different in the case when F is a finite field. We will give the argument in Exercise 2. $\qquad\square$

Here are two situations when the hypotheses of Theorem 5.4.1 are satisfied.

Corollary 5.4.2. *Let $F \subset L$ be a finite extension.*
(a) *If $F \subset L$ is separable, then there is $\alpha \in L$ such that $L = F(\alpha)$.*
(b) *If F has characteristic 0, then there is $\alpha \in L$ such that $L = F(\alpha)$. Furthermore, if $L = F(\alpha_1, \ldots, \alpha_n)$, then α can be chosen to be of the form*

$$\alpha = t_1\alpha_1 + \cdots + t_n\alpha_n$$

where $t_1, \ldots, t_n \in F$.

Proof. We know that $L = F(\alpha_1, \ldots, \alpha_n)$ since $F \subset L$ is finite. In part (a), each α_i is separable since $F \subset L$ is separable, so we are done by Theorem 5.4.1. For part (b), let F have characteristic 0. Then F is infinite and each α_i is separable by Proposition 5.3.7. Again, we are done by Theorem 5.4.1. $\qquad\square$

Every field of characteristic 0 contains a copy of $\mathbb{Z}$. In Exercise 3 you will use this to show that in the equation $\alpha = t_1\alpha_1 + \cdots + t_n\alpha_n$ in part (b) of Corollary 5.4.2, we can assume that $t_1, \ldots, t_n \in \mathbb{Z}$. This observation is due to Galois.

In some simple cases, one can explicitly find primitive elements.

Example 5.4.3. Consider $\mathbb{Q} \subset \mathbb{Q}(\sqrt{2}, \sqrt{3})$. In the notation of the proof of Theorem 5.4.1, we have $\beta_1 = \sqrt{2}$, $\beta_2 = -\sqrt{2}$ for $f = x^2 - 2$, and $\gamma_1 = \sqrt{3}$, $\gamma_2 = -\sqrt{3}$ for $g = x^2 - 3$. Then any $\lambda \neq 0$ in $\mathbb{Q}$ satisfies (5.16). Thus $\sqrt{2} + \lambda\sqrt{3}$ is a primitive element of $\mathbb{Q} \subset \mathbb{Q}(\sqrt{2}, \sqrt{3})$ for all $\lambda \in \mathbb{Q} \setminus \{0\}$. $\qquad\triangleleft\triangleright$

Not all finite extensions have primitive elements. By Corollary 5.4.2, such an extension cannot have characteristic 0. Here is an example in characteristic p.

Example 5.4.4. Let k be a field of characteristic p and let t, u be variables. Consider the extension field

$$(5.20) \qquad\qquad F = k(t, u) \subset L,$$

where L is the splitting field of $(x^p - t)(x^p - u) \in F[x]$. Thus there are $\alpha, \beta \in L$ with $\alpha^p = t$ and $\beta^p = u$. By Exercise 4, we have $L = F(\alpha, \beta)$ and $[L : F] = p^2$.

Let us show that (5.20) has no primitive element. Given $\gamma \in L$, we can use $L = F(\alpha, \beta) = F[\alpha, \beta]$ to write

$$\gamma = \sum_{i,j} a_{ij} \alpha^i \beta^j, \quad a_{ij} \in F,$$

where the sum is finite. Lemma 5.3.10 implies that

$$\gamma^p = \left(\sum_{i,j} a_{ij} \alpha^i \beta^j\right)^p = \sum_{i,j} a_{ij}^p \alpha^{ip} \beta^{jp},$$

and then $\alpha^p = t$ and $\beta^p = u$ give

$$\gamma^p = \sum_{i,j} a_{ij}^p t^i u^j \in F.$$

Hence γ is a root of $x^p - \gamma^p \in F[x]$, so that $[F(\gamma) : F] \leq p$. Since $[L : F] = p^2$, we have $L \neq F(\gamma)$ for all $\gamma \in L$. Thus $F \subset L$ has no primitive element.   ◁▷

In Exercise 4 you will show that the extension (5.20) is purely inseparable.

Mathematical Notes

Theorem 5.4.1 leads to the following question about primitive elements.

▪ **Existence of Primitive Elements.** Corollary 5.4.2 tells us that all finite separable extensions have primitive elements. But this is not the full story, since the extension $F \subset L = F(\alpha)$ discussed in Example 5.3.11 is not separable but has a primitive element. The following theorem of Steinitz characterizes *all* finite extensions that have primitive elements.

Theorem 5.4.5. *A finite extension $F \subset L$ has a primitive element if and only if there are only finitely many intermediate fields $F \subset K \subset L$.*   □

A proof can be found in Section 4.14 of [Jacobson, Vol. I]. As an example of this result, consider the extension $F \subset L$ from (5.20). Since this has no primitive element, there must be infinitely many fields K such that $F \subset K \subset L$. You will construct an infinite collection of such fields in Exercise 5.

Historical Notes

We will see in Chapter 12 that Lagrange and Galois knew special cases of the Theorem of the Primitive Element.

Exercises for Section 5.4

Exercise 1. Use the hints given in the text to prove that (5.18) has coefficients in F.

Exercise 2. Let F be a finite field, and let $F \subset L$ be a finite extension. We claim that there is $\alpha \in L$ such that $L = F(\alpha)$ and α is separable over F.
(a) Show that L is a finite field.
(b) The set $L^* = L \setminus \{0\}$ is a finite group under multiplication and hence is cyclic by Proposition A.5.3. Let $\alpha \in L^*$ be a generator. Prove that $L = F(\alpha)$.
(c) Let $m = |L| - 1$. Show that α^i is a root of $x^m - 1 \in F[x]$ for all $0 \le i \le m - 1$, and conclude that

$$x^m - 1 = (x - 1)(x - \alpha)(x - \alpha^2) \cdots (x - \alpha^{m-1}).$$

(d) Use part (c) to show that α is separable over F.

Exercise 3. In the equation $\alpha = t_1 \alpha_1 + \cdots + t_n \alpha_n$ in part (b) of Corollary 5.4.2, show that we can assume that $t_1, \ldots, t_n \in \mathbb{Z}$.

Exercise 4. In the extension $F \subset L$ of Example 5.4.4, we have $F = k(t, u)$, where k has characteristic p and L is the splitting field of $(x^p - t)(x^p - u) \in F[x]$. We also have $\alpha, \beta \in L$ satisfying $\alpha^p = t$, $\beta^p = u$. Prove the following properties of $F \subset L$:
(a) $L = F(\alpha, \beta)$ and $[L : F] = p^2$.
(b) $[F(\gamma) : F] = p$ for all $\gamma \in L \setminus F$.
(c) $F \subset L$ is purely inseparable.

Exercise 5. Let $F \subset L = F(\alpha, \beta)$ be as in Exercise 4, and consider the intermediate fields $F \subset F(\alpha + \lambda\beta) \subset L$ as λ varies over all elements of F. Suppose that $\lambda \ne \mu$ are two elements of F such that $F(\alpha + \lambda\beta) = F(\alpha + \mu\beta)$.
(a) Show that $\alpha, \beta \in F(\alpha + \lambda\beta)$.
(b) Conclude that $F(\alpha + \lambda\beta) = F(\alpha, \beta)$, and explain why this contradicts Example 5.4.4. It follows that the fields $F(\alpha + \lambda\beta)$, $\lambda \in F$, are all distinct. Since F is infinite, we see that there are infinitely many fields between F and L.

Exercise 6. Explain why the proof of Theorem 5.4.1 implies that $F(\beta + \lambda\gamma) = F(\beta, \gamma)$ when γ is separable over F, β is algebraic over F, and λ satisfies (5.17).

Exercise 7. Let $F \subset L = F(\alpha_1, \ldots, \alpha_n)$ be a finite extension, and suppose that $\alpha_1, \ldots, \alpha_{n-1}$ are separable over F. Prove that $F \subset L$ has a primitive element.

Exercise 8. Use Exercise 7 to find an explicit primitive element for $F = k(t, u) \subset L$, where k has characteristic 3 and L is the splitting field of $(x^2 - t)(x^3 - u)$. Note that this extension is not separable, by Exercise 12 of Section 5.3.

REFERENCES

1. D. Cox, J. Little, and D. O'Shea, *Ideals, Varieties and Algorithms*, Second Edition, Springer-Verlag, New York, Berlin, Heidelberg, 1997.

2. M. Kreuzer and L. Robbiano, *Computational Commutative Algebra 1*, Springer-Verlag, New York, Berlin, Heidelberg, 2000.

3. L. Weisner, *Introduction to the Theory of Equations*, Macmillan, New York, 1938.

6

The Galois Group

In this chapter we will define the Galois group of a finite extension $F \subset L$. We will then study the Galois group of the splitting field of a separable polynomial and give some examples of Galois groups.

6.1 DEFINITION OF THE GALOIS GROUP

If L is a field, then an *automorphism* of L is a field isomorphism $\sigma : L \to L$. We now define one of the central objects in Galois theory.

Definition 6.1.1. *Let* $F \subset L$ *be a finite extension. Then* $\operatorname{Gal}(L/F)$ *is the set*

$$\{\sigma : L \to L \mid \sigma \text{ is an automorphism}, \sigma(a) = a \text{ for all } a \in F\}.$$

In other words, $\operatorname{Gal}(L/F)$ consists of all automorphisms of L that are the identity on F. The basic structure of $\operatorname{Gal}(L/F)$ is as follows.

Proposition 6.1.2. $\operatorname{Gal}(L/F)$ *is a group under composition.*

Proof. First suppose that $\sigma, \tau \in \operatorname{Gal}(L/F)$. Then $\sigma\tau$ is the composition $\sigma \circ \tau$, which is an automorphism because σ, τ are. Also, if $a \in F$, then $\sigma \circ \tau(a) = \sigma(\tau(a)) = \sigma(a) = a$, since σ, τ are the identity on F. Hence composition gives an operation on $\operatorname{Gal}(L/F)$, which is associative by standard properties of composition.

The identity map $1_L : L \to L$ is an isomorphism that is the identity on F, so that $1_L \in \operatorname{Gal}(L/F)$. One easily checks that $\sigma \circ 1_L = 1_L \circ \sigma = \sigma$ for all $\sigma \in \operatorname{Gal}(L/F)$. Thus 1_L is the identity element of $\operatorname{Gal}(L/F)$.

Finally, any $\sigma \in \text{Gal}(L/F)$ is an automorphism, which means that its inverse $\sigma^{-1} : L \to L$ is also an automorphism. Also, if $a \in F$, then $a = \sigma(a)$, which implies $\sigma^{-1}(a) = \sigma^{-1}(\sigma(a)) = a$. This shows that $\sigma^{-1} \in \text{Gal}(L/F)$ and completes the proof that $\text{Gal}(L/F)$ is a group under composition. $\qquad \square$

Because of this proposition, we call $\text{Gal}(L/F)$ the *Galois group* of $F \subset L$. In order to compute Galois groups, we need to know how elements of $\text{Gal}(L/F)$ behave. We begin with the following simple observation.

Lemma 6.1.3. *Let $F \subset L$ be finite, and fix $\sigma \in \text{Gal}(L/F)$. Given $h \in F[x_1, \ldots, x_n]$ and $\beta_1, \ldots, \beta_n \in L$, then*

$$\sigma\big(h(\beta_1, \ldots, \beta_n)\big) = h\big(\sigma(\beta_1), \ldots, \sigma(\beta_n)\big).$$

In particular, if $h \in F[x]$ and $\beta \in L$, then

$$\sigma\big(h(\beta)\big) = h\big(\sigma(\beta)\big).$$

Proof. This follows immediately because σ preserves addition and multiplication and is the identity on the coefficients of h. $\qquad \square$

This lemma has some nice consequences concerning the Galois group.

Proposition 6.1.4. *Let $F \subset L$ be a finite extension and let $\sigma \in \text{Gal}(L/F)$. Then:*
(a) *If $h \in F[x]$ is a nonconstant polynomial with $\alpha \in L$ as a root, then $\sigma(\alpha)$ is another root of h lying in L.*
(b) *If $L = F(\alpha_1, \ldots, \alpha_n)$, then σ is uniquely determined by its values on $\alpha_1, \ldots, \alpha_n$.*

Proof. By Lemma 6.1.3, $h \in F[x]$ and $0 = h(\alpha)$ imply that

$$0 = \sigma(0) = \sigma\big(h(\alpha)\big) = h\big(\sigma(\alpha)\big),$$

which shows that $\sigma(\alpha) \in L$ is also a root of h. Part (a) follows.

Turning to part (b), note that $L = F[\alpha_1, \ldots, \alpha_n]$, since $L = F(\alpha_1, \ldots, \alpha_n)$ is a finite extension of F. Hence any $\beta \in L$ can be written

$$\beta = h(\alpha_1, \ldots, \alpha_n)$$

for some polynomial $h \in F[x_1, \ldots, x_n]$. By Lemma 6.1.3,

$$\sigma(\beta) = \sigma\big(h(\alpha_1, \ldots, \alpha_n)\big) = h\big(\sigma(\alpha_1), \ldots, \sigma(\alpha_n)\big).$$

It follows that $\sigma : L \to L$ is uniquely determined by $\sigma(\alpha_1), \ldots, \sigma(\alpha_n)$. $\qquad \square$

This proposition leads to our first result on the structure of $\text{Gal}(L/F)$.

Corollary 6.1.5. *Let $F \subset L$ be a finite extension. Then its Galois group $\text{Gal}(L/F)$ is finite.*

Proof. Since $F \subset L$ is finite, $L = F(\alpha_1, \ldots, \alpha_n)$, where each α_i is algebraic over F. Now suppose that $\sigma \in \mathrm{Gal}(L/F)$. By part (b) of Proposition 6.1.4, σ is uniquely determined by $\sigma(\alpha_1), \ldots, \sigma(\alpha_n)$. Furthermore, if $p_i \in F[x]$ is the minimal polynomial of α_i, then part (a) shows that there are at most $\deg(p_i)$ possibilities for $\sigma(\alpha_i)$. The finiteness of $\mathrm{Gal}(L/F)$ follows immediately. $\square$

In Exercise 1 you will show that in the situation of Corollary 6.1.5, one has $|\mathrm{Gal}(L/F)| \leq \deg(p_1) \cdots \deg(p_n)$.

Let us now use Proposition 6.1.4 to compute some Galois groups. We begin by observing that Galois groups are sometimes unexpectedly small.

Example 6.1.6. Consider the extension

$$\mathbb{Q} \subset L = \mathbb{Q}(\sqrt[3]{2})$$

studied in Example 5.2.2. The minimal polynomial of $\sqrt[3]{2}$ over $\mathbb{Q}$ is $x^3 - 2$, which has roots $\sqrt[3]{2}, \omega\sqrt[3]{2}, \omega^2\sqrt[3]{2}$, where $\omega = e^{2\pi i/3}$. The last two are not real and hence can't lie in L. Hence every $\sigma \in \mathrm{Gal}(L/\mathbb{Q})$ must satisfy $\sigma(\sqrt[3]{2}) = \sqrt[3]{2}$. Since σ is uniquely determined by $\sigma(\sqrt[3]{2})$, it must be the identity. Thus $\mathrm{Gal}(L/\mathbb{Q}) = \{1_L\}$. Do you see how this argument uses both parts of Proposition 6.1.4? ◁▷

Example 6.1.7. Let $F = k(t)$, where k is a field of characteristic p, and let $F \subset L$ be the splitting field of $f = x^p - t \in F[x]$. If $\alpha \in L$ is a root of f, then $L = F(\alpha)$ and $f = (x - \alpha)^p$ by Example 5.3.11. Thus α is the only root of f. Arguing as in the previous example, we see that $\mathrm{Gal}(L/F) = \{1_L\}$. ◁▷

Here are some examples where the Galois group is nontrivial.

Example 6.1.8. Let $\tau : \mathbb{C} \to \mathbb{C}$ be complex conjugation, that is, $\tau(z) = \overline{z}$ for $z \in \mathbb{C}$. By (A.5), we know that τ is a homomorphism of fields, and it is an automorphism because $\tau \circ \tau$ is the identity. Furthermore, we have $\tau(a) = a$ for all $a \in \mathbb{R}$, so that $\tau \in \mathrm{Gal}(\mathbb{C}/\mathbb{R})$. Thus $\mathrm{Gal}(\mathbb{C}/\mathbb{R})$ has *at least* two elements, since $1_{\mathbb{C}} \in \mathrm{Gal}(\mathbb{C}/\mathbb{R})$.

However, we also know that $\mathbb{C} = \mathbb{R}(i)$. Since the roots of $x^2 + 1$ are $\pm i$, Proposition 6.1.4 implies that $\sigma \in \mathrm{Gal}(\mathbb{C}/\mathbb{R})$ is determined uniquely by $\sigma(i) = \pm i$. Hence $\mathrm{Gal}(\mathbb{C}/\mathbb{R})$ has *at most* two elements. Combining this with the previous paragraph, we conclude that

$$\mathrm{Gal}(\mathbb{C}/\mathbb{R}) = \{1_{\mathbb{C}}, \tau\}.$$

It follows that $\mathrm{Gal}(\mathbb{C}/\mathbb{R}) \simeq \mathbb{Z}/2\mathbb{Z}$. ◁▷

Example 6.1.9. Next consider the extension $\mathbb{Q} \subset L = \mathbb{Q}(\sqrt{2})$. Arguing as in the previous example shows that $\sigma \in \mathrm{Gal}(L/\mathbb{Q})$ is determined uniquely by $\sigma(\sqrt{2}) = \pm\sqrt{2}$. Thus $|\mathrm{Gal}(L/\mathbb{Q})| \leq 2$. There are two ways to see that equality occurs:

- By explicit computation, one can show that $\sigma(a + b\sqrt{2}) = a - b\sqrt{2}$ is an automorphism of L.
- $L = \mathbb{Q}(\sqrt{2})$ is the splitting field of $x^2 - 2$ over $\mathbb{Q}$. Since $x^2 - 2$ is irreducible over $\mathbb{Q}$ and $\pm\sqrt{2} \in L$, Proposition 5.1.8 implies that there is an automorphism of L that takes $\sqrt{2}$ to $-\sqrt{2}$ and is the identity on $\mathbb{Q}$. ◁▷

Our last example will appear often in this chapter.

Example 6.1.10. For the extension $\mathbb{Q} \subset L = \mathbb{Q}(\sqrt{2}, \sqrt{3})$, Proposition 6.1.4 implies that $\sigma \in \mathrm{Gal}(L/\mathbb{Q})$ is determined uniquely by

$$(6.1) \qquad\qquad \sigma(\sqrt{2}) = \pm\sqrt{2}, \quad \sigma(\sqrt{3}) = \pm\sqrt{3}.$$

This gives the inequality $|\mathrm{Gal}(L/\mathbb{Q})| \leq 4$. The natural question is whether all possible sign combinations in (6.1) actually occur, that is, whether $|\mathrm{Gal}(L/\mathbb{Q})| = 4$. In Exercise 2 you will prove this using Proposition 5.1.8 as in the previous example. We will learn a much quicker method in Section 6.2.   ◁▷

Finally, we study what happens when we go to an isomorphic field.

Proposition 6.1.11. *Suppose that $F \subset L_1$ and $F \subset L_2$ are finite extensions, and let $\varphi : L_1 \to L_2$ be an isomorphism that is the identity on F. Then the map sending σ to $\varphi \circ \sigma \circ \varphi^{-1}$ defines a group isomorphism*

$$\mathrm{Gal}(L_1/F) \simeq \mathrm{Gal}(L_2/F).$$

Proof. You will prove this in Exercise 3.   □

Proposition 6.1.11 shows that isomorphic fields give isomorphic Galois groups. We use this as follows.

Definition 6.1.12. *Let $f \in F[x]$. The **Galois group of f over** F is $\mathrm{Gal}(L/F)$, where L is a splitting field of f over F.*

To check that Definition 6.1.12 makes sense, suppose that L_1 and L_2 are splitting fields of $f \in F[x]$. Corollary 5.1.7 implies $L_1 \simeq L_2$ via an isomorphism that is the identity on F, and hence $\mathrm{Gal}(L_1/F) \simeq \mathrm{Gal}(L_2/F)$ by Proposition 6.1.11. Thus the Galois group of f over F is well defined up to isomorphism.

Using this terminology, Example 6.1.8 tells us that the Galois group of $x^2 + 1$ over $\mathbb{R}$ is $\mathbb{Z}/2\mathbb{Z}$.

Historical Notes

The definition of Galois group given here is *very* different from the one given by Galois. He only dealt with splitting fields, and for him, the Galois group consisted of certain permutations of the roots. We will give Galois's definition and explore its relation to Definition 6.1.1 in Chapter 12.

Isomorphisms of fields were first defined by Richard Dedekind in 1877 under the name "permutations." Here is his definition from [1, pp. 108–109]:

> Now let Ω be *any* field. By a *permutation* of Ω we mean a substitution which changes each number
>
> $$\alpha, \quad \beta, \quad \alpha + \beta, \quad \alpha - \beta, \quad \alpha\beta, \quad \alpha/\beta$$

of Ω into a corresponding number

$$\alpha', \quad \beta', \quad (\alpha + \beta)', \quad (\alpha - \beta)', \quad (\alpha\beta)', \quad (\alpha/\beta)'$$

in such a way that

$$(\alpha + \beta)' = \alpha' + \beta'$$
$$(\alpha\beta)' = \alpha'\beta'$$

are satisfied and the substitute numbers $\alpha', \beta', \ldots$ are not all zero. We shall see that the set Ω' of the latter numbers forms a new field, ...

In Exercise 4 you will show that this implies that the map $\Omega \to \Omega'$ given by $\alpha \mapsto \alpha'$ is an isomorphism of fields.

By 1894 Dedekind was also aware of the relevance of automorphisms to Galois theory. Dedekind's influence can be seen in the work of his student, Heinrich Weber, who gave a careful account of group theory and Galois theory in the first volume of his *Lehrbuch der Algebra*, which appeared in 1894. In this book Weber begins with Galois's definition of the Galois group and shows how this leads to automorphisms of the splitting field.

The final step in the evolution of the Galois group is due to Emil Artin, who during the 1920s made Definition 6.1.1 the starting point of Galois theory. The first exposition of this approach appeared in the 1930 edition of [van der Waerden]. Artin published his own account of Galois theory in 1938 and 1942. The latter was enormously influential and is still in print as [Artin]. See [2] for more details.

Exercises for Section 6.1

Exercise 1. Let $L = F(\alpha_1, \ldots, \alpha_n)$, and let $p_i \in F[x]$ be a nonzero polynomial vanishing at α_i. Explain why the proof of Corollary 6.1.5 implies that $|\mathrm{Gal}(L/F)| \leq \deg(p_1) \cdots \deg(p_n)$.

Exercise 2. Consider the extension $\mathbb{Q} \subset L = \mathbb{Q}(\sqrt{2}, \sqrt{3})$. In Exercise 13 of Section 5.1, you used Proposition 5.1.8 to construct an automorphism of L that takes $\sqrt{3}$ to $-\sqrt{3}$ and is the identity on $\mathbb{Q}(\sqrt{2})$. By interchanging the roles of 2 and 3 in this construction, explain why all possible signs in (6.1) can occur. This shows that $|\mathrm{Gal}(L/\mathbb{Q})| = 4$.

Exercise 3. This exercise will prove a generalized form of Proposition 6.1.11.
  (a) Let $\varphi : L_1 \simeq L_2$ be an isomorphism of fields. Given a subfield $F_1 \subset L_1$, set $F_2 = \varphi(F_1)$, which is a subfield of L_2. Prove that the map sending $\sigma \in \mathrm{Gal}(L_1/F_1)$ to $\varphi \circ \sigma \circ \varphi^{-1}$ induces an isomorphism $\mathrm{Gal}(L_1/F_1) \simeq \mathrm{Gal}(L_2/F_2)$.
  (b) Explain why Proposition 6.1.11 follows from part (a).

Exercise 4. In the Historical Notes, we saw that Dedekind defined a "permutation" $\alpha \mapsto \alpha'$ to be a map $\Omega \to \Omega'$ satisfying $(\alpha + \beta)' = \alpha' + \beta'$ and $(\alpha\beta)' = \alpha'\beta'$ for all $\alpha, \beta \in \Omega$. Dedekind also assumes that $\Omega' = \{\alpha' \mid \alpha \in \Omega\}$ and that the α' are not all zero.
  (a) Show that $1 \in \Omega$ maps to $1 \in \Omega'$. Once this is proved, it follows that $\alpha \mapsto \alpha'$ is a ring homomorphism. (Recall that sending 1 to 1 is part of the definition of ring homomorphism given in Appendix A.)
  (b) Show that the map $\alpha \mapsto \alpha'$ is one to one.
This shows that Dedekind's definition of field isomorphism is equivalent to ours.

Exercise 5. Prove the following inequalities:
  (a) $|\mathrm{Gal}(\mathbb{Q}(\sqrt{2}, \sqrt{3}, \sqrt{5})/\mathbb{Q})| \leq 8$.
  (b) $|\mathrm{Gal}(\mathbb{Q}(\sqrt{p_1}, \ldots, \sqrt{p_n})/\mathbb{Q})| \leq 2^n$, where $p_1, \ldots, p_n$ are the first n primes.
In each case, one can show that these are actually equalities.

Exercise 6. If we apply Exercise 1 to the extension $\mathbb{Q} \subset L = \mathbb{Q}(\sqrt{6}, \sqrt{10}, \sqrt{15})$, we get the inequality $|\mathrm{Gal}(L/\mathbb{Q})| \leq 8$. Show that $|\mathrm{Gal}(L/\mathbb{Q})| \leq 4$.

Exercise 7. Let $F \subset L$ be a finite extension, and let $\sigma : L \to L$ be a ring homomorphism that is the identity on F. This exercise will show that σ is an automorphism.
  (a) Show that σ is one to one.
  (b) Show that σ is onto.

6.2 GALOIS GROUPS OF SPLITTING FIELDS

In this section we will study the Galois group of the splitting field of a separable polynomial. Recall from Section 5.3 that $f \in F[x]$ is separable if it has distinct roots in a splitting field. This is the situation considered by Galois.

We now prove the first main theorem of Galois theory.

Theorem 6.2.1. *If L is the splitting field of a separable polynomial in $F[x]$, then the Galois group of $F \subset L$ has order $|\mathrm{Gal}(L/F)| = [L : F]$.*

Proof. Our hypothesis implies that $L = F(\alpha_1, \ldots, \alpha_n)$, where $\alpha_1, \ldots, \alpha_n$ are the roots of a separable polynomial $f \in F[x]$. Then each α_i is separable over F (be sure you can explain why). By the Theorem of the Primitive Element (Theorem 5.4.1), we can find $\beta \in L$ separable over F such that $L = F(\beta)$. Let $h \in F[x]$ be the minimal polynomial of β. Note that h is separable, since β is.

Since $L = F(\beta)$, Proposition 4.3.4 implies that $[L : F] = m$, where $m = \deg(h)$. To prove the theorem, we need to show that $\mathrm{Gal}(L/F)$ has m elements. We will use the following ideas from Chapter 5:

- **Normality** (Section 5.2): If an irreducible polynomial has one root in a splitting field, then all of its roots lie in the splitting field.

- **Separability** (Section 5.3): Separability means that a polynomial has distinct roots in its splitting field.

- **Isomorphisms** (Proposition 5.1.8): If two elements in a splitting field L are roots of the same irreducible polynomial over F, then there is an automorphism of L that is the identity on F and takes one root to the other.

As we will now explain, the theorem follows easily from these ideas.

The above polynomial $h \in F[x]$ is separable and has a root $\beta \in L$. Since L is a splitting field over F, the bullets for normality and separability imply that h has distinct roots $\beta = \beta_1, \beta_2, \ldots, \beta_m$, $m = \deg(h)$, all of which lie in L. Now fix one of the roots, say β_i. Then β and β_i are roots of the irreducible polynomial h. Since L is a splitting field over F, the bullet for isomorphisms implies that there is an automorphism σ_i of L such that $\sigma_i(\beta) = \beta_i$ and σ_i is the identity on F.

It follows that $\sigma_1, \ldots, \sigma_m \in \mathrm{Gal}(L/F)$. Note that $\sigma_i \neq \sigma_j$ for $i \neq j$, since $\sigma_i(\beta) = \beta_i \neq \beta_j = \sigma_j(\beta)$. Thus $\mathrm{Gal}(L/F)$ has at least m distinct elements. But given *any* $\sigma \in \mathrm{Gal}(L/F)$, Proposition 6.1.4 and $L = F(\beta)$ imply that σ is uniquely determined by $\sigma(\beta) \in \{\beta_1, \ldots, \beta_m\}$. It follows that $\sigma = \sigma_i$ for some i. This completes the proof of the theorem. $\qquad\square$

The following example illustrates the power of the theorem just proved.

Example 6.2.2. Consider $\mathbb{Q} \subset L = \mathbb{Q}(\sqrt{2}, \sqrt{3})$. In Example 6.1.10, we saw that $\sigma \in \mathrm{Gal}(L/\mathbb{Q})$ is uniquely determined by

$$\sigma(\sqrt{2}) = \pm\sqrt{2}, \quad \sigma(\sqrt{3}) = \pm\sqrt{3},$$

which implies that $|\mathrm{Gal}(L/\mathbb{Q})| \leq 4$. We also asked whether equality occurs.

This is now easy to decide, for $[L:\mathbb{Q}] = 4$ by Example 4.3.9 and L is the splitting field of the separable polynomial $(x^2 - 2)(x^2 - 3)$. Hence all of the above sign combinations must occur. In particular, we can find $\sigma, \tau \in \mathrm{Gal}(L/\mathbb{Q})$ such that

(6.2)
$$\begin{aligned} \sigma(\sqrt{2}) &= \sqrt{2}, & \sigma(\sqrt{3}) &= -\sqrt{3}, \\ \tau(\sqrt{2}) &= -\sqrt{2}, & \tau(\sqrt{3}) &= \sqrt{3}. \end{aligned}$$

In Exercise 1 you will show that $\mathrm{Gal}(L/\mathbb{Q}) = \{1_L, \sigma, \tau, \sigma\tau\} \simeq \mathbb{Z}/2\mathbb{Z} \times \mathbb{Z}/2\mathbb{Z}$ (this is usually called the *Klein four-group*). $\qquad\triangleleft\triangleright$

It is important to understand why the hypotheses *splitting field* and *separable* are necessary in the proof of Theorem 6.2.1. We can see this in the first two examples considered in Section 6.1:

- Consider $\mathbb{Q} \subset \mathbb{Q}(\sqrt[3]{2})$. The Galois group is trivial by Example 6.1.6. This extension is not a splitting field, by Example 5.2.2.
- Consider $F = k(t) \subset L$, where k has characteristic p and L is the splitting field of $f = x^p - t$. The Galois group is trivial by Example 6.1.7. This polynomial is not separable by Example 5.3.11.

In both of these examples, note that $|\mathrm{Gal}(L/F)| < [L:F]$. This is no accident, for in Section 7.1 we will prove that $|\mathrm{Gal}(L/F)| \leq [L:F]$, with equality if and only if L is the splitting field of a separable polynomial in $F[x]$. Such extensions will be called *Galois extensions* in Chapter 7.

Exercises for Section 6.2

Exercise 1. Complete Example 6.2.2 by showing that $\mathrm{Gal}(L/\mathbb{Q}) = \{1_L, \sigma, \tau, \sigma\tau\}$ and that $\mathrm{Gal}(L/\mathbb{Q}) \simeq \mathbb{Z}/2\mathbb{Z} \times \mathbb{Z}/2\mathbb{Z}$.

Exercise 2. Consider $\mathbb{Q} \subset L = \mathbb{Q}(\omega, \sqrt[3]{2})$, where $\omega = e^{2\pi i/3}$.
  (a) Explain why $\sigma \in \mathrm{Gal}(L/\mathbb{Q})$ is uniquely determined by $\sigma(\omega) \in \{\omega, \omega^2\}$ and $\sigma(\sqrt[3]{2}) \in \{\sqrt[3]{2}, \omega\sqrt[3]{2}, \omega^2\sqrt[3]{2}\}$.
  (b) Explain why all possible combinations for $\sigma(\omega)$ and $\sigma(\sqrt[3]{2})$ actually occur.
In the next section we will show that $\mathrm{Gal}(L/\mathbb{Q}) \simeq S_3$.

Exercise 3. Consider $\mathbb{Q} \subset L = \mathbb{Q}(\zeta_5, \sqrt[5]{2})$, where $\zeta_5 = e^{2\pi i/5}$. By Proposition 4.2.5, the minimal polynomial of ζ_5 over $\mathbb{Q}$ is $x^4 + x^3 + x^2 + x + 1$.
(a) Show that $[L:\mathbb{Q}] = 20$.
(b) Show that L is the splitting field of $x^5 - 2$ over $\mathbb{Q}$, and conclude that $\mathrm{Gal}(L/\mathbb{Q})$ is a group of order 20.
We will describe the structure of this Galois group in Section 6.4.

Exercise 4. Consider the nth root of unity $\zeta_n = e^{2\pi i/n}$. We call $\mathbb{Q} \subset \mathbb{Q}(\zeta_n)$ a *cyclotomic extension* of $\mathbb{Q}$.
(a) Show that $\mathbb{Q} \subset \mathbb{Q}(\zeta_n)$ is a splitting field of a separable polynomial.
(b) Given $\sigma \in \mathrm{Gal}(\mathbb{Q}(\zeta_n)/\mathbb{Q})$, show that $\sigma(\zeta_n) = \zeta_n^i$ for some integer i.
(c) Show that the integer i in part (b) is relatively prime to n.
(d) The set of congruence classes modulo n relatively prime to n form a group under multiplication, denoted $(\mathbb{Z}/n\mathbb{Z})^*$. Show that the map $\sigma \mapsto [i]$, where $\sigma(\zeta_n) = \zeta_n^i$, defines a one-to-one group homomorphism $\mathrm{Gal}(\mathbb{Q}(\zeta_n)/\mathbb{Q}) \to (\mathbb{Z}/n\mathbb{Z})^*$.
(e) The order of $(\mathbb{Z}/n\mathbb{Z})^*$ is $|(\mathbb{Z}/n\mathbb{Z})^*| = \phi(n)$, where $\phi(n)$ is the Euler ϕ-function from number theory. Prove that the homomorphism of part (d) is an isomorphism if and only if $[\mathbb{Q}(\zeta_n):\mathbb{Q}] = \phi(n)$.
(f) Let p be prime. Use part (e) and Proposition 4.2.5 to show that $\mathrm{Gal}(\mathbb{Q}(\zeta_p)/\mathbb{Q}) \simeq (\mathbb{Z}/p\mathbb{Z})^*$.
In Chapter 9 we will prove that $[\mathbb{Q}(\zeta_n):\mathbb{Q}] = \phi(n)$. By part (e), this will imply that there is an isomorphism $\mathrm{Gal}(\mathbb{Q}(\zeta_n)/\mathbb{Q}) \simeq (\mathbb{Z}/n\mathbb{Z})^*$ for all n.

Exercise 5. Let F have characteristic p, and assume that $f = x^p - x + a \in F[x]$ is irreducible over F. Then let $L = F(\alpha)$, where α is a root of f in some splitting field. In Exercise 15 of Section 5.3, you showed that $F \subset L$ is a normal separable extension.
(a) Show that $|\mathrm{Gal}(L/F)| = p$, and use this to prove that $\mathrm{Gal}(L/F) \simeq \mathbb{Z}/p\mathbb{Z}$.
(b) Exercise 15 of Section 5.3 showed that $\alpha + 1$ is a root of f. For $i = 0, \ldots, p - 1$, show that there is a unique element of $\mathrm{Gal}(L/F)$ that takes α to $\alpha + i$.
(c) Use part (b) to describe an explicit isomorphism $\mathrm{Gal}(L/F) \simeq \mathbb{Z}/p\mathbb{Z}$.

Exercise 6. Let $f \in F[x]$ be irreducible and separable of degree n, and let $F \subset L$ be a splitting field of f. Prove that n divides $|\mathrm{Gal}(L/F)|$.

6.3 PERMUTATIONS OF THE ROOTS

In Chapter 1 we saw that permutations of the roots of a cubic arise naturally from Cardan's formulas. In fact, the title of Section 1.2 was "Permutations of the Roots." We now explain more generally how Galois groups relate to permutations. As in the previous section, we assume that L is the splitting field of a separable polynomial $f \in F[x]$. Our goal is to interpret $\mathrm{Gal}(L/F)$ in terms of permutations of roots of f.

Let $n = \deg(f)$. Then in $L[x]$ we can write f as the product

$$f = c(x - \alpha_1) \cdots (x - \alpha_n),$$

where $c \neq 0$ and $\alpha_1, \ldots, \alpha_n \in L$ are distinct. In this situation we get a map

(6.3) $$\mathrm{Gal}(L/F) \longrightarrow S_n$$

as follows. Given $\sigma \in \text{Gal}(L/F)$, Proposition 6.1.4 implies that $\sigma(\alpha_i)$ is a root of f (since α_i is), so that $\sigma(\alpha_i) = \alpha_{\tau(i)}$ for some $\tau(i) \in \{1, \ldots, n\}$. Note that $\tau(i)$ is uniquely determined, since $\alpha_1, \ldots, \alpha_n$ are distinct. Also,

$$\tau : \{1, \ldots, n\} \longrightarrow \{1, \ldots, n\}$$

is one to one since σ is (be sure you see why). It follows that τ is a permutation, that is, $\tau \in S_n$. This defines the map (6.3).

Proposition 6.3.1. *The map* $\text{Gal}(L/F) \to S_n$ *described in* (6.3) *is a one-to-one group homomorphism.*

Proof. Suppose that $\sigma_1, \sigma_2 \in \text{Gal}(L/F)$ correspond to $\tau_1, \tau_2 \in S_n$ via (6.3). This means that $\sigma_1(\alpha_i) = \alpha_{\tau_1(i)}$, and similarly for σ_2 and τ_2. Then

$$\sigma_1 \circ \sigma_2(\alpha_i) = \sigma_1(\sigma_2(\alpha_i)) = \sigma_1(\alpha_{\tau_2(i)}) = \alpha_{\tau_1(\tau_2(i))} = \alpha_{\tau_1 \tau_2(i)}.$$

This shows that $\sigma_1 \circ \sigma_2$ corresponds to $\tau_1 \tau_2$, so that (6.3) is a group homomorphism.

It remains to show that (6.3) is one to one. This follows immediately from Proposition 6.1.4, since $L = F(\alpha_1, \ldots, \alpha_n)$. The proof is now complete. $\square$

Proposition 6.3.1 shows that for the splitting field of a separable polynomial of degree n, we can regard the Galois group as a subgroup of S_n. By Lagrange's Theorem, it follows that $|\text{Gal}(L/F)|$ divides $n!$. Combining this with $[L:F] = |\text{Gal}(L/F)|$ from Theorem 6.2.1, we get the following corollary.

Corollary 6.3.2. *If L is the splitting field of a separable polynomial $f \in F[x]$, then $[L:F]$ divides $n!$, where $n = \deg(f)$.* $\square$

Theorem 5.1.5 states that $[L:F] \le n!$ when L is the splitting field of $f \in F[x]$ of degree n. Do you see how Corollary 6.3.2 refines this result when f is separable?

Example 6.3.3. We know that the splitting field of $f = (x^2 - 2)(x^2 - 3)$ over $\mathbb{Q}$ is $L = \mathbb{Q}(\sqrt{2}, \sqrt{3})$. Example 6.2.2 shows that $\text{Gal}(L/\mathbb{Q}) = \{1_L, \sigma, \tau, \sigma\tau\}$, where σ and τ satisfy

$$\sigma(\sqrt{2}) = \sqrt{2}, \ \sigma(\sqrt{3}) = -\sqrt{3} \quad \text{and} \quad \tau(\sqrt{2}) = -\sqrt{2}, \ \tau(\sqrt{3}) = \sqrt{3}.$$

Let $\alpha_1 = \sqrt{2}, \alpha_2 = -\sqrt{2}, \alpha_3 = \sqrt{3}$, and $\alpha_4 = -\sqrt{3}$. Then $\text{Gal}(L/\mathbb{Q})$ is isomorphic to a subgroup of S_4 by Proposition 6.3.1. The automorphism σ clearly fixes α_1, α_2 and interchanges α_3, α_4. It follows that $\sigma \mapsto (34) \in S_4$. One similarly shows that $\tau \mapsto (12)$, so that $\sigma\tau \mapsto (34)(12) = (12)(34)$. Hence $\text{Gal}(L/\mathbb{Q}) \simeq \{e, (12), (34), (12)(34)\} \subset S_4$. ◁▷

Example 6.3.4. Consider the extension $\mathbb{Q} \subset L = \mathbb{Q}(\omega, \sqrt[3]{2})$, $\omega = e^{2\pi i/3}$. Since L is the splitting field of $x^3 - 2$ over $\mathbb{Q}$ (Exercise 1 of Section 5.1), we get a one-to-one group homomorphism $\text{Gal}(L/\mathbb{Q}) \to S_3$. However, we learned in Example 4.3.10 that $[L:\mathbb{Q}] = 6$. Since $[L:\mathbb{Q}] = |\text{Gal}(L/\mathbb{Q})|$, it follows that $\text{Gal}(L/\mathbb{Q}) \simeq S_3$. You will work out the details of this isomorphism in Exercise 1. ◁▷

When one thinks of Galois groups in terms of permutations, it makes sense to ask how properties of the permutations relate to properties of the corresponding field extension. One nice example of this involves the following subgroups of S_n.

Definition 6.3.5. *A subgroup $H \subset S_n$ is **transitive** if for every pair of elements $i, j \in \{1, \ldots, n\}$, there is $\tau \in H$ such that $\tau(i) = j$.*

For example, S_n is a transitive subgroup of itself, since the transposition $(i\,j)$ takes i to j. But not all subgroups of S_n are transitive.

Example 6.3.6. The subgroup $\{e, (12), (34), (12)(34)\} \subset S_4$ from Example 6.3.3 is not transitive, since no element of the subgroup takes 1 to 3.   ◁▷

It is natural to ask if the subgroup of S_n corresponding to $\mathrm{Gal}(L/F)$ is transitive. This question was answered by Camille Jordan in 1870 as follows.

Proposition 6.3.7. *Let L be the splitting field of a separable polynomial $f \in F[x]$ of degree n. Then the subgroup of S_n corresponding to $\mathrm{Gal}(L/F)$ is transitive if and only if f is irreducible over F.*

Proof. First suppose that f is irreducible with distinct roots $\alpha_1, \ldots, \alpha_n \in L$. As in the proof of Theorem 6.2.1, we can use Proposition 5.1.8 to construct an automorphism $\sigma : L \simeq L$ that takes α_i to α_j and is the identity on F. Then $\sigma \in \mathrm{Gal}(L/F)$, and the corresponding permutation in S_n clearly takes i to j. Thus $\mathrm{Gal}(L/F)$ gives a transitive subgroup of S_n.

Conversely, suppose that $\mathrm{Gal}(L/F)$ corresponds to a transitive subgroup of S_n, and let h be an irreducible factor of f. We will show that $\deg(h) \geq n$, which easily implies that f is irreducible (do you see why?).

For this purpose, let the roots of f be $\alpha_1, \ldots, \alpha_n \in L$. Since h is a nonconstant factor of f, we can find i such that $h(\alpha_i) = 0$. Now pick *any* $j \in \{1, \ldots, n\}$. By our transitivity assumption, there is $\sigma \in \mathrm{Gal}(L/F)$ such that $\sigma(\alpha_i) = \alpha_j$. Since h has coefficients in F, part (a) of Proposition 6.1.4 implies that $\sigma(\alpha_i) = \alpha_j$ is also a root of h. Since j was arbitrary and $\alpha_1, \ldots, \alpha_n$ are distinct, it follows that h has at least n roots, which implies that $\deg(h) \geq n$. $\square$

Mathematical Notes

Here are two topics for further discussion.

■ **The Galois Group of a Polynomial.** In Section 6.1, the Galois group of $f \in F[x]$ was defined to be $\mathrm{Gal}(L/F)$, where $F \subset L$ is a splitting field of f over F. But when f is separable of degree n, the Galois group of f has extra structure given by its action on the roots of f. Hence one can argue that the correct definition of "the Galois group of f" is the homomorphism $\mathrm{Gal}(L/F) \to S_n$ studied in this section.

■ **Transitive Group Actions.** Definition 6.3.5 defines a transitive subgroup of S_n. This can be generalized to any group action: if a group G acts on a set X (as defined in Section A.4), then the action is *transitive* if for all $x, y \in X$, there is $g \in G$ such that $g \cdot x = y$. For example, if L is the splitting field of a separable polynomial $f \in F[x]$, then $\mathrm{Gal}(L/F)$ acts on the roots of f. Thus Proposition 6.3.7 can be restated as saying that f is irreducible if and only if $\mathrm{Gal}(L/F)$ acts transitively on the roots of f.

Historical Notes

Proposition 6.3.1 shows that for the splitting field of a separable polynomial of degree n, $\mathrm{Gal}(L/F)$ is isomorphic to a subgroup of S_n. The permutations in this subgroup correspond to those permutations that respect the algebraic structure of the roots, that is, those that come from automorphisms of the splitting field.

Galois defined his "group" to consist of certain arrangements of the roots of the given polynomial. In Chapter 12 we will show that this set of permutations in S_n is the image of $\mathrm{Gal}(L/F) \to S_n$. Hence his group agrees with ours up to isomorphism. What is interesting is that Galois had no notion of automorphism, although automorphisms are implicit in his development of the theory.

In [Galois, p. 79] Galois defines transitive subgroups of S_n and gives an example of a nontransitive subgroup that in modern terms is written $\langle (12), (345) \rangle \subset S_5$. However, he terminology is different: he writes "irreducible" instead of "transitive." This shows that he also knew Proposition 6.3.7, though Jordan was the first to state the result explicitly.

Exercises for Section 6.3

Exercise 1. Consider $\mathrm{Gal}(L/\mathbb{Q})$, where $L = \mathbb{Q}(\omega, \sqrt[3]{2})$, $\omega = e^{2\pi i/3}$. By Exercise 2 of Section 6.2, there are $\sigma, \tau \in \mathrm{Gal}(L/\mathbb{Q})$ such that

$$\sigma(\sqrt[3]{2}) = \omega \sqrt[3]{2}, \ \sigma(\omega) = \omega \quad \text{and} \quad \tau(\sqrt[3]{2}) = \sqrt[3]{2}, \ \tau(\omega) = \omega^2.$$

Find the permutations in S_3 corresponding to σ and τ.

Exercise 2. For each of the following Galois groups, find an explicit subgroup of S_4 that is isomorphic to the group. Also explain which known group the Galois group is isomorphic to. (By "known groups," we mean cyclic groups, dihedral groups, the quaternion group, symmetric groups, alternating groups, products of these groups, etc. You may need to look up some of these in your abstract algebra text.)
  (a) $\mathrm{Gal}(\mathbb{Q}(i, \sqrt{2})/\mathbb{Q})$.
  (b) $\mathrm{Gal}(\mathbb{Q}(i, \sqrt[4]{2})/\mathbb{Q})$.

Exercise 3. In the terminology of Exercise 2, what known group is $\mathrm{Gal}(\mathbb{Q}(i, \sqrt{2}, \sqrt{3})/\mathbb{Q})$ isomorphic to? Explain your reasoning in detail.

Exercise 4. Consider the extension $\mathbb{Q} \subset L = \mathbb{Q}(\alpha)$, where $\alpha = \sqrt{2 + \sqrt{2}}$. In Exercise 6 of Section 5.1, you showed that $f = x^4 - 4x^2 + 2$ is the minimal polynomial of α over $\mathbb{Q}$ and that L is the splitting field of f over $\mathbb{Q}$. Show that $\mathrm{Gal}(L/\mathbb{Q}) \simeq \mathbb{Z}/4\mathbb{Z}$.

Exercise 5. Let $f \in F[x]$ be separable, where $f = g_1 \cdots g_s$ for $g_i \in F[x]$ of degree $d_i > 0$, and let L be the splitting field of f over F. Show that $\mathrm{Gal}(L/F)$ is isomorphic to a subgroup of the product group $S_{d_1} \times \cdots \times S_{d_s}$.

Exercise 6. Let H be a transitive subgroup of S_n. Prove that $|H|$ is a multiple of n.

Exercise 7. Let $f \in F[x]$ be irreducible and separable of degree n and let $F \subset L$ be a splitting field of f. Use Exercise 6 and Proposition 6.3.7 to prove that n divides $|\mathrm{Gal}(L/F)|$. This gives an alternate proof of Exercise 6 of Section 6.2.

6.4   EXAMPLES OF GALOIS GROUPS

In this section we will give some interesting examples of Galois groups.

A.   The pth Roots of 2. Let $\zeta_p = e^{2\pi i/p}$ be a pth root of unity, where p is prime. By Section A.2 the roots of $x^p - 2$ are $\zeta_p^j \sqrt[p]{2}$ for $0 \le j \le p - 1$, so that

$$L = \mathbb{Q}\big(\sqrt[p]{2}, \zeta_p \sqrt[p]{2}, \zeta_p^2 \sqrt[p]{2}, \ldots, \zeta_p^{p-1} \sqrt[p]{2}\big) = \mathbb{Q}\big(\zeta_p, \sqrt[p]{2}\big)$$

is the splitting field of $x^p - 2$ over $\mathbb{Q}$. Our goal is to describe $\mathrm{Gal}(L/\mathbb{Q})$.

The minimal polynomial of ζ_p over $\mathbb{Q}$ is $x^{p-1} + \cdots + 1$ by Proposition 4.2.5 and the roots of this polynomial are ζ_p^i for $1 \le i \le p - 1$ by Section A.2. Furthermore, the minimal polynomial of $\sqrt[p]{2}$ over $\mathbb{Q}$ is $x^p - 2$ by the Schönemann–Eisenstein criterion, and its roots are listed above. Since p and $p - 1$ are relatively prime, the method used in Exercise 5 of Section 4.3 implies that $[L : \mathbb{Q}] = p(p - 1)$. (See also Exercise 8 of Section 5.1.)

It follows from Theorem 6.2.1 that $\mathrm{Gal}(L/\mathbb{Q})$ is a group of order $p(p - 1)$. To see what group this is, let $\sigma \in \mathrm{Gal}(L/\mathbb{Q})$. Then Proposition 6.1.4 implies that σ is uniquely determined by

$$\sigma(\zeta_p) \in \big\{\zeta_p, \ldots, \zeta_p^{p-1}\big\}, \quad \sigma(\sqrt[p]{2}) \in \big\{\sqrt[p]{2}, \zeta_p \sqrt[p]{2}, \zeta_p^2 \sqrt[p]{2}, \ldots, \zeta_p^{p-1} \sqrt[p]{2}\big\}.$$

In other words, there are integers $1 \le i \le p - 1$ and $0 \le j \le p - 1$ such that

(6.4) $$\sigma(\zeta_p) = \zeta_p^i, \quad \sigma(\sqrt[p]{2}) = \zeta_p^j \sqrt[p]{2}.$$

We will denote this σ by $\sigma_{i,j}$. The number of possible pairs (i, j) is $(p - 1) \cdot p = p(p - 1)$. Since this is also the order of $\mathrm{Gal}(L/\mathbb{Q})$, it follows that all possible pairs

(6.5) $$(i, j) \in \{1, \ldots, p - 1\} \times \{0, \ldots, p - 1\}$$

must occur in (6.4).

To determine the group structure, we need to compute the composition of $\sigma_{i,j}$ and $\sigma_{r,s}$. This is done as follows:

$$\begin{aligned}
\sigma_{i,j} \circ \sigma_{r,s}(\zeta_p) &= \sigma_{i,j}(\zeta_p^r) = \big(\sigma_{i,j}(\zeta_p)\big)^r = (\zeta_p^i)^r \\
&= \zeta_p^{ir},
\end{aligned}$$

$$\begin{aligned}
\sigma_{i,j} \circ \sigma_{r,s}(\sqrt[p]{2}) &= \sigma_{i,j}(\zeta_p^s \sqrt[p]{2}) = \big(\sigma_{i,j}(\zeta_p)\big)^s \sigma_{i,j}(\sqrt[p]{2}) = (\zeta_p^i)^s (\zeta_p^j \sqrt[p]{2}) \\
&= \zeta_p^{is+j} \sqrt[p]{2}.
\end{aligned}$$

This computation suggests that

$$\sigma_{i,j} \circ \sigma_{r,s} = \sigma_{ir,is+j}.$$

Unfortunately, the pair $(ir, is + j)$ need not lie in (6.5). We can resolve this difficulty by realizing that for $i \in \mathbb{Z}$, ζ_p^i depends only on the congruence class of i modulo p. In other words, for $a = [i] \in \mathbb{F}_p = \mathbb{Z}/p\mathbb{Z}$, the number $\zeta_p^a = \zeta_p^i$ is well defined.

If we set $\mathbb{F}_p^* = \mathbb{F}_p \setminus \{0\}$, then for

$$(a, b) \in \mathbb{F}_p^* \times \mathbb{F}_p,$$

we can define $\sigma_{a,b}$ to be the element of $\mathrm{Gal}(L/\mathbb{Q})$ such that

$$\sigma_{a,b}(\zeta_p) = \zeta_p^a, \quad \sigma_{a,b}(\sqrt[p]{2}) = \zeta_p^b \sqrt[p]{2}.$$

Then the above computation shows that $\sigma_{a,b} \circ \sigma_{c,d} = \sigma_{ac,ad+b}$.

This composition formula leads to a geometric description of the Galois group $\mathrm{Gal}(L/\mathbb{Q})$. Given $a, b \in \mathbb{F}_p$, the function $\gamma_{a,b} : \mathbb{F}_p \to \mathbb{F}_p$ defined by $\gamma_{a,b}(u) = au + b$ is an *affine linear transformation*. By Exercise 1, $\gamma_{a,b}$ is one to one and onto if and only if $a \neq 0$, and all such $\gamma_{a,b}$ form a group of order $p(p-1)$ under composition. This group is called $\mathrm{AGL}(1, \mathbb{F}_p)$, the *one-dimensional affine linear group modulo p*. To understand its structure, we take $u \in \mathbb{F}_p$ and compute

$$\gamma_{a,b} \circ \gamma_{c,d}(u) = \gamma_{a,b}(\gamma_{c,d}(u)) = \gamma_{a,b}(cu + d)$$
$$= a(cu + d) + b = acu + (ad + b) = \gamma_{ac,ad+b}(u).$$

Thus $\gamma_{a,b} \circ \gamma_{c,d} = \gamma_{ac,ad+b}$, so that the map $\sigma_{a,b} \mapsto \gamma_{a,b}$ gives an isomorphism

$$\mathrm{Gal}(L/\mathbb{Q}) \simeq \mathrm{AGL}(1, \mathbb{F}_p).$$

Another way to understand the structure of $\mathrm{AGL}(1, \mathbb{F}_p)$ is via the subgroup

$$T = \{\gamma_{1,b} \mid b \in \mathbb{F}_p\}.$$

In Exercise 2 you will show that there is a group isomorphism $T \simeq \mathbb{F}_p$. You will also prove that T is a normal subgroup of $\mathrm{AGL}(1, \mathbb{F}_p)$ with quotient

$$(6.6) \qquad\qquad \mathrm{AGL}(1, \mathbb{F}_p)/T \simeq \mathbb{F}_p^*.$$

As a group, $\mathbb{F}_p$ is cyclic of order p, and Proposition A.5.3 implies that $\mathbb{F}_p^*$ is cyclic of order $p - 1$. In the Mathematical and Historical Notes we will say more about how $\mathrm{AGL}(1, \mathbb{F}_p)$ is built from these cyclic groups.

B. The Universal Extension. In Chapter 2 we studied the elementary symmetric polynomials $\sigma_1, \ldots, \sigma_n$ in variables $x_1, \ldots, x_n$. Recall from Proposition 2.1.4 that

$$(x - x_1) \cdots (x - x_n) = x^n - \sigma_1 x^{n-1} + \cdots + (-1)^r \sigma_r x^{n-r} + \cdots + (-1)^n \sigma_n.$$

This is the *universal polynomial of degree n* introduced in Section 2.2 and denoted by $\tilde{f}$. Note that $\tilde{f}$ is a polynomial in x with coefficients in the field

$$K = F(\sigma_1, \ldots, \sigma_n).$$

Since the roots of $\tilde{f}$ are $x_1, \ldots, x_n$, it follows easily that

$$L = F(x_1, \ldots, x_n)$$

is the splitting field of $\tilde{f}$ over K. We call $K \subset L$ the *universal extension in degree n*.

Since $\tilde{f}$ has distinct roots, Section 6.3 gives a one-to-one group homomorphism $\mathrm{Gal}(L/K) \to S_n$. We now prove that this map is an isomorphism.

Theorem 6.4.1. *The universal extension $K = F(\sigma_1, \ldots, \sigma_n) \subset L = F(x_1, \ldots, x_n)$ in degree n is the splitting field of a separable polynomial. The action of the Galois group on the roots of the universal polynomial of degree n gives an isomorphism*

$$\mathrm{Gal}(L/K) \simeq S_n.$$

Proof. We showed above that L is the splitting field of the universal polynomial $\tilde{f}$. Notice also that $\tilde{f}$ is separable, since its roots $x_1, \ldots, x_n$ are distinct.

To prove the final assertion of the theorem, we will use the action of S_n on $F[x_1, \ldots, x_n]$ discussed in Section 2.4. Recall that for $f \in F[x_1, \ldots, x_n]$ and $\tau \in S_n$, $\tau \cdot f$ is the polynomial obtained by permuting the variables according to τ. This action has the properties

$$
\begin{aligned}
\tau \cdot (f + g) &= \tau \cdot f + \tau \cdot g, \\
(6.7) \qquad\qquad \tau \cdot (fg) &= (\tau \cdot f)(\tau \cdot g), \\
\tau \cdot (\gamma \cdot f) &= (\tau\gamma) \cdot f,
\end{aligned}
$$

where $\tau, \gamma \in S_n$ and $f, g \in F[x_1, \ldots, x_n]$. You will prove this in Exercise 3.

In Exercises 4 and 5 you will also show that $f \mapsto \tau \cdot f$ is a ring isomorphism from $F[x_1, \ldots, x_n]$ to itself and hence extends to an isomorphism of its field of fractions. It follows that permuting the variables according to τ gives an automorphism of $L = F(x_1, \ldots, x_n)$. Since the elementary symmetric polynomials are fixed by the action of $\tau \in S_n$, this automorphism is the identity on $F(\sigma_1, \ldots, \sigma_n)$.

We have thus shown that $f \mapsto \tau \cdot f$ is an element of $\mathrm{Gal}(L/K)$. Under the map $\mathrm{Gal}(L/K) \to S_n$ of Proposition 6.3.1, this automorphism obviously maps to τ. Since τ was an arbitrary element of S_n, we see that $\mathrm{Gal}(L/K) \to S_n$ is onto, which completes the proof of the theorem. $\qquad\square$

Chapter 7 will describe the Galois theory of the universal extension.

C. A Polynomial of Degree 5.

Consider the polynomial $f = x^5 - 6x + 3$, and let L be the splitting field of f over $\mathbb{Q}$. The Schönemann–Eisenstein criterion implies that f is irreducible and hence separable. Thus $\mathrm{Gal}(L/\mathbb{Q})$ is isomorphic to a subgroup $H \subset S_5$. We will show that $H = S_5$, so that

$$(6.8) \qquad\qquad \mathrm{Gal}(L/\mathbb{Q}) \simeq S_5.$$

We will sketch the proof and leave the details for Exercise 6.

By Exercise 6 of Section 6.2, $|\text{Gal}(L/\mathbb{Q})| = |H|$ is divisible by 5, since f is irreducible. By Cauchy's Theorem (Theorem A.1.5) from group theory, H must have an element g of order 5. Recall that g is a product of disjoint cycles whose order is the least common multiple of the lengths of the cycles. Since g is in S_5 and has order 5, one easily sees that g is in fact a 5-cycle. Thus H contains a 5-cycle.

By the Fundamental Theorem of Algebra, we can assume that $L \subset \mathbb{C}$, so that the roots of f can be regarded as complex numbers. Furthermore, using curve graphing techniques from calculus, one also sees that f has exactly three real roots. It follows that complex conjugation gives an element $\tau \in \text{Gal}(L/\mathbb{Q})$ that interchanges two of the roots and fixes the other three. Since τ maps to a transposition in S_5, we conclude that H contains a transposition.

Hence, relabeling the roots appropriately, we may assume that H contains (12345) and $(1i)$ for some $i \in \{2, 3, 4, 5\}$. Since $(12345)^{i-1}$ is a 5-cycle beginning $(1i \ldots)$, we can relabel the roots again so that H contains (12345) and (12). It is a classic result in group theory that these two permutations generate S_5. You probably studied this in your abstract algebra course (if not, you should do Exercise 7). This shows that $H = S_5$ and completes the proof of (6.8).

This example is taken from [Stewart, Chapter 14]. In Chapter 8 we will see that $f = x^5 - 6x + 3$ is not solvable by radicals, since $\text{Gal}(L/\mathbb{Q}) \simeq S_5$. Different proofs of (6.8) will be given in Examples 13.2.8 and 13.4.7 of Chapter 13.

Mathematical Notes

This section has several topics of interest to discuss.

■ **Specialization of Galois Groups.** For the fifth roots of 2, we have

the splitting field of $x^5 - 2$ over $\mathbb{Q}$ has Galois group $\text{AGL}(1, \mathbb{F}_5)$,

while for the universal extension in degree 5 over $F = \mathbb{Q}$, we have

the splitting field of $x^5 - \sigma_1 x^4 + \sigma_2 x^3 - \sigma_3 x^2 + \sigma_4 x - \sigma_5$

over $\mathbb{Q}(\sigma_1, \sigma_2, \sigma_3, \sigma_4, \sigma_5)$ has Galois group S_5.

Since the second polynomial is the universal polynomial of degree 5, the first can be regarded as the specialization of the second obtained by the mapping

$$\sigma_1 \mapsto 0, \quad \sigma_2 \mapsto 0, \quad \sigma_3 \mapsto 0, \quad \sigma_4 \mapsto 0, \quad \sigma_5 \mapsto 2.$$

However, the Galois groups are not the same, which implies that *specialization does not always preserve the Galois group*. This is part of what makes Galois theory so hard—polynomials of the same degree may have different Galois groups.

On the other hand, most specializations of the universal polynomial of degree n over $\mathbb{Q}$ have S_n as their Galois group. This follows from the *Hilbert Irreducibility Theorem*, which is discussed in [Hadlock]. For example, one can prove that the Galois group of $x^n - x - 1 \in \mathbb{Q}[x]$ is S_n for all $n \geq 2$ (see [4, p. 42]).

■ **Semidirect Products.** In the text we noted that the one-dimensional affine group $\mathrm{AGL}(1, \mathbb{F}_p)$ has a normal subgroup $T \simeq \mathbb{F}_p$ such that $\mathrm{AGL}(1, \mathbb{F}_p)/T \simeq \mathbb{F}_p^*$. We will now explain how $\mathrm{AGL}(1, \mathbb{F}_p)$ is the *semidirect product* of $\mathbb{F}_p$ and $\mathbb{F}_p^*$ via the action of $\mathbb{F}_p^*$ on $\mathbb{F}_p$ given by $a \cdot u = au$.

In general, let G and H be groups, and assume that G acts on H (as defined in Section A.4) in the following special way: for any $g \in G$, the map $h \mapsto g \cdot h$ given by the action on g on H is a group homomorphism from H to itself. Then define a binary operation on the set $H \times G$ by

(6.9) $$(h, g) \cdot (h', g') = (h \, (g \cdot h'), gg')$$

where $h \, (g \cdot h')$ is the product of $h, g \cdot h' \in H$. The intuition behind this formula is that when we multiply (h, g) and (h', g'), we "twist" by the action of g, since g is between h and h'. In Exercise 8 you will show that this defines a group, called the *semidirect product $H \rtimes G$*.

For example, the action of $\mathbb{F}_p^*$ on $\mathbb{F}_p$ gives the semidirect product $\mathbb{F}_p \rtimes \mathbb{F}_p^*$. In this group, the product is given by

$$(b, a) \cdot (d, c) = (b + a \cdot d, ac) = (ad + b, ac),$$

since the group operation is addition in $\mathbb{F}_p$ and multiplication in $\mathbb{F}_p^*$. It follows that $\gamma_{a,b} \mapsto (b, a)$ gives an isomorphism

(6.10) $$\mathrm{AGL}(1, \mathbb{F}_p) \simeq \mathbb{F}_p \rtimes \mathbb{F}_p^*.$$

For any semidirect product $H \rtimes G$, the map $(h, g) \mapsto g$ is a group homomorphism that is clearly onto. In Exercise 8 you will check that the kernel of this map is

$$\{(h, e) \mid h \in H\} = H \times \{e\} \simeq H.$$

Then the Fundamental Theorem of Group Homomorphisms implies that

$$(H \rtimes G)/(H \times \{e\}) \simeq G.$$

In Exercise 9 you will explore how this relates to (6.6) and (6.10).

■ **The Extension Problem.** Given groups G and H, a third group G_1 is an *extension of H by G* if G_1 contains a normal subgroup $H_1 \simeq H$ such that $G_1/H_1 \simeq G$. For example, when G acts on H by group homomorphisms as above, the semidirect product $H \rtimes G$ is an extension of H by G.

An important observation is that the same groups can have nonisomorphic extensions. For example, Exercise 10 will show that the product $\mathbb{F}_p \times \mathbb{F}_p^*$ and the semidirect product $\mathbb{F}_p \rtimes \mathbb{F}_p^*$ are nonisomorphic extensions of $\mathbb{F}_p$ by $\mathbb{F}_p^*$ when $p \geq 3$.

The *extension problem* in group theory asks whether it is possible to classify all extensions of H by G. This is a difficult problem and is one of the reasons why groups are hard to classify. The extension problem is also related to *group cohomology*.

Historical Notes

For the extension $\mathbb{Q} \subset L = \mathbb{Q}(\zeta_p, \sqrt[p]{2})$, one can also describe $\mathrm{Gal}(L/\mathbb{Q}) \simeq$ $\mathrm{AGL}(1, \mathbb{F}_p)$ in terms of permutations in S_p. We do this by replacing $\{1, \dots, p\}$ with the congruence classes $\{[1], \dots, [p]\} = \mathbb{F}_p$. Then the affine linear transformation $\gamma_{a,b}$ defined in the text becomes the element of S_p represented by the permutation

$$\begin{pmatrix} 1 & 2 & \cdots & p \\ a1 + b & a2 + b & \cdots & ap + b \end{pmatrix}$$

provided we think of the entries as congruence classes modulo p. This can be expressed more succinctly in the form

(6.11)
$$\begin{pmatrix} i \\ ai + b \end{pmatrix}.$$

We will see in Chapter 12 that the permutations (6.11) are implicit in the work of Lagrange. These permutations also appear explicitly in Galois's study of irreducible polynomials of prime degree that are solvable by radicals. We will have more to say about this in Chapter 14.

In the late nineteenth and early twentieth centuries the subgroup of S_p consisting of the permutations (6.11) was called *the metacyclic group*. These days the term *metacyclic* is used more generally to mean any group G possessing a normal subgroup H such that both H and G/H are cyclic. In Exercise 11 you will show that the group of permutations of the form (6.11) is metacyclic in this sense.

As for the Galois group of the universal extension computed in Theorem 6.4.1, Galois states this result as follows [Galois, p. 51]:

> In the case of algebraic equations, the group is nothing other than the collection of $1.2.3 \dots m$ possible permutations on the m letters…

Here, "algebraic equation" refers to the universal case where the "m letters" are the roots of the universal polynomial of degree m. However, we will see in Chapter 12 that Galois's use of the word "permutation" is different from ours.

Exercises for Section 6.4

Exercise 1. Given $a, b \in \mathbb{F}_p$, define $\gamma_{a,b} : \mathbb{F}_p \to \mathbb{F}_p$ by $\gamma_{a,b}(u) = au + b$.
  (a) Prove that $\gamma_{a,b}$ is one-to-one and onto if and only if $a \neq 0$.
  (b) Suppose that $a \neq 0$. Prove that the inverse function of $\gamma_{a,b}$ is $\gamma_{a^{-1},-a^{-1}b}$.
  (c) Show that
$$\mathrm{AGL}(1, \mathbb{F}_p) = \{\gamma_{a,b} \mid (a, b) \in \mathbb{F}_p^* \times \mathbb{F}_p\}$$
  is a group under composition.

Exercise 2. Consider the map $\mathrm{AGL}(1, \mathbb{F}_p) \to \mathbb{F}_p^*$ defined by $\gamma_{a,b} \mapsto a$.
  (a) Show that this map is an onto group homomorphism with kernel $T = \{\gamma_{1,b} \mid b \in \mathbb{F}_p\}$. Then use this to prove (6.6).
  (b) Show that $T \simeq \mathbb{F}_p$.

Exercise 3. This exercise is concerned with the proof of (6.7). Given $\tau \in S_n$, observe that $f \mapsto \tau \cdot f$ can be regarded as the evaluation map from $F[x_1, \ldots, x_n]$ to itself that evaluates $f(x_1, \ldots, x_n)$ at $(x_{\tau(1)}, \ldots, x_{\tau(n)})$.

(a) Explain why Theorem 2.1.2 implies that $f \mapsto \tau \cdot f$ is a ring homomorphism. This proves the first two bullets of (6.7).

(b) Prove the third bullet of (6.7).

Exercise 4. Let $\tau \in S_n$. Prove that $f \mapsto \tau \cdot f$ is a ring isomorphism from $F[x_1, \ldots, x_n]$ to itself.

Exercise 5. Let R be an integral domain, and let K be its field of fractions. Prove that every ring isomorphism $\phi : R \to R$ extends uniquely to an automorphism $\tilde{\phi} : K \to K$.

Exercise 6. As in the text, let $f = x^5 - 6x + 3$.

(a) Use the hints given in the text to show that every element of S_5 of order 5 is a 5-cycle.

(b) Use curve graphing from calculus to show that f has exactly three real roots.

Exercise 7. Show that S_n is generated by the transposition (12) and the n-cycle $(12 \ldots n)$.

Exercise 8. Let G and H be groups where G acts on H by group homomorphisms. As in the text, we let $H \rtimes G$ denote the set $H \times G$ with the binary operation given by (6.9).

(a) Prove that $H \rtimes G$ is a group.

(b) Prove that the map $H \rtimes G \to G$ defined by $(h, g) \mapsto g$ is an onto group homomorphism with kernel $H \times \{e\}$.

(c) Prove that $h \mapsto (h, e)$ defines an isomorphism $H \simeq H \times \{e\}$ (where the group structure on $H \times \{e\}$ comes from $H \rtimes G$).

Exercise 9. Explain how (6.6) and (6.10) relate to the last paragraph of the discussion of semidirect products in the Mathematical Notes.

Exercise 10. Let $p \geq 3$ be prime, and let $\mathbb{F}_p \rtimes \mathbb{F}_p^*$ be the semidirect product described in the Mathematical Notes.

(a) Show that $\mathbb{F}_p \rtimes \mathbb{F}_p^*$ is not Abelian.

(b) Show that the product group $\mathbb{F}_p \times \mathbb{F}_p^*$ is Abelian.

(c) Show that $\mathbb{F}_p \times \mathbb{F}_p^*$ is an extension of $\mathbb{F}_p$ by $\mathbb{F}_p^*$.

Since we already know that $\mathbb{F}_p \rtimes \mathbb{F}_p^*$ is an extension of $\mathbb{F}_p$ by $\mathbb{F}_p^*$, we see that (a) and (b) give nonisomorphic extensions.

Exercise 11. The goal of this exercise is to show that the group G of permutations (6.11) is metacyclic in the sense that G has a normal subgroup H such that H and G/H are cyclic. Show that this follows from $G \simeq \mathrm{AGL}(1, \mathbb{F}_p)$ together with (6.6) and Proposition A.5.3.

Exercise 12. Let p be prime. Generalize part (a) of Exercise 6 by showing that every element of S_p of order p is a p-cycle.

Exercise 13. Let L be the splitting field of $2x^5 - 10x + 5$ over $\mathbb{Q}$. Prove that $\mathrm{Gal}(L/\mathbb{Q}) \simeq S_5$.

Exercise 14. let $L = \mathbb{Q}(\zeta_p, \sqrt[p]{2})$. Prove that $L = \mathbb{Q}(\sqrt[p]{2}, \zeta_p \sqrt[p]{2})$, that is, the splitting field of $x^p - 2$ over $\mathbb{Q}$ can be generated by two of its roots. Chapter 14 will show that this follows from Galois's criterion for an irreducible polynomial of prime degree to be solvable by radicals.

Exercise 15. Let $L = \mathbb{Q}(\zeta_p, \sqrt[p]{2})$. The description of $\mathrm{Gal}(L/\mathbb{Q})$ given in the text enables one to construct some elements of $\mathrm{Gal}(L/\mathbb{Q}(\zeta_p))$. Use these automorphisms and Proposition 6.3.7 to prove that $x^p - 2$ is irreducible over $\mathbb{Q}(\zeta_p)$.

6.5 ABELIAN EQUATIONS (OPTIONAL)

In this section we will discuss the following theorem of Abel:

> If the roots of an equation of arbitrary degree are related among themselves in such a way that *all* the roots can be expressed rationally by means of one of them, which we denote by x; if in addition whenever one denotes by θx, $\theta_1 x$ two other arbitrary roots, one has
>
> $$\theta\theta_1 x = \theta_1\theta x,$$
>
> then the equation to which they belong will always be solvable algebraically.

(See [Abel, p. 479].) Our goal is to interpret this theorem in terms of Galois theory.

We begin by translating Abel's theorem into modern terminology. First observe that Abel talks about an equation $f = 0$ rather than a polynomial f. This is typical for the early nineteenth century. We will assume that f is a nonconstant polynomial whose coefficients lie in a field F. Since we prefer x to be a variable, we will replace Abel's x with α. So α will be a root of f in some extension field.

Now let $\alpha_1 = \alpha, \alpha_2, \ldots, \alpha_n$ be the roots of f in a splitting field L. Then, when Abel says that the roots can be "expressed rationally" in terms of α, he means that there are rational functions θ_i with coefficients in F such that $\alpha_i = \theta_i(\alpha)$. In Exercise 1 you will show that this is equivalent to

$$(6.12) \qquad\qquad L = F(\alpha).$$

Here is an example of this from Chapter 4.

Example 6.5.1. If we let $\alpha = \sqrt{2} + \sqrt{3}$, then (4.4) implies that $f = x^4 - 10x^2 + 1$ factors as

$$(x - \alpha)(x + \alpha)(x - 10\alpha + \alpha^3)(x + 10\alpha - \alpha^3).$$

If we set $\theta_1(x) = x$, $\theta_2(x) = -x$, $\theta_3(x) = 10x - x^3$, and $\theta_4(x) = -10x + x^3$, then the roots of f are $\theta_1(\alpha)$, $\theta_2(\alpha)$, $\theta_3(\alpha)$, and $\theta_4(\alpha)$. ◁▷

The rational functions θ_i in Abel's theorem give functions from L to L that usually fail to be automorphisms. For instance, the function $\theta_2(x) = -x$ in the above example does not preserve multiplication, since $\theta_2(ab) = -ab$ differs from $\theta_2(a)\theta_2(b) = (-a)(-b) = ab$ whenever $ab \neq 0$.

In our notation, the displayed equation in the quote from Abel becomes

$$(6.13) \qquad\qquad \theta_i(\theta_j(\alpha)) = \theta_j(\theta_i(\alpha)), \quad 1 \leq i, j \leq n.$$

In Exercise 2 you will show that the rational functions of Example 6.5.1 satisfy this condition. Following Kronecker and Jordan, we call $f = 0$ an *Abelian equation* if f is a nonconstant polynomial with a root α satisfying (6.12) and (6.13).

The conclusion of Abel's theorem states that f is "solvable algebraically." In modern terms this means "solvable by radicals," which will be defined carefully in Chapter 8. Thus Abel's theorem can be restated as follows.

Theorem 6.5.2. *In characteristic* 0, *every Abelian equation is solvable by radicals.*

The hypothesis about characteristic 0 is not in Abel's original statement but is needed since the theory of solvability developed in Chapter 8 only applies to fields of characteristic 0. Abel always worked in characteristic 0.

Theorem 6.5.2 is a consequence of the following two theorems. Recall from Section 6.1 that the Galois group of $f \in F[x]$ is $\text{Gal}(L/F)$, where L is a splitting field of f over L.

Theorem 6.5.3. *The Galois group of an Abelian equation is an Abelian group.*

Theorem 6.5.4. *In characteristic* 0, *a polynomial with Abelian Galois group is solvable by radicals.*

We will prove Theorem 6.5.4 in Chapter 8. The proof will follow from Galois's criterion for solvability by radicals together with the fact that every finite Abelian group is solvable. (All of these terms will be defined in Chapter 8.)

We now prove Theorem 6.5.3.

Proof of Theorem 6.5.3. If the Abelian equation is $f = 0$, where $f \in F[x]$, then f has a root α such that $L = F(\alpha)$ is the splitting field of f. In particular, we have rational functions $\theta_i(x) \in F(x)$, $1 \le i \le n$, such that the $\theta_i(\alpha)$ are the roots of f. Now let $\sigma, \tau \in \text{Gal}(L/F)$. In Exercise 3 you will prove the following:

- $\sigma(\alpha) = \theta_i(\alpha)$ and $\tau(\alpha) = \theta_j(\alpha)$ for some i and j.
- $\sigma\tau = \tau\sigma$ in $\text{Gal}(L/F)$ if and only if $\sigma(\tau(\alpha)) = \tau(\sigma(\alpha))$ in L.
- $\sigma(\tau(\alpha)) = \theta_j(\theta_i(\alpha))$ and $\tau(\sigma(\alpha)) = \theta_i(\theta_j(\alpha))$.

Since $f = 0$ is Abelian, the theorem follows easily from these bullets. $\square$

In the 1880s, Weber applied the term "Abelian" to commutative groups because of this theorem.

Historical Notes

The story of Abelian equations begins with Gauss, who showed in 1801 that $x^n - 1 = 0$ is solvable by radicals. We will study Gauss's work in Chapter 9, and in Chapter 10 we will explore the surprising geometric consequences of his results. In Exercise 4 you will show that $x^n - 1 = 0$ is an Abelian equation over $\mathbb{Q}$.

In his 1829 paper *Mémoire sur une classe particulière d'équations résolubles algébriquement* [Abel, Vol. I, pp. 478–507], Abel states the theorem quoted at the beginning of the section and goes on to say

> After having explained this theory [the solvability of Abelian equations] in
> general, I will apply it to circular and elliptic functions.

(See [Abel, Vol. I, p. 479].) In this passage, "circular functions" refer to the work of Gauss just mentioned, and "elliptic functions" refer to Abel's deep results on elliptic functions and complex multiplication. We will discuss a special case of this involving the lemniscate in Chapter 15. Abel died at age 26 before he could publish the full details of his work.

Kronecker introduced the term "Abelian equation" in 1853 in the special case when the Galois group was cyclic. The general sense of the term, as defined here, is due to Jordan in 1870. Kronecker's interest in Abelian equations is related to his amazing conjecture that *the roots of an Abelian equation over $\mathbb{Q}$ can be expressed rationally in terms of a root of unity.* This was proved in 1886 by Weber and is now called the Kronecker–Weber Theorem. The modern version of this theorem is stated as follows.

Theorem 6.5.5. *Suppose that $\mathbb{Q} \subset L$ is a finite extension such that $L \subset \mathbb{C}$. Then the following conditions are equivalent:*

(a) *$\mathbb{Q} \subset L$ is normal and $\mathrm{Gal}(L/\mathbb{Q})$ is Abelian.*

(b) *There is a root of unity $\zeta_n = e^{2\pi i/n}$ such that $L \subset \mathbb{Q}(\zeta_n)$.*   □

In the next chapter, you will prove (b) $\Rightarrow$ (a) in Exercise 14 of Section 7.3, and a proof of (a) $\Rightarrow$ (b) can be found in [3, pp. 125–129]. The proof of (a) $\Rightarrow$ (b) uses ideas from algebraic number theory and is beyond the scope of this book.

The early history of group theory and Galois theory are closely related—after all, Galois was the person who introduced the term "group" into mathematics. So it is not surprising that notions like Abelian equations from Galois theory influenced the terminology of group theory. We will see many more examples of this phenomenon in the next chapter.

Exercises for Section 6.5

Exercise 1. Assume that $f \in F[x]$ is nonconstant and has roots $\alpha_1 = \alpha, \alpha_2, \ldots, \alpha_n$ in a splitting field L. Prove that $L = F(\alpha)$ if and only if there are rational functions $\theta_i \in F(x)$ such that $\alpha_i = \theta_i(\alpha)$. Can we assume that the θ_i are polynomials?

Exercise 2. Show that the equation $x^4 - 10x^2 + 1 = 0$ discussed in Example 6.5.1 is Abelian.

Exercise 3. Complete the proof of Theorem 6.5.3.

Exercise 4. Show that $x^n - 1 = 0$ is an Abelian equation over $\mathbb{Q}$.

Exercise 5. Let f be the minimal polynomial of $\sqrt{2 + \sqrt{2}}$ over $\mathbb{Q}$. Show that $f = 0$ is an Abelian equation.

Exercise 6. In this exercise, you will prove a partial converse to Theorem 6.5.3. Suppose that a finite extension $F \subset L$ is normal and separable and has Abelian Galois group.

(a) Explain why $F \subset L$ has a primitive element.

(b) By part (a), we can find $\alpha \in L$ such that $L = F(\alpha)$. Let f be the minimal polynomial of α. Prove that $f = 0$ is an Abelian equation over F.

Exercise 7. Show that the implication (a) $\Rightarrow$ (b) of Theorem 6.5.5 is equivalent to Kronecker's assertion that the roots of an Abelian equation over $\mathbb{Q}$ can be expressed rationally in terms of a root of unity.

REFERENCES

1. R. Dedekind, *Theory of Algebraic Integers*, English translation by J. Stillwell, Cambridge U. P., Cambridge, 1996. (Translation of 1877 French edition.)

2. B. M. Kiernan, *The development of Galois theory from Lagrange to Artin*, Arch. Hist. Exact Sci. **8** (1971), 40–154.

3. D. A. Marcus, *Number Fields*, Springer-Verlag, New York, Berlin, Heidelberg, 1977.

4. J.-P. Serre, *Topics in Galois Theory*, Jones and Bartlett, Boston, 1992.

7

The Galois Correspondence

This chapter will draw on the work we did in Chapters 4, 5, and 6 to state and prove the main theorems of Galois theory. We will also give some applications.

7.1 GALOIS EXTENSIONS

In Section 6.2 we learned that splitting fields of separable polynomials are especially nice from the point of view of Galois theory. The main goal of this section is to characterize such extensions in terms of normality and separability. We will also apply this theory to study separable extensions.

A. Splitting Fields of Separable Polynomials. Before stating our main result, we introduce the idea of a *fixed field*. Suppose that we have a finite extension $F \subset L$ with Galois group $\text{Gal}(L/F)$. Given a subgroup $H \subset \text{Gal}(L/F)$, we call

$$L_H = \{\alpha \in L \mid \sigma(\alpha) = \alpha \text{ for all } \sigma \in H\}$$

the *fixed field of H*. This terminology is justified by Exercise 1, where you will show that L_H is a subfield of L containing F.

Here is one of the important theorems of Galois theory.

Theorem 7.1.1. *Let $F \subset L$ be a finite extension. Then the following are equivalent:*
  (a) *L is the splitting field of a separable polynomial in $F[x]$.*
  (b) *F is the fixed field of $\text{Gal}(L/F)$ acting on L.*
  (c) *$F \subset L$ is a normal separable extension.*

Proof. (a) ⇒ (b): Let K be the fixed field of $\text{Gal}(L/F)$. By Exercise 1 we have $F \subset K \subset L$, and the goal is to show $K = F$. For this purpose, note that since L is the splitting field of a separable polynomial $f \in F[x]$ over F, the same is true over the larger field K, since f also lies in $K[x]$. By Theorem 6.2.1 it follows that

$$[L:F] = |\text{Gal}(L/F)| \quad \text{and} \quad [L:K] = |\text{Gal}(L/K)|.$$

Next observe that $\text{Gal}(L/K) \subset \text{Gal}(L/F)$, since if an automorphism of L is the identity on K, then it is also the identity on the smaller field F. The reverse inclusion also holds, since every $\sigma \in \text{Gal}(L/F)$ is the identity on K, for K is the fixed field of $\text{Gal}(L/F)$. It follows that $\text{Gal}(L/K) = \text{Gal}(L/F)$. Combining this with the above equations, we see that

$$[L:F] = [L:K].$$

Since $[L:F] = [L:K][K:F]$, we have $[K:F] = 1$, and $K = F$ follows.

   (b) ⇒ (c): Now suppose that F is the fixed field of $\text{Gal}(L/F)$ and let $\alpha \in L$. We will find the minimal polynomial of α over F using a construction due to Lagrange. Let $\alpha_1 = \alpha, \alpha_2, \ldots, \alpha_r$ be the distinct elements of L obtained by applying the elements of $\text{Gal}(L/F)$ to α. Then consider the polynomial

(7.1)
$$h(x) = \prod_{i=1}^{r}(x - \alpha_i) \in L[x].$$

We claim that $h \in F[x]$ and that h is irreducible over F.

   We first show that $\sigma \in \text{Gal}(L/F)$ permutes the α_i. By definition, $\alpha_i = \tau(\alpha)$ for some $\tau \in \text{Gal}(L/F)$. Then $\sigma(\alpha_i) = \sigma(\tau(\alpha)) = (\sigma\tau)(\alpha_i)$, which is α_j for some j. Thus σ maps $\{\alpha_1, \ldots, \alpha_r\}$ to itself, which gives a permutation, since σ is one to one.

   Since σ permutes the α_i, it also permutes the factors $x - \alpha_i$ of h. This shows that the coefficients of h are fixed by $\text{Gal}(L/F)$ and hence lie in the fixed field, which is F by assumption. Hence $h \in F[x]$, as claimed.

   Next let $g \in F[x]$ be the irreducible factor of h that vanishes at α. Then Proposition 6.1.4 shows that $\sigma(\alpha)$ is also a root of g for all $\sigma \in \text{Gal}(L/F)$. Since the α_i are the distinct elements of L obtained in this way, (7.1) shows that $h|g$. It follows that h is irreducible over F, since g is an irreducible factor of h.

   We conclude that $h \in F[x]$ is the minimal polynomial of α over F, since h is irreducible over F and has α as a root. The above formula for h also shows that h is separable and splits completely over L. Hence:

- **Normality:** If $f \in F[x]$ is irreducible and has a root $\alpha \in L$, then f is the polynomial h defined in (7.1) (up to a constant factor). Thus f splits completely over L, which proves normality.

- **Separability:** If $\alpha \in L$, then its minimal polynomial is the polynomial h. Then α is separable over F because h is, and separability follows.

This shows that $F \subset L$ is normal and separable, as claimed.

(c) $\Rightarrow$ (a): Finally, suppose that $F \subset L$ is normal and separable. We can write $L = F(\alpha_1, \ldots, \alpha_n)$, where the minimal polynomial p_i of α_i over F is separable. Let $q_1, \ldots, q_r$ be the distinct elements of the set $\{p_1, \ldots, p_n\}$, and set

$$f = q_1 \cdots q_r.$$

By Lemma 5.3.4, f is separable (the lemma applies because the q_i are monic—do you see why?). Furthermore, the proof of Theorem 5.2.4 shows that L is the splitting field of f over F (you will check this in Exercise 2). Thus L is the splitting field over F of a separable polynomial in $F[x]$, as claimed. $\square$

In light of this theorem, we make the following definition.

Definition 7.1.2. *An extension $F \subset L$ is called a **Galois extension** if it is a finite extension satisfying any of the equivalent conditions of Theorem 7.1.1.*

To see how Definition 7.1.2 works, consider the following extensions.

- The extension $\mathbb{Q} \subset \mathbb{Q}(\sqrt{2}, \sqrt{3})$ is Galois, since $\mathbb{Q}(\sqrt{2}, \sqrt{3})$ is the splitting field of $(x^2 - 2)(x^2 - 3)$ over $\mathbb{Q}$. This uses part (a) of Theorem 7.1.1.
- The extension $\mathbb{Q} \subset \mathbb{Q}(\sqrt[3]{2})$ is not Galois, since $x^3 - 2$ is irreducible over $\mathbb{Q}$, has a root in $\mathbb{Q}(\sqrt[3]{2})$, but does not split completely over $\mathbb{Q}(\sqrt[3]{2})$. This uses part (c) of Theorem 7.1.1.

Here is one case where being a Galois extension is automatic.

Proposition 7.1.3. *Suppose that $F \subset L$ is a Galois extension and that we have an intermediate field $F \subset K \subset L$. Then $K \subset L$ is a Galois extension.*

Proof. We will use part (a) of Theorem 7.1.1. If $F \subset L$ is Galois, then L is the splitting field of a separable polynomial in $f \in F[x]$. By regarding f as an element of $K[x]$, it follows immediately that the same is true over the larger field K. (This is the argument used in the proof of (a) $\Rightarrow$ (b) from Theorem 7.1.1.) $\square$

While the proof of Proposition 7.1.3 seems easy, notice that it is *much* less obvious if we think in terms of parts (b) and (c) of Theorem 7.1.1.

The reader should also note that in the situation of Proposition 7.1.3, $F \subset K$ need not be Galois. Here is a simple example to illustrate this.

Example 7.1.4. By Example 4.1.10, $\mathbb{Q} \subset \mathbb{Q}(i, \sqrt[4]{2})$ is the splitting field of $x^4 - 2$ and hence is a Galois extension. Consider the intermediate fields $\mathbb{Q}(i)$ and $\mathbb{Q}(\sqrt[4]{2})$. Then $\mathbb{Q} \subset \mathbb{Q}(i)$ is Galois (it is the splitting field of $x^2 + 1$), while $\mathbb{Q} \subset \mathbb{Q}(\sqrt[4]{2})$ is not ($x^4 - 2$ is the minimal polynomial of $\sqrt[4]{2}$ but doesn't split completely).   ◁▷

In Section 7.2 we will learn how to recognize exactly when $F \subset K$ is Galois in the situation of Proposition 7.1.3.

Definition 7.1.2 and Theorem 6.2.1 imply that $|\mathrm{Gal}(L/F)| = [L:F]$ whenever $F \subset L$ is Galois. For an arbitrary finite extension, the relation between the order of the Galois group and the degree of the extension can be described as follows.

Theorem 7.1.5. *Let* $F \subset L$ *be a finite extension. Then:*
(a) $|\mathrm{Gal}(L/F)|$ *divides* $[L:F]$.
(b) $|\mathrm{Gal}(L/F)| \leq [L:F]$.
(c) $F \subset L$ *is a Galois extension if and only if* $|\mathrm{Gal}(L/F)| = [L:F]$.

Proof. To prove part (a), let K be the fixed field of $\mathrm{Gal}(L/F)$. Then $F \subset K \subset L$, and the proof of (a) $\Rightarrow$ (b) from Theorem 7.1.1 implies that $\mathrm{Gal}(L/K) = \mathrm{Gal}(L/F)$ (be sure you understand why). Thus K is the fixed field of $\mathrm{Gal}(L/K)$, so that $K \subset L$ is a Galois extension by Theorem 7.1.1. Hence

$$[L:F] = [L:K][K:F] = |\mathrm{Gal}(L/K)|[K:F] = |\mathrm{Gal}(L/F)|[K:F],$$

where the first equality uses Theorem 4.3.8, the second uses Theorem 6.2.1 ($K \subset L$ is Galois), and the third uses $\mathrm{Gal}(L/K) = \mathrm{Gal}(L/F)$. It follows that the order of $\mathrm{Gal}(L/F)$ group divides $[L:F]$, as claimed.

Part (b) is an immediate consequence of part (a). As for part (c), note that one direction follows from Theorem 6.2.1. For the converse, suppose that $F \subset L$ is a finite extension with $|\mathrm{Gal}(L/F)| = [L:F]$, and let K be the fixed field of $\mathrm{Gal}(L/F)$. If we can prove that $K = F$, then Theorem 7.1.1 will imply that $F \subset L$ is a Galois extension.

To show that $K = F$, first observe that the proof of part (a) given above implies that $K \subset L$ is a Galois extension and that $\mathrm{Gal}(L/K) = \mathrm{Gal}(L/F)$. Then

$$[L:F] = |\mathrm{Gal}(L/F)| = |\mathrm{Gal}(L/K)| = [L:K],$$

where the first equality is by assumption, the second uses $\mathrm{Gal}(L/K) = \mathrm{Gal}(L/F)$, and the third holds because $K \subset L$ is a Galois extension. We conclude that $K = F$ just as in the proof of (a) $\Rightarrow$ (b) from Theorem 7.1.1. $\qquad\square$

Theorem 7.1.1 gave three ways to characterize Galois extensions, and part (c) of Theorem 7.1.5 gives a fourth. Putting these together, we see that a finite extension $F \subset L$ is Galois if and only if any of the following equivalent conditions is satisfied:

- L is the splitting field of a separable polynomial in $F[x]$.
- F is the fixed field of $\mathrm{Gal}(L/F)$ acting on L.
- $F \subset L$ is a normal separable extension.
- $|\mathrm{Gal}(L/F)| = [L:F]$.

B. Finite Separable Extensions. The theory of Galois extensions implies the following characterization of finite separable extensions.

Proposition 7.1.6. *Let* $F \subset L$ *be a finite extension. Then L is separable over F if and only if* $L = F(\alpha_1, \ldots, \alpha_n)$, *where each α_i is separable over F.*

Proof. First assume that $F \subset L$ is separable. Since it is also finite, Theorem 4.4.3 implies that L has the desired form. For the converse, let $L = F(\alpha_1, \ldots, \alpha_n)$, where each α_i is separable over F. Our strategy will be to embed L in a larger field that is separable over F.

Let p_i be the minimal polynomial of α_i over F, and let $q_1, \ldots, q_r$ be the distinct elements of the set $\{p_1, \ldots, p_n\}$. Then Lemma 5.3.4 implies that $f = q_1 \cdots q_r$ is separable, since each q_i is. Let M be the splitting field of f, regarded as a polynomial in $L[x]$. Thus $M = L(\beta_1, \ldots, \beta_m)$, where $\beta_1, \ldots, \beta_m$ are the roots of f.

We claim that $M = F(\beta_1, \ldots, \beta_m)$. To see why, note that we have the obvious inclusion

$$(7.2) \qquad F(\beta_1, \ldots, \beta_m) \subset L(\beta_1, \ldots, \beta_m) = M.$$

However, the roots $\beta_1, \ldots, \beta_m$ include $\alpha_1, \ldots, \alpha_n$, so that

$$L = F(\alpha_1, \ldots, \alpha_n) \subset F(\beta_1, \ldots, \beta_m).$$

Thus $F(\beta_1, \ldots, \beta_m)$ contains both L and $\beta_1, \ldots, \beta_m$, which gives the inclusion

$$M = L(\beta_1, \ldots, \beta_m) \subset F(\beta_1, \ldots, \beta_m).$$

Combining this with (7.2), we see that $M = F(\beta_1, \ldots, \beta_m)$, as claimed.

This shows that M is the splitting field over F of the separable polynomial f. Then $F \subset M$ is Galois and hence separable by Theorem 7.1.1. Since $L \subset M$, every element of L is separable over F, so that $F \subset L$ is separable. $\square$

Proposition 7.1.6 has some nice consequences. For example, if $F \subset L$ and $\alpha, \beta \in L$ are separable over F, then so are $\alpha + \beta$, $\alpha\beta$, and α/β (assuming $\beta \neq 0$). This in turn implies that in characteristic p, any finite extension can be written as a separable extension followed by a purely inseparable one. You will be asked to prove these assertions in Exercises 3 and 4.

C. Galois Closures. The proof of Proposition 7.1.6 shows how to embed a finite separable extension $F \subset L$ into a larger Galois extension. This leads to the idea of *Galois closure*, which roughly speaking is the smallest extension of L that is Galois over F. More precisely, we have the following result.

Proposition 7.1.7. *Let $F \subset L$ be a finite separable extension. Then there is an extension $L \subset M$ such that:*
(a) *M is Galois over F, that is, $F \subset M$ is a Galois extension.*
(b) *Given any other extension $L \subset M'$ such that M' is Galois over F, there is a field homomorphism $\varphi : M \to M'$ that is the identity on L.*

Proof. Since $F \subset L$ is finite and separable, we can write $L = F(\alpha_1, \ldots, \alpha_n)$, where α_i is separable over F. Following the proof of Proposition 7.1.6, we get an extension $L \subset M$ such that M is a splitting field over L of the separable polynomial $f = q_1 \cdots q_r$, where $q_1, \ldots, q_r$ are the distinct elements of $\{p_1, \ldots, p_n\}$ and p_i is the minimal polynomial of α_i over F. As in the proof of Proposition 7.1.6, we see that $F \subset M$ is a Galois extension.

To show that $L \subset M$ satisfies part (b) of the proposition, let $L \subset M'$ be an extension where M' is Galois over F. By Theorem 7.1.1, $F \subset M'$ is normal, so that each p_i splits completely over M'. It follows that f splits completely in M'. Let $M'' \subset M'$ be the subfield obtained by adjoining the roots of f to F. Furthermore, since $\alpha_i \in L \subset M'$ is a root of f, we have $L \subset M''$. Thus we can regard M'' as a splitting field of f over L. By the uniqueness of splitting fields (Corollary 5.1.7), there is an isomorphism $\varphi : M \to M''$ that is the identity on L. Since $M'' \subset M'$, we can regard φ as a field homomorphism $\varphi : M \to M'$. This completes the proof of the proposition.   □

Be sure you understand why part (b) of the proposition implies that $L \subset M$ can be thought of as the smallest extension of L that is Galois over F. The field constructed in Proposition 7.1.7 is called the *Galois closure* of L over F. In Exercise 5 you will show that the Galois closure of $F \subset L$ is unique up to an isomorphism that is the identity on L.

Related to the idea of Galois closure is the *normal closure* of a finite extension $F \subset L$. Roughly speaking, this is the smallest extension of L that is normal over F. The theory of normal closures is worked out in Exercises 6 and 7.

Historical Notes

Of the criteria for $F \subset L$ to be a Galois extension given in Theorem 7.1.1, the most elegant is the one involving the fixed field of $\mathrm{Gal}(L/F)$. For Galois, this was his Proposition I, which was the first of his main results [Galois, p. 51]:

PROPOSITION I

THEOREM. For a given equation, let $a, b, c, \ldots$ be the m roots. There is always a group of permutations on the letters $a, b, c, \ldots$ that enjoys the following property:

1° that every function of the roots that is invariant** under the substitutions of the group, is rationally known;

2° conversely, that every function of the roots that is rationally determined, is invariant under these substitutions*.

For Galois, "rationally known" and "rationally determined" refer to elements of a field F containing the coefficients of the given equation. Adjoining the roots of this equation gives the splitting field $L = F(a, b, c, \ldots)$. Furthermore, Galois assumes that the given polynomial "does not have equal roots." Hence the polynomial is separable, so that L is a Galois extension of F. Since every element of L is a "function of the roots," parts 1° and 2° of Galois's Proposition I say that F is the fixed field of the Galois group acting on L. Thus we recover part (b) of Theorem 7.1.1.

In Galois's manuscript, Proposition I includes two notes, marked with ** and * above. We will explain these notes when we discuss Galois's work in Chapter 12.

We should also mention that Galois knew the formula (7.1) for the minimal polynomial given in the proof of Theorem 7.1.1 (see [Galois, p. 85]). In Chapter 12, we will see that this formula is a generalization of the *resolvent polynomial* defined by Lagrange in 1770.

Exercises for Section 7.1

Exercise 1. Given a finite extension $F \subset L$ and a subgroup $H \subset \mathrm{Gal}(L/F)$, prove that $L_H = \{\alpha \in L \mid \sigma(\alpha) = \alpha \text{ for all } \sigma \in H\}$ is a subfield of L containing F.

Exercise 2. In the proof of (c) $\Rightarrow$ (a) in Theorem 7.1.1, give the details of how the proof of Theorem 5.2.4 shows that L is the splitting field of f over F.

Exercise 3. Suppose that $F \subset L$ and that $\alpha, \beta \in L$ are separable over F. Prove that $\alpha + \beta$, $\alpha\beta$, and α/β (assuming $\beta \neq 0$) are also separable over F.

Exercise 4. Let $F \subset L$ be a finite extension, and assume F has characteristic p. Then consider the set $K = \{\alpha \in L \mid \alpha \text{ is separable over } F\}$.
(a) Use Proposition 7.1.6 to show that K is a subfield of L containing F. Thus $F \subset K$ is a separable extension.
(b) Use part (c) of Theorem 5.3.15 to show that $K \subset L$ is purely inseparable.

Exercise 5. Prove that the Galois closure of a finite separable extension $F \subset L$ is unique up to an isomorphism that is the identity on L.

Exercise 6. In analogy with the Galois closure of a finite separable extension, every finite extension $F \subset L$ has a *normal closure*, which is essentially the smallest extension of L that is normal over F. State and prove the analog of Proposition 7.1.7 for normal closures.

Exercise 7. Prove that the normal closure of a finite extension $F \subset L$ is unique up to an isomorphism that is the identity on L.

Exercise 8. Let h be the polynomial (7.1) used in the proof of (b) $\Rightarrow$ (c) from Theorem 7.1.1. Show that there is an integer m such that

$$\prod_{\sigma \in \mathrm{Gal}(L/F)} \left(x - \sigma(\alpha)\right) = h^m.$$

Exercise 9. For each of the following extensions, say whether it is a Galois extension. Be sure to say which of our four criteria (the three parts of Theorem 7.1.1 and part (c) of Theorem 7.1.5) you are using.
(a) $\mathbb{Q} \subset \mathbb{Q}(\sqrt{2}, \sqrt[3]{2})$.
(b) $\mathbb{Q} \subset \mathbb{Q}(\alpha, \beta)$, α, β distinct roots of $x^3 + x^2 + 2x + 1$.
(c) $\mathbb{F}_p(t^p) \subset \mathbb{F}_p(t)$, t a variable.
(d) $\mathbb{C}(t + t^{-1}) \subset \mathbb{C}(t)$, t a variable.
(e) $\mathbb{C}(t^n) \subset \mathbb{C}(t)$, t a variable, n a positive integer.
The ideas underlying the extensions given in parts (d) and (e) will be discussed in Section 7.5.

Exercise 10. Prove that $\mathbb{Q}(\omega, \sqrt[3]{2})$ is the Galois closure of $\mathbb{Q}(\sqrt[3]{2})$ over $\mathbb{Q}$.

Exercise 11. Construct the Galois closure of $\mathbb{Q} \subset \mathbb{Q}(\sqrt[4]{2})$.

Exercise 12. Let $F \subset L$ be an extension of degree 2, where F has characteristic $\neq 2$.
(a) Show that $L = F(\alpha)$, where α is a root of an irreducible polynomial of degree 2.
(b) Show that the minimal polynomial of α over F is separable.
(c) Conclude that $F \subset L$ is a Galois extension with $\mathrm{Gal}(L/F) \simeq \mathbb{Z}/2\mathbb{Z}$.
(d) By completing the square, show that there is $\beta \in L$ such that $L = F(\beta)$ and $\beta^2 \in F$.
For β as in part (d), let $a = \beta^2 \in F$. Then we can write $\beta = \sqrt{a}$. This shows that if F has characteristic $\neq 2$, then every degree 2 extension of F is obtained by taking a square root.

7.2   NORMAL SUBGROUPS AND NORMAL EXTENSIONS

In Chapter 5 we introduced normal extensions, and in abstract algebra you learned about normal subgroups. This section will explain why it is no accident that these concepts have the same name.

A. Conjugate Fields. In high school algebra one calls $2 - \sqrt{3}$ the *conjugate* of $2 + \sqrt{3}$. This terminology is used for subfields as follows.

Definition 7.2.1. *Suppose that we have finite extensions $F \subset K \subset L$. Then, for an automorphism $\sigma \in \mathrm{Gal}(L/F)$, we call*

$$\sigma K = \{\sigma(\alpha) \mid \alpha \in K\}$$

a conjugate field of K.

We should write $\sigma(K)$ instead of σK, but we prefer the latter because it is less cumbersome. Note that σK is a subfield of L, since σ is a field isomorphism.

We can compute the degree of a conjugate field as follows.

Lemma 7.2.2. *Let $F \subset K \subset L$ and $\sigma \in \mathrm{Gal}(L/F)$ be as in Definition 7.2.1. Then $F \subset \sigma K \subset L$ and $[K:F] = [\sigma K:F]$.*

Proof. The inclusion $F \subset \sigma K$ is obvious, since $F \subset K$ and σ is the identity on F. Also, $\sigma(a\alpha) = \sigma(a)\sigma(\alpha) = a\sigma(\alpha)$ when $a \in F$ and $\alpha \in K$. It follows that $\sigma|_K : K \to \sigma K$ is linear over F in the sense of linear algebra. Hence $\sigma|_K$ is an isomorphism of vector spaces over F, so that $[K:F] = \dim_F K = \dim_F \sigma K = [\sigma K:F]$ by the definition of degree given in Section 4.3.   □

Here is an example of conjugate fields.

Example 7.2.3. Consider the extension $\mathbb{Q} \subset \mathbb{Q}(\omega, \sqrt[3]{2})$, where $\omega = e^{2\pi i/3}$. Then we have the following intermediate fields:

(7.3)

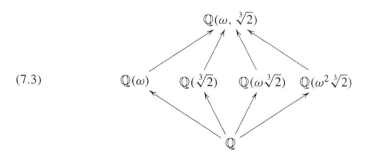

Recall that $\sigma \in \mathrm{Gal}(\mathbb{Q}(\omega, \sqrt[3]{2})/\mathbb{Q})$ is determined uniquely by

(7.4) $\sigma(\omega) \in \{\omega, \omega^2\}$ and $\sigma(\sqrt[3]{2}) \in \{\sqrt[3]{2}, \omega\sqrt[3]{2}, \omega^2\sqrt[3]{2}\}$.

In Exercise 2 of Section 6.2 we showed that all possible combinations of $\sigma(\omega)$ and $\sigma(\sqrt[3]{2})$ actually occur. In Exercise 1 below you will check the following:

- $\mathbb{Q}(\sqrt[3]{2})$ has conjugate fields $\mathbb{Q}(\sqrt[3]{2})$, $\mathbb{Q}(\omega\sqrt[3]{2})$, and $\mathbb{Q}(\omega^2\sqrt[3]{2})$.
- $\mathbb{Q}(\omega)$ equals all of its conjugates.

Later in the section we will explain the second bullet using Galois theory.   ◁▷

We next relate intermediate fields to subgroups of the Galois group.

Lemma 7.2.4. *Suppose that we have finite extensions $F \subset K \subset L$. Then:*
(a) $\mathrm{Gal}(L/K)$ *is a subgroup of* $\mathrm{Gal}(L/F)$.
(b) *If $\sigma \in \mathrm{Gal}(L/F)$, then $\mathrm{Gal}(L/\sigma K) = \sigma \mathrm{Gal}(L/K)\sigma^{-1}$ in $\mathrm{Gal}(L/F)$.*

Proof. Each $\sigma \in \mathrm{Gal}(L/K)$ is an automorphism of L that is the identity on K. Then $\sigma \in \mathrm{Gal}(L/F)$ follows from $F \subset K$; hence $\mathrm{Gal}(L/K) \subset \mathrm{Gal}(L/F)$. Since both are groups under composition, we conclude that $\mathrm{Gal}(L/K)$ is a subgroup.

To prove part (b), let $\gamma \in \sigma\mathrm{Gal}(L/K)\sigma^{-1}$ and $\beta \in \sigma K$. Then $\gamma = \sigma\tau\sigma^{-1}$ for some $\tau \in \mathrm{Gal}(L/K)$, and $\beta = \sigma(\alpha)$ for some $\alpha \in K$. Thus

$$
\begin{aligned}
\gamma(\beta) &= \sigma\tau\sigma^{-1}\big(\sigma(\alpha)\big) \\
&= \sigma\big(\tau(\alpha)\big) \\
&= \sigma(\alpha) = \beta,
\end{aligned}
$$

where the third equality follows because τ is the identity on K. Hence γ is the identity on σK, which implies that $\sigma\mathrm{Gal}(L/K)\sigma^{-1} \subset \mathrm{Gal}(L/\sigma K)$. The opposite inclusion is equally straightforward (see Exercise 2), and the lemma follows.   □

In group theory, a *conjugate* of a subgroup $H \subset G$ is a subgroup of the form gHg^{-1} for some $g \in G$. Thus part (b) of Lemma 7.2.4 tells us that conjugate fields correspond to conjugate subgroups.

B. Normal Subgroups. The first main theorem of this section explains how normal subgroups relate to normal extensions.

Theorem 7.2.5. *Suppose that we have fields $F \subset K \subset L$, where $F \subset L$ is a Galois extension. Then the following conditions are equivalent:*
(a) $K = \sigma K$ *for all $\sigma \in \mathrm{Gal}(L/F)$, that is, K equals all of its conjugates.*
(b) $\mathrm{Gal}(L/K)$ *is a normal subgroup of $\mathrm{Gal}(L/F)$.*
(c) $F \subset K$ *is a Galois extension.*
(d) $F \subset K$ *is a normal extension.*

Proof. We first show that (a) and (b) are equivalent. Proving (a) $\Rightarrow$ (b) is especially easy, for $K = \sigma K$ and Lemma 7.2.4 imply that

$$
\mathrm{Gal}(L/K) = \mathrm{Gal}(L/\sigma K) = \sigma\mathrm{Gal}(L/K)\sigma^{-1}.
$$

Thus $\text{Gal}(L/K)$ is normal in $\text{Gal}(L/F)$ since this holds for all $\sigma \in \text{Gal}(L/F)$. To prove (b) $\Rightarrow$ (a), first note that if $\text{Gal}(L/K)$ is normal and $\sigma \in \text{Gal}(L/F)$, then using Lemma 7.2.4 a second time implies that

$$\text{Gal}(L/K) = \sigma \text{Gal}(L/K)\sigma^{-1} = \text{Gal}(L/\sigma K).$$

However, $K \subset L$ and $\sigma K \subset L$ are Galois extensions by Proposition 7.1.3. Hence

$$K = \text{fixed field of } \text{Gal}(L/K) = \text{fixed field of } \text{Gal}(L/\sigma K) = \sigma K,$$

where the first and third equalities use Theorem 7.1.1.

We next observe that (c) and (d) are equivalent. The implication (c) $\Rightarrow$ (d) is trivial, since every Galois extension is normal and separable. For (d) $\Rightarrow$ (c), note that since $F \subset L$ is Galois, it is also separable, and then any intermediate field $F \subset K \subset L$ is also separable over F. If in addition K is normal over F, then it is normal and separable, and hence Galois.

Finally, we prove that (a) $\Leftrightarrow$ (d). For (a) $\Rightarrow$ (d), let $f \in F[x]$ be irreducible over F with a root $\alpha \in K$. We need to show that f splits completely over K. In the proof of Theorem 7.1.1 we showed that up to a constant, f is the polynomial

$$h(x) = \prod_{i=1}^{r} (x - \alpha_i)$$

from (7.1). Recall that $\alpha_1 = \alpha, \alpha_2, \ldots, \alpha_r$ are the distinct elements of L obtained by applying the elements of $\text{Gal}(L/F)$ to α. Since $\alpha \in K$, each α_i lies in a conjugate field of K. Using (a), we conclude that $\alpha_i \in K$ for all i, so that h and hence f split completely over K.

It remains to show that (d) $\Rightarrow$ (a). Take $\alpha \in K$ and $\sigma \in \text{Gal}(L/F)$, and let p be the minimal polynomial of α over F. By Proposition 6.1.4, $\sigma(\alpha)$ is also a root of p. Since $F \subset K$ is normal, p splits completely over K, which implies that $\sigma(\alpha) \in K$. Thus $\sigma K \subset K$, and then equality follows by Lemma 7.2.2. $\qquad\square$

Here is an example of how this theorem works.

Example 7.2.6. Consider $\mathbb{Q} \subset L = \mathbb{Q}(\omega, \sqrt[3]{2})$ studied in (7.3). By the discussion following (7.4), there are automorphisms $\sigma, \tau \in \text{Gal}(L/\mathbb{Q})$ such that

$$(7.5) \qquad \sigma(\omega) = \omega, \ \sigma(\sqrt[3]{2}) = \omega\sqrt[3]{2} \quad \text{and} \quad \tau(\omega) = \omega^2, \ \tau(\sqrt[3]{2}) = \sqrt[3]{2}.$$

Label the roots of $x^3 - 2$ as $\alpha_1 = \sqrt[3]{2}, \alpha_2 = \omega\sqrt[3]{2}$, and $\alpha_3 = \omega^2\sqrt[3]{2}$, and consider the isomorphism $\text{Gal}(L/\mathbb{Q}) \simeq S_3$ given by the action of the automorphisms on the roots $\alpha_1, \alpha_2, \alpha_3$. Then it is easy to see that

$$\sigma \mapsto (123), \quad \tau \mapsto (23).$$

Since these permutations generate S_3, it follows that σ and τ generate $\text{Gal}(L/\mathbb{Q})$.

Now consider the fields in the diagram (7.3). Each such field K gives a subgroup $\text{Gal}(L/K) \subset \text{Gal}(L/\mathbb{Q})$. Furthermore, in Exercise 3 you will show that

$$(7.6) \qquad K_1 \subset K_2 \subset L \Longrightarrow \text{Gal}(L/K_1) \supset \text{Gal}(L/K_2).$$

In other words, larger fields correspond to smaller Galois groups. Then we claim that for the fields K of (7.3), the map $K \mapsto \mathrm{Gal}(L/K)$ gives the following diagram of subgroups of $\mathrm{Gal}(L/\mathbb{Q})$:

(7.7)

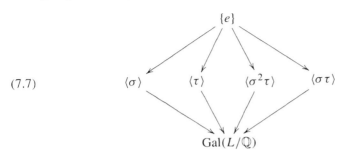

In this diagram, $\langle \sigma \rangle$ is the subgroup generated by σ. Thus $\langle \sigma \rangle = \{e, \sigma, \sigma^2\}$, since σ has order 3. Similarly, $\langle \tau \rangle$, $\langle \sigma^2 \tau \rangle$, $\langle \sigma \tau \rangle$ are subgroups of order 2.

To see how (7.3) gives (7.7), consider the case of $\mathbb{Q}(\omega)$. We know that

$$|\mathrm{Gal}(L/\mathbb{Q}(\omega))| = [L : \mathbb{Q}(\omega)] = [\mathbb{Q}(\omega, \sqrt[3]{2}) : \mathbb{Q}(\omega)] = \frac{[\mathbb{Q}(\omega, \sqrt[3]{2}) : \mathbb{Q}]}{[\mathbb{Q}(\omega) : \mathbb{Q}]} = \frac{6}{2} = 3.$$

Furthermore, (7.5) shows that σ is the identity on $\mathbb{Q}(\omega)$, since $\sigma(\omega) = \omega$. Thus $\sigma \in \mathrm{Gal}(L/\mathbb{Q}(\omega))$, and it follows easily that $\mathrm{Gal}(L/\mathbb{Q}(\omega)) = \langle \sigma \rangle$. In Exercise 4 you will give similar arguments for the other fields in (7.3) to verify that applying $K \mapsto \mathrm{Gal}(L/K)$ to (7.3) gives (7.7).

We can relate (7.3) and (7.7) to Lemma 7.2.4 and Theorem 7.2.5 as follows. First consider $\mathbb{Q}(\omega)$. This is the splitting field of $x^2 + x + 1$ over $\mathbb{Q}$, so that $\mathbb{Q} \subset \mathbb{Q}(\omega)$ is Galois. By Theorem 7.2.5, this implies:

- $\mathbb{Q}(\omega)$ coincides with its conjugates in L, as we saw in Example 7.2.3.
- $\mathrm{Gal}(L/\mathbb{Q}(\omega)) = \langle \sigma \rangle$ is normal in $\mathrm{Gal}(L/\mathbb{Q})$.

We can also go the other way. Under the isomorphism $\mathrm{Gal}(L/\mathbb{Q}) \simeq S_3$, $\langle \sigma \rangle$ maps to the normal subgroup A_3. Thus $\langle \sigma \rangle$ is normal in $\mathrm{Gal}(L/\mathbb{Q})$, so that $\mathbb{Q} \subset \mathbb{Q}(\omega)$ is a Galois extension by Theorem 7.2.5.

We can do a similar analysis for $\mathbb{Q}(\sqrt[3]{2})$. Example 7.2.3 shows that this field has three conjugates in L. Hence

$$(7.8) \qquad \mathrm{Gal}\big(L/\mathbb{Q}(\sqrt[3]{2})\big) = \langle \tau \rangle$$

is not normal in $\mathrm{Gal}(L/\mathbb{Q})$, by Theorem 7.2.5. We can relate the conjugate fields of $\mathbb{Q}(\sqrt[3]{2})$ to the conjugates of $\langle \tau \rangle$ as follows. By Exercise 1, the conjugate fields of $\mathbb{Q}(\sqrt[3]{2})$ are itself, $\sigma \mathbb{Q}(\sqrt[3]{2})$, and $\sigma^2 \mathbb{Q}(\sqrt[3]{2})$. Then Lemma 7.2.4 implies that the Galois groups of L over these fields are the conjugate subgroups

$$\langle \tau \rangle, \ \sigma \langle \tau \rangle \sigma^{-1}, \ \sigma^2 \langle \tau \rangle \sigma^{-2}.$$

One easily checks that these are the subgroups $\langle \tau \rangle$, $\langle \sigma^2 \tau \rangle$, $\langle \sigma \tau \rangle$ from (7.7). ◁▷

In group theory, normal subgroups are important because they lead to quotient groups. Recall that if N is normal in G, then left cosets of N coincide with right cosets, and the set G/N consisting of all cosets of N in G becomes a group under multiplication, the *quotient group*. Theorem 7.2.5 shows that normal subgroups arise naturally in Galois theory.

When $\text{Gal}(L/K) \subset \text{Gal}(L/F)$ is normal, the second main theorem of this section explains how to interpret the quotient group.

Theorem 7.2.7. *Suppose that we have extension fields $F \subset K \subset L$, where $F \subset K$ and $F \subset L$ are Galois. Then $\text{Gal}(L/K)$ is a normal subgroup of $\text{Gal}(L/F)$, and there is a natural isomorphism of groups*

$$\text{Gal}(L/F)/\text{Gal}(L/K) \simeq \text{Gal}(K/F).$$

Proof. If $F \subset K$ is Galois, then $\text{Gal}(L/K)$ is normal in $\text{Gal}(L/F)$ by Theorem 7.2.5. It remains to relate $\text{Gal}(L/F)/\text{Gal}(L/K)$ to $\text{Gal}(K/F)$.

For a fixed $\sigma \in \text{Gal}(L/F)$, the restriction of σ to K gives the isomorphism $\sigma|_K : K \simeq \sigma K$. But Theorem 7.2.5 tells us that $\sigma K = K$, since $F \subset K$ is Galois. It follows that $\sigma|_K$ is an automorphism of K. Since σ is the identity on F, the same is true for $\sigma|_K$ (do you see why?), so that $\sigma|_K \in \text{Gal}(K/F)$.

The mapping $\sigma \mapsto \sigma|_K$ gives a function

$$\Phi : \text{Gal}(L/F) \longrightarrow \text{Gal}(K/F).$$

Furthermore, in Exercise 5 you will verify that for $\sigma, \tau \in \text{Gal}(L/F)$, we have

$$(7.9) \qquad \sigma\tau|_K = (\sigma \circ \tau)|_K = \sigma|_K \circ \tau|_K = \sigma|_K \tau|_K,$$

where the first composition takes place in $\text{Gal}(L/F)$ and the second in $\text{Gal}(K/F)$. This shows that Φ is a group homomorphism.

The kernel of Φ is easy to determine, for if $\sigma \in \text{Gal}(L/F)$, then

$$\sigma \in \text{Ker}(\Phi) \;\Leftrightarrow\; \sigma|_K = 1_K \;\Leftrightarrow\; \sigma \text{ is the identity on } K \;\Leftrightarrow\; \sigma \in \text{Gal}(L/K).$$

Thus $\text{Ker}(\Phi) = \text{Gal}(L/K)$, and then the Fundamental Theorem of Group Homomorphisms implies that Φ induces an isomorphism

$$\text{Gal}(L/F)/\text{Gal}(L/K) \simeq \text{Im}(\Phi) \subset \text{Gal}(K/F).$$

The final step is to show that $\text{Im}(\Phi) = \text{Gal}(K/F)$. The key point is that since all of the extensions involved are Galois extensions, their degrees equal the order of the corresponding Galois groups. Thus

$$\begin{aligned}
|\text{Im}(\Phi)| &= |\text{Gal}(L/F)/\text{Gal}(L/K)| \\
&= \frac{|\text{Gal}(L/F)|}{|\text{Gal}(L/K)|} = \frac{[L:F]}{[L:K]} = [K:F] = |\text{Gal}(K/F)|.
\end{aligned}$$

This shows that $\text{Im}(\Phi) = \text{Gal}(K/F)$ and completes the proof of the theorem. $\quad\square$

Here is a simple example of Theorem 7.2.7.

Example 7.2.8. Consider $\mathbb{Q} \subset \mathbb{Q}(\omega) \subset L = \mathbb{Q}(\omega, \sqrt[3]{2})$. Since $\mathbb{Q} \subset \mathbb{Q}(\omega)$ is Galois and $\mathrm{Gal}(L/\mathbb{Q}(\omega)) = \langle \sigma \rangle$, where σ is as in (7.5), the theorem implies that

$$\mathrm{Gal}(\mathbb{Q}(\omega)/\mathbb{Q}) \simeq \mathrm{Gal}(L/\mathbb{Q})/\langle \sigma \rangle \simeq S_3/A_3 \simeq \mathbb{Z}/2\mathbb{Z}.$$

Note that if τ is as in (7.5), then $\mathrm{Gal}(\mathbb{Q}(\omega)/\mathbb{Q}) = \{1_{\mathbb{Q}(\omega)}, \tau|_{\mathbb{Q}(\omega)}\}$.   ◁▷

Mathematical Notes

There are two ideas in this section to comment on.

■ **The Galois Correspondence.** In Section 7.3 we will see that (7.3) and (7.7) give an example of the Galois correspondence. It is easy to check that (7.7) gives all subgroups of $\mathrm{Gal}(L/\mathbb{Q})$ for $L = \mathbb{Q}(\omega, \sqrt[3]{2})$ (see Exercise 6). Then Corollary 7.3.2 will tell us that (7.3) gives *all* fields between $\mathbb{Q}$ and $\mathbb{Q}(\omega, \sqrt[3]{2})$. This is not obvious—while the subfields in (7.3) are easy to find, how do we know that they are *all* the subfields? This is a good illustration of the power of the Galois correspondence.

We should also mention that (7.6) is also part of the Galois correspondence. The idea behind (7.6) is that as the field K gets larger, the Galois group $\mathrm{Gal}(L/K)$ gets smaller. This explains why the arrows in (7.3) go up while those in (7.7) go down.

■ **Conjugate Fields.** If $F \subset K$ is not Galois, then K will have a certain number of conjugate fields in L (assuming $F \subset K \subset L$ and L is Galois over F). We claim that the *number* of such conjugate fields is related to the *normalizer* of a subgroup.

To see this, we first analyze when a conjugate equals the given field. Suppose that $F \subset K \subset L$, where L is Galois over F, and let $\sigma \in \mathrm{Gal}(L/F)$. In Exercise 7 you will show that

(7.10) $K = \sigma K \iff \mathrm{Gal}(L/K) = \sigma \mathrm{Gal}(L/K)\sigma^{-1}$ in $\mathrm{Gal}(L/F)$.

In group theory, the *normalizer* of a subgroup $H \subset G$ is the set

$$N_G(H) = \{g \in G \mid gHg^{-1} = H\}.$$

One can show that $N_G(H)$ is a subgroup of G, H is a normal subgroup of $N_G(H)$, and $N_G(H)$ is the largest subgroup of G in which H is normal. (You should do Exercise 8 if you're not familiar with normalizers.) From (7.10), it follows that for $\sigma \in \mathrm{Gal}(L/F)$, we have

$$K = \sigma K \iff \sigma \text{ is in the normalizer of } \mathrm{Gal}(L/K) \text{ in } \mathrm{Gal}(L/F).$$

Using this and standard facts about group actions, one can prove that if $F \subset L$ is Galois and K is an intermediate field, then the number of conjugate fields of K in L is given by the index

$$[\mathrm{Gal}(L/F):N] = \frac{|\mathrm{Gal}(L/F)|}{|N|},$$

where N is the normalizer of $\mathrm{Gal}(L/K)$ in $\mathrm{Gal}(L/F)$ (see Exercise 9 for the details).

Historical Notes

In a letter written the night before his fatal duel, Galois describes the concept of normal subgroup as follows [Galois, pp. 173–175]:

> In other words, when a group G contains another group H, the group G can be divided into groups that are obtained by performing the same substitution on the permutations of H, so that $G = H + HS + HS' + \cdots$ and it can also be divided into groups with the same substitutions so that $G = H + TH + T'H + \cdots$. These two decompositions do not ordinarily coincide. When they do coincide, the decomposition is said to be proper.

In modern terms, equality of the decompositions

$$G = H + HS + HS' + \cdots = H + TH + T'H + \cdots$$

implies that the left cosets of H coincide with the right cosets, which is equivalent to the usual definition of normal subgroup. However, we will see in Chapter 12 that Galois's "groups" are not quite what you might think.

Galois was also aware of Theorem 7.2.5, though again his terminology takes some explanation. The details of what Galois knew can be found in [Edwards, pp. 47–66], [2, pp. 80–84], and [7].

The second main theorem of this section, Theorem 7.2.7, concerns quotient groups. Quotient groups weren't defined until much later in the nineteenth century, though hints can be found in the examples worked out in Galois's memoir (see [2, p. 82]). When discussing Galois's work in 1852, Betti made some further progress toward a definition of quotient group, and by the 1880s quotient groups were well established. For us, the key point is that both normality and quotient groups first arose in the context of Galois theory.

Exercises for Section 7.2

Exercise 1. In the diagram (7.3), verify the following.
 (a) $\mathbb{Q}(\sqrt[3]{2})$ has conjugate fields $\mathbb{Q}(\sqrt[3]{2})$, $\mathbb{Q}(\omega\sqrt[3]{2})$, and $\mathbb{Q}(\omega^2\sqrt[3]{2})$.
 (b) $\mathbb{Q}(\omega)$ equals all of its conjugates.

Exercise 2. Complete the proof of Lemma 7.2.4 by showing that

$$\mathrm{Gal}(L/\sigma K) \subset \sigma\,\mathrm{Gal}(L/K)\sigma^{-1}.$$

Exercise 3. Prove (7.6).

Exercise 4. Verify that applying $K \mapsto \mathrm{Gal}(L/K)$ to (7.3) gives (7.7). Don't forget to include the extreme cases $K = \mathbb{Q}$ and $K = L$.

Exercise 5. Prove (7.9) in the proof of Theorem 7.2.7.

Exercise 6. For the extension $\mathbb{Q} \subset L = \mathbb{Q}(\omega, \sqrt[3]{2})$, we listed some subgroups of $\mathrm{Gal}(L/\mathbb{Q})$ in diagram (7.7). Prove that this gives *all* subgroups of $\mathrm{Gal}(L/\mathbb{Q})$.

Exercise 7. Suppose that $F \subset K \subset L$, where L is Galois over F, and let $\sigma \in \mathrm{Gal}(L/F)$. Show that

$$K = \sigma K \iff \mathrm{Gal}(L/K) = \sigma\,\mathrm{Gal}(L/K)\sigma^{-1} \text{ in } \mathrm{Gal}(L/F).$$

Exercise 8. Let H be a subgroup of a group G, and let $N_G(H) = \{g \in G \mid gHg^{-1} = H\}$ be the normalizer of H in G, as defined in the Mathematical Notes.
(a) Prove that $N_G(H)$ is a subgroup of G containing H.
(b) Prove that H is normal in $N_G(H)$.
(c) Let N be a subgroup of G containing H. Prove that H is normal in N if and only if $N \subset N_G(H)$. Do you see why this shows that $N_G(H)$ is the largest subgroup of G in which H is normal?
(d) Prove that H is normal in G if and only if $N_G(H) = G$.

Exercise 9. Let $F \subset L$ be Galois, and suppose that $F \subset K \subset L$ is an intermediate field. The goal of this exercise is to show that the number of conjugates of K in L is

$$[\mathrm{Gal}(L/F) : N] = \frac{|\mathrm{Gal}(L/F)|}{|N|},$$

where N is the normalizer of $\mathrm{Gal}(L/K)$ in $\mathrm{Gal}(L/F)$. More precisely, suppose that the distinct conjugates of K are

$$K = \sigma_1 K, \sigma_2 K, \dots, \sigma_r K,$$

where $\sigma_1 = e$. Then we need to show that $r = [\mathrm{Gal}(L/F) : N]$.
(a) Show that $\mathrm{Gal}(L/F)$ acts on the set of conjugates $\{\sigma_1 K, \sigma_2 K, \dots, \sigma_r K\}$.
(b) Show that the isotropy subgroup of K is the normalizer subgroup N.
(c) Explain how $r = [\mathrm{Gal}(L/F) : N]$ follows from the Fundamental Theorem of Group Actions (Theorem A.4.9 from Appendix A).

Exercise 10. In (7.5), explain why τ is complex conjugation restricted to $\mathbb{Q}(\omega, \sqrt[3]{2})$.

Exercise 11. Consider the extension $\mathbb{Q} \subset L = \mathbb{Q}(\sqrt{2}, \sqrt{3})$.
(a) Show that $\mathrm{Gal}(L/\mathbb{Q}) = \{e, \sigma, \tau, \sigma\tau\}$, where

$$\sigma(\sqrt{2}) = \sqrt{2}, \qquad \sigma(\sqrt{3}) = -\sqrt{3},$$
$$\tau(\sqrt{2}) = -\sqrt{2}, \qquad \tau(\sqrt{3}) = \sqrt{3}.$$

(b) Find all subgroups of $\mathrm{Gal}(L/\mathbb{Q})$, and use this to draw a picture similar to (7.7).
(c) For each subgroup of part (b), determine the corresponding subfield of L and use this to draw a picture similar to (7.3).
(d) Explain why all of the subgroups in part (b) are normal. What does this imply about the subfields in part (c)?
In the next section, we will see that the Galois correspondence implies that the subfields you found in part (c) give *all* subfields of L.

7.3   THE FUNDAMENTAL THEOREM OF GALOIS THEORY

We can now state the main result of this chapter, which describes precisely the relation between subgroups and subfields. Recall that if we are given a finite extension $F \subset L$ and a subgroup $H \subset \mathrm{Gal}(L/F)$, then we have the fixed field

$$L_H = \{\alpha \in L \mid \sigma(\alpha) = \alpha \text{ for all } \sigma \in H\}.$$

In Exercise 1 of Section 7.1 you showed that L_H is a subfield of L containing F. The first part of the *Fundamental Theorem of Galois Theory* goes as follows.

Theorem 7.3.1. *Let $F \subset L$ be a Galois extension.*

(a) *For an intermediate field $F \subset K \subset L$, its Galois group $\mathrm{Gal}(L/K) \subset \mathrm{Gal}(L/F)$ has fixed field*

$$L_{\mathrm{Gal}(L/K)} = K.$$

Furthermore, $|\mathrm{Gal}(L/K)| = [L:K]$ and $[\mathrm{Gal}(L/F):\mathrm{Gal}(L/K)] = [K:F]$.

(b) *For a subgroup $H \subset \mathrm{Gal}(L/F)$, its fixed field $F \subset L_H \subset L$ has Galois group*

$$\mathrm{Gal}(L/L_H) = H.$$

Furthermore, $[L:L_H] = |H|$ and $[L_H:F] = [\mathrm{Gal}(L/F):H]$.

Proof. Part (a) follows easily from earlier results. We are assuming that $F \subset L$ is Galois, so that $K \subset L$ is also Galois by Proposition 7.1.3. Then $K = L_{\mathrm{Gal}(L/K)}$ follows from Theorem 7.1.1 and the definition of Galois extension.

Since $K \subset L$ and $F \subset L$ are both Galois, we have $|\mathrm{Gal}(L/K)| = [L:K]$ and $|\mathrm{Gal}(L/F)| = [L:F]$ by Theorem 6.2.1. Using these equalities and the Tower Theorem (Theorem 4.3.8), we obtain

$$[\mathrm{Gal}(L/F):\mathrm{Gal}(L/K)] = \frac{|\mathrm{Gal}(L/F)|}{|\mathrm{Gal}(L/K)|} = \frac{[L:F]}{[L:K]} = [K:F].$$

This completes the proof of part (a).

To prove part (b), let H be a subgroup of $\mathrm{Gal}(L/F)$. This gives $F \subset L_H \subset L$, and since every $\sigma \in H$ is the identity on L_H, we have

$$(7.11) \qquad H \subset \mathrm{Gal}(L/L_H).$$

To prove that equality occurs, we will give a classic proof using the Theorem of the Primitive Element. Observe that $L_H \subset L$ is a finite separable extension (since $F \subset L$ is), so that $L = L_H(\alpha)$ for some $\alpha \in L$ by Corollary 5.4.2. Then consider

$$h(x) = \prod_{\sigma \in H} (x - \sigma(\alpha)).$$

By standard arguments, the coefficients of h are fixed by H (be sure you can prove this carefully). Thus $h \in L_H[x]$ satisfies $h(\alpha) = 0$. It follows that if $p \in L_H[x]$ is the minimal polynomial of α over L_H, then $p \mid h$. This implies that

$$(7.12) \qquad |H| = \deg(h) \geq \deg(p) = [L_H(\alpha):L_H] = [L:L_H],$$

where the second equality follows because p is the minimal polynomial of α over L_H. Combining this with (7.11), we obtain

$$[L:L_H] \leq |H| \leq |\mathrm{Gal}(L/L_H)|.$$

However, Proposition 7.1.3 implies that $L_H \subset L$ is Galois, so that we also have $|\text{Gal}(L/L_H)| = [L : L_H]$. Then the above inequalities easily imply that

$$|H| = |\text{Gal}(L/L_H)|,$$

and $H = \text{Gal}(L/L_H)$ follows immediately. As in part (a), we conclude that $[\text{Gal}(L/F) : H] = [L_H : F]$. We leave the details as Exercise 1.   □

Here is the second part of the Fundamental Theorem of Galois Theory.

Theorem 7.3.2. *Let $F \subset L$ be a Galois extension. Then the maps between intermediate fields $F \subset K \subset L$ and subgroups $H \subset \text{Gal}(L/F)$ given by*

$$K \mapsto \text{Gal}(L/K),$$

$$H \mapsto L_H$$

reverse inclusions and are inverses of each other. Furthermore, if a subfield K corresponds to a subgroup H under these maps, then K is Galois over F if and only if H is normal in $\text{Gal}(L/F)$, and when this happens, there is a natural isomorphism

$$\text{Gal}(L/F)/H \simeq \text{Gal}(K/F).$$

Proof. Composing the maps one way gives

$$K \mapsto \text{Gal}(L/K) \mapsto L_{\text{Gal}(L/K)} = K$$

by part (a) of Theorem 7.3.1, and going the other way gives

$$H \mapsto L_H \mapsto \text{Gal}(L/L_H) = H$$

by part (b) of the theorem. This proves that the maps $K \mapsto \text{Gal}(L/K)$ and $H \mapsto L_H$ are inverses of each other. The map $K \mapsto \text{Gal}(L/K)$ is inclusion-reversing by (7.6), and $H_1 \subset H_2 \Rightarrow L_{H_1} \supset L_{H_2}$ follows from the definition of fixed field.

The final assertions of the theorem follow from Theorems 7.2.5 and 7.2.7.   □

We next give two examples of the Galois correspondence.

Example 7.3.3. Consider the extension $\mathbb{Q} \subset L = \mathbb{Q}(\omega, \sqrt[3]{2})$, $\omega = e^{2\pi i/3}$. Recall from (7.7) that $\text{Gal}(L/\mathbb{Q}) \simeq S_3$ has subgroups

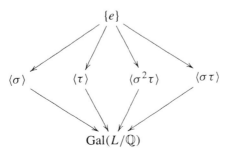

Here, $\sigma, \tau \in \mathrm{Gal}(L/\mathbb{Q})$ are as in (7.5), and Exercise 6 of Section 7.2 shows that these are *all* subgroups of $\mathrm{Gal}(L/\mathbb{Q})$.

By (7.3), the corresponding fixed fields are

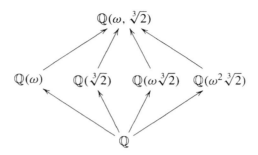

The key point is that according to Theorem 7.3.2, these are *all* subfields of $L = \mathbb{Q}(\omega, \sqrt[3]{2})$ containing $\mathbb{Q}$. Furthermore, note that the discussion of conjugate extensions, normal subgroups, etc., given in Example 7.2.6 verifies the fine details of Theorem 7.3.2.   ◁▷

Here is a slightly more complicated example.

Example 7.3.4. We get a similar picture for the extension $\mathbb{Q} \subset L = \mathbb{Q}(i, \sqrt[4]{2})$. In Exercise 2 you will describe $\mathrm{Gal}(L/\mathbb{Q})$ as follows.

- $\mathrm{Gal}(L/\mathbb{Q})$ is generated by elements σ, τ such that

$$\sigma(i) = i, \ \sigma(\sqrt[4]{2}) = i\sqrt[4]{2} \quad \text{and} \quad \tau(i) = -i, \ \tau(\sqrt[4]{2}) = \sqrt[4]{2}.$$

We also have $o(\sigma) = 4$ and $o(\tau) = 2$.

- $\mathrm{Gal}(L/\mathbb{Q}) \simeq D_8$, where D_8 is the dihedral group of order 8.

We next work out the correspondence between subfields of L and subgroups of $\mathrm{Gal}(L/\mathbb{Q})$.

In Exercise 3 you will show that all subgroups of $\mathrm{Gal}(L/\mathbb{Q})$ are given by

(7.13)

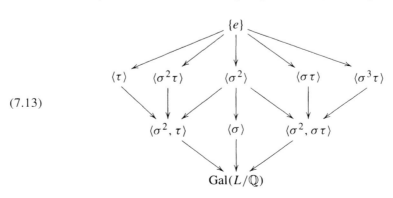

and that the corresponding fixed fields are given by

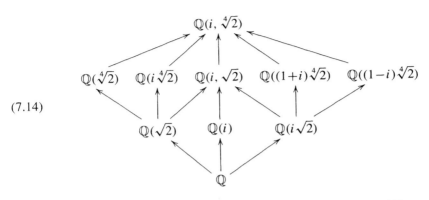

(7.14)

Again, Theorem 7.3.2 implies that this gives *all* subfields of $L = \mathbb{Q}(i, \sqrt[4]{2})$ containing $\mathbb{Q}$. Exercise 3 will work out the details of the Galois correspondence.   ◁▷

Finally, let us give an interesting application of the Galois correspondence.

Proposition 7.3.5. *Let $F \subset L$ be a finite separable extension. Then there are only finitely many intermediate fields $F \subset K \subset L$.*

Proof. By Proposition 7.1.7 there is an extension $L \subset M$ such that $F \subset M$ is Galois. Then Theorem 7.3.2 implies that subfields of M containing F correspond to subgroups of $\mathrm{Gal}(M/F)$. Since $\mathrm{Gal}(M/F)$ is finite, it has only finitely many subgroups, so that there are only finitely many subfields of M containing F. Since $F \subset L \subset M$, it follows in particular that there are only finitely many intermediate fields between F and L.   □

In contrast, there are finite purely inseparable extensions that have infinitely many intermediate fields. Here is a classic example.

Example 7.3.6. Let k be a field of characteristic p, and consider the extension

$$F = k(t, u) \subset L,$$

where L is the splitting field of $(x^p - t)(x^p - u) \in F[x]$. This extension was studied in Example 5.4.4, where we showed that it has no primitive element. Furthermore, Exercise 4 of Section 5.4 showed that $F \subset L$ is purely inseparable and $L = F(\alpha, \beta)$, where $\alpha^p = t$ and $\beta^p = u$.

In Exercise 5 of Section 5.4 you proved that the intermediate fields

(7.15) $F \subset F(\alpha + \lambda \beta) \subset L$

are all distinct as λ ranges over the distinct elements of F. Since F is infinite, we see that there are infinitely many intermediate fields.

In Exercise 4, you will show that $\mathrm{Gal}(L/F) = \{1_L\}$. This means that in particular, $\mathrm{Gal}(L/F)$ has only one subgroup, namely $\{e\}$, yet $F \subset L$ has the infinitely many intermediate fields given by (7.15). ◁▷

This example shows that the Galois correspondence can break down spectacularly for purely inseparable splitting fields.

Exercises for Section 7.3

Exercise 1. Complete the proof of Theorem 7.3.1 by showing that $[\mathrm{Gal}(L/F) : H] = [L_H : F]$ for all subgroups $H \subset \mathrm{Gal}(L/F)$.

Exercise 2. Consider $\mathbb{Q} \subset L = \mathbb{Q}(i, \sqrt[4]{2})$.
(a) Show that there are $\sigma, \tau \in \mathrm{Gal}(L/\mathbb{Q})$ such that

$$\sigma(i) = i, \ \sigma(\sqrt[4]{2}) = i\sqrt[4]{2} \quad \text{and} \quad \tau(i) = -i, \ \tau(\sqrt[4]{2}) = \sqrt[4]{2}.$$

(b) Prove that $o(\sigma) = 4$, $o(\tau) = 2$ and that τ is complex conjugation restricted to L.
(c) Prove that σ and τ generate $\mathrm{Gal}(L/\mathbb{Q})$.
(d) Show that $\mathrm{Gal}(L/\mathbb{Q}) \simeq D_8$, where D_8 is the dihedral group of order 8.

Exercise 3. Let $L = \mathbb{Q}(i, \sqrt[4]{2})$ and $\sigma, \tau \in \mathrm{Gal}(L/\mathbb{Q})$ be as in Exercise 2 and Example 7.3.4.
(a) Show that all subgroups of $\mathrm{Gal}(L/\mathbb{Q})$ are given by (7.13).
(b) Show that the corresponding fixed fields are given by (7.14).
(c) Determine which subgroups in part (a) are normal in $\mathrm{Gal}(L/\mathbb{Q})$, and for those that are normal, construct a polynomial whose splitting field is the corresponding fixed field.
(d) For the subfields in part (b) that are not Galois over $\mathbb{Q}$, find all of their conjugate fields. Also describe the conjugates of their corresponding groups.

Exercise 4. Prove that the extension $F \subset L$ of Example 7.3.6 has $\mathrm{Gal}(L/F) = \{1_L\}$.

Exercise 5. Consider the extension $F = \mathbb{C}(t^4) \subset L = \mathbb{C}(t)$, where t is a variable.
(a) Show that L is the splitting field of $x^4 - t^4 \in F[x]$ over F.
(b) Show that $x^4 - t^4$ is irreducible over F.
(c) Show that $\mathrm{Gal}(L/F) \simeq \mathbb{Z}/4\mathbb{Z}$.
(d) Similar to what you did in Exercise 3, determine all subgroups of $\mathrm{Gal}(L/F)$ and the corresponding intermediate fields between F and L.
We will say more about this type of extension in Section 7.5.

Exercise 6. This exercise will work out the Galois correspondence for the splitting field L of $x^4 - 4x^2 + 2$ over $\mathbb{Q}$. In Exercise 6 of Section 5.1 you showed that $L = \mathbb{Q}(\sqrt{2 + \sqrt{2}})$ and that $\mathrm{Gal}(L/\mathbb{Q}) \simeq \mathbb{Z}/4\mathbb{Z}$. Now, similar to Example 7.3.4, determine all subgroups of $\mathrm{Gal}(L/\mathbb{Q})$ and the corresponding intermediate fields of $\mathbb{Q} \subset L$.

Exercise 7. Let $\zeta_7 = e^{2\pi i/7}$, and consider the extension $\mathbb{Q} \subset L = \mathbb{Q}(\zeta_7)$.
(a) Show that L is the splitting field of $f = x^6 + x^5 + x^4 + x^3 + x^2 + x + 1$ over $\mathbb{Q}$ and that f is the minimal polynomial of ζ_7.
(b) Let $(\mathbb{Z}/7\mathbb{Z})^*$ be the group of nonzero congruence classes modulo 7 under multiplication. By Exercise 4 of Section 6.2 there is a group isomorphism $\mathrm{Gal}(L/\mathbb{Q}) \simeq (\mathbb{Z}/7\mathbb{Z})^*$. Let $H \subset (\mathbb{Z}/7\mathbb{Z})^*$ be the subgroup generated by the congruence class of -1. Prove that $\mathbb{Q}(\zeta_7 + \zeta_7^{-1})$ is the fixed field of the subgroup of $\mathrm{Gal}(L/\mathbb{Q})$ corresponding to H.

Exercise 8. Let $\alpha = \zeta_7 + \zeta_7^{-1}$, where $\zeta_7 = e^{2\pi i/7}$.
(a) Show that the minimal polynomial of α over $\mathbb{Q}$ is $x^3 + x^2 - 2x - 1$.
(b) Use Exercise 7 to show that the splitting field of $x^3 + x^2 - 2x - 1$ over $\mathbb{Q}$ is a Galois extension of degree 3 with Galois group isomorphic to $\mathbb{Z}/3\mathbb{Z}$.

Exercise 9. Let F be a field of characteristic different from 2, and let $F \subset L$ be a finite extension. Prove that the following are equivalent:
(a) L is a Galois extension of F with $\text{Gal}(L/F) \simeq \mathbb{Z}/2\mathbb{Z} \times \mathbb{Z}/2\mathbb{Z}$.
(b) L is the splitting field of a polynomial of the form $(x^2 - a)(x^2 - b)$, where $a, b \in F$ but $\sqrt{a}, \sqrt{b}, \sqrt{ab}$ do not lie in F.

Exercise 10. Suppose that $\alpha, \beta \in \mathbb{C}$ are algebraic of degree 2 over $\mathbb{Q}$ (i.e., they are both roots of irreducible quadratic polynomials in $\mathbb{Q}[x]$). Prove that the following are equivalent:
(a) $\mathbb{Q}(\alpha) = \mathbb{Q}(\beta)$.
(b) $\alpha = a + b\beta$ for some $a, b \in \mathbb{Q}, b \neq 0$.
(c) $\alpha + \beta$ is the root of a quadratic polynomial in $\mathbb{Q}[x]$.

Exercise 11. Let $F \subset L$ be a Galois extension, and let $F \subset K \subset L$ be an intermediate field. Then let N be the normalizer (as defined in the Mathematical Notes to Section 7.2) of $\text{Gal}(L/K) \subset \text{Gal}(L/F)$. Prove that the fixed field L_N is the smallest subfield of K such that K is Galois over the subfield.

Exercise 12. Let H be a subgroup of a group G, and let $N = \bigcap_{g \in G} gHg^{-1}$.
(a) Show that N is a normal subgroup of G.
(b) Show that N is the largest normal subgroup of G contained in H.

Exercise 13. Let $F \subset L$ be a Galois extension, and let $F \subset K \subset L$ be an intermediate field. If we apply the construction of Exercise 12 to $\text{Gal}(L/K) \subset \text{Gal}(L/F)$, then we obtain a normal subgroup $N \subset \text{Gal}(L/F)$. Prove that the fixed field L_N is the Galois closure of K.

Exercise 14. Prove the implication (b) $\Rightarrow$ (a) of Theorem 6.5.5.

Exercise 15. Let p be prime. Consider the extension $\mathbb{Q} \subset L = \mathbb{Q}(\zeta_p, \sqrt[p]{2})$ discussed in Section 6.4. There, we showed that $\text{Gal}(L/\mathbb{Q}) \simeq \text{AGL}(1, \mathbb{F}_p)$. The group $\text{AGL}(1, \mathbb{F}_p)$ has two subgroups defined as follows:

$$T = \{\gamma_{1,b} \mid b \in \mathbb{F}_p\} \quad \text{and} \quad D = \{\gamma_{a,0} \mid a \in \mathbb{F}_p^*\},$$

where $\gamma_{a,b}(u) = au + b, u \in \mathbb{F}_p$. Let T' and D' be the corresponding subgroups of $\text{Gal}(L/\mathbb{Q})$.
(a) Show that the fixed field of T' is $\mathbb{Q}(\zeta_p)$.
(b) What is the fixed field of D'?

7.4 FIRST APPLICATIONS

This section is devoted to three applications of the Galois correspondence.

A. The Discriminant. The discriminant $\Delta(f) \in F$ of a nonconstant monic polynomial $f \in F[x]$ was defined in Section 2.4. There, we showed that if f has degree $n \geq 2$ and $f = (x - \alpha_1) \cdots (x - \alpha_n)$ in a splitting field L of f, then

$$\Delta(f) = \prod_{i < j}(\alpha_i - \alpha_j)^2 \in F.$$

In Section 5.3, we saw that f is separable if and only if $\Delta(f) \neq 0$. We define

$$\sqrt{\Delta(f)} = \prod_{i<j}(\alpha_i - \alpha_j) \in L.$$

Note that while $\Delta(f)$ is uniquely determined by f, the above square root depends on how the roots are labeled.

If $f \in F[x]$ is separable, then by Section 6.3 the action of the Galois group on $\alpha_1, \ldots, \alpha_n$ gives a one-to-one group homomorphism

$$\mathrm{Gal}(L/F) \longrightarrow S_n.$$

In S_n we also have the alternating group $A_n \subset S_n$. Our first result shows that $\sqrt{\Delta(f)}$ controls the relation between A_n and $\mathrm{Gal}(L/F)$.

Theorem 7.4.1. *Let f and $F \subset L$ be as above, and assume that the characteristic of F is different from* 2.
(a) *If $\sigma \in \mathrm{Gal}(L/F)$ corresponds to $\tau \in S_n$, then*

$$\sigma\left(\sqrt{\Delta(f)}\right) = \mathrm{sgn}(\tau)\sqrt{\Delta(f)}.$$

(b) *The image of $\mathrm{Gal}(L/F)$ lies in A_n if and only if $\sqrt{\Delta(f)} \in F$ (i.e., $\Delta(f)$ is the square of an element of F).*

Proof. The result is trivial if $n = 1$. Hence we may assume that $n \geq 2$. Recall from Proposition 2.4.1 that $\sqrt{\Delta} = \prod_{i<j}(x_i - x_j) \in F[x_1, \ldots, x_n]$ has the property that

$$(7.16) \qquad\qquad \tau \cdot \sqrt{\Delta} = \mathrm{sgn}(\tau)\sqrt{\Delta}$$

for all $\tau \in S_n$. This gives the identity

$$\prod_{i<j}(x_{\tau(i)} - x_{\tau(j)}) = \mathrm{sgn}(\tau)\prod_{i<j}(x_i - x_j)$$

in $F[x_1, \ldots, x_n]$. Since the evaluation map $F[x_1, \ldots, x_n] \to L$ sending x_i to α_i is a ring homomorphism (Theorem 2.1.2), it follows that

$$\prod_{i<j}(\alpha_{\tau(i)} - \alpha_{\tau(j)}) = \mathrm{sgn}(\tau)\prod_{i<j}(\alpha_i - \alpha_j) = \mathrm{sgn}(\tau)\sqrt{\Delta(f)}.$$

However, we also have $\sigma(\alpha_i) = \alpha_{\tau(i)}$, which implies that

$$\prod_{i<j}(\alpha_{\tau(i)} - \alpha_{\tau(j)}) = \sigma\left(\sqrt{\Delta(f)}\right).$$

This completes the proof of part (a). For part (b), observe that $F \subset L$ is Galois, so that F is the fixed field of $\mathrm{Gal}(L/F)$. Combining this with (a), we obtain

$$\sqrt{\Delta(f)} \in F \iff \sigma\left(\sqrt{\Delta(f)}\right) = \sqrt{\Delta(f)} \text{ for all } \sigma \in \mathrm{Gal}(L/F)$$
$$\iff \mathrm{sgn}(\tau)\sqrt{\Delta(f)} = \sqrt{\Delta(f)} \text{ for all } \sigma \in \mathrm{Gal}(L/F),$$

where τ is the image of σ under the map $\mathrm{Gal}(L/F) \to S_n$. Since $\Delta(f) \neq 0$ and F has characteristic $\neq 2$, the last condition is equivalent to $\mathrm{sgn}(\tau) = 1$ for all τ coming from $\mathrm{Gal}(L/F)$. Then we are done, since $\mathrm{sgn}(\tau) = 1$ if and only if $\tau \in A_n$. $\qquad\square$

This result allows us to compute the Galois group of an irreducible cubic.

Proposition 7.4.2. *Let $f \in F[x]$ be a monic irreducible separable cubic, where F has characteristic $\neq 2$. If L is the splitting field of f over F, then*

$$\mathrm{Gal}(L/F) \simeq \begin{cases} \mathbb{Z}/3\mathbb{Z}, & \text{if } \Delta(f) \text{ is a square in } F, \\ S_3, & \text{otherwise.} \end{cases}$$

Proof. Exercise 6 of Section 6.2 implies that $|\mathrm{Gal}(L/F)|$ is a multiple of 3, since f is irreducible and separable. We also have the one-to-one map $\mathrm{Gal}(L/F) \to S_3$. Since the only subgroups of S_3 of order divisible by 3 are S_3 and A_3, the proposition follows easily from Theorem 7.4.1. We leave the details as Exercise 1. $\qquad\square$

Here is a simple example of this proposition.

Example 7.4.3. Consider $f = x^3 + x^2 - 2x - 1 \in \mathbb{Q}[x]$. It is easy to see that f is irreducible over $\mathbb{Q}$ and hence is separable, since we are in characteristic 0. Using the method discussed in Section 5.3 one computes that

$$\Delta(f) = 49 = 7^2.$$

By Proposition 7.4.2 the Galois group of f over $\mathbb{Q}$ is cyclic of order 3.

In Exercise 2 you will compute the Galois groups of some other cubics, and in Chapter 13 we will compute the Galois groups of quartics and quintics.

B. The Universal Extension.

Consider the universal extension in degree n,

$$(7.17) \qquad K = F(\sigma_1, \ldots, \sigma_n) \subset L = F(x_1, \ldots, x_n),$$

where as usual $\sigma_1, \ldots, \sigma_n$ are the elementary symmetric polynomials. Recall from Section 6.4 that this is the splitting field of the universal polynomial of degree n,

$$\tilde{f} = x^n - \sigma_1 x^{n-1} + \cdots + (-1)^n \sigma_n = \prod_{i=1}^n (x - x_i).$$

Theorem 6.4.1 implies that $K \subset L$ is Galois with Galois group $\mathrm{Gal}(L/K) \simeq S_n$. Furthermore, if we identify $\mathrm{Gal}(L/K)$ with S_n, then $\sigma \in S_n$ becomes the automorphism of $L = F(x_1, \ldots, x_n)$ that permutes the x_i according to σ.

Then the Fundamental Theorem of Galois Theory implies the following facts about symmetric functions. As above, set

$$\sqrt{\Delta} = \prod_{i<j} (x_i - x_j).$$

Theorem 7.4.4. *Let $f \in F(x_1, \ldots, x_n)$.*

(a) *f is invariant under S_n if and only if $f \in F(\sigma_1, \ldots, \sigma_n)$.*

(b) *Assume that F has characteristic $\neq 2$. Then f is invariant under A_n if and only if there are $A, B \in F(\sigma_1, \ldots, \sigma_n)$ such that*

$$f = A + B\sqrt{\Delta}.$$

Proof. Since (7.17) is a Galois extension, Theorem 7.3.1 implies that K is the fixed field of $\text{Gal}(L/K) = S_n$ acting on L. This proves part (a).

To prove part (b), let $M = L_{A_n}$ be the fixed field of A_n acting on L. Since A_n has index 2 in $\text{Gal}(L/K) = S_n$, Theorem 7.3.1 implies that $K \subset M$ is an extension of degree 2. However, (7.16) shows that $\sqrt{\Delta} \in M$, so that we have

$$K \subset K(\sqrt{\Delta}) \subset M.$$

By the Tower Theorem, it follows that

$$2 = [M : K] = [M : K(\sqrt{\Delta})][K(\sqrt{\Delta}) : K].$$

But (7.16) also shows that $\sqrt{\Delta} \notin K$ (since F and hence K have characteristic $\neq 2$). We conclude that $K(\sqrt{\Delta}) = M$. Finally, since $\sqrt{\Delta}$ is a primitive element of the degree 2 extension $K \subset M$, Proposition 4.3.4 implies that

$$M = \{A + B\sqrt{\Delta} \mid A, B \in K\}.$$

This completes the proof of part (b). $\square$

In Chapter 2 we proved part (a) of Theorem 7.4.4 by first considering the case when f is a polynomial in $x_1, \ldots, x_n$ (Theorem 2.2.2) and then doing the case when f is a rational function (Exercise 11 of Section 2.2). The proof given above is much shorter. This illustrates nicely the power of Galois theory.

On the other hand, if f is a polynomial in $x_1, \ldots, x_n$, then part (a) of Theorem 7.4.4 does not assert that f is a polynomial in the σ_i—the theorem only tells us that f is in the field $F(\sigma_1, \ldots, \sigma_n)$. The point is that Galois theory deals with *fields* rather than *rings*. In Exercise 3 you will study what happens in part (b) of Theorem 7.4.4 when f is a polynomial in $x_1, \ldots, x_n$.

C. The Inverse Galois Problem.

The Galois group $\text{Gal}(L/F)$ of a finite extension $F \subset L$ is a finite group. But what finite groups can arise in this way? We will discuss two aspects of this question.

We first note that a finite group G of order n is isomorphic to a subgroup of S_n. The easiest way to see this is to label the elements of G as $g_1, \ldots, g_n$. Then the group operation on G can be represented by its *Cayley table*, where the entry in row

i and column j is $g_i g_j$. For example, the Cayley table of S_3 is

	e	(123)	(132)	(12)	(13)	(23)
e	e	(123)	(132)	(12)	(13)	(23)
(123)	(123)	(132)	e	(13)	(23)	(12)
(132)	(132)	e	(123)	(23)	(12)	(13)
(12)	(12)	(23)	(13)	e	(132)	(123)
(13)	(13)	(12)	(23)	(123)	e	(132)
(23)	(23)	(13)	(12)	(132)	(123)	e

(7.18)

In 1854 Cayley observed that every row of the Cayley table of G is a permutation of the elements of G (do Exercise 4 if you didn't prove this in your abstract algebra course). It follows that for the ith row, there is a permutation $\sigma_i \in S_n$ (where $n = |G|$) such that

(7.19) $$g_i g_j = g_{\sigma_i(j)}.$$

In Exercise 5 you will compute the six elements of S_6 given by the rows of (7.18), and in Exercise 6 you will show that in general, the map $G \to S_n$ given by $g_i \mapsto \sigma_i$ is a one-to-one group homomorphism. It follows that G is isomorphic to a subgroup of S_n. Combining this with the Galois correspondence for the universal extension in degree n gives the following nice result.

Theorem 7.4.5. *Given a finite group G, there is a Galois extension whose Galois group is isomorphic to G.*

Proof. Let G be a finite group of order n, and let F be an arbitrary field. We know that the universal extension in degreee n,

$$K = F(\sigma_1, \ldots, \sigma_n) \subset L = F(x_1, \ldots, x_n),$$

is a Galois extension with Galois group $\mathrm{Gal}(L/K) \simeq S_n$.

Since G is isomorphic to a subgroup of S_n, it follows that G is also isomorphic to a subgroup $H \subset \mathrm{Gal}(L/K)$. Then the fixed field of H is an intermediate field $K \subset L_H \subset L$, and the Fundamental Theorem of Galois Theory tells us that $L_H \subset L$ is a Galois extension with Galois group

$$\mathrm{Gal}(L/L_H) = H \simeq G.$$

This shows that $L_H \subset L$ is the desired extension. $\square$

However, this is not the end of the story, for in the extension $L_H \subset L$ constructed in Theorem 7.4.5, the smaller field L_H depends on the group. In explicit examples, one is often interested in Galois groups of polynomials over $\mathbb{Q}$. Thus the question is: which finite groups can occur as the Galois group of a finite extension of $\mathbb{Q}$? This is called *the inverse Galois problem for* $\mathbb{Q}$.

There has been a lot of work on this problem, starting with Hilbert, who used his irreducibility theorem (mentioned in the Mathematical Notes to Section 6.4) to show that for every $n \geq 1$, both S_n and A_n can occur as Galois groups of Galois extensions of $\mathbb{Q}$. In Section 6.4 we also gave the example of $x^n - x - 1$, whose Galois group over $\mathbb{Q}$ is S_n for $n \geq 2$. Another example is the polynomial

$$p_n(x) = 1 + x + \tfrac{1}{2}x^2 + \tfrac{1}{6}x^3 + \cdots + \tfrac{1}{n!}x^n$$

obtained by truncating the power series for e^x. In 1930, Schur proved that the Galois group of p_n over $\mathbb{Q}$ is A_n when $n \equiv 0 \bmod 4$ and S_n otherwise (see [Chebotarev, p. 398] for references and further examples). In the case of a prime p, the paper [6] uses elementary methods to construct a polynomial of degree p whose Galois group over $\mathbb{Q}$ is S_p. The book *Inverse Galois Theory* [4] discusses some of the powerful methods used to study this unsolved problem in general.

Historical Notes

The universal extension in degree n,

$$K = F(\sigma_1, \ldots, \sigma_n) \subset L = F(x_1, \ldots, x_n),$$

plays a central (though somewhat implicit) role in Lagrange's 1770 study of solving equations by radicals (see the Historical Notes to Section 1.2). In this treatise, Lagrange proves many interesting results, including a theorem that (in modern terminology) says that if $f \in L$, then

$$K(f) = L_{\mathrm{Gal}(L/K(f))}.$$

For us, this is part of the Galois correspondence, yet Lagrange proved this result sixty years before Galois. We will discuss Lagrange's work in more detail in Chapter 12.

Exercises for Section 7.4

Exercise 1. Give a detailed proof of Proposition 7.4.2.

Exercise 2. Compute the Galois groups of the following cubic polynomials:
(a) $x^3 - 4x + 2$ over $\mathbb{Q}$.
(b) $x^3 - 4x + 2$ over $\mathbb{Q}(\sqrt{37})$.
(c) $x^3 - 3x + 1$ over $\mathbb{Q}$.
(d) $x^3 - t$ over $\mathbb{C}(t)$, t a variable.
(e) $x^3 - t$ over $\mathbb{Q}(t)$, t a variable.

Exercise 3. This exercise will study part (b) of Theorem 7.4.4 when f is a *polynomial* in $x_1, \ldots, x_n$ that is invariant under A_n. The theorem implies that $f = A + B\sqrt{\Delta}$ for some $A, B \in F(\sigma_1, \ldots, \sigma_n)$. You will prove that A and B are polynomials in the σ_i.
(a) Show that $f + (12) \cdot f = 2A$.
(b) In part (a), the left-hand side is a polynomial while the right-hand side is a symmetric rational function. Use Theorem 2.2.2 to conclude that A is a polynomial in the σ_i.
(c) Let P denote the product of $f - A$ and $(12) \cdot (f - A)$. Show that $P = -B^2 \Delta$.
(d) Let $B = u/v$, where $u, v \in F[\sigma_1, \ldots, \sigma_n]$ are relatively prime (recall that $F[\sigma_1, \ldots, \sigma_n]$ is a UFD). In Exercise 8 of Section 2.4 you showed that Δ is irreducible in $F[\sigma_1, \ldots, \sigma_n]$.

Use this and the equation $v^2 P = -u^2 \Delta$ to show that v must be constant. This will prove that $B \in F[\sigma_1, \ldots, \sigma_n]$.

Exercise 4. Let G be a group of order n, and fix $g \in G$.
(a) Show that the map $G \to G$ defined by $h \mapsto gh$ is one to one and onto.
(b) Explain why part (a) implies that each row of the Cayley table of G is a permutation of the elements of G.
(c) Write $G = \{g_1, \ldots, g_n\}$, and fix $g_i \in G$. Use part (a) to show the existence of $\sigma_i \in S_n$ satisfying $g_i g_j = g_{\sigma_i(j)}$ as in (7.19).

Exercise 5. Label the elements of S_3 as $g_1 = e$, $g_2 = (123)$, $g_3 = (132)$, $g_4 = (12)$, $g_5 = (13)$, and $g_6 = (23)$. Write down the six permutations $\sigma_i \in S_6$ defined by the rows of the Cayley table (7.18).

Exercise 6. In the situation of Exercise 4, let $G = \{g_1, \ldots, g_n\}$, and assume that $g_i g_j = g_k$. Let $\sigma_i, \sigma_j, \sigma_k \in S_n$ be the corresponding permutations determined by (7.19).
(a) Prove that $\sigma_i \sigma_j = \sigma_k$.
(b) Prove that the map $G \to S_n$ defined by $g_i \mapsto \sigma_i$ is a one-to-one group homomorphism.

Exercise 7. Let f and $F \subset L$ satisfy the hypothesis of Proposition 7.4.2, and assume that $\sqrt{\Delta(f)} \notin F$. Prove that $\mathrm{Gal}\left(L/F(\sqrt{\Delta(f)})\right) = \mathbb{Z}/3\mathbb{Z}$ and that f is irreducible over $F(\sqrt{\Delta(f)})$.

7.5 AUTOMORPHISMS AND GEOMETRY (OPTIONAL)

This optional section will explore some unexpected connections between geometry and Galois theory.

A. Groups of Automorphisms.
The theory developed in Chapters 6 and 7 begins with an extension $F \subset L$ and then considers its Galois group $\mathrm{Gal}(L/F)$. We will now change our point of view and instead begin with a field L and a finite group G of automorphisms of L.

Here are two simple examples.

Example 7.5.1. Let $L = \mathbb{Q}(\zeta_7)$, and consider the automorphism σ of L that maps ζ_7 to ζ_7^2. Since $2^3 \equiv 1 \bmod 7$, we see that $G = \langle \sigma \rangle = \{1_L, \sigma, \sigma^2\} \simeq \mathbb{Z}/3\mathbb{Z}$ is a group of automorphisms of L.   ◁▷

Example 7.5.2. Let $L = \mathbb{C}(t)$, where t is a variable. It is easy to see that $t \mapsto 1/t$ induces an automorphism σ of L. This gives the group $G = \langle \sigma \rangle = \{1_L, \sigma\} \simeq \mathbb{Z}/2\mathbb{Z}$ of automorphisms of L.   ◁▷

Given a finite group G of automorphisms of a field L, we get the fixed field

$$L_G \subset L.$$

Furthermore, the definition of L_G easily implies that $G \subset \mathrm{Gal}(L/L_G)$. However, much more is true, as we will now prove.

Theorem 7.5.3. *Let G be a finite group of automorphisms of a field L. Then:*
(a) $[L : L_G] = |G|$.
(b) $L_G \subset L$ *is a Galois extension.*
(c) $\mathrm{Gal}(L/L_G) = G$.

Proof. Let $n = |G|$. We first claim that

$$(7.20) \qquad\qquad [L : L_G] \le n.$$

If (7.20) is false, then we can find $\alpha_1, \ldots, \alpha_{n+1} \in L$ that are linearly independent over L_G. Also let the elements of G be $\sigma_1 = 1_L, \sigma_2, \ldots, \sigma_n$. Then, given unknowns $x_1, \ldots, x_{n+1}$, consider the equations

$$
\begin{aligned}
x_1 \sigma_1(\alpha_1) + \cdots + x_{n+1} \sigma_1(\alpha_{n+1}) &= 0, \\
x_1 \sigma_2(\alpha_1) + \cdots + x_{n+1} \sigma_2(\alpha_{n+1}) &= 0,
\end{aligned}
$$

$$(7.21) \qquad\qquad\qquad\qquad\qquad \vdots$$

$$x_1 \sigma_n(\alpha_1) + \cdots + x_{n+1} \sigma_n(\alpha_{n+1}) = 0.$$

This is a system of n homogeneous equations in $n + 1$ unknowns with coefficients in the field L.

Since the number of unknowns exceeds the number of equations, (7.21) must have a nontrivial solution $(x_1, \ldots, x_{n+1}) = (\beta_1, \ldots, \beta_{n+1})$ in L^{n+1}. Among all nontrivial solutions in L^{n+1}, pick one that has the fewest nonzero β_i's. Relabeling, we can write this solution as $(\beta_1, \ldots, \beta_r, 0, \ldots, 0)$, where $\beta_1, \ldots, \beta_r$ are nonzero. Then being a solution of (7.21) means that

$$\beta_1 \sigma_i(\alpha_1) + \cdots + \beta_r \sigma_i(\alpha_r) = 0, \quad i = 1, \ldots, n.$$

Observe that $r > 1$ since $\beta_1 \ne 0$ and $\alpha_1 \ne 0$. Furthermore, we may divide by β_1 and relabel $\beta_2, \ldots, \beta_r$ to obtain

$$(7.22) \qquad \sigma_i(\alpha_1) + \beta_2 \sigma_i(\alpha_2) + \cdots + \beta_r \sigma_i(\alpha_r) = 0, \quad i = 1, \ldots, n.$$

Since $\sigma_1 = 1_L$, setting $i = 1$ in (7.22) gives

$$\alpha_1 + \beta_2 \alpha_2 + \cdots + \beta_r \alpha_r = 0.$$

Hence $\beta_2, \ldots, \beta_r$ cannot all lie in L_G since $\alpha_1, \ldots, \alpha_r$ are linearly independent over L_G by assumption. Relabeling, we may assume that $\beta_r \notin L_G$, so that $\sigma(\beta_r) \ne \beta_r$ for some $\sigma \in G$. Now apply this σ to (7.22) to obtain

$$\sigma\sigma_i(\alpha_1) + \sigma(\beta_2) \sigma\sigma_i(\alpha_2) + \cdots + \sigma(\beta_r) \sigma\sigma_i(\alpha_r) = 0, \quad i = 1, \ldots, n.$$

Since $G = \{\sigma_1, \ldots, \sigma_n\}$ is a group under composition, the product $\sigma\sigma_i$ gives all elements of G as we vary σ_i. Thus we obtain

$$(7.23) \qquad \sigma_i(\alpha_1) + \sigma(\beta_2) \sigma_i(\alpha_2) + \cdots + \sigma(\beta_r) \sigma_i(\alpha_r) = 0, \quad i = 1, \ldots, n.$$

Now multiply the equations of (7.22) by $\sigma(\beta_r)$ and the equations of (7.23) by β_r and subtract. This choice of multipliers causes the coefficients of $\sigma_i(\alpha_r)$ to cancel. Hence we are left with the equations

$$\left(\sigma(\beta_r) - \beta_r\right)\sigma_i(\alpha_1) + \left(\sigma(\beta_r)\beta_2 - \beta_r\sigma(\beta_2)\right)\sigma_i(\alpha_2) + \cdots$$
$$+ \left(\sigma(\beta_r)\beta_{r-1} - \beta_r\sigma(\beta_{r-1})\right)\sigma_i(\alpha_{r-1}) = 0, \quad i = 1, \ldots, n.$$

Thus the $(n+1)$-tuple

$$\left(\sigma(\beta_r) - \beta_r, \sigma(\beta_r)\beta_2 - \beta_r\sigma(\beta_2), \ldots, \sigma(\beta_r)\beta_{r-1} - \beta_r\sigma(\beta_{r-1}), 0, \ldots, 0\right)$$

is a solution of (7.21). It has at most $r - 1$ nonzero entries and is nontrivial, since $\sigma(\beta_r) \neq \beta_r$. This contradicts our choice of r and completes the proof of (7.20).

It follows that $L_G \subset L$ is a finite extension. Furthermore, we have

$$L_G \subset L_{\mathrm{Gal}(L/L_G)} \subset L_G,$$

where the first inclusion follows because elements of $\mathrm{Gal}(L/L_G)$ are the identity on L_G, and the second follows from $G \subset \mathrm{Gal}(L/L_G)$. We conclude that L_G is the fixed field of $\mathrm{Gal}(L/L_G)$, and then Theorem 7.1.1 implies that $L_G \subset L$ is normal and separable and hence is a Galois extension. This proves part (b) of the theorem.

Since $L_G \subset L$ is Galois, Theorem 7.1.5 implies that $[L : L_G] = |\mathrm{Gal}(L/L_G)|$. Combining this with (7.20), we have

$$|\mathrm{Gal}(L/L_G)| = [L : L_G] \leq n = |G| \leq |\mathrm{Gal}(L/L_G)|,$$

since $G \subset \mathrm{Gal}(L/L_G)$. From here, parts (a) and (c) follow easily. $\qquad\square$

Here is an example of how to use this theorem.

Example 7.5.4. In Example 7.5.2 we considered the group of automorphisms of $L = \mathbb{C}(t)$ given by $G = \langle\sigma\rangle$, where $\sigma(t) = 1/t$. Since $t + t^{-1}$ is obviously fixed by G, we have

$$(7.24) \qquad\qquad \mathbb{C}(t + t^{-1}) \subset L_G \subset L = \mathbb{C}(t).$$

However, t is a root of $(x - t)(x - t^{-1}) = x^2 - (t + t^{-1})x + 1 \in \mathbb{C}(t + t^{-1})[x]$. Furthermore, the inclusions

$$\mathbb{C}(t) \subset \mathbb{C}(t + t^{-1})(t) \subset \mathbb{C}(t)$$

show that $\mathbb{C}(t + t^{-1})(t) = \mathbb{C}(t)$. Thus $\mathbb{C}(t)$ is obtained by adjoining t to $\mathbb{C}(t + t^{-1})$. Since t is a root of a quadratic equation with coefficients in $\mathbb{C}(t + t^{-1})$, we have

$$[\mathbb{C}(t) : \mathbb{C}(t + t^{-1})] \leq 2.$$

Theorem 7.5.3 implies that $[L : L_G] = |G| = 2$. Using (7.24) and the Tower Theorem, it follows easily that $L_G = \mathbb{C}(t + t^{-1})$. ◁▷

B. Function Fields in One Variable. Example 7.5.4 is a Galois extension constructed from the field $\mathbb{C}(t)$ of rational functions in the variable t with coefficients in $\mathbb{C}$. More generally, the function fields $F(t)$, where F is any field, have some interesting subfields as follows.

Proposition 7.5.5. *Assume that $\alpha \in F(t)$ is a rational function not in F, and write $\alpha = a(t)/b(t)$, where $a(t), b(t) \in F[t]$ are relatively prime. Then:*
(a) α *is transcendental over F.*
(b) *The polynomial $a(x) - \alpha b(x) \in F(\alpha)[x]$ is irreducible over $F(\alpha)$.*
(c) $F(\alpha) \subset F(t)$ *is a finite extension of degree*

$$[F(t):F(\alpha)] = \max\big(\deg(a), \deg(b)\big).$$

Proof. If α is algebraic over F, then α satisfies an equation

$$\alpha^n + a_1\alpha^{n-1} + \cdots + a_n = 0,$$

where $n \geq 1$ and $a_1, \ldots, a_n \in F$. Substituting $\alpha = a(t)/b(t)$ into the above equation and multiplying by $b(t)^n$ gives

(7.25) $$a(t)^n + a_1 a(t)^{n-1} b(t) + \cdots + a_n b(t)^n = 0$$

in the polynomial ring $F[t]$. This implies that

$$a(t)^n = b(t) \cdot \big(- a_1 a(t)^{n-1} - \cdots - a_n b(t)^{n-1} \big).$$

Since $a(t)$ and $b(t)$ are relatively prime, $b(t)$ must be constant, say $b_0 \in F$. Then substituting $b(t) = b_0$ into (7.25) gives

$$a(t)^n + a_1 a(t)^{n-1} b_0 + \cdots + a_n b_0^n = 0$$

in $F[t]$. This implies that $a(t)$ is also constant (can you explain why?), say $a_0 \in F$. Then $\alpha = a(t)/b(t) = a_0/b_0 \in F$, which is a contradiction. Part (a) follows.

For parts (b) and (c), first observe that since α is a rational function of t, we have

$$F(\alpha) \subset F(\alpha)(t) = F(\alpha, t) = F(t).$$

In other words, $F(t)$ is obtained by adjoining t to $F(\alpha)$. Thus $[F(t):F(\alpha)]$ is the degree of the minimal polynomial of t over $F(\alpha)$. To find the minimal polynomial, we will use the relatively prime polynomials $a(t), b(t) \in F[t]$ appearing in $\alpha = a(t)/b(t)$. Consider the polynomial in x defined by $a(x) - \alpha b(x)$. Then:
- $a(x) - \alpha b(x)$ is a polynomial in x with coefficients in $F(\alpha)$.
- t is a root of $a(x) - \alpha b(x)$ since $a(t) - \alpha b(t) = a(t) - \frac{a(t)}{b(t)}b(t) = 0$.
- If $a(x) = a_0 x^n + \cdots$ and $b(x) = b_0 x^m + \cdots$ have degrees n and m, then

$$a(x) - \alpha\, b(x) = (a_0 x^n + \cdots) - \alpha\,(b_0 x^m + \cdots).$$

None of the coefficients can cancel because $\alpha \notin F$. Hence the degree of x in $a(x) - \alpha\, b(x)$ is $\max(n, m) = \max\big(\deg(a), \deg(b)\big)$.

Now suppose that $a(x) - \alpha b(x)$ is irreducible over $F(\alpha)$. Then the above bullets imply that

$$[F(t) : F(\alpha)] = \text{the degree of } x \text{ in } a(x) - \alpha b(x) = \max\big(\deg(a), \deg(b)\big).$$

Thus part (c) of the proposition follows from part (b).

To prove part (b), we begin in the polynomial ring $F[x, y]$ for variables x, y. By Theorem 2.1.1, $F[x, y]$ is a UFD. We first claim that $a(x) - y\, b(x)$ is irreducible in $F[x, y]$. This is easy to see, for if

$$a(x) - y\, b(x) = AB, \quad A, B \in F[x, y],$$

then A and B can't both have positive degree in y. We may assume that $A \in F[x]$. In Exercise 1 you will show that this implies that A divides $a(x)$ and $b(x)$. Hence A is constant, since $a(x), b(x)$ are relatively prime. This proves that $a(x) - y\, b(x)$ is irreducible in $F[x, y]$.

Now consider $a(x) - y\, b(x)$ as a polynomial in $F(y)[x]$, that is, as a polynomial in x with coefficients in $F(y)$. We claim that it is irreducible over $F(y)$ because it is irreducible in $F[x, y]$. This can be proved several ways—Exercise 2 uses Gauss's Lemma, and Exercise 3 gives a more elementary proof.

To apply this to our situation, recall that α is transcendental over F by part (a). This means that α can be regarded as a variable over F. Hence $y \mapsto \alpha$ induces a ring isomorphism $F(y)[x] \simeq F(\alpha)[x]$ that takes $a(x) - y\, b(x)$ to $a(x) - \alpha b(x)$. Then the previous paragraph implies that $a(x) - \alpha b(x) \in F(\alpha)[x]$ is irreducible over $F(\alpha)$. This completes the proof of the proposition. $\qquad\square$

Here is an example that illustrates Theorem 7.5.3 and Proposition 7.5.5.

Example 7.5.6. Consider the automorphisms σ and τ of $\mathbb{C}(t)$ defined by

$$\sigma\big(\alpha(t)\big) = \alpha(\zeta_n t) \quad \text{and} \quad \tau\big(\alpha(t)\big) = \alpha(t^{-1}),$$

where $\zeta_n = e^{2\pi i / n}$ and $\alpha(t)$ is an arbitrary rational function in $\mathbb{C}(t)$. It is easy to see that σ has order n and τ has order 2. Furthermore, the computation

$$\tau \circ \sigma \circ \tau\big(\alpha(t)\big) = \tau \circ \sigma\big(\alpha(t^{-1})\big) = \tau\big(\alpha((\zeta_n t)^{-1})\big) = \tau\big(\alpha(\zeta_n^{-1} t^{-1})\big)$$
$$= \alpha\big(\zeta_n^{-1} t\big) = \sigma^{-1}\big(\alpha(t)\big)$$

shows that $\tau \circ \sigma \circ \tau = \sigma^{-1}$. It follows that σ and τ generate a group G of automorphisms of $\mathbb{C}(t)$ isomorphic to the dihedral group D_{2n} of order $2n$. If we let $L = \mathbb{C}(t)$, then Theorem 7.5.3 implies that

$$L_G \subset \mathbb{C}(t)$$

is a Galois extension of degree $2n$ with Galois group isomorphic to D_{2n}.

To describe this extension more explicitly, let

$$t^n + t^{-n} = \frac{t^{2n} + 1}{t^n} \in \mathbb{C}(t).$$

Proposition 7.5.5 implies that $\mathbb{C}(t^n + t^{-n}) \subset \mathbb{C}(t)$ also has degree $2n$. Since $t^n + t^{-n}$ is invariant under the action of σ and τ, we have extensions

$$\mathbb{C}(t^n + t^{-n}) \subset L_G \subset \mathbb{C}(t)$$

where $\mathbb{C}(t)$ has degree $2n$ over both smaller fields. Thus $\mathbb{C}(t^n + t^{-n}) = L_G$, and

$$\mathbb{C}(t^n + t^{-n}) \subset \mathbb{C}(t)$$

is a Galois extension with Galois group isomorphic to D_{2n}. ◁▷

C. Linear Fractional Transformations.

Given a field F and a variable t,

$$F \subset F(t)$$

is an extension of infinite degree, so that the theory developed in previous sections doesn't apply. But we can still define $\mathrm{Gal}(F(t)/F)$ to consist of all automorphisms of $F(t)$ that are the identity on F. We will describe this group using matrices.

Let $\mathrm{GL}(2, F)$ be the group of 2×2 invertible matrices with entries in F. Then

$$\begin{pmatrix} a & b \\ c & d \end{pmatrix} \in \mathrm{GL}(2, F),$$

gives a rational function

$$\frac{at + b}{ct + d} \in F(t).$$

This relates to $\mathrm{Gal}(F(t)/F)$ as follows.

Theorem 7.5.7. *Let $F \subset F(t)$ be as defined above.*

(a) *For $\gamma = \begin{pmatrix} a & b \\ c & d \end{pmatrix} \in \mathrm{GL}(2, F)$, the function $\sigma_\gamma : F(t) \to F(t)$ defined by $\alpha(t) \mapsto$ $\alpha\left(\frac{at+b}{ct+d}\right)$ is an automorphism that is the identity on F. Thus $\sigma_\gamma \in \mathrm{Gal}(F(t)/F)$.*

(b) *The map $\gamma \mapsto \sigma_{\gamma^{-1}}$ defines a group homomorphism $\mathrm{GL}(2, F) \to \mathrm{Gal}(F(t)/F)$.*

(c) *The homomorphism of part* (b) *is onto, and its kernel consists of all nonzero multiples of the identity matrix.*

Proof. For part (a), we first observe that the evaluation map sending t to $\frac{at+b}{ct+d} \in F(t)$ induces a ring homomorphism

$$F[t] \longrightarrow F(t).$$

This map is one to one, since $\frac{at+b}{ct+d}$ is transcendental over F by Proposition 7.5.5. Thus it extends to the field of fractions of $F[t]$, which gives the map

$$\sigma_\gamma : F(t) \longrightarrow F(t)$$

described in the statement of the theorem. Hence σ_γ is a one-to-one homomorphism (see Exercise 2 of Section 3.1). Furthermore, the image of σ_γ is the subfield $F\left(\frac{at+b}{ct+d}\right)$. By Proposition 7.5.5, the extension $F\left(\frac{at+b}{ct+d}\right) \subset F(t)$ has degree

$$\max\left(\deg(at + b), \deg(ct + d)\right) = 1,$$

since a or c is nonzero. Thus $F\left(\frac{at+b}{ct+d}\right) = F(t)$, so that σ_γ is onto. It follows that $\sigma_\gamma \in \mathrm{Gal}(F(t)/F)$, since it is obviously the identity on F.

This proves part (a) and shows that we have a map

$$\Phi : \mathrm{GL}(2, F) \longrightarrow \mathrm{Gal}(F(t)/F)$$

defined by $\Phi(\gamma) = \sigma_{\gamma^{-1}}$. The inverse in the definition of Φ is needed to make it a group homomorphism, as you will prove in Exercise 4. Part (b) follows.

For part (c), let $\sigma \in \mathrm{Gal}(F(t)/F)$. Then $\sigma(\alpha(t)) = \alpha(\sigma(t))$ for all $\alpha(t) \in F(t)$, since σ is the identity on F. This implies that $F(\sigma(t)) = \sigma(F(t)) = F(t)$. Setting $\sigma(t) = A(t)/B(t)$, where $A(t), B(t) \in F[t]$ are relatively prime, we can write this as $F\big(A(t)/B(t)\big) = F(t)$. By Proposition 7.5.5, it follows that

$$\max\big(\deg(A), \deg(B)\big) = 1.$$

Thus $A(t) = at + b$ and $B(t) = ct + d$, where $a, b, c, d \in F$. In Exercise 5 you will show that

(7.26)
$$\begin{pmatrix} a & b \\ c & d \end{pmatrix} \in \mathrm{GL}(2, F).$$

Since $\sigma(t) = \frac{at+b}{ct+d}$, it follows that $\sigma = \sigma_{\gamma^{-1}}$, where γ^{-1} is the matrix (7.26). This proves that Φ is onto.

Finally, take $\gamma \in \mathrm{GL}(2, F)$ in the kernel of Φ. Then $\sigma_{\gamma^{-1}} = 1_{F(t)}$, so that

$$t = \sigma_{\gamma^{-1}}(t) = \frac{at + b}{ct + d}$$

in $F(t)$, where $\gamma^{-1} = \begin{pmatrix} a & b \\ c & d \end{pmatrix}$. Clearing denominators and collecting terms gives

$$ct^2 + (d - a)t - b = 0$$

in $F[t]$. Thus $c = b = 0$ and $a = d$, which shows that γ is a nonzero multiple of the identity matrix. Since all nonzero multiples of the identity are in the kernel (check this), we get the desired description of the kernel. $\qquad\square$

If I_2 is the 2×2 identity matrix, then $F^* I_2$ consists of all nonzero multiples of the identity. Then the Fundamental Theorem of Group Homomorphisms and Theorem 7.5.7 imply that we have an isomorphism

$$\mathrm{GL}(2, F)/F^* I_2 \simeq \mathrm{Gal}(F(t)/F).$$

The quotient group $\mathrm{GL}(2, F)/F^* I_2$ is denoted $\mathrm{PGL}(2, F)$. Thus

$$\mathrm{PGL}(2, F) \simeq \mathrm{Gal}(F(t)/F).$$

The group $\mathrm{PGL}(2, F)$ will play an important role in what follows. The "PGL" in $\mathrm{PGL}(2, F)$ stands for *projective linear group*. We will learn more about projective linear groups in Chapters 13 and 14.

D. Stereographic Projection. Let F be a field. Then define

$$\widehat{F} = F \cup \{\infty\},$$

where ∞ is a formal symbol that stands for the "point at infinity" of F. Given

$$\gamma = \begin{pmatrix} a & b \\ c & d \end{pmatrix} \in GL(2, F) \quad \text{and} \quad \alpha \in \widehat{F},$$

we get $[\gamma] \in PGL(2, F) = GL(2, F)/F^* I_2$ and set

(7.27)
$$[\gamma] \cdot \alpha = \frac{a\alpha + b}{c\alpha + d}.$$

In Exercise 6 you will show that this gives a well-defined action of $PGL(2, F)$ on $\widehat{F}$. Furthermore, in Exercise 7 you will prove the following.

Proposition 7.5.8. *Let $(\alpha_1, \alpha_2, \alpha_3)$ and $(\beta_1, \beta_2, \beta_3)$ be triples of distinct points of $\widehat{F}$. Then there is a unique $[\gamma] \in PGL(2, F)$ such that*

$$[\gamma] \cdot \alpha_i = \beta_i$$

for $i = 1, 2, 3$.   □

We will assume that $F = \mathbb{C}$ for the rest of the section. This will allow geometry to give some interesting Galois groups. It is customary to call $\widehat{\mathbb{C}} = \mathbb{C} \cup \{\infty\}$ the *Riemann sphere* because we can map the unit sphere $S^2 \subset \mathbb{R}^3$ to $\widehat{\mathbb{C}}$ by *stereographic projection*.

The unit sphere S^2 is defined by $x^2 + y^2 + z^2 = 1$ in $\mathbb{R}^3$, and we identify $a + bi \in \mathbb{C}$ with $(a, b, 0) \in \mathbb{R}^3$, so that $\mathbb{C}$ becomes the plane $z = 0$ in $\mathbb{R}^3$. Then define

$$\pi : S^2 \setminus \{(0, 0, 1)\} \longrightarrow \mathbb{C}$$

as follows. Given $P \in S^2 \setminus \{(0, 0, 1)\}$, we draw the line connecting P to the north pole $(0, 0, 1)$ and define $\pi(P)$ to be the point where the line intersects the xy-plane. Here is the picture:

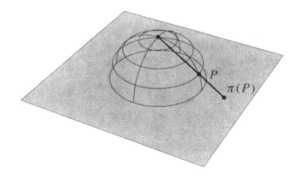

In this picture, the xy-plane is shaded gray and the top half of S^2 is shown as a wire frame. In Exercise 8 you will show that

$$(7.28) \qquad \pi(a, b, c) = \left(\frac{a}{1-c}, \frac{b}{1-c}, 0 \right) = \frac{a}{1-c} + i \frac{b}{1-c},$$

where the last equality uses the above identification of $\mathbb{C}$ with the xy-plane in $\mathbb{R}^3$. Under the map π, the south pole $(0, 0, -1)$ maps to $0 \in \mathbb{C}$ and the equator of the sphere maps to the unit circle $\{z \in \mathbb{C} \mid |z| = 1\}$.

We then extend π to *stereographic projection*

$$\hat{\pi} : S^2 \longrightarrow \widehat{\mathbb{C}}$$

by defining $\hat{\pi}(0, 0, 1) = \infty$ and $\hat{\pi}(P) = \pi(P)$ for $P \in S^2 \setminus \{(0, 0, 1)\}$. Note that $\hat{\pi}$ is one to one and onto.

The key geometric property of $\hat{\pi}$ is the following. Consider a rotation r of S^2 about some axis through the origin. This gives a map

$$r : S^2 \longrightarrow S^2.$$

We then obtain the map

$$\hat{\pi} \circ r \circ \hat{\pi}^{-1} : \widehat{\mathbb{C}} \longrightarrow \widehat{\mathbb{C}}$$

by composition. The remarkable fact is that this map is given by a linear fractional transformation. To state our result more precisely, let

$$\mathrm{Rot}(S^2)$$

be the group of all rotations of the sphere. A careful description of this group can be found in [8, Sec. 2.2], together with a proof of the following important result.

Theorem 7.5.9. *Given $r \in \mathrm{Rot}(S^2)$, there is a unique $[\gamma] \in \mathrm{PGL}(2, \mathbb{C})$ such that*

$$\hat{\pi} \circ r \circ \hat{\pi}^{-1}(z) = [\gamma] \cdot z$$

for all $z \in \widehat{\mathbb{C}}$. Furthermore, the map

$$\mathrm{Rot}(S^2) \longrightarrow \mathrm{PGL}(2, \mathbb{C})$$

defined by $r \mapsto [\gamma]$ is a one-to-one group homomorphism. $\square$

Here is an example of this result.

Example 7.5.10. Consider the octahedron with vertices at $(\pm 1, 0, 0)$, $(0, \pm 1, 0)$ and $(0, 0, \pm 1)$ in $\mathbb{R}^3$. This is inscribed in the sphere S^2 and when combined with stereographic projection gives the picture

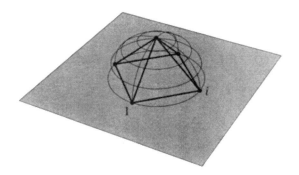

This shows the top half of the octahedron. Note that $(1, 0, 0)$ becomes $1 \in \mathbb{C}$ and $(0, 1, 0)$ becomes $i \in \mathbb{C}$, so that the six vertices of the octahedron map to the points

$$0, \pm 1, \pm i, \infty \in \widehat{\mathbb{C}}.$$

Now consider the rotation r_1 of S^2 by $180°$ about the x-axis. By Theorem 7.5.9, this gives an element of $\mathrm{PGL}(2, \mathbb{C})$ that takes 1 to itself and interchanges 0 and ∞. However, note that

$$\gamma_1 = \begin{pmatrix} 0 & 1 \\ 1 & 0 \end{pmatrix}$$

satisfies

$$[\gamma_1] \cdot z = \frac{1}{z}$$

for $z \in \widehat{\mathbb{C}}$. This also takes 1 to itself and interchanges 0 and ∞. Since $[\gamma_1]$ is uniquely determined by its values on $0, 1, \infty$ (this is Proposition 7.5.8), it follows that r_1 corresponds to $[\gamma_1]$.

Similarly, consider the rotation r_2 of S^2 by $90°$ counterclockwise about the z-axis. In Exercise 9 you will show that r_2 corresponds to $[\gamma_2]$, where

$$\gamma_2 = \begin{pmatrix} i & 0 \\ 0 & 1 \end{pmatrix}.$$

Note that

$$[\gamma_2] \cdot z = iz$$

for $z \in \widehat{\mathbb{C}}$.

Finally, consider the rotation r_3 of S^2 about the y-axis that takes $(0, 0, 1)$ to $(1, 0, 0)$. Under stereographic projection, this corresponds to an element $[\gamma_3] \in \mathrm{PGL}(2, \mathbb{C})$ that fixes $\pm i$ and maps

$$\infty \to 1 \to 0 \to -1 \to \infty \to \cdots .$$

Be sure that you can see this in the above picture. In Exercise 9 you will show that

$$\gamma_3 = \begin{pmatrix} 1 & -1 \\ 1 & 1 \end{pmatrix}.$$

Thus

$$[\gamma_3] \cdot z = \frac{z-1}{z+1}$$

for $z \in \widehat{\mathbb{C}}$.

In Exercise 10 you will show that the rotation group of the octahedron is isomorphic to S_4 and is generated by the rotations r_1, r_2 and r_3. By the isomorphism of Theorem 7.5.9, it follows that the subgroup

$$G = \langle [\gamma_1], [\gamma_2], [\gamma_3] \rangle \subset \mathrm{PGL}(2, \mathbb{C})$$

is isomorphic to S_4. By Theorem 7.5.7, we can regard G as a group of automorphisms of $L = \mathbb{C}(t)$, and then Theorem 7.5.3 shows that

$$L_G \subset L = \mathbb{C}(t)$$

is a Galois extension with Galois group $G \simeq S_4$.   ◁▷

Example 7.5.10 will have an unexpected application in Chapter 14.

Mathematical Notes

There are many ideas in this section to discuss.

■ **Finite Subgroups of Linear Fractional Transformations.** Theorems 7.5.3 and 7.5.7 show that any finite subgroup $G \subset \mathrm{PGL}(2, F)$ gives a Galois extension

$$L_G \subset L = F(t)$$

with Galois group G. The remarkable fact is that for many fields F, the finite subgroups $G \subset \mathrm{PGL}(2, F)$ have been classified.

For example, when $F = \mathbb{C}$, G is isomorphic to one of the following groups:

> A cyclic group C_n of order n, $n \geq 1$
>
> A dihedral group D_{2n} of order $2n$, $n \geq 2$
>
> The alternating group A_4 of order 12
>
> The symmetric group S_4 of order 24
>
> The alternating group A_5 of order 60

(7.29)

Furthermore, two finite subgroups of $\mathrm{PGL}(2, \mathbb{C})$ are conjugate in $\mathrm{PGL}(2, \mathbb{C})$ if and only if they are isomorphic as abstract groups. Proofs of these assertions can be found in [1, Sec. 2.13] and [8, Secs. 2.2, 2.6].

Even more remarkable is the geometric origin of these subgroups of $\mathrm{PGL}(2, \mathbb{C})$. We saw in Example 7.5.10 how the octahedron gave a subgroup $G \subset \mathrm{PGL}(2, \mathbb{C})$ isomorphic to S_4. In a similar way, the tetrahedron and icosahedron give subgroups isomorphic to A_4 and A_5 respectively. Furthermore, you will show in Exercise 12 how to realize C_n and D_{2n} as the symmetry groups of polyhedra. Thus polyhedra give Galois groups!

Chapter 14 will discuss the subgroups of $\mathrm{PGL}(2, F)$ when F is a finite field.

■ **Invariant Theory.** Examples 7.5.6 and 7.5.10 gave extensions with Galois groups D_{2n} and S_4 respectively. For D_{2n}, the extension was given explicitly as

$$\mathbb{C}(t^n + t^{-n}) \subset \mathbb{C}(t),$$

while for S_4, we merely wrote

(7.30) $$L_G \subset L = \mathbb{C}(t), \quad G \simeq S_4.$$

In fact, one can show that for the group $G = \langle [\gamma_1], [\gamma_2], [\gamma_3] \rangle \subset \mathrm{PGL}(2, \mathbb{C})$ of Example 7.5.10, the extension (7.30) is given by

$$\mathbb{C}\left(\tfrac{(t^8+14t^4+1)^3}{t^4(t^4-1)^4} \right) \subset \mathbb{C}(t).$$

While the invariance of $t^n + t^{-n}$ under D_{2n} is obvious, it is not at all clear that

$$\alpha = \frac{(t^8 + 14t^4 + 1)^3}{t^4(t^4 - 1)^4}$$

is invariant under $G = \langle [\gamma_1], [\gamma_2], [\gamma_3] \rangle$. But even if α is invariant, how does one find α in the first place? The answer involves *invariant theory* and is beyond the scope of this book. But as a small hint of where α comes from, we note that $t^8 + 14t^4 + 1$ has the following geometric description. The octahedron has eight faces that are equilateral triangles. If we project from the center of the sphere, the center of each face gives a point on S^2. Under stereographic projection, these give eight points of $\mathbb{C}$. Then the roots of $t^8 + 14t^4 + 1$ are precisely these eight points.

Invariant theory gives similar formulas for the extensions coming from the tetrahedron and icosahedron. Complete details can be found in Chapter 3 of [8].

■ **Lüroth's Theorem.** The above paragraph suggests that for any finite subgroup $G \subset \mathrm{PGL}(2, \mathbb{C})$, there is $\alpha \in \mathbb{C}(t)$ such that if $L = \mathbb{C}(t)$, then

$$L_G = \mathbb{C}(\alpha).$$

In fact, more is true: given *any* field F and *any* finite extension $K \subset F(t)$, there is $\alpha \in F(t)$ such that $K = F(\alpha)$. This result is known as *Lüroth's Theorem*. Proofs of Lüroth's Theorem may be found in [5, Sec. 9.5] and [8, Sec. 6.3].

■ **The Quintic and the Icosahedron.** When we use the rotations of the icosahedron, we get a Galois extension

$$L_G \subset L = \mathbb{C}(t)$$

with Galois group $G \simeq A_5$. By [8, Sec. 4.8], there is $W \in \mathbb{C}(t)$ such that

$$L_G = \mathbb{C}(W)$$

and $\mathbb{C}(W) \subset \mathbb{C}(t)$ is the splitting field of the irreducible quintic

$$x^5 - 10Wx^3 + 45W^2x^2 - W^2 \in \mathbb{C}(W)[x].$$

This is called a *Brioschi quintic*.

The results of Chapter 8 will imply that the Brioschi quintic is not solvable by radicals over $\mathbb{C}(W)$. However, in the nineteenth century it was discovered that (roughly speaking) the solution of any quintic can be reduced to solving the Brioschi resolvent. This involves a rich collection of ideas described by Klein in [3] and more recently in the book [8] and the poster [9].

The Galois group of an irreducible quintic will be computed in Chapter 13.

Exercises for Section 7.5

Exercise 1. Let $P, Q \in F[x, y]$ be polynomials such that $P|Q$ and $P \in F[x]$, and write $Q = a_0(x) + a_1(x)y + a_2(x)y^2 + \cdots + a_m(x)y^m$. Prove that $P|a_i$ for $i = 0, \ldots, m$.

Exercise 2. In the proof of Proposition 7.5.5, we showed that $a(x) - y\,b(x)$ is irreducible in $F[x, y]$ and we want to conclude that it is also irreducible in $F(y)[x]$. Prove this using the version of Gauss's Lemma stated in Theorem A.5.8.

Exercise 3. The proof of Proposition 7.5.5 shows that $a(x) - y\,b(x)$ is irreducible in $F[x, y]$. In this exercise, you will give an elementary proof that $a(x) - y\,b(x)$ is irreducible over $F(y)[x]$. Suppose that

$$a(x) - y\,b(x) = AB, \quad A, B \in F(y)[x].$$

You need to prove that A or B is constant, which in this case means that A or B lies in $F(y)$.
(a) Show that there are nonzero polynomials $g(y), h(y) \in F[y]$ that clear the denominators of A and B, i.e., $g(y)A = A_1$ and $h(y)B = B_1$ for some $A_1, B_1 \in F[x, y]$.
(b) Show that $g(y)h(y)\big(a(x) - y\,b(x)\big) = A_1 B_1$ in $F[x, y]$ and explain why $a(x) - y\,b(x)$ must divide either A_1 or B_1 in $F[x, y]$.
(c) Assume that $A_1 = \big(a(x) - y\,b(x)\big)A_2$, where $A_2 \in F[x, y]$. Show that this implies that $g(y)h(y) = A_2 B_1$, and then conclude that $B_1 \in F[y]$.
(d) Show that $B \in F(y)$.

Exercise 4. Prove that the map $\Phi : GL(2, F) \to \text{Gal}(F(t)/F)$ defined in the proof of Theorem 7.5.7 is a group homomorphism.

Exercise 5. Prove (7.26).

Exercise 6. In this exercise, you will prove that $PGL(2, F)$ acts on $\widehat{F} = F \cup \{\infty\}$.
(a) First show that

$$\gamma \cdot \alpha = \frac{a\alpha + b}{c\alpha + d}, \quad \gamma = \begin{pmatrix} a & b \\ c & d \end{pmatrix}$$

defines an action of $GL(2, F)$ on $\widehat{F}$. Explain carefully what happens when $\alpha = \infty$.

(b) Show that nonzero multiples of the identity matrix act trivially on $\widehat{F}$, and use this to give a careful proof that (7.27) gives a well-defined action of PGL(2, F) on $\widehat{F}$.

Exercise 7. Proposition 7.5.8 asserts that we can map any triple of distinct points of $\widehat{F}$ to any other such triple via a unique element $[\gamma] \in$ PGL(2, F). We will defer the proof of existence of $[\gamma]$ until Exercise 24 in Section 14.3. However, we will prove the uniqueness part of the proposition, since this is what is used in Example 7.5.10.
  (a) First suppose that $[\gamma] \in$ PGL(2, F) fixes ∞ and also fixes two points $\alpha \neq \beta$ of F. Prove that γ is a nonzero multiple of the identity matrix.
  (b) Now suppose that $[\gamma] \in$ PGL(2, F) fixes three distinct points of F, and let α be one of these points. Show that there is $[\delta] \in$ PGL(2, F) such that $[\delta] \cdot \alpha = \infty$. Then prove that γ is a nonzero multiple of the identity matrix by applying part (a) to $[\delta \gamma \delta^{-1}]$.
  (c) Show that the desired uniqueness follows from parts (a) and (b).

Exercise 8. Prove the formula (7.28) for stereographic projection.

Exercise 9. In Example 7.5.10, we considered rotations r_1, r_2, r_3 of the octahedron and defined matrices $\gamma_1, \gamma_2, \gamma_3 \in$ GL(2, $\mathbb{C}$). We also proved carefully that r_1 corresponds to $[\gamma_1]$ under the homomorphism of Theorem 7.5.9. In a similar way, prove that r_2 corresponds to $[\gamma_2]$ and r_3 corresponds to $[\gamma_3]$.

Exercise 10. The goal of this exercise is to prove that the symmetry group G of the octahedron is isomorphic to S_4. By *symmetry group*, we mean the group of rotations that carry the octahedron to itself. We think of G as acting on the octahedron.
  (a) Let v be a vertex of the octahedron. Use the action of G on v and the Fundamental Theorem of Group Actions to prove that $|G| = 24$.
  (b) The eight face centers of the octahedron form the vertices of an inscribed cube. Explain why the octahedron and its inscribed cube have the same symmetry group.
  (c) The cube has four long diagonals that connect a vertex to an opposite vertex. Explain why the action of G on these diagonals gives a group homomorphism $G \to S_4$.
  (d) Let $r_1, r_2, r_3 \in G$ be the rotations described in Example 7.5.10. Explain how each rotation acts on the inscribed cube and describe its corresponding permutation in S_4.
  (e) Prove that the three permutations constructed in part (d) generate S_4.
  (f) Use parts (a) and (e) to show that $G \simeq S_4$. Also prove that G is generated by r_1, r_2, r_3.
See Section 14.4 for a different approach to proving that a group is isomorphic to S_4.

Exercise 11. In Section 6.4, we defined the one-dimensional affine linear group AGL(1, $\mathbb{F}_p$) over the finite field $\mathbb{F}_p$. More generally, if F is any field, then AGL(1, F) consists of all functions $\gamma_{a,b} : F \to F$ defined by

$$\gamma_{a,b}(\alpha) = a\alpha + b, \quad \alpha \in F,$$

where $a \in F^*, b \in F$, and the group structure is given by composition. In this exercise, you will represent AGL(1, F) as a subgroup of PGL(2, F).
  (a) Show that the map $\gamma_{a,b} \mapsto \begin{bmatrix} a & b \\ 0 & 1 \end{bmatrix}$ defines a one-to-one group homomorphism

$$\text{AGL}(1, F) \longrightarrow \text{PGL}(2, F).$$

  (b) Consider the action of PGL(2, F) on $\widehat{F}$. Show that the isotropy subgroup of PGL(2, F) acting on ∞ is the image of the homomorphism of part (a).

Exercise 12. In this exercise, you will construct polyhedra whose symmetry groups are isomorphic to C_n and D_{2n}. For D_{2n}, consider the polyhedron whose vertices are the north and south poles of S^2 together with the nth roots of unity along the equator. For $n = 8$, this gives the following picture:

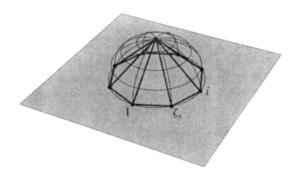

As usual, this shows only the top half of the polyhedron. Note that to obtain a three-dimensional object, we must assume $n \geq 3$.

   (a) Show that the symmetry group of this polyhedron is isomorphic to D_{2n} when $n \neq 4$ and S_4 when $n = 4$.

   (b) Now take the vertices on the equator and move them up in S^2 so that they become the vertices of a regular n-gon lying in the plane $z = c$, where $c > 0$ is small. Prove that the symmetry group of this polyhedron is isomorphic to C_n.

   (c) Find polyhedra inscribed in S^2 whose symmetry groups are C_1 (the trivial group), C_2, D_4 (the Klein four-group), and D_8 respectively.

Notice that the symmetry groups of these polyhedra, together with those of the tetrahedron, octahedron, and icosahedron, give *all* of the groups listed in (7.29).

Exercise 13. Consider the automorphism of $L = \mathbb{C}(t)$ defined by $\alpha(t) \mapsto \alpha(\zeta_n t)$. This generates a cyclic group G of automorphisms such that $|G| = n$. Adapt the methods of Example 7.5.6 to show that $L_G = \mathbb{C}(t^n)$ and conclude that $\mathbb{C}(t^n) \subset \mathbb{C}(t)$ is a Galois extension whose Galois group is cyclic of order n.

Exercise 14. Consider the automorphisms of $L = F(t)$ defined by

$$\sigma\big(\alpha(t)\big) = \alpha(t^{-1}) \quad \text{and} \quad \tau\big(\alpha(t)\big) = \alpha(1 - t).$$

   (a) Prove that σ and τ generate a group G of automorphisms of $F(t)$ isomorphic to S_3.

   (b) Show that G corresponds to the subgroup of $\mathrm{PGL}(2, F)$ consisting of all elements that map the subset $\{0, 1, \infty\} \subset \widehat{F}$ to itself.

   (c) Prove that

$$L_G = F\Big(\tfrac{(t^2 - t + 1)^3}{t^2(t-1)^2}\Big),$$

and conclude that

$$F\Big(\tfrac{(t^2 - t + 1)^3}{t^2(t-1)^2}\Big) \subset F(t)$$

is a Galois extension with Galois group $G \simeq S_3$.

REFERENCES

1. G. A. Jones and D. Singerman, *Complex Functions*, Cambridge U. P., Cambridge, 1987.

2. B. M. Kiernan, *The development of Galois theory from Lagrange to Artin*, Arch. Hist. Exact Sci. **8** (1971), 40–154.

3. F. Klein, *Lectures on the ikosahedron and the solution of equations of the fifth degree*, English translation by George Gavin Morrice, Trübner & Co., London, 1888. Reprint by Chelsea, New York, 1956.

4. G. Malle and B. H. Matzat, *Inverse Galois Theory*, Springer-Verlag, New York, Berlin, Heidelberg, 1999.

5. T. Mora, *Solving Polynomial Equations Systems I*, Cambridge U. P., Cambridge, 2003.

6. B. Osofsky, *Nice polynomials for introductory Galois theory*, Math. Mag. **72** (1999), 218–222.

7. I. Radloff, *Évariste Galois: principles and applications*, Historia Math. **29** (2002), 114–137.

8. J. Shurman, *Geometry of the Quintic*, Wiley, New York, 1997.

9. Wolfram Research, *Solving the Quintic*, Poster, Wolfram Research, Champaign, IL, 1995. Available at http://library.wolfram.com/examples/quintic/.

Part III

Applications

The next four chapters give classic applications of Galois theory.

The search for generalizations of Cardan's formulas leads to the notion of *solvability by radicals*. In Chapter 8, we relate this to the notion of a *solvable group* and use Galois theory to show that in general, polynomials of degree ≥ 5 cannot be solved by radicals.

Chapter 9 discusses the Galois theory of the *cyclotomic extension* $\mathbb{Q} \subset \mathbb{Q}(\zeta_n)$, where ζ_n is a primitive nth root of unity. We also explain how Gauss analyzed this extension when n is a prime.

In Chapter 10, we study straightedge-and-compass constructions from the point of view of Galois theory. This includes the classic Greek problems (trisecting the angle, duplicating the cube, squaring the circle) as well as the construction of regular polygons. We also explore what happens when we go beyond straightedge and compass to allow constructions using origami.

Finally, Chapter 11 explores the theory of finite fields. We consider the structure of finite fields and compute the Galois groups involved. We also describe irreducible polynomials and cyclotomic polynomials over a finite field.

8

Solvability by Radicals

In this chapter, we will use the Galois theory developed in Chapter 7 to determine when a polynomial equation can be solved by radicals. The idea is to translate the problem into group theory. Hence we begin with the group-theoretic concept of *solvable group*.

8.1 SOLVABLE GROUPS

Here is the basic definition of this section.

Definition 8.1.1. *A finite group G is **solvable** if there are subgroups*

$$\{e\} = G_n \subset G_{n-1} \subset \cdots \subset G_1 \subset G_0 = G$$

such that for $i = 1, \ldots, n$ we have:
(a) G_i *is normal in* G_{i-1}.
(b) $[G_{i-1} : G_i]$ *is prime.*

Since G_i is normal in G_{i-1}, part (b) of the definition can be replaced by the assertion that G_{i-1}/G_i is a cyclic group of prime order.

We will show below that every finite Abelian group is solvable. Here is an example of a non-Abelian solvable group.

Example 8.1.2. The subgroups

$$\{e\} \subset A_3 \subset S_3$$

show that S_3 is solvable, since each subgroup is normal in the next and the indices $[A_3 : \{e\}] = 3, [S_3 : A_3] = 2$ are prime. ◁▷

In Exercise 1 you will prove similarly that A_4 and S_4 are solvable. On the other hand, we will see in Section 8.4 that A_n and S_n are nonsolvable for $n \geq 5$.

Here is our first result about solvability.

Proposition 8.1.3. *Every subgroup of a solvable finite group is solvable.*

Proof. Let G be finite and solvable with subgroups G_i as in Definition 8.1.1. Given a subgroup $H \subset G$, set $H_i = G_i \cap H$ and note that

$$H_0 = G_0 \cap H = G \cap H = H,$$
$$H_n = G_n \cap H = \{e\} \cap H = \{e\}.$$

Then consider the group homomorphism

$$\pi : H_{i-1} \longrightarrow G_{i-1}/G_i$$

that sends $h \in H_{i-1}$ to the coset $hG_i \in G_{i-1}/G_i$. Observe that $h \in H_{i-1}$ is in the kernel of π if and only if $hG_i = G_i$, which happens if and only if

$$h \in H_{i-1} \cap G_i = (G_{i-1} \cap H) \cap G_i = H \cap G_i = H_i,$$

where the second equality follows from $G_i \subset G_{i-1}$. This shows that $\text{Ker}(\pi) = H_i$. Thus H_i is normal in H_{i-1}. By the Fundamental Theorem of Group Homomorphisms, we get an isomorphism

$$H_{i-1}/H_i = H_{i-1}/\text{Ker}(\pi) \simeq \text{Im}(\pi) \subset G_{i-1}/G_i.$$

Since G_{i-1}/G_i is cyclic of prime order, it follows that H_{i-1}/H_i is either trivial or isomorphic to G_{i-1}/G_i. Thus either $H_{i-1} = H_i$ or $[H_{i-1} : H_i]$ is prime. Hence, by discarding duplicates, the subgroups

$$\{e\} = H_n \subset \cdots \subset H_i \subset H_{i-1} \subset \cdots \subset H_0 = H$$

show that H is solvable. □

Here is one of the main theoretical tools for dealing with solvable groups.

Theorem 8.1.4. *Let G be a finite group and H a normal subgroup. Then G is solvable if and only if H and G/H are.*

Proof. First suppose that G is solvable. Then H is solvable by Proposition 8.1.3. To show that G/H is solvable, suppose that G_i are subgroups of G as in Definition 8.1.1, and let $\pi : G \to G/H$ be the group homomorphism $g \mapsto gH$. Then let $\tilde{G}_i = \pi(G_i)$. In Exercise 2 you will show the following:

(a) $G_0 = G$ implies that $\widetilde{G}_0 = G/H$.

(b) $G_n = \{e\}$ implies that $\widetilde{G}_n = \{eH\}$ (where eH is the identity of G/H).

(c) G_i normal in G_{i-1} implies that $\widetilde{G}_i$ is normal in $\widetilde{G}_{i-1}$.

(d) The map $G_{i-1}/G_i \to \widetilde{G}_{i-1}/\widetilde{G}_i$ given by $gG_i \mapsto \pi(g)\widetilde{G}_i$ is a well-defined onto group homomorphism.

By assumption G_{i-1}/G_i is a group of prime order. In Exercise 2 you will show that this fact and (d) imply that $\widetilde{G}_{i-1}/\widetilde{G}_i$ is either trivial or also has prime order. Thus either $\widetilde{G}_{i-1} = \widetilde{G}_i$ or $[\widetilde{G}_{i-1} : \widetilde{G}_i]$ is prime. Then, as in the proof of Proposition 8.1.3, discarding duplicates among the subgroups

$$\{eH\} = \widetilde{G}_n \subset \cdots \subset \widetilde{G}_i \subset \widetilde{G}_{i-1} \subset \cdots \subset \widetilde{G}_0 = G/H$$

shows that G/H is solvable.

Conversely suppose that H and G/H are solvable. Let H_i, $i = 0, \ldots, \ell$ be subgroups of H satisfying Definition 8.1.1, and similarly let $\widetilde{G}_j$, $j = 0, \ldots, m$ be subgroups of G/H satisfying the definition.

As above, we have $\pi : G \to G/H$. Given a subgroup $K \subset G/H$, set

(8.1) $$\pi^{-1}(K) = \{g \in G \mid \pi(g) \in K\}.$$

In Exercise 3 you will verify that $\pi^{-1}(K)$ is a subgroup of G. You will also check that the kernel of π is

$$H = \pi^{-1}(\{eH\})$$

and that

$$G = \pi^{-1}(G/H).$$

If we apply this to the subgroups

$$\{eH\} = \widetilde{G}_m \subset \cdots \subset \widetilde{G}_0 = G/H,$$

then we obtain the subgroups

$$H = \pi^{-1}(\{eH\}) = \pi^{-1}(\widetilde{G}_m) \subset \cdots \subset \pi^{-1}(\widetilde{G}_0) = \pi^{-1}(G/H) = G.$$

However, we also have the subgroups

$$\{e\} = H_\ell \subset \cdots \subset H_i \subset H_{i-1} \subset \cdots \subset H_0 = H.$$

We "glue together" these sequences of subgroups by defining $G_i \subset G$ to be

$$G_i = \begin{cases} \pi^{-1}(\widetilde{G}_i), & 0 \leq i \leq m, \\ H_{i-m}, & m \leq i \leq \ell + m. \end{cases}$$

Note that the sequences are "joined" at $G_m = \pi^{-1}(\widetilde{G}_m) = H_0 = H$. It remains to show that G_i is normal in G_{i-1} of prime index. For $m \leq i \leq \ell + m$, this is obvious, since $G_i = H_{i-m}$ in this range and the H_i satisfy Definition 8.1.1. For $0 \leq i \leq m$, we leave it as Exercise 4 to show that for this range of indices, G_i is normal in G_{i-1} and that

(8.2)
$$G_{i-1}/G_i \simeq \widetilde{G}_{i-1}/\widetilde{G}_i.$$

Since the $\widetilde{G}_i$ satisfy Definition 8.1.1, it follows that G is solvable. This completes the proof of the theorem. $\qquad\square$

We next use Theorem 8.1.4 to show that Abelian groups are solvable.

Proposition 8.1.5. *Every finite Abelian group G is solvable.*

Proof. We will prove the proposition by complete induction on $n = |G|$. The case $n = 1$ being trivial, assume that G is an Abelian group of order $n > 1$ and that the result is true for all Abelian groups of order $< n$.

Let p be prime divisor of $|G|$. If $p = |G|$, then G is cyclic of order p and hence solvable. If $p < |G|$, then by Cauchy's Theorem (Theorem A.1.5), we can find $g \in G$ of order p. Now let $H = \langle g \rangle$ be the subgroup generated by g. Then H is normal, since G is Abelian and $|H| = p < |G|$. It follows that the orders of H and G/H are strictly smaller than $|G| = n$. Hence H and G/H are solvable (by our inductive assumption), and then Theorem 8.1.4 implies that G is solvable. $\qquad\square$

Here is an interesting non-Abelian solvable group.

Example 8.1.6. The one-dimensional affine linear group $\mathrm{AGL}(1, \mathbb{F}_p)$ over $\mathbb{F}_p$ was introduced in Section 6.4. There, the discussion leading up to (6.6) showed that $\mathrm{AGL}(1, \mathbb{F}_p)$ has a normal subgroup $T \simeq \mathbb{F}_p$ with quotient $\mathrm{AGL}(1, \mathbb{F}_p)/T \simeq \mathbb{F}_p^*$.

Since $\mathbb{F}_p$ and $\mathbb{F}_p^*$ are Abelian, they are solvable by Proposition 8.1.5, so that $\mathrm{AGL}(1, \mathbb{F}_p)$ is solvable by Theorem 8.1.4. We also know that $\mathrm{AGL}(1, \mathbb{F}_p)$ non-Abelian for $p \geq 3$ by part (a) of Exercise 10 of Section 6.4. This example will be important in Chapter 14. $\qquad\triangleleft\triangleright$

Mathematical Notes

The definition of solvability is related to the ideas of *simple groups, composition series*, and the *Jordan–Hölder Theorem*. We will say more about these topics in Section 8.4. However, some standard results used to study solvable groups need to be mentioned here.

■ **Solvability and the Order of a Group.** In some cases the solvability of a group is determined by its order. For example, in Exercise 5 you will prove the following.

Theorem 8.1.7. *If p is prime, then every group of order p^n, $n \geq 0$, is solvable.* $\quad\square$

In 1904, Burnside [4] generalized Theorem 8.1.7 as follows.

Theorem 8.1.8. *If p and q are distinct primes, then every group of order $p^n q^m$, $n, m \geq 0$, is solvable.* $\qquad\square$

In 1963, Feit and Thompson [5] proved the following surprising result.

Theorem 8.1.9. *Every group of odd order is solvable.* $\square$

In spite of its simple statement, the proof of Theorem 8.1.9 uses some very sophisticated mathematics and takes 255 pages.

▪ **Solvability and the Sylow Theorems.** The Sylow Theorems imply some nice results about solvability. As stated in Theorem A.5.1, we have:

- (First Sylow Theorem) If p^n is the highest power of a prime dividing the order of a finite group G, then G has a subgroup of order p^n, called a *p-Sylow subgroup.*
- (Second Sylow Theorem) All p-Sylow subgroups of G are conjugate in G.
- (Third Sylow Theorem) If G has N p-Sylow subgroups, then $N \equiv 1$ mod p and N divides $|G|$.

Here are two examples of how the Third Sylow Theorem can be used to prove that a given group is solvable.

Example 8.1.10. Let G be a group of order 14, and let N be the number of 7-Sylow subgroups of G. Then $N \equiv 1$ mod 7 and $N|14$ by the Third Sylow Theorem. It follows easily that $N = 1$, so that G has a unique 7-Sylow subgroup H. Since any conjugate of H is also a 7-Sylow subgroup, H coincides with its conjugates. Thus H is normal, and then Theorem 8.1.4 easily implies that G is solvable.   ◁▷

Example 8.1.11. Let G have order 42. Arguing as in Example 8.1.10, one sees that G has a normal 7-Sylow subgroup H. Then G/H has order 6, so that $G/H \simeq \mathbb{Z}/6\mathbb{Z}$ or S_6, both of which are solvable. Hence G is solvable by Theorem 8.1.4.   ◁▷

In Exercises 6 and 7 you will combine similar arguments with Burnside's Theorem (Theorem 8.1.8) to show that all groups of order < 60 are solvable.

Exercises for Section 8.1

Exercise 1. Consider the groups A_4 and S_4.
(a) Show that $\{(12)(34), (13)(24), (14)(23)\}$ is a normal subgroup of S_4.
(b) Show that A_4 and S_4 are solvable.

Exercise 2. This exercise is concerned with the first part of the proof of Theorem 8.1.4.
(a) Prove assertions (a)–(d) made in the proof of the theorem.
(b) Suppose that $\phi : M_1 \to M_2$ is an onto group homomorphism. If $|M_1| = p$, where p is prime, then prove that $|M_2| = 1$ or p.
(c) Explain how part (b) proves the assertion made in the text that $\widetilde{G}_{i-1}/\widetilde{G}_i$ either is trivial or has prime order.

Exercise 3. Consider the map $\pi : G \to G/H$ used in the proof of Theorem 8.1.4. Given a subgroup $K \subset G/H$, define $\pi^{-1}(K)$ as in (8.1).
(a) Show that $\pi^{-1}(K)$ is a subgroup of G containing H.
(b) Show that H is the kernel of π and that $H = \pi^{-1}(\{eH\})$.
(c) Show that $G = \pi^{-1}(G/H)$.

Exercise 4. In the situation of (8.2), prove that G_i is normal in G_{i-1} and that $gG_i \mapsto \pi(g)\widetilde{G}_i$ gives the isomorphism (8.2).

Exercise 5. In this exercise, you will prove Theorem 8.1.7. We begin with a classic result from group theory about the center of a group of prime power order. Recall that the *center* of a group G is the subset

$$Z(G) = \{g \in G \mid gh = hg \text{ for all } h \in G\}.$$

Most courses in abstract algebra prove that $Z(G) \neq \{e\}$ when $|G| = p^n$, p prime (see, for example, [Herstein, Thm. 2.11.2]). You may assume this result.
(a) In any group, show that $\langle g \rangle$ is normal for all $g \in Z(G)$.
(b) Prove Theorem 8.1.7 using induction on n, where $|G| = p^n$ and p is prime.

Exercise 6. In this exercise you will prove that groups of order 30 are solvable.
(a) Use the method of Example 8.1.10 to prove that groups of order 10 or 15 are solvable.
(b) Show that a group of order 30 is solvable if and only if it has a proper normal subgroup different from $\{e\}$.
(c) Let G be a group of order 30. Use the Third Sylow Theorem to show that G has one or ten 3-Sylow subgroups and one or six 5-Sylow subgroups.
(d) Show that the group G of part (c) can't simultaneously have ten 3-Sylow subgroups and six 5-Sylow subgroups. Conclude that G must be solvable.
See [Herstein, Sec. 2.12, Ex. 7] for further details on the structure of groups of order 30.

Exercise 7. Use Burnside's $p^n q^m$ Theorem (Theorem 8.1.8) to show that groups of order < 60 are solvable, with the possible exception of groups of order 30 or 42. When combined with the previous exercise and Example 8.1.11, this implies that groups of order < 60 are solvable. In Section 8.4 we will prove that A_5 is not solvable. Since A_5 has order 60, it is the smallest nonsolvable group. (One can show that A_5 is the only nonsolvable group of order 60 up to isomorphism.)

Exercise 8. Let G be a finite group, and suppose that we have subgroups

$$\{e\} = G_n \subset \cdots \subset G_0 = G$$

such that G_i is normal in G_{i-1} for $i = 1, \ldots, n$.
(a) Prove that G is solvable if G_{i-1}/G_i is Abelian for $i = 1, \ldots, n$.
(b) Prove that G is solvable if G_{i-1}/G_i is solvable for $i = 1, \ldots, n$.

8.2   RADICAL AND SOLVABLE EXTENSIONS

The purpose of this section is to introduce the field theory needed to study solvability by radicals.

A. Definitions and Examples. The naive idea of solvability by radicals arises from polynomials such as $x^3 + 3x + 1$, whose unique real root is

$$\sqrt[3]{\tfrac{1}{2}\left(-1 + \sqrt{5}\right)} + \sqrt[3]{\tfrac{1}{2}\left(-1 - \sqrt{5}\right)}$$

by Example 1.1.1. This algebraic number is built by taking successive radicals. When we cast this in terms of fields, we are led to the following definition.

Definition 8.2.1. *A field extension $F \subset L$ is **radical** if there are fields*

$$F = F_0 \subset F_1 \subset \cdots \subset F_{n-1} \subset F_n = L$$

where for $i = 1, \ldots, n$, there is $\gamma_i \in F_i$ with $F_i = F_{i-1}(\gamma_i)$, $\gamma_i^{m_i} \in F_{i-1}$, $m_i > 0$.

Notice that if we let $b_i = \gamma_i^{m_i} \in F_{i-1}$, then γ_i is an m_ith root of b_i. This allows us to write $\gamma_i = \sqrt[m_i]{b_i}$, so that

$$F_i = F_{i-1}\left(\sqrt[m_i]{b_i}\right), \quad b_i \in F_{i-1}.$$

This shows that radical extensions are obtained by adjoining successive radicals.

Here is our first example of a radical extension.

Example 8.2.2. For the field extension $\mathbb{Q} \subset \mathbb{Q}(\sqrt{2 + \sqrt{2}})$, let $\gamma_1 = \sqrt{2}$ and $\gamma_2 = \sqrt{2 + \sqrt{2}}$. Then we have the extensions

$$\mathbb{Q} \subset \mathbb{Q}(\gamma_1) = \mathbb{Q}(\sqrt{2}) \subset \mathbb{Q}(\sqrt{2})(\gamma_2) = \mathbb{Q}(\sqrt{2})(\sqrt{2 + \sqrt{2}}),$$

where $\gamma_1^2 = \sqrt{2}^2 = 2 \in \mathbb{Q}$ and $\gamma_2^2 = (\sqrt{2 + \sqrt{2}})^2 = 2 + \sqrt{2} \in \mathbb{Q}(\sqrt{2})$. Since

$$\mathbb{Q}(\sqrt{2})(\sqrt{2 + \sqrt{2}}) = \mathbb{Q}(\sqrt{2 + \sqrt{2}})$$

(be sure you can prove this), $\mathbb{Q} \subset \mathbb{Q}(\sqrt{2 + \sqrt{2}})$ is a radical extension.   ◁▷

An important observation is some extensions are not radical but are contained in larger radical extensions. Here is an example.

Example 8.2.3. Let $\mathbb{Q} \subset L$ be a splitting field of $f = x^3 + x^2 - 2x - 1 \in \mathbb{Q}[x]$. In Example 7.4.3, we showed that f is irreducible over $\mathbb{Q}$ with discriminant

$$\Delta(f) = 49 = 7^2 > 0.$$

By Theorem 1.3.1, the roots of f are real, which allows us to assume that $L \subset \mathbb{R}$. Furthermore, since $\Delta(f)$ is a perfect square, Proposition 7.4.2 implies that $\mathbb{Q} \subset L$ is a Galois extension of degree 3. Cardan's formulas imply that $\mathbb{Q} \subset L$ is contained in a radical extension (see also Exercise 1).

However, the extension $\mathbb{Q} \subset L$ is not radical itself. We prove this as follows. If $\mathbb{Q} \subset L$ were radical, then $[L : \mathbb{Q}] = 3$ would imply that $L = \mathbb{Q}(\gamma)$, where $\gamma^m \in \mathbb{Q}$ for some $m \geq 3$ (see Exercise 2 for the details). Then the minimal polynomial f of γ over $\mathbb{Q}$ would divide $x^m - \gamma^m$ and have degree $[L : \mathbb{Q}] = 3$. Since $\mathbb{Q} \subset L$ is Galois, f would split completely over $\mathbb{Q}(\gamma)$, so that three of $\gamma, \zeta_m \gamma, \zeta_m^2 \gamma, \ldots \zeta_m^{m-1}\gamma$ would lie in L. This is impossible, since $L \subset \mathbb{R}$. Hence $\mathbb{Q} \subset L$ is not radical.   ◁▷

This example motivates the following definition.

Definition 8.2.4. *A field extension $F \subset L$ is **solvable** (sometimes called **solvable by radicals**) if there is a field extension $L \subset M$ such that $F \subset M$ is radical.*

In this terminology, the extension $\mathbb{Q} \subset L$ considered in Example 8.2.3 is solvable, since it is contained in a radical extension.

B. Compositums and Galois Closures. In order to understand radical and solvable extensions, we need to define the *compositum* of two or more subfields.

Definition 8.2.5. *Suppose that we have a field L and two subfields $K_1 \subset L$ and $K_2 \subset L$. Then the* **compositum** *of K_1 and K_2 in L is the smallest subfield of L containing K_1 and K_2. We denote the compositum by $K_1 K_2$.*

In Exercise 3 you will show that the compositum always exists and that the compositum of $K_1 = F(\alpha_1, \ldots, \alpha_n) \subset L$ and $K_2 = F(\beta_1, \ldots, \beta_m) \subset L$ is

$$(8.3) \qquad K_1 K_2 = F(\alpha_1, \ldots, \alpha_n, \beta_1, \ldots, \beta_m).$$

For example, the compositum of $\mathbb{Q}(\sqrt{2})$ and $\mathbb{Q}(\sqrt{3})$ in $\mathbb{R}$ is $\mathbb{Q}(\sqrt{2}, \sqrt{3})$.

We next consider Galois closures. Proposition 7.1.7 proves that every finite separable extension $F \subset L$ has a Galois closure, which may be thought of as the smallest Galois extension of F containing L. We can express the Galois closure as a compositum as follows.

Proposition 8.2.6. *Suppose that $F \subset L \subset M$ where $F \subset M$ is Galois. Then the compositum of all conjugate fields of L in M is the Galois closure of $F \subset L$.*

Proof. The Theorem of the Primitive Element implies that $L = F(\alpha)$ for some $\alpha \in L$. Since $F \subset M$ is Galois, the minimal polynomial h of α over F is separable and splits completely over M, say $h(x) = (x - \alpha_1) \cdots (x - \alpha_r)$, where $\alpha_1 = \alpha$. It follows that

$$K = F(\alpha_1, \ldots, \alpha_r)$$

is a Galois extension of F containing L. In Exercise 4 you will prove that $F \subset K$ is the Galois closure of $F \subset L$. In Exercise 4 you will also show that the conjugate fields of L in M are $F(\alpha_i)$ for $i = 1, \ldots, r$. Then (8.3) implies that

$$F(\alpha_1) \cdots F(\alpha_r) = F(\alpha_1, \ldots, \alpha_r) = K.$$

This proves that K is the compositum of the conjugate fields of L in M. $\qquad \square$

C. Properties of Radical and Solvable Extensions. We begin with the following useful lemma.

Lemma 8.2.7.
(a) *If $F \subset L$ and $L \subset M$ are radical, then so is $F \subset M$.*
(b) *If we have $F \subset K_1 \subset L$ and $F \subset K_2 \subset L$ such that $F \subset K_1$ is radical, then $K_2 \subset K_1 K_2$ is radical.*
(c) *If we have $F \subset K_1 \subset L$ and $F \subset K_2 \subset L$ such that $F \subset K_1$ and $F \subset K_2$ are radical, then $F \subset K_1 K_2$ is radical.*

Proof. Part (a) follows easily from the definition of radical extension by combining the sequences of fields used for $F \subset L$ and $L \subset M$. We omit the details.

For part (b), we have fields $F = F_0 \subset F_1 \subset \cdots \subset F_{n-1} \subset F_n = K_1$ such that $F_i = F_{i-1}(\gamma_i)$, where $\gamma_i^{m_i} \in F_{i-1}$ for $1 \leq i \leq n$. Then define fields

(8.4)
$$
\begin{aligned}
F_0' &= K_2, \\
F_1' &= F_0'(\gamma_1), \\
&\vdots \\
F_n' &= F_{n-1}'(\gamma_n).
\end{aligned}
$$

In Exercise 5 you will use $F \subset K_2$ and induction to show that

(8.5)
$$F_i \subset F_i'$$

for $i = 0, \dots, n$. This in turn implies that

$$\gamma_i^{m_i} \in F_{i-1} \subset F_{i-1}'$$

for $i = 1, \dots, n$. It follows easily that $K_2 = F_0' \subset \cdots \subset F_n'$ is radical. In Exercise 5 you will show that F_n' is the compositum $K_1 K_2$, which will prove part (b).

Finally, for part (c), note that $K_2 \subset K_1 K_2$ is radical by part (b). Then we are done by part (a), since $F \subset K_2$ is radical by assumption. $\square$

We next use Proposition 8.2.6 to study the Galois closure of a radical extension.

Theorem 8.2.8. *If an extension $F \subset L$ is separable and radical, then its Galois closure is also radical.*

Proof. Find an extension $L \subset M$ such that $F \subset M$ is Galois (such an extension exists by the existence of Galois closures). Given $\sigma \in \mathrm{Gal}(M/F)$, we get the conjugate field $F \subset \sigma L \subset M$. Exercise 6 shows that $F \subset \sigma L$ is radical because $F \subset L$ is radical.

But once we know that each conjugate field is radical over F, Lemma 8.2.7 tells us that their compositum is also radical over F. Then we are done, since the compositum is the Galois closure by Proposition 8.2.6. $\square$

The following corollary of Theorem 8.2.8 will be used in Section 8.5.

Corollary 8.2.9. *If a finite extension $F \subset L$ of characteristic 0 is solvable, then so is its Galois closure.*

Proof. Since $F \subset L$ is solvable, we have $F \subset L \subset L'$ such that $F \subset L'$ is radical. Furthermore, $F \subset L'$ is separable (we are in characteristic 0) and hence has a Galois closure $F \subset L' \subset M$. Then $F \subset M$ is radical by Theorem 8.2.8.

Now consider $F \subset L \subset M$. Since $F \subset M$ is Galois, it contains the Galois closure of $F \subset L$ by Proposition 8.2.6. Thus the Galois closure lies in the radical extension $F \subset M$, so that the Galois closure is solvable by definition. $\square$

In the next section, we will see how the solvable extensions defined here relate to the solvable groups studied in Section 8.1.

Historical Notes

In 1824 Abel proved that the general quintic cannot be solved by radicals. He presented his proof in the privately printed *Memoir on algebraic equations, in which is demonstrated the impossibility of solving the general equation of the fifth degree* [Abel, pp. 28–33] that he sent to the leading mathematicians of Europe. In this memoir, Abel begins his proof as follows:

> Let
> $$y^5 - a\,y^4 + b\,y^3 - c\,y^2 + d\,y - e = 0$$
> be the general equation of the fifth degree and let us suppose that it is solvable algebraically, that is, one can express y by a function formed by radicals of the quantities a, b, c, d, and e.
>
> It is clear that in this case we can express y in the form
> $$y = p + p_1 R^{\frac{1}{m}} + p_2 R^{\frac{2}{m}} + \cdots + p_{m-1} R^{\frac{m-1}{m}},$$
> m being a prime number and R, p, p_1, p_2, etc. functions of the same form as y, and so on until we come to rational functions of the quantities a, b, c, d, and e.

(This is from the English translation in [9, pp. 155–169].) Abel's description of y in terms a radicals is a "top-down" version of the definition of radical extension. Definition 8.2.1 is a "bottom-up" approach that begins with the smallest field (here containing the coefficients a, b, c, d, e) and successively adds radicals. Abel instead begins with the largest field (here containing y) and successively strips away radicals. Another difference is that Abel focuses on individual elements rather than the fields in which they lie. Nevertheless, the above quotation contains a clear description of a radical extension. Be sure you understand this.

The reader may wonder why Abel assumes that m is prime in $R^{\frac{1}{m}}$. We will see in Lemma 8.6.2 that this is no restriction. Also note that Abel's "solvable algebraically" means "solvable by radicals" in modern terms. We will say more about Abel's proof in Sections 8.5 and 12.1.

Exercises for Section 8.2

Exercise 1. As in Example 8.2.3, let L be a splitting field of $x^3 + x^2 - 2x - 1$ over $\mathbb{Q}$. Also let $\zeta_7 = e^{2\pi i/7}$.
(a) Show that the roots of $x^3 + x^2 - 2x - 1$ are $2\cos(2j\pi/7) = \zeta_7^j + \zeta_7^{-j}$ for $j = 1, 2, 3$.
(b) Show that $\mathbb{Q} \subset L \subset \mathbb{Q}(\zeta_7)$, and explain why $\mathbb{Q} \subset \mathbb{Q}(\zeta_7)$ is radical.

Exercise 2. In the situation of Example 8.2.3, assume that $\mathbb{Q} \subset L$ is radical. Prove that $L = \mathbb{Q}(\gamma)$ where $\gamma^m \in \mathbb{Q}$ for some $m \geq 3$.

Exercise 3. Here you will prove two properties of compositums.
(a) Prove that the compositum $K_1 K_2$ exists.
(b) Prove (8.3).

Exercise 4. This exercise is concerned with the proof of Propostion 8.2.6
(a) Show that $K = F(\alpha_1, \ldots, \alpha_r)$ is the Galois closure of $F \subset L$.
(b) Prove that the conjugates of L in M are the fields $F(\alpha_i)$ for $i = 1, \ldots, r$.

Exercise 5. This exercise will complete the proof of part (b) of Lemma 8.2.7.
(a) Prove (8.5).
(b) Prove that the field F_n' defined in (8.4) is the compositum $K_1 K_2$.

Exercise 6. Suppose we have finite extensions $F \subset L \subset M$ and $\sigma \in \mathrm{Gal}(M/F)$, and assume that $F \subset L$ is radical. Prove that $F \subset \sigma L$ is also radical.

Exercise 7. Suppose that we have extensions $F \subset K_1 \subset L$ and $F \subset K_2 \subset L$ such that $F \subset K_1$ and $F \subset K_2$ are Galois. Prove that $F \subset K_1 K_2$ is Galois. This will show that the compositum of two Galois extensions is again Galois.

8.3 SOLVABLE EXTENSIONS AND SOLVABLE GROUPS

The main question we will answer in this section is: when is a finite extension $F \subset L$ solvable? Because of subtleties that can occur in characteristic p, we will make the following simplifying assumption:

All fields appearing in this section will have characteristic 0.

See Section 8.6 for an example to show what can go wrong in characteristic p.

A. Roots of Unity and Lagrange Resolvents. Section A.2 shows that given one mth root of a complex number, we get the others by multiplying by the mth roots of unity. Since radical extensions involve taking mth roots, it makes sense that roots of unity will play an important role. However, the roots of unity in Section A.2 are complex numbers, while the fields considered here need not be subfields of $\mathbb{C}$ (even though they have characteristic 0). For this reason, we need to study roots of unity for arbitrary fields of characteristic 0.

Given a positive integer m and a field L of characteristic 0, consider the splitting field of $x^m - 1$ over L. In Exercise 1 you will show that $x^m - 1$ has m distinct roots in its splitting field. These roots form a group under multiplication, which is cyclic by Proposition A.5.3. A generator ζ of this group has the following two properties:

- The m distinct roots of $x^m - 1$ are $1, \zeta, \ldots, \zeta^{m-1}$.
- The splitting field of $x^m - 1$ over L is $L(1, \zeta, \ldots, \zeta^{m-1}) = L(\zeta)$.

We call ζ a *primitive mth root of unity* in this situation. We claim that

(8.6) $L \subset L(\zeta)$ is Galois and $\mathrm{Gal}(L(\zeta)/L)$ is Abelian.

To prove this, note that $L \subset L(\zeta)$ is Galois, since $L(\zeta)$ is the splitting field of the separable polynomial $x^m - 1 \in L[x]$. Now suppose that $\sigma, \tau \in \mathrm{Gal}(L(\zeta)/L)$. Then σ, τ are determined by their values on ζ, and since the roots of $x^m - 1$ are $1, \zeta, \ldots, \zeta^{m-1}$, it follows that $\sigma(\zeta) = \zeta^i$ and $\tau(\zeta) = \zeta^j$ for integers i, j. Thus

$$\sigma\tau(\zeta) = \sigma(\zeta^j) = (\sigma(\zeta))^j = (\zeta^i)^j = \zeta^{ij}.$$

A similar computation shows that $\tau\sigma(\zeta) = \zeta^{ji} = \zeta^{ij}$. Then $\sigma\tau = \tau\sigma$, since $\sigma\tau$ and $\tau\sigma$ are uniquely determined by their values on ζ. Hence $\mathrm{Gal}(L(\zeta)/L)$ is Abelian.

Given a Galois extension $F \subset L$ and a primitive mth root of unity ζ, we get the extensions

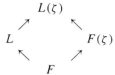

We can relate the solvability of the various Galois groups as follows.

Lemma 8.3.1. *Let $F \subset L$ be a Galois extension, and ζ be a primitive mth root of unity. Then $F \subset L(\zeta)$ and $F(\zeta) \subset L(\zeta)$ are also Galois, and*

$$\mathrm{Gal}(L/F) \text{ is solvable} \iff \mathrm{Gal}(L(\zeta)/F) \text{ is solvable}$$
$$\iff \mathrm{Gal}(L(\zeta)/F(\zeta)) \text{ is solvable.}$$

Proof. In Exercise 2 you will prove that $F \subset L(\zeta)$ is Galois, which implies that $F(\zeta) \subset L(\zeta)$ are also Galois. To prove the first equivalence, we use the extensions

$$F \subset L \subset L(\zeta).$$

Since $F \subset L(\zeta)$ and $F \subset L$ are Galois, Theorem 7.2.7 implies that $\mathrm{Gal}(L(\zeta)/L)$ is a normal subgroup of $\mathrm{Gal}(L(\zeta)/F)$ such that

$$\mathrm{Gal}(L/F) \simeq \mathrm{Gal}(L(\zeta)/F)/\mathrm{Gal}(L(\zeta)/L).$$

But $\mathrm{Gal}(L(\zeta)/L)$ is Abelian by (8.6) and hence solvable by Proposition 8.1.5. Then Theorem 8.1.4 implies that $\mathrm{Gal}(L(\zeta)/F)$ is solvable if and only if $\mathrm{Gal}(L/F)$ is. This proves the first equivalence of the lemma.

For the second equivalence, consider the extensions

$$F \subset F(\zeta) \subset L(\zeta).$$

Here, $F \subset F(\zeta)$ is Galois by (8.6) (applied with F in place of L), so that, arguing as above, we get a group isomorphism

$$\mathrm{Gal}(F(\zeta)/F) \simeq \mathrm{Gal}(L(\zeta)/F)/\mathrm{Gal}(L(\zeta)/F(\zeta)).$$

Also as above, $\mathrm{Gal}(F(\zeta)/F)$ is Abelian and hence solvable, and then Theorem 8.1.4 implies that $\mathrm{Gal}(L(\zeta)/F)$ is solvable if and only if $\mathrm{Gal}(L(\zeta)/F(\zeta))$ is. This proves the second equivalence of the lemma. $\square$

The following result will play a crucial role in our analysis of solvable extensions. The proof uses a clever construction due to Lagrange.

Lemma 8.3.2. *Suppose that $K \subset M$ is a Galois extension with $\mathrm{Gal}(M/K) \simeq \mathbb{Z}/p\mathbb{Z}$, p prime. If K contains a primitive pth root of unity ζ, then there is $\alpha \in M$ such that $M = K(\alpha)$ and $\alpha^p \in K$.*

Proof. By hypothesis, $\text{Gal}(M/K)$ is cyclic of order p. Let $\sigma \in \text{Gal}(M/K)$ be a generator, and fix $\beta \in M \setminus K$. Then, for each $i = 0, \ldots, p - 1$, consider the *Lagrange resolvent* defined by

$$(8.7) \qquad \alpha_i = \beta + \zeta^{-i} \sigma(\beta) + \zeta^{-2i} \sigma^2(\beta) + \cdots + \zeta^{-i(p-1)} \sigma^{p-1}(\beta).$$

This easily implies that

$$\zeta^{-i} \sigma(\alpha_i) = \zeta^{-i} \sigma(\beta) + \zeta^{-2i} \sigma^2(\beta) + \cdots + \zeta^{-i(p-1)} \sigma^{p-1}(\beta) + \zeta^{-ip} \sigma^p(\beta).$$

Since $\zeta^p = 1$ and σ^p is the identity, the final term on the right-hand side of the above equation simplifies to β, and then the equation becomes

$$\zeta^{-i} \sigma(\alpha_i) = \alpha_i,$$

so that

$$(8.8) \qquad\qquad\qquad\qquad \sigma(\alpha_i) = \zeta^i \alpha_i.$$

Since $\zeta \in K$ and $\zeta^p = 1$, (8.8) easily implies that

$$\sigma(\alpha_i^p) = \alpha_i^p.$$

But σ generates $\text{Gal}(M/K)$, so that the above equation shows that α_i^p is fixed by the Galois group. Hence $\alpha_i^p \in K$, since $K \subset M$ is Galois. Also, when $i = 0$, (8.8) becomes $\sigma(\alpha_0) = \alpha_0$, and then the argument just given shows that $\alpha_0 \in K$.

Suppose for the moment there is some i between 1 and $p - 1$ such that $\alpha_i \neq 0$. For these i's, we also have $\zeta^i \neq 1$, and it follows that $\zeta^i \alpha_i \neq \alpha_i$. Combining this with (8.8), we conclude that $\sigma(\alpha_i) \neq \alpha_i$, so that $\alpha_i \notin K$. This implies $M = K(\alpha_i)$, since $[M : K]$ is prime. Then $\alpha = \alpha_i$ has the desired properties, since $\alpha_i^p \in K$.

It remains to consider what happens if $\alpha_i = 0$ for all $i = 1, \ldots, p - 1$. In this case, we add up the equations (8.7) for $i = 0, \ldots, p - 1$ to obtain

$$
\begin{aligned}
\alpha_0 &= \alpha_0 + \alpha_1 + \cdots + \alpha_{p-1} \\
&= \left(\beta + \sigma(\beta) + \sigma^2(\beta) + \cdots + \sigma^{p-1}(\beta) \right) \\
&\quad + \left(\beta + \zeta^{-1} \sigma(\beta) + \zeta^{-2} \sigma^2(\beta) + \cdots + \zeta^{-(p-1)} \sigma^{p-1}(\beta) \right) \\
&\quad + \left(\beta + \zeta^{-2} \sigma(\beta) + \zeta^{-4} \sigma^2(\beta) + \cdots + \zeta^{-2(p-1)} \sigma^{p-1}(\beta) \right) + \cdots \\
&\quad + \left(\beta + \zeta^{-(p-1)} \sigma(\beta) + \zeta^{-2(p-1)} \sigma^2(\beta) + \cdots + \zeta^{-(p-1)(p-1)} \sigma^{p-1}(\beta) \right) \\
&= p\beta + (1 + \zeta^{-1} + \zeta^{-2} + \cdots + \zeta^{-(p-1)})\sigma(\beta) \\
&\quad + (1 + \zeta^{-2} + \zeta^{-4} + \cdots + \zeta^{-2(p-1)})\sigma^2(\beta) + \cdots \\
&\quad + (1 + \zeta^{-(p-1)} + \zeta^{-2(p-1)} + \cdots + \zeta^{-(p-1)(p-1)})\sigma^{p-1}(\beta).
\end{aligned}
$$

In Exercise 3 you will show that

$$(8.9) \qquad\qquad 1 + \zeta^{-i} + \zeta^{-2i} + \cdots + \zeta^{-(p-1)i} = 0$$

for $i = 1, \ldots, p - 1$. It follows that the above formula for α_0 simplifies to

$$\alpha_0 = p\beta,$$

so that $\beta = \alpha_0/p$ (remember that we are in characteristic 0). However, we proved above that $\alpha_0 \in K$, yet $\beta \notin K$ by assumption. This contradiction shows that at least one of $\alpha_1, \ldots, \alpha_{p-1}$ is nonzero, which completes the proof of the lemma. $\qquad\square$

B. Galois's Theorem. In Section 8.2, we showed that if $F \subset L$ is solvable, then we can find an extension $L \subset M$ such that $F \subset M$ is Galois and solvable. For an arbitrary Galois extension, the wonderful fact is that the Galois group determines whether or not the extension is solvable. The following theorem due to Galois is one of the most important applications of Galois theory.

Theorem 8.3.3. *Let $F \subset L$ be a Galois extension. Then the following are equivalent:*
(a) $F \subset L$ *is a solvable extension.*
(b) $\mathrm{Gal}(L/F)$ *is a solvable group.*

Proof. We prove (a) $\Rightarrow$ (b) in three steps.

Reduction to the Radical Case. Since $F \subset L$ is solvable, it lies in a radical extension $F \subset L'$. By Theorem 8.2.8, the Galois closure $F \subset M$ of $F \subset L'$ is radical over F. Thus we have $F \subset L \subset M$ where M is radical and Galois over F.

Suppose for the moment that $\mathrm{Gal}(M/F)$ is a solvable group. Since $F \subset L$ is Galois, Theorem 7.2.7 implies that we have an isomorphism

$$\mathrm{Gal}(L/F) \simeq \mathrm{Gal}(M/F)/\mathrm{Gal}(M/L).$$

Then Theorem 8.1.4 implies that $\mathrm{Gal}(L/F)$ is also solvable. Hence it suffices to prove that $\mathrm{Gal}(M/F)$ is solvable. In other words, we can assume that $F \subset L$ is radical and Galois.

Adjunction of Roots of Unity. Suppose that $F \subset L$ is radical and Galois. If we adjoin a primitive mth root of unity ζ to both F and L, then part (b) of Lemma 8.2.7 implies that the resulting extension $F(\zeta) \subset L(\zeta)$ is radical, since $L(\zeta)$ is the compositum of $F(\zeta)$ and L. This extension is also Galois by Lemma 8.3.1. If we can show that $\mathrm{Gal}(L(\zeta)/F(\zeta))$ is solvable, then Lemma 8.3.1 will imply that $\mathrm{Gal}(L/F)$ is solvable. Hence we can assume without loss of generality that F contains any mth root of unity we want.

Proof of Solvability. Since $F \subset L$ is radical, we have subfields

$$(8.10) \qquad F = F_0 \subset F_1 \subset \cdots \subset F_{n-1} \subset F_n = L$$

such that for $i = 1, \ldots, n$, we have $F_i = F_{i-1}(\gamma_i)$, where $\gamma_i^{m_i} \in F_{i-1}$ for some $m_i > 0$. By the previous step, we can also assume that F contains a primitive m_ith root of unity ζ_i for $i = 1, \ldots, n$. In this situation, we claim that

$$(8.11) \qquad F_{i-1} \subset F_i \text{ is Galois with cyclic Galois group.}$$

To prove this, note that $1, \zeta_i, \ldots, \zeta_i^{m_i-1}$ are the distinct m_ith roots of unity, which means that

$$\gamma_i, \zeta_i\gamma_i, \ldots, \zeta_i^{m_i-1}\gamma_i$$

are the distinct roots of $x^{m_i} - \gamma_i^{m_i} \in F_{i-1}[x]$. Since $\zeta_i \in F \subset F_{i-1}$, we have

$$F_{i-1}(\gamma_i, \zeta_i\gamma_i, \ldots, \zeta_i^{m_i-1}\gamma_i) = F_{i-1}(\gamma_i).$$

This shows that $F_{i-1} \subset F_i = F_{i-1}(\gamma_i)$ is Galois. The proof that the Galois group is cyclic is similar to the proof of (8.6) and is left to the reader as Exercise 4. This completes the proof of (8.11)

We now prove solvability. Given the subfields (8.10), consider the subgroups

$$G_i = \mathrm{Gal}(L/F_i) \subset \mathrm{Gal}(L/F).$$

Since the Galois correspondence is inclusion-reversing, (8.10) implies that

$$\{1_L\} = \mathrm{Gal}(L/L) = \mathrm{Gal}(L/F_n) = G_n \subset G_{n-1} \subset \cdots$$
$$\subset G_1 \subset G_0 = \mathrm{Gal}(L/F_0) = \mathrm{Gal}(L/F).$$

Consider the extensions $F_{i-1} \subset F_i \subset L$. Then $F_{i-1} \subset L$ is Galois, since F_{i-1} is an intermediate field of the Galois extension $F \subset L$. Furthermore, $F_{i-1} \subset F_i$ is also Galois by (8.11). Hence Theorem 7.2.7 implies that G_i is normal in G_{i-1} with

$$G_{i-1}/G_i = \mathrm{Gal}(L/F_{i-1})/\mathrm{Gal}(L/F_i) \simeq \mathrm{Gal}(F_i/F_{i-1}).$$

By (8.11), we conclude that G_{i-1}/G_i is Abelian. Since this is true for all $i = 1, \ldots, n$, part (a) of Exercise 8 from Section 8.1 implies that $\mathrm{Gal}(L/F)$ is solvable. This completes the proof of (a) $\Rightarrow$ (b).

It remains to prove (b) $\Rightarrow$ (a). We do this in two steps.

A Special Case. Let $F \subset L$ be Galois with solvable Galois group. Assume in addition that F satisfies the following special hypothesis:

(8.12) F has a primitive pth root of unity for every prime p dividing $|\mathrm{Gal}(L/F)|$.

We will prove that $F \subset L$ is radical in this situation. Since $\mathrm{Gal}(L/F)$ is solvable, we have subgroups $\{1_L\} = G_n \subset \cdots \subset G_0 = \mathrm{Gal}(L/F)$ as in Definition 8.1.1. Then consider the fixed fields

$$F_i = L_{G_i} \subset L.$$

Since the Galois correspondence is inclusion-reversing, this gives the fields

$$F = L_{\mathrm{Gal}(L/F)} = L_{G_0} = F_0 \subset F_1 \subset \cdots$$
$$\subset F_{n-1} \subset F_n = L_{G_n} = L_{\{1_L\}} = L.$$

Furthermore, since G_i is normal in G_{i-1}, the Galois correspondence together with Theorem 7.2.7 implies that

$$G_{i-1}/G_i \simeq \mathrm{Gal}(F_i/F_{i-1}).$$

Since $[G_{i-1} : G_i]$ is prime, $\mathrm{Gal}(F_i/F_{i-1}) \simeq \mathbb{Z}/p\mathbb{Z}$ for a prime p. In Exercise 5 you will prove that p divides $|\mathrm{Gal}(L/F)|$. By (8.12), F and hence F_{i-1} contain a primitive pth root of unity.

It follows that $F_{i-1} \subset F_i$ satisfies the conditions of Lemma 8.3.2. Thus F_i is obtained from F_{i-1} by adjunction of a pth root of an element of F_{i-1}. This proves that $F \subset L$ is a radical extension when F satisfies (8.12).

The General Case. Finally, we consider what happens when we only assume that $F \subset L$ is a Galois extension with solvable Galois group. In this situation, let ζ be a primitive mth root of unity, where $m = |\mathrm{Gal}(L/F)|$. By Lemma 8.3.1, $\mathrm{Gal}(L(\zeta)/F(\zeta))$ is solvable since $\mathrm{Gal}(L/F)$ is.

We relate the orders of these groups as follows. As in the proof of Lemma 8.3.2, we have an isomorphism

$$\mathrm{Gal}(L/F) \simeq \mathrm{Gal}(L(\zeta)/F)/\mathrm{Gal}(L(\zeta)/L).$$

If you look back at the proof of Theorem 7.2.7, you will see that this isomorphism comes from the homomorphism

$$\mathrm{Gal}(L(\zeta)/F) \longrightarrow \mathrm{Gal}(L/F)$$

given by restricting an automorphism of $L(\zeta)$ to L. Since $\mathrm{Gal}(L(\zeta)/F(\zeta))$ is a subgroup of $\mathrm{Gal}(L(\zeta)/F)$, we have a homomorphism

(8.13) $$\mathrm{Gal}(L(\zeta)/F(\zeta)) \longrightarrow \mathrm{Gal}(L/F)$$

also given by restriction to L. But the kernel of this map is the identity, since elements of the kernel are the identity on both L and $F(\zeta)$. Thus (8.13) is one to one, which by Lagrange's Theorem implies that

(8.14) $\qquad m = |\mathrm{Gal}(L/F)|$ is a multiple of $|\mathrm{Gal}(L(\zeta)/F(\zeta))|$.

Now let p be a prime dividing $|\mathrm{Gal}(L(\zeta)/F(\zeta))|$. Then p divides m by (8.14). Since ζ is a primitive mth root of unity, $\zeta^{m/p}$ is a primitive pth root of unity (see Exercise 6). Since $\zeta^{m/p} \in F(\zeta)$, we conclude that $F(\zeta) \subset L(\zeta)$ satisfies (8.12) with F and L replaced by $F(\zeta)$ and $L(\zeta)$ respectively. It follows that $F(\zeta) \subset L(\zeta)$ is radical by the Special Case. But $F \subset F(\zeta)$ is obviously radical ($\zeta^m = 1 \in F$), so that $F \subset L(\zeta)$ is radical by part (a) of Proposition 8.2.7.

Since $F \subset L(\zeta)$ is radical, the obvious inclusion $L \subset L(\zeta)$ implies that L lies in a radical extension of F. Hence $F \subset L$ is solvable by definition. This completes the proof of the theorem. $\qquad\square$

The proof of Theorem 8.3.3 implies that a solvable Galois extension becomes radical after adjoining a suitable root of unity. Here is the precise result.

Corollary 8.3.4. *Let $F \subset L$ be Galois and solvable, and let ζ be a primitive mth root of unity, where $m = [L : F]$. Then $F \subset L(\zeta)$ is radical.*

Proof. If $F \subset L$ is Galois and solvable, then $\mathrm{Gal}(L/F)$ is solvable. The General Case of the proof of (b) $\Rightarrow$ (a) in Theorem 8.3.3 shows that $F \subset L(\zeta)$ is radical, where ζ is a primitive mth root of unity for $m = |\mathrm{Gal}(L/F)|$. Then we are done, since $|\mathrm{Gal}(L/F)| = [L : F]$ for Galois extensions. $\qquad\square$

Exercise 7 will give a more refined version of this result.

C. Cardan's Formulas. We conclude this section with a surprising application of Lagrange resolvents. Let $F = \mathbb{Q}(\omega)$, where $\omega = e^{2\pi i/3}$ is our usual cube root of unity. Note that ω is primitive as defined at the beginning of the section.

We will study the universal cubic

$$\tilde{f} = x^3 - \sigma_1 x^2 + \sigma_2 x - \sigma_3 = (x - x_1)(x - x_2)(x - x_3).$$

If we regard this as a polynomial with coefficients in $K = F(\sigma_1, \sigma_2, \sigma_3)$, then the splitting field of $\tilde{f}$ is the universal extension in degree 3,

$$K = F(\sigma_1, \sigma_2, \sigma_3) \subset L = F(x_1, x_2, x_3),$$

with Galois group $\mathrm{Gal}(L/K) = S_3$ (we identify an automorphism with the permutation it induces on the roots). As noted in Example 8.1.2, the subgroups

$$\{e\} \subset A_3 \subset S_3$$

show that S_3 is solvable. Hence $K \subset L$ is solvable.

A more interesting picture emerges when we apply the proof of Theorem 8.3.3 to this situation. Since (8.12) is satisfied, the Special Case tells us to take the fixed fields of the above subgroups. By Theorem 7.4.4, these fixed fields are

$$K \subset K(\sqrt{\Delta}) \subset L,$$

where $\Delta \in K$ is the discriminant of $\tilde{f}$ and $\sqrt{\Delta} = (x_2 - x_1)(x_3 - x_2)(x_3 - x_1)$. (This differs by a sign from the formula for $\sqrt{\Delta}$ used in Theorem 7.4.4. However, it gives the same field $K(\sqrt{\Delta})$ and leads to nicer formulas below.)

Since $K \subset K(\sqrt{\Delta})$ is clearly radical, we turn our attention to $K(\sqrt{\Delta}) \subset L$. Here, the Galois group is $A_3 \simeq \mathbb{Z}/3\mathbb{Z}$ (be sure you know why). Since K contains the primitive cube root of unity ω, Lemma 8.3.2 implies that there is $\alpha \in L$ such that

$$L = K(\sqrt{\Delta})(\alpha), \quad \alpha^3 \in K(\sqrt{\Delta}).$$

To get an explicit formula for α, we use the Lagrange resolvent α_i defined in (8.7) for the generator $\sigma = (123)$ of A_3. Setting $\zeta = \omega$, $\beta = x_1$, and $i = 1$ gives

$$\alpha_1 = x_1 + \omega^{-1}\sigma \cdot x_1 + \omega^{-2}\sigma^2 \cdot x_1 = x_1 + \omega^2 x_2 + \omega x_3,$$

since $\omega^{-1} = \omega^2$. This formula for α_1 relates nicely to Section 1.2:

- In (1.10) of Section 1.2 we showed that

$$z_1 = \tfrac{1}{3}(x_1 + \omega^2 x_2 + \omega x_3)$$

is a root of the cubic resolvent (1.9). Up to the factor of $\tfrac{1}{3}$, this is precisely α_1. So Galois theory explains where z_1 in (1.10) comes from!

- Furthermore, recall that the roots of the cubic resolvent listed in (1.10) were obtained from z_1 by applying elements of S_3. Thus (1.10) comes from z_1 by applying elements of $\mathrm{Gal}(L/K)$. It follows from (7.1) that the cubic resolvent is the minimal polynomial of z_1. Galois theory explains the cubic resolvent!

This is nice, but things get even better when we use our methods for computing symmetric polynomials. Namely, $\alpha_1^3 \in K(\sqrt{\Delta})$ implies that $\alpha_1^3 = A + B\sqrt{\Delta}$ for some $A, B \in F(\sigma_1, \sigma_2, \sigma_3)$. Since α_1 is a polynomial in the x_i, Exercise 3 of Section 7.4 implies that $A, B \in F[\sigma_1, \sigma_2, \sigma_3]$. In Exercise 8 you will show that

$$\alpha_1^3 = \tfrac{-27}{2}q + \tfrac{3\sqrt{3}i}{2}\sqrt{\Delta} = \tfrac{27}{2}\left(-q + \sqrt{\tfrac{-\Delta}{27}}\right),$$

where $q = -2\sigma_1^3/27 + \sigma_1\sigma_2/3 - \sigma_3$. This allows us to write

$$\alpha_1 = x_1 + \omega^2 x_2 + \omega x_3 = 3\sqrt[3]{\tfrac{1}{2}\left(-q + \sqrt{\tfrac{-\Delta}{27}}\right)}.$$

In Exercise 8 you will also show that if we set $\beta_1 = (23) \cdot \alpha_1$, then

$$\beta_1 = x_1 + \omega^2 x_3 + \omega x_2 = 3\sqrt[3]{\tfrac{1}{2}\left(-q - \sqrt{\tfrac{-\Delta}{27}}\right)}$$

and

(8.15)
$$\begin{aligned}
x_1 &= \tfrac{1}{3}(\sigma_1 + \alpha_1 + \beta_1), \\
x_2 &= \tfrac{1}{3}(\sigma_1 + \omega\alpha_1 + \omega^2\beta_1), \\
x_3 &= \tfrac{1}{3}(\sigma_1 + \omega^2\alpha_1 + \omega\beta_1).
\end{aligned}$$

If you compare this with (1.8), (1.18), and (1.19) in Section 1.2, you will see we have derived Cardan's formulas using Galois theory.

We will say more about solving polynomials by radicals in Section 8.5.

Historical Notes

Solvable groups first appeared in Galois's version of Theorem 8.3.3. Here is an extract from his statement of the theorem [Galois, pp. 57–59]:

> I first observe that to solve an equation, it is necessary to reduce its group until it contains only a single permutation ...

> Given this, we will try to find the condition satisfied by the group of an equation for which it is possible to reduce the group [to a single permutation] by adjunction of radical quantities ...

The second part of the quotation refers to a radical extension. Furthermore, since the radicals need not lie in the splitting field of the polynomial, Galois is describing a solvable extension.

The first part of the quotation explains the strategy used by Galois: as more radicals are adjoined, the field gets bigger, so that under the Galois correspondence the group gets smaller. Furthermore, if the splitting field is $K \subset L$, then the fixed field of $\{e\} \subset \mathrm{Gal}(L/K)$ is L. This means that when the group is reduced to "a single permutation," we have found the splitting field.

Galois wants to know the "condition satisfied by the group" in this situation. His method is to study how the Galois group changes under the adjunction of a pth root for some prime p. In [Galois, p. 59], he says the following:

> We can always suppose ... that included among the quantities adjoined earlier to the equation is a pth root of unity α ...
>
> Consequently, by theorems II and III, the group of the equation should decompose into p groups that have in relation to one another the following double property: 1° that one passes from one to the other by a single permutation; 2° that they all contain the same substitutions.

The first part of this quotation refers to adjoining roots of unity, just as we did in the proof of Theorem 8.3.3. The second part seems more obscure until one realizes that the "double property" 1° and 2° is Galois's awkward way of saying normal subgroup. Then decomposing into "p groups" refers to cosets of the subgroup, so that we have a normal subgroup of prime index p. Since this happens for the radical adjunctions that reduce the Galois group to the identity, we see that the Galois group is solvable. This is the condition that Galois sought and is the first appearance of solvable groups in mathematics.

Galois also asserts that the converse is true. The main point is Lemma 8.3.2, which Galois states as follows [Galois, pp. 59–61]:

> I say reciprocally that if the group of the equation can be partitioned into p groups that have this double property, one can, by a simple extraction of a pth root, and by adjunction of this pth root, reduce the group of the equation to one of the partial groups

The key step in his proof is a Lagrange resolvent, which Galois writes as

$$\theta + \alpha\theta_1 + \alpha^2\theta_2 + \cdots + \alpha^{p-1}\theta_{p-1},$$

where $\theta, \theta_1, \ldots, \theta_{p-1}$ correspond to $\beta, \sigma(\beta), \ldots, \sigma^{p-1}(\beta)$ in (8.7) and α is a pth root of unity. Students usually find the proof of Theorem 8.3.3 to be straightforward, with the exception of the Lagrange resolvent (8.7)—this seems to come out of the blue. Yet here is Galois using essentially the same resolvent with no explanation whatsoever. As we will learn in Chapter 12, Galois didn't need to say anything, for Lagrange had worked out the theory of such resolvents in detail in 1770.

One observation is that when Galois says "partitioned into p groups," he seems to be using the term "group" for both a subgroup and its cosets. In fact, the situation is even more complicated, as we will see in Chapter 12. Given that we are at the birth of group theory, some confusion about terminology is understandable.

Exercises for Section 8.3

Exercise 1. Let m be a positive integer, and let L be a field of characteristic 0. Then let $L \subset M$ be the splitting field of $x^m - 1 \in L[x]$.
(a) Prove that $x^m - 1$ is separable.
(b) Prove that the roots of $x^m - 1$ lying in M form a group under multiplication.

Exercise 2. Assume that $F \subset L$ is a Galois extension and that F has characteristic 0. Also, consider the extension $L \subset L(\zeta)$ obtained by adjoining a primitive mth root of unity. Prove that $F \subset L(\zeta)$ is Galois.

Exercise 3. Prove (8.9), where ζ is a primitive pth root of unity and $1 \le i \le p - 1$.

Exercise 4. Consider the extension $F_{i-1} \subset F_i$ of (8.11). In the discussion following (8.11), we showed that this extension is Galois. We now describe its Galois group.
(a) Let $\sigma \in \mathrm{Gal}(F_i/F_{i-1})$. Show that there is a unique integer $0 \le \ell \le m_i - 1$ such that $\sigma(\gamma_i) = \zeta_i^\ell \gamma_i$.
(b) Show that $\sigma \mapsto [\ell]$ defines a one-to-one homomorphism $\mathrm{Gal}(F_i/F_{i-1}) \to \mathbb{Z}/m_i\mathbb{Z}$, where $[\ell]$ is the congruence class of ℓ modulo m_i.
(c) Conclude that $\mathrm{Gal}(F_i/F_{i-1})$ is cyclic.

Exercise 5. Suppose that we have extensions $F \subset F_{i-1} \subset F_i \subset L$ such that L is Galois over F and F_i is Galois over F_{i-1}. Prove that $|\mathrm{Gal}(F_i/F_{i-1})|$ divides $|\mathrm{Gal}(L/F)|$.

Exercise 6. Let L be a field containing a primitive mth root of unity ζ and let n be a positive divisor of m. Prove that $\zeta^{m/n}$ is a primitive nth root of unity.

Exercise 7. Let $F \subset L$ be Galois and solvable (with F of characteristic 0). This exercise will consider a variation of Corollary 8.3.4. Let $p_1, \ldots, p_r$ be the distinct primes dividing $[L : F]$.
(a) Show that F contains a primitive $(p_1 \cdots p_r)$th root of unity if and only if F contains a primitive p_ith root of unity for $i = 1, \ldots, r$.
(b) Prove that $F \subset L$ is radical when F contains a primitive $(p_1 \cdots p_r)$th root of unity.
(c) Prove that $F \subset L(\zeta)$ is radical, where ζ is a primitive $(p_1 \cdots p_r)$th root of unity.

Exercise 8. This exercise concerns the details of our derivation of Cardan's formulas.
(a) Use the computational methods of Section 2.3 to obtain the formulas for α_1^3 and β_1 stated in the text.
(b) Prove (8.15).

8.4 SIMPLE GROUPS

Here is the key definition of this section.

Definition 8.4.1. *A group G is **simple** if its only normal subgroups are $\{e\}$ and G.*

Some simple groups are easy to find.

Example 8.4.2. If p is prime, then Lagrange's Theorem implies that the cyclic group $\mathbb{Z}/p\mathbb{Z}$ is simple. In Exercise 1 you will prove that these are the only nontrivial Abelian finite simple groups. ◁▷

Here are some more interesting simple groups.

Theorem 8.4.3. *The alternating group A_n is simple for all $n \geq 5$.*

Proof. Our argument will use the following properties of A_n:

- An l-cycle $(i_1 \, i_2 \cdots i_l)$ lies in A_n if and only if l is odd.
- If $n \geq 3$, then A_n is generated by 3-cycles.

The first property follows from the identity (A.2)

$$(i_1 \, i_2 \cdots i_l) = (i_1 \, i_l) \cdots (i_1 \, i_3)(i_1 \, i_2)$$

of Section A.1, which shows that an l-cycle is a product of $l - 1$ transpositions. The second is less obvious and will be proved in Exercise 2.

Suppose that $H \neq \{e\}$ is a normal subgroup of A_n. It suffices to show that $H = A_n$. To prove this, we first show that H contains a 3-cycle. By assumption H contains a nontrivial permutation σ. We will create a 3-cycle in H by considering the decomposition of σ into disjoint cycles.

Since A_n contains the 3-cycle $(j_1 \, j_2 \, j_3)$ and $H \subset A_n$ is normal, it follows that

(8.16) $$\sigma^{-1}(j_1 \, j_2 \, j_3)^{-1}\sigma(j_1 \, j_2 \, j_3) \in H.$$

This will be useful because it will allow us to create some interesting elements of H. In Exercise 3 you will prove that the permutation (8.16) has the following property:

(8.17) If neither j nor $\sigma(j)$ lies in $\{j_1, j_2, j_3\}$, then $\sigma^{-1}(j_1 \, j_2 \, j_3)^{-1}\sigma(j_1 \, j_2 \, j_3)$ fixes j.

This is important because the given permutation $\sigma \in H$ might be very complicated, especially if n is large. But (8.17) shows that $\sigma^{-1}(j_1 \, j_2 \, j_3)\sigma(j_1 \, j_2 \, j_3)$ is a simpler permutation, since it moves at most six elements of $\{1, \ldots, n\}$. Furthermore, this simpler permutation lies in H by (8.16). We will exploit this by making careful choices of the 3-cycle $(j_1 \, j_2 \, j_3)$.

We now prove that H contains a 3-cycle by considering the following cases.

Case 1. First suppose that one of the cycles in σ has length ≥ 4, say

$$\sigma = (i_1 \, i_2 \, i_3 \, i_4 \cdots)(\cdots) \cdots .$$

In this case, we claim that

(8.18) $$\sigma^{-1}(i_2 \, i_3 \, i_4)^{-1}\sigma(i_2 \, i_3 \, i_4) = (i_1 \, i_3 \, i_4).$$

By (8.17), this permutation fixes all $j \notin \{i_1, i_2, i_3, i_4\}$, and from here it is easy to verify (8.18). We leave the details as part of Exercise 3. Since (8.18) and (8.16) imply that $(i_1 \, i_3 \, i_4) \in H$, we have the desired 3-cycle.

Case 2. Next suppose that σ has a 3-cycle. If σ is a 3-cycle, then we are done. Hence we may assume that

$$\sigma = (i_1 \, i_2 \, i_3)(i_4 \, i_5 \cdots) \cdots .$$

We claim that

(8.19) $$\sigma^{-1}(i_2 \, i_3 \, i_5)^{-1}\sigma(i_2 \, i_3 \, i_5) = (i_1 \, i_4 \, i_2 \, i_3 \, i_5).$$

The proof is similar to (8.18) and is part of Exercise 3. As in Case 1, it follows that $(i_1 \, i_4 \, i_2 \, i_3 \, i_5) \in H$. This shows that H contains a 5-cycle. Then H contains a 3-cycle by Case 1.

Case 3. Finally suppose that σ is a product of disjoint 2-cycles. There must be at least two since $\sigma \in H \subset A_n$, so that

$$\sigma = (i_1 \, i_2)(i_3 \, i_4)(\cdots)(\cdots)\cdots.$$

This time, we have

(8.20) $$\sigma^{-1}(i_2 \, i_3 \, i_4)^{-1}\sigma(i_2 \, i_3 \, i_4) = (i_1 \, i_3)(i_2 \, i_4)$$

(see Exercise 3). As usual, this shows that $(i_1 \, i_3)(i_2 \, i_4) \in H$. To turn this into a 3-cycle, let i_5 be distinct from i_1, i_2, i_3, i_4 (this is where we use $n \geq 5$). Then we compute directly that

$$\big((i_1 \, i_3)(i_2 \, i_4)\big)^{-1}(i_1 \, i_3 \, i_5)^{-1}\big((i_1 \, i_3)(i_2 \, i_4)\big)(i_1 \, i_3 \, i_5) = (i_1 \, i_5 \, i_3).$$

Again we get a 3-cycle in H.

Since every $\sigma \neq e$ in H must satisfy one of these three cases, we conclude that H contains some 3-cycle, say $(i \, j \, k)$. We next claim that H contains *all* 3-cycles, since it is normal. To prove this, suppose that i', j', k' are distinct, and let θ be a permutation that satisfies

$$\theta(i) = i', \qquad \theta(j) = j', \qquad \theta(k) = k'.$$

An important property of permutations is that for any cycle $(i_1 \, i_2 \, \cdots \, i_l)$, we have the identity

(8.21) $$\theta(i_1 \, i_2 \, \cdots \, i_l)\theta^{-1} = \big(\theta(i_1)\,\theta(i_2)\cdots\theta(i_l)\big)$$

(see Exercise 3). This implies that

$$\theta(i \, j \, k)\theta^{-1} = (i' \, j' \, k').$$

If $\theta \in A_n$, then $(i' \, j' \, k') \in H$, since H is normal in A_n. On the other hand, if $\theta \notin A_n$, then $\theta' = \theta(i \, j) \in A_n$. The above computation, performed using θ' instead of θ, shows that $(j' \, i' \, k') \in H$ (you should verify this carefully). Then $(i' \, j' \, k') = (j' \, i' \, k')^{-1} \in H$, as claimed.

Thus H contains all 3-cycles. At the beginning of the proof, we noted that A_n is generated by 3-cycles. It follows immediately that $H = A_n$, and we are done. □

We next observe that non-Abelian finite simple groups are not solvable.

Lemma 8.4.4. *Let G be a non-Abelian finite simple group. Then G is not solvable.*

Proof. Suppose that G is solvable. Then we can find a normal subgroup $G_1 \subset G_0 = G$ such that $[G:G_1] = [G_0:G_1]$ is prime. Since G is simple, we must have $G_1 = \{e\}$, since $G_1 \neq G$. Thus

$$|G| = [G:G_1]|G_1| = [G:G_1]|\{e\}| = [G:G_1],$$

so that G has prime order. But this implies that G is cyclic and hence Abelian. The lemma follows by contradiction. □

Combining Lemma 8.4.4 with earlier results gives us infinitely many nonsolvable groups as follows.

Theorem 8.4.5. *The alternating group A_n and the symmetric group S_n are solvable if and only if $n \leq 4$.*

Proof. The cases $n = 1, 2$ are trivial, and we saw in Example 8.1.2 and Exercise 1 of Section 8.1 that S_3 and S_4 are solvable. By Proposition 8.1.3 it follows that A_3 and A_4 are solvable (this is also easy to prove directly).

Now suppose that $n \geq 5$. Then A_n is non-Abelian (the 3-cycles (123) and (124) don't commute) and simple (by Theorem 8.4.3). By Lemma 8.4.4 we conclude that A_n is not solvable for $n \geq 5$. Then Proposition 8.1.3 shows that S_n is also not solvable for $n \geq 5$. □

For later purposes, we determine the normal subgroups of S_n.

Proposition 8.4.6. *If $n \geq 5$ and $H \subset S_n$ is a normal subgroup, then either $H = \{e\}$, $H = A_n$, or $H = S_n$.*

Proof. If H is normal in S_n, then $H \cap A_n$ is normal in A_n (see Exercise 4). Since $n \geq 5$, Theorem 8.4.3 implies that $H \cap A_n$ is $\{e\}$ or A_n. In the latter case, we have $A_n \subset H$, which easily implies that $H = A_n$ or S_n, since $[S_n : A_n] = 2$.

Finally suppose that $H \cap A_n = \{e\}$. If $H \neq \{e\}$, then Exercise 5 will show that $H = \{e, \sigma\}$, where

$$\sigma = (i\ j)(\cdots)\cdots$$

is the product of an odd number of disjoint 2-cycles. Now pick k different from i and j, and let $\theta = (j\ k)$. Then (8.21) implies that

$$\theta\sigma\theta^{-1} = \theta(i\ j)\theta^{-1}\theta(\cdots)\theta^{-1}\theta\cdots\theta^{-1}$$
$$= (i\ k)(\cdots)\cdots.$$

This is still a product of disjoint 2-cycles. Since one of the cycles is $(i\ k)$, it can't equal $\sigma = (i\ j)(\cdots)\cdots$. Thus $\theta\sigma\theta^{-1} \notin H$, which is impossible, since H is normal. This contradiction shows that $H = \{e\}$ and completes the proof of the proposition. □

Mathematical Notes

The relation between simple and solvable groups is more interesting than indicated by Lemma 8.4.4. The key observation is that all groups are "built" out of simple groups by means of what are called *composition series*.

■ **Composition Series and the Jordan–Hölder Theorem.** Definition 8.1.1 says that a group G is solvable if we can find subgroups

$$(8.22) \qquad \{e\} = G_n \subset G_{n-1} \subset \cdots \subset G_1 \subset G_0 = G$$

such that G_i is normal in G_{i-1} and $[G_{i-1} : G_i]$ is prime for $i = 1, \ldots, n$. This implies in particular that the quotient G_{i-1}/G_i is simple, since it has prime order.

More generally, if G is a finite group, then a *composition series* of G consists of subgroups (8.22) such that G_i is normal in G_{i-1} and the quotient G_{i-1}/G_i is simple for all i. We call the G_{i-1}/G_i the *composition factors* of G.

Example 8.4.7. Let $n \geq 5$. Since A_n is simple, a composition series of S_n is

$$\{e\} \subset A_n \subset S_n$$

The composition factors are $A_n/\{e\} \simeq A_n$ and $S_n/A_n \simeq \mathbb{Z}/2\mathbb{Z}$.   ◁▷

It is straightforward to show that any finite group has a composition series (see Exercise 6). However, a given group may have more than one composition series. For example, the cyclic group $\mathbb{Z}/6\mathbb{Z} = \langle [1] \rangle$ has the composition series

$$\{e\} \subset \langle [2] \rangle \subset \mathbb{Z}/6\mathbb{Z} \quad \text{and} \quad \{e\} \subset \langle [3] \rangle \subset \mathbb{Z}/6\mathbb{Z}.$$

The factors of the first composition series are $\mathbb{Z}/2\mathbb{Z}$ and $\mathbb{Z}/3\mathbb{Z}$, while the factors for the second are $\mathbb{Z}/3\mathbb{Z}$ and $\mathbb{Z}/2\mathbb{Z}$. The *Jordan–Hölder Theorem* asserts that any two composition series of a given group have the same length and that the corresponding composition factors can be permuted so that they become isomorphic. Hence the composition factors of a group are the simple groups from which the group is built. Here "built" refers to the extension problem discussed in the Mathematical Notes to Section 6.4. For more on composition series, see [Jacobson, Vol. I, Sec. 4.6].

In particular, a finite group is solvable if and only if its composition factors are the "simplest" simple groups, namely the Abelian ones. This shows that solvable groups form a very special class of groups.

Historical Notes

The term "simple group" is due to Jordan. He was the first to prove that A_n is simple for $n \geq 5$. However, concerning A_5, Galois noted in 1832 that "the smallest number of permutations for which there is an indecomposable group is $5 \cdot 4 \cdot 3$ when the number is not prime" [Galois, p. 175]. The simplicity of A_5 is also implicit in the work of Ruffini and Abel on the unsolvability of the quintic equation.

The idea of a composition series is due to Jordan. He proved that any two composition series have the same length and that the indices $[G_{i-1} : G_i]$ are unique up to a permutation. Later, once the concept of quotient group was better understood, Hölder proved the Jordan–Hölder Theorem mentioned above.

Exercises for Section 8.4

Exercise 1. Let G be a nontrivial finite Abelian group. Prove that G is simple if and only if $G \simeq \mathbb{Z}/p\mathbb{Z}$ for some prime p.

Exercise 2. Prove that A_n is generated by 3-cycles when $n \geq 3$.

Exercise 3. This exercise is concerned with the proof of Theorem 8.4.3.
(a) Prove (8.17).
(b) Verify the identities (8.18), (8.19), and (8.20).
(c) Verify the conjugation identity (8.21).

Exercise 4. Let H_1 and H_2 be subgroups of a group G and assume that H_1 is normal in G. Prove that $H_1 \cap H_2$ is normal in H_2.

Exercise 5. Suppose that $H \subset S_n$ is a subgroup such that $H \neq \{e\}$ and $H \cap A_n = \{e\}$. Prove that $H = \{e, \sigma\}$, where σ is a product of an odd number of disjoint 2-cycles.

Exercise 6. Let G be a finite group.
(a) Among all normal subgroups of G different from G itself, pick one of maximal order and call it H. Prove that G/H is a simple group.
(b) Use part (a) and complete induction on $|G|$ to prove that G has a composition series.

Exercise 7. Show that the Feit–Thompson Theorem (Theorem 8.1.9) is equivalent to the assertion that every non-Abelian finite simple group has even order.

Exercise 8. Prove that $\mathbb{Z}/4\mathbb{Z}$ and $\mathbb{Z}/2\mathbb{Z} \times \mathbb{Z}/2\mathbb{Z}$ are nonisomorphic groups with the same composition factors.

8.5 SOLVING POLYNOMIALS BY RADICALS

As in Section 8.3 we will assume the following:

All fields appearing in this section will have characteristic 0.

The goal of this section is to study the roots of polynomials using Galois theory.

A. Roots and Radicals. So far, our discussion of solvability by radicals has focused on field extensions. We now shift our attention to polynomials and their roots.

Definition 8.5.1. *Let $f \in F[x]$ be nonconstant with splitting field $F \subset L$.*
(a) *A root $\alpha \in L$ of f is **expressible by radicals** over F if α lies in some radical extension of F.*
(b) *The polynomial f is **solvable by radicals** over F if $F \subset L$ is a solvable extension.*

In Exercise 1 you will show that part (b) of this definition doesn't depend on which splitting field of f over F we use.

Definition 8.5.1 implies that if a nonconstant polynomial in $F[x]$ is solvable by radicals, then *all* of its roots are expressible by radicals. However, for an irreducible polynomial, it turns out that solvability by radicals is satisfied as soon as *one* root is expressible by radicals. Here is the precise result.

Proposition 8.5.2. *Let $f \in F[x]$ be irreducible. Then f is solvable by radicals over F if and only if f has a root expressible by radicals over F.*

Proof. One direction is obvious. Going the other way, suppose that f has a root α in some radical extension of F. This means that $F \subset F(\alpha)$ is solvable, so that by Corollary 8.2.9, its Galois closure $F \subset F(\alpha) \subset M$ is also solvable. (Remember that we are in characteristic 0.)

Since a Galois extension is normal and f is irreducible over F with a root in M, we see that f splits completely over M. Thus M contains the splitting field of f over F (in fact, M is the splitting field—see Exercise 2). The proposition follows, since $F \subset M$ is solvable. □

We can now apply the theory developed in Sections 8.3 and 8.4. Recall from Definition 6.1.12 that the Galois group of $f \in F[x]$ is $\mathrm{Gal}(L/F)$, where L is a splitting field of f over F. Then Theorem 8.3.3 implies the following.

Theorem 8.5.3. *A polynomial $f \in F[x]$ is solvable by radicals over F if and only if the Galois group of f over F is solvable.* □

We can apply this to polynomials of low degree as follows.

Proposition 8.5.4. *If $f \in F[x]$ has degree $n \leq 4$, then f is solvable by radicals.*

Proof. If f is separable, then the Galois group of f is isomorphic to a subgroup of S_n by Proposition 6.3.1, and we are done by Theorem 8.5.3, since S_n is solvable for $n \leq 4$. See Exercise 3 for the case when f is not separable. □

Once we get to degree 5, a different picture emerges.

Example 8.5.5. In Section 6.4, we showed that $f = x^5 - 6x + 3$ has S_5 as Galois group over $\mathbb{Q}$. But S_5 is not solvable by Theorem 8.4.5, so that f is not solvable by radicals over $\mathbb{Q}$ by Theorem 8.5.3. Furthermore, f is irreducible, so that by Proposition 8.5.2, no root of f is expressible by radicals over $\mathbb{Q}$.   ◁▷

This example requires that we revise how we think about the roots of a polynomial. Most students come into a course on Galois theory thinking that the roots of a polynomial $f \in \mathbb{Q}[x]$ are numbers like

$$\sqrt{2} + \sqrt{3}, \quad \sqrt{2 + \sqrt{2}}, \quad \sqrt[7]{12 + 7i}, \quad \text{etc.}$$

The English word "root" comes from the Latin "radix," and the radical symbol $\sqrt{}$ is a modified version of the first letter "r" of "radix." Historically, "root" came to refer to a solution of $f(x) = 0$ because of the intuition that roots are built from radicals. But the above example shows that this intuition is wrong. Roots of polynomials are intrinsically more complicated than just radicals.

B. The Universal Polynomial. The quadratic formula shows that the universal quadratic

$$\tilde{f} = x^2 - \sigma_1 x + \sigma_2 = (x - x_1)(x - x_2).$$

is solvable by radicals, and Cardan's formulas imply that the universal cubic

$$\tilde{f} = x^3 - \sigma_1 x^2 + \sigma_2 x - \sigma_3 = (x - x_1)(x - x_2)(x - x_3)$$

is solvable by radicals. Furthermore, once we these formulas in the universal case, then they apply to *all* polynomials of degree 2 and 3.

This discussion shows that asking if the quadratic formula and Cardan's formulas generalize to polynomials of degree n is equivalent to asking if the universal polynomial of degree n,

(8.23) $$\tilde{f} = x^n - \sigma_1 x^{n-1} + \cdots + (-1)^n \sigma_n = (x - x_1) \cdots (x - x_n),$$

is solvable by radicals. By Section 6.4, the splitting field of $\tilde{f}$ is the universal extension in degree n

$$K = F(\sigma_1, \ldots, \sigma_n) \subset L = F(x_1, \ldots, x_n),$$

whose Galois group $\mathrm{Gal}(L/K)$ is isomorphic to S_n. Then Theorem 8.5.3 implies that the existence of radical formulas generalizing the quadratic formula or Cardan's formulas is equivalent to the solvability of S_n.

In particular, the solvability of S_4 implies the existence of radical formulas for polynomials of degree 4. These are *Ferrari's formulas*, to be discussed in Chapter 12. However, when $n \geq 5$, we have the following.

Theorem 8.5.6. *If $n \geq 5$, then the universal polynomial $\tilde{f} \in K[x]$ of degree n is not solvable by radicals over K, and no root of $\tilde{f}$ is expressible by radicals over K.*

Proof. The first assertion follows from Theorem 8.5.3, since S_n is not solvable when $n \geq 5$ by Theorem 8.4.5, and the second assertion follows from Proposition 8.5.2, since $\tilde{f}$ is irreducible over K. $\quad\square$

Thus, while we have the quadratic formula for polynomials of degree 2, Cardan's formulas for degree 3, and Ferrari's formulas for degree 4, it is impossible to find radical formulas that apply to all polynomials of degree n when $n \geq 5$.

It is important to keep in mind that for every $n \geq 5$, there are always *some* polynomials of degree n, such as $x^n - 2 \in \mathbb{Q}[x]$, that are solvable by radicals. It is only when we try to solve *all* polynomials by radicals that we run into problems.

C. Abelian Equations. In 1829 Abel considered separable polynomials $f \in F[x]$ that have a root α such that the roots of f are $\theta_1(\alpha), \ldots, \theta_n(\alpha)$, where $\theta_1, \ldots, \theta_n$ are rational functions with coefficients in F satisfying

$$\theta_i(\theta_j(\alpha)) = \theta_j(\theta_i(\alpha))$$

for all i, j. Following Kronecker and Jordan, we call $f = 0$ an *Abelian equation* in this situation. Abel showed that Abelian equations are solvable by radicals. We can prove Abel's theorem as follows.

Theorem 8.5.7. *Let $f \in F[x]$. If $f = 0$ is an Abelian equation, then f is solvable by radicals over F.*

Proof. Theorem 6.5.3 states that the Galois group of an Abelian equation is Abelian. (As noted in Section 6.5, this result is the origin of the term "Abelian group.") Since Abelian groups are solvable by Proposition 8.1.5, we are done by Theorem 8.5.3. $\square$

This shows that Abel's theorem on the solvability of Abelian equations follows from Galois theory and basic facts about solvable groups.

We studied Abelian equations in the optional Section 6.5 of Chapter 6. For those who read that section, note that Theorem 6.5.2 is simply a restatement of Theorem 8.5.7 and that Theorem 6.5.4 follows from Theorem 8.5.3 because Abelian groups are solvable.

D. The Fundamental Theorem of Algebra Revisited. In Chapter 3 we proved the Fundamental Theorem of Algebra using the following two facts:

• Every polynomial of odd degree in $\mathbb{R}[x]$ has a root in $\mathbb{R}$ (Proposition 3.2.2).

• Every quadratic polynomial in $\mathbb{C}[x]$ splits completely over $\mathbb{C}$ (Lemma 3.2.3).

The proof given in Section 3.2 used induction on the power of 2 in the degree of $f \in \mathbb{R}[x]$. Artin gave an elegant version of this argument using the Galois correspondence, the solvability of groups of prime power order (Theorem 8.1.7), and the First Sylow Theorem (see the Mathematical Notes to Section 8.1). Here is Artin's proof.

Theorem 8.5.8. *Every nonconstant polynomial in $\mathbb{C}[x]$ splits completely over $\mathbb{C}$, that is, $\mathbb{C}$ is algebraically closed.*

Proof. By Proposition 3.2.1, it suffices to prove that every nonconstant polynomial in $\mathbb{R}[x]$ splits completely over $\mathbb{C}$. Given such a polynomial f, let $\mathbb{R} \subset L$ be its splitting field. Since $\mathbb{R}$ has characteristic 0, this extension is separable and hence Galois. Let $G = \mathrm{Gal}(L/\mathbb{R})$, and define $H \subset G$ as follows: if $|G|$ is odd, then $H = \{e\}$, and if $|G|$ is even, then H is a 2-Sylow subgroup of G. Hence H is a subgroup of G such that $|H|$ is the highest power of 2 dividing $|G|$.

By the Galois correspondence, the fixed field $\mathbb{R} \subset L_H$ has degree $[L_H : \mathbb{R}] = [G : H] = |G|/|H|$. This is odd by the definition of H, so that $\mathbb{R} \subset L_H$ has odd degree. It follows that if $\alpha \in L_H$ is a primitive element over $\mathbb{R}$, then the minimal polynomial $f \in \mathbb{R}[x]$ of α has odd degree. But by the first bullet above, f has a root in $\mathbb{R}$. Since minimal polynomials are irreducible, this means that f must have degree 1, which implies $L_H = \mathbb{R}$.

Then the Galois correspondence implies $H = G$, so that $|G|$ is a power of 2, say $|G| = 2^n$. If $n = 0$, then G is trivial, which implies that $L = \mathbb{R}$. Hence f splits completely over $\mathbb{R}$ in this case. Now suppose that $n \geq 1$. By Theorem 8.1.7, G is solvable, which by $|G| = 2^n$ and Definition 8.1.1 means that we have subgroups

$$\{e\} = G_n \subset G_{n-1} \subset \cdots \subset G_1 \subset G_0 = G$$

such that G_i is normal in G_{i-1} of index 2 for $1 \leq i \leq n$. This gives the fixed fields

$$\mathbb{R} = L_{G_0} \subset L_{G_1} \subset L_{G_2} \subset \cdots$$

such that $L_{G_{i-1}} \subset L_{G_i}$ has degree 2 for every i.

Since $n \geq 1$, we have the degree 2 extension $\mathbb{R} \subset L_{G_1}$. The minimal polynomial of a primitive element of this extension is a quadratic polynomial with no real roots. It follows easily that $L_{G_1} \simeq \mathbb{C}$.

Now suppose that $n \geq 2$. Since $L_{G_1} \subset L_{G_2}$, we have a degree 2 extension of $\mathbb{C}$. By the second bullet above, this is impossible, since every quadratic polynomial in $\mathbb{C}$ splits completely over $\mathbb{C}$. Hence we must have $n = 1$, which implies that $|G| = 2$ and $L = L_{G_1} \simeq \mathbb{C}$. It follows that f splits completely over $\mathbb{C}$, as claimed. □

Notice that our proof of Theorem 8.5.8 translates the above two bullets into the following field-theoretic facts about $\mathbb{R}$ and $\mathbb{C}$:

- $\mathbb{R}$ has no extensions of odd degree > 1.
- $\mathbb{C}$ has no extensions of degree 2.

The essence of Artin's argument is that these facts combine with Galois theory and results from group theory (the First Sylow Theorem and the solvability of groups of order 2^n) to prove the Fundamental Theorem of Algebra.

Historical Notes

The universal polynomial $\tilde{f}$ considered in (8.23) is sometimes called the "general polynomial" of degree n. We will see in Chapter 12 that Lagrange tried hard to solve the general quintic in 1770. Using these methods, Ruffini proved the impossibility of solving the general quintic by radicals in 1799, though his proof was difficult to follow (see [2]). In 1824, Abel, also using the methods of Lagrange, found a proof that came to be more generally accepted. The general quintic is discussed in [3], [8], and [11]. Also, one can prove the unsolvability the general equation of degree $n \geq 5$ by radicals (Theorem 8.5.6) without using Galois theory (see [1]). See [9] for an account of Abel's proof and [10] for more on his life.

A tantalizing comment in [Galois, p. 72] suggests that Galois may have known the First Sylow Theorem. We don't know whether he had a proof or simply conjectured the result.

However, there is no doubt that Galois knew a *lot* about solvability by radicals. Chapter 14 will explore Galois's amazing insights about when an irreducible polynomial of degree p or p^2, p prime, is solvable by radicals.

Exercises for Section 8.5

Exercise 1. Let $F \subset L_1$ and $F \subset L_2$ be splitting fields of $f \in F[x]$. Prove that $F \subset L_1$ is solvable if and only if $F \subset L_2$ is solvable.

Exercise 2. Let $f \in F[x]$ be separable and irreducible, and assume that we have an extension $F \subset F(\alpha)$ where α is a root of f. Prove that the Galois closure of this extension (as defined in Section 7.1) is the splitting field of f over F.

Exercise 3. Let F have characteristic 0 and suppose that $f \in F[x]$ has degree ≤ 4 and is not separable. Prove that f is solvable by radicals over F.

Exercise 4. Let f be the minimal polynomial of $\sqrt[5]{\sqrt[3]{17} + \sqrt[4]{37}}$ over $\mathbb{Q}$, where all of the indicated radicals are real. Prove that f is solvable by radicals over $\mathbb{Q}$.

Exercise 5. Let F have characteristic 0, and assume that we have fields $F \subset K \subset L$. Also suppose that $\alpha \in L$ is expressible by radicals over K and that the extension $F \subset K$ is a solvable extension. Prove carefully that the minimal polynomial of α over F is solvable by radicals over F.

Exercise 6. The proof of Theorem 8.5.8 used the Theorem of the Primitive Element to show that $\mathbb{R}$ has no extensions of odd degree > 1. Prove this without using primitive elements.

8.6   THE *CASUS IRREDUCIBILIS* (OPTIONAL)

In this optional section we will complete our discussion of the *casus irreducibilis* begun in Chapter 1. We will also give an example to show how solvability by radicals can fail in characteristic p.

A. Real Radicals. By Section 1.3, a monic separable cubic polynomial $f \in \mathbb{R}[x]$ with real roots has discriminant $\Delta(f) > 0$. Then Cardan's formulas (8.15) imply that the complex number

$$\sqrt{\frac{-\Delta(f)}{27}} = i\sqrt{\frac{\Delta(f)}{27}}$$

appears in the formulas for the roots of f, even though the roots are real.

It is natural to ask whether it is possible to express the roots of f in terms of real radicals in this situation. In some cases, such as $f = x^3 + x^2 - 5x - 5$, the answer is yes, since $f = (x + 1)(x^2 - 5)$ has roots $-1, \pm\sqrt{5}$, which are expressible using real radicals. However, we will show below that the answer is no whenever the cubic f is irreducible. This is the *casus irreducibilis* from Section 1.3.

We first give a careful definition of what it means for a real number to be expressible by real radicals.

Definition 8.6.1. *Let F be a subfield of $\mathbb{R}$. Then:*
(a) $F \subset L$ *is a **real radical extension** if $F \subset L$ is radical and $L \subset \mathbb{R}$.*

(b) $\alpha \in \mathbb{R}$ *is **expressible by real radicals** over F if there is a real radical extension $F \subset L$ such that $\alpha \in L$.*

Before proving our main theorem, we need to study radical extensions. Our first result allows us to limit ourselves to prime radicals.

Lemma 8.6.2. *If $F \subset K$ is a radical extension, then there are fields*

$$F = F_0 \subset F_1 \subset \cdots \subset F_{n-1} \subset F_n = K$$

where for $i = 1, \ldots, n$, there is $\gamma_i \in F_i$ such that $F_i = F_{i-1}(\gamma_i)$ and $\gamma_i^{m_i} \in F_{i-1}$ for some prime m_i.

Proof. We first show that the lemma is true for an extension $F \subset F(\gamma)$ with $\gamma^m \in F$ for some $m > 1$. If m is prime, then we are done, and if m is not prime, then let p be a prime dividing m and set $\delta = \gamma^p$. This gives extensions

$$F \subset F(\delta) \subset F(\delta)(\gamma) = F(\gamma)$$

such that $\gamma^p = \delta \in F(\delta)$ and $\delta^{m/p} = \gamma^m \in F$. If m/p is prime, then we are done, and if not, pick a prime dividing m/p and continue as above. Thus the lemma holds for $F \subset F(\gamma)$. Since any radical extension is a sequence of such extensions, the lemma follows. $\qquad\square$

We next study extensions obtained by adjoining real prime radicals.

Lemma 8.6.3. *Let E be a subfield of $\mathbb{R}$, and suppose that $\gamma \in \mathbb{R}$ satisfies $\gamma \notin E$ and $\gamma^m \in E$, where m is prime. Then $g = x^m - \gamma^m$ is irreducible over E, and $[E(\gamma) : E] = m$.*

Proof. By Proposition 4.2.6, it suffices to show that g has no roots in E. If $\beta \in E$ is a root of g, then $\beta^m = \gamma^m$, so that $\beta = \zeta\gamma$ for some mth root of unity ζ. It follows that $\beta = \pm\gamma$ since β and γ are real and nonzero and the only real roots of unity are ± 1. Thus $\gamma = \pm\beta \in E$. This contradiction proves the lemma. $\qquad\square$

The following result will be a key tool in our analysis of the *casus irreducibilis*.

Proposition 8.6.4. *Suppose that $M \subset L$ is a Galois extension with $L \subset \mathbb{R}$ and $[L : M] = p$ for an odd prime p. Then L cannot lie in a real radical extension of M.*

Proof. Suppose that we have an extension

$$M \subset M(\gamma)$$

where $\gamma \notin M$, $\gamma \in \mathbb{R}$, and $\gamma^m \in M$ for some prime m. Then $M \subset \mathbb{R}$ implies that $[M(\gamma) : M] = m$ by Lemma 8.6.3. We will relate $[L(\gamma) : M(\gamma)]$ to $[L : M]$ by considering the following diagram:

(8.24)

$$
\begin{array}{ccc}
 & L(\gamma) & \\
 \nearrow & & \nwarrow \\
L & & M(\gamma) \\
 \nwarrow & & \nearrow \\
 & M &
\end{array}
$$

If $\gamma \in L$, then $M(\gamma) = L$, since $\gamma \notin M$ and $[L:M]$ is prime. It follows that

$$(8.25) \qquad m = [M(\gamma):M] = [L:M] = p,$$

so that m is odd. Furthermore, Lemma 8.6.3 implies that $x^m - \gamma^m$ is the minimal polynomial of γ over M. Since $M \subset L$ is normal, $x^m - \gamma^m$ splits completely over L. The roots of this polynomial are $\zeta_m^\ell \gamma$ for $\ell = 0, \ldots, m-1$, where $\zeta_m = e^{2\pi i/m}$. Since $\gamma \neq 0$, it follows that $\zeta_m \in L$. This is impossible, since m is odd and $L \subset \mathbb{R}$.

Hence $\gamma \notin L$, so that $[L(\gamma):L] = m$ by Lemma 8.6.3. Using $m = [M(\gamma):M]$ and the Tower Theorem, (8.24) easily implies that

$$(8.26) \qquad [L(\gamma):M(\gamma)] = [L:M] = p.$$

Thus adjoining a real prime radical doesn't change the degree.

By Lemma 8.6.2, a real radical extension $M \subset K$ is obtained by adjoining successive real prime radicals. Each time we do this, (8.26) shows that the degree is unchanged. In Exercise 1 you will use this to prove that

$$(8.27) \qquad [KL:K] = [L:M] = p,$$

where KL is the compositum of K and L. It follows that $KL \neq K$, which in turn implies that $L \not\subset K$ (see Exercise 1). Since $M \subset K$ is an arbitrary real radical extension of M, we conclude that L cannot lie in such an extension, as claimed. $\qquad\square$

B. Irreducible Polynomials with Real Radical Roots. We can now state a generalized version of the *casus irreducibilis* proved by Hölder in 1891 [6] and independently by Isaacs in 1985 [7]. Their result shows that when an irreducible polynomial has all real roots, the roots are expressible by real radicals only in very special cases.

Theorem 8.6.5. *Let F be a subfield of $\mathbb{R}$ and let $f \in F[x]$ be irreducible with splitting field $F \subset L \subset \mathbb{R}$. Then the following conditions are equivalent:*
(a) *Some root of f is expressible by real radicals over F.*
(b) *All roots of f are expressible by real radicals over F in which only square roots appear.*
(c) *$F \subset L$ is a radical extension.*
(d) *$[L:F]$ is a power of 2.*

Proof. Some implications of the proof are easy. For example, (b) $\Rightarrow$ (a) is trivial and (c) $\Rightarrow$ (a) follows from $L \subset \mathbb{R}$.

Now suppose that (d) holds. This implies that $|\mathrm{Gal}(L/F)|$ is a power of 2. As in the proof of Theorem 8.5.8, this leads to subfields

$$F = L_{G_0} \subset L_{G_1} \subset \cdots \subset L_{G_{n-1}} \subset L_{G_n} = L$$

where each field has degree 2 over the previous. Since the characteristic is different from 2, each field is obtained from the previous by adjoining a square root. This shows that $F \subset L$ is radical, so that (d) $\Rightarrow$ (c) follows. We also obtain (d) $\Rightarrow$ (b) since $L \subset \mathbb{R}$.

It remains to prove (a) $\Rightarrow$ (d). We have $f \in F[x]$ with splitting field $F \subset L \subset \mathbb{R}$. Now assume that some root α of f lies in a real radical extension $F \subset K$ and that $[L : F]$ is a not power of 2. Our goal is to derive a contradiction.

We will use a clever idea from [7] to reduce to the situation of Proposition 8.6.4. Let p be an odd prime dividing $[L : F]$. We claim the following:

(8.28)   there is $\sigma \in \mathrm{Gal}(L/F)$ of order p such that $\sigma(\alpha) \neq \alpha$.

Let's first use (8.28) to get our desired contradiction. Given σ as in (8.28), let $M = L_{\langle\sigma\rangle}$ be the fixed field of the cyclic group generated by σ. Then the Galois correspondence implies that $M \subset L$ is a Galois extension such that

$$[L : M] = |\mathrm{Gal}(L/M)| = |\mathrm{Gal}(L/L_{\langle\sigma\rangle})| = |\langle\sigma\rangle| = p.$$

By Proposition 8.6.4, L lies in no real radical extension of M.

On the other hand, $\alpha \in L$ because L is the splitting field of f, yet $\alpha \notin M$ because $\sigma(\alpha) \neq \alpha$. Since $[L : M]$ is prime, $L = M(\alpha)$ follows. Furthermore, we are assuming that $\alpha \in K$, where $F \subset K$ is real radical. Hence:

- $L = M(\alpha) \subset MK$.
- $M \subset MK$ is real radical, since $F \subset K$ is (see Exercise 2).

These bullets imply that L lies in a real radical extension of M, which contradicts the previous paragraph.

It remains to prove (8.28). Since p divides $[L : K] = |\mathrm{Gal}(L/K)|$, Cauchy's Theorem implies that $\mathrm{Gal}(L/K)$ has an element τ of order p. Let the roots of f be $\alpha_1 = \alpha, \ldots, \alpha_m$, $m = \deg(f)$. Using $L = F(\alpha_1, \ldots, \alpha_m)$ and $\tau \neq 1_L$, we see that $\tau(\alpha_i) \neq \alpha_i$ for some i. However, f is irreducible, so that by Proposition 5.1.8, there is $\sigma_i \in \mathrm{Gal}(L/F)$ such that $\sigma_i(\alpha) = \alpha_i$. Then $\sigma_i^{-1}\tau\sigma_i$ has order p and

$$\sigma_i^{-1}\tau\sigma_i(\alpha) = \sigma_i^{-1}\tau(\alpha_i) \neq \sigma_i^{-1}(\alpha_i) = \alpha.$$

It follows easily that $\sigma = \sigma_i^{-1}\tau\sigma_i$ satisfies the conditions of (8.28). The proof of theorem is complete. $\square$

Theorem 8.6.5 has the following useful corollary.

Corollary 8.6.6. *Let F be a subfield of $\mathbb{R}$, and assume that $f \in F[x]$ is irreducible and $\deg(f)$ is not a power of 2. If f splits completely over $\mathbb{R}$, then no root of f is expressible by real radicals over F.*

Proof. Let $F \subset L \subset \mathbb{R}$ be the splitting field of f over F, and let $\alpha \in L$ be a root of f. Then the extensions

$$F \subset F(\alpha) \subset L$$

and the Tower Theorem imply that $[L : F]$ is a multiple of $\deg(f)$. Since $\deg(f)$ is not a power of 2, the same is true for $[L : F]$. Then the corollary follows from the equivalence (a) $\Leftrightarrow$ (d) of Theorem 8.6.5. $\square$

In concrete terms, Corollary 8.6.6 says that if a polynomial with real roots is irreducible over a subfield $F \subset \mathbb{R}$ and has degree not a power of 2, then it is impossible to express any of its roots using real radicals over F. In particular, this is true for any irreducible cubic with real roots, which is the *casus irreducibilis*.

Here is an example Theorem 8.6.5.

Example 8.6.7. Consider the polynomial

$$f = x^4 - 4x^2 + x + 1.$$

By Exercise 3, f is irreducible over $\mathbb{Q}$ and all of its roots are real. In Chapter 13 we will show that the Galois group of f over $\mathbb{Q}$ is isomorphic to S_4. If L is the splitting field of f over $\mathbb{Q}$, it follows that $[L:\mathbb{Q}] = 24$. This is not a power of 2, so that by Theorem 8.6.5, no root of f is expressible in terms of real radicals. Yet f is solvable by radicals since S_4 is solvable. ◁▷

We can use Corollary 8.6.6 to construct solvable extensions that are not radical.

Example 8.6.8. Consider $f = x^3 + x^2 - 2x - 1$. In Example 8.2.3, we showed that the splitting field $\mathbb{Q} \subset L$ of f is solvable but not radical. This follows immediately from Corollary 8.6.6 since f is irreducible of degree 3 with all real roots. ◁▷

We will also see that Theorem 8.6.5 has applications to the geometric constructions considered in the Mathematical Notes to Section 10.1.

C. The Failure of Solvability in Characteristic p. One surprise is that the methods used to study real radicals can also shed light on solvability by radicals for fields of characteristic p. When we considered solvable extensions in Sections 8.3 and 8.5, we explicitly assumed that we were working in characteristic 0. As we will see, this is necessary because of the lack of pth roots of unity in characteristic p.

We begin with a result on adjoining prime radicals. We say that a field F *contains all roots of unity* if $x^m - 1$ splits completely over F for all integers $m \geq 1$.

Lemma 8.6.9. *Let F be a field that contains all roots of unity. Also assume that F has an extension containing an element γ such that $\gamma \notin F$ but $\gamma^m \in F$ for some prime m. Then $g = x^m - \gamma^m$ is irreducible over F and $[F(\gamma):F] = m$.*

Proof. If g has a root $\beta \in F$, then $\beta^m = \gamma^m$, which implies that $\beta = \zeta\gamma$ for some mth root of unity ζ. Then $\zeta, \beta \in F$ imply that $\gamma \in F$, which is a contradiction. Hence g has no roots in F and the lemma follows from Proposition 4.2.6. □

We now state our main result.

Proposition 8.6.10. *Let M be a field of characteristic p that contains all roots of unity. Then any Galois extension $M \subset L$ of degree p is not solvable.*

Proof. We need to prove that L cannot lie in a radical extension of M. The argument is remarkably similar to the proof of Proposition 8.6.4. Consider an extension

$$M \subset M(\gamma)$$

such that $\gamma \notin M$ and $\gamma^m \in M$ for some prime m. Lemma 8.6.9 implies that $g = x^m - \gamma^m$ is irreducible over M and $[M(\gamma):M] = m$.

Following the proof of Proposition 8.6.4, we have the diagram (8.24). If $\gamma \in L$, then $M(\gamma) = L$ since $[L:M]$ is prime. Using (8.25), we see that $m = p$. Then the minimal polynomial of γ over M is $g = x^p - \gamma^p$. However, $M \subset L$ is Galois and hence separable, so that γ would be separable over M. Yet $g = x^p - \gamma^p = (x - \gamma)^p$ is clearly inseparable.

This contradiction shows that $\gamma \notin L$, so that $[L(\gamma):L] = m$ by Lemma 8.6.9. Since $[M(\gamma):M] = m$, the Tower Theorem and (8.24) imply that

$$[L(\gamma):M(\gamma)] = [L:M] = p.$$

Thus adjoining a prime radical doesn't change the degree. Also, the extension $M(\gamma) \subset L(\gamma)$ is Galois (you will prove this in Exercise 4), and $M(\gamma)$ contains all roots of unity because M does.

From here, it is straightforward to show that L lies in no radical extension of M. We leave the details as Exercise 4. $\square$

We construct an extension $M \subset L$ that satisfies of Proposition 8.6.10 as follows.

Example 8.6.11. Let k be an algebraically closed field of characteristic p (the existence of such a field is proved in [Jacobson, Vol. II, Sec. 8.1]), and let $M = k(t)$, where t is a variable. Since $x^m - 1 \in k[x]$ splits completely over k for any m, it follows that M contains all roots of unity.

Following Artin and Schreier, we consider the polynomial

$$f = x^p - x + t \in M[x].$$

Let L be a splitting field of f over M. We know that $M \subset L$ is a Galois extension by Exercise 15 of Section 5.3. In Exercise 5 you will show that there is a one-to-one group homomorphism

(8.29) $\mathrm{Gal}(L/M) \longrightarrow \mathbb{Z}/p\mathbb{Z}.$

Since $[L:M] = |\mathrm{Gal}(L/M)|$, it follows that $[L:M] = 1$ or p. The former would imply that $L = M$, which would mean that f splits completely over M. However, Exercise 6 will show that f has no roots in M. Hence $[L:M] = p$.

It follows that $M \subset L$ is a Galois extension of degree p. Since M contains all roots of unity, Proposition 8.6.10 implies that $M \subset L$ is not solvable.   ◁▷

In Example 8.6.11, note that $\mathrm{Gal}(L/M) \simeq \mathbb{Z}/p\mathbb{Z}$ since $M \subset L$ is a Galois extension of degree p. So the Galois group is Abelian and hence solvable, yet the extension is not solvable. This shows that the relation between solvable extensions and solvable Galois groups breaks down in characteristic p.

To see where the problem is, note that a key step in the proof of Proposition 8.6.10 is the observation that in characteristic p, the polynomial $g = x^p - \gamma^p = (x - \gamma)^p$ is not separable and hence γ can't lie in a nontrivial Galois extension. The inseparability of g can be explained by the small number of pth roots of unity in characteristic p, for any two roots of $g = x^p - \gamma^p$ differ by a pth root of unity, but $x^p - 1 = (x - 1)^p$ implies that the only pth root of unity is 1.

It turns out that if one avoids extensions whose degree is divisible by p, then the relation between solvable extensions and solvable Galois groups works out nicely in characteristic p. See Exercise 7 for a proof.

Historical Notes

Although the *casus irreducibilis* for the cubic equation dates back to the sixteenth century, rigorous proofs didn't appear until late in the nineteenth century. Mollame gave a proof in 1890, followed a year later by the more general result of Hölder proved in the text. A quick proof for cubics can be found in [van der Waerden].

We should also mention the following result of Loewy from the 1920s.

Theorem 8.6.12. *If $F \subset \mathbb{R}$ and $f \in F[x]$ is irreducible of degree $2^m n$, n odd, then f has at most 2^m roots expressible by real radicals over F.* □

When f is irreducible of odd degree, Theorem 8.6.12 implies that at most one root can be expressible by real radicals. For cubics, this is consistent with Cardan's formulas (see Example 1.1.1). References and a proof of Theorem 8.6.12 when the degree is odd can be found in [Chebotarev, p. 350].

Exercises for Section 8.6

Exercise 1. Here are some details from the proof of Proposition 8.6.4.
(a) Prove (8.27).
(b) Prove that $KL = K$ if and only if $L \subset K$.

Exercise 2. Let $F \subset K$ be a real radical extension and that $F \subset M \subset \mathbb{R}$. Prove that $M \subset MK$ is a real radical extension.

Exercise 3. Show that the polynomial $f = x^4 - 4x^2 + x + 1$ of Example 8.6.7 is irreducible over $\mathbb{Q}$ and has four real roots.

Exercise 4. Complete the proof of Proposition 8.6.10.

Exercise 5. This exercise will consider the polynomial $f = x^p - x + t$ from Example 8.6.11. Let $\alpha \in L$ be a root of f.
(a) Show that the roots of f are $\alpha, \alpha + 1, \ldots, \alpha + p - 1$.
(b) Let $\sigma \in \mathrm{Gal}(L/M)$. By part (a), $\sigma(\alpha) = \alpha + i$ for some i. Prove that $\sigma \mapsto [i]$ gives the desired one-to-one homomorphism (8.29).

Exercise 6. Let k be a field and let $M = k(t)$, where t is a variable. The goal of this exercise is to prove that if $n > 1$, then there is no element $\beta \in M$ such that $\beta^n - \beta + t = 0$.
(a) Write $\beta = A/B$, where $A, B \in k[t]$ are relatively prime polynomials. Prove that $\beta^n - \beta + t = 0$ implies that $B|A$ and hence that B is constant.
(b) Show that $A^n - A + t \neq 0$ for all polynomials $A \in k[t]$.

Exercise 7. Suppose the F is a field of characteristic p and that $F \subset L$ is a Galois extension. Also assume that $\mathrm{Gal}(L/F)$ is solvable and that $p \nmid [L : F]$. Prove that $F \subset L$ is solvable.

REFERENCES

1. R. G. Ayoub, *On the nonsolvability of the general polynomial*, Amer. Math. Monthly **89** (1982), 397–401.

2. R. G. Ayoub, *Paolo Ruffini's contributions to the quintic*, Arch. Hist. Exact Sci. **23** (1980), 253–277.

3. I. G. Bashmakova and G. S. Smirnova, *The Beginnings and Evolution of Algebra*, English translation by A. Shenitzer, MAA, Washington, DC, 1999.

4. W. Burnside, *On groups of order $p^a q^b$*, Proc. London Math. Soc. **2** (1904), 388–392.

5. W. Feit and J. Thompson, *Solvability of groups of odd order*, Pacific J. Math. **13** (1963), 775–1029.

6. O. L. Hölder, *Ueber den Casus Irreducibilis bei der Gleichung dritten Grades*, Math. Annalen **38** (1891), 307–312.

7. I. M. Isaacs, *Solution of polynomials by real radicals*, Amer. Math. Monthly **92** (1985), 571–575.

8. B. M. Kiernan, *The development of Galois theory from Lagrange to Artin*, Arch. Hist. Exact Sci. **8** (1971), 40–154.

9. P. Pesic, *Abel's Proof*, MIT Press, Cambridge, MA, 2003.

10. A. Stubhaug, *Niels Henrik Abel and his Times*, Springer-Verlag, New York, Berlin, Heidelberg, 2000.

11. H. Wussing, *The Genesis of the Abstract Group Concept*, English translation by A. Shenitzer, MIT Press, Cambridge, MA, 1984.

9

Cyclotomic Extensions

In this chapter we will explore the Galois theory of *cyclotomic extensions*, which are extensions of the form $\mathbb{Q} \subset \mathbb{Q}(\zeta_n)$, $\zeta_n = e^{2\pi i/n}$. This will involve a study of cyclotomic polynomials and Gauss's theory of periods. In the next chapter we will apply these results to determine which regular polygons are constructible by straightedge and compass.

9.1 CYCLOTOMIC POLYNOMIALS

In Section 4.2 we showed that if p is prime, then

$$\Phi_p(x) = x^{p-1} + x^{p-2} + \cdots + x + 1$$

is the minimal polynomial of $\zeta_p = e^{2\pi i/p}$ over $\mathbb{Q}$. In this section, we will describe the minimal polynomial of

$$\zeta_n = e^{2\pi i/n}$$

over $\mathbb{Q}$, where n is now an arbitrary integer ≥ 1. We will also compute the Galois group $\mathrm{Gal}(\mathbb{Q}(\zeta_n)/\mathbb{Q})$. But first, we need two facts from elementary number theory.

A. Some Number Theory. We begin with the Euler ϕ-function. Given a positive integer n, we define $\phi(n)$ to be the number of integers i such that $0 \leq i < n$ and $\gcd(i, n) = 1$. We can interpret $\phi(n)$ in terms of the ring $\mathbb{Z}/n\mathbb{Z}$ as follows. The invertible elements of this ring form the set

$$(\mathbb{Z}/n\mathbb{Z})^* = \{[i] \in \mathbb{Z}/n\mathbb{Z} \mid [i][j] = [1] \text{ for some } [j] \in \mathbb{Z}/n\mathbb{Z}\}.$$

One easily sees that $(\mathbb{Z}/n\mathbb{Z})^*$ is a group under multiplication. In Exercise 1 you will show that $(\mathbb{Z}/n\mathbb{Z})^*$ has order $\phi(n)$. Thus

$$(9.1) \qquad\qquad \phi(n) = |(\mathbb{Z}/n\mathbb{Z})^*|.$$

Our first lemma gives the basic properties of the ϕ-function.

Lemma 9.1.1. *Let ϕ be defined as above.*
(a) *If n and m are relatively prime positive integers, then $\phi(nm) = \phi(n)\phi(m)$.*
(b) *If $n > 1$ is an integer, then*

$$\phi(n) = n \prod_{p|n} \left(1 - \frac{1}{p}\right),$$

where the product is over all primes p dividing n.

Proof. Since $\gcd(n, m) = 1$, Lemma A.5.2 implies that there is a ring isomorphism $\alpha : \mathbb{Z}/nm\mathbb{Z} \simeq \mathbb{Z}/n\mathbb{Z} \times \mathbb{Z}/m\mathbb{Z}$. In Exercise 2 you will show that α induces a group isomorphism

$$(\mathbb{Z}/nm\mathbb{Z})^* \simeq (\mathbb{Z}/n\mathbb{Z})^* \times (\mathbb{Z}/m\mathbb{Z})^*.$$

Then $\phi(nm) = \phi(n)\phi(m)$ follows immediately from (9.1).

Next observe that if p is prime and $a \geq 1$, then $\phi(p^a)$ counts the number of integers i such that $0 \leq i < p^a$ and $p \nmid i$. In other words, if

$$S = \{j \in \mathbb{Z} \mid 0 \leq j < p^a \text{ and } p|j\},$$

then $\phi(p^a) = p^a - |S|$. However, $p|j$ for some $0 \leq j < p^a$ if and only if $j = p\ell$ for some $0 \leq \ell < p^{a-1}$. Thus $|S| = p^{a-1}$, so that $\phi(p^a) = p^a - p^{a-1}$.

For arbitrary $n > 1$, write $n = p_1^{a_1} \cdots p_s^{a_s}$, where the p_i are distinct primes and $a_i \geq 1$ for all i. Using part (a) and the formula $\phi(p^a) = p^a - p^{a-1}$, we obtain

$$\phi(n) = \phi(p_1^{a_1} \cdots p_s^{a_s}) = \phi(p_1^{a_1}) \cdots \phi(p_s^{a_s}) = (p_1^{a_1} - p_1^{a_1-1}) \cdots (p_s^{r_s} - p_s^{r_s-1})$$
$$= p_1^{a_1}\left(1 - \frac{1}{p_1}\right) \cdots p_s^{a_s}\left(1 - \frac{1}{p_s}\right) = n \prod_{p|n}\left(1 - \frac{1}{p}\right).$$

This completes the proof. $\qquad\qquad\qquad\qquad\qquad\qquad\qquad\qquad\qquad\square$

Our second lemma is sometimes called *Fermat's Little Theorem*.

Lemma 9.1.2. *If p is prime, then $a^p \equiv a \bmod p$ for all integers a.*

Proof. Since the congruence is true when $p|a$, we may assume that $p \nmid a$. Then $[a] \in (\mathbb{Z}/p\mathbb{Z})^*$, so that $[a]^{p-1} = [1]$, since $(\mathbb{Z}/p\mathbb{Z})^*$ is a group of order $p - 1$ under multiplication. In congruence notation, this means that $a^{p-1} \equiv 1 \bmod p$. The desired congruence follows by multiplying each side by a. $\qquad\qquad\square$

B. Definition of Cyclotomic Polynomials. Our next task is to define the cyclotomic polynomial $\Phi_n(x)$ for $n \geq 1$ and show that it has integer coefficients. We begin with the factorization

$$(9.2) \qquad x^n - 1 = \prod_{0 \leq i < n} (x - \zeta_n^i).$$

Then define the *nth cyclotomic polynomial* $\Phi_n(x)$ to be the product

$$(9.3) \qquad \Phi_n(x) = \prod_{\substack{0 \leq i < n \\ \gcd(i,n)=1}} (x - \zeta_n^i).$$

Thus the roots of $\Phi_n(x)$ are ζ_n^i for those $0 \leq i < n$ relatively prime to n. It follows that $\Phi_n(x)$ has degree $\phi(n)$. Combining this with (9.1), we see that

$$\phi(n) = \deg\left(\Phi_n(x)\right) = |(\mathbb{Z}/n\mathbb{Z})^*|.$$

This link between $\Phi_n(x)$ and $(\mathbb{Z}/n\mathbb{Z})^*$ will be used to determine $\mathrm{Gal}(\mathbb{Q}(\zeta_n)/\mathbb{Q})$.

In Section 8.3 we defined a root of $x^n - 1$ to be a *primitive nth root of unity* if its powers give all roots of $x^n - 1$. In Exercise 3 you will prove that in our situation, the primitive nth roots of unity are ζ_n^i for $0 \leq i < n$ and $\gcd(i, n) = 1$. Thus the roots of $\Phi_n(x)$ are the primitive nth roots of unity in $\mathbb{C}$.

Here are some examples of cyclotomic polynomials.

Example 9.1.3. When $n = 2$, the only primitive square root of unity is -1, so that $\Phi_2(x) = x + 1$. When $n = 4$, the primitive fourth roots of unity are i and $i^3 = -i$, so that

$$\Phi_4(x) = (x - i)(x + i) = x^2 + 1.$$

Since $\Phi_1(x) = x - 1$, we get the factorization

$$x^4 - 1 = (x - 1)(x + 1)(x^2 + 1) = \Phi_1(x)\Phi_2(x)\Phi_4(x).$$

Proposition 9.1.5 will show that $x^n - 1$ has a similar factorization. ◁▷

Example 9.1.4. Let p be prime. Since $1, \ldots, p - 1$ are relatively prime to p, it follows that

$$\Phi_p(x) = (x - \zeta_p)(x - \zeta_p^2) \cdots (x - \zeta_p^{p-1}) = \frac{x^p - 1}{x - 1}.$$

Using $x^p - 1 = (x - 1)(x^{p-1} + \cdots + x + 1)$, we obtain $\Phi_p(x) = x^{p-1} + \cdots + x + 1$, which agrees with the definition of $\Phi_p(x)$ given in Section 4.2. ◁▷

In the following discussion we will write $d \mid n$ to indicate that d is a positive divisor of n. We now state some elementary properties of cyclotomic polynomials.

Proposition 9.1.5. $\Phi_n(x)$ *is a monic polynomial with integer coefficients and has degree $\phi(n)$. Furthermore, these polynomials satisfy the identity*

$$(9.4) \qquad\qquad x^n - 1 = \prod_{d\mid n} \Phi_d(x).$$

Proof. $\Phi_n(x)$ is monic by definition and has degree $\phi(n)$ as shown above. Next we prove the factorization (9.4). The basic idea is that every number i in the range $0 \le i < n$ gives a divisor $d = \gcd(i, n)$ of n. Since different values of i can give the same d, we can organize the factorization (9.2) according to d. This gives

$$x^n - 1 = \prod_{d\mid n} \prod_{\substack{0 \le i < n \\ \gcd(i,n)=d}} (x - \zeta_n^i).$$

For a fixed positive divisor d of n, the corresponding part of this factorization is

$$(9.5) \qquad\qquad \prod_{\substack{0 \le i < n \\ \gcd(i,n)=d}} (x - \zeta_n^i).$$

But $\gcd(i, n) = d$ implies that $i = dj$ and $n = d\frac{n}{d}$, where $\gcd(j, \frac{n}{d}) = 1$. Also:
- $0 \le i < n$ becomes $0 \le dj < d\frac{n}{d}$, which is equivalent to $0 \le j < \frac{n}{d}$.
- $\zeta_n^d = \zeta_{\frac{n}{d}}$, so that $x - \zeta_n^i = x - \zeta_n^{dj} = x - (\zeta_{\frac{n}{d}})^j$.

It follows that (9.5) can be written

$$\prod_{\substack{0 \le j < \frac{n}{d} \\ \gcd(j, \frac{n}{d})=1}} \left(x - (\zeta_{\frac{n}{d}})^j\right),$$

which by (9.3) is the cyclotomic polynomial $\Phi_{\frac{n}{d}}(x)$. Thus the above factorization of $x^n - 1$ becomes

$$x^n - 1 = \prod_{d\mid n} \Phi_{\frac{n}{d}}(x).$$

Then (9.4) follows since d is a positive divisor of n if and only if $\frac{n}{d}$ is.

It remains to show that $\Phi_n(x)$ has integer coefficients. We prove this by complete induction on n. The base case $n = 1$ is trivial, since $\Phi_1(x) = x - 1$. Furthermore, if $n > 1$, then (9.4) and our inductive hypothesis imply that

$$x^n - 1 = \Phi_n(x) \cdot \prod_{d\mid n, d < n} \Phi_d(x)$$

$$= \Phi_n(x) \cdot \text{a monic polynomial } g(x) \text{ with integer coefficients.}$$

Hence $\Phi_n(x)$ is the quotient of $x^n - 1$ by $g(x)$. Since $x^n - 1$ and $g(x)$ lie in $\mathbb{Z}[x]$ and $g(x)$ is monic, the refinement of the division algorithm presented in Exercise 4 implies that $\Phi_n(x) \in \mathbb{Z}[x]$. This completes the proof. $\qquad\square$

Here are some examples of how to use the identity (9.4).

Example 9.1.6. Let p be prime. Proposition 9.1.5 implies that

$$x^p - 1 = \Phi_1(x)\Phi_p(x) \quad \text{and} \quad x^{p^2} - 1 = \Phi_1(x)\Phi_p(x)\Phi_{p^2}(x).$$

Thus

$$x^{p^2} - 1 = (x^p - 1)\Phi_{p^2}(x).$$

It follows that

$$\Phi_{p^2}(x) = \frac{x^{p^2} - 1}{x^p - 1} = x^{p(p-1)} + x^{p(p-2)} + \cdots + x^{2p} + x^p + 1,$$

where the second equality follows from

$$\frac{x^p - 1}{x - 1} = x^{p-1} + x^{p-2} + \cdots + x^2 + x + 1$$

by replacing x with x^p. ◁▷

Example 9.1.7. In the examples of cyclotomic polynomials given so far, the coefficients are always 0 or ± 1. This is true for all $n < 105$. You will show in Exercise 5 that $\Phi_{105}(x)$ is the polynomial

$$\begin{aligned}
1 &+ x + x^2 - x^5 - x^6 - 2x^7 - x^8 - x^9 + x^{12} + x^{13} + x^{14} + x^{15} \\
&+ x^{16} + x^{17} - x^{20} - x^{22} - x^{24} - x^{26} - x^{28} + x^{31} + x^{32} + x^{33} + x^{34} \\
&+ x^{35} + x^{36} - x^{39} - x^{40} - 2x^{41} - x^{42} - x^{43} + x^{46} + x^{47} + x^{48}.
\end{aligned}$$

As n increases, the coefficients of $\Phi_n(x)$ can get arbitrarily large (see [1]). ◁▷

C. The Galois Group of a Cyclotomic Extension.

The first step in computing $\text{Gal}(\mathbb{Q}(\zeta_n)/\mathbb{Q})$ is to prove that $\Phi_n(x)$ is irreducible. For this, we will need the following application of symmetric polynomials and Lemma 9.1.2.

Lemma 9.1.8. Let $f \in \mathbb{Z}[x]$ be monic of positive degree, and let p be prime. If f_p is the monic polynomial whose roots are the pth powers of the roots of f, then:
(a) $f_p \in \mathbb{Z}[x]$.
(b) The coefficients of f and f_p are congruent modulo p.

Proof. If f has roots $\gamma_1, \ldots, \gamma_r$, $r = \deg(f)$, then

$$f_p(x) = \prod_{i=1}^{r}(x - \gamma_i^p) = x^r - \sigma_1(\gamma_1^p, \ldots, \gamma_r^p)x^{r-1} + \cdots + (-1)^r\sigma_r(\gamma_1^p, \ldots, \gamma_r^p).$$

Similarly, $f(x) = x^r - \sigma_1(\gamma_1, \ldots, \gamma_r)x^{r-1} + \cdots + (-1)^r\sigma_r(\gamma_1, \ldots, \gamma_r)$. In these formulas, $\sigma_1, \ldots, \sigma_r$ are the elementary symmetric polynomials from Chapter 2.

Observe that $\sigma_i(x_1^p, \ldots, x_r^p)$ is a symmetric polynomial. In Exercise 6 you will show that the algorithm of Theorem 2.2.2 implies that

(9.6) $$\sigma_i(x_1^p, \ldots, x_r^p) = \sigma_i^p + S(\sigma_1, \ldots, \sigma_r),$$

where $S(\sigma_1, \ldots, \sigma_r)$ is a polynomial in $\sigma_1, \ldots, \sigma_r$ with integer coefficients. However, if we reduce modulo p, then Lemma 5.3.10 implies that

$$\sigma_i^p = \sigma_i(x_1, \ldots, x_r)^p = \sigma_i(x_1^p, \ldots, x_r^p)$$

as polynomials with coefficients in $\mathbb{F}_p$ (see Exercise 6 for details). Combining this with (9.6), we see that the coefficients of $S(\sigma_1, \ldots, \sigma_r)$ are all divisible by p.

Now substitute $\gamma_1, \ldots, \gamma_r$ for $x_1, \ldots, x_r$ in (9.6). Since $\sigma_i(\gamma_1, \ldots, \gamma_r) \in \mathbb{Z}$ for all i and S has integer coefficients, we conclude that $\sigma_i(\gamma_1^p, \ldots, \gamma_r^p) \in \mathbb{Z}$. Since the coefficients of S are all divisible by p, we also have

$$\sigma_i(\gamma_1^p, \ldots, \gamma_r^p) \equiv \sigma_i(\gamma_1, \ldots, \gamma_r)^p \equiv \sigma_i(\gamma_1, \ldots, \gamma_r) \bmod p,$$

where the second congruence follows from Lemma 9.1.2. Thus the coefficients of f and f_p are congruent modulo p. $\square$

We now show that $\Phi_n(x)$ is the minimal polynomial of ζ_n over $\mathbb{Q}$.

Theorem 9.1.9. *The cyclotomic polynomial $\Phi_n(x)$ is irreducible over $\mathbb{Q}$.*

Proof. Let $f \in \mathbb{Q}[x]$ be an irreducible factor of $\Phi_n(x)$. Then Gauss's Lemma, in the form of Corollary 4.2.1, allows us to assume that $f \in \mathbb{Z}[x]$ and that

(9.7) $$\Phi_n(x) = f(x)g(x),$$

for some $g \in \mathbb{Z}[x]$. We can also assume that f and g are monic, since $\Phi_n(x)$ is.

Let p be a prime not dividing n. The first step in the proof is to show that

(9.8) If ζ is a root of f, then so is ζ^p.

We will prove (9.8) by contradiction, so suppose that $f(\zeta) = 0$ and $f(\zeta^p) \neq 0$.

As in Lemma 9.1.8, let $f_p \in \mathbb{Z}[x]$ be the monic polynomial whose roots are the pth powers of the roots of f. In Exercise 7 you will show that the roots of f_p are distinct primitive nth roots of unity, which implies that f_p divides $\Phi_n(x)$. If f and f_p had a common root, then f would divide f_p, since f is irreducible. This would force $f = f_p$, since they are monic of the same degree. But $f = f_p$ is impossible, since $f(\zeta^p) \neq 0$ and $f_p(\zeta^p) = 0$ (the latter follows from $f(\zeta) = 0$ by the definition of f_p). Thus they have no common roots, so that (9.7) can be written

$$\Phi_n(x) = f(x)f_p(x)h(x).$$

Since $\Phi_n(x)$, $f(x)$, and $f_p(x)$ are monic with integer coefficients, the refined division algorithm of Exercise 4 implies that the same is true for $h(x)$.

Consider the map sending $q(x) \in \mathbb{Z}[x]$ to the polynomial $\overline{q}(x) \in \mathbb{F}_p[x]$ obtained by reducing the coefficients of $q(x)$ modulo p. Since $\overline{f}(x) = \overline{f}_p(x)$ by Lemma 9.1.8, the above factorization implies that $\overline{f}^2(x)$ divides $\overline{\Phi}_n(x)$ in $\mathbb{F}_p[x]$. Thus $\overline{f}^2(x)$ divides $x^n - 1$, so that $x^n - 1$ is not separable in $\mathbb{F}_p[x]$. But $x^n - 1$ is separable, since $p \nmid n$. This contradiction completes the proof of (9.8).

Now let ζ be a fixed root of f and let ζ' be any primitive nth root of unity. In Exercise 7 you will show that $\zeta' = \zeta^j$ for some j relatively prime to n. Let $j = p_1 \cdots p_r$ be the prime factorization of j, and note that each p_i is relatively prime to n. Then successive application of (9.8) shows that

$$\zeta, \zeta^{p_1}, \zeta^{p_1 p_2}, \zeta^{p_1 p_2 p_3}, \ldots, \zeta^{p_1 \cdots p_r} = \zeta'$$

are roots of f. Hence every primitive nth root of unity is a root of f. Since f divides $\Phi_n(x)$, we conclude that $f = \Phi_n(x)$. Thus $\Phi_n(x)$ is irreducible, since f is. $\square$

Theorem 9.1.9 implies that $\Phi_n(x)$ is the minimal polynomial of ζ_n over $\mathbb{Q}$. Thus $[\mathbb{Q}(\zeta_n) : \mathbb{Q}] = \deg\big(\Phi_n(x)\big) = \phi(n)$, which proves the following corollary.

Corollary 9.1.10. $[\mathbb{Q}(\zeta_n) : \mathbb{Q}] = \phi(n)$. $\square$

This makes it easy to compute the Galois group of a cyclotomic extension.

Theorem 9.1.11. *There is an isomorphism* $\mathrm{Gal}(\mathbb{Q}(\zeta_n)/\mathbb{Q}) \simeq (\mathbb{Z}/n\mathbb{Z})^*$ *such that* $\sigma \in \mathrm{Gal}(\mathbb{Q}(\zeta_n)/\mathbb{Q})$ *maps to* $[\ell] \in (\mathbb{Z}/n\mathbb{Z})^*$ *if and only if* $\sigma(\zeta_n) = \zeta_n^\ell$.

Proof. We know from (8.6) that $\mathbb{Q} \subset \mathbb{Q}(\zeta n)$ is a Galois extension. Furthermore, an element $\sigma \in \mathrm{Gal}(\mathbb{Q}(\zeta_n)/\mathbb{Q})$ is uniquely determined by $\sigma(\zeta_n)$, which is a root of $\Phi_n(x)$ because ζ_n is. Thus $\sigma(\zeta_n) = \zeta_n^\ell$ for some ℓ relatively prime to n. By Exercise 4 of Section 6.2, the map $\sigma \mapsto [\ell]$ is a well-defined one-to-one group homomorphism $\mathrm{Gal}(\mathbb{Q}(\zeta_n)/\mathbb{Q}) \to (\mathbb{Z}/n\mathbb{Z})^*$. Then Corollary 9.1.10 implies that

$$|\mathrm{Gal}(\mathbb{Q}(\zeta_n)/\mathbb{Q})| = [\mathbb{Q}(\zeta_n) : \mathbb{Q}] = \phi(n) = |(\mathbb{Z}/n\mathbb{Z})^*|.$$

It follows that $\mathrm{Gal}(\mathbb{Q}(\zeta_n)/\mathbb{Q}) \to (\mathbb{Z}/n\mathbb{Z})^*$ is an isomorphism. $\square$

In the next chapter we will use Corollary 9.1.10 to characterize those n for which a regular polygon with n sides is constructible by straightedge and compass.

Historical Notes

While both Lagrange and Vandermonde made significant use of roots of unity, the first systematic study of cyclotomic extensions is due to Gauss. Most of Gauss's results appear in Section VII of *Disquisitiones Arithmeticae* [4], published in 1801. This amazing book covers a wide range of topics in number theory. In particular, Gauss introduces the congruence notation $a \equiv b \bmod n$ and proves a version of Gauss's Lemma (Theorem A.3.2).

In Section VII Gauss studies the extension $\mathbb{Q} \subset \mathbb{Q}(\zeta_p)$, where p is prime. As we will see in the next section, Gauss constructs primitive elements for intermediate fields and essentially describes the Galois correspondence. In Article 365 of [4] he applies his results to the constructibility of regular polygons by straightedge and compass. We will discuss this in the next chapter.

To study $\mathbb{Q} \subset \mathbb{Q}(\zeta_p)$, Gauss needed to know that $\Phi_p(x) = x^{p-1} + \cdots + 1$ is irreducible over $\mathbb{Q}$. Not surprisingly, he proves this using Gauss's Lemma. For general $n \geq 1$, the entry dated June 12, 1808 of Gauss's mathematical diary (see [5]) reads as follows:

> The equation ... that contains all primitive roots of the equation $x^n - 1 = 0$ cannot be decomposed into factors with rational coefficients, proved for composite values of n.

Unfortunately, Gauss's proof has been lost. The first published proof that $\Phi_n(x)$ is irreducible (Theorem 9.1.9) appeared in 1854 and is due to Kronecker. Our proof is based on arguments of Dedekind, as presented by Jordan in 1870. The key step is (9.8), which we proved using Lemma 9.1.8. Schönemann's proof of this lemma dates from 1846, though Gauss proved it much earlier in an unpublished continuation of [4]. A modern proof of (9.8) is sketched in Exercise 8.

Exercises for Section 9.1

Exercise 1. Prove that a congruence class $[i] \in \mathbb{Z}/n\mathbb{Z}$ has a multiplicative inverse if and only if $\gcd(i, n) = 1$. Conclude that $(\mathbb{Z}/n\mathbb{Z})^*$ has order $\phi(n)$. Be sure that you understand what happens when $n = 1$.

Exercise 2. Assume that $\gcd(n, m) = 1$. By Lemma A.5.2, we have a ring isomorphism $\alpha : \mathbb{Z}/nm\mathbb{Z} \simeq \mathbb{Z}/n\mathbb{Z} \times \mathbb{Z}/m\mathbb{Z}$ that sends $[a]_{nm}$ to $([a]_n, [a]_m)$. Prove that α induces a group isomorphism $(\mathbb{Z}/nm\mathbb{Z})^* \simeq (\mathbb{Z}/n\mathbb{Z})^* \times (\mathbb{Z}/m\mathbb{Z})^*$.

Exercise 3. Let $\zeta_n = e^{2\pi i/n} \in \mathbb{C}$. Prove that ζ_n^i for $0 \leq i < n$ and $\gcd(i, n) = 1$ are the primitive nth roots of unity in $\mathbb{C}$.

Exercise 4. Let R be an integral domain, and let $f, g \in R[x]$, where $f \neq 0$. If K is the field of fractions of R, then we can divide g by f in $K[x]$ using the division algorithm of Theorem A.1.14. This gives $g = qf + r$, though $q, r \in K[x]$ need not lie in $R[x]$.
(a) Show that dividing x^2 by $2x + 1$ in $\mathbb{Q}[x]$ gives $x^2 = q \cdot (2x + 1) + r$, where $q, r \in \mathbb{Q}[x]$ are not in $\mathbb{Z}[x]$, even though x^2 and $2x + 1$ lie in $\mathbb{Z}[x]$.
(b) Show that if f is monic, then the division algorithm gives $g = qf + r$, where $q, r \in R[x]$. Hence the division algorithm works over R provided we divide by *monic* polynomials.

Exercise 5. Verify the formula for $\Phi_{105}(x)$ given in Example 9.1.7.

Exercise 6. This exercise is concerned with the proof of Lemma 9.1.8.
(a) Let $f \in \mathbb{Z}[x_1, \ldots, x_n]$ be symmetric. Prove that f is a polynomial in $\sigma_1, \ldots, \sigma_n$ with integer coefficients.
(b) Let p be prime and let $h \in \mathbb{F}_p[x_1, \ldots, x_n]$. Prove that $h(x_1, \ldots, x_n)^p = h(x_1^p, \ldots, x_n^p)$.

Exercise 7. This exercise is concerned with the proof of Theorem 9.1.9.
(a) Let ζ be a primitive nth root of unity, and let i be relatively prime to n. Prove that ζ^i is a primitive nth root of unity and that every primitive nth root of unity is of this form.
(b) Let $\gamma_1, \ldots, \gamma_r$ be distinct primitive nth roots of unity and let i be relatively prime to n. Prove that $\gamma_1^i, \ldots, \gamma_r^i$ are distinct.

Exercise 8. This exercise will present an alternate proof of (9.8) that doesn't use symmetric polynomials. Assume that ζ is root of f such that $f(\zeta^p) \neq 0$. As in the text, $q(x) \in \mathbb{Z}[x]$ maps to the polynomial $\bar{q}(x) \in \mathbb{F}_p[x]$. Let $g(x)$ be as in (9.7).
  (a) Prove that ζ is a root of $g(x^p)$, and conclude that $f(x)|g(x^p)$.
  (b) Use Gauss's Lemma to explain why $f(x)$ divides $g(x^p)$ in $\mathbb{Z}[x]$, and conclude that $\bar{f}(x)$ divides $\bar{g}(x^p)$ in $\mathbb{F}_p[x]$.
  (c) Use Exercise 7 to prove that $\bar{g}(x)^p = \bar{g}(x^p)$, and conclude that $\bar{f}(x)$ divides $\bar{g}(x)^p$.
  (d) Now let $h(x) \in \mathbb{F}_p[x]$ be an irreducible factor of $\bar{f}(x)$. Show that $h(x)$ divides $\bar{g}(x)$, so that $h(x)^2$ divides $\bar{f}(x)\bar{g}(x)$.
  (e) Conclude that $h(x)^2$ divides $x^n - 1 \in \mathbb{F}_p[x]$.
  (f) Use separability to obtain a contradiction.

Exercise 9. In proving Fermat's Little Theorem $a^p \equiv a \bmod p$, recall from the proof of Lemma 9.1.2 that we first proved $a^{p-1} \equiv 1 \bmod p$ when a is relatively prime to p. For general $n > 1$, Euler showed that $a^{\phi(n)} \equiv 1 \bmod n$ when a is relatively prime to n. Prove this. What basic fact from group theory do you use?

Exercise 10. Prove that a cyclic group of order n has $\phi(n)$ generators.

Exercise 11. Prove that $n = \sum_{d|n} \phi(d)$.

Exercise 12. Here are some further properties of cyclotomic polynomials.
  (a) Given n, let $m = \prod_{p|n} p$. Prove that $\Phi_n(x) = \Phi_m(x^{n/m})$. This shows that we can reduce computing $\Phi_n(x)$ to the case when n is squarefree.
  (b) Let n be an odd integer. Prove that $\Phi_{2n}(x) = \Phi_n(-x)$.
  (c) Let p be a prime not dividing an integer $n > 1$. Prove that $\Phi_{pn}(x) = \Phi_n(x^p)/\Phi_n(x)$.

Exercise 13. We know $\Phi_p(x)$ when p is prime. Use this and Exercise 12 to compute $\Phi_{15}(x)$ and $\Phi_{105}(x)$.

Exercise 14. The Möbius function is defined for integers $n \geq 1$ by

$$\mu(n) = \begin{cases} 1, & \text{if } n = 1, \\ (-1)^s, & \text{if } n = p_1 \cdots p_s \text{ for distinct primes } p_1, \ldots, p_s, \\ 0, & \text{otherwise.} \end{cases}$$

Prove that $\sum_{d|n} \mu\left(\frac{n}{d}\right) = 0$ when $n > 1$.

Exercise 15. Let μ be the Möbius function defined in Exercise 14. Prove that

$$\Phi_n(x) = \prod_{d|n} (x^d - 1)^{\mu(n/d)}.$$

This representation of $\Phi_n(x)$ is useful when studying the size of its coefficients.

Exercise 16. Let n and m be relatively prime positive integers.
  (a) Prove that $\mathbb{Q}(\zeta_n, \zeta_m) = \mathbb{Q}(\zeta_{nm})$.
  (b) Prove that $\Phi_n(x)$ is irreducible over $\mathbb{Q}(\zeta_m)$.

9.2   GAUSS AND ROOTS OF UNITY (OPTIONAL)

In this section we will explore how Gauss studied $\mathbb{Q} \subset \mathbb{Q}(\zeta_p)$, where p is an odd prime. Working 30 years before Galois, Gauss described the intermediate fields of this extension and used his results to show that $x^p - 1 = 0$ is solvable by radicals.

A. The Galois Correspondence. If p is an odd prime, then Proposition 9.1.11 implies that

$$\mathrm{Gal}(\mathbb{Q}(\zeta_p)/\mathbb{Q}) \simeq (\mathbb{Z}/p\mathbb{Z})^*.$$

Let's recall what we know about this group:

- $(\mathbb{Z}/p\mathbb{Z})^*$ is cyclic of order $p - 1$ by Proposition A.5.3.
- For every positive divisor f of $p - 1$, $(\mathbb{Z}/p\mathbb{Z})^*$ has a unique subgroup H_f of order f by Theorem A.1.4.

Following Gauss, we let $e = \frac{p-1}{f}$. Thus

$$ef = p - 1,$$

and H_f has index e in $(\mathbb{Z}/p\mathbb{Z})^*$. We will use this notation throughout the section. One further fact not mentioned earlier is the following:

- If f and f' are positive divisors of $p - 1$, then $H_f \subset H_{f'}$ if and only if $f | f'$.

You will prove this in Exercise 1. Hence we can easily check when one subgroup is contained in another.

By the isomorphism $\mathrm{Gal}(\mathbb{Q}(\zeta_p)/\mathbb{Q}) \simeq (\mathbb{Z}/p\mathbb{Z})^*$ and the Galois correspondence, the intermediate fields of $\mathbb{Q} \subset \mathbb{Q}(\zeta_p)$ are the fixed fields

$$L_f = \left\{ \alpha \in \mathbb{Q}(\zeta_p) \mid \sigma(\alpha) = \alpha \text{ for all } \sigma \text{ with } \sigma(\zeta_p) = \zeta_p^i, \, [i] \in H_f \right\}$$

as f ranges over all positive divisors of $p - 1$. These fixed fields have the following nice properties.

Proposition 9.2.1. *The intermediate fields $\mathbb{Q} \subset L_f \subset \mathbb{Q}(\zeta_p)$ satisfy:*
(a) *L_f is a Galois extension of $\mathbb{Q}$ of degree e.*
(b) *If f and f' are positive divisors of $p - 1$, then $L_f \supset L_{f'}$ if and only if $f | f'$.*
(c) *If f and f' are positive divisors of $p - 1$ such that $f | f'$, then $\mathrm{Gal}(L_f/L_{f'})$ is cyclic of order f'/f.*

Proof. You will supply the straightforward proof in Exercise 2.   □

In particular, if $p - 1 = q_1 q_2 \cdots q_r$ is the prime factorization of $p - 1$, then we get subfields

$$(9.9) \qquad \mathbb{Q} = L_{q_1 \cdots q_r} \subset L_{q_2 \cdots q_r} \subset \cdots \subset L_{q_{r-1} q_r} \subset L_{q_r} \subset L_1 = \mathbb{Q}(\zeta_p)$$

where $[L_{q_{i+1} \cdots q_r} : L_{q_i q_{i+1} \cdots q_r}] = q_i$. Thus every element of $L_{q_{i+1} \cdots q_r}$ is the root of a polynomial of degree q_i over $L_{q_i q_{i+1} \cdots q_r}$.

All of this is a simple application of Galois theory. The surprise is that Gauss understood most of this, including (9.9). Before discussing Gauss's results, let's do an example.

Example 9.2.2. Let $p = 7$. Then (9.9) with $p - 1 = 6 = 3 \cdot 2$ becomes

$$\mathbb{Q} = L_6 \subset L_2 \subset L_1 = \mathbb{Q}(\zeta_7),$$

where L_2 is the fixed field of the unique subgroup of order 2 of $\mathrm{Gal}(\mathbb{Q}(\zeta_7)/\mathbb{Q})$.

To make this more explicit, consider $\eta_1 = \zeta_7 + \zeta_7^{-1} = \zeta_7 + \overline{\zeta_7} = 2\cos(2\pi/7)$. In Exercise 3 you will show that $\mathbb{Q}(\eta_1)$ corresponds to the subgroup $\{e, \tau\}$ of $\mathrm{Gal}(\mathbb{Q}(\zeta_7)/\mathbb{Q})$, where τ is complex conjugation. This subgroup has order 2, which implies that

$$L_2 = \mathbb{Q}(\eta_1).$$

In Exercise 3 you will also show that the conjugates of η_1 over $\mathbb{Q}$ are

$$\eta_2 = \zeta_7^2 + \zeta_7^{-2} = 2\cos(4\pi/7) \quad \text{and} \quad \eta_3 = \zeta_7^3 + \zeta_7^{-3} = 2\cos(6\pi/7),$$

and that η_1, η_2, η_3 are roots of the cubic equation

$$y^3 + y^2 - 2y - 1 = 0.$$

It is easy to check that ζ_7 is a root of $x^2 - \eta_1 x + 1 \in L_2[x]$. From here we can express ζ_7 in terms of radicals as follows. Applying Cardan's formulas to the above cubic, one sees that

$$(9.10) \qquad \eta_1 = -\frac{1}{3} + \frac{1}{3}\sqrt[3]{\frac{7}{2}(1 + 3i\sqrt{3})} + \frac{1}{3}\sqrt[3]{\frac{7}{2}(1 - 3i\sqrt{3})},$$

provided that the cube roots are chosen correctly (see Exercise 3). Then applying the quadratic formula to $x^2 - \eta_1 x + 1 = 0$ gives

$$(9.11) \qquad \begin{aligned} \zeta_7 = &-\frac{1}{6} + \frac{1}{6}\sqrt[3]{\frac{7}{2}(1 + 3i\sqrt{3})} + \frac{1}{6}\sqrt[3]{\frac{7}{2}(1 - 3i\sqrt{3})} \\ &+ \frac{i}{2}\sqrt{1 - \left(\frac{1}{3} - \frac{1}{3}\sqrt[3]{\frac{7}{2}(1 + 3i\sqrt{3})} - \frac{1}{3}\sqrt[3]{\frac{7}{2}(1 - 3i\sqrt{3})}\right)^2}, \end{aligned}$$

where we use the same cube roots as in (9.10).

Notice how (9.11) is similar to the formula

$$\zeta_5 = \frac{-1 + \sqrt{5}}{4} + \frac{i}{2}\sqrt{\frac{5 + \sqrt{5}}{2}}$$

from Exercise 8 of Section A.2. These formulas were known to Lagrange and Vandermonde in the 1770s. Vandermonde also worked out a similar formula for ζ_{11}, which is more surprising in that it required solving an equation of degree 5 by radicals (see [Tignol, Ch. 11]).

B. Periods. In Section VII of *Disquisitiones*, Gauss proves the existence of radical formulas for ζ_p for any odd prime p. His proof uses *periods*, which for positive divisors f of $p - 1$ are carefully chosen primitive elements of L_f over $\mathbb{Q}$.

Let $ef = p - 1$, and let $H_f \subset (\mathbb{Z}/p\mathbb{Z})^*$ be the unique subgroup of order f. Given an element $a = [i] \in (\mathbb{Z}/p\mathbb{Z})^*$, set $\zeta_p^a = \zeta_p^i$. This is well defined, since $\zeta_p^p = 1$. Hence we can make the following definition.

Definition 9.2.3. *Let* $\lambda \in \mathbb{Z}$ *be relatively prime to* p. *This gives* $[\lambda] \in (\mathbb{Z}/p\mathbb{Z})^*$ *and the coset* $[\lambda]H_f$ *of* H_f *in* $(\mathbb{Z}/p\mathbb{Z})^*$. *Then we define an* **f-period** *to be the sum*

$$(f, \lambda) = \sum_{a \in [\lambda]H_f} \zeta_p^a.$$

Here are some simple properties of f-periods.

Lemma 9.2.4. *Let* $ef = p - 1$, *and let* (f, λ) *be defined as above. Then:*
(a) *Two* f-*periods either are identical or have no terms in common.*
(b) *There are* e *distinct* f-*periods.*
(c) *The* f-*periods are linearly independent over* $\mathbb{Q}$.
(d) *Let* $\sigma \in \mathrm{Gal}(\mathbb{Q}(\zeta_p)/\mathbb{Q})$ *satisfy* $\sigma(\zeta_p) = \zeta_p^i$. *Then, for any* f-*period* (f, λ),

$$\sigma\big((f, \lambda)\big) = (f, i\lambda).$$

Proof. Recall that $1, \zeta_p, \ldots, \zeta_p^{p-2} \in \mathbb{Q}(\zeta_p)$ are linearly independent over $\mathbb{Q}$, since $[\mathbb{Q}(\zeta_p) : \mathbb{Q}] = p - 1$. Multiplying by ζ_p shows that the same is true for $\zeta_p, \ldots, \zeta_p^{p-1}$. This implies that two f-periods coincide if and only if the corresponding cosets of H_f are equal. Then part (a) follows because cosets are either identical or disjoint, and part (b) because the number of cosets is the index of H_f in $(\mathbb{Z}/p\mathbb{Z})^*$, which is $e = \frac{p-1}{f}$. Then part (c) is a consequence of part (a) together with the linear independence of $\zeta_p, \ldots, \zeta_p^{p-1}$ over $\mathbb{Q}$.

For part (d), observe that $\zeta_p^i = \zeta_p^{[i]}$. Thus $\sigma(\zeta_p) = \zeta_p^i$ implies that

$$\sigma\big((f, \lambda)\big) = \sum_{a \in [\lambda]H_f} \left(\zeta_p^i\right)^a = \sum_{a \in [\lambda]H_f} \zeta_p^{[i]a} = \sum_{b \in [i\lambda]H_f} \zeta_p^b = (f, i\lambda),$$

where the third equality follows via the substitution $b = [i]a$. $\square$

Here are some particularly simple periods.

Example 9.2.5. Since p is odd, the unique subgroup of $(\mathbb{Z}/p\mathbb{Z})^*$ of order 2 is $H_2 = \{[1], [-1]\}$. The cosets of this subgroup are $[\lambda]H_2 = \{[\lambda], [-\lambda]\}$, so that the 2-periods are

$$(2, \lambda) = \zeta_p^\lambda + \zeta_p^{-\lambda} = 2\cos(2\pi\lambda/p).$$

The number of 2-periods is $e = \frac{p-1}{2}$.

In particular, when $p = 7$, the distinct 2-periods are $(2, 1)$, $(2, 2)$, and $(2, 3)$. These were denoted η_1, η_2, and η_3 in Example 9.2.2.   ◁▷

We now prove that f-periods give the desired primitive elements.

Proposition 9.2.6. *Let L_f be the fixed field of H_f. Then:*

(a) *Let $(f, \lambda_1), \ldots, (f, \lambda_e)$ be the distinct f-periods. Then*

$$g(x) = \big(x - (f, \lambda_1)\big) \cdots \big(x - (f, \lambda_e)\big)$$

is in $\mathbb{Q}[x]$ and is the minimal polynomial of any f-period over $\mathbb{Q}$.

(b) *Any f-period is a primitive element of L_f over $\mathbb{Q}$.*

Proof. An f-period $\eta = (f, \lambda)$ corresponds to a coset $[\lambda]H_f$. If $[i] \in (\mathbb{Z}/p\mathbb{Z})^*$, then the f-period $(f, i\lambda)$ corresponding to $[i\lambda]H_f$ is a conjugate of η over $\mathbb{Q}$, by Lemma 9.2.4. Since $[i\lambda]H_f$ gives all cosets of H_f as we vary $[i]$, the conjugates of η over $\mathbb{Q}$ are the e distinct f-periods $(f, \lambda_1), \ldots, (f, \lambda_e)$. Then part (a) follows from the formula for the minimal polynomial given in equation (7.1) of Chapter 7.

It follows that $\mathbb{Q} \subset \mathbb{Q}(\eta)$ and $\mathbb{Q} \subset L_f$ are extensions of degree e. Since $\mathrm{Gal}(\mathbb{Q}(\zeta_p)/\mathbb{Q}) \simeq (\mathbb{Z}/p\mathbb{Z})^*$ has a unique subgroup of index e, the Galois correspondence implies that $\mathbb{Q}(\eta) = L_f$. This proves part (b). $\qquad \square$

As a corollary, we get the following interesting basis of L_f over $\mathbb{Q}$.

Corollary 9.2.7. *The f-periods form a basis of L_f over $\mathbb{Q}$.*

Proof. The f-periods lie in L_f by Proposition 9.2.6. Furthermore, Lemma 9.2.4 tells us that the e such periods are linearly independent over $\mathbb{Q}$. The corollary follows, since $[L_f : \mathbb{Q}] = e$ by Proposition 9.2.1. $\qquad \square$

Our next task is to describe the extension $L_{f'} \subset L_f$ in terms of periods, where f and f' are positive divisors of $p - 1$ satisfying $f | f'$. Set $d = f'/f$, so that $[L_f : L_{f'}] = d$. Any f-period (f, λ) is a primitive element of L_f over $L_{f'}$. We need to describe its minimal polynomial over $L_{f'}$.

This is done as follows. Observe that H_f is a subgroup of index $d = f'/f$ in $H_{f'}$. Hence every coset of $H_{f'}$ in $(\mathbb{Z}/p\mathbb{Z})^*$ is a disjoint union of d cosets of H_f. (Do Exercise 4 if you are unsure of this.) In particular, $[\lambda]H_{f'}$ is a disjoint union

(9.12) $$[\lambda]H_{f'} = [\lambda_1]H_f \cup \cdots \cup [\lambda_d]H_f,$$

where we may assume $\lambda_1 = \lambda$, since $[\lambda]H_f \subset [\lambda]H_{f'}$. This leads to the following description of the desired minimal polynomial.

Proposition 9.2.8. *Let f and f' be positive divisors of $p - 1$ such that $f | f'$, and set $d = f'/f$. Given an f-period (f, λ), let $\lambda_1 = \lambda, \lambda_2, \ldots, \lambda_d$ be as in (9.12). Then*

$$h(x) = \big(x - (f, \lambda_1)\big) \cdots \big(x - (f, \lambda_d)\big)$$

is in $L_{f'}[x]$ and is the minimal polynomial of (f, λ) over $L_{f'}$.

Proof. The proof is similar to what we did in part (b) of Proposition 9.2.6. Setting $\eta = (f, \lambda)$, we need to show that as σ varies over $\mathrm{Gal}(\mathbb{Q}(\zeta_p)/L_{f'})$, the elements $\sigma(\eta)$ give the f-periods $(f, \lambda_1), \ldots, (f, \lambda_d)$.

To prove this, let $\sigma \in \mathrm{Gal}(\mathbb{Q}(\zeta_p)/L_{f'})$, so that $\sigma(\zeta_p) = \zeta_p^i$ for $[i] \in H_{f'}$. Then

$$\sigma(\eta) = \sigma\big((f, \lambda)\big) = (f, i\lambda).$$

This f-period corresponds to the coset $[i\lambda]H_f$. However,

$$[i\lambda]H_f \subset [i\lambda]H_{f'} = [\lambda][i]H_{f'} = [\lambda]H_{f'},$$

where the final equality uses $[i] \in H_{f'}$. By (9.12), it follows that $[i\lambda]H_f = [\lambda_j]H_f$ for some j, so that $\sigma(\eta) = (f, i\lambda) = (f, \lambda_j)$. Since every (f, λ_j) arises in this way (see Exercise 5), the proposition is proved. $\square$

We will give an example of Proposition 9.2.8 below.

C. Explicit Calculations. The above results are pretty but somewhat abstract. To compute specific examples, we need a concrete way to work with periods. The key idea, due to Gauss, is to pick a generator $[g]$ of the cyclic group $(\mathbb{Z}/p\mathbb{Z})^*$. Since this group has order $p - 1$, it follows that

$$(\mathbb{Z}/p\mathbb{Z})^* = \big\{[1], [g], [g^2], \ldots, [g^{p-2}]\big\}.$$

In other words, the $p - 1$ numbers $1, g, g^2, \ldots, g^{p-2}$ represent the nonzero congruence classes modulo p. We call g a *primitive root* modulo p.

Given a primitive root g and $ef = p - 1$ as usual, Exercise 1 implies that H_f is generated by g^e, that is,

$$H_f = \big\{[1], [g^e], [g^{2e}], \ldots, [g^{(f-1)e}]\big\}.$$

It follows that the coset $[\lambda]H_f$ gives the f-period

$$(9.13) \qquad (f, \lambda) = \zeta_p^\lambda + \zeta_p^{\lambda g^e} + \zeta_p^{\lambda g^{2e}} + \cdots + \zeta_p^{\lambda g^{(f-1)e}} = \sum_{j=0}^{f-1} \zeta_p^{\lambda g^{je}}.$$

So far, we have assumed that $[\lambda] \in (\mathbb{Z}/p\mathbb{Z})^*$, that is, $p \nmid \lambda$. However, (9.13) makes sense for *any* integer λ. Since $\zeta_p^p = 1$, one easily sees that

$$(f, \lambda) = f \quad \text{when} \quad p|\lambda.$$

For an arbitrary $\lambda \in \mathbb{Z}$, we call (f, λ) a *generalized period*. Thus a generalized period is an ordinary period if $p \nmid \lambda$ and is equal to f if $p|\lambda$.

In order to compute the minimal polynomials appearing in Propositions 9.2.6 and 9.2.8, we need to know how to multiply f-periods. Gauss expressed the product of two f-periods in terms of generalized periods as follows.

Proposition 9.2.9. *If (f, λ) and (f, μ) are f-periods with $p \nmid \lambda$ and $p \nmid \mu$, then*

$$(f, \lambda)(f, \mu) = \sum_{[\lambda'] \in [\lambda]H_f} (f, \lambda' + \mu) = \sum_{j=0}^{f-1} (f, \lambda g^{je} + \mu).$$

Proof. Following [4, Art. 345], we set $h = g^e$, so that

$$(f, \mu) = \sum_{\ell=0}^{f-1} \zeta_p^{\mu h^\ell}$$

since $[h]$ generates H_f. We also have $[\lambda]H_f = [\lambda h^\ell]H_f$ for any ℓ, which implies that $(f, \lambda) = (f, \lambda h^\ell)$. Thus

$$(f, \lambda)(f, \mu) = \sum_{\ell=0}^{f-1} (f, \lambda) \zeta_p^{\mu h^\ell} = \sum_{\ell=0}^{f-1} (f, \lambda h^\ell) \zeta_p^{\mu h^\ell}$$

$$= \sum_{\ell=0}^{f-1} \left(\sum_{j=0}^{f-1} \zeta_p^{\lambda h^\ell h^j} \right) \zeta_p^{\mu h^\ell} = \sum_{j=0}^{f-1} \left(\sum_{\ell=0}^{f-1} \zeta_p^{(\lambda h^j + \mu) h^\ell} \right)$$

$$= \sum_{j=0}^{f-1} (f, \lambda h^j + \mu).$$

This gives the desired formula, since $h = g^e$. $\qquad\qquad\square$

Here is an example from [4, Art. 346].

Example 9.2.10. In this example and three that follow, we will consider the 6-periods for $p = 19$. In Exercise 7 you will show that $g = 2$ is a primitive root modulo 19. Since $f = 6$ implies $e = 3$, the unique subgroup of order 6 in $(\mathbb{Z}/19\mathbb{Z})^*$ is generated by $[2]^3 = [8]$. Thus

$$H_6 = \{[1], [8], [8]^2, [8]^3, [8]^4, [8]^5\} = \{[1], [8], [7], [18], [11], [12]\} \subset (\mathbb{Z}/19\mathbb{Z})^*.$$

For simplicity, we will write $[n]$ as n, so that

$$H_6 = \{1, 7, 8, 11, 12, 18\}.$$

The $e = 3$ cosets of H_6 in $(\mathbb{Z}/19\mathbb{Z})^*$ are H_6 together with

$$2H_6 = \{2, 14, 16, 22, 24, 36\} = \{2, 3, 5, 14, 16, 17\},$$
$$4H_6 = \{4, 28, 32, 44, 48, 72\} = \{4, 6, 9, 10, 13, 15\}.$$

(Remember that we are working modulo 19.)

According to Proposition 9.2.9,

$$(6, 1)^2 = (6, 1+1) + (6, 7+1) + (6, 8+1) + (6, 11+1) + (6, 12+1) + (6, 18+1)$$
$$= (6, 2) + (6, 8) + (6, 9) + (6, 12) + (6, 13) + 6,$$

where the second equality uses $(6, 19) = 6$. This shows that generalized periods can arise when we multiply ordinary periods. However,

$$(6, 8) = (6, 1)$$

since 8 and 1 lie in the same coset of H_6. Using similar simplifications, we get

$$(6, 1)^2 = 2(6, 1) + (6, 2) + 2(6, 4) + 6.$$

By Exercise 6 we also have

$$(6, 1) + (6, 2) + (6, 4) = -1.$$

Then the formula for $(6, 1)^2$ simplifies to

$$(6, 1)^2 = 4 - (6, 2).$$

You will work out similar formulas in Exercise 7.   ◁▷

Example 9.2.11. Still assuming $p = 19$, our next task is to compute the minimal polynomial of the 6-periods over $\mathbb{Q}$. We will use the notation of the previous example. By Proposition 9.2.6, the minimal polynomial is

$$(9.14) \qquad \big(x - (6, 1)\big)\big(x - (6, 2)\big)\big(x - (6, 4)\big).$$

In Exercise 7 you will use the methods of Example 9.2.10 to show that

$$(9.15) \qquad \begin{aligned} (6, 1)(6, 2) &= (6, 1) + 2(6, 2) + 3(6, 4), \\ (6, 1)(6, 4) &= 2(6, 1) + 3(6, 2) + (6, 4), \\ (6, 2)(6, 4) &= 3(6, 1) + (6, 2) + 2(6, 4). \end{aligned}$$

Note that these sum to $6(6, 1) + 6(6, 2) + 6(6, 4) = -6$, since (as noted above) $(6, 1) + (6, 2) + (6, 4) = -1$.

Using (9.15) and $(6, 1)^2 = 4 - (6, 2)$ (from Example 9.2.10), we have

$$\begin{aligned} (6, 1)(6, 2)(6, 4) &= (6, 1)\big(3(6, 1) + (6, 2) + 2(6, 4)\big) \\ &= 3(6, 1)^2 + (6, 1)(6, 2) + 2(6, 1)(6, 4) \\ &= 12 + 5(6, 1) + 5(6, 2) + 5(6, 4) = 7 \end{aligned}$$

(see Exercise 7 for the details). It follows that multiplying out (9.14) gives

$$(9.16) \qquad x^3 + x^2 - 6x - 7.$$

This is the minimal polynomial of the 6-periods over $\mathbb{Q}$. Its splitting field is $\mathbb{Q} \subset L_6$, the extension generated by the 6-periods.   ◁▷

Example 9.2.12. Now consider the 3-periods for $p = 19$. Since $6/3 = 2$, we see that $L_6 \subset L_3$ is an extension of degree 2. Hence 3-periods have quadratic minimal polynomials over L_6.

Since 2 is a primitive root modulo 19, the subgroup $H_3 \subset (\mathbb{Z}/19\mathbb{Z})^*$ is generated by $[2]^6 = [8]^2 = [7]$. Using the notation of Example 9.2.10, we have

$$H_6 = \{1, 7, 8, 11, 12, 18\} = \{1, 7, 11\} \cup \{8, 12, 18\} = H_3 \cup 8H_3.$$

This shows that

$$(6, 1) = (3, 1) + (3, 8),$$

and in a similar way, one obtains

$$(6, 2) = (3, 2) + (3, 16),$$
$$(6, 4) = (3, 4) + (3, 13).$$

However, Proposition 9.2.9 implies that

$$(3, 1)(3, 8) = (3, 1 + 8) + (3, 7 + 8) + (3, 11 + 8) = (3, 9) + (3, 15) + 3,$$

and since $(3, 9) = (3, 4)$ and $(3, 15) = (3, 13)$ (do you see why?), we get

$$(3, 1)(3, 8) = (3, 4) + (3, 13) + 3 = (6, 4) + 3.$$

By Proposition 9.2.8, the minimal polynomial of $(3, 1)$ and $(3, 8)$ over L_6 is

(9.17) $$\big(x - (3, 1)\big)\big(x - (3, 8)\big) = x^2 - (6, 1)x + (6, 4) + 3.$$

Exercise 7 will consider the minimal polynomials of the other 3-periods.   ◁▷

Example 9.2.13. The 1-periods for $p = 19$ are the primitive 19th roots of unity $(1, \lambda) = \zeta_{19}^\lambda$ for $\lambda = 1, \ldots, 18$. In Example 9.2.12, we noted that $H_3 = \{1, 7, 11\}$, which means that

$$(3, 1) = \zeta_{19} + \zeta_{19}^7 + \zeta_{19}^{11}.$$

By Exercise 7 the minimal polynomial of ζ_{19}, ζ_{19}^7 and ζ_{19}^{11} over L_3 is

(9.18) $$(x - \zeta_{19})(x - \zeta_{19}^7)(x - \zeta_{19}^{11}) = x^3 - (3, 1)x^2 + (3, 8)x - 1.$$

Combining this with (9.17) and (9.16), one can write an explicit formula for ζ_{19} that involves only square and cube roots.   ◁▷

In Exercises 8 and 9 you will use similar methods to derive the formula

(9.19) $$\cos(2\pi/17) = -\tfrac{1}{16} + \tfrac{1}{16}\sqrt{17} + \tfrac{1}{16}\sqrt{34 - 2\sqrt{17}}$$
$$+ \tfrac{1}{8}\sqrt{17 + 3\sqrt{17} - \sqrt{34 - 2\sqrt{17}} - 2\sqrt{34 + 2\sqrt{17}}},$$

due to Gauss. In Chapter 10, we will see that this leads immediately to a straightedge-and-compass construction of a regular 17-gon.

One reason these methods work so well is that the f-periods are linearly independent over $\mathbb{Q}$ by Lemma 9.2.4. Hence any linear combination of f-periods with coefficients in $\mathbb{Z}$ or $\mathbb{Q}$ is unique. However, we've seen cases where generalized periods (f, λ), $p|\lambda$, also occur. But this is no problem, since $(f, \lambda) = f$ in such a case, and we also know that the distinct f-periods sum to -1 (see Exercise 6). Thus a generalized f-period can be expressed in terms of ordinary f-periods. Hence we can always reduce to an expression involving only f-periods, where we know that the representation is unique.

D. Solvability by Radicals. When studying $\mathbb{Q} \subset \mathbb{Q}(\zeta_p)$, we saw in (9.9) that a prime factorization $p - 1 = q_1 q_2 \cdots q_r$ gives intermediate fields

$$\mathbb{Q} = L_{q_1 \cdots q_r} \subset L_{q_2 \cdots q_r} \subset \cdots \subset L_{q_{r-1} q_r} \subset L_{q_r} \subset L_1 = \mathbb{Q}(\zeta_p)$$

such that $[L_{q_{i+1} \cdots q_r} : L_{q_i q_{i+1} \cdots q_r}] = q_i$. If we focus on one of these fields and the next larger one, then we get an extension of the form

(9.20) $L_{fq} \subset L_f$

where fq divides $p - 1$ and q is prime. The theory of periods shows that $(f, 1)$ is a primitive element of L_f and the examples given above make it clear that in any particular case we can compute the minimal polynomial of $(f, 1)$ over L_{fq}.

When $p = 19$, the minimal polynomials found in Examples 9.2.10 to 9.2.13 have degrees 2 or 3. Hence their roots can be found by known formulas. But when $p = 11$, the period $(2, 1) = 2 \cos(2\pi/11)$ has minimal polynomial

$$y^5 + y^4 - 4y^3 - 3y^2 + 3y + 1$$

(see Exercise 10). Is this polynomial solvable by radicals? More generally, are the minimal polynomials of periods solvable by radicals?

For a theoretical point of view, this question is trivial, since $\mathbb{Q} \subset \mathbb{Q}(\zeta_p)$ is a radical extension ($\zeta_p^p = 1 \in \mathbb{Q}$). It follows by definition that any f-period (f, λ) is expressible by radicals over $\mathbb{Q}$, since $(f, \lambda) \in \mathbb{Q}(\zeta_p)$. Things become even more trivial if you recall that when we studied solvability by radicals in Chapter 8, we felt free to adjoin any roots of unity we needed, including ζ_p.

Hence it appears that solving the minimal polynomials of periods by radicals is completely uninteresting. The problem is that this ignores the *inductive* nature of what's going on. The real goal, which goes back to Lagrange's strategy for solving equations, is to construct pth roots of unity using only radicals and roots of unity of *lower degree* (we will discuss Lagrange's strategy in Chapter 12). This is what Gauss does in *Disquisitiones*.

Thus, when studying $\mathbb{Q} \subset \mathbb{Q}(\zeta_p)$, we may assume inductively that we know all mth roots of unity for $m < p$. Furthermore, as explained in the discussion preceding (9.20), it suffices to consider the extension $L_{fq} \subset L_f$, where we may assume that the fq-periods are known. The idea is to express an f-period in terms of radicals that are qth roots involving fq-periods and qth roots of unity. These roots of unity are known, since $q < p$.

To do this in practice, we will use Lagrange resolvents. Let ω be a primitive qth root of unity. In Exercise 11 you will prove that

$$\mathrm{Gal}(L_f(\omega)/L_{fq}(\omega)) \simeq \mathrm{Gal}(L_f/L_{fq}) \simeq \mathbb{Z}/q\mathbb{Z}.$$

Since $L_{fq}(\omega)$ contains a primitive qth root of unity, Lemma 8.3.2 implies that $L_f(\omega)$ is obtained from $L_{fq}(\omega)$ by adjoining a qth root. Furthermore, the proof of Lemma 8.3.2 shows that the element adjoined is a Lagrange resolvent. Recall from (8.7) that if σ is a generator of $\mathrm{Gal}(L_f(\omega)/L_{fq}(\omega))$ and $\beta \in L_f(\omega)$, then we get the *Lagrange resolvents*

$$\alpha_i = \beta + \omega^{-i}\sigma(\beta) + \cdots + \omega^{-i(q-1)}\sigma^{q-1}(\beta)$$

for $i = 0, \ldots, q - 1$. We will use $\beta = (f, 1) \in L_f \subset L_f(\omega)$. In Exercise 11 you will show that we can pick the generator σ so that for any f-period (f, λ),

(9.21)
$$\sigma\big((f, \lambda)\big) = (f, g^{e/q}\lambda)$$

(note that $q|e$, since $fq|p - 1$). Thus the above Lagrange resolvents can be written

(9.22)
$$\alpha_i = (f, 1) + \omega^{-i}(f, g^{e/q}) + \cdots + \omega^{-i(q-1)}(f, g^{(q-1)e/q}).$$

If we set $A_i = \alpha_i^q$, then we can define $\sqrt[q]{A_i} = \alpha_i$. Then the f-periods in (9.22) can be expressed in terms of radicals as follows.

Theorem 9.2.14. *Let α_i and $A_i = \alpha_i^q$ be defined as above.*
(a) *$\alpha_0 \in L_{fq}(\omega)$ and $A_i = \alpha_i^q \in L_{fq}(\omega)$ for $1 \leq i \leq q - 1$.*
(b) *For $0 \leq \ell \leq q - 1$,*

$$(f, g^{\ell e/q}) = \frac{1}{q}\left(\alpha_0 + \omega^{\ell}\sqrt[q]{A_1} + \omega^{2\ell}\sqrt[q]{A_2} + \cdots + \omega^{(q-1)\ell}\sqrt[q]{A_{q-1}}\right).$$

Before beginning the proof, let's explain the f-periods appearing in the theorem. The extension $L_{fq} \subset L_f$ corresponds to the subgroups $H_f \subset H_{fq}$ of $(\mathbb{Z}/p\mathbb{Z})^*$. Since $e = \frac{p-1}{f}$, Exercise 1 shows that these subgroups are generated by $[g^e]$ and $[g^{e/q}]$ respectively. In Exercise 11 you will use this to prove that

(9.23)
$$H_{fq} = H_f \cup g^{e/q}H_f \cup g^{2e/q}H_f \cup \cdots \cup g^{(q-1)e/q}H_f.$$

By Proposition 9.2.8, the f-periods $(f, g^{\ell e/q})$ are the conjugates of $(f, 1)$ over L_{fq}.

Proof of Theorem 9.2.14. Part (a) follows easily from the properties of Lagrange resolvents presented in the proof of Lemma 8.3.2. For part (b), let $\lambda_\ell = g^{\ell e/q}$. Then for any integer m we have

$$\sum_{i=0}^{q-1}\omega^{mi}\alpha_i = \sum_{i=0}^{q-1}\omega^{mi}\left(\sum_{\ell=0}^{q-1}\omega^{-i\ell}(f, \lambda_\ell)\right)$$

$$= \sum_{\ell=0}^{q-1}\left(\sum_{i=0}^{q-1}\omega^{(m-\ell)i}\right)(f, \lambda_\ell) = q\,(f, \lambda_m),$$

where the last equality follows from Exercise 9 of Section A.2. This gives the desired formula for (f, λ_m), since $\alpha_i = \sqrt[q]{A_i}$ for $i > 0$.   □

From a computational point of view, the results of this section give a systematic method for expressing $A_i = \alpha_i^q$ in terms of fq-periods and qth roots of unity. This works because f-periods and fq-periods are linearly independent not only over $\mathbb{Q}$ but also over $\mathbb{Q}(\omega)$, where ω is a primitive qth root of unity (you will prove this in Exercise 12). Thus the radical formula for $(f, g^{\ell e/q})$ given in Theorem 9.2.14 is explicitly computable.

Mathematical Notes

Here are comments on two topics relevant to what we did in this section.

■ **Primitive Roots Modulo p.** The formulas presented in this section illustrate the usefulness of knowing primitive roots modulo p. Gauss explains a method for finding primitive roots in [4, Art. 73–74]. See also [9, p. 163].

Let g_p denote the smallest positive primitive root modulo p. For example, 2 is a primitive root modulo 19, which implies that $g_{19} = 2$. In 1962 Burgess [3] proved that for any $\varepsilon > 0$ there is a positive constant $C(\varepsilon)$ such that

$$g_p \le C(\varepsilon) p^{\frac{1}{4}+\varepsilon}$$

for all odd primes p. This says that g_p can't be too big relative to p. On the other hand, Kearnes [8] proved in 1984 that given any integer $m > 0$ there are infinitely many primes $p > m$ such that $g_p > m$. So g_p can still get large.

If we fix a primitive root g modulo p, then the *discrete log* problem asks the following: given an integer a not divisible by p, find i such that $a \equiv g^i \bmod p$. We write this as $i = \log_g a$. It is easy to describe an algorithm for finding $\log_g a$ (divide $a - g^i$ by p for $i = 0, 1, 2, \ldots$, and stop when the remainder is zero). But finding an *efficient* algorithm for $\log_g a$ is much more difficult. Several modern encryption schemes, including the Pohig–Hellman symmetric key exponentiation cipher (described in [9, Sec. 3.1]) and the Diffie–Hellman key exchange protocol (described in [2, Sec. 7.4] and [9, Sec. 3.1]), would be easy to break if discrete logs were easy to compute.

Primitive roots modulo p are also used in the *Digital Signature Algorithm* suggested by the National Institute of Standards and Technology. A description can be found in [2, Sec. 11.5]. As above, one could forge digital signatures if discrete logs were easy to compute.

There are also purely mathematical questions about primitive roots modulo p. A list of unsolved problems can be found in [6, Sec. F.9].

■ **Periods and Gauss Sums.** Let $p = 17$. By Exercise 9 we have

$$(8, 1) = \tfrac{1}{2}\left(-1 + \sqrt{17}\right),$$
$$(8, 3) = \tfrac{1}{2}\left(-1 - \sqrt{17}\right),$$

which easily implies
$$(8, 1) - (8, 3) = \sqrt{17}.$$

In Exercise 13 you will show that this can be written

(9.24)
$$\sum_{a=0}^{17} \left(\frac{a}{17}\right) \zeta_{17}^a = \sqrt{17},$$

where, for an odd prime p, the *Legendre symbol* $\left(\frac{a}{p}\right)$ is defined by

$$\left(\frac{a}{p}\right) = \begin{cases} 0, & \text{if } p \mid a, \\ +1, & \text{if } p \nmid a, \ x^2 \equiv a \bmod p \text{ has a solution}, \\ -1, & \text{if } p \nmid a, \ x^2 \equiv a \bmod p \text{ has no solution}. \end{cases}$$

More generally, for an odd prime p, a *quadratic Gauss sum* is defined to be

$$g_\ell = \sum_{a=0}^{p} \left(\frac{a}{p}\right) \zeta_p^{\ell a}.$$

Gauss used these sums to prove quadratic reciprocity. He also proved the remarkable formula

$$g_1 = \begin{cases} \sqrt{p}, & \text{if } p \equiv 1 \bmod 4, \\ i\sqrt{p}, & \text{if } p \equiv 3 \bmod 4. \end{cases}$$

Notice how this generalizes (9.24). A careful discussion of quadratic Gauss sums can be found in [7, Ch. 6].

Historical Notes

Most results of this section are implicit in Section VII of *Disquisitiones*. The main difference is that we have stated things in terms of the Galois correspondence, which to each divisor f of $p - 1$ associates the subgroup H_f of $(\mathbb{Z}/p\mathbb{Z})^*$ and the subfield L_f of $\mathbb{Q}(\zeta_p)$. For Gauss, on the other hand, each divisor f gets associated to the collection of f-periods (f, λ). In general, he considers elements rather than the fields in which they lie. For example, consider [4, Art. 346], which asserts that given (f, λ), any other f-period (f, μ) can be expressed as

$$(f, \mu) = \alpha_0 + \alpha_1(f, \lambda) + \alpha_2(f, \lambda)^2 + \cdots + \alpha_{e-1}(f, \lambda)^{e-1}$$

for some uniquely determined integers $\alpha_0, \ldots, \alpha_{e-1}$. For us, this gives the unique representation of (f, μ) as an element of $L_f = \mathbb{Q}\big((f, \lambda)\big)$.

Another difference is our use of cosets. For example, if g is a primitive root modulo p and f divides $p - 1$, then Gauss notes that the distinct f-periods are

$$(f, 1), (f, g), (f, g^2), \ldots, (f, g^{e-1}),$$

where $e = \frac{p-1}{f}$. For us, this follows from Lemma 9.2.4, since $H_f \subset (\mathbb{Z}/p\mathbb{Z})^*$ is generated by $[g^e]$, so that its cosets in $(\mathbb{Z}/p\mathbb{Z})^*$ are

$$[1]H_f, [g]H_f, [g^2]H_f, \ldots, [g^{e-1}]H_f.$$

Cosets give a conceptual basis for what Gauss is doing, and the same is true for the minimal polynomials computed in Proposition 9.2.8.

It is also interesting to note that Gauss makes implicit use of the Galois group $\mathrm{Gal}(\mathbb{Q}(\zeta_p)/\mathbb{Q})$. We saw in Lemma 9.2.4 that $\sigma(\zeta_p) = \zeta_p^k$ implies that $\sigma\big((f, \lambda)\big) = (f, k\lambda)$. Now consider the following quote from [4, Art. 345]:

> IV. It follows that if in any rational integral algebraic function $F = \phi(t, u, v, \ldots)$ we substitute for the unknowns t, u, v, etc. respectively the similar periods (f, λ), (f, μ), (f, v), etc., its value will be reducible to the form
>
> $$A + B(f, 1) + B'(f, g) + B''(f, g^2) \ldots + B^\varepsilon(f, g^{e-1})$$
>
> and the coefficients A, B, B', etc. will all be integers if all the coefficients in F are integers. But if afterward we substitute $(f, k\lambda)$, $(f, k\mu)$, (f, kv), etc. for t, u, v, etc. respectively, the value of F will be reduced to $A + B(f, k) + B'(f, kg) + $etc.

A "rational integral algebraic function" is a polynomial with coefficients in $\mathbb{Q}$. Here is an example of what this means.

Example 9.2.15. In Example 9.2.10, we showed that

$$(6, 1)^2 = 4 - (6, 2)$$

when $p = 19$. Using $k = 2$, the above quotation from Gauss tells us that

$$(6, 2)^2 = 4 - (6, 4).$$

In modern terms, this follows by applying the automorphism $\sigma \in \mathrm{Gal}(\mathbb{Q}(\zeta_{19})/\mathbb{Q})$ that takes ζ_{19} to ζ_{19}^2. So the Galois action is implicit in Gauss's theory! ◁▷

Gauss's result that $x^p - 1$ is solvable by radicals is less compelling from the modern perspective, though it is still interesting when one thinks inductively. But historically, being able to solve special but nontrivial equations of high degree was important. Here is what Gauss says in [4, Art. 359]:

> Everyone knows that the most eminent geometers have been unsuccessful in the search for a general solution of equations higher than the fourth degree, or (to define the search more accurately) for the THE REDUCTION OF MIXED EQUATIONS TO PURE EQUATIONS. ... Nevertheless, it is certain that there are innumerable mixed equations of every degree that admit a reduction to pure equations, and we trust that geometers will find it gratifying if we show that our equations are always of this kind.

For Gauss, an equation is "pure" if it is of the form $x^m - A = 0$ and "mixed" otherwise. Thus, reducing "mixed equations to pure equations" is what we call solvability by radicals. Of course, in saying "our equations," Gauss is referring to the minimal polynomials satisfied by the periods, as constructed in Proposition 9.2.8.

Gauss's study of the pth roots of unity is an important midpoint in the development leading from Lagrange to the emergence of Galois theory. Gauss uses Lagrange's inductive strategy to work out the Galois correspondence for $\mathbb{Q} \subset \mathbb{Q}(\zeta_p)$, and his theory of periods makes everything explicit and computable. He also shows that Lagrange resolvents are the correct tool for studying solvability by radicals, paving the way for Galois's analysis of the general case.

In spite of its beauty, what Gauss does in Section VII of [4] is not perfect. Some proofs are omitted and others have gaps. For example, Gauss does not prove the assertion about the Galois action made in the quotation before Example 9.2.15. Also, as noted in [Tignol, p. 195], Gauss's study of solvability assumes without proof that when fq divides $p - 1$, the f-periods are linearly independent over the field generated by the qth roots of unity. (You will prove this in Exercise 12.)

Galois was very aware of Section VII of *Disquisitiones*. For example, Galois describes the "group" of $\mathbb{Q} \subset \mathbb{Q}(\zeta_n)$, n prime, as follows [Galois, pp. 51–53]:

> In the case of the equation $\frac{x^n - 1}{x - 1} = 0$, if one supposes $a = r, b = r^g, c = r^{g^2}$,
> ..., g being a primitive root, the group of permutations will simply be as follows:

$$
\begin{array}{cccccc}
a & b & c & d & \ldots & k \\
b & c & d & \ldots & k & a \\
c & d & \ldots & k & a & b \\
\ldots & \ldots & \ldots & \ldots & \ldots & \ldots \\
k & a & b & c & \ldots & i
\end{array}
$$

> in this particular case, the number of permutations is equal to the degree of the equation, and the same will be true for equations where all of the roots are rational functions of each other.

Here, r is a primitive nth root of unity. Each line is a cyclic permutation of the one above it, which leads to a cyclic group of order $n - 1$. This quotation also reveals that for Galois, a "permutation" was an arrangement of the roots and that the permutations (in the modern sense) are obtained by mapping the first arrangement in the table to the others. You will work out the details of this in Exercise 14. We will say more about how Galois thought about Galois groups in Chapter 12.

Exercises for Section 9.2

Exercise 1. Let G be a cyclic group of order n and let g be a generator of G.
  (a) Let f be a positive divisor of n and set $e = n/f$. Prove that $H_f = \langle g^e \rangle$ has order f and hence is the unique subgroup of order f.
  (b) Let f and f' be positive divisors of $p - 1$. Prove that $H_f \subset H_{f'}$ if and only if $f | f'$.

Exercise 2. Prove Proposition 9.2.1.

Exercise 3. Let η_1, η_2, η_3 be as in Example 9.2.2.
  (a) We know that ζ_7 is a root of $x^6 + x^5 + x^4 + x^3 + x^2 + x + 1 = 0$. Dividing by x^3 gives

$$x^3 + x^2 + x + 1 + x^{-1} + x^{-2} + x^{-3} = 0.$$

Use this to show that η_1, η_2, η_3 are roots of $y^3 + y^2 - 2y - 1$.

(b) Prove that $[\mathbb{Q}(\eta_1):\mathbb{Q}] = 3$, and conclude that $\mathbb{Q}(\eta_1)$ is the fixed field of the subgroup $\{e, \tau\} \subset \mathrm{Gal}(\mathbb{Q}(\zeta_7)/\mathbb{Q})$, where τ is complex conjugation.

(c) Prove (9.10).

Exercise 4. Let $A \subset B$ be subgroups of a group G, and assume that A has index d in B. Prove that every left coset of B in G is a disjoint union of d left cosets of A in G.

Exercise 5. Complete the proof of Proposition 9.2.8.

Exercise 6. Prove that the sum of the distinct f-periods equals -1.

Exercise 7. This exercise is concerned with the details of Examples 9.2.10, 9.2.11, 9.2.12, and 9.2.13.

(a) Show that 2 is a primitive root modulo 19.

(b) Use the methods of Example 9.2.10 to obtain formulas for $(6, 2)^2$ and $(6, 4)^2$.

(c) Show that the formulas of part (b) follow from $(6, 1)^2 = 4 - (6, 2)$ and part (d) of Lemma 9.2.4.

(d) Prove (9.15) and use this and Exercise 6 to show that $(6, 1)(6, 2)(6, 4) = 7$.

(e) Find the minimal polynomials of $(3, 2)$ and $(3, 4)$ over the field L_6 considered in Example 9.2.12.

(f) Show that (9.18) is the minimal polynomial of ζ_{19} over the field L_3 considered in Example 9.2.13.

Exercise 8. In this exercise and the next, you will derive Gauss's radical formula (9.19) for $\cos(2\pi/17)$.

(a) Show that 3 is a primitive root modulo 17.

(b) Show that

$$H_8 = \{1, 2, 4, 8, 9, 13, 15, 16\},$$
$$H_4 = \{1, 4, 13, 16\},$$
$$H_2 = \{1, 16\},$$

where we write the congruence class $[n]$ modulo 17 as n.

(c) Use Propositions 9.2.8 and 9.2.9 to compute the following minimal polynomials:

Extension	Primitive Elements	Minimal Polynomial
$\mathbb{Q} \subset L_8$	$(8, 1), (8, 3)$	$x^2 + x - 4$
$L_8 \subset L_4$	$(4, 1), (4, 2)$	$x^2 - (8, 1)x - 1$
	$(4, 3), (4, 6)$	$x^2 - (8, 3)x - 1$
$L_4 \subset L_2$	$(2, 1), (2, 4)$	$x^2 - (4, 1)x + (4, 3)$

The resulting quadratic equations are easy to solve using the quadratic formula. But how do the roots correspond to the periods? For example, the roots $(8, 1)$, $(8, 3)$ of $x^2 + x - 4$ are $(-1 \pm \sqrt{17})/2$. How do these match up? The answer will be given in the next exercise.

Exercise 9. In this exercise, you will use numerical computations and the previous exercise to find radical expressions for various f-periods when $p = 17$.

(a) Show that

$$(8, 1) = 2\cos(2\pi/17) + 2\cos(4\pi/17) + 2\cos(8\pi/17) + 2\cos(16\pi/17),$$
$$(4, 1) = 2\cos(2\pi/17) + 2\cos(8\pi/17),$$
$$(4, 3) = 2\cos(6\pi/17) + 2\cos(10\pi/17),$$
$$(2, 1) = 2\cos(2\pi/17).$$

Then compute each of these periods to five decimal places.
(b) Use the numerical computations of part (a) and the quadratic polynomials of Exercise 8 to show that

$$(8, 1) = \tfrac{1}{2}\left(-1 + \sqrt{17}\right),$$
$$(8, 3) = \tfrac{1}{2}\left(-1 - \sqrt{17}\right),$$
$$(4, 1) = \tfrac{1}{4}\left(-1 + \sqrt{17} + \sqrt{34 - 2\sqrt{17}}\right),$$
$$(4, 2) = \tfrac{1}{4}\left(-1 + \sqrt{17} - \sqrt{34 - 2\sqrt{17}}\right),$$
$$(4, 3) = \tfrac{1}{4}\left(-1 - \sqrt{17} + \sqrt{34 + 2\sqrt{17}}\right).$$

(c) Use the quadratic polynomial $x^2 - (4, 1)x - (4, 3)$ and part (b) to derive (9.19).

Exercise 10. Let $p = 11$. Prove that $y^5 + y^4 - 4y^3 - 3y^2 + 3y + 1$ is the minimal polynomial of the 2-period $(2, 1) = 2\cos(2\pi/11)$.

Exercise 11. Let $L_{fq} \subset L_f$ be the extension studied in Theorem 9.2.14. Thus f and fq divide $p - 1$, and q is prime. As usual, $ef = p - 1$ and g is a primitive root modulo p. Finally, let ω be a primitive qth root of unity.
(a) Let $\tau \in \mathrm{Gal}(\mathbb{Q}(\zeta_p)/\mathbb{Q})$ satisfy $\tau(\zeta_p) = \zeta_p^{g^{e/q}}$, and let $\sigma' = \tau|_{L_f}$ be the restriction of τ to L_f. Prove that σ' generates $\mathrm{Gal}(L_f/L_{fq})$.
(b) Prove that $\mathrm{Gal}(L_f(\omega)/L_{fq}(\omega)) \simeq \mathrm{Gal}(L_f/L_{fq})$, where the isomorphism is defined by restriction to L_f.
(c) Let $\sigma \in \mathrm{Gal}(L_f(\omega)/L_{fq}(\omega))$ map to the element $\sigma' \in \mathrm{Gal}(L_f/L_{fq})$ constructed in part (a). Prove that σ satisfies (9.21).
(d) Prove the coset decomposition of H_{fq} given in (9.23).

Exercise 12. Let p be an odd prime, and let m be a positive integer relatively prime to p.
(a) Prove that $1, \zeta_p, \ldots, \zeta_p^{p-2}$ are linearly independent over $\mathbb{Q}(\zeta_m)$.
(b) Explain why part (a) implies that $\zeta_p, \ldots, \zeta_p^{p-1}$ are linearly independent over $\mathbb{Q}(\zeta_m)$.
(c) Let $f | p - 1$. Prove that the f-periods are linearly independent over $\mathbb{Q}(\zeta_m)$.

Exercise 13. Prove (9.24).

Exercise 14. Consider the quotation from Galois given at the end of the Historical Notes.
(a) Show that the permutations obtained by mapping the first line in the displayed table to the other lines give a cyclic group of order $n - 1$. Also explain how these permutations relate to the Galois group.
(b) Explain what Galois is saying in the last sentence of the quotation.

Exercise 15. What are the 1-periods?

Exercise 16. Redo Exercise 3 using periods.

Exercise 17. Let f be an even divisor of $p - 1$, where p is an odd prime. Prove that every f-period (f, λ) lies in $\mathbb{R}$.

REFERENCES

1. G. Bachman, *On the coefficients of cyclotomic polynomials*, Mem. Amer. Math. Soc. **106** (1993).

2. J. A. Buchmann, *Introduction to Cryptography*, Springer-Verlag, New York, Berlin, Heidelberg, 2001.

3. D. A. Burgess, *On character sums and L-series*, Proc. London Math. Soc. **12** (1962), 193-206.

4. C. F. Gauss, *Disquisitiones Arithmeticae*, Leipzig, 1801. Republished in 1863 as Volume I of [Gauss]. French translation, *Recherches Arithmétiques*, Paris, 1807. Reprint by Hermann, Paris, 1910. German translation, *Untersuchungen über Höhere Arithmetik*, Berlin, 1889. Reprint by Chelsea, New York, 1965. English translation, Yale U. P., New Haven, 1966. Reprint by Springer-Verlag, New York, Berlin, Heidelberg, 1986.

5. J. J. Gray, *A commentary on Gauss's mathematical diary, 1796–1814, with an English translation*, Expo. Math. **2** (1984), 97–130. (The Latin original of Gauss's diary is reprinted in [Gauss, Vol. X.1].)

6. R. K. Guy, *Unsolved Problems in Number Theory*, Springer-Verlag, New York, Berlin, Heidelberg, 1994.

7. K. Ireland and M. Rosen, *A Classical Introduction to Modern Number Theory*, Springer-Verlag, New York, Berlin, Heidelberg, 1982.

8. K. Kearnes, *Solution of Problem 6420*, Amer. Math. Monthly **91** (1984), 521.

9. R. A. Mollin, *Introduction to Cryptography*, Chapman & Hall/CRC, Boca Raton, FL, 2001.

10

Geometric Constructions

The idea of geometric constructions using straightedge and compass goes back to the ancient Greeks. This chapter will explore the surprising connection between geometric constructions and Galois theory. Topics covered include classic problems from Greek geometry, the work of Gauss described in Chapter 9, and the use of origami to solve cubic and quartic equations.

10.1 CONSTRUCTIBLE NUMBERS

Recall that a straightedge is an unmarked ruler. We also assume that you remember how to do standard straightedge-and-compass constructions such as bisecting a given angle, erecting a perpendicular to a given line at a given point, and dropping a perpendicular from a given point to a given line.

To prove theorems about geometric constructions, we need a careful description of what a construction is. The basic idea is that we begin with some known points. From these points, we use straightedge and compass to construct lines and circles:

C1. From $\alpha \neq \beta$, we can draw the line ℓ that goes through α and β.
C2. From $\alpha \neq \beta$ and γ, we can draw the circle C with center γ whose radius is the distance from α to β.

In this labeling, "C" stands for "Construct." Then the intersections of these lines and circles give new points:

P1. The point of intersection of distinct lines ℓ_1 and ℓ_2 constructed as above.
P2. The points of intersection of a line ℓ and circle C constructed as above.
P3. The points of intersection of distinct circles C_1 and C_2 constructed as above.

Here, "P" stands for "Point." We regard these newly constructed points as known. Then we can apply C1, C2, P1, P2, P3 to our enlarged collection of known points. We continue this process until the construction is completed.

For us, the plane will be the field of complex numbers $\mathbb{C}$, so that constructing a point means constructing a complex number. Our constructions will all start from the same two numbers, 0 and 1. This leads to the following definition.

Definition 10.1.1. *A complex number α is **constructible** if there is a finite sequence of straightedge and compass constructions using* C1, C2, P1, P2, P3 *that begins with* 0 *and* 1 *and ends with* α.

Here are some examples of constructible numbers.

Example 10.1.2.

(a) From 0 and 1, we get the x-axis using C1 and the circle of radius 1 centered at 1 using C2. These intersect in the numbers 0 and 2. By P2, 2 is constructible. Iterating this shows that every $n \in \mathbb{Z}$ is constructible.
(b) In Exercise 1 you will use standard methods from high school geometry to construct the line perpendicular to the x-axis at 0. This will show that the y-axis is constructible. Then use C2 to construct the circle of radius 1 centered at 0. These intersect in $\pm i$. By P2, $i \in \mathbb{C}$ is constructible.

These constructions will be useful later.   ◁▷

Example 10.1.3. Suppose that we can construct a regular polygon with n sides somewhere in the plane (we call this a *regular n-gon*). Using two consecutive vertices and the center of the n-gon, we get the triangle shown here:

(We may have constructed the center in the process of constructing the n-gon. If not, then the center is constructible, since it is the intersection of the bisectors of the interior angles.) One easily sees that the angle at the center is $\theta = 2\pi/n$. In Exercise 2 you will show how to copy this angle to the origin:

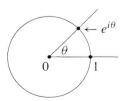

Intersecting this with the unit circle shows that the nth root of unity $\zeta_n = e^{2\pi i/n}$ is constructible. In Exercise 2 you will show that this process can be reversed. Hence ζ_n is constructible if and only if a regular n-gon can be constructed by straightedge and compass. Section 10.2 will determine those n's for which this is possible. ◁▷

The set of constructible numbers has the following properties.

Theorem 10.1.4. *The set $\mathscr{C} = \{\alpha \in \mathbb{C} \mid \alpha \text{ is constructible}\}$ is a subfield of $\mathbb{C}$. Furthermore:*

(a) *Let $\alpha = a + ib \in \mathbb{C}$, where $a, b \in \mathbb{R}$. Then $\alpha \in \mathscr{C}$ if and only if $a, b \in \mathscr{C}$.*
(b) *$\alpha \in \mathscr{C}$ implies that $\sqrt{\alpha} \in \mathscr{C}$.*

Proof. We first show that $\mathscr{C}$ is a subgroup of $\mathbb{C}$ under addition. Given $\alpha \in \mathscr{C} \setminus \{0\}$, construct the line connecting 0 and α by C1 and the circle of radius $|\alpha|$ centered at the origin by C2. These intersect in $\pm\alpha$, so that $-\alpha$ is constructible by P2.

Now suppose that α and β are constructible. If α, β, and 0 are not collinear, then use C2 twice to construct the circle of radius $|\alpha|$ with center β and the circle of radius $|\beta|$ with center α. One of the points of intersection is $\alpha + \beta$:

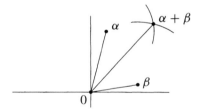

By P3, we conclude that $\alpha + \beta$ is constructible. In Exercise 3 you will show that this is also true when α, β, and 0 are collinear. Since $0 \in \mathscr{C}$ by definition, it follows that $\mathscr{C}$ is a subgroup of $\mathbb{C}$ under addition.

We next prove part (a). Given $\alpha = a + ib \in \mathscr{C}$, we can drop perpendiculars from α to the x-axis and y-axis constructed in Example 10.1.2. This shows that $a, ib \in \mathscr{C}$. Since the circle of radius $|ib|$ centered at 0 intersects the x-axis at b, C2 and P2 imply that $b \in \mathscr{C}$. Conversely, given $a, b \in \mathscr{C} \cap \mathbb{R}$, applying C2 and P2 to the circle of radius $|b|$ centered at 0 shows that ib is constructible. By the previous paragraph, $a + ib$ is constructible and part (a) follows.

Now let $a, b \in \mathscr{C} \cap \{x \in \mathbb{R} \mid x > 0\}$ and consider the following two figures:

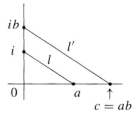

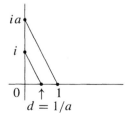

Figure 1   Figure 2

Recall that i was constructed in Example 10.1.2. In Figure 1 we construct ib as above and then use C1 to draw the line l containing i and a. By high school geometry, we can draw the line l' through ib that is parallel to l. Then P1 shows that l' and the x-axis intersect at a constructible real number c. But $c = ab$ follows easily by similar triangles, so that ab is constructible. In a similar way Figure 2 shows that $1/a$ is constructible. We leave the details as Exercise 3. This exercise will also show that $\mathscr{C} \cap \mathbb{R}$ is a subfield of $\mathbb{R}$.

To show that $\mathscr{C}$ is closed under multiplication and taking reciprocals of nonzero elements, let $\alpha = a + ib$ and $\beta = c + id$ be constructible numbers. Then

$$\alpha\beta = (a + ib)(c + id) = ac - bd + i(ad + bc).$$

However, $a, b, c, d \in \mathscr{C} \cap \mathbb{R}$ by part (a), so that $ac - bd, ad + bc \in \mathscr{C} \cap \mathbb{R}$, since the latter is a subfield of $\mathbb{R}$. Using part (a) again, we conclude that $\alpha\beta \in \mathscr{C}$. Furthermore, if $\alpha \neq 0$, then

$$\frac{1}{\alpha} = \frac{1}{a + ib}\frac{a - ib}{a - ib} = \frac{a}{a^2 + b^2} + i\frac{-b}{a^2 + b^2}.$$

Using part (a) and the fact that $\mathscr{C} \cap \mathbb{R}$ is a subfield of $\mathbb{R}$, we easily see that $1/\alpha \in \mathscr{C}$. Thus $\mathscr{C}$ is a subfield of $\mathbb{C}$.

Finally, we show that $\sqrt{\alpha}$ is constructible when α is. We can assume that $\alpha \neq 0$. If we write $\alpha = re^{i\theta}, r = |\alpha| > 0$, then it suffices to show that $\sqrt{r}e^{i\theta/2}$ is constructible. To prove this, note that the constructibility of α implies the following:

- First, using the x-axis and the line containing 0 and α (by C1), we can construct the angle θ, which we can then bisect by the usual straightedge-and-compass construction. Thus the angle $\theta/2$ is constructible.
- Second, the circle of radius $r = |\alpha|$ centered at 0 (by C2) intersects the x-axis at $\pm r$. By P2, we see that r is constructible.
- Third, if we can construct $\sqrt{r}$, then we can construct the circle of radius $\sqrt{r}$ centered at the origin by C2. Then P2, applied to this circle and the angle $\theta/2$ constructed above, implies that $\sqrt{r}e^{i\theta/2}$ is constructible.

It remains to prove that $\sqrt{r}$ is constructible whenever $r > 0$ is. Consider the diagram

(10.1)

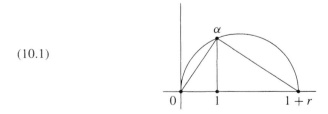

In Exercise 3 you will show that α is constructible. By Euclidean geometry, the triangle with vertices $1, \alpha$, and $1 + r$ is a right triangle. The triangles that share the side determined by 1 and α are similar, so that

$$\frac{1}{d} = \frac{d}{r},$$

where d is the distance from 1 to α. Thus $d^2 = r$ and hence $d = \sqrt{r}$. Since d is easily seen to be constructible, we conclude that $\sqrt{r}$ is constructible. $\square$

Here is an example of how to use Theorem 10.1.4.

Example 10.1.5. By Exercise 8 of Section A.2, $\zeta_5 = e^{2\pi i/5}$ is given by the formula

$$\zeta_5 = \frac{-1 + \sqrt{5}}{4} + \frac{i}{2}\sqrt{\frac{5 + \sqrt{5}}{2}}.$$

Since the field $\mathscr{C}$ is closed under the operation of taking square roots, it follows easily that ζ_5 is constructible. By Example 10.1.3, we conclude that a regular pentagon can be constructed by straightedge and compass.   ◁▷

We call $\mathscr{C}$ the *field of constructible numbers*. We next study the structure of $\mathscr{C}$.

Theorem 10.1.6. *Let α be a complex number. Then $\alpha \in \mathscr{C}$ if and only if there are subfields*

$$\mathbb{Q} = F_0 \subset F_1 \subset \cdots \subset F_{n-1} \subset F_n \subset \mathbb{C}$$

such that $\alpha \in F_n$ and $[F_i : F_{i-1}] = 2$ for $1 \leq i \leq n$.

Proof. First suppose that we have $\mathbb{Q} = F_0 \subset \cdots \subset F_n \subset \mathbb{C}$ where $[F_i : F_{i-1}] = 2$. By Exercise 12 of Section 7.1, $F_i = F_{i-1}(\sqrt{\alpha_i})$ for some $\alpha_i \in F_i$. We will prove that $F_i \subset \mathscr{C}$ by induction on $0 \leq i \leq n$. The case $F_0 = \mathbb{Q} \subset \mathscr{C}$ follows because $\mathscr{C}$ is a subfield of $\mathbb{C}$. Now suppose that $F_{i-1} \subset \mathscr{C}$. Then $\alpha_i \in F_{i-1}$ is constructible, which implies $\sqrt{\alpha_i} \in \mathscr{C}$ by Theorem 10.1.4. Thus $F_i = F_{i-1}(\sqrt{\alpha_i}) \subset \mathscr{C}$, as claimed. This shows that $F_n \subset \mathscr{C}$, so that in particular, any $\alpha \in F_n$ is constructible.

Conversely, given $\alpha \in \mathscr{C}$, we need to create successive quadratic extensions that start from $\mathbb{Q}$ and eventually contain α. We will prove that there are extensions $\mathbb{Q} = F_0 \subset \cdots \subset F_n \subset \mathbb{C}$ where $[F_i : F_{i-1}] = 2$ such that F_n contains the real and imaginary parts of all numbers constructed during the course of constructing α. The theorem will follow, since $\alpha = a + ib$ will imply that $a, b \in F_n$, so that $\alpha \in F_n(i)$.

We will prove this by induction on the number N of times we use P1, P2, P3 in the construction of α. When $N = 0$, we must have $\alpha = 0$ or 1, in which case we let $F_n = F_0 = \mathbb{Q}$. Now suppose that α is constructed in $N > 1$ steps, where the last step uses P1, the intersection of distinct lines ℓ_1 and ℓ_2. But then ℓ_1 was constructed from distinct points α_1 and β_1 using C1, and similarly ℓ_2 was constructed from distinct points α_2 and β_2. By our inductive assumption, there are extensions $\mathbb{Q} = F_0 \subset \cdots \subset F_n \subset \mathbb{C}$ where $[F_i : F_{i-1}] = 2$ such that F_n contains the real and imaginary parts of $\alpha_1, \beta_1, \alpha_2, \beta_2$. We will prove that F_n contains the real and imaginary parts of α.

The line ℓ_1 has an equation of the form $a_1 x + b_1 y = c_1$ and goes through $\alpha_1 \neq \beta_1$. Since the real and imaginary parts of α_1, β_1 lie in F_n, Exercise 4 implies that we can assume that the coefficients a_1, b_1, c_1 lie in F_n. Similarly, ℓ_2 has an equation of the form $a_2 x + b_2 y = c_2$ where $a_2, b_2, c_2 \in F_n$. Hence the real and imaginary parts of

α give the unique solution of the equations

$$a_1x + b_1y = c_1,$$
$$a_2x + b_2y = c_2.$$

In this situation, linear algebra tells us that the matrix

$$\begin{pmatrix} a_1 & b_1 \\ a_2 & b_2 \end{pmatrix}$$

is invertible (be sure you can explain why), so that the unique solution is

$$\begin{pmatrix} x \\ y \end{pmatrix} = \begin{pmatrix} a_1 & b_1 \\ a_2 & b_2 \end{pmatrix}^{-1} \begin{pmatrix} c_1 \\ c_2 \end{pmatrix}.$$

It follows immediately that the real and imaginary parts of α lie in F_n.

Next suppose that the last step in the construction of α uses P2, the intersection of a line ℓ and a circle C. Thus ℓ is the line through $\alpha_1 \neq \beta_1$ (from C1), and C is the circle with center γ_2 and radius $|\alpha_2 - \beta_2|$ (from C2). The five points $\alpha_1, \beta_1, \alpha_2, \beta_2, \gamma_2$ come from earlier steps in the construction, so that by our inductive assumption, there are extensions $\mathbb{Q} = F_0 \subset \cdots \subset F_n \subset \mathbb{C}$ where $[F_i : F_{i-1}] = 2$ such that F_n contains the real and imaginary parts of these five points. We will show that the real and imaginary parts of α lie in F_n or in a quadratic extension of F_n.

As above, ℓ is given by an equation

$$(10.2) \qquad a_1x + b_1y = c_1,$$

where $a_1, b_1, c_1 \in F_n$. In Exercise 4 you will show that C is given by an equation

$$(10.3) \qquad x^2 + y^2 + a_2x + b_2y + c_2 = 0,$$

where $a_2, b_2, c_2 \in F_n$. Now suppose that $a_1 \neq 0$. Then dividing (10.2) by a_1 and relabeling, we can assume that the line ℓ is given by $x + b_1y = c_1$. Substituting $x = -b_1y + c_1$ into (10.3) gives the quadratic equation

$$(-b_1y + c_1)^2 + y^2 + a_2(-b_1y + c_1) + c_2 = 0.$$

By the quadratic formula, the values of y involve the square root of an expression in F_n. If this lies in F_n, then so do y and $x = -b_1y + c_1$, and it follows that the real and imaginary parts of α lie in F_n. On the other hand, if this square root does not lie in F_n, then it lies in a quadratic extension $F_n \subset F_{n+1}$. Then y and $x = -b_1y + c_1$ also lie in F_{n+1}, which shows that the real and imaginary parts of α lie in a quadratic extension of F_n. When $a_1 = 0$, the argument is similar and is left as part of Exercise 4.

Finally, suppose that the last step in the construction of α uses P3, the intersection of distinct circles C_1 and C_2. As above, we can find $\mathbb{Q} = F_0 \subset \cdots \subset F_n \subset \mathbb{C}$ where $[F_i : F_{i-1}] = 2$ such that the circles are given by equations

$$(10.4) \qquad \begin{aligned} x^2 + y^2 + a_1x + b_1y + c_1 &= 0, \\ x^2 + y^2 + a_2x + b_2y + c_2 &= 0, \end{aligned}$$

where all of the coefficients lie in F_n. Furthermore, we know that the real and imaginary parts of α give a solution of (10.4).

If we subtract these equations, we get the equation

$$(10.5) \qquad (a_1 - a_2)x + (b_1 - b_2)y + (c_1 - c_2) = 0.$$

Since the circles (10.4) are distinct but not disjoint, one easily sees that the coefficients of x and y in (10.5) don't vanish simultaneously. Thus (10.5) defines a line. Furthermore, if we combine this equation with the first equation of (10.4), then we are in the previous case of the intersection of a circle and a line. We conclude that the real and imaginary parts of α lie in F_n or in a quadratic extension of F_n. This completes the proof. $\qquad\qquad\square$

Corollary 10.1.7. *$\mathscr{C}$ is the smallest subfield of $\mathbb{C}$ that is closed under the operation of taking square roots.*

Proof. By Theorem 10.1.4, we know that $\alpha \in \mathscr{C}$ implies that $\sqrt{\alpha} \in \mathscr{C}$. Now let F be any subfield of $\mathbb{C}$ closed under taking square roots, and suppose that $\alpha \in \mathscr{C}$. By Theorem 10.1.6, we have $\mathbb{Q} = F_0 \subset \cdots \subset F_n \subset \mathbb{C}$ where $[F_i : F_{i-1}] = 2$ and $\alpha \in F_n$. The first paragraph of the proof of Theorem 10.1.6 shows that $F_n \subset F$. Thus $\alpha \in F_n \subset F$, and $\mathscr{C} \subset F$ follows as desired. $\qquad\qquad\square$

Theorem 10.1.6 also has the following useful consequence.

Corollary 10.1.8. *If $\alpha \in \mathscr{C}$, then $[\mathbb{Q}(\alpha) : \mathbb{Q}] = 2^m$ for some $m \geq 0$. Thus every constructible number is algebraic over $\mathbb{Q}$, and the degree of its minimal polynomial over $\mathbb{Q}$ is a power of 2.*

Proof. If $\alpha \in \mathscr{C}$, then Theorem 10.1.6 gives extensions $\mathbb{Q} = F_0 \subset \cdots \subset F_n \subset \mathbb{C}$ where $[F_i : F_{i-1}] = 2$ and $\alpha \in F_n$. Hence

$$[F_n : \mathbb{Q}] = [F_n : F_0] = [F_n : F_{n-1}] \cdots [F_1 : F_0] = 2^n$$

by the Tower Theorem. However, we also have $\mathbb{Q} \subset \mathbb{Q}(\alpha) \subset F_n$. Using the Tower Theorem again, we conclude that $[\mathbb{Q}(\alpha) : \mathbb{Q}]$ divides $[F_n : \mathbb{Q}] = 2^n$. $\qquad\square$

Some of the most famous problems in Greek geometry are trisection of the angle, duplication of the cube, and squaring the circle. Using Corollary 10.1.8, we can solve these as follows.

Example 10.1.9. *Trisection of the Angle.* As we know, every angle can be bisected using straightedge and compass. We will prove that not every angle can be trisected by straightedge and compass. In particular, suppose that we could trisect a $120°$ angle in this way. Since we can construct a $120°$ angle from 0 and 1 by straightedge and compass (see Exercise 5), a trisection of this angle would imply that we could construct a $40°$ angle from 0 and 1 by straightedge and compass. Intersecting this with the unit circle centered at the origin, it would follow that the 9th root of unity $\zeta_9 = e^{2\pi i/9}$ would be a constructible number (since $40° = 2\pi/9$ radians).

However, Theorem 9.1.9 implies that the minimal polynomial of ζ_9 is the cyclotomic polynomial $\Phi_9(x)$, and the factorization

$$x^9 - 1 = \Phi_1(x)\Phi_3(x)\Phi_9(x) = (x - 1)(x^2 + x + 1)(x^6 + x^3 + 1)$$

from Proposition 9.1.5 shows that $x^6 + x^3 + 1$ is the minimal polynomial of ζ_9. By Corollary 10.1.8, ζ_9 is not constructible. This contradiction proves that we cannot trisect $120°$ using straightedge and compass. In Exercise 6 you will show that it is also impossible to trisect $60°$ by straightedge and compass.

In Section 10.2, we will use the results of Section 9.1 to determine all n for which ζ_n is constructible. ◁▷

Example 10.1.10. *Duplication of the Cube.* Here, the problem is to take a given cube and construct one with exactly twice the volume. We can pick our units of measurement so that the given cube has edges of length 1. In these units, the volume is also 1, which means that we need to construct a cube of volume 2. Since volume is edge length cubed, it follows that if we could duplicate the cube, then we could construct a number s such that $s^3 = 2$, that is, $s = \sqrt[3]{2}$. Furthermore, since the cube has edge length 1, we can assume that the construction begins with 0 and 1. It follows that duplicating the cube by straightedge and compass implies that $s = \sqrt[3]{2}$ is constructible. But $x^3 - 2$ is the minimal polynomial of $\sqrt[3]{2}$ over $\mathbb{Q}$, so that $\sqrt[3]{2}$ is not constructible, by Corollary 10.1.8. This contradiction proves that we cannot duplicate the cube by straightedge and compass. ◁▷

Example 10.1.11. *Squaring the Circle.* This is the problem of constructing a square whose area is equal to that of a given circle. If we pick our units of measurement so that the given circle has radius 1, then the circle has area π. Since a square of area π has side $\sqrt{\pi}$, it follows that if we could square the circle, then we could construct $\sqrt{\pi}$. Furthermore, since the circle has radius 1, we can assume that the construction begins with 0 and 1. It follows that squaring the circle by straightedge and compass would imply that $\sqrt{\pi}$ is constructible.

Since $\mathscr{C}$ is a field, the constructibility of $\sqrt{\pi}$ would imply that $\pi = \sqrt{\pi}^2$ is also constructible. Then Corollary 10.1.8 would imply that π is algebraic over $\mathbb{Q}$. However, in 1882 Lindemann proved that π is transcendental over $\mathbb{Q}$. A self-contained proof can be found in [Hadlock, Sec. 1.7]. This contradiction shows that we cannot square the circle by straightedge and compass. ◁▷

One could also ask whether the converse of Corollary 10.1.8 is true. In other words, if $\alpha \in \mathbb{C}$ is algebraic over $\mathbb{Q}$ and the degree of its minimal polynomial is a power of 2, is α constructible? The following result will answer this question.

Theorem 10.1.12. *Let $\alpha \in \mathbb{C}$ be algebraic over $\mathbb{Q}$, and let $\mathbb{Q} \subset L$ be the splitting field of the minimal polynomial of α over $\mathbb{Q}$. Then α is constructible if and only if $[L : \mathbb{Q}]$ is a power of 2.*

Proof. First suppose that $[L : \mathbb{Q}]$ is a power of 2. We will follow the proof of the Fundamental Theorem of Algebra given in Section 8.5. Since $\mathbb{Q} \subset L$ is Galois, it follows that $|\text{Gal}(L/\mathbb{Q})| = [L:\mathbb{Q}]$ is a power of 2, say $|\text{Gal}(L/\mathbb{Q})| = 2^m$. By Theorem 8.1.7, $\text{Gal}(L/\mathbb{Q})$ is solvable, which by Definition 8.1.1 means that we have subgroups

$$\{e\} = G_m \subset G_{m-1} \subset \cdots \subset G_1 \subset G_0 = \text{Gal}(L/\mathbb{Q})$$

such that G_i is normal in G_{i-1} of index 2 (since $|\text{Gal}(L/\mathbb{Q})| = 2^m$). This gives

$$\mathbb{Q} = L_{G_0} \subset L_{G_1} \subset \cdots \subset L_{G_m} = L,$$

where $[L_{G_i} : L_{G_{i-1}}] = 2$ for all i. By Theorem 10.1.6, $\alpha \in L$ is constructible.

Turning to the converse, we first show that $\mathbb{Q} \subset \mathscr{C}$ is a normal extension. For this, let $\alpha \in \mathscr{C}$, and let f be the minimal polynomial of f over $\mathbb{Q}$. We need to prove that f splits completely over $\mathscr{C}$. Since α is constructible, Theorem 10.1.6 gives extensions $\mathbb{Q} = F_0 \subset \cdots \subset F_n \subset \mathbb{C}$ where $[F_i : F_{i-1}] = 2$ and $\alpha \in F_n$. Then let $\mathbb{Q} \subset M$ be the Galois closure of $\mathbb{Q} \subset F_n$, as constructed in Proposition 7.1.7. In Exercise 7 you will show that we may assume that $M \subset \mathbb{C}$.

Note that f splits completely in M, since M is normal over $\mathbb{Q}$, f is irreducible over $\mathbb{Q}$, and $\alpha \in F_n \subset M$ is a root of f. Now let $\beta \in M$ be any root of f. By Proposition 5.1.8, there is $\sigma \in \text{Gal}(M/\mathbb{Q})$ such that $\sigma(\alpha) = \beta$. Applying σ to the fields $\mathbb{Q} = F_0 \subset \cdots \subset F_n \subset M$ gives

$$\mathbb{Q} = \sigma(\mathbb{Q}) = \sigma(F_0) \subset \cdots \subset \sigma(F_n)$$

such that $[\sigma(F_i) : \sigma(F_{i-1})] = [F_i : F_{i-1}] = 2$ for all i. By Theorem 10.1.6, $\beta = \sigma(\alpha) \in \sigma(F_n)$ is constructible. This shows that f splits completely over $\mathscr{C}$.

It follows that $\mathscr{C}$ contains a splitting field L of f over $\mathbb{Q}$. By the Theorem of the Primitive Element, we have $L = \mathbb{Q}(\gamma)$ for some $\gamma \in L$. Since $\gamma \in \mathscr{C}$, Corollary 10.1.8 implies that $[\mathbb{Q}(\gamma):\mathbb{Q}] = [L:\mathbb{Q}]$ is a power of 2, as claimed. This completes the proof of the theorem. $\qquad\square$

We can use Theorem 10.1.12 to show that the converse of Corollary 10.1.8 is false. Here is an example.

Example 10.1.13. Let α be a root of the polynomial

$$f = x^4 - 4x^2 + x + 1.$$

One easily checks that f is irreducible over $\mathbb{Q}$, so that $[\mathbb{Q}(\alpha):\mathbb{Q}] = 4$. However, in Chapter 13 we will show that the splitting field L of f over $\mathbb{Q}$ satisfies $[L:\mathbb{Q}] = 24$. By Theorem 10.1.12 we conclude that α is not constructible. $\qquad\triangleleft\triangleright$

Mathematical Notes

There are several issues that are worthy of further comment.

■ **Starting Configurations.** According to Definition 10.1.1, a construcible number is constructed by a sequence of constructions that always begins with 0 and 1. It is possible to begin constructions with different starting configurations. For example, three noncollinear points α, β, γ determine an angle with vertex α and rays $\overrightarrow{\alpha\beta}$ and $\overrightarrow{\alpha\gamma}$. This angle can be bisected by straightedge and compass, even though α, β, γ need not be constructible.

We will not develop the theory of such constructions beyond the comment that the trisection of the angle is most naturally stated in this context: given an angle determined by α, β, γ as above, one seeks a construction that trisects this angle by straightedge and compass. In Example 10.1.9 we showed that this is impossible by finding a particular case of α, β, γ to which we could apply Corollary 10.1.8.

■ **Compasses.** In C2, the compass uses points $\alpha \neq \beta$ to give the radius, with the center given by a third point γ. This is slightly different from what Euclid does, for he uses the compass with points $\alpha \neq \beta$ where the center is either α or β. One can prove that this more restrictive notion of compass (called the *Euclidean compass* in Martin's book [12]) gives the same set of constructible points.

More surprising is the fact that we can dispense with the straightedge entirely. The Mohr–Mascheroni Theorem states that $\alpha \in \mathbb{C}$ is constructible if and only if there is a sequence of Euclidean compass constructions that starts with 0 and 1 and ends with α. A proof can be found in [12, Ch. 3].

■ **Straightedge and Dividers.** A set of *dividers* is a tool that can copy line segments. In other words, given points $\alpha \neq \beta$ and a point γ on a line l, dividers allow us to construct points δ_1, $\delta_2 \in l$ such that the distance from δ_1 or δ_2 to γ equals the distance from α to β, as in the following picture:

Let $\mathscr{P}$ denote the set of real numbers that can be obtained from 0, 1, and i by a sequence of straightedge-and-dividers constructions. By [12, Thm. 5.6], $\mathscr{P}$ is a subfield of $\mathbb{R}$. A more interesting property of $\mathscr{P}$ is that if $a, b \in \mathscr{P}$, then $\sqrt{a^2 + b^2} \in \mathscr{P}$. To prove this, note that the y-axis is constructible using 0, i, and the straightedge, so that given $b \in \mathscr{P}$, we can construct ib using our dividers. Combining this with $a \in \mathscr{P}$, we get the diagram

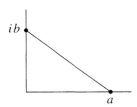

Now use the dividers to transfer the line segment from a and ib to the positive x-axis, starting from 0. The Pythagorean Theorem implies that $\sqrt{a^2 + b^2} \in \mathscr{P}$, as claimed.

In general, a subfield of $\mathbb{R}$ that contains $\sqrt{a^2 + b^2}$ whenever it contains a and b is called *Pythagorean*. Thus $\mathscr{P}$ is Pythagorean, and in Exercise 8 you will show that $\mathscr{P}$ is the smallest Pythagorean subfield of $\mathbb{R}$. This is an analog of Corollary 10.1.7. We call $\mathscr{P}$ the field of *Pythagorean numbers*.

The most interesting result about $\mathscr{P}$ is the following analog of Theorem 10.1.12.

Theorem 10.1.14. *Let $\alpha \in \mathbb{R}$ be algebraic over $\mathbb{Q}$, and let f be the minimal polynomial of α over $\mathbb{Q}$ with splitting field L. Then the following are equivalent:*

(a) *$\alpha \in \mathscr{P}$.*

(b) *All roots of f are real, and α is constructible.*

(c) *All roots of f are real, and $[L : \mathbb{Q}]$ is a power of 2.*

Proof. The equivalence (a) $\Leftrightarrow$ (b) is proved by Auckly and Cleveland in [3, p. 225]. Then (b) $\Leftrightarrow$ (c) follows from Theorem 10.1.12. $\square$

For those who read the discussion of the *casus irreducibilis* in Section 8.6, we note the following corollary of Theorems 10.1.14 and 8.6.5.

Theorem 10.1.15. *Let $\alpha \in \mathbb{R}$ be algebraic over $\mathbb{Q}$. Then $\alpha \in \mathscr{P}$ if and only if α is expressible by real radicals and all conjugates of α over $\mathbb{Q}$ are real.* $\square$

This shows an unexpected relation between geometric constructions and solvability by radicals.

Numbers in $\mathscr{P}$ are constructed using straightedge and dividers. Since a compass can be used as a pair of dividers, one sees easily that

$$\mathscr{P} \subset \mathscr{C} \cap \mathbb{R}.$$

Using Theorem 10.1.14, we prove that these fields are not equal as follows.

Example 10.1.16. Consider

$$\alpha = \sqrt{2 + 2\sqrt{2}}.$$

Then $\alpha \in \mathscr{C} \cap \mathbb{R}$, since $\mathscr{C}$ is closed under taking square roots. However, the minimal polynomial of α is

$$f = x^4 - 4x^2 - 4,$$

which has roots $\pm\sqrt{2 \pm 2\sqrt{2}}$. Two of these roots are not real, so that $\alpha \notin \mathscr{P}$ by Theorem 10.1.14.   ◁▷

■ **Marked Rulers.** The straightedge we've been using is an unmarked ruler. But suppose instead that we have a *marked ruler*, which is a straightedge with two marks on it one unit apart. Such a ruler allows the construction of some interesting lines and points. Provided one starts from the points $0, 1, i$ in $\mathbb{C}$, one can prove that all straightedge-and-compass constructions can be done using *only* a marked ruler. We will see in Section 10.3 that there are also marked-ruler constructions for trisecting angles and taking cube roots.

Historical Notes

There are two versions of where the problem of duplicating the cube first arose. In one version, King Minos was unhappy with the cubical tomb of his son Glaucus and ordered its size doubled. In the other, a delegation from Athens asked the oracle at Delos for advice about a plague in Athens. The Athenians were told to double the size of the cubical altar of Apollo. In both versions, the first attempt was to double the sides, which multiplies the volume by a factor of eight. The point, of course, is that what was required was to double the volume, which means that the sides must be multiplied by $\sqrt[3]{2}$. Because of the popularity of the second version of the story, the duplication of the cube is sometimes called the Delian problem.

Less is known about the origin of the other two Greek problems. It may have been something like the following. A line segment is easily bisected or trisected by straightedge and compass, and bisecting angles is equally easy. Hence it is natural to ask about trisecting angles. Similarly, a rectangle or triangle is easily squared by straightedge and compass. Since the next most basic geometric figure is a circle, it is natural to ask whether it is also squarable.

Greek geometers never solved these problems, though the search for solutions led to some wonderful mathematics. For example, consider the lune indicated by the shaded region:

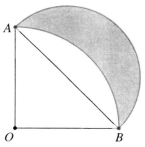

Around 440 B.C., Hippocrates of Chios discovered this lune in the course of his work on squaring the circle. The above figure has a quarter circle with radius $OA = OB$ and a semicircle with diameter AB. The lune is the region outside the quarter circle but inside the semicircle. Hippocrates showed that the shaded region has the same area as the triangle AOB (you will verify this in Exercise 9). It follows that this lune is squarable, so that at least part of a circle is squarable. Hippocrates found other squarable lunes, and in the twentieth century, Chebotarev and Dorodnov showed that there are only five squarable lunes. A proof can be found in [14].

Somewhat later, Hippias of Elis (ca. 425 B.C.) discovered a curve called the *quadratrix*, which he used to trisect angles and square the circle. In modern terms, this curve is given by

$$(10.6) \qquad y = x \cot\left(\frac{\pi x}{2}\right).$$

In Exercise 10 you will use this curve to trisect angles and square the circle. As for duplication of the cube, Menaechmus introduced parabolas around 350 B.C. as a by-product of his work on duplication. In Exercise 11 you will show that duplication of the cube can be solved by intersecting two parabolas. We will see in Section 10.3 that many other constructions can be done by intersecting conic sections.

There is a lot more to say about Greek work on these problems. For example, the spiral of Archimedes $r = \theta$ has some nice applications that you will study in Exercise 12. Numerous other examples can be found in Chapter 4 of [8]. Greek geometry is more interesting than what you learned in high school, in part because the Greeks had these great problems to inspire them. In modern mathematics, unsolved problems play a similar role of inspiring research. For example, the inverse Galois problem mentioned in Section 7.4 is still unsolved and is being actively studied by many researchers.

The three Greek problems were solved in the nineteenth century. In 1837 Wantzel showed that duplication of the cube and trisection of the angle cannot be done by straightedge and compass. His argument used the irreducibility of certain cubic polynomials. This is similar to what we did in Exercise 6 (for trisection of the angle) and Example 10.1.10 (for duplication of the cube). The first page of Wantzel's paper is reproduced on page 84 of [Escofier]. Finally, as noted in Example 10.1.11, the problem of squaring the circle was solved in 1882 when Lindemann showed that π is transcendental over $\mathbb{Q}$.

We should also note that the variants on straightedge-and-compass constructions mentioned in the Mathematical Notes have a long and interesting history. This is discussed in [12]. See also the Historical Notes to Section 10.3.

Exercises for Section 10.1

Exercise 1. In part (a) of Example 10.1.2 we constructed the x-axis. In a similar way show that the y-axis is constructible. For each step in your construction be sure to say which of C1, C2, P1, P2, P3 you are using.

Exercise 2. Suppose that α, β, γ are noncollinear and consider the rays $\overrightarrow{\alpha\beta}$ and $\overrightarrow{\alpha\gamma}$ emanating from α that go through β and γ respectively. We call this the angle formed by α, β, γ. Also assume that α, β, γ are constructible.
  (a) Prove that there is a constructible number δ with positive y-coordinate such that the angle formed by α, β, γ is congruent to the angle formed by $0, 1, \delta$. As in Exercise 1, each step in the construction should be justified by C1, C2, P1, P2, or P3.
  (b) Prove the claim made in Example 10.1.3 that $\zeta_n = e^{2\pi i/n}$ is constructible if and only if a regular n-gon can be constructed by straightedge and compass.

Exercise 3. This exercise covers the details omitted in the proof of Theorem 10.1.4.
  (a) Let α, β be constructible numbers such that $0, \alpha, \beta$ are collinear. Prove that $\alpha + \beta$ is constructible.
  (b) Let $a \in \mathscr{C} \cap \{x \in \mathbb{R} \mid x > 0\}$. Use Figure 2 in the proof of Theorem 10.1.4 to show that $1/a$ is construcible.
  (c) In the proof of Theorem 10.1.4, we showed that $\mathscr{C} \cap \{x \in \mathbb{R} \mid x > 0\}$ is closed under addition, multiplication, and multiplicative inverses. Use this to prove that $\mathscr{C} \cap \mathbb{R}$ is a subfield of $\mathbb{R}$.
  (d) Prove that the number α pictured in (10.1) is constructible (assuming that r is constructible).

Exercise 4. This exercise covers the details omitted in the proof of Theorem 10.1.6.

(a) Suppose that a line ℓ_1 goes through distinct points $\alpha_1 = u_1 + iv_1$ and $\beta_1 = u_2 + iv_2$, where u_1, v_1, u_2, v_2 lie in a subfield $F \subset \mathbb{R}$. Prove that ℓ_1 is defined by an equation of the form $a_1 x + b_1 y = c_1$ where $a_1, b_1, c_1 \in F$.

(b) Suppose that $\alpha_2 \neq \beta_2$ and γ_2 are complex numbers whose real and imaginary parts lie in a subfield $F \subset \mathbb{R}$. Prove that the circle C with center γ_2 and radius $|\alpha_2 - \beta_2|$ has an equation of the form (10.3) with $a_2, b_2, c_2 \in F$.

(c) In the proof of Theorem 10.1.6, we considered the equations (10.2) and (10.3) when $a_1 \neq 0$. Explain what happens when $a_1 = 0$ in (10.2).

Exercise 5. In this exercise you will give two proofs that $\zeta_3 = e^{2\pi i/3}$ is constructible.

(a) Give a direct geometric construction of ζ_3 with each step justified by citing C1, C2, P1, P2, or P3.

(b) Use Theorem 10.1.6 to show that ζ_3 is constructible.

Exercise 6. Show that it is impossible to trisect a $60°$ angle by straightedge and compass.

Exercise 7. Suppose we have extensions $\mathbb{Q} \subset F \subset \mathbb{C}$ where $[F : \mathbb{Q}]$ is finite. Prove that there is a field M such that $F \subset M \subset \mathbb{C}$ and M is a Galois closure of F over $\mathbb{Q}$.

Exercise 8. In the Mathematical Notes we defined the field $\mathscr{P} \subset \mathbb{R}$ and what it means for a subfield $F \subset \mathbb{R}$ to be Pythagorean.

(a) Let α be a real number. Prove that $\alpha \in \mathscr{P}$ if and only if there is a sequence of fields $\mathbb{Q} = F_0 \subset \cdots \subset F_n \subset \mathbb{R}$ such that $\alpha \in F_n$, and for $i = 1, \ldots, n$ there are $a_i, b_i \in F_{i-1}$ such that $F_i = F_{i-1}\left(\sqrt{a_i^2 + b_i^2}\right)$.

(b) Prove that $\mathscr{P}$ is the smallest Pythagorean subfield of $\mathbb{R}$.

Exercise 9. Show that the lune illustrated in the Historical Notes has the same area as the triangle AOB in the illustration.

Exercise 10. The quadratrix is the curve $y = x \cot(\pi x/2)$ for $0 < x \leq 1$. In this problem, you will use this curve to square the circle and trisect the angle.

(a) Show that $2/\pi = \lim_{x \to 0^+} x \cot(\pi x/2)$, that is, the quadratrix meets the y axis at $y = 2/\pi$. We will follow Hippias and include this point in the curve.

(b) Show that we can square the circle starting from 0 and 1 and constructing new points using C1, C2, P1, P2, P3, and the intersections of constructible lines with the quadratrix.

(c) A point (a, b) on the quadratrix determines an angle θ as pictured below. Prove that $\theta = \pi a/2$.

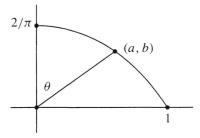

(d) Suppose that we are given an angle $0 < \theta < \pi/2$. Prove that we can trisect θ starting from 0, 1, and θ and constructing new points using C1, C2, P1, P2, P3, and the intersections of constructible lines with the quadratrix.

(e) Explain how the method of part (d) can be adapted to trisect arbitrary angles.

(f) Using the quadratrix, what else can you do to angles besides trisecting them?

Exercise 11. Explain how the points of intersection of the parabolas $y = x^2$ and $x = 2y^2$ enable one to duplicate the cube. Your explanation should include a picture.

Exercise 12. The spiral of Archimedes is the curve whose polar equation is $r = \theta$:

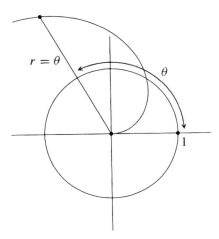

(a) Explain how the spiral and $\theta = \pi/2$ enable one to square the circle.

(b) Given an angle θ_0, explain how the spiral enables one to trisect θ_0.

10.2 REGULAR POLYGONS AND ROOTS OF UNITY

Our next task is to apply the theory developed in Section 10.1 to the question of which regular polygons can be constructed by straightedge and compass. Our main tool will be the cyclotomic extension $\mathbb{Q} \subset \mathbb{Q}(\zeta_n)$ studied in Chapter 9.

Before stating our main result, we need some terminology: an odd prime p is a *Fermat prime* if it can be written in the form

$$p = 2^{2^m} + 1$$

for some integer $m \geq 0$. Following Gauss, we can now characterize constructible regular polygons as follows.

Theorem 10.2.1. *Let $n > 2$ be an integer. Then a regular n-gon can be constructed by straightedge and compass if and only if*

$$n = 2^s p_1 \cdots p_r,$$

where $s \geq 0$ is an integer and $p_1, \ldots, p_r$ are $r \geq 0$ distinct Fermat primes.

Proof. In Example 10.1.3 we saw that a regular n-gon is constructible by straightedge and compass if and only if ζ_n is constructible. Using results proved earlier, we can determine when ζ_n is constructible as follows:

- By (8.6), $\mathbb{Q} \subset \mathbb{Q}(\zeta_n)$ is a Galois extension.
- By Theorem 10.1.12, it follows that ζ_n is constructible if and only if $[\mathbb{Q}(\zeta_n):\mathbb{Q}]$ is a power of 2.
- By Corollary 9.1.10, $[\mathbb{Q}(\zeta_n):\mathbb{Q}] = \phi(n)$, where $\phi(n)$ is the Euler ϕ-function defined in Section 9.1.

We conclude that ζ_n is constructible if and only if $\phi(n)$ is a power of 2.

First suppose that $n = 2^s p_1 \cdots p_r$, where $p_1, \ldots, p_r$ are distinct Fermat primes. Then part (b) of Lemma 9.1.1 gives the formula

$$\phi(n) = n \prod_{p\mid n} \left(1 - \frac{1}{p}\right) = \begin{cases} 2^{s-1}(p_1 - 1) \cdots (p_r - 1), & s > 0, \\ (p_1 - 1) \cdots (p_r - 1), & s = 0. \end{cases}$$

It follows that $\phi(n)$ is a power of 2, since each p_i is a Fermat prime.

Conversely, suppose that $\phi(n)$ is a power of 2, and let the factorization of n be $n = q_1^{a_1} \cdots q_s^{a_s}$, where $q_1, \ldots, q_s$ are distinct primes and the exponents $a_1, \ldots, a_s$ are all ≥ 1. Then part (b) of Lemma 9.1.1 gives the formula

$$\phi(n) = n \prod_{i=1}^{s} \left(1 - \frac{1}{q_i}\right) = q_1^{a_1-1}(q_1 - 1) \cdots q_s^{a_s-1}(q_s - 1).$$

If q_i is odd, then we must have $a_i = 1$, since $\phi(n)$ is a power of 2, and we also conclude that $q_i - 1$ must be a power of 2. However, in Exercise 1 you will show that if an odd prime p is of the form $2^k + 1$, then k must be a power of 2, that is, p is a Fermat prime. It follows that the odd primes dividing n have exponent 1 and are Fermat primes. This completes the proof of the theorem. $\square$

Notice that the power of 2 in Theorem 10.2.1 makes sense, for if we can construct a regular n-gon, then we get a regular $2n$-gon by bisecting the sides of the n-gon.

The proof of Theorem 10.2.1 is short and elegant because of the work we did in Chapter 9 to prove that the cyclotomic polynomial $\Phi_n(x)$ is irreducible over $\mathbb{Q}$. This is not the only way to prove the theorem. In Exercise 2 you will use the Schönemann–Eisenstein criterion to give a direct proof of the irreducibility of $\Phi_{p^2}(x)$, p prime. Exercises 3–6 will use this to give an alternate proof of Theorem 10.2.1.

The mth *Fermat number* is $F_m = 2^{2^m} + 1$, and a Fermat prime is a Fermat number that is prime. The five known Fermat primes are

$$F_0 = 3,$$
$$F_1 = 5,$$
$$F_2 = 17,$$
$$F_3 = 257,$$
$$F_4 = 65537$$

(see Exercise 7). It is also known that $F_5, F_6, \ldots, F_{31}, F_{32}$ are composite, but the status of F_{33} is still uncertain as of 2003. This is a very large number—in Exercise 8 you will estimate the number of digits in the decimal expansion of F_{33}. In addition, F_m is known to be composite for other scattered values of m. For example, in 2003 it was shown that $F_{2478782}$ is divisible by $3 \cdot 2^{2478785} + 1$. These are extremely large numbers.

Many people suspect that F_0, F_1, F_2, F_3, F_4 are the only Fermat primes, though this has not been proved. If true, then Theorem 10.2.1 would imply that a regular n-gon is constructible by straightedge and compass if and only if

$$n = 2^s \cdot 3^a \cdot 5^b \cdot 17^c \cdot 257^d \cdot 65537^e,$$

where $s \geq 0$ and a, b, c, d, e are 0 or 1.

Historical Notes

Beginning around 1640, numbers of the form $2^{2^m} + 1$ first appeared in Fermat's correspondence. He knew that they were prime for $0 \leq m \leq 4$ and conjectured that this was also true for $m \geq 5$, though he never found a rigorous proof.

In 1729, Goldbach's first letter to the young Euler mentions Fermat's conjecture about $2^{2^m} + 1$. Euler was sufficiently intrigued to start reading Fermat's letters. His first paper on number theory, published in 1732, shows that $F_5 = 2^{32} + 1$ is divisible by 641, thereby disproving Fermat's claim. Encouraged by this success, he went on to study other problems posed by Fermat over the course of the next fifty years. For example, he defined $\phi(n)$ in his attempt to understand Fermat's Little Theorem (Lemma 9.1.2). You proved Euler's wonderful generalization of this result in Exercise 9 of Section 9.1.

Because of Euler's negative result, there was little interest in Fermat primes until Gauss discovered their relation to the constructibility of regular polygons. The first entry in his famous mathematical diary, dated March 30, 1796, reads as follows:

> The principles upon which the division of the circle depend, and geometrical divisibility of the same into seventeen parts, etc.

(See [10].) Gauss wrote this one month before his 19th birthday.

The details of what Gauss proved about regular polygons appear in Section VII of *Disquisitiones Arithmeticae* [9]. As explained in Section 9.2, Gauss studied the equations satisfied by periods (special primitive elements of intermediate fields) of the extension $\mathbb{Q} \subset \mathbb{Q}(\zeta_p)$, where p is prime. Then, in Article 365 of [9], he applies his results to show that ζ_p is constructible when p is a Fermat prime. Though he asserts that the converse is true, the first published proof of this is due to Wantzel in 1837. In Article 366 Gauss describes which ζ_n are constructible when n is arbitrary (Theorem 10.2.1), though his proof is again incomplete.

Gauss knew that a straightedge-and-compass construction of a regular 17-gon was a big deal. As he says in Article 365 of [9]:

> It is certainly astonishing that although the geometric divisibility of the circle into three and five parts was already known in Euclid's time, nothing was added to this discovery for 2000 years.

But rather than give an explicit construction, Gauss shows that

$$\cos(2\pi/17) = -\tfrac{1}{16} + \tfrac{1}{16}\sqrt{17} + \tfrac{1}{16}\sqrt{34 - 2\sqrt{17}}$$
$$+ \tfrac{1}{8}\sqrt{17 + 3\sqrt{17} - \sqrt{34 - 2\sqrt{17}} - 2\sqrt{34 + 2\sqrt{17}}}.$$

Exercises 8 and 9 of Section 9.2 explain how this formula follows from Gauss's theory of periods. From here one can design a construction for the regular 17-gon, though it is not very efficient. A more elegant construction can be found in [Stewart, Ch. 17], along with a reference for Richelot's construction of the regular 257-gon in 1832. There is also the story of Professor Hermes of Lingen, who late in the nineteenth century worked 10 years on the construction of the regular 65537-gon.

We conclude with some remarks about arc length. This was an important topic in the seventeenth and eighteenth centuries. For example, by inscribing a regular n-gon in the unit circle, one easily sees that constructing the n-gon by straightedge and compass is equivalent to dividing a circle into n equal arcs by straightedge and compass. Another example involves the lemniscate, which is the curve in the plane defined by the polar equation $r^2 = \cos 2\theta$:

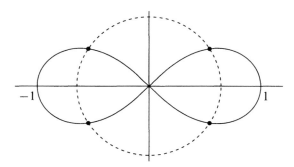

In 1716 Fagnano discovered a method for doubling and halving an arc of the lemniscate. In particular, he showed that the circle of radius $\sqrt{\sqrt{2} - 1}$ (drawn in dashes above) divides each quadrant of the lemniscate into arcs of equal length. Hence the lemniscate can be divided into eight equal arcs by straightedge and compass. In Chapter 15 we will explore a remarkable generalization of this discovered by Abel.

Exercises for Section 10.2

Exercise 1. Suppose that $2^k + 1$ is an odd prime. Prove that k is a power of 2.

Exercise 2. Let p be prime. In Example 9.1.6, we showed that

$$\Phi_{p^2}(x) = x^{p(p-1)} + x^{p(p-2)} + \cdots + x^{2p} + x^p + 1.$$

The goal of this exercise is prove that $\Phi_{p^2}(x)$ is irreducible over $\mathbb{Q}$ using only the Schönemann–Eisenstein criterion.

(a) Explain how the formulas of Example 9.1.6 imply that

$$(x + 1)^{p^2} - 1 = \left((x + 1)^p - 1\right)\Phi_{p^2}(x + 1).$$

(b) Let $\overline{\Phi}_{p^2}(x + 1)$ be the reduction of $\Phi_{p^2}(x + 1)$ modulo p. Show that

$$x^{p^2} = x^p \overline{\Phi}_{p^2}(x + 1).$$

(c) Show that $\Phi_{p^2}(x + 1)$ is irreducible over $\mathbb{Q}$ by the Schönemann–Eisenstein criterion. As in the proof of Proposition 4.2.5, this will imply that the same is true for $\Phi_{p^2}(x)$.

Exercise 3. Using only Proposition 4.2.5, Theorem 10.1.12, and Exercise 1, show that ζ_p is constructible if and only if p is a Fermat prime,

Exercise 4. Prove that

$$(\zeta_n)^{\frac{n}{m}} = \zeta_m$$

when $m \mid n$, $m > 0$, and use this to conclude that if ζ_n is constructible and $m \mid n$, $m > 0$, then ζ_m is constructible.

Exercise 5. Suppose that $n = 2^s p_1 \cdots p_r$, where $p_1, \ldots, p_r$ are distinct Fermat primes. Then ζ_{p_i} is constructible by Exercise 3.
(a) Show that ζ_{2^s} is constructible.
(b) Assume that ζ_a, ζ_b are constructible and $\gcd(a, b) = 1$. Prove that ζ_{ab} is constructible.
(c) Conclude that ζ_n is constructible, since $\zeta_{2^s}, \zeta_{p_1}, \ldots, \zeta_{p_r}$ are.

Exercise 6. Now suppose that ζ_n is constructible for some $n > 2$. The goal of this exercise is to prove that if p is an odd prime dividing n, then p is a Fermat prime and $p^2 \nmid n$. This and Exercise 5 will give a proof of Theorem 10.2.1 that doesn't require knowing that $\Phi_n(x)$ is irreducible for arbitrary n.
(a) Let p be an odd prime dividing n. Use Exercises 3 and 4 to show that p is a Fermat prime.
(b) Now assume that p is an odd prime and $p^2 \mid n$. Use Exercise 4 to show that ζ_{p^2} is constructible. Then use Theorem 10.1.12 and Exercise 2 to obtain a contradiction.
In Chapter 15 we will use a similar strategy to prove Abel's theorem about straightedge-and-compass constructions on the lemniscate.

Exercise 7. Prove that 3, 5, 17, 257, and 65537 are Fermat primes.

Exercise 8. Use $\log_{10}(F_{33}) \approx 2^{33} \log_{10}(2)$ to estimate the number of digits in the decimal expansion of F_{33}. Then do the same for $F_{2478782}$.

10.3   ORIGAMI (OPTIONAL)

In this optional section, we will use *origami*—the art of Japanese paper folding—to do some constructions not possible by straightedge and compass. We will also give a careful description of *origami numbers* and explain what they mean from the point of view of Galois theory.

A. Origami Constructions. We begin with a classic origami construction. Take an arbitrary angle θ between $\pi/4$ and $\pi/2$, and put it in the bottom left corner of a square sheet of paper. This gives the picture on the left:

(10.7)

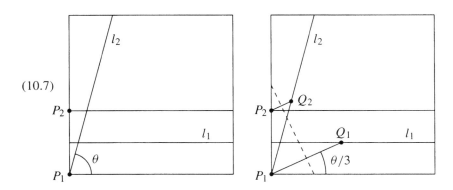

Thus θ is the angle between the line l_2 and the bottom of the sheet. Then, as indicated on the left, fold the sheet twice to obtain two lines parallel to the bottom such that the line l_1 is equidistant to the parallel lines through the points P_1 and P_2.

Now, turning to the picture on the right, do a classic origami move that folds the sheet so that P_1 moves to a point Q_1 on l_1 and P_2 moves to a point Q_2 on l_2. You should try this on a sheet of paper (a rectangular sheet will work fine). In Exercise 1 you will prove that the angle made by the bottom and the segment $\overline{P_1 Q_1}$ is $\theta/3$. Thus we have trisected an arbitrary angle $\pi/4 \le \theta \le \pi/2$ using origami!

Origami also makes it easy to double or halve angles. From this and the above construction, it follows that one can trisect *any* angle using origami (see Exercise 2).

We can also solve cubic equations using origami. But before explaining this, we need to think about the underlying geometry of the trisection given in (10.7). The surprise is that we are dealing with *simultaneous tangents to parabolas*. To see how this works, consider the geometric description of a parabola, which is defined as the locus of all points P equidistant from a fixed point P_1 (the *focus*) and a fixed line l_1 (the *directrix*). This gives the following picture:

(10.8)

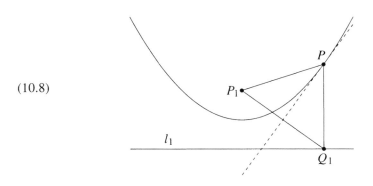

In this picture, the segments $\overline{P_1 P}$ and $\overline{P Q_1}$ have equal length, and $\overline{P Q_1}$ is perpendicular to the directrix l_1. The key point, which you will prove in Exercise 3, is that Q_1 is the reflection of P_1 about the tangent line at P (the dashed line in the picture). You will also prove the converse. Thus we have the following result.

Lemma 10.3.1. *In the plane, let P_1 be a point not on a line l_1. Then, given another line ℓ, the reflection of P_1 about ℓ lies on l_1 if and only if ℓ is tangent to the parabola with focus P_1 and directrix l_1.* ☐

To see how this relates to origami, look back at (10.7). The origami move we used took P_1 to $Q_1 \in l_1$ and P_2 to $Q_2 \in l_2$ and was done by folding along the dashed line. This means that the reflection of P_1 about the dashed line lies on l_1, so that the dashed line is tangent to the parabola with focus P_1 and directrix l_1 by Lemma 10.3.1. The same argument shows that the dashed line is also tangent to the parabola with focus P_2 and directrix l_2. We conclude that *using origami, one can find the simultaneous tangents to two given parabolas.*

Here is an example of how to use this.

Example 10.3.2. Let's find the real roots of the cubic equation $x^3 + ax + b = 0$, where $a, b \in \mathbb{R}$ and $b \neq 0$. Following the paper [1], we consider the parabolas

(10.9)
$$(y - \tfrac{1}{2}a)^2 = 2bx \quad \text{and} \quad y = \tfrac{1}{2}x^2.$$

Let ℓ be a line with slope m that is simultaneously tangent to these parabolas, say at points (x_1, y_1) on the first and (x_2, y_2) on the second. In Exercise 4 you will use calculus to show that the slope of the tangent line to the first parabola at (x_1, y_1) is

$$m = \frac{b}{y_1 - \tfrac{1}{2}a}.$$

This implies that $m \neq 0$ and $y_1 - \tfrac{1}{2}a = \frac{b}{m}$, from which we easily conclude that

(10.10)
$$x_1 = \frac{(y - \tfrac{1}{2}a)^2}{2b} = \frac{(\tfrac{b}{m})^2}{2b} = \frac{b}{2m^2},$$
$$y_1 = \frac{b}{m} + \frac{a}{2}.$$

Computing the slope of the tangent line to the second parabola at (x_2, y_2) gives

(10.11)
$$x_2 = m,$$

which easily implies that

$$y_2 = \frac{m^2}{2}.$$

If we substitute these values into $m = (y_2 - y_1)/(x_2 - x_1)$, then we obtain

$$m = \frac{y_2 - y_1}{x_2 - x_1} = \frac{\frac{m^2}{2} - (\frac{b}{m} + \frac{a}{2})}{m - \frac{b}{2m^2}} = \frac{m^4 - 2bm - am^2}{2m^3 - b}.$$

Since $m \neq 0$, it follows without difficulty that m satisfies the equation

(10.12) $$m^3 + am + b = 0.$$

Hence the slopes of the simultaneous tangents to the parabolas (10.9) are roots of the cubic $m^3 + am + b$. In Exercise 5 you will do this using origami. ◁▷

B. Origami Numbers. Our next task is to give a careful description of the numbers we get when we add the origami move used in (10.7) to the constructions C1 and C2 defined in Section 10.1. More precisely, consider the following origami construction:

C3. From $\alpha_1 \neq \alpha_2$ not lying on lines $\ell_1 \neq \ell_2$, we can draw a line ℓ that reflects α_1 to a point on ℓ_1 and α_2 to a point on ℓ_2.

The dashed line in (10.7) is an example of C3. There are situations where no line ℓ satisfies the conditions of C3 (see Exercise 6). Hence what C3 really says is that we are allowed to use such a line ℓ whenever it exists.

By Lemma 10.3.1, C3 enables us to draw a simultaneous tangent to two given parabolas (assuming there is such a tangent). Notice that C3 constructs only the line ℓ. This is because in origami, a line is a fold and a point is an intersection of folds. Of course, once we have ℓ, we can construct the reflections of α_1 and α_2 about ℓ by further straightedge and compass constructions.

The constructions C1, C2, C3 create circles and lines, and intersecting them using P1, P2, P3 from Section 10.1 gives new points that can be used for further constructions. We define origami numbers as follows.

Definition 10.3.3. *A complex number α is an **origami number** if there is a finite sequence of constructions using* C1, C2, C3, P1, P2, P3 *that begins with 0 and 1 and ends with α.*

This definition appears to involve compass, straightedge, and origami. However, in Chapter 10 of Martin's book [12], it is shown that all straightedge-and-compass constructions can be done using origami (called "paperfolding" by Martin). In particular, one can replace C1, C2, C3, P1, P2, P3 with constructions that involve only origami and give the same set of origami numbers.

The set of all origami numbers has the following structure.

Theorem 10.3.4. *The set $\mathcal{O} = \{\alpha \in \mathbb{C} \mid \alpha \text{ is an origami number}\}$ is a subfield of $\mathbb{C}$. Furthermore:*

(a) *Let $\alpha = a + ib$, where $a, b \in \mathbb{R}$. Then $\alpha \in \mathcal{O}$ if and only if $a, b \in \mathcal{O}$.*
(b) *$\alpha \in \mathcal{O}$ implies that $\sqrt{\alpha}, \sqrt[3]{\alpha} \in \mathcal{O}$.*
(c) *A complex number α lies in $\mathcal{O}$ if and only if there are subfields*

$$\mathbb{Q} = F_0 \subset F_1 \subset \cdots \subset F_{n-1} \subset F_n \subset \mathbb{C}$$

such that $\alpha \in F_n$ and $[F_i : F_{i-1}] = 2$ or 3 for $1 \leq i \leq n$.

Proof. We refer to [12, Ch. 10] for the proof that $\mathcal{O}$ is a subfield of $\mathbb{C}$. The proof of part (a) is similar to what we did in Theorem 10.1.4 and is omitted.

To prove part (b), write α in polar form as $\alpha = re^{i\theta}$. We may assume $r > 0$. Using the compass, we can transfer r to the x-axis, and then the straightedge-and-compass construction given in (10.1) shows that $\sqrt{r} \in \mathcal{O}$. Since we can also bisect θ by straightedge and compass, it follows that

$$\sqrt{\alpha} = \pm\sqrt{r}e^{i\theta/2} \in \mathcal{O}.$$

For the cube root, we can trisect θ using (10.7) and Exercise 2. To construct $\sqrt[3]{r}$, consider the parabolas (10.9) with $a = 0$ and $b = -r$. By Exercise 7 the foci α_1, α_2 and directrices l_1, l_2 of these parabolas are defined over any subfield of $\mathbb{R}$ containing r and hence can be constructed from r by straightedge and compass. Applying C3 to α_1, α_2 and l_1, l_2, we can construct a simultaneous tangent ℓ to these parabolas. By (10.12), ℓ has slope $m = \sqrt[3]{r}$. This easily implies that $\sqrt[3]{r} \in \mathcal{O}$. Since $\omega = e^{2\pi i/3} \in \mathcal{O}$ (do you see why?), it follows that

$$\sqrt[3]{\alpha} = \omega^i \sqrt[3]{r}e^{i\theta/3} \in \mathcal{O}, \quad i = 0, 1, 2.$$

In the proof of part (c) we will say that fields $\mathbb{Q} = F_0 \subset \cdots \subset F_n \subset \mathbb{C}$ form a *2-3 tower* if $[F_i : F_{i-1}] = 2$ or 3 for $1 \le i \le n$ (this differs slightly from the terminology used by Videla in [16]). Now suppose that $\mathbb{Q} = F_0 \subset \cdots \subset F_n$ is a 2-3 tower. We will prove that $F_n \subset \mathcal{O}$ by induction on n. Since the case $n = 0$ is obvious, we may assume that $F_{n-1} \subset \mathcal{O}$. Given $\alpha \in F_n$, we know that α is a root of a polynomial $f \in \mathcal{O}[x]$ of degree at most 3, since $[F_n : F_{n-1}] = 2$ or 3. If f has degree 1, then $\alpha \in \mathcal{O}$ is immediate, and if f has degree 2 or 3, then, by the quadratic formula or Cardan's formula, α can be expressed in terms of square roots, cube roots, and elements of $\mathcal{O}$. By part (b), it follows that $\alpha \in \mathcal{O}$.

Going the other way, let α be an origami number. We will show that there is a 2-3 tower $\mathbb{Q} = F_0 \subset \cdots \subset F_n \subset \mathbb{C}$ such that F_n contains the real and imaginary parts of all numbers constructed in the course of constructing α. The theorem will follow, since $\alpha = a + ib$ will imply that $a, b \in F_n$, so that $\alpha \in F_n(i)$. (We used the same strategy in the proof of Theorem 10.1.6.)

We will prove this by induction on the number N of times we use P1, P2, P3 in the construction of α. First suppose that α is constructed in $N > 1$ steps and that the last step uses P1. Thus α is the intersection of distinct lines ℓ_1 and ℓ_2 created earlier in the construction. If both lines come from C1, then we are done, as in the proof of Theorem 10.1.6. However, if we used C3 to construct either of the lines, then more work is needed.

If ℓ_1 was created using C3, then ℓ_1 is simultaneously tangent to two parabolas whose foci and directrices were created earlier in the construction. We claim that ℓ_1 has an equation whose coefficients lie in a 2-3 tower. To prove this, first consider the special case when the parabolas are of the form (10.9) for some $a, b \in \mathbb{R}$. Here, our inductive assumption and Exercise 7 imply that a, b lie in a 2-3 tower. Then the slope m of ℓ_1 satisfies the cubic equation (10.12), so that we can extend the 2-3 tower to get one that contains a, b, m. By (10.10), the point $(x_1, y_1) \in \ell_1$ has coordinates in the 2-3 tower. It follows that ℓ_1 has an equation $Ax + By = C$ whose coefficients lie in the same 2-3 tower. In the general case when ℓ_1 was created using

C3 for two arbitrary parabolas, one can argue similarly that ℓ_1 has an equation whose coefficients lie in a 2-3 tower. We omit the details.

It follows that if ℓ_1 or ℓ_2 (or both) were created using C3, then there is a 2-3 tower containing the coefficients of their defining equations. As in the proof of Theorem 10.1.6, we conclude that the coordinates of the intersection of ℓ_1 and ℓ_2 lie in the same tower.

Next suppose that we use P2 to create α, so that α comes from the intersection of a circle and a line. By our inductive assumption and the above argument, we may assume that the circle and line are defined by equations whose coefficients lie in a 2-3 tower, and then we are done by the argument of Theorem 10.1.6.

Finally, when we use P3, the argument is identical to what we did in Theorem 10.1.6. This completes the proof of the theorem. $\qquad \square$

Here is a nice example of Theorem 10.3.4.

Example 10.3.5. The 2-3 tower

$$\mathbb{Q} \subset \mathbb{Q}(2\cos(2\pi/7)) \subset \mathbb{Q}(\zeta_7)$$

shows that ζ_7 is an origami number. It follows that a regular heptagon (7-gon) can be constructed by origami. $\qquad \triangleleft\triangleright$

We can also characterize origami numbers using Galois theory.

Theorem 10.3.6. *Let $\alpha \in \mathbb{C}$ be algebraic over $\mathbb{Q}$ and let $\mathbb{Q} \subset L$ be the splitting field of the minimal polynomial of α over $\mathbb{Q}$. Then α is an origami number if and only if $[L : \mathbb{Q}] = 2^a 3^b$ for some integers $a, b \geq 0$.*

Proof. The argument is similar to the proof of Theorem 10.1.12. If $\alpha \in \mathcal{O}$, one first proves that $\mathbb{Q} \subset \mathcal{O}$ is a normal extension, so that $L \subset \mathcal{O}$. Then the formula for $[L : \mathbb{Q}]$ follows by applying Theorem 10.3.4 to a primitive element of $\mathbb{Q} \subset L$. For the converse one uses Burnside's $p^n q^m$ Theorem (Theorem 8.1.8) to show that $\mathrm{Gal}(L/\mathbb{Q})$ is solvable because $|\mathrm{Gal}(L/\mathbb{Q})| = [L : \mathbb{Q}] = 2^a 3^b$. The desired 2-3 tower is then easily constructed using the Galois correspondence and the definition of solvable group. We leave the details as Exercise 8. $\qquad \square$

Here are two examples of Theorem 10.3.6.

Example 10.3.7. The results of Section 9.1 imply that $\mathbb{Q} \subset \mathbb{Q}(\zeta_{11})$ is a Galois extension of degree 10. This is not of the form $2^a 3^b$, so that ζ_{11} cannot be constructed by origami. $\qquad \triangleleft\triangleright$

Example 10.3.8. Let $\alpha \in \mathbb{C}$ be a root of $f = x^6 + x + 1$. Using *Maple* or *Mathematica*, one easily checks that f is irreducible over $\mathbb{Q}$, so that $\mathbb{Q} \subset \mathbb{Q}(\alpha)$ is an extension of degree 6. However, even though $6 = 2 \cdot 3$, α is not an origami number. This follows from the `galois` command in *Maple*, which shows that the splitting field $\mathbb{Q} \subset L$ of f has Galois group $\mathrm{Gal}(L/\mathbb{Q}) \simeq S_6$. Then Theorem 10.3.6 implies that $\alpha \notin \mathcal{O}$, since $[L : \mathbb{Q}] = |\mathrm{Gal}(L/\mathbb{Q})| = 6! = 2^4 \cdot 3^2 \cdot 5$. $\qquad \triangleleft\triangleright$

You will prove the following corollary of Theorem 10.3.6 in Exercise 9.

Corollary 10.3.9. *Let $f(x) \in \mathbb{Q}[x]$ be a polynomial of degree ≤ 4. Then the roots of $f(x)$ are origami numbers, that is, we can solve $f(x) = 0$ by origami.*   □

See [7] for a description of how to solve quartics by origami.

C. Marked Rulers and Intersections of Conics. Origami numbers can be also constructed with a *marked ruler* or *intersections of conics*. We will discuss these methods briefly (without proofs), beginning with the marked ruler.

A *marked ruler* is a straightedge with two marks on it one unit apart. This is sometimes called a *twice-notched straightedge*. A marked ruler can construct a line in two ways: first, by connecting two known points, and second, by *verging*, which given a known point P and known lines l_1 and l_2 draws a line through P that meets l_1 at Q_1 and l_2 at Q_2 such that the segment $\overline{Q_1 Q_2}$ has length 1:

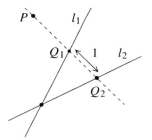

A marked-ruler construction begins with the points 0, 1, and i. At each step, one constructs a new line by applying either of the two operations just described to the already constructed lines and points and then intersecting the new line with other already constructed lines to get new points in $\mathbb{C}$.

Here is a quartic equation that comes from verging with a marked ruler.

Example 10.3.10. Let l_1 be the line $y = x$ and l_2 be the line $y = -\frac{1}{2}x$, and let $P = (\frac{1}{2}, 0)$. Then let ℓ be a line with slope m through P. In Exercise 10 you will show that ℓ meets l_1 at the point

$$(10.13) \qquad Q_1 = (x_1, x_1) \in l_1, \quad \text{where } x_1 = \frac{m}{2m - 2},$$

and meets l_2 at the point

$$(10.14) \qquad Q_2 = (x_2, -\tfrac{1}{2}x_2) \in l_2, \quad \text{where } x_2 = \frac{m}{2m + 1}.$$

If we think of ℓ as the marked ruler, then verging from P with l_1 and l_2 means that the distance from Q_1 to Q_2 is 1. This gives the equation

$$\left(\frac{m}{2m + 1} - \frac{m}{2m - 2}\right)^2 + \left(\frac{-m}{2(2m + 1)} - \frac{m}{2m - 2}\right)^2 = 1,$$

which simplifies to the quartic equation

$$(10.15) \qquad 7m^4 - 16m^3 - 21m^2 + 8m + 4 = 0.$$

The roots of this equation are all real and represent the slopes of the four lines through P that are constructed with a marked ruler by verging with the lines l_1 and l_2.

In Exercise 9 of Section 13.1, we will see that the Galois group of (10.15) is S_4 (this can also be done using the `galois` command in *Maple*). Hence the splitting field is an extension of $\mathbb{Q}$ of degree 24. This is not a power of 2, so that these lines are not constructible with straightedge and compass. ◁▷

See Exercise 11 for an example where verging leads to a cubic equation. Given that origami also solves cubic and quartic equations, the following result proved in [12, Ch. 10] is not surprising.

Theorem 10.3.11. *Let $\alpha \in \mathbb{C}$. Then α can be constructed using a marked ruler if and only if α is an origami number, that is, $\alpha \in \mathcal{O}$.* □

We next consider conics. These can be defined geometrically in terms of foci, directrices, and eccentricities, or one can work algebraically, giving separate treatments for ellipses, hyperbolas, and parabolas. We will use a third approach, which defines a conic to be a curve in the plane defined by an equation of the form

$$(10.16) \qquad F(x, y) = ax^2 + bxy + cy^2 + dx + ey + f = 0$$

where $a, b, c, d, e, f \in \mathbb{R}$ and $(a, b, c) \neq (0, 0, 0)$. We also assume that (10.16) has at least one solution with x, y real. This excludes equations like $x^2 + y^2 + 1 = 0$.

We write the equation (10.16) in matrix form as follows. Let

$$A = \begin{pmatrix} a & \frac{1}{2}b & \frac{1}{2}d \\ \frac{1}{2}b & c & \frac{1}{2}e \\ \frac{1}{2}d & \frac{1}{2}e & f \end{pmatrix},$$

and let

$$x = \begin{pmatrix} x \\ y \\ 1 \end{pmatrix}.$$

Then one easily checks that

$$(10.17) \qquad F(x, y) = x^t A x,$$

where x^t is the transpose of x. Then the conic C defined by (10.16) is *nondegenerate* if $\det(A) \neq 0$. As shown in [2], if C is nondegenerate, then

$$b^2 - 4ac < 0 \quad \Longleftrightarrow \quad C \text{ is an ellipse,}$$
$$b^2 - 4ac = 0 \quad \Longleftrightarrow \quad C \text{ is a parabola,}$$
$$b^2 - 4ac > 0 \quad \Longleftrightarrow \quad C \text{ is a hyperbola.}$$

To do constructions by intersecting conics, start with 0 and 1, and construct either a line connecting two already constructed points or a conic whose coefficients are previously constructed real numbers. Then we get new points by intersecting these lines and conics. This gives the following set of complex numbers.

Theorem 10.3.12. *Let* $\alpha \in \mathbb{C}$. *Then* α *can be constructible by intersecting conics if and only if* α *is an origami number, that is,* $\alpha \in \mathcal{O}$. ☐

Proof. This is proved by Alperin in [1]. Alternatively, Videla shows in [16] that α is constructible by conics if and only if α lies in a 2-3 tower. When we combine this with part (c) of Theorem 10.3.4, the theorem follows immediately. ☐

Putting together the results from this section, we have the following equivalences for a complex number α:

α is an origami number $\iff$ α is constructible by marked ruler

$\iff$ α is constructible by intersecting conics

$\iff$ α lies in a 2-3 tower $\mathbb{Q} = F_0 \subset \cdots \subset F_n$

$\iff$ α is algebraic over $\mathbb{Q}$, and the Galois group of its minimal polynomial has order $2^a 3^b$.

Mathematical Notes

Here are two topics for further discussion.

▪ **Marked Ruler and Compass.** By using a marked ruler *and* a compass, one can do constructions beyond what is possible by marked ruler alone. A marked ruler allows us to verge using a point and two lines, but with a compass to draw circles, we can also verge using a point and two circles or a point, a circle, and a line. An example of the latter is Archimedes' angle trisection:

(10.18)

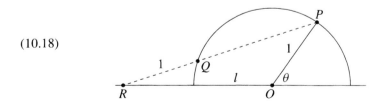

Here, we have a unit circle centered at O and a point P on the circle that makes the indicated angle θ with the line l. Then verging from P with l and the circle gives the dashed line containing points Q on the circle and R on l that are one unit apart. In Exercise 12 you will prove that $\angle PRO = \theta/3$. (Note that we can trisect angles using only the marked ruler. Exercise 13 gives such a trisection due to Pappus.)

Using a marked ruler and compass also enables us to construct points not possible by marked ruler alone. An example in Baragar's paper [4] shows that the real roots of the polynomial

$$x^5 - 4x^4 + 2x^3 + 4x^2 + 2x - 6$$

can be constructed using a marked ruler and compass (what Baragar calls a "compass and twice-notched ruler construction"). This polynomial is irreducible over $\mathbb{Q}$, and the methods used to analyze $x^5 - 6x + 3$ in Section 6.4 imply that its splitting field $\mathbb{Q} \subset L$ has Galois group $\mathrm{Gal}(L/\mathbb{Q}) \simeq S_5$. It follows that the roots of this polynomial are not expressible in terms of radicals and are not origami numbers, but can be constructed using marked ruler and compass. However, it is not known exactly which numbers can be constructed in this way.

In the exercises you will show that marked-ruler-and-compass constructions can be interpreted in terms of intersecting conchoids and limaçons with lines and circles. Further details may be found in [4].

■ **Origami and Dual Conics.** Origami and intersections of conics lead to the same set of complex numbers. Since origami involves simultaneous tangents to parabolas, it is reasonable to ask if origami has an intrinsic connection to intersections of conics. The answer involves the *dual conic* of a parabola, whose points correspond to tangent lines of the parabola. Then simultaneous tangents to two parabolas correspond to intersections of their dual conics. To make these ideas precise, one needs to work in the *projective plane*, which is beyond the scope of this book. See [1] for a discussion of the ideas involved.

Historical Notes

What we call "conics" are more properly called "conic sections," for they were defined by the Greeks as the intersections of a cone with a plane. One of the first Greek geometers to consider conic sections was Menaechmus (ca. 350 B.C.). He was a student of Plato and Eudoxus. He showed how to duplicate the cube by intersecting two parabolas (Exercise 11 of Section 10.1). Thus the idea of solving cubic equations by intersecting conics goes back to the very beginning of the study of conic sections.

In his book *On the Heptagon in the Circle*, Archimedes (287–212 B.C.) may have constructed a regular heptagon using the intersections of conics. Although this book no longer exists, works by Islamic geometers such as Thābit ibn Qurra (826–901) on the same problem mention Archimedes' book and use these methods to construct the regular heptagon.

One of the major works of Greek geometry is the *Conic Sections* by Apollonius (ca. 262–190 B.C.). This treatise introduced the terms *ellipse*, *parabola*, and *hyperbola*. A description of the *Conic Sections* can be found in [8, Ch. 6].

A later writer was Pappus (ca. 300), who wrote extensive commentaries on various aspects of Greek geometry. His work contains the first known description of a conic section in terms of focus, directrix, and eccentricity, though his description probably appeared in earlier but now lost works. Pappus gave a nice angle trisection using intersections of conics (see Exercise 14).

There is also a large Islamic literature on constructions using conic sections. As noted by Martin [12, p. 135], over a dozen conic constructions of the regular heptagon were found by Islamic geometers during the Middle Ages. Besides Thābit ibn Qurra mentioned earlier, another prominent geometer is Abu Ali Hasan ibn al-Haytham (ca. 965–1039), known in the West as Alhazen. He is best known for the problem of describing reflections in a circular mirror, which he solved by intersecting a hyperbola and a circle.

The emergence of equations for conics, such as (10.16), took a while. The standard equations for the ellipse, hyperbola, and parabola are implicit in many of the results proved the Greeks, but it wasn't until the work of L'Hôspital in 1707 that they were written down in their modern form. Much more on the history of the conic sections can be found in [5] and [6].

The marked ruler was first used by Nicomedes (ca. 240 B.C.) to construct cube roots using a marked ruler (see Exercise 15). Pappus, in one of his commentaries on Greek geometry, described verging as moving a ruler "about a fixed point until by trial the intercept [the portion of the ruler lying between the given lines] was found to be equal to the given length." We've already seen that Archimedes and Pappus used a marked ruler to trisect angles. Nicomedes also introduced the *conchoid*, which is a curve created by verging with a line (see Exercise 16). In 1593, Viète proposed that verging with a marked ruler be allowed for geometric constructions. See [4] and [12] for more details on the history of the marked ruler.

The connection with paperfolding or origami seems to be more recent. One of the earliest references is *Geometric Exercises in Paper Folding* by T. Sundara Row [15], published in Madras in 1893. The origami trisection given at the beginning of the section was discovered in the 1970s by Hisashi Abe and is taken from Hull's paper [11]. More references on origami can be found in [1], [3], and [11].

Exercises for Section 10.3

Exercise 1. This exercise will use the diagram

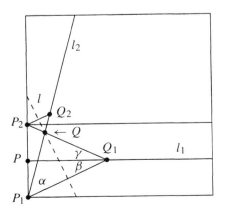

to prove that the origami construction described at the beginning of the section trisects the angle θ formed by the line l_2 and the bottom of the square.

(a) Let Q be the intersection of the line segments $\overline{P_1 Q_2}$ and $\overline{P_2 Q_1}$. Prove that Q lies on the dashed line l.

(b) Prove that θ is congruent to $\alpha + \beta$.

(c) Use triangles $\triangle P_1 P Q_1$ and $\triangle P_2 P Q_1$ to prove that β and γ are congruent.

(d) Use triangle $\triangle P_1 Q Q_1$ to prove that α is congruent to $\beta + \gamma$.

(e) Conclude that α is congruent to $2\theta/3$ and that the angle formed by $\overline{P_1 Q_1}$ and the bottom of the square is $\theta/3$.

Exercise 2. In the text we showed how to trisect an angle between $\pi/4$ and $\pi/2$ by origami.

(a) Explain how to bisect and double angles by origami.

(b) Explain how to trisect an arbitrary angle by origami.

Exercise 3. Let P_1 be a point not lying on a line l_1 in the plane. Drop a perpendicular from P_1 to l_1 that meets l_1 at a point S. Then choose rectangular coordinates such that P_1 lies on the positive y-axis and the x-axis is the perpendicular bisector of the segment $\overline{P_1 S}$. In this coordinate system, $P_1 = (0, a)$ and l_1 is defined by $y = -a$, where $a > 0$.

(a) The parabola with focus P_1 and directrix l_1 is defined to be the set of all points Q that are equidistant from P_1 and l_1. Prove that it is defined by the equation $4ay = x^2$.

(b) Let $Q = (x_0, y_0)$ be a point on the parabola. Prove that the y-intercept of its tangent line is $-y_0$.

(c) Let $Q = (x_0, y_0)$ be a point on the parabola, and let $Q_1 \in l_1$ be obtained by dropping a perpendicular from Q. Prove that Q_1 is the reflection of P_1 about the tangent line to the parabola at Q.

(d) Part (c) proves one direction of Lemma 10.3.1. Prove the other direction to complete the proof of the lemma.

Exercise 4. Show that the tangent line at a point (x_1, y_1) on the first parabola in (10.9) has slope given by

$$m = \frac{b}{y_1 - \frac{1}{2}a}.$$

Exercise 5. In the text we showed that the slopes of the simultaneous tangents to the parabolas in (10.9) are roots of (10.12). In this exercise, you will give an origami version of this in the special case when $a = 2$ and $b = 1$. Begin with a square sheet of paper folded so that the bottom edge touches the top. This fold will be the positive x-axis, and the left edge of the sheet will be the directrix for the first parabola in (10.9).

(a) Describe the origami moves one would use to construct the foci and directrices of the parabolas in (10.9) when $a = 2$ and $b = 1$. Also construct the y-axis. Exercise 7 will be helpful.

(b) Now perform an origami move that takes the focus of each parabola to a point on the corresponding directrix. Explain why there is only one way to do this.

(c) Part (b) gives a line whose slope m is the real root of $x^3 + 2x + 1$. Explain what origami moves you would use to find the point on the x-axis whose coordinates are $(m, 0)$.

Exercise 6. Suppose that in the situation of C3, we have points $\alpha_1 \neq \alpha_2$ not lying on lines $\ell_1 \neq \ell_2$. Also assume that ℓ_1 and ℓ_2 are parallel and that there is a line ℓ satisfying C3 (i.e., ℓ reflects α_i to a point on ℓ_i for $i = 1, 2$). Prove that the distance between ℓ_1 and ℓ_2 is at most the distance between α_1 and α_2. This makes it easy to find examples where the line described in C3 does not exist.

Exercise 7. Consider the parabolas $(y - \frac{1}{2}a)^2 = 2bx$ and $y = \frac{1}{2}x^2$ from (10.9).

(a) Show that the first parabola has focus $(\frac{1}{2}b, \frac{1}{2}a)$ and directrix $x = -\frac{1}{2}b$.

(b) Show that the second parabola has focus $(\frac{1}{2}, 0)$ and directrix $y = -\frac{1}{2}$.

Hence the focus and directrix of the first parabola are defined over any subfield of $\mathbb{R}$ containing a and b. For the second, this is true over any subfield of $\mathbb{R}$.

Exercise 8. Complete the proof of Theorem 10.3.6 sketched in the text.

Exercise 9. Prove Corollary 10.3.9.

Exercise 10. In Example 10.3.10, prove that ℓ meets l_1 and l_2 at the points Q_1 and Q_2 given in (10.13) and (10.14). Also draw the four lines whose slopes are the roots of (10.15).

Exercise 11. This exercise will give an example of a cubic equation that arises from verging. Consider the lines l_1 defined by $y = 0$ and l_2 defined by $y = x$ and verge from $P = (1, \frac{1}{2})$ using a marked ruler. Show that this gives the vertical line $x = 1$ together with three nonvertical lines whose slopes m satisfy the cubic equation

$$4m^3 + m^2 - 4m + 1 = 0.$$

Also show that the nonvertical lines cannot be constructed by straightedge and compass.

Exercise 12. Prove that $\angle PRO = \theta/3$ in the construction (10.18).

Exercise 13. According to [12], Pappus used a marked ruler to trisect angles as follows. Given an angle $0 < \theta < \pi/2$, write it as $\theta = \angle POA$, where:

- The distance between P and O is $1/2$.
- The line l_1 determined by P and A is perpendicular to the line determined by O and A.

Any angle $0 < \theta < \pi/2$ can be put in this form by a marked-ruler construction. Finally, let l_2 be the line through P that is perpendicular to l_1. Then verging with O and the lines l_1 and l_2 gives points $Q \in l_1$ and $R \in l_2$ such that Q and R are one unit apart:

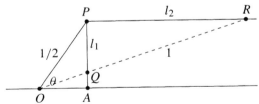

Prove that $\angle QOA = \theta/3$.

Exercise 14. As explained in [16], Pappus used intersections of conics to trisect angles as follows. Consider the unit circle centered at the origin, and let θ satisfy $0 < \theta < \pi/2$. Then $P = (\cos\theta, \sin\theta)$ is the corresponding point on the unit circle. We assume that P is known. Also let $O = (0, 0)$ be the origin, and set $A = (1, 0)$. Thus $\theta = \angle POA$.

(a) Consider the curve C consisting of all points $Q = (x, y)$ such that the distance from P to Q is twice the distance from Q to the x-axis. The curve C intersects the unit circle at a point R lying in the interior of $\angle POA$. Prove that $\angle ROA = \theta/3$.

(b) Show that the curve C is a hyperbola. It follows that we have trisected an angle using the intersection of a hyperbola and a circle, that is, an intersection of conics.

Exercise 15. In this exercise, we discuss a marked-ruler construction of cube roots due to Nicomedes and taken from [12]. Let k be a real number such that $0 < k < 8$, and consider an isosceles triangle $\triangle ABC$ such that AC and BC have length 1 and AB has length $k/4$. Then extend AC and AB as indicated in the picture below, and choose D on the extension of AC so that AD also has length 1. Finally, draw the line through D and B.

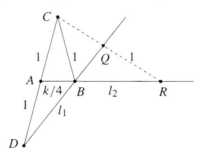

Verging from C with the lines l_1 and l_2 indicated above gives points $Q \in l_1$ and $R \in l_2$ that are one unit apart. Assume that $Q \neq A$.

(a) Explain why the restriction $0 < k < 8$ is necessary.

(b) Prove that the distance between B and R is $\sqrt[3]{k}$.

(c) Explain how to give a marked-ruler construction of $\sqrt[3]{k}$ for any $k > 0$.

Exercise 16. Let P be a point distance $b > 0$ from a line l. Put a marked ruler though P with one mark at $R \in l$. When R moves along l, the other mark Q_1 or Q_2 (depending on which side of l it is on) traces out the *conchoid of Nicomedes*. When $b < 1$ we get the picture

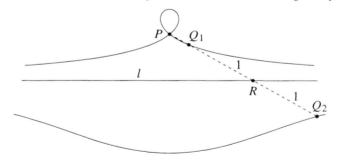

We can relate the conchoid to construction problems as follows.

(a) Suppose we are given a point P and lines l_1, l_2, and assume that $P \notin l_1$. Prove that a point Q is obtained by verging with P and l_1, l_2 if and only if Q is one of the points of intersection of l_2 with the conchoid determined by P and l_1.

(b) Prove that the angle trisection of (10.18) can be interpreted as the intersection of the unit circle with the conchoid determined by P and l.

(c) Suppose that $P = (0, 0)$ and l is the horizontal line $y = -b$. Prove that the polar equation of the conchoid is

$$r = b \csc \theta \pm 1,$$

where the minus sign gives the portion of the curve above l and the plus sign gives the portion below.

(d) Under the assumptions of part (c), show that the Cartesian equation of the conchoid is

$$(x^2 + y^2)(y - b)^2 = y^2.$$

By part (a), verging is the same as intersecting the conchoid with a line. Since the above equation has degree 4, this explains why verging leads to an equation of degree 4.

Exercise 17. Let P be a point on a circle, and consider a marked ruler that goes through P. If we place one mark on a point Q on the circle, then the other mark R_1 or R_2 (depending on whether it is inside or outside the circle) traces out a curve called the *limaçon of Pascal*:

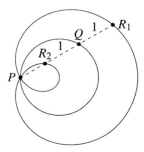

This curve was known to Jordanus Nemorarius (1225–1260) and Albrecht Dürer (1471–1528) and possibly the ancient Greeks. It was rediscovered by Étienne Pascal (father of Blaise Pascal) about a century after Dürer. In 1650 Roberval, unaware of the earlier work, named the curve in Pascal's honor.

(a) Show that the angle trisection (10.18) can be interpreted as the intersection of the line l with the limaçon determined by the circle and the point P.

(b) Let $P = (0, 0)$ and let C be the circle of radius a and center $(a, 0)$. Show that the corresponding limaçon has polar equation

$$r = 1 + 2a \cos \theta.$$

(c) In the situation of part (b), show that the Cartesian equation of the limaçon is

$$(x^2 + y^2 - 2ax)^2 = x^2 + y^2.$$

Exercise 18. A *Pierpont prime* is a prime $p > 3$ of the form $p = 2^k 3^l + 1$. Prove that a regular n-gon can be constructed by origami (or by marked ruler or by intersections of conics) if and only if $n = 2^a 3^b p_1 \cdots p_s$, where $a, b \geq 0$ and $p_1, \ldots, p_s$ are distinct Pierpont primes. This was first proved by Pierpont in [13].

REFERENCES

1. R. C. Alperin, *A mathematical theory of origami constructions and numbers*, New York J. Math. **6** (2000), 119–133.

2. J. W. Archbold, *Introduction to the Algebraic Geometry of the Plane*, Edward Arnold, London, 1948.

3. D. Auckly and J. Cleveland, *Totally real origami and impossible paper folding*, Amer. Math. Monthly **102** (1995), 215–226.

4. A. Baragar, *Constructions using a compass and twice-notched straightedge*, Amer. Math. Monthly **109** (2002), 151–164.

5. J. L. Coolidge, *A History of the Conic Sections and Quadric Surfaces*, Oxford U. P., 1945. Reprint by Dover, New York, 1968.

6. J. L. Coolidge, *A History of Geometrical Methods*, Oxford U. P., 1940. Reprint by Dover, New York, 1963.

7. B. Carter Edwards and J. Shurman, *Folding quartic roots*, Math. Mag. **74** (2001), 19–25.

8. H. Eves, *An Introduction to the History of Mathematics*, Sixth Edition, Brooks/Cole, Pacific Grove, CA, 1990.

9. C. F. Gauss, *Disquisitiones Arithmeticae*, Leipzig, 1801. Republished in 1863 as Volume I of [Gauss]. French translation, *Recherches Arithmétiques*, Paris, 1807. Reprint by Hermann, Paris, 1910. German translation, *Untersuchungen über Höhere Arithmetik*, Berlin, 1889. Reprint by Chelsea, New York, 1965. English translation, Yale U. P., New Haven, 1966. Reprint by Springer-Verlag, New York, Berlin, Heidelberg, 1986.

10. J. J. Gray, *A commentary on Gauss's mathematical diary, 1796–1814, with an English translation*, Expo. Math. **2** (1984), 97–130. (The Latin original of Gauss's diary is reprinted in [Gauss, Vol. X.1].)

11. T. Hull, *A note on "impossible" paper folding*, Amer. Math. Monthly **103** (1996), 240–241.

12. G. E. Martin, *Geometric Constructions*, Springer-Verlag, New York, Berlin, Heidelberg, 1998.

13. J. Pierpont, *On an undemonstrated theorem of the Disquisitiones Arithmeticæ*, Bull. Amer. Math. Soc. **2** (1895–1896), 77–83.

14. M. M. Postnikov, *The problem of squarable lunes*, Amer. Math. Monthly **107** (2000), 645–651.

15. T. S. Row, *Geometric Exercises in Paper Folding*, Addison & Co., Madras, 1893. Also published by Open Court, Chicago 1901. Reprint by Dover, New York, 1966.

16. C. R. Videla, *On points constructible from conics*, Math. Intelligencer **19** (1997), 53–57.

11

Finite Fields

The main topic of this chapter is the theory of finite fields. We will study their existence and uniqueness and compute their Galois groups. We will also consider irreducible polynomials over finite fields.

11.1 THE STRUCTURE OF FINITE FIELDS

In this section we discuss the basic properties of finite fields.

A. Existence and Uniqueness. The simplest examples of finite fields are $\mathbb{F}_p$, the integers modulo a prime p. These relate to arbitrary finite fields as follows.

Proposition 11.1.1. *Let F be a finite field. Then:*
(a) *There is a unique prime p such that F contains a subfield isomorphic to $\mathbb{F}_p$.*
(b) *F is a finite extension of $\mathbb{F}_p$, and*

$$|F| = p^n, \qquad \text{where } n = [F : \mathbb{F}_p].$$

Proof. Every field of characteristic 0 contains a subfield isomorphic to $\mathbb{Q}$ and hence is infinite. Thus F has characteristic p for some prime p. Furthermore, the discussion of characteristic in Section A.1 shows that $p\mathbb{Z} \subset \mathbb{Z}$ is the kernel of the ring homomorphism that sends $m \in \mathbb{Z}$ to $m \cdot 1 \in F$. By the Fundamental Theorem of Ring Homomorphisms, F contains a subfield isomorphic to $\mathbb{Z}/p\mathbb{Z} = \mathbb{F}_p$.

The map $\mathbb{F}_p \to F$ makes F an extension field of $\mathbb{F}_p$. Following our usual practice (see Definition 3.1.2), we identify $\mathbb{F}_p$ with its image and write $\mathbb{F}_p \subset F$. Now consider F as a vector space over $\mathbb{F}_p$. The elements of F give finitely many

289

vectors in F, whose span over $\mathbb{F}_p$ is obviously F. It follows that F is a finite-dimensional vector space over $\mathbb{F}_p$. As in Section 4.3, this means that F is a finite extension of $\mathbb{F}_p$. Furthermore, if $n = [F : \mathbb{F}_p]$, then we can find a basis $\alpha_1, \ldots, \alpha_n$ of F over $\mathbb{F}_p$. Hence every element of $\beta \in F$ can be written uniquely as

$$\beta = a_1\alpha_1 + \cdots + a_n\alpha_n, \quad a_i \in \mathbb{F}_p.$$

Since the a_i can be any of the p elements of $\mathbb{F}_p$, there are p^n possibilities for β. Thus $|F| = p^n$. This completes the proof. $\qquad\square$

For the rest of this chapter, we will assume as in the above proof that a finite field F contains $\mathbb{F}_p$ as a subfield. Our first major result is that F is the splitting field over $\mathbb{F}_p$ of a particularly simple polynomial.

Theorem 11.1.2. *Let F be a finite field with $q = p^n$ elements. Then:*
(a) $\alpha^q = \alpha$ *for all $\alpha \in F$.*
(b) $x^q - x = \prod_{\alpha \in F}(x - \alpha)$.
(c) F *is a splitting field over $\mathbb{F}_p$ of $x^q - x \in \mathbb{F}_p[x]$.*

Proof. Since F has q elements, its multiplicative group $F^* = F \setminus \{0\}$ is a group with $q - 1$ elements. It follows that $\alpha^{q-1} = 1$ for all $\alpha \in F^*$, so that $\alpha^q = \alpha$ for all $\alpha \in F$. This proves part (a) and shows that the q elements of F are roots of $x^q - x$. Then part (b) follows since $x^q - x$ is monic of degree q. Hence $x^q - x$ splits completely over F. Since every element of F is a root, $x^q - x$ can't split completely over any strictly smaller field. Hence F is a splitting field of $x^q - x \in \mathbb{F}_p[x]$. $\quad\square$

Using this theorem, we obtain the following uniqueness result for finite fields.

Corollary 11.1.3. *Two finite fields with the same number of elements are isomorphic.*

Proof. Corollary 5.1.7 implies that any two splitting fields of $x^q - x \in \mathbb{F}_p[x]$ are isomorphic. Then Corollary 11.1.3 follows immediately from Theorem 11.1.2. $\quad\square$

We next show that a finite field of order p^n exists for any p and n.

Theorem 11.1.4. *Given any prime p and any positive integer n, there is a finite field with p^n elements.*

Proof. Let $q = p^n$, and let L be an extension of $\mathbb{F}_p$ such that $x^q - x$ splits completely over L. Since we are in characteristic p, the derivative of $x^q - x$ is -1, so that $\gcd(x^q - x, (x^q - x)') = 1$. Thus $x^q - x$ is separable and hence has distinct roots in L. This means that $F = \{\alpha \in L \mid \alpha^q = \alpha\}$ is a subset of L consisting of q elements. In Exercise 1 you will show that F is a subfield of L. It follows that F is a finite field with $q = p^n$ elements. $\qquad\square$

Given any $q = p^n$ as in Theorem 11.1.4, the finite field of order q constructed in the theorem is unique up to isomorphism by Corollary 11.1.3. Hence we can speak of "the" finite field with q elements. We will denote this field as $\mathbb{F}_q$. Since these fields were first described by Galois (see the Historical Notes), $\mathbb{F}_q$ is sometimes denoted as GF(q), where "GF" stands for "Galois Field."

One can use Theorem 11.1.2 to count the number of roots of a polynomial in a finite field as follows.

Proposition 11.1.5. *If $f \in \mathbb{F}_p[x]$ is nonconstant and $n \geq 1$, then the number of roots of f in $\mathbb{F}_{p^n}$ is the degree of the polynomial* $\gcd(f, x^{p^n} - x)$.

Proof. Let $g = \gcd(f, x^{p^n} - x)$, where the gcd is computed in $\mathbb{F}_p[x]$. A useful observation is that if one replaces $\mathbb{F}_p$ with any larger field, then one gets the same polynomial g (you will prove this in Exercise 2). Thus we may compute the gcd in $\mathbb{F}_{p^n}[x]$. If we denote the elements of this field by α_i for $i = 1, \ldots, p^n$, then

$$x^{p^n} - x = (x - \alpha_1) \cdots (x - \alpha_{p^n}).$$

by part (b) of Theorem 11.1.2. This is the irreducible factorization of $x^{p^n} - x$ in $\mathbb{F}_{p^n}[x]$. Hence g is the product of those $x - \alpha_i$ that divide f. Since $x - \alpha_i$ divides f if and only if $f(\alpha_i) = 0$, we obtain the product formula

$$g = \prod_{f(\alpha_i)=0} (x - \alpha_i).$$

The proposition now follows immediately. $\qquad\qquad\qquad\qquad\qquad\square$

Here is an example to illustrate Proposition 11.1.5.

Example 11.1.6. Consider the polynomial

$$f = x^{11} + x^5 + 2x + 1 \in \mathbb{F}_7[x].$$

To compute the number of roots in $\mathbb{F}_{7^3}$, we need to compute $\gcd(f, x^{7^3} - x)$. In *Maple*, we do this using the command

$$\texttt{Gcd(f, x\^{}(7\^{}3)-x) mod 7;}$$

In *Mathematica*, we would type

$$\texttt{PolynomialGCD[f, x\^{}(7\^{}3)-x, Modulus -> 7]}$$

In both cases, the output is the polynomial

$$x^3 + 4x^2 + x + 6.$$

By Proposition 11.1.5, f has three roots in $\mathbb{F}_{7^3}$. Furthermore, replacing 7^3 with 7^4 in the above computation gives a gcd of 1, so that f has no roots in $\mathbb{F}_{7^4}$.

One drawback of this method is that the degree of $x^{7^n} - x$ increases rapidly. For example, if we replace 7^3 with 7^8 in the above computation, then *Maple* gives

$$x^8 + 3x^7 + x^6 + x^5 + 5x^4 + x^3 + 4x + 6$$

(thus f has eight roots in $\mathbb{F}_{7^8}$), whereas *Mathematica* gives an error message because the degree of $x^{7^8} - x = x^{5764801} - x$ is too large for $\texttt{PolynomialGCD}$. $\quad\triangleleft\triangleright$

B. Galois Groups. We next compute the Galois group of the extension $\mathbb{F}_p \subset \mathbb{F}_q$.

Theorem 11.1.7. *If $q = p^n$, then:*

(a) $\mathbb{F}_p \subset \mathbb{F}_q$ *is a Galois extension.*

(b) *The map* $\mathrm{Frob}_p : \mathbb{F}_q \to \mathbb{F}_q$ *defined by* $\mathrm{Frob}_p(\alpha) = \alpha^p$ *is an automorphism of* $\mathbb{F}_q$ *that is the identity on* $\mathbb{F}_p$.

(c) *There is a group isomorphism*

$$\mathrm{Gal}(\mathbb{F}_q / \mathbb{F}_p) \simeq \mathbb{Z}/n\mathbb{Z}$$

that sends $\mathrm{Frob}_p \in \mathrm{Gal}(\mathbb{F}_q / \mathbb{F}_p)$ *to* $[1] \in \mathbb{Z}/n\mathbb{Z}$.

Proof. In the proof of Theorem 11.1.4, we noted that $x^q - x$ is separable. Then Theorem 11.1.2 implies that $\mathbb{F}_q$ is the splitting field of a separable polynomial. Hence $\mathbb{F}_p \subset \mathbb{F}_q$ is Galois.

Turning to part (b), observe that by Lemma 5.3.10,

$$\mathrm{Frob}_p(\alpha + \beta) = (\alpha + \beta)^p = \alpha^p + \beta^p = \mathrm{Frob}_p(\alpha) + \mathrm{Frob}_p(\beta).$$

Since we also have $\mathrm{Frob}_p(1) = 1^p = 1$ and

$$\mathrm{Frob}_p(\alpha\beta) = (\alpha\beta)^p = \alpha^p \beta^p = \mathrm{Frob}_p(\alpha)\mathrm{Frob}_p(\beta),$$

it follows that Frob_p is a ring homomorphism. By Exercise 2 of Section 3.1, Frob_p is also one to one and hence onto, since it maps the finite set $\mathbb{F}_q$ to itself. Thus Frob_p is an automorphism of $\mathbb{F}_q$. Since it is the identity on $\mathbb{F}_p$ by Lemma 9.1.2, we conclude that $\mathrm{Frob}_p \in \mathrm{Gal}(\mathbb{F}_q / \mathbb{F}_p)$.

For part (c), we first note that since $\mathbb{F}_p \subset \mathbb{F}_q$ is Galois, we have

$$|\mathrm{Gal}(\mathbb{F}_q / \mathbb{F}_p)| = [\mathbb{F}_q : \mathbb{F}_p] = n,$$

where the second equality uses $q = p^n$ and Proposition 11.1.1. It follows that the order of Frob_p divides n. Suppose that $(\mathrm{Frob}_p)^r$ is the identity, where $0 < r < n$. Here, $(\mathrm{Frob}_p)^r$ denotes the r-fold composition of Frob_p with itself, so that

$$(\mathrm{Frob}_p)^r(\alpha) = \overbrace{\mathrm{Frob}_p(\cdots \mathrm{Frob}_p(\mathrm{Frob}_p(\alpha))\cdots)}^{r \text{ times}} = (\cdots (\alpha \overbrace{{}^p)^p \cdots)^p}^{r \text{ times}} = \alpha^{p^r}.$$

Thus, if $(\mathrm{Frob}_p)^r$ is the identity element of $\mathrm{Gal}(\mathbb{F}_q / \mathbb{F}_p)$, then

$$\alpha^{p^r} = \alpha$$

for all $\alpha \in \mathbb{F}_q$. Since $0 < r < n$, this implies that the polynomial $x^{p^r} - x$ of degree $p^r < p^n = q$ has q roots, which is clearly impossible. Hence Frob_p has order n, which easily gives the desired isomorphism $\mathrm{Gal}(\mathbb{F}_q / \mathbb{F}_p) \simeq \mathbb{Z}/n\mathbb{Z}$. $\qquad\square$

We call Frob_p the *Frobenius automorphism* of $\mathbb{F}_q$. We next use Theorem 11.1.7 to determine when one finite field is contained in another.

Corollary 11.1.8. *Let $\mathbb{F}_{p^m}$ and $\mathbb{F}_{p^n}$ be finite fields. Then $\mathbb{F}_{p^m}$ is isomorphic to a subfield of $\mathbb{F}_{p^n}$ if and only if $m \mid n$.*

Proof. First suppose that $\mathbb{F}_{p^m}$ is isomorphic to a subfield of $\mathbb{F}_{p^n}$. Writing this as an inclusion, we obtain

$$\mathbb{F}_p \subset \mathbb{F}_{p^m} \subset \mathbb{F}_{p^n}.$$

Proposition 11.1.1 and the Tower Theorem imply that

$$n = [\mathbb{F}_{p^n} : \mathbb{F}_p] = [\mathbb{F}_{p^n} : \mathbb{F}_{p^m}][\mathbb{F}_{p^m} : \mathbb{F}_p] = [\mathbb{F}_{p^n} : \mathbb{F}_{p^m}] \, m.$$

This shows that m divides n.

Conversely, suppose that $m \mid n$. Since $\mathrm{Gal}(\mathbb{F}_{p^n}/\mathbb{F}_p)$ is cyclic of order n by Theorem 11.1.7, we know that $\mathrm{Gal}(\mathbb{F}_{p^n}/\mathbb{F}_p)$ has a subgroup H of order $\frac{n}{m}$. By the Galois correspondence of Section 7.3, the fixed field F of H is an extension

$$\mathbb{F}_p \subset F \subset \mathbb{F}_{p^n}$$

satisfying

$$[F : \mathbb{F}_p] = [\mathrm{Gal}(\mathbb{F}_{p^n}/\mathbb{F}_p) : H] = \frac{n}{n/m} = m.$$

Using Proposition 11.1.1, we see that F has order p^m. By Corollary 11.1.3, it follows that F is a subfield of $\mathbb{F}_{p^n}$ isomorphic to $\mathbb{F}_{p^m}$. $\qquad\square$

In Exercise 3 you will prove Corollary 11.1.8 using neither Theorem 11.1.7 nor the Galois correspondence.

When $m \mid n$, Corollary 11.1.8 gives $\mathbb{F}_{p^m} \simeq F \subset \mathbb{F}_{p^n}$. As usual, we identify $\mathbb{F}_{p^m}$ with F, which gives the inclusion

$$\mathbb{F}_{p^m} \subset \mathbb{F}_{p^n}.$$

Then we can generalize Theorem 11.1.7 as follows.

Theorem 11.1.9. *Let $m \mid n$ and $\mathbb{F}_{p^m} \subset \mathbb{F}_{p^n}$. Then there is a group isomorphism*

$$\mathrm{Gal}(\mathbb{F}_{p^n}/\mathbb{F}_{p^m}) \simeq \mathbb{Z}/\tfrac{n}{m}\mathbb{Z}$$

that sends $(\mathrm{Frob}_p)^m \in \mathrm{Gal}(\mathbb{F}_{p^n}/\mathbb{F}_{p^m})$ *to* $[1] \in \mathbb{Z}/\tfrac{n}{m}\mathbb{Z}$.

Proof. You will prove this in Exercise 4. $\qquad\square$

This result makes it easy to work out the Galois correspondence of $\mathbb{F}_p \subset \mathbb{F}_{p^n}$. The key point is that subfields of $\mathbb{F}_{p^n}$ correspond to subgroups of $\mathbb{Z}/n\mathbb{Z}$, yet subgroups of $\mathbb{Z}/n\mathbb{Z}$ correspond to positive divisors of n. Here is an example inspired by the classic book [12] by Lidland Niederreiter.

Example 11.1.10. For $\mathbb{F}_{2^{30}}$, the above remarks show that subfields of $\mathbb{F}_{2^{30}}$ correspond to positive divisors of 30. This gives the following Galois correspondence:

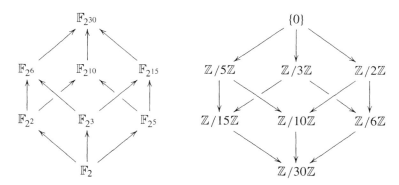

To understand the diagram, recall that an intermediate field $\mathbb{F}_2 \subset F \subset \mathbb{F}_{2^{30}}$ corresponds to the subgroup $\text{Gal}(\mathbb{F}_{2^{30}}/F) \subset \text{Gal}(\mathbb{F}_{2^{30}}/\mathbb{F}_2) \simeq \mathbb{Z}/30\mathbb{Z}$. Combining this with Theorem 11.1.9, we see that $F = \mathbb{F}_{2^m}$ corresponds to $\text{Gal}(\mathbb{F}_{2^{30}}/\mathbb{F}_{2^m}) \simeq \mathbb{Z}/\frac{30}{m}\mathbb{Z}$. When we do this for each divisor of 30, we obtain the above diagram.   ◁▷

Mathematical Notes

Finite fields are used in many different areas of mathematics. Here are three of particular importance.

▪ **Finite Groups.** If F is any field, then the set $\text{GL}(n, F)$ of invertible $n \times n$ matrices with entries in F is a group under matrix multiplication. In particular, if F is a finite field, then $\text{GL}(n, F)$ is a finite group. These groups play an important role in both Galois theory and the theory of finite groups. We will say more about this in Section 14.3.

▪ **Equations over Finite Fields.** Given a nonzero polynomial $f \in \mathbb{R}[x, y]$, the solutions of $f(x, y) = 0$ lying in $\mathbb{R}^2$ form a curve in the plane. For instance, the unit circle is defined by $x^2 + y^2 - 1 = 0$. Similarly, given a polynomial $f \in \mathbb{F}_p[x, y]$, we can consider solutions of $f(x, y) = 0$ lying in $\mathbb{F}_p^2$. Such equations were of interest to Gauss. For instance, the equation $x^2 + y^2 + x^2y^2 \equiv 1 \bmod p$ appears in the last entry in his mathematical diary [9].

Things get more interesting when one realizes that for $f \in \mathbb{F}_p[x, y]$, we can also consider solutions of $f(x, y) = 0$ lying in $\mathbb{F}_{p^n}^2$ for any $n \geq 1$. Since $\mathbb{F}_{p^n}$ is finite, there are only finitely many such solutions, though as n gets larger, the number of solutions increases. This is all controlled by the *zeta function* of the equation. See [8, Ch. 5], [10, §11.5], and [11, pp. 158–160] for more about zeta functions.

▪ **Coding Theory.** Information in a computer is often represented as a string of 0's and 1's. A good example is the ASCII code, which uses numbers between 0 and 127 to represent letters, digits, punctuation marks, and common special characters on computers. The ASCII value of "g" is 103 (decimal) = 1100111 (binary), while "G" is "071" (decimal) = 1000111 (binary). Thus the ASCII code uses $\mathbb{F}_2^7$, which is a vector space of dimension 7 over $\mathbb{F}_2$.

In algebraic coding theory, a code consists of a subset of $\mathbb{F}_q^m$ (the set of *code words*), where q is usually a power of 2. In the study of error-correcting codes, one wants the code words to be widely spaced so that errors in transmission can be detected, yet if they are too widely spaced, the code is not very efficient. Finding good codes is an active area of research. An introduction to coding theory can be found in Lidl and Niederreiter [12, Sec. 9.1]. One surprise is that an important class of codes (the so-called *Goppa codes*) involve equations $f(x, y) = 0$ over finite fields. See Moreno's book [13] for an introduction.

Historical Notes

The finite field $\mathbb{F}_p$ first arose in the study of congruences modulo the prime p. For an arbitrary modulus n, congruences are implicit in the work of Fermat, Euler, Lagrange, and Legendre, though Gauss was the first to give an explicit definition. All of the these people knew that congruences modulo a prime have special properties. For example, in *Disquisitiones*, Gauss states the following result:

> *A congruence of the mth degree*
>
> $$Ax^m + Bx^{m-1} + Cx^{m-2} + \text{etc.} + Mx + N \equiv 0$$
>
> *whose modulus is a prime number p that does not divide A, cannot be solved in more than m different ways, that is, it cannot have more than m noncongruent roots relative to p ...*

(See [7, Art. 43].) In modern terms, this says that a polynomial in $\mathbb{F}_p[x]$ of degree m has at most m roots in $\mathbb{F}_p$, which is true because $\mathbb{F}_p$ is a field. Exercise 5 will give an example of how this can fail when the modulus is not a prime.

In 1830, Galois published *Sur la théorie des nombres* in the *Bulletin des sciences mathématiques de Ferussac* (see [Galois, pp. 113–127]). Galois begins the paper by noting that for congruences $F(x) \equiv 0 \bmod p$ as above, "one customarily considers only integer solutions." If instead one considers "incommensurable" solutions, Galois states that "I have arrived at certain results that I think are new." Essentially everything in this section can be traced back to these results of Galois.

His construction goes as follows. Consider a polynomial $F \in \mathbb{Z}[x]$ of degree $\nu > 1$ whose reduction modulo p in $\mathbb{F}_p[x]$ is irreducible of degree ν (Exercise 6 explains how Galois stated this). Since $\nu > 1$, it follows that the congruence

(11.1) $F(x) \equiv 0 \bmod p$

has no integer solutions. Thus, as stated in [Galois, p. 113], "One must regard the roots of this congruence as a kind of imaginary symbol." In analogy with the usual symbol for $\sqrt{-1}$, Galois uses i to denote a root of the congruence (11.1). Then he considers expressions of the form

$$\alpha = a + a_1 i + a_2 i^2 + \cdots + a_{\nu-1} i^{\nu-1},$$

where $a, a_1, \ldots, a_{\nu-1}$ are integers modulo p. There are p^ν different choices for α, which give the finite field with p^ν elements. Galois proves the following facts about this field:

- $\alpha^{p^\nu - 1} = 1$ when $\alpha \neq 0$.
- The elements of this field are the roots of $x^{p^\nu} - x$ (Theorem 11.1.2).
- All irreducible polynomials of degree ν lead to the same field (Corollary 11.1.3).
- Primitive roots exist, that is, the nonzero elements of the field form a cyclic group under multiplication (Proposition A.5.3).

Galois also knew Proposition 11.1.5, which he states as follows:

> Next, to get the integer solutions, it suffices, as M. Libri appears to have been the first to remark, to find the greatest factor common to $Fx = 0$ and $x^{p-1} = 1$.
>
> If now one wants to have imaginary solutions of the second degree, one finds the greatest factor common to $Fx = 0$ and $x^{p^2 - 1} = 1$, and in general, the solutions of order ν will be given by the greatest factor common to $Fx = 0$ and $x^{p^\nu - 1} = 1$.

(See [Galois, pp. 123–125].) Note that Galois uses an equal sign $=$ to denote congruence modulo p. Do you see how his version of Proposition 11.1.5 counts only nonzero solutions?

Galois's arguments are sketchy and assume the existence of a root i of (11.1). Nevertheless, his account is remarkably complete. For instance, given an arbitrary $\nu > 1$, he uses a splitting field of $x^{p^\nu} - x$ to prove the existence of a polynomial F of degree ν that is irreducible modulo p. This is his way of proving Theorem 11.1.4. As for the Galois group $\mathrm{Gal}(\mathbb{F}_{p^\nu}/\mathbb{F}_p)$, Galois doesn't state Theorem 11.1.7 directly. However, if i is a root of (11.1), then Galois knew that the other roots are given by $i^p, i^{p^2}, \ldots, i^{p^{\nu-1}}$, which is the usual way we use the Galois group to find the other roots of an irreducible polynomial (see Exercise 7). Also, in Section 14.3 we will see how Galois used $\mathrm{Gal}(\mathbb{F}_{p^\nu}/\mathbb{F}_p)$ to construct some interesting matrix groups.

However, Galois was not the only person to discover finite fields. Gauss described a theory of "higher congruences" in an unpublished manuscript from around 1800, and Schönemann described a theory of finite fields based on congruences in 1846. Here is Schönemann's approach. Let $f \in \mathbb{Z}[x]$ be monic of degree n whose reduction modulo p is irreducible in $\mathbb{F}_p[x]$. Also let $\alpha \in \mathbb{C}$ be a root of f. Then Schönemann considers expressions of the form $\varphi(\alpha)$, where $\varphi \in \mathbb{Z}[x]$, and he writes

$$(11.2) \qquad \varphi(\alpha) \equiv \psi(\alpha) \ (\mathrm{mod}. \ p, \alpha)$$

to mean $\varphi(\alpha) = \psi(\alpha) + pR(\alpha)$, where $\varphi, \psi, R \in \mathbb{Z}[x]$. Using this, he shows that every $\varphi(\alpha)$ is congruent modulo (p, α) to an expression of the form

$$a_{n-1}\alpha^{n-1} + \cdots + a_1\alpha + a_0,$$

where $0 \leq a_i \leq p - 1$ for $i = 0, \ldots, n - 1$. There are p^n such expressions.

From a modern perspective, $\varphi(\alpha)$ lies in the ring $\mathbb{Z}[\alpha]$, and the congruence classes of (11.2) give the quotient ring

$$(11.3) \qquad \mathbb{Z}[\alpha]/p\mathbb{Z}[\alpha],$$

where $p\mathbb{Z}[\alpha]$ is the ideal of $\mathbb{Z}[\alpha]$ generated by p. It follows that Schönemann's construction leads to a ring with p^n elements.

We prove that this ring is a field as follows. In Exercises 8 and 9 you will show that since f is monic, the evaluation map $q(x) \mapsto q(\alpha)$ induces a ring isomorphism

$$\mathbb{Z}[x]/f\mathbb{Z}[x] \simeq \mathbb{Z}[\alpha].$$

Furthermore, you will also show that this isomorphism induces an isomorphism

$$\mathbb{Z}[x]/\langle p, f\rangle \simeq \mathbb{Z}[\alpha]/p\mathbb{Z}[\alpha],$$

where

(11.4) $\langle p, f\rangle = p\mathbb{Z}[x] + f\mathbb{Z}[x] = \{pR(x) + f(x)S(x) \mid R(x), S(x) \in \mathbb{Z}[x]\}.$

Thus we can interpret Schönemann's construction as taking the quotient of $\mathbb{Z}[x]$ by $\langle p, f\rangle$ in two steps: first quotient out by f to get $\mathbb{Z}[\alpha]$, and then quotient out by p to get (11.3). However, we can reverse the order of the steps: first quotient out by p to get $\mathbb{Z}[x]/p\mathbb{Z}[x] \simeq \mathbb{F}_p[x]$, which sends f to $\bar{f} \in \mathbb{F}_p[x]$, and then quotient out by $\bar{f}$ to obtain the isomorphism

$$\mathbb{Z}[x]/\langle p, f\rangle \simeq \mathbb{F}_p[x]/\langle \bar{f}\rangle$$

(see Exercise 9). Since $\bar{f} \in \mathbb{F}_p[x]$ is irreducible, the results of Chapter 3 show that $\mathbb{F}_p[x]/\langle \bar{f}\rangle$ is an extension field of $\mathbb{F}_p$ in which $\bar{f}$ has a root (namely, the coset of x).

Combining these isomorphisms gives

(11.5) $\mathbb{Z}[\alpha]/p\mathbb{Z}[\alpha] \simeq \mathbb{Z}[x]/\langle p, f\rangle \simeq \mathbb{F}_p[x]/\langle \bar{f}\rangle,$

which proves that Schönemann's construction gives a finite field with p^n elements. Also, this isomorphism sends the coset of α to the coset of x. If we let i denote this coset, then we recover Galois's construction of the finite field. So everything fits together nicely.

Besides the constructions of Schönemann and Galois, the isomorphisms (11.5) show that we can represent a finite field with p^n elements as

(11.6) $\mathbb{Z}[x]/\langle p, f\rangle$

whenever f is monic of degree n and irreducible modulo p. This construction is implicit in Schönemann's work and was made explicit by Dedekind in 1857. Unlike Galois (who assumed the existence of i) and Schönemann (who used the Fundamental Theorem of Algebra to find α), Dedekind's construction is purely algebraic and became the standard way to define finite fields (see, for example, Dickson's book [5] from 1901). Initially, the notation $\mathrm{GF}(p^n)$ was reserved for the finite field with p^n elements constructed in (11.6). It was only in 1893 that E. H. Moore showed that that *any* finite field is isomorphic to one of the form (11.6).

For more details on the history of finite fields, we refer the reader to [6, Vol. I, pp. 233–252] and [12, pp. 73–78].

Exercises for Section 11.1

Exercise 1. Let $\mathbb{F}_p \subset L$ be an extension such that $x^q - x$ splits completely over L, where $q = p^n$, and let F be the set of roots of this polynomial. Prove that F is a subfield of L.

Exercise 2. Suppose that $f, g \in F[x]$ are polynomials, not both zero, and let h be their greatest common divisor as computed in $F[x]$. Now let L be an extension field of F. Prove that h is greatest common divisor of f, g when considered as polynomials in $L[x]$.

Exercise 3. Give a proof of Corollary 11.1.8 that uses neither Theorem 11.1.7 nor the Galois correspondence.

Exercise 4. Prove Theorem 11.1.9.

Exercise 5. As noted in the text, if $f \in \mathbb{Z}[x]$ has degree n and its leading coefficient is not divisible by a prime p, then $f(x) \equiv 0 \bmod p$ has at most n solutions modulo p. Here are two questions that explore what happens when $n = 2$ and the modulus is arbitrary.
(a) How many solutions does the congruence $x^2 - 1 \equiv 0 \bmod 8$ have modulo 8?
(b) Fix an integer $m > 1$, and assume that if every polynomial of degree 2 in $\mathbb{Z}/m\mathbb{Z}[x]$ has at most two roots in $\mathbb{Z}/m\mathbb{Z}$. Is m prime?

Exercise 6. Let $F \in \mathbb{Z}[x]$ have degree n, and assume that the leading coefficient of F is not divisible by p. Prove that the reduction of F modulo p is irreducible over $\mathbb{F}_p$ if and only if it is not possible to find polynomials $\varphi, \psi, \chi \in \mathbb{Z}[x]$, where $\deg(\varphi), \deg(\psi) < n$, such that

$$\varphi(x)\psi(x) = F(x) + p\chi(x).$$

This is how Galois defines irreducibility modulo p in [Galois, p. 113].

Exercise 7. Let $f \in \mathbb{F}_p[x]$ be irreducible of degree v. Use (7.1) and Theorem 11.1.7 to prove Galois's observation that if i is one root of f in a splitting field, then the other roots are given by $i^p, i^{p^2}, \ldots, i^{p^{v-1}}$.

Exercise 8. Let I and J be ideals in a ring R, and let $I + J = \{r + s \mid r \in I, s \in J\}$ be their sum. Also let $\overline{I} = \{r + J \mid r \in I\}$. This is a subset of the quotient ring R/J.
(a) Prove that $I + J$ is an ideal of R and that $\overline{I}$ is an ideal of R/J.
(b) Show that the map $r + (I + J) \mapsto (r + J) + \overline{I}$ defines a well-defined ring isomorphism $R/(I + J) \simeq (R/J)/\overline{I}$.

Exercise 9. Let $f \in \mathbb{Z}[x]$ be monic and irreducible, and let $\alpha \in \mathbb{C}$ be a root of f. Then let $\bar{f} \in \mathbb{F}_p[x]$ be the reduction of f modulo the prime p, and let $\langle p, f \rangle$ be as in (11.4).
(a) Prove that the map $q(x) + f\mathbb{Z}[x] \mapsto q(\alpha)$ defines a well-defined ring isomorphism $\mathbb{Z}[x]/f\mathbb{Z}[x] \simeq \mathbb{Z}[\alpha]$.
(b) Use Exercise 8 to prove that $\mathbb{Z}[x]/\langle p, f \rangle \simeq \mathbb{Z}[\alpha]/p\mathbb{Z}[\alpha]$.
(c) Similarly prove that $\mathbb{Z}[x]/\langle p, f \rangle \simeq \mathbb{F}_p[x]/\langle \bar{f} \rangle$.

Exercise 10. Let $f = 2 + 2x + 2x^2 + 2x^3 + 2x^4 + 2x^5 + 2x^6 + 2x^7 + x^8 + x^9 + x^{10} \in \mathbb{F}_3[x]$.
(a) Use the method of Example 11.1.6 to determine the number of roots of f in $\mathbb{F}_{3^3}$ and $\mathbb{F}_{3^7}$.
(b) Explain why the splitting field of f over $\mathbb{F}_3$ is $\mathbb{F}_{3^{21}}$.

Exercise 11. Let $f \in \mathbb{F}_p[x]$ be an irreducible polynomial of degree n. Prove that f splits completely in $\mathbb{F}_{p^n}$.

11.2 IRREDUCIBLE POLYNOMIALS OVER FINITE FIELDS (OPTIONAL)

Our presentation of finite fields in Section 11.1 was very abstract—it wasn't until the Historical Notes that we explicitly constructed $\mathbb{F}_{p^n}$ as $\mathbb{F}_p[x]/\langle f\rangle$, where $f \in \mathbb{F}_p[x]$ is irreducible of degree n. Yet whenever finite fields are implemented on a computer, such a representation is essential. It follows that we need a good understanding of irreducible polynomials in $\mathbb{F}_p[x]$.

A. Irreducible Polynomials of Fixed Degree. We begin with the following easy result concerning irreducible polynomials in $\mathbb{F}_p[x]$.

Proposition 11.2.1. *Let $f \in \mathbb{F}_p[x]$ be irreducible of degree m. Then:*

(a) *f divides $x^{p^m} - x$.*

(b) *f is separable.*

(c) *Given an integer $n \geq 1$, f divides $x^{p^n} - x \Leftrightarrow f$ has a root in $\mathbb{F}_{p^n} \Leftrightarrow m \mid n$.*

Proof. We begin with part (c). Let α be a root of f in a splitting field over $\mathbb{F}_p$. Since f is irreducible, $\mathbb{F}_p \subset \mathbb{F}_p(\alpha)$ has degree m. Then $\mathbb{F}_p(\alpha) \simeq \mathbb{F}_{p^m}$ by Proposition 11.1.1 and Corollary 11.1.3. From here, the second equivalence of part (c) follows directly from Corollary 11.1.8. Also note that $f \mid \gcd(f, x^{p^n} - x)$ if and only if $\deg\big(\gcd(f, x^{p^n} - x)\big) > 0$, since f is irreducible over $\mathbb{F}_p$. Then the first equivalence follows from Proposition 11.1.5.

We get part (a) by taking $n = m$ in part (c), and part (b) follows immediately, since $x^{p^m} - x$ is separable by the proof of Theorem 11.1.4.   □

More generally, one can show that if F is any finite field and $f \in F[x]$ is irreducible, then f is separable (see Exercise 1). Since irreducible polynomials in characteristic 0 are always separable, it follows that nonseparable irreducible polynomials can occur only over infinite fields of characteristic p. Do you see how this relates to the examples of such polynomials presented in Example 5.3.11?

We next want to count the number of irreducible polynomials of fixed degree in $\mathbb{F}_p[x]$. This number is finite, since we are working over a finite field. Furthermore, since any irreducible polynomial becomes monic after multiplying by a suitable constant, it suffices to compute the number

$$N_m = \big|\{f \in \mathbb{F}_p[x] \mid f \text{ is monic irreducible of degree } m\}\big|.$$

These numbers are related as follows.

Theorem 11.2.2. *Let N_m be as defined above. Then, for any $n \geq 1$, we have*

$$\sum_{m \mid n} m N_m = p^n,$$

where the sum is over all positive divisors of n.

Proof. Since $x^{p^n} - x$ is separable, we know that it factors as a product of distinct irreducible polynomials in $\mathbb{F}_p[x]$. Furthermore, since it is monic, we can assume that the polynomials in the factorization are also monic. Finally, part (c) of Proposition 11.2.1 shows that the polynomials in the factorization are *all* monic irreducible polynomials of $\mathbb{F}_p[x]$ whose degree m divides n.

This allows us to write $x^{p^n} - x$ as follows. Let

$$\mathcal{N}_m = \{f \in \mathbb{F}_p[x] \mid f \text{ is monic irreducible of degree } m\},$$

so that $N_m = |\mathcal{N}_m|$. Then the above paragraph implies that

$$(11.7) \qquad\qquad x^{p^n} - x = \prod_{f \in \mathcal{N}_m, \, m \mid n} f$$

(be sure you understand why). Since every $f \in \mathcal{N}_m$ has degree m, taking the degree of each side of (11.7) gives the desired formula. $\qquad\square$

Here is an example of Theorem 11.2.2.

Example 11.2.3. The monic irreducible polynomials of degree 1 in $\mathbb{F}_p[x]$ are of the form $x - a$ for $a \in \mathbb{F}_p$. Thus $N_1 = p$. Then the theorem implies that

$$p^2 = 2N_2 + N_1 = 2N_2 + p,$$

so that $N_2 = \frac{1}{2}(p^2 - p)$. In Exercise 2 you will use this to prove that

$$(11.8) \qquad\qquad N_4 = \tfrac{1}{4}(p^4 - p^2).$$

These formulas show that N_2 and N_4 are positive, which implies that we can find irreducible polynomials of degrees 2 and 4. In particular, this proves the existence of finite fields of orders p^2 and p^4. $\qquad\qquad\triangleleft\triangleright$

To generalize the formulas in this example, we will use the *Möbius function* $\mu(n)$ from Exercise 14 of Section 9.1. This function is defined by

$$\mu(n) = \begin{cases} 1, & \text{if } n = 1, \\ (-1)^s, & \text{if } n = p_1 \cdots p_s \text{ for distinct primes } p_1, \ldots, p_s, \\ 0, & \text{otherwise.} \end{cases}$$

Then we have the following formula for N_n.

Theorem 11.2.4. *The number of monic irreducible polynomials of degree n in $\mathbb{F}_p[x]$ is given by*

$$N_n = \frac{1}{n} \sum_{m \mid n} \mu(m) \, p^{\frac{n}{m}}.$$

Proof. Let F be a complex-valued function defined on the set of positive integers. Then we get another such function G defined by

$$G(n) = \sum_{m|n} F(m),$$

where as usual the sum is over all positive divisors of n. The *Möbius inversion formula* asserts that in this situation, we can express F in terms of G as follows:

$$F(n) = \sum_{m|n} \mu(m) G\left(\tfrac{n}{m}\right).$$

A proof of the Möbius inversion formula can be found in most books on number theory. See [10, Sec. 2.2] or [14, Sec. 4.3].

In particular, if $F(n) = nN_n$, then Theorem 11.2.2 implies that

$$G(n) = \sum_{m|n} F(m) = \sum_{m|n} mN_m = p^n.$$

By the inversion formula, we obtain

$$nN_n = F(n) = \sum_{m|n} \mu(m) G\left(\tfrac{n}{m}\right) = \sum_{m|n} \mu(m) p^{\frac{n}{m}}.$$

The desired formula follows immediately. $\qquad\square$

Here are some examples of how this theorem works.

Example 11.2.5. When $n = 4$, Theorem 11.2.4 implies that

$$
\begin{aligned}
N_4 &= \tfrac{1}{4}\left(\mu(1)p^{\frac{4}{1}} + \mu(2)p^{\frac{4}{2}} + \mu(4)p^{\frac{4}{4}}\right) \\
&= \tfrac{1}{4}\left(1 \cdot p^4 + (-1) \cdot p^2 + 0 \cdot p\right) \\
&= \tfrac{1}{4}\left(p^4 - p^2\right),
\end{aligned}
$$

which agrees with (11.8). Similarly, when $n = 6$, you will show that

$$N_6 = \tfrac{1}{6}\left(p^6 - p^3 - p^2 + p\right)$$

in Exercise 3. ◁▷

In Exercise 4 you will use Theorem 11.2.4 to prove that $N_n > 0$ for all $n \geq 1$.

B. Cyclotomic Polynomials Modulo p.

Theorem 11.1.2 tells us that $\alpha^q = \alpha$ for all $\alpha \in \mathbb{F}_q$. It follows that $\alpha^{q-1} = 1$ when $\alpha \neq 0$, so that every nonzero element of $\mathbb{F}_q$ is a root of unity. In characteristic 0, the minimal polynomial of a primitive dth root of unity in $\mathbb{C}$ is the cyclotomic polynomial $\Phi_d(x)$. We will now explore what happens when we reduce these polynomials modulo p.

By Section 9.1, $\Phi_d(x)$ is the monic polynomial whose roots are the primitive dth roots of unity in $\mathbb{C}$. Furthermore, $\Phi_d(x)$ has integer coefficients, is irreducible of degree $\phi(d)$, and has the factorization

(11.9)
$$x^n - 1 = \prod_{d|n} \Phi_d(x)$$

in $\mathbb{Z}[x]$. The reduction of $\Phi_d(x)$ modulo p should be denoted $\overline{\Phi}_d(x)$, but for simplicity we will denote it by $\Phi_d(x)$. Thus (11.9) becomes an identity over in $\mathbb{F}_p[x]$. We will concentrate on the case when $\gcd(d, p) = 1$. This restriction is explained in Exercise 5.

We begin by describing the roots of $\Phi_d(x)$. Recall that a dth root of unity α is *primitive* if d is the smallest positive integer such that $\alpha^d = 1$.

Proposition 11.2.6. *If* $\gcd(d, p) = 1$ *and* $q = p^n$, *then the following are equivalent:*
(a) $d | q - 1$.
(b) $\Phi_d(x)$ *splits completely in* $\mathbb{F}_q$.
(c) $\Phi_d(x)$ *has a root in* $\mathbb{F}_q$.
Furthermore, when these conditions are satisfied, the roots of $\Phi_d(x)$ in $\mathbb{F}_q$ consist of the primitive dth roots of unity.

Proof. We first study the primitive dth roots of unity in characteristic p. Observe that $x^d - 1$ is separable, since $\gcd(d, p) = 1$. Hence $x^d - 1$ has d roots in a splitting field. These roots form a group under multiplication, which is cyclic of order d by Proposition A.5.3. Such a group has $\phi(d)$ generators (Exercise 10 of Section 9.1), so that there are $\phi(d)$ primitive dth roots of unity. By Exercise 11 of Section 9.1,

$$d = \sum_{\ell|d} \phi(\ell),$$

and by (11.9) for $n = d$,
$$x^d - 1 = \prod_{\ell|d} \Phi_\ell(x).$$

From these facts, it is straightforward to prove by complete induction on d that the roots of $\Phi_d(x)$ are the primitive dth roots of unity in characteristic p.

We now prove (a) $\Rightarrow$ (b). Applying (11.9) with $n = q - 1$, we obtain

$$x^{q-1} - 1 = \prod_{\ell|q-1} \Phi_\ell(x).$$

Since $x^{q-1} - 1$ splits completely in $\mathbb{F}_q$ and $d | q - 1$, we see that $\Phi_d(x)$ splits completely in $\mathbb{F}_q$. The implication (b) $\Rightarrow$ (c) is trivial. Finally, to prove (c) $\Rightarrow$ (a), note that a root $\alpha \in \mathbb{F}_q$ of $\Phi_d(x)$ is a primitive dth root of unity by the above analysis. Then $d | q - 1$ (since α has order d in $\mathbb{F}_q^*$), and (a) is proved.

The final assertion of the proposition now follows immediately. $\square$

We next compute the irreducible factors of $\Phi_d(x)$. Since $\gcd(d, p) = 1$, we have $[p] \in (\mathbb{Z}/d\mathbb{Z})^*$. Let m denote the order of $[p]$ in this group. Writing $[p]^m = [1]$ as a congruence, we see that m is the smallest positive integer such that $d \mid p^m - 1$. This number determines the degree of the irreducible factors of $\Phi_d(x)$ as follows.

Theorem 11.2.7. *Given d, let m be as above. Then $\Phi_d(x)$ is the product of $\phi(d)/m$ irreducible polynomials in $\mathbb{F}_p[x]$ of degree m.*

Proof. Let f be an irreducible factor of $\Phi_d(x)$. To show that $\deg(f) = m$, it suffices to show that the smallest positive integer ℓ such that $d \mid p^\ell - 1$ is $\ell = \deg(f)$. We prove this as follows. For an integer $\ell \geq 1$, observe that

$$d \mid p^\ell - 1 \iff \Phi_d(x) \text{ splits completely over } \mathbb{F}_{p^\ell}$$
$$\iff f \text{ splits completely over } \mathbb{F}_{p^\ell}$$
$$\iff f \text{ has a root in } \mathbb{F}_{p^\ell}$$
$$\iff \deg(f) \mid \ell.$$

The first equivalence is by (a) $\Leftrightarrow$ (b) of Proposition 11.2.6, the second follows from (b) $\Leftrightarrow$ (c) of the same proposition because $f \mid \Phi_d(x)$, the third follows because $\mathbb{F}_p \subset \mathbb{F}_{p^m}$ is Galois and hence normal, and the fourth is by part (c) of Proposition 11.2.1. These equivalences show that $\deg(f)$ has the desired property. $\square$

Here is an example of Theorem 11.2.7.

Example 11.2.8. Let $p = 2$. For $d = 5$, one easily sees that $[2]$ has order 4 in $(\mathbb{Z}/5\mathbb{Z})^*$. By Theorem 11.2.7, $\Phi_5(x)$ is the product of $\phi(5)/4 = 1$ irreducible polynomials of degree 4 in $\mathbb{F}_2[x]$. Thus $\Phi_5(x) = x^4 + x^3 + x^2 + x + 1$ is irreducible in $\mathbb{F}_2[x]$. Its roots are the primitive 5th roots of unity in $\mathbb{F}_{16}$.

When $d = 15$, $[2]$ also has order 4 in $(\mathbb{Z}/15\mathbb{Z})^*$. Thus $\Phi_{15}(x)$ is the product of $\phi(15)/4 = 8/4 = 2$ irreducible polynomials of degree 4. In Exercise 6 you will verify that the factorization in $\mathbb{F}_2[x]$ is

$$\Phi_{15}(x) = x^8 + x^7 + x^5 + x^4 + x^3 + x + 1 = (x^4 + x^3 + 1)(x^4 + x + 1).$$

The roots of these polynomials are the primitive 15th roots of unity in $\mathbb{F}_{16}$. ◁▷

C. Berlekamp's Algorithm. We conclude this section by explaining how to determine whether a given nonconstant polynomial $f \in \mathbb{F}_p[x]$ is irreducible. One method for doing this would be to list the finitely many nonconstant polynomials of degree $< \deg(f)$ and divide each into f. We will give a much more efficient method based on Berlekamp's factoring algorithm that uses a nice combination of linear algebra and the division algorithm.

Suppose that we want to test whether a given polynomial $f \in \mathbb{F}_p[x]$ is irreducible. We may assume that f has degree $n > 1$. Furthermore, since irreducible polynomials over $\mathbb{F}_p$ are separable by Proposition 11.2.1, we may also assume that f is separable. We will use the quotient ring

$$R = \mathbb{F}_p[x]/\langle f \rangle.$$

By the division algorithm, every element of the ring R can be written uniquely in the form

$$a_0 + a_1 x + \cdots + a_{n-1} x^{n-1} + \langle f \rangle, \quad a_i \in \mathbb{F}_p.$$

It follows that as a vector space over $\mathbb{F}_p$, R has dimension n. Now consider the map

$$T : R \longrightarrow R$$

defined by $T(g + \langle f \rangle) = g^p + \langle f \rangle$. This is well defined, for if $g + \langle f \rangle = h + \langle f \rangle$, then $g = h + fB$ for some $B \in \mathbb{F}_p[x]$. Since we are in characteristic p, we have

$$g^p = (h + fB)^p = h^p + f^p B^p = h^p + f \cdot f^{p-1} B^p,$$

which implies $g^p + \langle f \rangle = h^p + \langle f \rangle$. Furthermore, it is easy to see that T is linear over $\mathbb{F}_p$ (you will prove this in Exercise 7). The identity map $1_R : R \to R$ is also linear. Then we get the following unexpected result.

Theorem 11.2.9. *Let $f \in \mathbb{F}_p[x]$ be separable of degree $n > 1$, and let $R = \mathbb{F}_p[x]/\langle f \rangle$. Then f is irreducible if and only if the linear map $T - 1_R : R \to R$ has rank $n - 1$.*

Proof. If f is irreducible, then R is a field and T is the Frobenius automorphism. Hence the kernel of $T - 1_R$ consists of the solutions of $\alpha^p = \alpha$. This gives $\mathbb{F}_p$, so that the kernel has dimension 1. By the dimension formula from linear algebra, it follows that $T - 1_R$ has rank $n - 1$.

On the other hand, if f is reducible, then $f = gh$ where $g, h \in \mathbb{F}_p[x]$ have degree $< \deg(f)$. Furthermore, g and h must be relatively prime, since f is separable. Hence we can find $A, B \in \mathbb{F}_p[x]$ such that $Ag + Bh = 1$.

Observe that $Ag + \langle f \rangle$ is in the kernel of $T - 1_R$ if and only if $(Ag)^p - Ag$ is a multiple of f. Using the binomial theorem and $f = gh$, we have

$$\begin{aligned}
(Ag)^p &= Ag(1 - Bh)^{p-1} = Ag\bigl(1 - (p-1)Bh + \cdots + (-1)^{p-1}(Bh)^{p-1}\bigr) \\
&= Ag - gh \cdot (p-1)AB + \cdots + gh \cdot (-1)^{p-1}AB^{p-1}h^{p-2} \\
&\equiv Ag \bmod f.
\end{aligned}$$

It follows that $Ag + \langle f \rangle$ is in the kernel. Interchanging the roles of Ag and Bh, we see that $Bh + \langle f \rangle$ is also in the kernel.

If we can show that $Ag + \langle f \rangle$ and $Bh + \langle f \rangle$ are linearly independent elements of R, then $T - 1_R$ will have rank at most $n - 2$ and the theorem will follow. So suppose that some linear combination of these cosets is zero, that is, there are $a, b \in \mathbb{F}_p$ such that $aAg + bBh$ is a multiple of $f = gh$. Then

$$aAg + bBh = ghC$$

for some $C \in \mathbb{F}_p[x]$. Since g and h are relatively prime, it follows easily that $g | bB$ and $h | aA$. But $Ag + Bh = 1$ implies that $\gcd(g, B) = \gcd(h, A) = 1$, so that we must have $a = b = 0$. This proves the desired linear independence. $\qquad \square$

One useful observation is that from the vector $Ag + \langle f \rangle$ in the kernel of $T - 1_R$, we can recover the factorization of f, since $g = \gcd(f, Ag)$ follows from $f = gh$ and $Ag + Bh = 1$. In general, elements of the kernel can be more complicated, but it is still possible to use them to find the irreducible factorization of f. This is Berlekamp's algorithm, which is described in [12, Sec. 4.1].

Here is an example to show how Theorem 11.2.9 can be used.

Example 11.2.10. Let $f = x^5 + x^4 + 1 \in \mathbb{F}_2[x]$. Note that f is separable, since $\gcd(f, f') = 1$. Then $R = \mathbb{F}_2[x]/\langle f \rangle$ is a vector space over $\mathbb{F}_2$ of dimension 5 with basis $1 + \langle f \rangle, x + \langle f \rangle, x^2 + \langle f \rangle, x^3 + \langle f \rangle, x^4 + \langle f \rangle$, which for simplicity we write as $1, x, x^2, x^3, x^4$.

Note that $T : R \to R$ is the squaring map, since $p = 2$. To compute the matrix of T, we apply T to each basis element and represent the result in terms of the basis:

$$1 \mapsto 1,$$
$$x \mapsto x^2,$$
$$x^2 \mapsto x^4,$$
$$x^3 \mapsto x^6 = 1 + x + x^4,$$
$$x^4 \mapsto x^8 = 1 + x + x^2 + x^3 + x^4.$$

Here, $x^6 = 1 + x + x^4$ means that $1 + x + x^4$ is the remainder of x^6 on division by $f = x^5 + x^4 + 1$, and similarly for the last line.

It follows that the matrix of $T - 1_R$ with respect to the basis $1, x, x^2, x^3, x^4$ is

$$\begin{pmatrix} 1 & 0 & 0 & 1 & 1 \\ 0 & 0 & 0 & 1 & 1 \\ 0 & 1 & 0 & 0 & 1 \\ 0 & 0 & 0 & 0 & 1 \\ 0 & 0 & 1 & 1 & 1 \end{pmatrix} - \begin{pmatrix} 1 & 0 & 0 & 0 & 0 \\ 0 & 1 & 0 & 0 & 0 \\ 0 & 0 & 1 & 0 & 0 \\ 0 & 0 & 0 & 1 & 0 \\ 0 & 0 & 0 & 0 & 1 \end{pmatrix} = \begin{pmatrix} 0 & 0 & 0 & 1 & 1 \\ 0 & 1 & 0 & 1 & 1 \\ 0 & 1 & 1 & 0 & 1 \\ 0 & 0 & 0 & 1 & 1 \\ 0 & 0 & 1 & 1 & 0 \end{pmatrix}$$

(remember that we are in characteristic 2). One easily sees that this matrix has rank at most 3, since the first column is zero and the sum of the last three columns is zero. Since $3 < 4 = \deg(f) - 1$, Theorem 11.2.9 implies that f is reducible.   ◁▷

Historical Notes

In the early 1800s Gauss showed that the number of monic irreducible polynomials in $\mathbb{F}_p[x]$ of degree n is given by the formula

(11.10) $$N_n = \frac{1}{n}\left(p^n - \sum_a p^{\frac{n}{a}} + \sum_{ab} p^{\frac{n}{ab}} - \sum_{abc} p^{\frac{n}{abc}} + \cdots\right),$$

where $\sum_a$ is the sum over all distinct primes dividing n, $\sum_{ab}$ is the sum over all products of distinct primes dividing n, and so on. In Exercise 8 you will show

FINITE FIELDS

that (11.10) is equivalent to the formula given in Theorem 11.2.4. Schönemann discovered (11.10) independently in 1846. He and Gauss also knew the factorization of $x^{p^n} - x$ given in (11.7).

It is also possible to count the number of monic irreducible polynomials in $\mathbb{F}_q[x]$ of degree n. In Exercise 9 you will prove the analogs of Theorems 11.2.2 and 11.2.4 for arbitrary finite fields.

The Berlekamp algorithm is much more recent. The relation between irreducibility and the rank of $T - 1_R$ given in Theorem 11.2.9 is due to Butler [4] in 1954. He proved more generally that if $f \in \mathbb{F}_p[x]$ is separable of degree $n > 0$, then the rank of $T - 1_R$ is $n - k$, where k is the number of irreducible factors of f in $\mathbb{F}_p[x]$ (see Exercises 10 and 11). This theorem can be generalized to any finite field. In 1967 Berlekamp [2] rediscovered Butler's result and used it as the basis for his factoring algorithm. Some of the beginning steps of his method will be discussed in Exercise 12, and the details can be found in [12, Sec. 4.1]. Berlekamp's algorithm works best for small finite fields; other methods are used for larger ones.

We refer the reader to Chapters 3 and 4 of [12] for *much* more on the mathematics and history of polynomials over finite fields.

Exercises for Section 11.2

Exercise 1. Let $f \in F[x]$ be irreducible, where F is a finite field. Prove that f separable.

Exercise 2. This exercise concerns Theorem 11.2.2 and the factorization (11.7).
(a) Compute N_3 and N_4 using only Theorem 11.2.2.
(b) Write down the factorization (11.7) explicitly when $p^n = 4$ and 8.

Exercise 3. Use Theorem 11.2.4 to compute N_6 and N_{36}.

Exercise 4. In Theorem 11.1.4 we used splitting fields to show that a field of order p^n exists for any prime p and integer $n \geq 1$. When Galois and others considered this question in the nineteenth century, their approach was to prove the existence of an irreducible polynomial in $\mathbb{F}_p[x]$ of degree n. In other words, they needed to prove that $N_n > 0$.
(a) Prove that $N_n > 0$ using Theorem 11.1.4.
(b) Suppose that we have proved Theorem 11.2.4 but not Theorem 11.1.4. Use this to prove that $N_n > 0$.

Exercise 5. Let F be a field of characteristic p, and let $\alpha \in F$ be a root of unity. Prove that there is some $d \geq 1$ relatively prime to p such that α is a dth root of unity.

Exercise 6. This exercise is concerned with Example 11.2.8.
(a) Show that $N_4 = 3$ when $p = 2$. Then write down these three irreducible polynomials explicitly.
(b) Verify the factorization of $\Phi_{15}(x)$ given in the example.
(c) Show that the roots of $x^4 + x^3 + 1$ and $x^4 + x + 1$ are the reciprocals of each other.

Exercise 7. As in the discussion of Berlekamp's algorithm, let $R = \mathbb{F}_p[x]/\langle f \rangle$ and consider the pth-power map $T : R \to R$. Prove that T is a linear map when R is regarded as a vector space over $\mathbb{F}_p$.

Exercise 8. Prove that Gauss's formula (11.10) is equivalent to the formula given in Theorem 11.2.4.

Exercise 9. State and prove analogs of Theorems 11.2.2 and 11.2.4 that count monic irreducible polynomials of degree n in $\mathbb{F}_q[x]$, where q is now a power of the prime p.

Exercise 10. Suppose that a monic polynomial $f \in \mathbb{F}_p[x]$ has a factorization $f = f_1 \cdots f_k$, where $f_1, \ldots, f_k$ are distinct monic irreducible polynomials. Let $R = \mathbb{F}_p[x]/\langle f \rangle$, and let $R_i = \mathbb{F}_p[x]/\langle f_i \rangle$ for $i = 1, \ldots, k$. Then consider the map

$$\varphi : R \longrightarrow R_1 \times \cdots \times R_k.$$

defined by

$$\varphi(g + \langle f \rangle) = (g + \langle f_1 \rangle, \ldots, g + \langle f_k \rangle).$$

The goal of this exercise is to prove that φ is a ring isomorphism when we make $R_1 \times \cdots \times R_k$ into a ring using coordinatewise addition and multiplication.
  (a) Prove that φ is a well-defined ring homomorphism.
  (b) Prove that φ is one to one.
  (c) Show that R and $R_1 \times \cdots \times R_k$ have the same dimension when considered as vector spaces over $\mathbb{F}_p$.
  (d) Use the dimension theorem from linear algebra to conclude that φ is a ring isomorphism.

Exercise 11. In the situation of Theorem 11.2.9, let $T : R \to R$ be the pth-power map, where $R = \mathbb{F}_p[x]/\langle f \rangle$ and f is separable of degree n. The goal of this exercise is to prove that the rank of $T - 1_R$ is $n - k$, where k is the number of irreducible factors of f in $\mathbb{F}_p[x]$. We will use the isomorphism $\varphi : R \simeq R' = R_1 \times \cdots \times R_k$ constructed in Exercise 10.
  (a) Let $T' : R' \to R'$ be the map that is the pth power on each coordinate. Prove that φ induces an isomorphism between the kernel of $T - 1_R$ and the kernel of $T' - 1_{R'}$.
  (b) Prove that the kernel of $T' - 1_{R'}$ has dimension k as a vector space over $\mathbb{F}_p$.
  (c) Prove that $T - 1_R$ has rank $n - k$, and use this to give another proof of Theorem 11.2.9.

Exercise 12. Let $f \in \mathbb{F}_p[x]$ be monic and separable of degree $n > 1$, and assume that $T - 1_R$ has rank $\neq n - 1$. By Theorem 11.2.9, f is reducible. In this exercise, you will use the kernel of $T - 1_R$ to produce a nontrivial factorization of f.
  (a) Show that the constant polynomials in $\mathbb{F}_p[x]$ give a one-dimensional subspace of the kernel of $T - 1_R$.
  (b) Prove that there is a nonconstant polynomial $h \in \mathbb{F}_p[x]$ of degree $< n$ such that $f \mid h^p - p$. Parts (c), (d), and (e) will use h to produce a nontrivial factorization of f.
  (c) Explain why $h^p - h = \prod_{a \in \mathbb{F}_p} h - a$ in $\mathbb{F}_p[x]$.
  (d) Use parts (b) and (c) to show that $f = \prod_{a \in \mathbb{F}_p} \gcd(f, h - a)$.
  (e) Use $\deg(h) < n$ to show that $f \nmid \gcd(f, h - a)$ when $a \in \mathbb{F}_p$. Conclude that the factorization of part (d) is nontrivial, that is, $\gcd(f, h - a)$ is a nonconstant factor of f of degree $< n$ for at least two $a \in \mathbb{F}_p$.
The basic idea of Berlekamp's algorithm is that one can factor f into irreducibles by taking the gcd's of the nontrivial factors $\gcd(f, h - a)$ produced by part (e) as we vary h and a.

Exercise 13. Consider the polynomial $f = x^6 + x^4 + x + 1 \in \mathbb{F}_2[x]$.
  (a) Use Exercise 11 and the method of Example 11.2.10 to show that f is the product of three irreducible polynomials in $\mathbb{F}_2[x]$. Also find a basis of the kernel of $T - 1_R$.
  (b) One element of the kernel is $(0, 0, 1, 1, 0, 1)$. This corresponds to $h = x^2 + x^3 + x^5$, since we're using the basis of R given by the cosets of $1, x, \ldots, x^5$. Show that $\gcd(f, h)$ and $\gcd(f, h + 1)$ give a nontrivial factorization $f = g_1 g_2$ as in Exercise 12.

(c) Pick an element h' of the kernel not in the span of 1 and h. Compute $\gcd(g_i, h')$ and $\gcd(g_i, h' + 1)$ for $1 = 1, 2$.

(d) Part (c) should show that f is a product of three nonconstant polynomials. Why is this the irreducible factorization of f?

Exercise 14. In this exercise we will count the number of primitive elements of the extension $\mathbb{F}_p \subset \mathbb{F}_{p^n}$. This is the number

$$P_n = \left|\{\alpha \in \mathbb{F}_{p^n} \mid \mathbb{F}_{p^n} = \mathbb{F}_p(\alpha)\}\right|.$$

(a) Use Corollary 11.1.8 to prove that $p^n = \sum_{m|n} P_m$.

(b) Use the Möbius inversion formula to conclude that $P_n = \sum_{m|n} \mu(m) \, p^{\frac{n}{m}}$. This formula was first proved by Dedekind in 1857.

(c) Explain how the formula of part (b) relates to Theorem 11.2.4.

Exercise 15. This exercise will illustrate how the word "primitive" is sometimes overused in mathematics. In the previous problem, we computed the number of primitive elements of $\mathbb{F}_p \subset \mathbb{F}_{p^n}$. In this problem, we consider the *primitive roots* of $\mathbb{F}_{p^n}$, which are generators of the cyclic group $\mathbb{F}_{p^n}^*$. The minimal polynomial over $\mathbb{F}_p$ of a primitive root of $\mathbb{F}_{p^n}$ is called a *primitive polynomial* for $\mathbb{F}_{p^n}$. These are the minimal polynomials of the primitive $(p^n - 1)$st roots of unity in characteristic p.

(a) Prove that $\mathbb{F}_{p^n}$ has $\phi(p^n - 1)$ primitive roots, where ϕ is the Euler ϕ-function.

(b) Prove that every primitive polynomial for $\mathbb{F}_{p^n}$ has degree n.

(c) Prove that the product of the primitive polynomials for $\mathbb{F}_{p^n}$ is $\Phi_{p^n-1}(x)$.

Exercise 16. Consider the trinomial $f = x^r + x^s + 1 \in \mathbb{F}_2[x]$, where $r > s > 0$ and r is prime. Prove that f is irreducible over $\mathbb{F}_2$ if and only if $f \mid x^{2^r} - x$. If in addition r is large and f is primitive in the sense of Exercise 15, then one can use f to make a that takes a long time to repeat itself. For example, $x^{6972593} + x^{3037958} + 1$ is primitive trinomial of large degree. See [3] for more details.

REFERENCES

1. E. Berlekamp, *Algebraic Coding Theory*, McGraw-Hill, New York, 1968.

2. E. Berlekamp, *Factoring polynomials over finite fields*, Bell System Tech. J. **46** (1967), 1853–1859. See also Chapter 6 of [1].

3. R. P. Brent and P. Zimmerman, *Random number generators with period divisible by a Mersenne prime*, in *Proc. ICCSA (Montreal, 2003)*.

4. M. C. R. Butler, *On the reducibility of polynomials over a finite field*, Quart. J. Math. Oxford Ser. (2) **5** (1954), 102–107.

5. L. E. Dickson, *Linear Groups with an Exposition of the Galois Field Theory*, B. G. Teubner, Leipzig, 1901. Reprint by Dover, New York, 1958.

6. L. E. Dickson, *History of the Theory of Numbers*, Carnegie Institute, Washington, DC, 1919–1923. Reprint by Chelsea, New York, 1971.

7. C. F. Gauss, *Disquisitiones Arithmeticae*, Leipzig, 1801. Republished in 1863 as Volume I of [Gauss]. French translation, *Recherches Arithmétiques*, Paris, 1807. Reprint by Hermann, Paris, 1910. German translation, *Untersuchungen über Höhere Arithmetik*, Berlin, 1889. Reprint by Chelsea, New York, 1965. English translation, Yale U. P., New Haven, 1966. Reprint by Springer-Verlag, New York, Berlin, Heidelberg, 1986.

8. D. M. Goldschmidt, *Algebraic Functions and Projective Curves*, Springer-Verlag, New York, Berlin, Heidelberg, 2003.

9. J. J. Gray, *A commentary on Gauss's mathematical diary, 1796–1814, with an English translation*, Expo. Math. **2** (1984), 97–130. (The Latin original of Gauss's diary is reprinted in [Gauss, Vol. X.1].)

10. K. Ireland and M. Rosen, *A Classical Introduction to Modern Number Theory*, Springer-Verlag, New York, Berlin, Heidelberg, 1982.

11. N. Koblitz, *A Course in Number Theory and Cryptography*, Springer-Verlag, New York, Berlin, Heidelberg, 1987.

12. R. Lidl and H. Niederreiter, *Finite Fields*, Encyclopedia of Mathematics and its Applications **20**, Addison-Wesley, Reading, MA, 1983.

13. C. J. Moreno, *Algebraic Curves over Finite Fields*, Cambridge U. P., Cambridge, 1991.

14. I. Niven and H. S. Zuckerman, *An Introduction to the Theory of Numbers*, Third Edition, Wiley, New York, 1972.

Part IV

Further Topics

The final four chapters explore the further riches of Galois theory.

The goal of Chapter 12 is to deepen our understanding of where Galois theory came from. We begin with Lagrange, who proved studied an important special case of the Galois correspondence. We then explain how Galois thought about his theory and discuss Kronecker's approach to the existence of roots and splitting fields.

Given an arbitrary polynomial, how do we find its Galois group? In Chapter 13 we show that in principle this can always be done. We also explore various more efficient methods for dealing with polynomials of small degree.

Chapter 14 continues our study of solvability by radicals. We give Galois's wonderful criterion for when an irreducible polynomial of prime degree is solvable by radicals. Then, following Galois, we consider irreducible polynomials of prime-squared degree. This requires that we study the theory of permutation groups. Primitive and imprimitive permutation groups arise naturally in this context.

Finally, Chapter 15 discusses Abel's theorem on straightedge-and-compass constructions on the lemniscate. This involves some truly wonderful mathematics. In particular, we use certain elliptic functions constructed from the lemniscate to create extensions of $\mathbb{Q}(i)$ with Abelian Galois groups.

12

Lagrange, Galois, and Kronecker

This chapter will explore the contributions to Galois theory made by Lagrange, Galois, and Kronecker. Our account of these great mathematicians will touch on the high points of their work on the roots of polynomials.

As you read this chapter, you will be asked to look back at numerous arguments in previous chapters. The goal is to gain a better understanding of where these arguments came from and to see how the basic concepts of Galois theory evolved.

12.1 LAGRANGE

As we noted in the Historical Notes to Section 1.2, Lagrange's 1770 treatise *Réflexions sur la résolution algébrique des équations* studied the known methods for solving equations of degree ≤ 4 and analyzed these methods using permutations. Lagrange's hope was that these methods could be adapted to equations of degree ≥ 5. This section will discuss some of Lagrange's ideas and explain why his approach was doomed to failure for degree ≥ 5.

Lagrange's goal in 1770 was to understand the roots of an arbitrary polynomial. However, when dealing with expressions in the roots, Lagrange was concerned "only with the form" of such expressions and not "with their numerical quantity" [Lagrange, p. 385]. In modern terms, this means the following. Given a field F and roots $\alpha_1, \ldots, \alpha_n$ of a polynomial $f \in F[x]$ of degree n, then for Lagrange, an "expression" in the roots is a quotient of the form

$$\frac{A(\alpha_1, \ldots, \alpha_n)}{B(\alpha_1, \ldots, \alpha_n)},$$

where $A, B \in F[x_1, \ldots, x_n]$ are polynomials in variables $x_1, \ldots, x_n$ with coefficients in F. Hence the "form" of this expression is the rational function

$$\frac{A(x_1, \ldots, x_n)}{B(x_1, \ldots, x_n)} \in F(x_1, \ldots, x_n).$$

By using only the "form," Lagrange is dealing with the case when the roots are variables $x_1, \ldots, x_n$. These are the roots of the universal polynomial of degree n,

$$\tilde{f} = (x - x_1) \cdots (x - x_n) = x^n - \sigma_1 x^{n-1} + \cdots + (-1)^n \sigma_n,$$

where $\sigma_1, \ldots, \sigma_n$ are the elementary symmetric polynomials from Section 2.1.

The coefficients of $\tilde{f}$ lie in $K = F(\sigma_1, \ldots, \sigma_n)$, and the splitting field of $\tilde{f}$ over K is $L = F(x_1, \ldots, x_n)$. We will use the universal extension in degree n,

$$K \subset L,$$

throughout this section. We will also assume that F has characteristic 0 so that we can use the results on solvability by radicals proved in Chapter 8. Theorem 6.4.1 shows that $K \subset L$ is a Galois extension with Galois group

(12.1) $\mathrm{Gal}(L/K) \simeq S_n,$

where $\sigma \in S_n$ gives the automorphism that sends $f \in L$ to the rational function $\sigma \cdot f$ obtained from f by replacing x_i with $x_{\sigma(i)}$ for each i. We are thus in a rich mathematical context: we know explicitly how the Galois group acts, we have the Galois correspondence, and we understand solvability by radicals.

Lagrange, on the other hand, was working 60 years before Galois and 150 years before Artin formalized the Galois correspondence. Lagrange's main tools were the following results from the theory of symmetric functions:

- A polynomial in $F[x_1, \ldots, x_n]$ unchanged by all permutations in S_n lies in $F[\sigma_1, \ldots, \sigma_n]$.
- A rational function in $L = F(x_1, \ldots, x_n)$ unchanged by all permutations in S_n lies in $K = F(\sigma_1, \ldots, \sigma_n)$ (i.e., K is the fixed field of (12.1) acting on L).

We proved the first bullet in Theorem 2.2.2 and the second in Exercises 7 and 8 from Section 2.2. Using these facts, Lagrange discovered several important parts of Galois theory, even though the concept of "group" didn't exist in 1770. Lagrange didn't even have Cauchy's efficient notation for expressing permutations.

We will now explore some of what Lagrange did in his *Réflexions*.

A. Resolvent Polynomials. Fix a rational function $f \in L$, and consider the rational functions $\sigma \cdot f$ for all $\sigma \in S_n$. Let

(12.2) $f_1 = f, f_2, \ldots, f_r$

be the distinct rational functions we get in this way. The polynomial

(12.3) $$\theta(x) = \prod_{i=1}^{r} (x - f_i)$$

with roots $f_1, \ldots, f_r$ is a special case of the polynomial (7.1) used in the proof of Theorem 7.1.1. Reread the proof of Theorem 7.1.1, especially the part where we show that (7.1) is the minimal polynomial of an element in a Galois extension. Be sure you understand how this construction, applied to $f \in L$, gives the polynomial θ defined above. It follows that that θ has coefficients in K, is separable, and is the minimal polynomial of f over K. Hence we have proved the following.

Proposition 12.1.1. *The polynomial θ from (12.3) lies in $K[x]$ and is separable and irreducible.* $\square$

We call θ the *resolvent polynomial* of f. Hence the "Galois" construction of minimal polynomials given in (7.1) is actually due to Lagrange. In Exercise 1 you will follow Lagrange by proving that $\theta \in K[x]$ using the second of the above bullets.

Here is an example of a resolvent polynomial from Chapter 1.

Example 12.1.2. Let $n = 3$ and $z_1 = \frac{1}{3}(x_1 + \omega^2 x_2 + \omega x_3)$, $\omega = e^{2\pi i/3}$. You will check in Exercise 2 that S_3 acting on z_1 gives the elements $z_1, z_2 = (23) \cdot z_1$, $\omega z_1, \omega z_2, \omega^2 z_1, \omega^2 z_2$ listed in (1.10) and that the resulting resolvent is

$$\theta(z) = (z - z_1)(z - z_2)(z - \omega z_1)(z - \omega z_2)(z - \omega^2 z_1)(z - \omega^2 z_2)$$
$$= z^6 + q z^3 - p^3/27,$$

where

$$p = -\frac{\sigma_1^2}{3} + \sigma_2, \quad q = -\frac{2\sigma_1^3}{27} + \frac{\sigma_1 \sigma_2}{3} - \sigma_3.$$

These formulas become identical to those derived in (1.2) and (1.4) of Section 1.1 if we use $x^3 + bx^2 + cx + d$ in place of $x^3 - \sigma_1 x^2 + \sigma_2 x - \sigma_3$. ◁▷

One of Lagrange's main goals was to replace the clever substitutions used to derive Cardan's formulas in Section 1.1 with the following systematic process:

- Pick $z_1 \in L = F(x_1, x_2, x_3)$ whose resolvent $\theta(z)$ is easy to solve.
- Express the roots x_1, x_2, x_3 in terms of the roots of the resolvent.

Example 12.1.2 does the first step of this process, since $\theta(z) = z^6 + q z^3 - p^3/27 = 0$ can be solved by the quadratic formula. The second step is equally easy, since

$$z_1 = \frac{1}{3}(x_1 + \omega^2 x_2 + \omega x_3),$$
$$z_2 = \frac{1}{3}(x_1 + \omega x_2 + \omega^2 x_3),$$

together with $1 + \omega + \omega^2 = 0$ and $\sigma_1 = x_1 + x_2 + x_3$, imply that

$$x_1 = \frac{1}{3}\sigma_1 + z_1 + z_2,$$
$$x_2 = \frac{1}{3}\sigma_1 + \omega z_1 + \omega^2 z_2,$$
$$x_3 = \frac{1}{3}\sigma_1 + \omega^2 z_1 + \omega z_2.$$

Comparing this with (8.15) from Section 8.3 reveals that our "Galois" approach to Cardan's formulas is virtually identical to the "Lagrange" approach just described. The major difference is that in (8.15), we used the Lagrange resolvent

$$\alpha_1 = x_1 + \omega^{-1}x_2 + \omega^{-2}x_3 = x_1 + \omega^2 x_2 + \omega x_3$$

rather than $z_1 = \frac{1}{3}(x_1 + \omega^2 x_2 + \omega x_3)$ as above. The name "Lagrange resolvent" is no accident, as we will see later in the section.

Here is another example of a resolvent polynomial.

Example 12.1.3. Consider $y_1 = x_1x_2 + x_3x_4 \in L = F(x_1, x_2, x_3, x_4)$. In Exercise 3 you will show that the action of S_4 on y_1 gives the three polynomials

$$y_1 = x_1x_2 + x_3x_4, \quad y_2 = x_1x_3 + x_2x_4, \quad y_3 = x_1x_4 + x_2x_3,$$

and that the corresponding resolvent, as a polynomial in y, is

$$\begin{aligned}
\theta(y) &= \big(y - (x_1x_2 + x_3x_4)\big)\big(y - (x_1x_3 + x_2x_4)\big)\big(y - (x_1x_4 + x_2x_3)\big) \\
&= y^3 - \sigma_2 y^2 + (\sigma_1\sigma_3 - 4\sigma_4) y - \sigma_3^2 - \sigma_1^2\sigma_4 + 4\sigma_2\sigma_4.
\end{aligned}$$

This will be useful later in the section when we solve the quartic equation.   ◁▷

In Example 12.1.3, note that when the 24 permutations $\sigma \in S_4$ are substituted into $\sigma \cdot y_1$, the result is always one of the three polynomials y_1, y_2, y_3. Lagrange made the crucial observation that this happens because many permutations leave $y_1 = x_1x_2 + x_3x_4$ unchanged. From the modern point of view this is best stated using the language of *group actions*. Group actions are discussed in Section A.4 and have been used in several places in the text, most prominently in the proof of Theorem 6.4.1 in Section 6.4.

In terms of group actions, we can describe what Lagrange did as follows. In the proof of Theorem 6.4.1, we showed that S_n acts on $L = F(x_1, \ldots, x_n)$. For our chosen element $f \in L$, we have the *orbit*

$$S_n \cdot f = \{\sigma \cdot f \mid \sigma \in S_n\} = \{f_1 = f, f_2, \ldots, f_r\},$$

where the last equality uses the notation of (12.2). We also have the *isotropy subgroup*

$$H(f) = \{\sigma \in S_n \mid \sigma \cdot f = f\}.$$

Since every $\sigma \in H(f)$ satisfies $\sigma \cdot f = f$, we can write this symbolically as

$$H(f) \cdot f = f.$$

Now consider f_i in the orbit of f. This implies that $f_i = \sigma_i \cdot f$ for some $\sigma_i \in S_n$. One easily see that $\sigma \cdot f = f_i$ for every $\sigma \in \sigma_i H(f)$ (be sure you can show this). As above, we write this symbolically as

$$(12.4) \qquad\qquad \sigma_i H(f) \cdot f = f_i.$$

In this way, we partition S_n into r cosets of $H(f)$. Since each coset has $|H(f)|$ elements, we conclude that

$$|S_n| = r|H(f)|.$$

Although he didn't use this terminology, the above partition is implicit in what Lagrange did. Hence Lagrange in essence proved the following:

- $|H(f)|$ divides the order of S_n. Thus the index

$$[S_n : H(f)] = \frac{|S_n|}{|H(f)|} = \frac{n!}{|H(f)|}$$

  is an integer. This is a special case of *Lagrange's Theorem*.
- $r = |S_n|/|H(f)| = [S_n : H(f)]$. Thus the number of elements in the orbit of f is the index of the isotropy subgroup $H(f)$. This is a special case of the *Fundamental Theorem of Group Actions* from Appendix A.4.

These results are more than just special cases: they represent the first time these issues were considered in mathematics. The name "Lagrange's Theorem" was chosen in honor of Lagrange's analysis of this situation.

On the other hand, the details of Lagrange's arguments are quite different from ours. To see how he approached these matters, we need to think in terms of resolvent polynomials. For this purpose, consider the polynomial of degree $n!$ defined by

$$\Theta(x) = \prod_{\sigma \in S_n} (x - \sigma \cdot f).$$

To compare this with the resolvent

$$\theta(x) = \prod_{i=1}^{r} (x - f_i),$$

we organize the product formula for Θ according to cosets of $H(f)$ in S_n. The key observation is that (12.4) and $|\sigma_i H(f)| = |H(f)|$ imply that

$$\prod_{\sigma \in \sigma_i H(f)} (x - \sigma \cdot f) = (x - f_i)^{|\sigma_i H(f)|} = (x - f_i)^{|H(f)|}.$$

Since this holds for $i = 1, \ldots, r$, we obtain the following theorem of Lagrange.

Theorem 12.1.4. *Given $f \in L$, the polynomials Θ and θ are related by the formula*

$$\Theta(x) = \theta(x)^{|H(f)|}.$$

In particular, the degree of the resolvent polynomial θ is the index

$$[S_n : H(f)] = \frac{|S_n|}{|H(f)|} = \frac{n!}{|H(f)|}. \qquad \square$$

Here is an example to show how this result can be used.

Example 12.1.5. Example 12.1.3 shows that the resolvent of $y_1 = x_1x_2 + x_3x_4$ has degree 3. Thus, by Theorem 12.1.4, the isotropy group $H(y_1)$ has $\frac{24}{3} = 8$ elements. Also note that $H(y_1)$ contains (12) and (1324). In Exercise 3 you will show that

$$H(y_1) = \langle (12), (1324) \rangle \subset S_4$$

and that $H(y_1)$ is isomorphic to the dihedral group of order 8.   ◁▷

It is fun to read Lagrange's statement of Theorem 12.1.4:

> One can show in the same manner that, if the function
>
> $$f[(x')(x'')(x''')(x^{IV})\ldots]$$
>
> is by its own nature such that it conserves the same value when two, or three, or a greater number of different permutations are made among the roots x', x'', x''', x^{IV}, ..., the roots of the equation $\Theta = 0$ will be equal three by three, or four by four, or etc.; so that the quantity Θ will be equal to a cube θ^3, or a square-square θ^4, or etc., and consequently the equation $\Theta = 0$ will reduce to that of $\theta = 0$, whose degree will be equal to $\varpi/3$, or to $\varpi/4$, or etc.

(See [Lagrange, pp. 370–371].) Here, ϖ is Lagrange's notation for $n!$. In this statement, Lagrange says that if f is fixed by 2, 3, etc. permutations, then the resolvent has degree $\frac{n!}{3}$, $\frac{n!}{4}$, etc. At first glance, this seems wrong, for the denominator is one more than the number of permutations. The reason for the discrepancy is that Lagrange didn't count the identity permutation.

In most courses on group theory, students usually study cosets and Lagrange's Theorem in one part of the course and group actions in another. Pedagogically, this makes sense, but it is also important to remember that historically, things are often more complicated. In considering resolvent polynomials, Lagrange had to deal with many issues all at once. It is a testament to his power as a mathematician that Lagrange could see what was important and thereby enable his successors to sort out the details of what he did.

B. Similar Functions. In [Lagrange, pp. 358–359] Lagrange says that "one calls functions *similar* those that vary at the same time or remain the same when one makes the same permutations among the quantities of which they are composed." In modern terms this means that $f, g \in L = F(x_1, \ldots, x_n)$ are similar if for all $\sigma \in S_n$, we have

$$\sigma \cdot f = f \iff \sigma \cdot g = g.$$

Thus f and g are similar if and only if they have the same isotropy subgroup, that is, $H(f) = H(g)$.

Lagrange makes a careful study of similar functions, though his most amazing result concerns the more general situation where we have $f, g \in L$ with the property that g is fixed by every permutation that fixes f. In terms of isotropy groups, this condition can be written

$$H(f) \subset H(g)$$

(be sure you understand this). Assuming we know f, can we determine which g's satisfy the above condition? Here is Lagrange's remarkable answer.

Theorem 12.1.6. *Suppose that rational functions $f, g \in L = F(x_1, \ldots, x_n)$ have the property that g is fixed by every permutation fixing f. Then g is a rational function in f with coefficients in $K = F(\sigma_1, \ldots, \sigma_n)$.*

Proof. Our first proof of the theorem will use the Galois correspondence for $K \subset L$. Using $f \in L$, we get the intermediate field $K \subset K(f) \subset L$. Then:

- Under the group isomorphism (12.1), $\mathrm{Gal}(L/K(f)) \subset \mathrm{Gal}(L/K)$ corresponds to $H(f) \subset S_n$. Be sure you understand how this follows from Proposition 6.1.4.
- By hypothesis, g is fixed by $H(f)$, so that g is in the fixed field $L_{\mathrm{Gal}(L/K(f))}$.
- By the Galois correspondence, $K(f) = L_{\mathrm{Gal}(L/K(f))}$.

Thus $g \in K(f)$, and hence g is a rational function in f with coefficients in K.

Our second proof is taken from [3] and is much more in the spirit of Lagrange. As above, let $f_1 = f, \ldots, f_r$ be the different rational functions obtained by letting S_n act on f. Fix i between 1 and r. The proof of Theorem 12.1.4 shows that f_i corresponds to some left coset of $H(f)$, say $\sigma_i H(f)$. Since $H(f)$ doesn't affect g, it follows that every element of $\sigma_i H(f)$ takes g to $g_i = \sigma_i \cdot g$. In this way, we get elements $g_1 = g, \ldots, g_r$, which need not be distinct (do you see why?).

Using the f_i and g_i, consider the function

$$(12.5) \quad \psi(x) = \theta(x) \left(\frac{g_1}{x - f_1} + \cdots + \frac{g_r}{x - f_r} \right) = g_1 \frac{\theta(x)}{x - f_1} + \cdots + g_r \frac{\theta(x)}{x - f_r},$$

where $\theta(x)$ is the resolvent of f. In spite of the denominators, ψ is actually a polynomial in x, since $\theta(x)$ is by definition divisible by $x - f_1, \ldots, x - f_r$. An element $\sigma \in S_n$ permutes the $g_i/(x - f_i)$, so that $g_1/(x - f_1) + \cdots + g_r/(x - f_r)$ is unaffected by S_n. Since $\theta(x) \in K[x]$, the coefficients of $\psi(x)$ must be symmetric. Hence $\psi(x) \in K[x]$.

Next observe that if we evaluate the polynomial

$$\frac{\theta(x)}{x - f_i} = \prod_{j \neq i} (x - f_j)$$

at f_1, then we get $\prod_{j=2}^{r}(f_1 - f_j)$ when $i = 1$ and 0 when $2 \leq i \leq r$. Looking at the formula (12.5) for $\psi(x)$, we conclude that

$$(12.6) \qquad \psi(f_1) = g_1 \prod_{j=2}^{r} (f_1 - f_j) + 0 + \cdots + 0 = g_1 \prod_{j=2}^{r} (f_1 - f_j).$$

However, $\theta(x) = (x - f_1) \cdots (x - f_r)$ and (5.7) imply that

$$(12.7) \qquad \theta'(f_1) = (f_1 - f_2) \cdots (f_1 - f_r) = \prod_{j=2}^{r} (f_1 - f_j).$$

Since $f_1 = f$ and $g_1 = g$, equations (12.6) and (12.7) give the equation

(12.8)
$$g = \frac{\psi(f)}{\theta'(f)}.$$

This expression lies in $K(f)$ since $\psi(x)$ and $\theta'(x)$ have coefficients in K.   □

One advantage of the second proof of Theorem 12.1.6 is that it gives an explicit formula (12.8) for expressing g in terms of f, though in practice computing this formula can be unpleasant. While this proof differs from Lagrange's, it uses f_i and g_i in the same way, and Lagrange knew the formula (12.5), which is closely related to the Lagrange interpolation formula stated in Exercise 1 of Section 4.2.

Here are some simple applications of Theorem 12.1.6.

Example 12.1.7. Let $L = F(x_1, x_2, x_3, x_4)$, where F has characteristic $\neq 2$. The isotropy subgroup of $t_1 = x_1 + x_2 - x_3 - x_4$ is $\langle (12), (34) \rangle \subset S_4$. Since $y_1 = x_1 x_2 + x_3 x_4$ is fixed by these permutations, we conclude that $y_1 \in K(t_1)$.   ◁▷

Example 12.1.8. Since fields of characteristic 0 are infinite, we can pick distinct elements $t_1, \ldots, t_n \in F$. Now consider

$$\beta = t_1 x_1 + \cdots + t_n x_n \in L = F(x_1, \ldots, x_n).$$

If $\sigma \in S_n$, then $\sigma \cdot \beta = t_1 x_{\sigma(1)} + \cdots + t_n x_{\sigma(n)}$. Since $t_1, \ldots, t_n$ are distinct, it follows that $\sigma \cdot \beta = \beta$ if and only if σ is the identity. Thus $H(\beta) = \{e\}$. This means that any $g \in L$ is fixed by $H(\beta) = \{e\}$, so that $g \in K(\beta)$ by Theorem 12.1.6. Since g was an arbitrary element of L, we see that $L = K(\beta)$, that is, β is a primitive element of L. Furthermore, $H(\beta) = \{e\}$ and Theorem 12.1.4 imply that the resolvent of β has degree $n!$. Thus

$$\theta(y) = \prod_{\sigma \in S_n} (y - \sigma \cdot \beta) = \prod_{\sigma \in S_n} \left(y - (t_1 x_{\sigma(1)} + \cdots + t_n x_{\sigma(n)}) \right)$$

is the resolvent of β. This will be useful in the next section.   ◁▷

Lagrange was aware of Example 12.1.8, so that the idea of a primitive element dates back to the very beginnings of Galois theory.

From the point of view of Galois theory, Theorem 12.1.6 is exciting in that it reveals a strong connection with the Galois correspondence. It gets even better when we bring in Lagrange's similar functions. Here is the precise result.

Theorem 12.1.9. *Let $f, g \in L$. Then f and g are similar functions if and only if $K(f) = K(g)$.*

Proof. We have $K(f) = K(g) \iff \mathrm{Gal}(L/K(f)) = \mathrm{Gal}(L/K(g))$ by the Galois correspondence. Using (12.1), this gives $K(f) = K(g) \iff H(f) = H(g)$ (be sure you understand why). Since f and g are similar if and only if $H(f) = H(g)$, the theorem follows.   □

This theorem shows that the intrinsic object corresponding to similar functions is the field they generate when adjoined to K. For us, it is natural to think in terms of fields. Lagrange, on the other hand, was writing before set theory was developed, so that he and his contemporaries tended to think of individual elements rather than the sets in which they lie. At the same time, Lagrange knew that individual functions $f \in L$ weren't intrinsic, which is why he introduced the concept of similar functions. Taken together, similar functions and Theorem 12.1.6 show that Lagrange had an implicit understanding of the Galois correspondence for the extension $K \subset L$.

C. The Quartic. After analyzing the solutions of cubic and quartic equations, Lagrange states his strategy for solving equations as follows [Lagrange, p. 355]:

> As should be clear from this analysis that we have just given of the main known methods for the solution of equations, all these methods reduce to the same general principle, namely to find functions of the roots of the proposed equation such that: $1°$ the equation or equations by which they are given, i.e., of which they are the roots (equations that are usually called reduced equations), happen to be of a degree smaller than that of the proposed equation, or at least decomposable into other equations of a smaller degree than this one; $2°$ the values of the desired roots can be easily deduced from them.

Here, "functions of the roots of the proposed equation" are elements $f \in L$, and "reduced equations" are the corresponding resolvent polynomials. So Lagrange's idea is to look for resolvent polynomials that either have smaller degree or factor into polynomials of smaller degree.

To see what this means in practice, let's apply Lagrange's methods to Ferrari's solution of the universal quartic equation

$$x^4 - \sigma_1 x^3 + \sigma_2 x^2 - \sigma_3 x + \sigma_4 = 0.$$

We first describe what Ferrari did (with some of the algebra left to the exercises). Write the above equation as

$$x^4 - \sigma_1 x^3 = -\sigma_2 x^2 + \sigma_3 x - \sigma_4.$$

Since F has characteristic 0, we can add the quantity

$$y x^2 + \frac{1}{4}(-\sigma_1 x + y)^2 = \left(y + \frac{\sigma_1^2}{4}\right)x^2 - \frac{\sigma_1}{2}y x + \frac{y^2}{4}$$

to each side, where y is yet to be chosen. In Exercise 4 you will show that this leads to the equation

$$(12.9) \quad \left(x^2 - \frac{\sigma_1}{2}x + \frac{y}{2}\right)^2 = \left(y + \frac{\sigma_1^2}{4} - \sigma_2\right)x^2 + \left(\frac{-\sigma_1}{2}y + \sigma_3\right)x + \frac{y^2}{4} - \sigma_4.$$

We next choose y so that the right-hand side of (12.9) is also a perfect square. The right-hand side is quadratic in x. In general, if $A \neq 0$, then

$$Ax^2 + Bx + C = A\left(x + \frac{B}{2A}\right)^2 \iff B^2 = 4AC.$$

Applying this to the right-hand side of (12.9), you will show in Exercise 4 that $B^2 = 4AC$ leads to the cubic equation

$$(12.10) \qquad y^3 - \sigma_2 y^2 + (\sigma_1\sigma_3 - 4\sigma_4) y - \sigma_3^2 - \sigma_1^2\sigma_4 + 4\sigma_2\sigma_4 = 0.$$

This is the resolvent from Example 12.1.3 and is called the *Ferrari resolvent*.

If y_1 is a root of this resolvent, then the above formula for $Ax^2 + Bx + C$ shows that the right-hand side of (12.9) becomes

$$\left(y + \frac{\sigma_1^2}{4} - \sigma_2\right)\left(x + \frac{\frac{-\sigma_1}{2} y + \sigma_3}{2(y + \frac{\sigma_1^2}{4} - \sigma_2)}\right)^2.$$

It follows that (12.9) can be written as

$$\left(x^2 - \frac{\sigma_1}{2}x + \frac{y_1}{2}\right)^2 = \left(y_1 + \frac{\sigma_1^2}{4} - \sigma_2\right)\left(x + \frac{\frac{-\sigma_1}{2} y_1 + \sigma_3}{2(y_1 + \frac{\sigma_1^2}{4} - \sigma_2)}\right)^2,$$

so that

$$(12.11) \qquad x^2 - \frac{\sigma_1}{2}x + \frac{y_1}{2} = \pm\sqrt{y_1 + \frac{\sigma_1^2}{4} - \sigma_2}\left(x + \frac{\frac{-\sigma_1}{2} y_1 + \sigma_3}{2(y_1 + \frac{\sigma_1^2}{4} - \sigma_2)}\right).$$

Solving these two quadratic equations in x gives the four roots x_1, x_2, x_3, x_4 of our quartic equation. This is Ferrari's solution of the quartic.

One of Lagrange's main observations is that the auxiliary polynomials and radicals used in solving cubics or quartics come from expressions built from the roots and hence can be explained in terms of resolvent polynomials. For example, the Ferrari resolvent (12.10) is the resolvent polynomial of $y_1 = x_1x_2 + x_3x_4$ from Example 12.1.3. Exercise 5 will show how $y_1 = x_1x_2 + x_3x_4$ follows from (12.11).

We can also explain the square root in (12.11) in the same way. Using $y_1 = x_1x_2 + x_3x_4$ and setting

$$t_1 = x_1 + x_2 - x_3 - x_4,$$

one checks that

$$y_1 + \frac{\sigma_1^2}{4} - \sigma_2 = \frac{t_1^2}{4}.$$

This allows us to define

$$(12.12) \qquad \sqrt{y_1 + \frac{\sigma_1^2}{4} - \sigma_2} = \frac{t_1}{2}.$$

The isotropy subgroup of t_1 is easily seen to be $\langle(12), (34)\rangle$, which means that its resolvent polynomial $\theta(t)$ has degree 6. By (12.12), t_1 is a root of the quadratic polynomial

$$t^2 - (4y_1 + \sigma_1^2 - 4\sigma_2) = t^2 - 4y_1 - \sigma_1^2 + 4\sigma_2,$$

which has coefficients in $K(y_1)$. To get a polynomial with coefficients in K, we use the other roots y_2, y_3 of the quartic resolvent (12.10). This gives

$$(12.13) \quad \left(t^2 - 4y_1 - \sigma_1^2 + 4\sigma_2\right)\left(t^2 - 4y_2 - \sigma_1^2 + 4\sigma_2\right)\left(t^2 - 4y_3 - \sigma_1^2 + 4\sigma_2\right).$$

In Exercise 6 you will show that this polynomial lies in $K[t]$ and hence is the resolvent $\theta(t)$ since it has degree 6.

The passage from Lagrange quoted at the beginning of our discussion of the quartic states that we need resolvent polynomials "of a degree smaller than that of the proposed equation, or at least decomposable into other equations of a smaller degree than this one." In terms of what we just did for the quartic, this means the following:

- For $y_1 = x_1x_2 + x_3x_4$, the resolvent has degree 3, which is smaller than 4. So we can find y_1, y_2, y_3 by Cardan's formulas.
- For $t_1 = x_1 + x_2 - x_3 - x_4$, the resolvent has degree 6, but since we already know y_1, y_2, y_3, we can decompose the resolvent into quadratics as in (12.13). Then we get t_1 in terms of y_1 by extracting a square root.

Hence Ferrari's solution can be seen as a special case of Lagrange's strategy.

We can also describe the above derivation in terms of fields and Galois groups. Example 12.1.7 shows that we have fields

$$(12.14) \qquad\qquad K \subset K(y_1) \subset K(t_1) \subset L.$$

Since $K \subset K(y_1)$ has degree 3, we can use Cardan's formulas to express the splitting field, and (12.12) shows that $K(y_1) \subset K(t_1)$ is obtained by adjoining a square root. If we take the Galois groups of (12.14), then $\mathrm{Gal}(L/K) \simeq S_4$ gives the subgroups

$$H(t_1) = \langle (12), (34) \rangle \subset H(y_1) = \langle (1324), (12) \rangle \subset S_4.$$

This differs from what we did in Chapter 8. There, we wanted a chain of subgroups where each was normal in the next larger. In contrast, $H(y_1) \subset S_4$ is not normal (see Exercise 3). Hence Lagrange did not follow a strictly "Galois-theoretic" approach to solving the quartic. However, the way he built up the solution using extensions of smaller degree shows that he had the beginnings of a theory of solvability.

There is still more to say about the quartic, since the fields in (12.14) only give $t_1 = x_1 + x_2 - x_3 - x_4$. We need to explain how to go from here to the roots x_1, x_2, x_3, x_4. Rather than pursuing Ferrari's solution (12.11), we will switch to Euler's solution, which follows naturally from what we have done so far.

The key idea is to work simultaneously with the extensions (12.14) and their conjugates. For $K(y_1)$, this means using $K(y_2)$ and $K(y_3)$. As for $K(t_1)$, the six conjugates of t_1 are $\pm t_1$, $\pm t_2$, $\pm t_3$, where $t_2 = (23) \cdot t_1 = x_1 - x_2 + x_3 - x_4$ and $t_3 = (24) \cdot t_2 = x_1 - x_2 - x_3 + x_4$. By (12.12), we know that

$$t_1 = \sqrt{\sigma_1^2 - 4\sigma_2 + 4y_1}$$

and similarly one can show that

$$t_2 = \sqrt{\sigma_1^2 - 4\sigma_2 + 4y_2} \quad \text{and} \quad t_3 = \sqrt{\sigma_1^2 - 4\sigma_2 + 4y_3},$$

where $y_2 = (23) \cdot y_1$ and $y_3 = (24) \cdot y_1$. We can thus express t_1, t_2, t_3 in terms of radicals once we find y_1, y_2, y_3 using Cardan's formulas.

In Exercise 7 you will show that the equations

(12.15)
$$\begin{aligned}
\sigma_1 &= x_1 + x_2 + x_3 + x_4, \\
t_1 &= x_1 + x_2 - x_3 - x_4, \\
t_2 &= x_1 - x_2 + x_3 - x_4, \\
t_3 &= x_1 - x_2 - x_3 + x_4
\end{aligned}$$

imply that

$$\begin{aligned}
x_1 &= \tfrac{1}{4}(\sigma_1 + t_1 + t_2 + t_3), \\
x_2 &= \tfrac{1}{4}(\sigma_1 + t_1 - t_2 - t_3), \\
x_3 &= \tfrac{1}{4}(\sigma_1 - t_1 + t_2 - t_3), \\
x_4 &= \tfrac{1}{4}(\sigma_1 - t_1 - t_2 + t_3).
\end{aligned}$$

Thus the x_i can be expressed as sums involving three square roots. However, we can't make independent choices of signs, since this would lead to eight values for the roots. The point is that t_1, t_2, t_3 satisfy the identity

(12.16)
$$t_1 t_2 t_3 = \sigma_1^3 - 4\sigma_1\sigma_2 + 8\sigma_3$$

(see Exercise 8), so that knowing two of the square roots determines the third. Hence, if y_1, y_2, y_3 are the roots of the quartic resolvent (12.10), then the four roots of the quartic are

(12.17) $$\frac{1}{4}\left(\sigma_1 \pm \sqrt{4y_1 + \sigma_1^2 - 4\sigma_2} \pm \sqrt{4y_2 + \sigma_1^2 - 4\sigma_2} \pm \sqrt{4y_3 + \sigma_1^2 - 4\sigma_2}\right),$$

where the $\pm$ signs are chosen so that the product of the radicals is the right-hand side of (12.16). This is Euler's solution of the quartic.

Lagrange discusses other solutions of the quartic and interprets them in terms of resolvents. In general, his approach to solving equations anticipates many features of Galois theory, though there are important differences, especially in the appearance of nonnormal subgroups and the use of conjugate fields.

D. Higher Degrees. Although Lagrange's methods work wonderfully for equations of degree ≤ 4, they fail for degrees 5 and greater. One way to see this is by the theory of Chapter 8, which tells us that $K \subset L$ is not solvable by radicals for $n \geq 5$, since $\mathrm{Gal}(L/K) \simeq S_n$ is not a solvable group for $n \geq 5$.

It is also possible to describe this failure in terms of Lagrange's strategy. Since the degree of a resolvent polynomial is the index of the isotropy subgroup, finding resolvents of small degree is equivalent to finding subgroups of S_n of small index. However, as soon as $n \geq 5$, such subgroups are hard to find, as we will now prove in the following theorem.

Theorem 12.1.10. *Let* $n \geq 5$.

(a) *If* $H \subset S_n$ *is a subgroup of index* $[S_n : H] > 1$, *then either* $H = A_n$ *or* $[S_n : H] \geq n$.

(b) *If* $H \subset A_n$ *is a subgroup of index* $[A_n : H] > 1$, *then* $[A_n : H] \geq n$.

Proof. To prove part (a), we first note that there is $f \in L$ whose isotropy group is precisely the subgroup H, that is, $H = H(f)$. You will prove this in Exercise 9. Then write the distinct rational functions of the form $\sigma \cdot f$ for $\sigma \in S_n$ as

$$f_1 = f, f_2, \ldots, f_r.$$

By Theorem 12.1.4 we know that $r = [S_n : H]$. Now consider the set

$$N = \{\sigma \in S_n \mid \sigma \cdot f_i = f_i \text{ for all } i = 1, \ldots, r\}.$$

In Exercise 10 you will show that N is a subgroup of S_n. Note also that every $\sigma \in N$ fixes $f_1 = f$, which implies that $N \subset H$.

The key point of the proof is that N is a normal subgroup of S_n. To prove this, we must show that $\tau^{-1}\sigma\tau \in N$ for all $\sigma \in N$ and $\tau \in S_n$. Fix i between 1 and r. If $\tau \cdot f_i = f_j$ for some j, then $\tau^{-1} \cdot f_j = f_i$. Using $\sigma \in N$, this implies that

$$(\tau^{-1}\sigma\tau) \cdot f_i = (\tau^{-1}\sigma) \cdot f_j = \tau^{-1} \cdot (\sigma \cdot f_j) = \tau^{-1} \cdot f_j = f_i.$$

This is true for all i, so that $\tau^{-1}\sigma\tau \in N$. Thus N is normal in S_n.

Since $N \subset H \neq S_n$, Proposition 8.4.6 implies that either $N = \{e\}$ or $N = H = A_n$. To complete the proof, we must show that $N = \{e\}$ implies $[S_n : H] \geq n$.

We will show that $N = \{e\}$ and $r = [S_n : H] < n$ lead to a contradiction. First observe that every $\tau \in S_n$ permutes the f_i. The number of f_i's is r, so that they can be permuted in $r!$ ways. Yet the number of τ's is $n!$. Since $r < n$ implies $r! < n!$, there must be $\tau_1 \neq \tau_2$ in S_n that give the same permutation of the f_i. Thus

$$\tau_1 \cdot f_i = \tau_2 \cdot f_i \qquad \text{for all } i = 1, \ldots, r,$$

which easily implies that

$$\tau_2^{-1}\tau_1 \cdot f_i = f_i \qquad \text{for all } i = 1, \ldots, r.$$

Thus $\tau_2^{-1}\tau_1 \in N$, so that $N \neq \{e\}$, since $\tau_1 \neq \tau_2$. This contradicts $N = \{e\}$. You will prove part (b) in Exercise 11. $\square$

To see how this messes up Lagrange's strategy, suppose that $n \geq 5$ and that the resolvent $\theta(x)$ of $f \in L$ has degree > 1. Since $\deg(\theta) = [S_n : H(f)]$, part (a) of Theorem 12.1.10 tells us that either $\deg(\theta) \geq n$ or $H(f) = A_n$, in which case $f \in K(\sqrt{\Delta})$ by Theorem 7.4.4. Hence the only reasonable way to begin Lagrange's strategy is to pick $f = \sqrt{\Delta}$. But then continuing his strategy would entail finding a proper subgroup of A_n of index $< n$ (you will verify this in Exercise 12). Such subgroups don't exist (this is part (b) of Theorem 12.1.10), so that Lagrange's strategy fails for $n \geq 5$.

E. Lagrange Resolvents. To see where Lagrange resolvents come from, recall that the solution of the cubic used

$$z_1 = \tfrac{1}{3}(x_1 + \omega x_3 + \omega^2 x_2), \quad \omega = e^{2\pi i/3},$$

and the above solution of the quartic used

$$t_1 = x_1 + x_2 - x_3 - x_4 = x_1 + (-1)x_3 + (-1)^2 x_2 + (-1)^3 x_4.$$

Aside from the factor of $\tfrac{1}{3}$, both expressions involve the roots multiplied by roots of unity to increasing powers. Here's how Lagrange says this [Lagrange, p. 356]:

> As to equations that do not exceed the fourth degree, the simplest functions that yield their solution can be represented by the general formula
>
> $$x' + y x'' + y^2 x''' + \cdots + y^{\mu-1} x^{(\mu)},$$
>
> $x', x'', x''', \ldots, x^{(\mu)}$ being the roots of the proposed equation, which is assumed to have degree μ, and y being a root different from 1 of the equation
>
> $$y^\mu - 1 = 0$$

In Lemma 8.3.2 of Chapter 8, we used the name "Lagrange resolvent" for such expressions. There, we wanted to show that Galois extensions of prime degree p are obtained by adjoining a pth root when the smaller field contains a primitive pth root of unity. Our main tool was the Lagrange resolvent (8.7):

$$\alpha_i = \beta + \zeta^{-i} \sigma(\beta) + \zeta^{-2i} \sigma^2(\beta) + \cdots + \zeta^{-i(p-1)} \sigma^{p-1}(\beta).$$

You should reread the proof of Lemma 8.3.2, especially (8.7) and (8.8).

We can apply the formula for α_i to the extension $K \subset L$ as follows. Let $\sigma = (12 \ldots n) \in S_n \simeq \mathrm{Gal}(L/K)$, let $\beta = x_1$, and let ζ be an nth root of unity. Then replacing p with n in the above formula gives the Lagrange resolvent

(12.18)
$$\begin{aligned}
\alpha_i &= x_1 + \zeta^{-i} \sigma \cdot x_1 + \zeta^{-2i} \sigma^2 \cdot x_1 + \cdots + \zeta^{-i(n-1)} \sigma^{n-1} \cdot x_1 \\
&= x_1 + \zeta^{-i} x_2 + \zeta^{-2i} x_3 + \cdots + \zeta^{-i(n-1)} x_n.
\end{aligned}$$

This agrees with Lagrange's "general formula." One can prove the identity

$$\sigma \cdot \alpha_i = \zeta^i \alpha_i,$$

which easily implies that $\sigma = (12 \ldots n)$ fixes

$$\theta_i = \alpha_i^n = \left(x_1 + \zeta^{-i} x_2 + \zeta^{-2i} x_3 + \cdots + \zeta^{-i(n-1)} x_n\right)^n.$$

The proofs are identical to the arguments used in the proof of Lemma 8.3.2. One can also show that if ζ^{-i} is a primitive nth root of unity, then $(12 \ldots n)$ generates the isotropy subgroup of $\theta_i = \alpha_i^n$, that is, $H(\theta_i) = \langle (12 \ldots n) \rangle$ (see Exercise 13). It follows that the resolvent polynomial of θ_i has degree $n!/n = (n-1)!$. Lagrange states this as follows [Lagrange, pp. 332–333]:

one gets an equation in θ of degree $1.2.3\ldots(\mu-1)$, whose roots are the values of θ that come from the permutations of the $\mu-1$ roots $x'', x''', \ldots$ ignoring the root x'.

(As above, Lagrange uses μ instead of n for the degree.) In Exercise 14 you will work out how the final part of this statement relates to the proof of Theorem 12.1.4.

We conclude this section with an unexpected property of Lagrange resolvents. Let $n = p$ be prime and $\zeta_p = e^{2\pi i/p}$. Then (12.18) gives the Lagrange resolvent

$$(12.19) \qquad \alpha_i = x_1 + \zeta_p^i x_2 + \zeta_p^{2i} x_3 + \cdots + \zeta_p^{(p-1)i} x_p,$$

where $i = 0, 1, \ldots, p-1$. Ignoring $\alpha_0 = \sigma_1$, we set

$$\theta_i = \alpha_i^p = \left(x_1 + \zeta_p^i x_2 + \zeta_p^{2i} x_3 + \cdots + \zeta_n^{(p-1)i} x_p\right)^p, \quad 1 \le i \le p-1.$$

Lagrange forms the polynomial

$$(12.20) \quad (x - \theta_1)\cdots(x - \theta_{p-1}) = x^{p-1} - T x^{p-2} + U x^{p-3} - X x^{p-4} + \cdots,$$

so that $T, U, X, \ldots$ are the elementary symmetric polynomials of $\theta_1, \ldots, \theta_{p-1}$. Lagrange then asserts that "the coefficients $T, U, X, \ldots$ are each given by an equation of degree $1.2.3\ldots(\mu-2)$", that is, degree $(p-2)!$ [Lagrange, p. 333].

By claiming that these resolvents have degree $(p-2)!$, Lagrange is in effect saying that their isotropy subgroups have order $p!/(p-2)! = p(p-1)$. In fact, one can prove that the isotropy group of T is the subgroup $M_p \subset S_p$ consisting of the permutations

$$\begin{pmatrix} i \\ ai + b \end{pmatrix}, \quad 1 \le a \le p-1, \ 0 \le b \le p-1,$$

where everything is interpreted modulo p. In the Historical Notes to Section 6.4, we showed that M_p is isomorphic to the affine linear group $\mathrm{AGL}(1, \mathbb{F}_p)$. Furthermore, we will prove in Proposition 14.1.4 of Section 14.1 that M_p is a maximal solvable subgroup of S_p. So Lagrange essentially found a maximal solvable subgroup of S_p.

Lagrange concludes that his strategy fails in degree 5 and that

> if the algebraic resolution of equations of degrees greater than four is not impossible, then it must depend on some functions of the roots, different from the preceding.

(See [Lagrange, p. 357].) In spite of this failure, it is impressive to see how far Lagrange got. His 1770 treatise is one of the great works of algebra.

Historical Notes

In the sixteenth century del Ferro, Tartaglia, and Cardan solved the cubic, and Ferrari (a student of Cardan) solved the quartic. This was followed by mathematicians such as Viète, Hudde, Descartes, Tschirnhaus, Euler, and Bézout, who simplified and improved these solutions and found some entirely new solutions. Many of these methods are analyzed by Lagrange in his *Réflexions*.

The late eighteenth century was a time of active research on the roots of polynomials. Besides the work of Lagrange just discussed, we have Euler's solution of the quartic, which appeared in his 1770 algebra text [10, pp. 282–288] (a nice discussion of Euler's method can be found in [4, pp. 104–107]). Euler also found many examples of quintics that are solvable by radicals and gave an incomplete proof of the Fundamental Theorem of Algebra. In 1772 Lagrange used the methods of his *Réflexions* to fill most of the gaps in Euler's proof—though, as mentioned in the Historical Notes to Section 3.2, his argument is still incomplete.

Another important development in the early 1770s was Vandermonde's *Mémoire sur la résolution des équations*. This paper covers much of the same material as Lagrange's *Réflexions*, though Vandermonde's approach is different from Lagrange's. In particular, he considered permutations in more detail than Lagrange and understood how resolvents relate the action of permutations. He also used these methods to treat cubic and quartic equations and independently discovered Lagrange resolvents, though he didn't pursue the general theory to the same depth as Lagrange. One way in which he went significantly beyond Lagrange was his treatment of the equation $x^n - 1 = 0$. For example, Lagrange notes that $x^{11} - 1$ reduces to solving a quintic, but Vandermonde actually solved the resulting quintic by radicals. This may have been part of what inspired Gauss to investigate $x^p - 1 = 0$, p prime. His results are discussed in Section 9.2.

Lagrange hoped to solve equations by finding functions of the roots that gave a resolvent of small degree. In this section, we learned that

$$\begin{array}{ccc} \text{the degree of the} & \text{the number of distinct} & \text{the index of the} \\ \text{resolvent of } f \in L & = & \text{values } \sigma \cdot f, \ \sigma \in S_n & = & \text{isotropy subgroup of } f. \end{array}$$

In 1845, Cauchy studied "the problem of the number of values that can be assumed by functions," which by the last equality means the study of the index of subgroups of S_n. This was one of the important problems in the early history of group theory. The key result is Theorem 12.1.10, which we used to show the failure of Lagrange's strategy. In modern terminology, here are some highlights of how we got from Lagrange to Theorem 12.1.10:

- In 1799 Ruffini published a proof that the quintic is not solvable by radicals. His proof was hard to follow, but he did show that S_5 had no subgroups of index 3 or 4, which is part (a) of Theorem 12.1.10 for $n = 5$. He also proved the irreducibility of resolvent polynomials.

- In 1815 Cauchy generalized Ruffini's result by showing that the index of a subgroup $H \neq S_n$ is either 2 or at least the largest prime $\leq n$. Cauchy used the word "index" to denote the number of values of a function, which is where the modern term "index" comes from. The same paper also proved that A_n is generated by 3-cycles. Cauchy also emphasized the importance of the identity permutation and introduced the two-row notation for permutations.

- In 1824 Abel gave the first generally accepted proof that the general quintic is not solvable by radicals. He used Cauchy's (and Ruffini's) results on the index of subgroups of S_5.

- In 1832 Galois defined a normal subgroup of a group and asserted that a noncyclic simple group has order at least 60. Note that $|A_5| = 60$.

- In 1845 first Bertrand and then Cauchy proved part (a) of Theorem 12.1.10, though their proofs are quite different from the one given here. Cauchy also introduced the cycle notation now taught in introductory courses in abstract algebra.

- In 1869 Jordan defined the concept of simple group, and in 1870 he showed that A_n is simple for $n \geq 5$.

- In 1879 Kronecker proved Theorem 12.1.10 using the simplicity of A_n for $n \geq 5$. This is the proof used in the text.

Further results on the index of subgroups in S_n can be found in [15, pp. 138–139] or [16, p. 274, Note 120].

The Historical Notes to Section 8.5 mentioned the work of Ruffini and Abel on the unsolvability of the quintic. Now that we know Lagrange's *Réflexions*, we can get a better idea of what they did. Very roughly, both Ruffini and Abel tried to prove the unsolvability of the quintic by showing the nonexistence of the required resolvent polynomials, which in terms of group theory reduces to Theorem 12.1.10 for $n = 5$. But this alone is not enough, for Lagrange's theory only deals with rational functions of the roots. But suppose that there were formulas for the roots $x_1, \ldots, x_5$ that involved expressions like $\sqrt{x_1} + \cdots + \sqrt{x_5}$? Lagrange's methods would no longer apply. So the first thing Abel had to do was prove that if the quintic were solvable by radicals, then one could write the solution entirely within $L = F(x_1, \ldots, x_n)$, assuming that F had suitable roots of unity. This is discussed in [1]. See also [9] and [10] in the references to Chapter 8.

Then comes Galois, whose work in the early 1830s is the main topic of the next section. The important thing to say here is that Galois's analysis of solvability by radicals led to the concept of solvable group and gave a dramatically simpler approach to all of these questions. Namely, once one proves that S_n is not solvable for $n \geq 5$, then one *immediately* concludes that the general polynomial of degree $n \geq 5$ isn't solvable by radicals and that Lagrange's strategy for the quintic must fail. Results like Theorem 12.1.10 are simply not needed. This shows the power of good ideas. And the fact that Cauchy was still pursuing Lagrange's approach in 1845 shows how long it took to understand these ideas.

Further comments on eighteenth-century algebra can be found in Chapter 6 of [2]. The book [Tignol] has chapters on the work of Lagrange, Vandermonde, Ruffini, and Abel. A good description of Lagrange's *Réflexions* can be found in [11]. This paper also discusses the subsequent history of Galois theory.

Exercises for Section 12.1

Exercise 1. Let $\theta(x)$ be the resolvent polynomial defined in (12.3). Use the second bullet following (12.1) to show that $\theta(x) \in K[x]$.

Exercise 2. Work out the details of Example 12.1.2.

Exercise 3. This exercise concerns Examples 12.1.3 and 12.1.5.
(a) Compute the resolvent $\theta(y)$ of Example 12.1.3. This can be done using the methods of Section 2.3.
(b) Let $y_1 = x_1x_2 + x_3x_4$. Show that $H(y_1) = \langle (12), (1324) \rangle \subset S_4$.
(c) Show that $H(y_1)$ is not normal in S_4.
(d) Show that $H(y_1)$ is isomorphic to D_8, the dihedral group of order 8.

Exercise 4. Verify (12.9) and (12.10).

Exercise 5. This exercise will study the quadratic equations (12.11). Each quadratic has two roots, which together make up the four roots x_1, x_2, x_3, x_4 of our quartic.
(a) For the moment, forget all of the theory developed so far, and let y be some root of the Ferrari resolvent (12.10). Given only this, can we determine how y relates to the x_i? This is surprisingly easy to do. Suppose x_i, x_j are the roots of (12.11) for one choice of sign, and x_k, x_l are the roots for the other. Thus i, j, k, l are the numbers 1, 2, 3, 4 in some order. Prove that y is given by $y = x_ix_j + x_kx_l$.
(b) Now let $y_1 = x_1x_2 + x_3x_4$, and define the square root in (12.11) using (12.12). Show that the roots of (12.11) are x_1, x_2 for the plus sign and x_3, x_4 for the minus sign.
Historically, the Ferrari resolvent was just a tool for solving the quartic. Lagrange was the first to observe that the roots of (12.10) can be expressed in terms of the roots of the quartic. His argument [Lagrange, p. 262] is similar to what we did here.

Exercise 6. Explain why the polynomial (12.13) has coefficients in $K = F(\sigma_1, \sigma_2, \sigma_3, \sigma_4)$.

Exercise 7. Show that (12.15) implies the equations for x_1, x_2, x_3, x_4 given in the text.

Exercise 8. Let t_1, t_2, t_3 be defined as in (12.15).
(a) Lagrange noted that any transposition fixes exactly one of t_1, t_2, t_3 and interchanges the other two, possibly changing the sign of both. Prove this and use it to show that $t_1t_2t_3$ is fixed by all elements of S_4.
(b) Use the methods of Chapter 2 express $t_1t_2t_3$ in terms of the σ_i. The result should be the identity (12.16).

Exercise 9. Let H be a subgroup of S_n. In this exercise you will give two proofs that there is $f \in L$ such that $H = H(f)$.
(a) (First Proof.) The fixed field L_H gives an extension $K \subset L_H$. Explain why the Theorem of the Primitive Element applies to give $f \in L_H$ such that $L_H = K(f)$. Show that this f has the desired property.
(b) (Second Proof.) Let $m = x_1^{a_1} \cdots x_n^{a_n}$ be a monomial in $x_1, \ldots, x_n$ with distinct exponents $a_1, \ldots, a_n$. Then define

$$ f = \sum_{\sigma \in H} \sigma \cdot m = \sum_{\sigma \in H} x_{\sigma(1)}^{a_1} \cdots x_{\sigma(n)}^{a_n}. $$

Prove that $H(f) = H$.

Exercise 10. Prove that the subset $N \subset S_n$ defined in the proof of Theorem 12.1.10 is a subgroup of S_n.

Exercise 11. Let H be a proper subgroup of A_n with $n \geq 5$. Prove that $[A_n : H] \geq n$.

Exercise 12. The discussion following Theorem 12.1.10 shows that if we are going to use Lagrange's strategy when $n \geq 5$, then we need to begin with $f = \sqrt{\Delta}$, which has isotropy subgroup A_n. Suppose that $g \in L$ is our next choice, and let $\theta(x)$ be the resolvent of g. Since we regard $K(\sqrt{\Delta})$ as known, we may assume that $g \notin K(\sqrt{\Delta})$. The idea is to factor $\theta(x)$ over $K(\sqrt{\Delta})$, say $\theta = R_1 \cdots R_s$, where $R_i \in K(\sqrt{\Delta})[x]$ is irreducible. This is similar to how (12.13) factors the resolvent of t_1 over $K(y_1)$. Suppose that g enables us to continue Lagrange's inductive strategy. This means that some factor of θ, say R_j, has degree $< n$. Your goal is to prove that this implies the existence of a proper subgroup of A_n of index $< n$.
  (a) Prove that $\deg(R_j) \geq 2$.
  (b) Let g_j be a root of R_j and consider the fields

$$ K \subset K(\sqrt{\Delta}) \subset M = K(\sqrt{\Delta}, g_j) \subset L. $$

Let $H_j \subset S_n$ be the subgroup corresponding to $\mathrm{Gal}(L/M) \subset \mathrm{Gal}(L/K)$ under (12.1). Prove that $H_j \subset A_n$ and that $[A_n : H_j]$ is the degree of R_j.
  (c) Conclude that $\deg(R_j) < n$ implies that H_j is a proper subgroup of A_n of index $< n$.
With more work, one can show that $\deg(R_i) = [A_n : A_n \cap H(g)]$ for all i and that

$$ s = \frac{2}{[H(g) : A_n \cap H(g)]}. $$

It follows that $s = 1$ or 2.

Exercise 13. Let ζ be a primitive nth root of unity, and let $\alpha = (x_1 + \zeta x_2 + \cdots + \zeta^{n-1} x_n)^n$. Prove that $H(\alpha^n) = \langle (12 \ldots n) \rangle \subset S_n$.

Exercise 14. Let α_i be as in (12.18). The quotation given in the discussion following (12.18) can be paraphrased as saying that the roots of the resolvent of $\theta_i = \alpha_i^n$ come from the permutations of the $n - 1$ roots $x_2, \ldots, x_n$ that ignore the root x_1. What does this mean?
  (a) Show that each left coset of $\langle (12 \ldots n) \rangle$ in S_n can be written uniquely as $\sigma \langle (12 \ldots n) \rangle$, where σ fixes 1.
  (b) Explain how Lagrange's statement follows from part (a).
In general, we say that $g_1, \ldots, g_m \in G$ are *coset representatives* of a subgroup $H \subset G$ if $g_1 H, \ldots, g_m H$ are the distinct left cosets of G in H (so $m = [G : H]$). Thus Lagrange's quotation gives an explicit set of coset representatives for $\langle (12 \ldots n) \rangle \subset S_n$.

Exercise 15. Given the Lagrange resolvents $\alpha_1, \ldots, \alpha_{p-1}$ defined in (12.19), the goal of this exercise is to prove that

$$ x_i = \frac{1}{p} \Big(\sigma_1 + \sum_{j=1}^{p-1} \zeta_p^{-j(i-1)} \alpha_j \Big). $$

  (a) Write $\alpha_j = \sum_{l=1}^{p} \zeta_p^{j(l-1)} x_l$ for $1 \leq j \leq p$, so that $\alpha_p = \sigma_1$. Then show that

$$ \sum_{j=1}^{p} \zeta_p^{-j(i-1)} \alpha_j = \sum_{j,l=1}^{p} (\zeta_p^{l-i})^j x_l. $$

  (b) Given an integer m, use Exercise 9 of Section A.2 to prove that

$$ \sum_{j=1}^{p} (\zeta_p^m)^j = \begin{cases} p, & \text{if } m \equiv 0 \bmod p, \\ 0, & \text{otherwise.} \end{cases} $$

(c) Use parts (a) and (b) to prove the desired formula for x_i.

Exercise 16. Prove that Theorem 7.4.4 follows from Theorem 12.1.6 and Proposition 2.4.1.

Exercise 17. In Theorem 12.1.9, we used the Galois correspondence to show that f and g are similar if and only if $K(f) = K(g)$. Give another proof of this result that uses only Theorem 12.1.6.

Exercise 18. Consider the quartic polynomial $f = x^4 + 2x^2 - 4x + 2 \in \mathbb{Q}[x]$.
(a) Show that the Ferrari resolvent (12.10) is $y^3 - 2y^2 - 8y$.
(b) Using the root $y_1 = 0$ of the cubic of part (a), show that (12.11) becomes

$$x^2 = \pm\sqrt{-2}(x - 1)$$

and conclude that the four roots of f are

$$\frac{\sqrt{2}}{2}i \pm \frac{1}{2}\sqrt{-2 - 4i\sqrt{2}} \quad \text{and} \quad -\frac{\sqrt{2}}{2}i \pm \frac{1}{2}\sqrt{-2 + 4i\sqrt{2}}.$$

(c) Use Euler's solution (12.17) to find the roots of f. The formulas are surprisingly different. We will see in Chapter 13 that this quartic is especially simple. For most quartics, the formulas for the roots are much more complicated.

Exercise 19. This exercise will prove a version of Theorem 12.1.10 for a subgroup H of an arbitrary finite group G. When $G = S_n$, Theorem 12.1.10 used the action of S_n on L and chose $f \in L$ so that $H(f) = H$. In general, we use the action of G on the left cosets of H defined by $g \cdot hH = ghH$ for $g, h \in G$.
(a) Prove that $g \cdot hH = ghH$ is well defined, that is, $hH = h'H$ implies that $ghH = gh'H$.
(b) Prove that H is the isotropy subgroup of the identity coset eH.
(c) Let $m = [G : H]$, so that the left cosets of H can be labeled $g_1H, \ldots, g_mH$. Then, for $g \in G$, let $\sigma \in S_m$ be the permutation such that $g \cdot g_iH = g_{\sigma(i)}H$. Prove that the map $g \mapsto \sigma$ defines a group homomorphism $G \to S_m$.
(d) Let N be the kernel of the map of part (c). Thus N is a normal subgroup of G. Prove that $N \subset H$.
(e) Prove that $[G : N]$ divides $m!$.
(f) Explain why you have proved the following result: if H is a subgroup of a finite group G, then H contains a normal subgroup of G whose index divides $[G : H]!$.
(g) Use part (f) and Proposition 8.4.6 to give a quick proof of Theorem 12.1.10.

Exercise 20. Let G be a finite group and let p be the smallest prime dividing $|G|$. Prove that every subgroup of index p in G is normal.

Exercise 21. Part (a) of Theorem 12.1.10 implies that when $n \geq 5$, the index of a proper subgroup of S_n is either 2 or $\geq n$.
(a) Prove that S_n always has a subgroup H of index n. This means that equality can occur in the bound $[S_n : H] \geq n$.
(b) Give an example to show that Theorem 12.1.10 is false when $n = 4$.

12.2   GALOIS

In this section we will explore several aspects of Galois's work. Our discussion will be based on his 1831 memoir on Galois theory, entitled *Mémoire sur les conditions de résolubilité des équations par radicaux.* See [Galois, pp. 42–71] for the French original and [Edwards, pp. 101–113] for an English translation.

A. Beyond Lagrange. In Section 12.1 we saw that Lagrange studied the universal case where the roots are variables $x_1, \ldots, x_n$. In contrast, Galois created a theory that applies to arbitrary polynomials. To see the difference, recall the quotation from Galois given in the Historical Notes to Section 7.1 [Galois, p. 51]:

PROPOSITION I

THEOREM. For a given equation, let $a, b, c, \ldots$ be the m roots. There is always a group of permutations on the letters $a, b, c, \ldots$ that enjoys the following property:

1° that every function of the roots that is invariant** under the substitutions of the group, is rationally known;

2° conversely, that every function of the roots that is rationally determined, is invariant under these substitutions*.

As noted in Section 7.1, this asserts that in a Galois extension $F \subset L$, the field F is the fixed field of the Galois group $\mathrm{Gal}(L/F)$. For our purposes here, the most interesting part of the the proposition is the double asterisk **, which refers to the following marginal note in Galois's manuscript [Galois, p. 50]:

Here we call invariant not only a function whose form is invariant under the substitutions of the roots among themselves, but also those [functions] for which the numerical value does not vary under the substitutions.

In Section 12.1 we saw that Lagrange's concern was "only with the form" of expressions and not "with their numerical quantity." In this marginal note, we see that Galois is consciously going beyond Lagrange. We finally have a theory that applies to all polynomials, not just the universal one.

The single asterisk * above will be discussed in the Historical Notes.

B. Galois Resolvents. To understand the splitting field of a separable polynomial, Galois used a variation of Lagrange's notion of resolvent polynomial. Suppose that $f \in F[x]$ can be written $f = c(x - \alpha_1) \cdots (x - \alpha_n)$ in a splitting field L, where $\alpha_1, \ldots, \alpha_n$ are distinct. We also assume that F is infinite. Given $t_1, \ldots, t_n \in F$, consider the polynomial of degree $n!$ defined by

$$(12.21) \qquad s(y) = \prod_{\sigma \in S_n} \left(y - (t_1 \alpha_{\sigma(1)} + \cdots + t_n \alpha_{\sigma(n)}) \right).$$

The discussion following (5.4) in the proof of Proposition 5.2.1 in Section 5.2 shows that $s(y) \in F[y]$. You should reread this argument, which uses symmetric functions and is similar to Galois's.

In this situation, Galois asserts that since $\alpha_1, \ldots, \alpha_n$ are distinct, one can find $t_1, \ldots, t_n \in F$ so that the $n!$ elements

$$t_1 \alpha_{\sigma(1)} + \cdots + t_n \alpha_{\sigma(n)}, \qquad \sigma \in S_n,$$

are all distinct. In other words, $t_1, \ldots, t_n \in F$ may be chosen so that $s(y)$ is separable. When this happens, we call $s(y)$ a *Galois resolvent* of f. Exercises 1 and 2 will prove that such $t_1, \ldots, t_n$ exist. Galois uses the letter V to denote $t_1\alpha_1 + \cdots + t_n\alpha_n$. Following Lagrange, he refers to V as a "function of the roots."

Here is an example of a Galois resolvent.

Example 12.2.1. Let's compute a Galois resolvent of $f = (x^2 - 2)(x^2 - 3) = x^4 - 5x^2 + 6 \in \mathbb{Q}[x]$, which has roots $\sqrt{2}, -\sqrt{2}, \sqrt{3}$ and $-\sqrt{3}$. Let $(t_1, t_2, t_3, t_4) = (0, 1, 2, 4)$. In the notation used by Galois,

$$V = 0 \cdot \sqrt{2} + 1 \cdot (-\sqrt{2}) + 2 \cdot \sqrt{3} + 4 \cdot (-\sqrt{3}) = -\sqrt{2} - 2\sqrt{3}.$$

Using *Maple* or *Mathematica*, one can compute that (12.21) gives

$$\begin{aligned}
s(y) = {} & 731025000000 - 5765769000000\,y^2 + 13335274350000\,y^4 \\
& - 12343809230400\,y^6 + 5171341381036\,y^8 - 1110939359380\,y^{10} \\
& + 129730351909\,y^{12} - 8413645990\,y^{14} + 308394211\,y^{16} \\
& - 6392440\,y^{18} + 73339\,y^{20} - 430\,y^{22} + y^{24}.
\end{aligned}$$

A computer calculation also shows that $\gcd(s(y), s'(y)) = 1$, which implies that $s(y)$ is a Galois resolvent of f. Factoring $s(y)$ into irreducibles gives

$$\begin{aligned}
s(y) = {} & (900 - 132\,y^2 + y^4)(25 - 118\,y^2 + y^4)(361 - 70\,y^2 + y^4) \\
& (36 - 60\,y^2 + y^4)(100 - 28\,y^2 + y^4)(25 - 22\,y^2 + y^4).
\end{aligned}$$

Hence the Galois resolvent is reducible in this case.   ◁▷

As stated in [Galois, p. 49], the key property of V is the following:

> LEMMA III. If the function V is chosen as indicated in the preceding article, then it has the property that every root of the given equation [our f] can be expressed rationally as a function of V.

This lemma says the roots $\alpha_1, \ldots, \alpha_n$ lie in $F(V)$. It follows easily that

$$L = F(\alpha_1, \ldots, \alpha_n) = F(V),$$

since $V = t_1\alpha_1 + \cdots + t_n\alpha_n \in F(\alpha_1, \ldots, \alpha_n) = L$. Thus Lemma III implies that V is a primitive element of the splitting field of f over F.

Galois's proof of Lemma III is so terse that when Galois submitted his memoir to the French Academy in 1831, Poisson complained that the proof was insufficient but could be completed using Lagrange's methods [Galois, p. 50]. A discussion of Galois's proof is in [Edwards, §37] and [12]. In Exercise 3 you will use Lagrange's methods to prove Lemma III.

Let's compare Galois's Lemma III with Example 12.1.8, where we considered

$$\beta = t_1 x_1 + \cdots + t_n x_n$$

for $K = F(\sigma_1, \ldots, \sigma_n) \subset L = F(x_1, \ldots, x_n)$. There, we used Theorems 12.1.4 and 12.1.6 to show that $L = K(\beta)$ and that

$$\theta(y) = \prod_{\sigma \in S_n} \left(y - (t_1 x_{\sigma(1)} + \cdots + t_n x_{\sigma(n)}) \right)$$

is the resolvent of β. Thus $\theta(y)$ is irreducible over K by Proposition 12.1.1. So we have a primitive element β and an irreducible polynomial $\theta(y)$ of degree $n!$. In Galois's situation, we have V and $s(y)$, and although V is a primitive element (by Lemma III), Example 12.2.1 shows that $s(y)$ need not be irreducible. Thus, while some of Lagrange's results apply to arbitrary polynomials (such as the construction of primitive elements), others do not (such as the irreducibility of resolvents).

As we have defined things, V is a root of the Galois resolvent $s = s(y) \in F[y]$. Since s can be reducible over F, we let $h = h(y) \in F[y]$ be the minimal polynomial of V over F. Then h is an irreducible factor of s. We will let m denote the degree of h. Note that h is separable, since s is.

Galois makes the crucial observation that the roots of h interact with the roots of the original polynomial f as follows [Galois, pp. 49–51]:

> LEMMA IV. Suppose that one forms the equation of V [our s], and that one takes one of its irreducible factors, such that V is a root of an irreducible equation [our h]. Let $V, V', V'', \ldots$ be the roots of this irreducible equation. If $a = \phi(V)$ is one of the roots of the given equation [our f], then $\phi(V')$ will also be a root of the given equation.

(In the original, Galois wrote $a = f(V)$. We have changed f to ϕ because we use f for the given polynomial.)

Proof of Lemma IV. Since $L = F(V)$ contains the root a of f, we can write $a = \phi(V)$, where $\phi \in F[x]$. Also note that by normality, h splits completely over L. This shows that $V, V', V'', \cdots \in L$. In particular, L contains the roots V and V' of the irreducible polynomial h. By Proposition 5.1.8, we can find $\sigma \in \text{Gal}(L/F)$ such that $\sigma(V) = V'$. (This proposition played a crucial role in our development of Galois theory, especially in the proof of Theorem 6.2.1.) Then Lemma IV follows immediately from

$$0 = \sigma(0) = \sigma(f(a)) = \sigma(f(\phi(V))) = f(\phi(\sigma(V))) = f(\phi(V')),$$

where the fourth equality uses $f, \phi \in F[x]$ and $\sigma \in \text{Gal}(L/F)$. $\square$

Galois's argument is different from ours and doesn't mention automorphisms explicitly. But it should be clear that automorphisms and how they act on roots are implicit in the statement of Lemma IV. Galois's proof of Lemma IV is described in [Edwards, pp. 51–52].

C. Galois's Group. We next explore how Galois defined the Galois group. He considered only splitting fields of separable polynomials. We will show that in this situation, Galois's definition is equivalent to the one given in Section 6.1.

Consider the splitting field L of a separable polynomial $f \in F[x]$. As above, we assume that F is infinite. For us, the Galois group $\mathrm{Gal}(L/F)$ consists of automorphisms of L that are the identity on F. However, Proposition 6.3.1 shows that we can interpret $\mathrm{Gal}(L/F)$ in terms of permutations of the roots of f. Thus we can think of $\mathrm{Gal}(L/F)$ as consisting of all permutations of the roots that come from automorphisms. In other words, we consider only those permutations that preserve the field operations. Here is an example of what this means.

Example 12.2.2. Let $L = \mathbb{Q}(\pm\sqrt{2 \pm \sqrt{2}})$ be the splitting field of $f = x^4 - 4x + 2$ over $\mathbb{Q}$, and consider the permutation of the roots defined by

$$\sqrt{2 + \sqrt{2}} \mapsto \sqrt{2 - \sqrt{2}}, \qquad \sqrt{2 - \sqrt{2}} \mapsto \sqrt{2 + \sqrt{2}},$$
$$-\sqrt{2 + \sqrt{2}} \mapsto -\sqrt{2 + \sqrt{2}}, \qquad -\sqrt{2 - \sqrt{2}} \mapsto -\sqrt{2 - \sqrt{2}}.$$

This is not consistent with the field operations, since $\sqrt{2 + \sqrt{2}} \mapsto \sqrt{2 - \sqrt{2}}$ should imply that $-\sqrt{2 + \sqrt{2}} \mapsto -\sqrt{2 - \sqrt{2}}$. Hence this permutation doesn't come from an automorphism in $\mathrm{Gal}(L/\mathbb{Q})$. ◁▷

Galois did not use the notion of field automorphism. So how did he decide which permutations to use? His approach is based on the primitive element V and minimal polynomial h constructed above. We will use the following notation. Let

$$V, V', V'', \dots, V^{(m-1)}$$

denote the roots of h. Furthermore, since $L = F(V)$ by Lemma III, the roots $\alpha_1, \dots, \alpha_n$ of f can be written

$$(12.22) \qquad \varphi(V), \ \varphi_1(V), \ \varphi_2(V), \dots, \varphi_{n-1}(V),$$

where $\varphi, \varphi_1, \varphi_2, \dots, \varphi_{n-1}$ have coefficients in F. Then Galois describes his group as follows [Galois, p. 53]:

> No matter what the given equation [our f] is, one can find a rational function V of the roots such that all of the roots are rational functions of V. Given this V, let us consider the irreducible equation of which V is a root (lemmas III and IV) [our h]. Let $V, V', V'', \dots, V^{(m-1)}$ be the roots of this equation.
>
> Let $\varphi V, \varphi_1 V, \varphi_2 V, \dots, \varphi_{n-1} V$ be roots of the proposed equation.
>
> Write down the following m permutations of the roots:

$(V),$	$\varphi V,$	$\varphi_1 V,$	$\varphi_2 V,$	$\dots,$	$\varphi_{n-1} V,$
$(V'),$	$\varphi V',$	$\varphi_1 V',$	$\varphi_2 V',$	$\dots,$	$\dots,$
$(V''),$	$\varphi V'',$	$\varphi_1 V'',$	$\varphi_2 V'',$	$\dots,$	$\dots,$
$\dots,$	$\dots,$	$\dots,$	$\dots,$	$\dots,$	$\dots,$
$(V^{(m-1)}),$	$\varphi V^{(m-1)},$	$\varphi_1 V^{(m-1)},$	$\varphi_2 V^{(m-1)},$	$\dots,$	$\varphi_{n-1} V^{(m-1)}$

> I say that this group of permutations has the desired property.

(In the original, n and m are interchanged. We have switched them in order to be consistent with the notation used here.) In this table, the first entry of a row is a label for the row, and the remaining n entries of the row are roots of f by Lemma IV.

One complication is that for Galois, the word "permutation" has a different meaning than it has for us. We will discuss this in the Mathematical Notes below. For now, we will understand the above quote as saying that Galois's "group" consists of the m permutations obtained by mapping $\varphi(V)$, $\varphi_1(V)$, $\varphi_2(V), \ldots, \varphi_{n-1}(V)$ to the m rows displayed in the quote. These permutations are related to the Galois group $\mathrm{Gal}(L/F)$ as follows.

Theorem 12.2.3. *Let L be the splitting field of the separable polynomial f in $F[x]$ and let the roots of f be denoted as in* (12.22). *Proposition 6.3.1 gives a one-to-one group homomorphism* $\mathrm{Gal}(L/F) \to S_n$. *Then the image of this map consists of the m permutations described by Galois.*

Proof. As above, V is a primitive element of $F \subset L$, and h is the minimal polynomial of V over F. The m roots of h will be denoted $V^{(0)} = V$, $V^{(1)} = V'$, $V^{(2)} = V''$,..., $V^{(m-1)}$. Note also that h is separable. Then the proof of Theorem 6.2.1 implies that

$$\mathrm{Gal}(L/F) = \{\sigma_1, \ldots, \sigma_m\},$$

where σ_i is the automorphism of L that takes the primitive element V to the root $V^{(i-1)}$ of h. As in the proof of Lemma IV, it follows that

(12.23) $$\sigma_i(\psi(V)) = \psi(\sigma_i(V)) = \psi(V^{(i-1)})$$

for any polynomial ψ with coefficients in F.

In the homomorphism $\mathrm{Gal}(L/F) \to S_n$ from Proposition 6.3.1, σ_i maps to the permutation that takes $\varphi(V)$, $\varphi_1(V)$, $\varphi_2(V), \ldots, \varphi_{n-1}(V)$ to

$$\sigma_i(\varphi(V)), \ \sigma_i(\varphi_1(V)), \ \sigma_i(\varphi_2(V)), \ldots, \sigma_i(\varphi_{n-1}(V)).$$

Using (12.23), this can be rewritten as

(12.24) $$\varphi(V^{(i-1)}), \ \varphi_1(V^{(i-1)}), \ \varphi_2(V^{(i-1)}), \ldots, \varphi_{n-1}(V^{(i-1)}).$$

One easily sees that (12.24) is the ith row displayed in the above quote since V, V', V'', ..., $V^{(m-1)}$ are now $V^{(0)}$, $V^{(1)}$, $V^{(2)}$, ..., $V^{(m-1)}$. Hence the images of the σ_i are the m permutations described by Galois. $\qquad\square$

This theorem shows that for the splitting field of a separable polynomial, the definition used by Galois is equivalent to Definition 6.1.1. The Historical Notes to Section 6.1 give a brief description of how we got from Galois's group to the modern Galois group $\mathrm{Gal}(L/F)$. A more detailed explanation appears in [11].

D. Natural and Accessory Irrationalities. Before explaining Galois's strategy for solving equations, we need to discuss some classical terminology. Let $F \subset L$ be the splitting field of a separable polynomial $f \in F[x]$. Then adjoin a quantity β

to F, where β is a root of an auxiliary equation that we assume to be known. For example, β could be a radical or a root of a resolvent equation. If $\beta \notin F$, then we call β a *natural irrationality* when $\beta \in L$ and an *accessory irrationality* when $\beta \notin L$.

Example 12.2.4. Let $f \in F[x]$ be solvable by radicals with splitting field $F \subset L$.

- If $F \subset L$ is radical in the sense of Section 8.2, then we can obtain the roots of f by adjoining natural irrationalities.
- If $F \subset L$ is solvable but not radical, then at least one of the radicals adjoined must be an accessory irrationality. ◁▷

The quantity $\beta \notin F$ in the above discussion gives an extension $F \subset K = F(\beta)$. Then it is easy to see that

$$\beta \text{ is a natural irrationality} \iff K \subset L,$$
$$\beta \text{ is an accessory irrationality} \iff K \not\subset L.$$

When $K \not\subset L$, we can assume that K and L are both contained in some larger field (see Exercise 4). Then we get a diagram

(12.25)

$$
\begin{array}{ccc}
 & KL & \\
\nearrow & & \nwarrow \\
K & & L \\
\nwarrow & & \nearrow \\
 & F &
\end{array}
$$

where KL is the compositum of K and L as in Definition 8.2.5. The relation between $\mathrm{Gal}(KL/K)$ and $\mathrm{Gal}(L/F)$ is described by the following result.

Theorem 12.2.5. *Suppose that we have a diagram* (12.25) *where $F \subset L$ is a Galois extension and $F \subset K$ is finite. Then $K \subset KL$ is a Galois extension and the restriction map $\sigma \mapsto \sigma|_L$ defines an isomorphism*

$$\mathrm{Gal}(KL/K) \simeq \mathrm{Gal}(L/K \cap L) \subset \mathrm{Gal}(L/F).$$

Proof. In Exercise 5 you will show that $K \subset KL$ is Galois whenever $F \subset L$ is. Now pick $\sigma \in \mathrm{Gal}(KL/K)$ and $\alpha \in L$. We claim that $\sigma(\alpha) \in L$. To prove this, let g be the minimal polynomial of α over F. Since σ is the identity on K and $F \subset K$, we see that $\sigma(\alpha)$ is a root of g. But g splits completely over L, since $F \subset L$ is Galois (be sure you can supply the details). This shows that $\sigma(\alpha) \in L$. Hence the restriction of σ to L gives a ring homomorphism

$$\sigma|_L : L \longrightarrow L.$$

Since $\sigma^{-1}|_L$ is the inverse of $\sigma|_L$ (see Exercise 6), $\sigma|_L$ is an automorphism of L. Furthermore, we noted above that σ is the identity on F, which implies that the same is true for $\sigma|_L$. It follows that $\sigma|_L \in \mathrm{Gal}(L/F)$, so that we have a map

(12.26) $$\mathrm{Gal}(KL/K) \longrightarrow \mathrm{Gal}(L/F).$$

You will show that this is a group homomorphism in Exercise 6.

To see that (12.26) is one to one, suppose that $\sigma \in \mathrm{Gal}(KL/K)$ and $\sigma|_L$ is the identity on L. Then σ is the identity on both K and L, which easily implies that σ is the identity on KL (see Exercise 6). Thus (12.26) is one to one.

Finally, we need to show that image is $\mathrm{Gal}(L/K \cap L)$. For this purpose, let $H \subset \mathrm{Gal}(L/F)$ be the image of (12.26). By the Galois correspondence, it suffices to show that $L_H = K \cap L$. First suppose $\alpha \in L_H$. Then α is fixed by all $\sigma|_L \in H$, which means that α is fixed by all $\sigma \in \mathrm{Gal}(KL/K)$ and hence is in the fixed field of $\mathrm{Gal}(KL/K)$. Thus $\alpha \in K$ since $K \subset KL$ is Galois, and $\alpha \in K \cap L$ follows from $\alpha \in L_H \subset L$. This proves that $L_H \subset K \cap L$.

For the other inclusion, let $\alpha \in K \cap L$. Then in particular, α is in K and hence is fixed by all $\sigma \in \mathrm{Gal}(KL/K)$. Thus $\sigma|_L(\alpha) = \alpha$ for all $\sigma|_L \in H$, which implies $\alpha \in L_H$, as claimed. This completes the proof. $\qquad\square$

Theorem 12.2.5 is sometimes called the *Theorem on Natural Irrationalities*. To see why, suppose that $K \not\subset L$, that is, K is obtained from F by adjoining accessory irrationalities. Then the isomorphism

$$\mathrm{Gal}(KL/K) \simeq \mathrm{Gal}(L/K \cap L)$$

of Theorem 12.2.5 means that $K \subset KL$ and $K \cap L \subset L$ have the same Galois group. But $K \cap L$ lies inside L and hence is obtained from F by adjoining natural irrationalities. Thus, from the point of view of Galois theory, Theorem 12.2.5 implies that we don't need accessory irrationalities.

E. Galois's Strategy. In Section 12.1 we saw that Lagrange formulated his strategy for solving equations in terms of resolvents. However, there are groups lurking in the background. For instance, our discussion of the quartic used

$$y_1 = x_1 x_2 + x_3 x_4 \quad \text{and} \quad t_1 = x_1 + x_2 - x_3 - x_4,$$

whose isotropy subgroups are

$$H(y_1) = \langle (12), (1324) \rangle \supset H(t_1) = \langle (12), (34) \rangle.$$

So the idea of reducing to smaller groups is implicit in what Lagrange was doing.

Getting smaller groups is the main goal of Galois's strategy. In the Historical Notes to Section 8.3, we gave the following quote where Galois discusses his approach to solvability by radicals [Galois, pp. 57–59]:

> I first observe that to solve an equation, it is necessary to reduce its group until it contains only a single permutation …

> Given this, we will try to find the condition satisfied by the group of an equation for which it is possible to reduce the group [to a single permutation] by adjunction of radical quantities …

In the first sentence, Galois states the goal of reducing the Galois group to the identity, and in the second, he says that in the case of solvability by radicals, the goal is to reduce the Galois group by adjoining radicals.

In its most general form, Galois's strategy is to successively adjoin known quantities (radicals or roots of resolvents) in order to reduce the Galois group to the identity. This adjunction process gives an extension $F \subset K$ that we regard as known. The splitting field of f over K is easily seen to be $K \subset KL$, which is one of the extensions in the diagram (12.25). By Theorem 12.2.5, we have

$$(12.27) \qquad \text{Gal}(KL/K) \simeq \text{a subgroup of Gal}(L/F).$$

Thus going from F to K gives a subgroup of the original Galois group. Furthermore, if the new Galois group is the identity, then $KL = K$, which implies that $L \subset K$. Since K is known, it follows that the roots of f are also known.

Here is an example of how this works.

Example 12.2.6. One easily checks that $f = x^3 + 9x - 2 \in \mathbb{Q}[x]$ is irreducible over $\mathbb{Q}$ with real root given by

$$\alpha_1 = \sqrt[3]{1 + 2\sqrt{7}} + \sqrt[3]{1 - 2\sqrt{7}}.$$

The other roots of f are a complex conjugate pair $\alpha_2 = \overline{\alpha_3}$, since $\Delta(f) = -3024$ is negative. If L is the splitting field of f over $\mathbb{Q}$, then Proposition 7.4.2 implies that $\text{Gal}(L/\mathbb{Q}) \simeq S_3$, since $\sqrt{\Delta(f)} \notin \mathbb{Q}$.

To make the Galois group smaller, we adjoin $\beta = \sqrt[3]{1 + 2\sqrt{7}}$ to $\mathbb{Q}$, which gives $K = \mathbb{Q}(\beta)$. In Exercise 7 you will show that $\sqrt[3]{1 - 2\sqrt{7}}$ and hence α_1 lie in K. This means that f factors as $(x - \alpha_1)g$, where $g \in K[x]$ has roots α_2, α_3. Thus KL is obtained from K by adjoining α_2, α_3. Since $K \subset \mathbb{R}$ and α_2, α_3 are not real, it follows that $[KL : K] = 2$. This shows that $\text{Gal}(KL/K) \simeq \mathbb{Z}/2\mathbb{Z}$.

Furthermore, if we think of KL as the splitting field of f over K, then we still have the map $\text{Gal}(KL/K) \to S_3$. Given how we've labeled the roots, the image of this map is clearly $\langle (23) \rangle \subset S_3$. So adjoining β reduces the group of permutations from S_3 to the smaller group $\langle (23) \rangle$.

In Exercise 7 you will show that if we adjoin $\omega = e^{2\pi i/3}$ to K, then $K' = K(\omega)$ contains all roots of f, so that $K'L = K'$. Hence the Galois group has been reduced to $\text{Gal}(K'L/K') = \{e\}$, which completes Galois's strategy.   ◁▷

To fully understand Galois's strategy, we need to think in terms of permutations. If our separable polynomial f has degree n, then the action of the Galois group on the roots gives a map

$$(12.28) \qquad \text{Gal}(L/F) \longrightarrow S_n$$

whose image is Galois's group by Theorem 12.2.3. Now let $F \subset K$ be a finite extension. Since $K \subset KL$ is the splitting field of f over K, we get a similar map

$$(12.29) \qquad \text{Gal}(KL/K) \longrightarrow S_n.$$

In Exercise 8 you will show that these maps are compatible with the isomorphism given in (12.27). Hence, when we regard $\text{Gal}(KL/K)$ and $\text{Gal}(L/F)$ as subgroups of S_n, the former is contained in the latter.

This makes Galois's strategy easy to understand. He works with a fixed separable polynomial f of degree n. For him, the group of f lies in S_n, but the field he works over keeps changing. Furthermore, each time he enlarges the field by adjoining something known (a radical or a root of a resolvent), he passes from the group to a subgroup (which may be the whole group). This leads to extensions

$$F \subset K_1 \subset K_2 \subset \cdots$$

where each K_i is regarded as consisting of things that are known. If at some point, say for K_m, the group reduces to the identity, then $\text{Gal}(K_m L / K_m) = \{e\}$, which as noted above implies that $L \subset K_m$. This allows us to express the roots of f in terms of known quantities.

Historical Notes

When reading Galois, one must keep in mind the distinction between *arrangements* of roots and *permutations* of roots. If you look back at the quotation giving Galois's definition of his group, you will see that he lists m arrangements of the roots. The corresponding permutations come from mapping the first arrangement to the others. To complicate matters, Galois uses different terminology from us:

Us	Galois
Arrangement	Permutation
Permutation	Substitution

So when Galois says "group of permutations," he really means "group of arrangements." But later in the memoir, we find the following [Galois, pp. 53–55]:

> It is evident that in the group of permutations considered here, the order of the letters is not of importance, but rather only the SUBSTITUTIONS of the letters by which one passes from one permutation to another.

By Theorem 12.2.3, these substitutions form a subgroup of S_n isomorphic to the Galois group.

Galois knew the difference between arrangements and substitutions, and was aware that the latter formed a group in the modern sense [Galois, p. 47]:

> ... if one has substitutions S and T within such a group, one is sure to have the substitution ST.

From the modern point of view, substitutions are more important. But this was not clear to Galois, especially given the vivid visual image provided by groups of arrangements. This is evident from Galois's definition of his group, and other examples can be found in [12]. Galois's memoir is written in terms of arrangements, although changes made shortly before his death in 1832 indicate that Galois was thinking about switching to substitutions. For example, we quoted Galois's Proposition I at the beginning of the section. This quotation includes an asterisk * that refers to a marginal note where Galois says "Put everywhere in place of the word permutation the word substitution" [Galois, p. 50]. But then Galois crosses this out!

It took a while for the mathematical community to understand Galois's ideas. In 1866 the third edition of Serret's *Cours d'algèbre supérieure* included a partial account of Galois theory. As quoted in [11, p. 110], Serret comments that "Galois used the notion of groups of permutations [our arrangements] ..., but it seems better for us to keep to substitutions." This quote also shows that "substitution" was the common name for elements of S_n in the nineteenth century. Another example of this is Jordan's 1870 text *Traité des substitutions et des équations algébriques* [Jordan1], which gave the first complete account of Galois theory.

Our discussion of Galois's strategy did not state his version of Theorem 12.2.5 (the Theorem on Natural Irrationalities). The reason is that one needs to understand the distinction between arrangements and permutations before reading Galois's version, which goes as follows [Galois, p. 55]:

PROPOSITION II

>   THEOREM. If one adjoins to a given equation [our f] the root r of an auxiliary irreducible equation *,
>
> $1°$ one of two things will occur: either the group of the equation will not change; or it will be partitioned into p groups each belonging to the original equation when one adjoins each of the roots of the auxiliary equation;
>
> $2°$ these groups have the remarkable property, that one passes from one to another by applying to all of the permutations of the first the same substitution of the letters.

The asterisk * indicates that r was a root of an auxiliary equation "of prime degree p" in an earlier version of Proposition II [Galois, p. 54]. This is the "p" that appears in $1°$. In Exercise 9 you will use the Galois correspondence and Theorem 12.2.5 to show that if $[K : F] = p$, then $\text{Gal}(KL/K)$ is isomorphic either to $\text{Gal}(L/F)$ or to a subgroup of index p in $\text{Gal}(L/F)$. The latter corresponds to "partitioned into p groups" in the above quotation.

It appears that Galois first proved Proposition II in the prime-degree case. The night before his fatal duel, he realized that his proof applied in greater generality. Writing in haste, he changed part but not all of the statement of Proposition II. He also knew that his proof was incomplete—this is where he writes "Je n'ai pas le temps" ("I don't have time") [Galois, p. 54].

This explains Proposition II up to the appearance of p. But what about the remainder of $1°$? The idea is that instead of adjoining one root r of the auxiliary equation, one could adjoin a different root r' of the same equation. This gives a different extension $F \subset K' = F(r')$. Then going from F to K' will reduce the group, but possibly in a different way. In modern terms, $F \subset K$ and $F \subset K'$ are conjugate extensions. By Theorem 12.2.5, $\text{Gal}(KL/K)$ and $\text{Gal}(K'L/K')$ are isomorphic to subgroups of $\text{Gal}(L/F)$. Then Galois's observation in $1°$ is that these subgroups are conjugate in $\text{Gal}(L/F)$. You will prove this in Exercise 10.

The precise meaning of $2°$ of Proposition II will be explored in Exercises 11 and 12. Galois's proof of Proposition II can be found in [5] and [12].

This concludes our discussion of Galois. However, to fully appreciate what Galois did, the reader should keep in mind Galois's other contributions to Galois theory, many of which were discussed earlier in the book:

- Extension fields (Historical Notes to Section 4.1).
- The Galois correspondence (Historical Notes to Section 7.1).
- Normal subgroups (Historical Notes to Section 7.2).
- Solvable groups and solvability by radicals (Historical Notes to Section 8.3).
- Finite fields (Historical Notes to Section 11.1).

This is an impressive list for someone who died at age 20. There is also Galois's amazing work on irreducible polynomials of degree p and p^2, where p prime. This will be described in Chapter 14.

For a fuller account of Galois's mathematical work, the reader should consult [5], [11, pp. 80–84], [12], [Edwards], or [Tignol, Ch. 14]. The recent biography of Galois [13] describes his short but intense life.

Exercises for Section 12.2

Exercise 1. Let F be an infinite field and let V be a finite-dimensional vector space over F. The goal of this exercise is to prove that V cannot be the union of a finite number of proper subspaces. This will be used in Exercise 2 to prove the existence of Galois resolvents.

Suppose that $W_1, \ldots, W_m$ are proper subspaces of V. We will prove by complete induction on $\dim(V)$ that $V \neq W_1 \cup \cdots \cup W_m$. The case $\dim(V) = 0$ is obvious since $V = \{0\}$ has no proper subspaces. Now assume that $\dim(V) > 0$ and that the assertion is true for all vector spaces of dimension $< \dim(V)$. Your job is to prove that $V = W_1 \cup \cdots \cup W_m$ leads to a contradiction.

(a) Suppose that $W_1 \subset W_i$ for some $i > 1$. Prove that $V = W_2 \cup \cdots \cup W_m$, and explain why this allows us to assume that $W_1 \not\subset W_i$ for all $i > 1$.
(b) Show that $W_1 \not\subset W_i$ for all $i > 1$ implies that $W_1 \not\subset W_2 \cup \cdots \cup W_m$.
(c) There is $v \in W_1 \setminus (W_2 \cup \cdots \cup W_m)$ by part (b), and there is $w \in V \setminus W_1$, since W_1 is a proper subspace. Then $\lambda v + w \in V = W_1 \cup \cdots \cup W_m$ for all $\lambda \in F$. Explain why this implies that there are $\lambda_1 \neq \lambda_2$ in F such that $\lambda_1 v + w, \lambda_2 v + w \in W_i$ for some i.
(d) Now derive the desired contradiction.

Exercise 2. Suppose that we have an extension $F \subset L$, where F is infinite. The goal of this exercise is to show that if $\alpha_1, \ldots, \alpha_n \in L$ are distinct, then $t_1, \ldots, t_n \in F$ can be chosen so that the polynomial $s(y)$ defined in (12.21) has distinct roots. Given $\sigma \neq \tau$ in S_n, let

$$W_{\sigma,\tau} = \left\{ (t_1, \ldots, t_n) \in F^n \mid \textstyle\sum_{i=1}^n (\alpha_{\sigma(i)} - \alpha_{\tau(i)})t_i = 0 \text{ in } L \right\}.$$

(a) Prove that $W_{\sigma,\tau}$ is a subspace of F^n and that $W_{\sigma,\tau} \neq F^n$.
(b) Show that parts (a) and (b) and Exercise 1 imply that there are $t_1, \ldots, t_n \in F$ such that the polynomial $s(y)$ from (12.21) has distinct roots.

Exercise 3. This exercise will prove Galois's Lemma III using the methods of Lagrange. Let $V = t_1 \alpha_1 + \cdots + t_n \alpha_n$, where $t_1, \ldots, t_n$ are chosen so that the Galois resolvent $s(y)$ from (12.21) is separable. Also let $V_\sigma = t_1 \alpha_{\sigma(1)} + \cdots + t_n \alpha_{\sigma(n)}$ for $\sigma \in S_n$. Prove that each α_j can be written as a rational function in V with coefficients in F by adapting the second proof of Theorem 12.1.6.

Exercise 4. In the discussion preceding (12.25), we have extensions $F \subset L$, which is a splitting field of $f \in F[x]$, and $F \subset K = F(\beta)$, where β is a root of an irreducible polynomial in $F[x]$. Given the many ways in which extension fields can be constructed, these extensions might not have much to do with each other. Prove that there is an extension $F \subset M$ that contains subfields $F \subset L_1 \subset M$ and $F \subset K_1 \subset M$ such that L_1, K_1 are isomorphic to L, K respectively, where the isomorphisms are the identity on F. Thus, by replacing L, K with the isomorphic fields L_1, K_1, we can assume that L, K lie in a larger field, as claimed in the text.

Exercise 5. Suppose that $F \subset L$ is the splitting field of a separable polynomial $f \in F[x]$. Also suppose that we have another finite extension $F \subset K$ such that the compositum KL is defined. Prove that $K \subset KL$ is the splitting field of f over K.

Exercise 6. This exercise will complete the proof of Theorem 12.2.5. Given $\sigma \in \text{Gal}(KL/K)$, we showed in the text that $\sigma|_L$ maps L to L.
(a) Show that $(\sigma\tau)|_L = \sigma|_L \tau|_L$.
(b) Use part (a) to show that $\sigma^{-1}|_L$ is the inverse function of $\sigma|_L$.
(c) Use part (a) to show that (12.26) is a group homomorphism.
(d) Let σ be an automorphism of KL that is the identity on both K and L. Prove that σ is the identity on KL.

Exercise 7. This exercise is concerned with the details of Example 12.2.6. As in the example, let L be the splitting field of $f = x^3 + 9x - 2$ over $\mathbb{Q}$ and set $K = \mathbb{Q}(\beta)$, where $\beta = \sqrt[3]{1 + 2\sqrt{7}}$.
(a) Show that $\sqrt[3]{1 - 2\sqrt{7}} \in K$.
(b) Show that $K' = K(\omega)$, $\omega = e^{2\pi i/3}$, contains all roots of f.

Exercise 8. In Theorem 12.2.5, we have the map (12.26) defined by $\sigma \mapsto \sigma|_L$. However, if $F \subset L$ is the splitting field of a separable polynomial $f \in F[x]$ of degree n, then we also have maps (12.28) and (12.29). Prove that these maps are compatible, that is, that $\sigma \in \text{Gal}(KL/K)$ and $\sigma|_L \in \text{Gal}(L/F)$ map to the same element of S_n under (12.28) and (12.29).

Exercise 9. In the situation of Theorem 12.2.5, suppose that $F \subset K$ is an extension of prime degree p. Prove that $\text{Gal}(KL/K)$ is isomorphic to either $\text{Gal}(L/F)$ or a subgroup of index p in $\text{Gal}(L/F)$.

Exercise 10. Suppose that we have a diagram (12.25) as in Theorem 12.2.5. Also assume that $K = F(\beta)$, and let $K' = F(\beta')$, where β' and β have the same minimal polynomial over F. You will show that $\text{Gal}(KL/K)$ and $\text{Gal}(K'L/K')$ give conjugate subgroups of $\text{Gal}(L/F)$. This is the modern version of what Galois says in 1° of Proposition II.
(a) Let $F \subset M'$ be the Galois closure of the extension $F \subset M$ be the field constructed in Exercise 4. Explain why we can regard L, K, and K' as subfields of M'.
(b) Explain why we can find $\tau \in \text{Gal}(M'/F)$ such that $\tau(K) = K'$.
(c) Show that $\tau|_L \in \text{Gal}(L/F)$ maps $K \cap L$ to $K' \cap L$. Thus $K \cap L$ and $K' \cap L$ are conjugate subfields of L.
(d) Use Lemma 7.2.4 to show that in Theorem 12.2.5, $\text{Gal}(KL/K)$ and $\text{Gal}(K'L/K')$ map to conjugate subgroups of $\text{Gal}(L/F)$.

Exercise 11. Let A denote the set of arrangements described by Galois. This is Galois's "group." For simplicity, we write the first arrangement on Galois's list as $\alpha_1 \cdots \alpha_n$. Then let G be the set of permutations that take the first element of A to the others. Theorem 12.2.3 implies that G is a subgroup of S_n isomorphic to $\text{Gal}(L/F)$.

We also have the action of S_n on the set of all $n!$ arrangements of roots by

$$\sigma \cdot \alpha_{i_1} \cdots \alpha_{i_n} = \alpha_{\sigma(i_1)} \cdots \alpha_{\sigma(i_n)}.$$

This induces an action of G on the set of arrangements.
  (a) Explain why A is the orbit of $\alpha_1 \cdots \alpha_n$ under the G action.
  (b) Show that the map $G \to A$ defined by $\sigma \mapsto \sigma \cdot \alpha_1 \cdots \alpha_n$ is one to one and onto.

Exercise 12. In the situation of Theorem 12.2.5, let $G \subset S_n$ correspond to $\mathrm{Gal}(L/F)$, and $H \subset S_n$ correspond to $\mathrm{Gal}(KL/K)$. By Exercise 8, we know that $H \subset G$. Also let A be the set of arrangements studied in Exercise 11. Then a left coset $\sigma H \subset G$ gives a subset $\sigma H \cdot \alpha_1 \cdots \alpha_n \subset A$, and since the map $\sigma \mapsto \sigma \cdot \alpha_1 \cdots \alpha_n$ is one to one and onto, the sets $\sigma H \cdot \alpha_1 \cdots \alpha_n$ partition A into disjoint subsets. We claim that these are the "groups" that appear in 1° and 2° of Galois's Proposition II.
  (a) Given any two such "groups" $\sigma H \cdot \alpha_1 \cdots \alpha_n$ and $\tau H \cdot \alpha_1 \cdots \alpha_n$, prove that there is $\gamma \in G$ such that (as Galois says in 2°) one passes from one to the other by applying γ to all arrangements in the first.
  (b) So far, it seems like Galois is describing cosets. However, as pointed out in [12], Galois thought of these "groups" differently. This is seen by explaining how they relate to 1° of Galois's proposition. Let M' be the field used in Exercise 10, and let $\tau \in \mathrm{Gal}(M'/F)$. Then $K' = \tau(K)$ is a conjugate of K. Let $\sigma \in G$ be the permutation corresponding to $\tau|_L \in \mathrm{Gal}(L/F)$. Show that $\sigma H \sigma^{-1}$ is the subgroup of S_n corresponding to $\mathrm{Gal}(K'L/K')$.
  (c) Using the setup of part (b), consider the "group" $\sigma H \cdot \alpha_1 \cdots \alpha_n \subset A$. Prove that $\sigma H \sigma^{-1} \subset S_n$ is the set of all permutations of S_n that map the first element of this "group," namely $\sigma \cdot \alpha_1 \cdots \alpha_n$, to another element of the "group." (Remember that this is the process for turning a "group" of arrangements into a subgroup of S_n.)
Combining parts (b) and (c), we see that what Galois says in 1° of Proposition II is fully consistent with what we did in Exercise 10.

Exercise 13. This exercise will show that not all choices of the t_i in (12.21) give Galois resolvents. As in Example 12.2.1, $f = (x^2 - 2)(x^2 - 3)$ has roots $\sqrt{2}, -\sqrt{2}, \sqrt{3}$ and $-\sqrt{3}$. This time we will use $(t_1, t_2, t_3, t_4) = (0, 1, 2, 3)$. Show that (12.21) gives the polynomial

$$\begin{aligned}
s(y) = {} & 1679616 - 45722880\, y^2 + 445417056\, y^4 - 1935550800\, y^6 \\
& + 4169468065\, y^8 - 4504515400\, y^{10} + 2268233020\, y^{12} - 432170200\, y^{14} \\
& + 36781990\, y^{16} - 1483000\, y^{18} + 29596\, y^{20} - 280\, y^{22} + y^{24} \\
= {} & (81 - 90\, y^2 + y^4)^2 (16 - 40\, y^2 + y^4)^2 (1 - 10\, y^2 + y^4)^2.
\end{aligned}$$

This does not have distinct roots, so that $s(y)$ is not a Galois resolvent.

Exercise 14. Use Theorem 12.2.5 and standard results about Galois extensions to prove that $|\mathrm{Gal}(KL/K)| = [L : K \cap L]$. Then explain why this implies that $|\mathrm{Gal}(KL/K)| < |\mathrm{Gal}(L/F)|$ if and only if F is a proper subfield of $K \cap L$.

Exercise 15. Let $F \subset L$ and $F \subset K$ be Galois extensions such that KL is defined. We will also assume that $K \cap L = F$. The goal of this exercise is to prove that $F \subset KL$ is a Galois extension with Galois group

$$\mathrm{Gal}(KL/F) \simeq \mathrm{Gal}(L/F) \times \mathrm{Gal}(K/F).$$

(a) Prove that $F \subset KL$ is Galois and that $\sigma \in \text{Gal}(KL/F)$ implies that $\sigma|_L \in \text{Gal}(L/F)$ and $\sigma|_K \in \text{Gal}(L/K)$.

(b) Use part (d) of Exercise 6 to show that there is a one-to-one group homomorphism

$$\text{Gal}(KL/F) \longrightarrow \text{Gal}(L/F) \times \text{Gal}(K/F).$$

(c) Use Exercise 14 and the Tower Theorem to show that $[KL : F] = [K : F][L : F]$.

(d) Conclude that the map of part (b) is an isomorphism.

12.3   KRONECKER

In this section we will explore how Kronecker combined ideas of Lagrange, Gauss, and Galois to give a powerful construction of the splitting field of a separable polynomial over a field of characteristic 0.

A.   Algebraic Quantities.   In 1882 Kronecker published the important paper *Grundzüge einer arithmetischen Theorie der algebraischen Grössen* [Kronecker, Vol. II, pp. 237–387]. In English, the title is "Foundations of an arithmetic theory of algebraic quantities," which signals Kronecker's intention to create a general foundation for dealing with algebraic objects.

Kronecker begins his *Grundzüge* by describing the fields that he will work over, although he deliberately avoids using Dedekind's terminology of "fields." The quotation from Dedekind given in the Historical Notes to Section 4.1 shows that Dedekind's definition is very abstract: anything that satisfies the field axioms is a field. Kronecker, on the other hand, wants to emphasize that the objects he deals with are very concrete. We will use the term "field" when discussing what Kronecker does, though Kronecker would not be entirely comfortable with this practice.

Kronecker's basic objects of study are elements of a *Rationalitäts-Bereich*, denoted $(\mathfrak{R}', \mathfrak{R}'', \mathfrak{R}''', \dots)$. Here, $\mathfrak{R}', \mathfrak{R}'', \mathfrak{R}''', \dots$ are finitely many *algebraic quantities* (*algebraischen Grössen*), which can be variables or roots of polynomials (we will say more about this below). Then an element of the Rationalitäts-Bereich $(\mathfrak{R}', \mathfrak{R}'', \mathfrak{R}''', \dots)$ is a rational function with integer coefficients in these quantities. In modern terms, this is the field

$$(12.30) \qquad\qquad L = \mathbb{Q}(\mathfrak{R}', \mathfrak{R}'', \mathfrak{R}''', \dots),$$

since by clearing denominators, every element of L can be written as a quotient of polynomials with integer coefficients. For Kronecker, however, the emphasis is more on the elements than on the field.

The basic operation on such fields is adjunction. Given L as in (12.30), consider an irreducible polynomial with coefficients in L. Then a root of this polynomial is an algebraic quantity $\mathfrak{G}$ that gives a new Rationalitäts-Bereich when adjoined to L. In §2 of the *Grundzüge*, Kronecker assumes without comment that $\mathfrak{G}$ exists. As we will see, he eventually explains why this assumption is valid.

Here is a simple example of Kronecker's adjunction process.

Example 12.3.1. Given a variable x, the field $\mathbb{Q}(x)$ is an example of a Rationalitäts-Bereich. Furthermore, in Exercise 1 you will verify that $y^2 - 4x^3 - x$ is irreducible as a polynomial in y with coefficients in $\mathbb{Q}(x)$. Thus, by adjoining a root y_1 of this polynomial to $\mathbb{Q}(x)$, we get a new Rationalitäts-Bereich

$$\mathbb{Q}(x, y_1) = \mathbb{Q}\left(x, \sqrt{4x^3 + x}\right).$$

In particular, $y_1 = \sqrt{4x^3 + x}$ is an algebraic quantity.   ◁▷

Kronecker then studies the structure of the fields (12.30). His main result is that any such field can be written as an extension

(12.31) $\mathbb{Q} \subset \mathbb{Q}(\mathfrak{R}_1, \mathfrak{R}_2, \mathfrak{R}_3, \dots) \subset \mathbb{Q}(\mathfrak{G}, \mathfrak{R}_1, \mathfrak{R}_2, \mathfrak{R}_3, \dots)$

where $\mathfrak{R}_1, \mathfrak{R}_2, \mathfrak{R}_3, \dots$ can be regarded as variables over $\mathbb{Q}$, and $\mathfrak{G}$ is algebraic over $\mathbb{Q}(\mathfrak{R}_1, \mathfrak{R}_2, \mathfrak{R}_3, \dots)$. In Exercise 2 you will show that this follows from the result of Steinitz discussed in the Mathematical Notes to Section 4.1, together with the Theorem of the Primitive Element.

In contrast to modern presentations, the fields considered by Kronecker are constructed explicitly. This constructive attitude runs very deep. For example, rather than just defining what it means for a polynomial to be irreducible, Kronecker gives a method for deciding whether or not a polynomial with coefficients in a field of the form (12.31) is irreducible. To do this, he first discusses polynomials in $\mathbb{Q}[x]$ and describes the algorithm presented in Proposition 4.2.2 for factoring polynomials over $\mathbb{Q}$. He then gives a terse explanation of how to factor in the general case. The missing details can be found in Edwards's book [7].

B. Module Systems. Besides developing a theory of fields, the *Grundzüge* also considers rings, ideals, and quotient rings. This begins in §5 of the *Grundzüge*, where Kronecker introduces the *Integritäts-Bereich* $[\mathfrak{R}', \mathfrak{R}'', \mathfrak{R}''', \dots]$, which in modern terms is the integral domain

$$R = \mathbb{Z}[\mathfrak{R}', \mathfrak{R}'', \mathfrak{R}''', \dots]$$

consisting of all polynomials in $\mathfrak{R}', \mathfrak{R}'', \mathfrak{R}''', \dots$ with integer coefficients. The field of fractions of R is the Rationalitäts-Bereich $\mathbb{Q}(\mathfrak{R}', \mathfrak{R}'', \mathfrak{R}''', \dots)$.

The next step is to define certain ideals of R. In §21 of the *Grundzüge*, Kronecker takes finitely many elements $M_1, M_2, M_3, \dots \in R$ and defines the *module system* $(M_1, M_2, M_3, \dots)$ to consist of all linear combinations with coefficients in R

$$A_1 M_1 + A_2 M_2 + A_3 M_3 + \cdots, \quad A_1, A_2, A_3, \dots \in R.$$

Given $M, M' \in R$, Kronecker then defines

(12.32) $M \equiv M' \ (\text{modd. } M_1, M_2, M_3, \dots)$

to mean that $M - M'$ is contained in $(M_1, M_2, M_3, \dots)$. These days, we say "ideal" rather than "module system" and we write the ideal as

$$I = \langle M_1, M_2, M_3, \dots \rangle = \left\{ A_1 M_1 + A_2 M_2 + A_3 M_3 + \cdots \mid A_1, A_2, A_3, \dots \in R \right\}.$$

(In Exercise 3 you will prove that this is an ideal of R.) Then (12.32) means that $M - M' \in I$, which is equivalent to the equality $M + I = M' + I$ of cosets in R/I.

It follows that Kronecker is developing the basic language of ideals and quotient rings. However, Kronecker didn't use Dedekind's term "ideal," because Dedekind allowed his ideals to be very abstract, while Kronecker was only interested in the explicitly constructed ideals described above.

For us, the most important application of these ideas came in Section 3.1. Recall that in the proof of Proposition 3.1.3, we showed that if $f \in F[x]$ is an irreducible polynomial, then $L = F[x]/\langle f \rangle$ is an extension field of F such that $\alpha = x + \langle f \rangle \in L$ is a root of f. This is how we proved the existence of roots.

This is equally important for Kronecker, for he uses a similar construction to give a precise meaning to the term algebraic quantity. The idea is as follows: in (12.31), let $G(x)$ is the minimal polynomial of $\mathfrak{G}$ over $\mathbb{Q}(\mathfrak{R}_1, \mathfrak{R}_2, \mathfrak{R}_3, \dots)$. Then one can replace the extension field $\mathbb{Q}(\mathfrak{G}, \mathfrak{R}_1, \mathfrak{R}_2, \mathfrak{R}_3, \dots)$ with the quotient ring

$$(12.33) \qquad \mathbb{Q}(\mathfrak{R}_1, \mathfrak{R}_2, \mathfrak{R}_3, \dots)[x]/\langle G \rangle,$$

where the coset $x + \langle G \rangle$ plays the role of the root $\mathfrak{G}$. Since $\mathfrak{R}_1, \mathfrak{R}_2, \mathfrak{R}_3, \dots$ are variables in (12.31), we now have a rigorous construction of the algebraic quantity $\mathfrak{G}$ in terms of polynomials and ideals.

In this construction, Kronecker preferred to work with rings rather than fields since he wanted to avoid denominators as much as possible. So Kronecker would replace (12.33) with

$$\mathbb{Z}[\mathfrak{R}_1, \mathfrak{R}_2, \mathfrak{R}_3, \dots, x]/\langle G \rangle,$$

where one has now suitably cleared denominators so that G is an irreducible polynomial in $\mathbb{Z}[\mathfrak{R}_1, \mathfrak{R}_2, \mathfrak{R}_3, \dots, x]$. This quotient ring is an integral domain whose field of fractions is the corresponding field (12.33).

Here is a simple example of what these presentations look like.

Example 12.3.2. Consider the field $\mathbb{Q}(x, \sqrt{4x^3 + x})$ constructed in Example 12.3.1. This is $\mathbb{Q} \subset \mathbb{Q}(x) \subset \mathbb{Q}(x, \sqrt{4x^3 + x})$. As a polynomial in y, the minimal polynomial of $\sqrt{4x^3 + x}$ is $y^2 - 4x^3 - x$, so that (12.33) becomes

$$\mathbb{Q}(x)[y]/\langle y^2 - 4x^3 - x \rangle.$$

Kronecker's presentation, which uses $\mathbb{Z}$ rather than $\mathbb{Q}$, would be to take the field of fractions of the integral domain

$$\mathbb{Z}[x, y]/\langle y^2 - 4x^3 - x \rangle.$$

Notice how polynomials in several variables appear naturally in this example. ◁▷

As noted in [7], Kronecker was aware that this construction allows one to dispense with "algebraic quantities." Kronecker states this as follows:

> … the whole arithmetic theory of algebraic quantities can be reduced to a theory of entire functions of variables and unknowns with integer coefficients …

(see [Kronecker, Vol. II, p. 377]). In the nineteenth century, "entire function" meant polynomial. Thus Kronecker is saying that we can construct all algebraic quantities using congruences of polynomials in several variables with coefficients in $\mathbb{Z}$.

C. Splitting Fields. One of the points made in [8] is that Kronecker's conception of algebraic quantity evolved during the writing of the *Grundzüge*. The early sections of the *Grundzüge* don't give a precise definition of algebraic quantity, yet the later sections provide the language needed for this purpose (as noted in the above quotation). But to rewrite the *Grundzüge* from this new point of view would have been an overwhelming task. Hence one needs to look at Kronecker's subsequent papers to see how he worked out these ideas.

For us, Kronecker's 1887 paper *Ein Fundamentalsatz der allgemeinen Arithmetik* [Kronecker, Vol. III, pp. 209–240] is the most relevant. Here, he uses his new methods to give an explicit construction of splitting fields. His presentation and proof involve a deep understanding of the ideas of Lagrange, Gauss, and Galois.

Let F be a field of characteristic 0, and let

$$(12.34) \qquad f(x) = x^n - c_1 x^{n-1} + c_2 x^{n-2} + \cdots + (-1)^n c_n \in F[x]$$

be a separable polynomial. In his *Fundamentalsatz* paper, Kronecker uses $F = \mathbb{Q}(\mathfrak{R}', \mathfrak{R}'', \ldots, \mathfrak{R}^{n-1})$, where $n \geq 1$ and $\mathfrak{R}', \mathfrak{R}'', \ldots, \mathfrak{R}^{n-1}$ are variables. So for Kronecker, the coefficients are explicitly known objects. His goal was to describe the roots of f using module systems and congruences.

Kronecker was inspired by Galois's approach to the Galois group. Galois assumed that the roots $\alpha_1, \ldots, \alpha_n$ of f lie in some extension of F. Let us recall some of the ideas developed in Section 12.2. The Galois resolvent (12.21) is the polynomial

$$s(y) = \prod_{\sigma \in S_n} \left(y - (t_1 \alpha_{\sigma(1)} + \cdots + t_n \alpha_{\sigma(n)}) \right),$$

where $t_1, \ldots, t_n \in F$ are chosen so that $s(y)$ is separable. Then the key player for Galois is the irreducible factor $h(y)$ of $s(y)$ that vanishes at $V = t_1 \alpha_1 + \cdots + t_n \alpha_n$. Recall from Section 12.2 that by Galois's Lemma III, V is a primitive element of the splitting field of f over F. Thus each root α_i is a polynomial $\varphi_i(V)$ in V with coefficients in F.

This implies that in the splitting field $F(V)$, we can write

$$f(x) = \prod_{i=1}^{n} (x - \alpha_i) = \prod_{i=1}^{n} \left(x - \varphi_i(V) \right).$$

Furthermore, since $h(y)$ is the minimal polynomial of V over F, we can replace $F(V)$ with the quotient ring $F[y]/\langle h \rangle$, where the coset $y + \langle h \rangle$ plays the role of V. If we substitute this into the above equation and use congruence notation $\equiv$, then we can write the above factorization as

$$(12.35) \qquad f(x) \equiv \prod_{i=1}^{n} \big(x - \varphi_i(y)\big) \bmod h(y).$$

This is close to what Kronecker states in his *Fundamentalsatz* paper. However, our derivation of (12.35) assumed knowledge of the roots $\alpha_1, \ldots, \alpha_n$ of f. Kronecker's goal is to compute this factorization *without* knowing the roots of the polynomial in advance.

How does Kronecker accomplish this? In reading Gauss's 1815 proof of the Fundamental Theorem of Algebra (discussed in the Historical Notes to Section 3.2), Kronecker learned the strategy of applying Lagrange's methods in the universal case and then specializing to the specific polynomial at hand. This works as follows.

In the universal situation, the variables $x_1, \ldots, x_n$ are roots of the universal polynomial of degree n,

$$\tilde{f} = (x - x_1) \cdots (x - x_n) = x^n - \sigma_1 x^{n-1} + \cdots + (-1)^n \sigma_n,$$

which has coefficients in $F(\sigma_1, \ldots, \sigma_n)$. Then the substitution $\sigma_i \mapsto c_i$ takes $\tilde{f}$ to $f = x^n - c_1 x^{n-1} + \cdots + (-1)^n c_n$.

Since F has characteristic 0, we can pick distinct integers $t_1, \ldots, t_n \in \mathbb{Z} \subset F$. This gives the universal Galois resolvent

$$(12.36) \qquad S(y) = \prod_{\sigma \in S_n} \big(y - (t_1 x_{\sigma(1)} + \cdots + t_n x_{\sigma(n)})\big) = \prod_{\sigma \in S_n} (y - \sigma \cdot \beta),$$

where $\beta = t_1 x_1 + \cdots + t_n x_n$. The theory of symmetric polynomials from Chapter 2 shows that $S(y)$ lies in $\mathbb{Q}[\sigma_1, \ldots, \sigma_n, y]$.

Now let $s(y) \in F[y]$ be the polynomial obtained from $S(y)$ by the substitution $\sigma_i \mapsto c_i$. Kronecker then claims that $t_1, \ldots, t_n \in \mathbb{Z}$ can be chosen so that $s(y)$ is separable. Once this is done, he will have constructed a Galois resolvent of f without knowing the roots of f. Exercises 1 and 2 of Section 12.2 proved the existence $t_1, \ldots, t_n$ in F (rather than in $\mathbb{Z}$) using the roots $\alpha_1, \ldots, \alpha_n$. Of course, Kronecker needs to use a different argument. Exercises 4 and 5 will show how to find the desired $t_1, \ldots, t_n \in \mathbb{Z}$ without using the roots.

In Section 3.2, we observed that Gauss used a similar method to compute the polynomial $Z(x, u) = \prod_{1 \le i < j \le n} (x - (\alpha_i + \alpha_j)u + \alpha_i \alpha_j)$ from (3.13) without knowing the roots of f. Do you see how this could have influenced Kronecker?

Given the Galois resolvent $s(y) \in F[y]$, Kronecker then factors $s(y)$ into a product of irreducible polynomials using the methods mentioned earlier in the section. Let $h(y)$ be one of the irreducible factors. This gives the polynomial $h(y)$ we need for (12.35). The important thing for Kronecker is that $h(y)$ was constructed without knowing the roots of f.

It remains to find explicit formulas for the polynomials $\varphi_i(y)$ in (12.35). For this purpose, we turn to the methods of Lagrange discussed in Section 12.1. Since $S(y)$ is the resolvent polynomial of $\beta = t_1 x_1 + \cdots + t_n x_n$, one can find explicitly computable polynomials $\Psi_i(y)$ in $F[\sigma_1, \ldots, \sigma_n, y]$ such that

$$(12.37) \qquad\qquad x_i = \frac{\Psi_i(\beta)}{\Delta(S)}, \qquad i = 1, \ldots, n,$$

where $\Delta(S)$ is the discriminant of S. You will prove this in Exercises 6–8.

The substitution $\sigma_i \mapsto c_i$ maps $S(y)$ to $s(y)$ and $\Delta(S)$ to $\Delta(s)$. Furthermore, $\Delta(s) \neq 0$, since $s(y)$ is separable. Thus $\Psi_i(y)/\Delta(S)$ maps to some polynomial $\varphi_i(y) \in F[y]$. We can now state the construction of the splitting field of f given in Kronecker's *Fundamentalsatz* paper [Kronecker, Vol. II, p. 216].

Theorem 12.3.3. *Let F have characteristic 0. Let $f \in F[x]$ be monic and separable of degree $n > 0$, and let $s(y)$ and $\varphi_i(y)$, $i = 1, \ldots, n$, be constructed as above. Then, for any irreducible factor $h(y) \in F[x]$ of $s(y)$, we have*

$$f(x) \equiv \big(x - \varphi_1(y)\big) \cdots \big(x - \varphi_n(y)\big) \bmod h(y).$$

This congruence means that each side is a polynomial in x whose coefficients are equal in the quotient ring $F[y]/\langle h(y)\rangle$. Furthermore, $F[y]/\langle h(y)\rangle$ is a splitting field of f over F. □

We will not prove this here. The reader should consult [7] for a modern (but fully constructive) proof of Theorem 12.3.3.

Kronecker remarks that because of Theorem 12.3.3, "one is then relieved of the introduction of algebraic quantities in many ... algebraic investigations" [Kronecker, Vol. III, p. 216]. This is his clearest statement of how to avoid algebraic quantities. Kronecker's choice of the word *Fundamentalsatz* ("Fundamental Theorem") in the title of his paper indicates the importance he attaches to this result.

Theorems 12.3.3 and 3.1.4 both prove that a splitting field of $f \in F[x]$ exists, though Kronecker's theorem differs from Theorem 3.1.4 in two ways:

- Rather than construct the splitting field L using a sequence of quotient rings (as in Theorem 3.1.4), Kronecker constructs all roots of f simultaneously using just one quotient ring $L = F[y]/\langle h(y)\rangle$.
- We will see in Chapter 13 that Kronecker's construction leads directly to an algorithm for computing the Galois group of f.

Hence Kronecker's construction of the splitting field contains a *lot* of information about the roots of the polynomial. It is harder than what we did in Theorem 3.1.4, but for a good reason.

Theorem 12.3.3 uses a lot of mathematics, including ideas of Lagrange (the universal case), Gauss (relating the universal to the specific), and Galois (the Galois resolvent). It is impressive how Kronecker was able to synthesize all of this mathematics into one theorem. One irony is that while Kronecker is given credit as the first to prove the existence of the roots of a polynomial, his version of this result is rarely mentioned, since most books use the proof of Theorem 3.1.4 given in Chapter 3. While the modern proof illustrates the power of abstract algebra, it does not reflect the richness of the historical context that led to Kronecker's proof of the existence of splitting fields.

Historical Notes

Congruences modulo a polynomial or module system play a central role in Kronecker's construction of algebraic quantities. When Kronecker wrote his *Grundzüge* in 1882, there were many known examples of congruences, including:

- $a \equiv b \bmod n$, for $a, b \in \mathbb{Z}$ (Gauss, 1801).
- $\phi(x) \equiv \chi(x) \bmod x^2 + 1$, for $\phi(x), \chi(x) \in \mathbb{R}[x]$ (Cauchy, 1845).
- $\varphi(\alpha) \equiv \psi(\alpha) \bmod p\mathbb{Z}[\alpha]$, for $\varphi(\alpha), \psi(\alpha) \in \mathbb{Z}[\alpha]$ (Schönemann, 1846).
- $P(x) \equiv Q(x) \bmod \langle p, f(x) \rangle$, for $P(x), Q(x) \in \mathbb{Z}[x]$ (Dedekind, 1857).

While Kronecker did not originate the use congruences in the polynomial setting, he was clearly the first to realize the full power of this construction.

Our discussion of Kronecker's work omitted several important topics covered in the *Grundzüge*. For example, Kronecker created a theory of *divisors* to generalize Kummer's *ideal numbers*. This differs from Dedekind's theory of ideals, which is another way to generalize ideal numbers. An exposition of divisor theory can be found in [6]. Kronecker also considered *discriminants* in detail.

We should also note that our discussion made liberal use of set theory. For example, when we gave Kronecker's definition of module system $(M_1, M_2, \dots)$ in a ring R, we immediately translated it into the ideal

$$\langle M_1, M_2, \dots \rangle = \left\{ A_1 M_1 + A_2 M_2 + \cdots \mid A_1, A_2, \dots \in R \right\}.$$

Kronecker, in contrast, says that $A_1 M_1 + A_2 M_2 + \cdots$ "contains the module system $(M_1, M_2, \dots)$" [Kronecker, Vol. II, p. 335]. To us, this seems backwards. As explained to me by Edwards, Kronecker's use of "contains" is similar to saying that 6 "contains" 2 as a divisor. This makes sense when one realizes the importance of divisors in Kronecker's mathematical thought. An introduction to Kronecker's views of set theory and the foundations of mathematics can be found in [9].

An important development in the late nineteenth century was the realization that one needs to study the foundations of mathematics. It was no longer sufficient to simply assume the existence of mathematical objects such as algebraic quantities. Rather, one had to give a rigorous proof of their existence. But instead of Kronecker's constructive vision of what this meant, the set-theoretic approach of Dedekind and Cantor came to dominate. This explains why modern mathematics is firmly based on sets and why abstract algebra is so different from high school algebra.

This chapter began with Lagrange's attempts to understand the roots of polynomials and ended with Kronecker's impressive construction of splitting fields. Along the way, we were able to witness the brilliance of Galois and the beginnings of modern algebra. It has been a remarkable odyssey.

Exercises for Section 12.3

Exercise 1. Prove that $y^2 - 4x^3 - x$ is irreducible when considered as an element of $\mathbb{Q}(x)[y]$.

Exercise 2. Show that (12.31) follows from the Theorem of the Primitive Element and the theorem of Steinitz mentioned in the Mathematical Notes to Section 4.1.

Exercise 3. Let R be a commutative ring and let $M_1, \ldots, M_s$ be elements of R. Prove that the set $\langle M_1, \ldots, M_s \rangle = \left\{ \sum_{i=1}^{s} A_i M_i \mid A_i \in R \right\}$ is an ideal of R.

Exercise 4. In the discussion leading up to Theorem 12.3.3, we have the polynomial $S(y) \in F[\sigma_1, \ldots, \sigma_n, y]$ defined in (12.36). Then $s(y) \in F[y]$ is obtained by $\sigma_i \mapsto c_i$, where c_i is as in (12.34). Both of these polynomials depend on $t_1, \ldots, t_n$. The goal of this exercise is to show that if f is separable, then $\Delta(s)$ is a nonzero polynomial when $t_1, \ldots, t_n$ are regarded as variables. Since F has characteristic 0, part (a) of Exercise 5 implies that $\Delta(s) \neq 0$ for some $t_1, \ldots, t_n \in \mathbb{Z}$.

To prove that $\Delta(s)$ is a nonzero polynomial in $t_1, \ldots, t_n$, let $F \subset L$ be the splitting field of f constructed in Theorem 3.1.4. Thus $f = (x - \alpha_1) \cdots (x - \alpha_n)$ in $L[x]$.

(a) If we regard the t_i as variables, explain why $S(y)$ becomes a polynomial in y with coefficients in $F[\sigma_1, \ldots, \sigma_n, t_1, \ldots, t_n]$. Conclude that $s(y) \in F[t_1, \ldots, t_n, y]$ and hence that $\Delta(s) \in F[t_1, \ldots, t_n]$.

(b) Explain why $s(y) = \prod_{\sigma \in S_n} \left(y - (t_1 \alpha_{\sigma(1)} + \cdots + t_n \alpha_{\sigma(n)}) \right)$ in $L[t_1, \ldots, t_n, y]$.

(c) Use part (b) and the separability of f to show that $s(y)$ has distinct roots, all of which lie in $L[t_1, \ldots, t_n]$. Conclude that $\Delta(s)$ is a nonzero element of $F[t_1, \ldots, t_n]$.

Exercise 5. Let F be a field, and let $g \in F[t_1, \ldots, t_n]$ be nonzero.

(a) Suppose that F has characteristic 0, so that $\mathbb{Q} \subset F$. For each i, pick a nonnegative integer N_i such that the highest power of t_i appearing in g is at most N_i, and let

$$A = \{(a_1, \ldots, a_n) \mid a_i \in \mathbb{Z}, \ 0 \le a_i \le N_i\}.$$

Prove that there is $(a_1, \ldots, a_n) \in A$ such that $g(a_1, \ldots, a_n) \neq 0$.

(b) Now suppose that F has characteristic p and is infinite. Modify the argument of part (a) to show that there are $a_1, \ldots, a_n \in F$ such that $g(a_1, \ldots, a_n) \neq 0$.

(c) Give an example to illustrate why the hypothesis "F is infinite" is needed in part (b).

Exercise 6. In $F[x_1, \ldots, x_n]$, consider the polynomial

$$\tilde{f} = (x - x_1) \cdots (x - x_n) = x^n - \sigma_1 x^{n-1} + \cdots + (-1)^n \sigma_n.$$

As noted in Section 2.2, we can regard $\tilde{f} \in F[\sigma_1, \ldots, \sigma_n]$ as the universal polynomial of degree n. The goal of this exercise is to show that if $\tilde{f}'$ denotes the derivative of $\tilde{f}$, then there are polynomials $\tilde{A}, \tilde{B} \in F[\sigma_1, \ldots, \sigma_n, x]$ such that $\deg(\tilde{A}) \le n - 2$, $\deg(\tilde{B}) \le n - 1$, and

$$\tilde{A} \tilde{f} + \tilde{B} \tilde{f}' = \Delta.$$

Here, Δ is the discriminant defined in Section 2.4. The proof given here is taken from Gauss's 1815 proof of the Fundamental Theorem of Algebra (see [14, pp. 293–295]).

(a) Show that

$$\tilde{B} = \frac{\Delta(x - x_2) \cdots (x - x_n)}{(x_1 - x_2)^2 \cdots (x_1 - x_n)^2} + \frac{\Delta(x - x_1)(x - x_3) \cdots (x - x_n)}{(x_2 - x_1)^2 (x_2 - x_3)^2 \cdots (x_2 - x_n)^2}$$
$$+ \cdots + \frac{\Delta(x - x_1) \cdots (x - x_{n-1})}{(x_n - x_1)^2 \cdots (x_n - x_{n-1})^2}$$

is a polynomial in x of degree at most $n - 1$ whose coefficients are symmetric polynomials in $x_1, \ldots, x_n$. Conclude that $\tilde{B} \in F[\sigma_1, \ldots, \sigma_n, x]$.

(b) Prove that $\Delta - \tilde{B} \tilde{f}'$ vanishes when $x = x_i$.

(c) Conclude that $\Delta - \widetilde{B} \tilde{f}'$ is divisible by $\tilde{f}$, and set

$$\widetilde{A} = \frac{\Delta - \widetilde{B} \tilde{f}'}{\tilde{f}}.$$

Show that $\widetilde{A}$ and $\widetilde{B}$ have the desired properties.

Exercise 7. Let $f \in F[x]$ be monic of degree $n > 0$ with discriminant $\Delta(f) \in F$. Use Exercise 6 to show that there are $A, B \in F[x]$ with $\deg(A) \leq n - 2$, $\deg(B) \leq n - 1$ such that the coefficients of A and B are polynomials in the coefficients of f and $A f + B f' = \Delta(f)$.

Exercise 8. This exercise is concerned with $\Psi_i(y)$ from (12.37). Let $S(y)$ be as in (12.36).
(a) Show that applying (12.5) and (12.8) from the proof of Theorem 12.1.6 with $f = \beta = t_1 x_1 + \cdots + t_n x_n$ and $g = x_i$ gives

$$x_i = \frac{\Phi_i(\beta)}{S'(\beta)},$$

where

$$\Phi_i(y) = \sum_{\sigma \in S_n} \frac{S(y) x_{\sigma(i)}}{y - \sigma \cdot \beta}.$$

Also prove that $\Phi_i(y) \in F[\sigma_1, \ldots, \sigma_n, y]$.
(b) Use Exercise 7 to show that there are polynomials $A, B \in F[\sigma_1, \ldots, \sigma_n, y]$ such that $A(y)S(y) + B(y)S'(y) = \Delta(S)$. Also show that $B(\beta)S'(\beta) = \Delta(S)$.
(c) Use part (b) to show that (12.37) holds with $\Psi_i(y) = B(y)\Phi_i(y)$.

REFERENCES

1. R. G. Ayoub, *Paolo Ruffini's contributions to the quintic*, Arch. Hist. Exact Sci. **23** (1980), 253–277.

2. I. G. Bashmakova and G. S. Smirnova, *The Beginnings and Evolution of Algebra*, English translation by A. Shenitzer, MAA, Washington, DC, 1999.

3. L. E. Dickson, *Introduction to the Theory of Algebraic Equations*, Wiley, New York, 1903. Reprinted in *Congruence of Sets and Other Monographs*, Chelsea, New York, 1967.

4. W. Dunham, *Euler: The Master of Us All*, MAA, Washington, DC, 1999.

5. H. M. Edwards, *A note on Galois theory*, Arch. Hist. Exact Sci. **41** (1990), 163–169.

6. H. M. Edwards, *Divisor Theory*, Birkhäuser, Boston, Basel, Berlin, 1990.

7. H. M. Edwards, *Essays in Constructive Mathematics*, Springer-Verlag, New York, Berlin, Heidelberg, 2004.

8. H. M. Edwards, *Kronecker's arithmetical theory of algebraic quantities*, Jber. d. Dt. Math.-Verein. **94** (1992), 130–139.

9. H. M. Edwards, *Kronecker's views of the foundations of mathematics*, in *The History of Modern Mathematics*, Volume I, edited by D. E. Rowe and J. Mc-Cleary, Academic Press, San Diego, CA, 1989.

10. L. Euler, *Elements of Algebra*, English translation by J. Hewlett, Springer-Verlag, New York, Berlin, Heidelberg, 1984. Reprint of 1840 English edition.

11. B. M. Kiernan, *The development of Galois theory from Lagrange to Artin*, Arch. Hist. Exact Sci. **8** (1971), 40–154.

12. I. Radloff, *Évariste Galois: principles and applications*, Historia Math. **29** (2002), 114–137.

13. L. Rigatelli, *Évariste Galois 1811–1832*, Springer-Verlag, New York, Berlin, Heidelberg, 2000.

14. D. E. Smith, *A Source Book in Mathematics*, Volume One, Ginn, Boston, New York, 1925. Reprint by Dover, New York, 1959.

15. B. L. van der Waerden, *History of Algebra*, Springer-Verlag, New York, Berlin, Heidelberg, 1985.

16. H. Wussing, *The Genesis of the Abstract Group Concept*, English translation by A. Shenitzer, MIT Press, Cambridge, MA, 1984.

13

Computing Galois Groups

Galois groups are not easy to compute. As Galois says in the "Discours préliminaire" to his first memoir on Galois theory [Galois, p. 39]:

> If now you give me an equation that you have chosen at will, and about which you want to know if it is or is not solvable by radicals, I cannot do any more than indicate the means for answering your question, without wanting to charge either myself or any other person with doing it. In a word, the calculations are impractical.

Even with the aid of modern computers, it is not easy to compute the Galois group of a polynomial of large degree (currently 23 or higher) unless the polynomial has some special structure.

This chapter will explore some (but not all) ways of computing Galois groups of arbitrary polynomials; Chapters 14 and 15 will describe special classes of polynomials for which it is possible to say more about the Galois group.

13.1   QUARTIC POLYNOMIALS

In Section 7.4 we explained how to compute the Galois group of a monic irreducible separable cubic polynomial f over a field F of characteristic different from 2. Recall that up to isomorphism the only possibilities for the Galois group are $\mathbb{Z}/3\mathbb{Z}$ and S_3, and that these cases are distinguished by whether or not the discriminant $\Delta(f)$ is the square of an element of F.

In this section we will prove a similar result for a monic irreducible quartic polynomial $f \in F[x]$, where F has characteristic $\neq 2$. Note that f is necessarily separable by Lemma 5.3.5. We will write f in the form

(13.1) $f = x^4 - c_1 x^3 + c_2 x^2 - c_3 x + c_4, \quad c_1, c_2, c_3, c_4 \in F.$

In computing the Galois group of f, the key players will be the discriminant

(13.2)
$$\begin{aligned}
\Delta(f) = {} & 144\, c_2\, c_1{}^2 c_4{}^2 + 18\, c_1\, c_3{}^3 c_2 - 192\, c_1\, c_3\, c_4{}^2 - 6\, c_1{}^2 c_3{}^2 c_4 \\
& + 144\, c_4\, c_3{}^2 c_2 - 4\, c_2{}^3 c_1{}^2 c_4 + c_2{}^2 c_1{}^2 c_3{}^2 + 256\, c_4{}^3 \\
& - 27\, c_3{}^4 + 18\, c_1{}^3 c_3\, c_2\, c_4 - 4\, c_1{}^3 c_3{}^3 - 128\, c_2{}^2 c_4{}^2 \\
& + 16\, c_2{}^4 c_4 - 4\, c_2{}^3 c_3{}^2 - 27\, c_1{}^4 c_4{}^2 - 80\, c_1\, c_3\, c_2{}^2 c_4,
\end{aligned}$$

computed by the methods of Section 5.3, and the Ferrari resolvent

(13.3) $\theta_f(y) = y^3 - c_2\, y^2 + (c_1 c_3 - 4 c_4)\, y - c_3^2 - c_1^2 c_4 + 4 c_2 c_4$

used in the solution of the quartic given in Section 12.1.

The Galois group of f is $\mathrm{Gal}(L/F)$, where L is a splitting field of f over F. Proposition 6.3.1 implies that there is a subgroup $G \subset S_4$ such that

(13.4) $\mathrm{Gal}(L/F) \simeq G \subset S_4.$

Since we only need $\mathrm{Gal}(L/F)$ up to isomorphism, we can focus on the subgroup G. But G depends on how we label the roots. In Exercise 1 you will show that if we change the labels, then G gets replaced by a conjugate subgroup in S_4. Since the roots are intrinsically unlabeled, our goal is to compute G up to conjugacy.

Theorem 13.1.1. *Let F have characteristic $\neq 2$, and $f \in F[x]$ be monic and irreducible of degree 4. Write f as in (13.1), and let $\Delta(f)$ and $\theta_f(y)$ be defined as in (13.2) and (13.3). Then the subgroup $G \subset S_4$ from (13.4) is determined as follows:*

(a) *If $\theta_f(y)$ is irreducible over F, then*

$$G = \begin{cases} S_4, & \text{if } \Delta(f) \notin F^2, \\ A_4, & \text{if } \Delta(f) \in F^2. \end{cases}$$

(b) *If $\theta_f(y)$ splits completely over F, then*

$$G = \langle (12)(34), (13)(24) \rangle \simeq \mathbb{Z}/2\mathbb{Z} \times \mathbb{Z}/2\mathbb{Z}.$$

Furthermore, $\theta_f(y)$ splits completely over F if and only if it is reducible over F and $\Delta(f) \in F^2$.

(c) *If $\theta_f(y)$ has a unique root β in F, then*

$$G \text{ is conjugate to } \begin{cases} \langle (1324), (12) \rangle \simeq D_8, & \text{if } 4\beta + c_1^2 - 4c_2 \neq 0 \text{ and} \\ & \Delta(f)\big(4\beta + c_1^2 - 4c_2\big) \notin (F^*)^2, \\ & \text{or } 4\beta + c_1^2 - 4c_2 = 0 \text{ and} \\ & \Delta(f)\big(\beta^2 - 4c_4\big) \notin (F^*)^2, \\ \langle (1324) \rangle \simeq \mathbb{Z}/4\mathbb{Z}, & \text{otherwise,} \end{cases}$$

where D_8 is the dihedral group of order 8. Furthermore, $\theta_f(y)$ has a unique root in F if and only if it is reducible over F and $\Delta(f) \notin F^2$.

Proof. In Section 12.1, we defined the universal Ferrari resolvent to be

$$\theta(y) = \left(y - (x_1x_2 + x_3x_4)\right)\left(y - (x_1x_3 + x_2x_4)\right)\left(y - (x_1x_4 + x_2x_3)\right)$$
$$= y^3 - \sigma_2 y^2 + (\sigma_1\sigma_3 - 4\sigma_4)\, y - \sigma_3^2 - \sigma_1^2\sigma_4 + 4\sigma_2\sigma_4.$$

If f has roots $\alpha_1, \alpha_2, \alpha_3, \alpha_4 \in L$, then the evaluation map $x_i \mapsto \alpha_i$ takes $\sigma_i \mapsto c_i$ and hence takes $\theta(y)$ to $\theta_f(y)$. It follows that the roots of $\theta_f(y)$ are

(13.5) $\qquad\qquad \alpha_1\alpha_2 + \alpha_3\alpha_4, \quad \alpha_1\alpha_3 + \alpha_2\alpha_4, \quad \alpha_1\alpha_4 + \alpha_2\alpha_3.$

In particular, $\theta_f(y)$ splits completely in L.

Using this, we can now prove part (a). Since f and θ_f are monic and irreducible over F, they are the minimal polynomials over F of α_1 and $\alpha_1\alpha_2 + \alpha_3\alpha_4$ respectively. By the Tower Theorem, we see that $[L : F]$ is divisible by 12, so that $|G| = |\mathrm{Gal}(L/F)|$ is also divisible by 12, since $F \subset L$ is Galois. In Exercise 2 you will show that A_4 is the only subgroup of S_4 with 12 elements. Thus the hypothesis of part (a) implies that $G = A_4$ or S_4. Then we are done by Theorem 7.4.1.

Before proving parts (b) and (c), we first observe that θ_f and f have the same discriminant, that is,

(13.6) $\qquad\qquad\qquad\qquad \Delta(\theta_f) = \Delta(f).$

In the universal case, this was proved in Exercise 9 of Section 2.4. In Exercise 3 you will explain why this implies (13.6). Since f is separable, we conclude that θ_f is a separable cubic.

Now suppose that θ_f is reducible over F. Since θ_f is a cubic, this implies that it has a root $\beta \in F$. By (13.5), we may relabel the roots of f so that

(13.7) $\qquad\qquad\qquad\qquad \beta = \alpha_1\alpha_2 + \alpha_3\alpha_4 \in F.$

As explained earlier, relabeling the roots of f replaces G with a conjugate subgroup.

We will analyze how (13.7) affects the Galois group. What follows is a special case of a general phenomenon that will play a central role in Sections 13.2 and 13.3. We claim that (13.7) implies that

$$G \subset \langle (1324), (12) \rangle \simeq D_8.$$

The rough idea is that the Galois group shrinks when a resolvent has a root in F.

To prove our claim, suppose that $\sigma \in \mathrm{Gal}(L/F)$ corresponds to $\tau \in G \subset S_4$. Since $\beta \in F$, one easily computes that

$$\beta = \sigma(\beta) = \sigma(\alpha_1\alpha_2 + \alpha_3\alpha_4) = \alpha_{\tau(1)}\alpha_{\tau(2)} + \alpha_{\tau(3)}\alpha_{\tau(4)}$$
$$= \begin{cases} \alpha_1\alpha_2 + \alpha_3\alpha_4, & \text{if } \tau \in \langle (1324), (12) \rangle, \\ \alpha_1\alpha_3 + \alpha_2\alpha_4, & \text{if } \tau \in (23)\langle (1324), (12) \rangle, \\ \alpha_1\alpha_4 + \alpha_2\alpha_3, & \text{if } \tau \in (24)\langle (1324), (12) \rangle, \end{cases}$$

where

$$S_4 = \langle(1324),(12)\rangle \cup (23)\langle(1324),(12)\rangle \cup (24)\langle(1324),(12)\rangle$$

is the decomposition of S_4 into left cosets of $\langle(1324),(12)\rangle$. Since θ_f is separable, this implies that $G \subset \langle(1324),(12)\rangle$ as claimed.

As in part (a), we know that 4 divides $|G|$, since f is irreducible. It follows that $|G| = 4$ or 8. Furthermore, we found all subgroups of D_8 when we worked out the Galois correspondence for $\mathbb{Q} \subset \mathbb{Q}(i, \sqrt[4]{2})$ in Section 7.3. In Exercise 4 you will use this to show that G is one of the four groups

(13.8) $\langle(12),(34)\rangle, \ \langle(12)(34),(13)(24)\rangle, \ \langle(1324)\rangle, \ \langle(1324),(12)\rangle.$

Since f is irreducible, Proposition 6.3.7 implies that G is a transitive subgroup of S_4 (reread Section 6.3 if you've forgotten what transitive means). The first group listed in (13.8) is not transitive, so that G is one of the remaining three groups. Parts (b) and (c) of the theorem describe how we distinguish among these possibilities.

We begin with part (b). Since F has characteristic $\neq 2$, Exercise 5 implies that a monic reducible cubic $g \in F[x]$ splits completely over F if and only if $\Delta(g) \in F^2$. By (13.6), we conclude that if θ_f is reducible over F, then it splits completely over F if and only if $\Delta(f) \in F^2$. This proves the final assertion of part (b). Also, when θ_f splits completely over F, Theorem 7.4.1 and $\Delta(f) \in F^2$ imply that $G \subset A_4$. Of the groups in (13.8), only $\langle(12)(34),(13)(24)\rangle \simeq \mathbb{Z}/2\mathbb{Z} \times \mathbb{Z}/2\mathbb{Z}$ lies in A_4. This proves part (b). No conjugacy is needed, since $\langle(12)(34),(13)(24)\rangle$ is normal in S_4.

The final assertion of part (c) follows from the final assertion of part (b). Now suppose that $\beta \in F$ is a root of θ_f and that $\Delta(f) \notin F^2$. The last condition implies $G \not\subset A_4$, so G must be one of the last two groups of (13.8). Our method for distinguishing these begins with Euler's formula (12.17) for the roots of the universal quartic $\tilde{f}$. This formula involves the square root

$$\sqrt{4y_1 + \sigma_1^2 - 4\sigma_2}, \quad y_1 = x_1 x_2 + x_3 x_4,$$

which is related to the roots x_1, x_2, x_3, x_4 of $\tilde{f}$ via

$$4y_1 + \sigma_1^2 - 4\sigma_2 = (x_1 + x_2 - x_3 - x_4)^2$$

(see the discussion leading up to (12.12)). If we apply the evaluation map $x_i \mapsto \alpha_i$ and use (13.7), then we obtain

$$4\beta + c_1^2 - 4c_2 = (\alpha_1 + \alpha_2 - \alpha_3 - \alpha_4)^2.$$

It follows that

(13.9) $$\sqrt{\Delta(f)(4\beta + c_1^2 - 4c_2)} = \sqrt{\Delta(f)} \cdot (\alpha_1 + \alpha_2 - \alpha_3 - \alpha_4) \in L.$$

Now suppose that $4\beta + c_1^2 - 4c_2 \neq 0$, that is, $4\beta + c_1^2 - 4c_2 \in F^*$. If $G = \langle(1324)\rangle$, then $\mathrm{Gal}(L/F)$ has a generator σ that maps to (1324). One easily computes that

$$\sigma(\sqrt{\Delta(f)}) = -\sqrt{\Delta(f)} \ \text{ and } \ \sigma(\alpha_1 + \alpha_2 - \alpha_3 - \alpha_4) = -(\alpha_1 + \alpha_2 - \alpha_3 - \alpha_4).$$

It follows that σ fixes (13.9), and since σ generates the Galois group, we conclude that $\sqrt{\Delta(f)(4\beta + c_1^2 - 4c_2)} \in F$. Thus $\Delta(f)(4\beta + c_1^2 - 4c_2) \in (F^*)^2$.

On the other hand, if $G = \langle(1324), (12)\rangle$, then some $\sigma \in \mathrm{Gal}(L/F)$ maps to (12). For this σ, we have

$$\sigma(\sqrt{\Delta(f)}) = -\sqrt{\Delta(f)} \quad \text{and} \quad \sigma(\alpha_1 + \alpha_2 - \alpha_3 - \alpha_4) = \alpha_1 + \alpha_2 - \alpha_3 - \alpha_4.$$

Hence σ takes (13.9) to its negative. Since F has characteristic $\neq 2$ and (13.9) is nonzero, we have $\sqrt{\Delta(f)(4\beta + c_1^2 - 4c_2)} \notin F$. Thus $\Delta(f)(4\beta + c_1^2 - 4c_2) \notin (F^*)^2$.

The above argument fails when $4\beta + c_1^2 - 4c_2 = 0$ (be sure you see why). In this case, we will use

$$\beta^2 - 4c_4 = (\alpha_1\alpha_2 + \alpha_3\alpha_4)^2 - 4\alpha_1\alpha_2\alpha_3\alpha_4 = (\alpha_1\alpha_2 - \alpha_3\alpha_4)^2.$$

In Exercise 6 you will show that $4\beta + c_1^2 - 4c_2 = 0$ implies that $\beta^2 - 4c_4 \in F^*$. Then, arguing as above, one easily sees that

$$\sqrt{\Delta(f)(\beta^2 - 4c_4)} = \sqrt{\Delta(f)}(\alpha_1\alpha_2 - \alpha_3\alpha_4) \notin F^*$$

if and only if $G = \langle(1324), (12)\rangle$ (see Exercise 6). This completes the proof. $\quad\square$

We now give some examples of Theorem 13.1.1.

Example 13.1.2. Consider $f = x^4 - 4x^2 + x + 1 \in \mathbb{Q}[x]$. One can show that f is irreducible of discriminant $\Delta(f) = 1957 = 19 \cdot 103$ and that its resolvent

$$\theta_f(y) = y^3 + 4x^2 - 4x - 17$$

is irreducible over $\mathbb{Q}$. By Theorem 13.1.1, the Galois group of f is S_4, so that the splitting field of f has degree 24 over $\mathbb{Q}$. This has the following consequences:

- In Example 8.6.7 of Section 8.6, we used f as an example of an irreducible polynomial of degree 4 whose roots are all real yet cannot be expressed by real radicals. This follows from Theorem 8.6.5, since the degree of the splitting field over $\mathbb{Q}$ is not a power of 2.
- In Example 10.1.13 of Section 10.1, we used f as an example of polynomial of degree 4 whose roots are not constructible. This follows from Theorem 10.1.12, since the degree of the splitting field over $\mathbb{Q}$ is not a power of 2. $\quad\triangleleft\triangleright$

In Exercises 7 and 8 you will apply Theorem 13.1.1 to other quartic polynomials from earlier in the text.

Our next example is taken from [Chebotarev, p. 253].

Example 13.1.3. Suppose that $f = x^4 + ax^3 + bx^2 + ax + 1 \in F[x]$ is irreducible, where F has characteristic $\neq 2$. In this case, the resolvent is

$$\theta_f(y) = y^3 - by^2 + (a^2 - 4)y - 2a^2 + 4b$$
$$= (y - 2)(y^2 + (2 - b)y + a^2 - 2b),$$

and the discriminant is

$$\Delta(f) = (4b - a^2 - 8)^2 (b - 2a + 2)(b + 2a + 2).$$

By Theorem 13.1.1, it follows that the Galois group of f is $\mathbb{Z}/2\mathbb{Z} \times \mathbb{Z}/2\mathbb{Z}$ if and only if $(b - 2a + 2)(b + 2a + 2) = (b + 2)^2 - 4a^2 \in F^2$.

The above factorization of $\theta_f(y)$ is easy to find using *Maple* or *Mathematica*. In Exercise 9 you will show that the factor $y - 2$ of $\theta_f(y)$ arises naturally from the symmetry of f.   ◁▷

Here is an example taken from [15] that illustrates part (c) of Theorem 13.1.1.

Example 13.1.4. Assume that $f = x^4 + bx^2 + d \in F[x]$ is irreducible, where F has characteristic $\neq 2$. Also assume that $d \notin F^2$. In Exercise 10 you will show that f has discriminant

$$\Delta(f) = 16d(b^2 - 4d)^2$$

and resolvent

$$\theta_f(y) = y^3 - by^2 - 4dy + 4bd = (y - b)(y^2 - 4d).$$

Since $d \notin F^2$ and $\theta_f(y)$ is reducible, part (c) of Theorem 13.1.1 applies. Using $4\beta + c_1^2 - 4c_2 = 4b + 0^2 - 4b = 0$ and $\beta^2 - 4c_4 = b^2 - 4d$, we see that the Galois group of f is D_8 if $d(b^2 - 4d) \notin F^2$ and $\mathbb{Z}/4\mathbb{Z}$ otherwise.

See [15] for an analysis of what happens when $d \in F^2$.   ◁▷

Mathematical Notes

The text contains several ideas that will be developed in subsequent sections. Here are some remarks to help us see what is involved.

■ **Transitive Permutation Groups.** We noted in the proof of Theorem 13.1.1 that G is a transitive subgroup of S_4, since f is irreducible. Transitive subgroups of S_n will play a prominent role in this chapter and the next. For example, implicit in Theorem 13.1.1 is the following classification of the transitive subgroups of S_4.

Theorem 13.1.5. *Up to conjugacy, the transitive subgroups of S_4 are*

$$S_4, \ A_4, \ \langle(1324), (12)\rangle, \ \langle(1324)\rangle, \ \langle(12)(34), (13)(24)\rangle.$$

Proof. Let $G \subset S_4$ be a transitive subgroup. If we can prove that G arises from the Galois action on the roots of some monic irreducible quartic over a field of characteristic $\neq 2$, then G is conjugate to one of the above five groups by Theorem 13.1.1.

We will find the desired quartic polynomial using the methods of Section 7.4. When $F = \mathbb{Q}$, the universal extension in degree 4,

$$K = \mathbb{Q}(\sigma_1, \sigma_2, \sigma_3, \sigma_4) \subset L = \mathbb{Q}(x_1, x_2, x_3, x_4),$$

is the splitting field of the universal quartic $\tilde{f} = x^4 - \sigma_1 x^3 + \sigma_2 x^2 - \sigma_3 x + \sigma_4$. Given a transitive subgroup $G \subset S_4 \simeq \mathrm{Gal}(L/K)$, the corresponding fixed field

$$K \subset M \subset L$$

satisfies $\text{Gal}(L/M) \simeq G$. Also observe that L is the splitting field over M of $\tilde{f}$ and that $\tilde{f}$ is irreducible over M since G is transitive (this is Proposition 6.3.7). As noted above, we are now done by Theorem 13.1.1. $\square$

In practice, a standard strategy for computing Galois groups is the reverse of what we did in this section. When considering irreducible polynomials $f \in F[x]$ of degree n, one *first* finds all transitive subgroups of S_n up to conjugacy and *then*, for each such subgroup, determines criteria for the Galois group of f to be conjugate to that subgroup. This is the approach we will use for the quintic in Section 13.2.

■ **Resolvents.** The general theory of resolvents is based on the ideas of Lagrange discussed in Section 12.1. For $\theta_f(y)$, recall from the proof of Theorem 13.1.1 that we began in the universal case with $x_1x_2 + x_3x_4$ and constructed the Ferrari resolvent $\theta(y)$ of the universal quartic $\tilde{f}$. Then specializing to f gave $\theta_f(y)$. All of the resolvents considered in this chapter will be constructed similarly.

Besides $\theta_f(y)$, Theorem 13.1.1 needs to know whether or not $\Delta(f) \in F^2$. This can be stated in terms of the resolvent polynomial $y^2 - \Delta(f)$, since $\Delta(f) \in F^2$ if and only if $y^2 - \Delta(f)$ has a root in F.

In general, if a resolvent has a root in F, then this puts strong restrictions on the Galois group. For example, if $y^2 - \Delta(f)$ has a root in F, then the group G lies in A_4 (be sure you understand why), and if $\theta_f(y)$ has a root in F, then the proof of Theorem 13.1.1 shows that some conjugate of G lies in $\langle (1324), (12) \rangle$. By combining information from different resolvents, we can obtain precise information about the Galois group. We will pursue these ideas in Section 13.3.

We can also use resolvents to explain part (c) of Theorem 13.1.1. This part of the theorem says that if $\beta \in F$ is a root of $\theta_f(y)$ and $\Delta(f) \notin F^2$, then the Galois group is D_8 when β satisfies the condition that either

$$4\beta + c_1^2 - 4c_2 \neq 0 \quad \text{and} \quad \Delta(f)(4\beta + c_1^2 - 4c_2) \notin (F^*)^2$$

or

$$4\beta + c_1^2 - 4c_2 = 0 \quad \text{and} \quad \Delta(f)(\beta^2 - 4c_4) \notin (F^*)^2.$$

The first part of the condition implies that $y^2 - \Delta(f)(4\beta + c_1^2 - 4c_2)$ has no roots in F. Because of the appearance of β, we will call $y^2 - \Delta(f)(4\beta + c_1^2 - 4c_2)$ a *relative resolvent* in Section 13.3. Similarly, the second part of the condition can be stated in terms of the relative resolvent $y^2 - \Delta(f)(\beta^2 - 4c_4)$.

Furthermore, $\Delta(f)(4\beta + c_1^2 - 4c_2)$ and $\Delta(f)(\beta^2 - 4c_4)$ are nonzero if and only if the corresponding relative resolvents have simple roots (as defined in Section 5.3). We will see in Section 13.3 that simple roots of resolvents or relative resolvents are needed in order to get useful information about the Galois group.

■ **Diophantine Equations.** So far, we have always had a fixed polynomial whose Galois group we wanted to determine. But if we let the coefficients of the polynomial vary, then the criteria of Theorem 13.1.1 lead to some interesting equations. We begin by revisiting an earlier example.

Example 13.1.6. Let $f = x^4 + ax^3 + bx^2 + ax + 1 \in \mathbb{Q}[x]$ and assume that f is irreducible over $\mathbb{Q}$. By Example 13.1.3, the Galois group of f over $\mathbb{Q}$ is $\mathbb{Z}/2\mathbb{Z} \times \mathbb{Z}/2\mathbb{Z}$ if and only if $(b + 2)^2 - 4a^2 \in \mathbb{Q}^2$. The latter is equivalent to saying that

(13.10) $$(b + 2)^2 - 4a^2 = c^2$$

for some $c \in \mathbb{Q}$. If we write this as

$$4a^2 + c^2 = (b + 2)^2,$$

then $f = x^4 + ax^3 + bx^2 + ax + 1 \in \mathbb{Q}[x]$ has $\mathbb{Z}/2\mathbb{Z} \times \mathbb{Z}/2\mathbb{Z}$ as Galois group if and only if there is $c \in \mathbb{Q}$ such that $(x, y, z) = (2a, c, b + 2)$ lies on the cone

$$x^2 + y^2 = z^2.$$

Hence we have an equation where we only want solutions whose coordinates all lie in $\mathbb{Q}$. This is an example of a *Diophantine equation*. Such equations are an important part of number theory.

In Exercise 11 you will show that if $f = x^4 + ax^3 + bx^2 + ax + 1$ is irreducible with positive integer coefficients, then the Galois group is $\mathbb{Z}/2\mathbb{Z} \times \mathbb{Z}/2\mathbb{Z}$ if and only if there is $c > 0$ in $\mathbb{Z}$ such that $(2a, c, b + 2)$ is a *Pythagorean triple*, that is, the integers $2a, c, b + 2$ are the sides of a right triangle with $b + 2$ as hypotenuse. ◁▷

Here is a polynomial that leads to a more sophisticated equation.

Example 13.1.7. Assume that $f = x^4 + x + b \in \mathbb{Z}[x]$ is irreducible over $\mathbb{Q}$. This polynomial has discriminant

$$\Delta(f) = 256b^3 - 27$$

and resolvent

$$\theta_f(y) = y^3 - 4by - 1.$$

It is easy to see that $\theta_f(y)$ is irreducible over $\mathbb{Q}$, since its only possible rational roots are ± 1 (be sure you can explain why). Thus the Galois group over $\mathbb{Q}$ is either S_4 or A_4, and the latter happens if and only if $256b^3 - 27 \in \mathbb{Q}^2$. In fact, we can replace $\mathbb{Q}^2$ with $\mathbb{Z}^2$ since $b \in \mathbb{Z}$. Thus the Galois group is A_4 if and only if there is $c \in \mathbb{Z}$ such that the point $(x, y) = (b, c)$ lies on the curve

$$y^2 = 256x^3 - 27.$$

This is an example of an *elliptic curve*. A famous theorem of Siegel asserts that such an equation has at most finitely many integer solutions (see [27]). So there are at most finitely many integers b such that $f = x^4 + x + b$ has Galois group A_4 over $\mathbb{Q}$.

This example can be extended in several ways. First, one could allow b to be a rational number. Then one seeks rational points on the above elliptic curve. Some of the deepest conjectures in number theory involve rational points on elliptic curves (see [27] for an introduction). Another direction would be to consider polynomials $x^4 + ax + b \in \mathbb{Z}[x]$ with Galois group A_4. This problem is solved in [31] using methods from algebraic number theory. ◁▷

Historical Notes

The first person to give a systematic method for finding the Galois group of a quartic was F. Hack, in his unpublished 1895 dissertation. In recent years, many books and papers have addressed this problem—see the references in [15], to which one can add [Escofier] and [Garling].

Our version of Theorem 13.1.1 is based on the recent paper [14], which, as pointed out in [19], is essentially equivalent to the method of [15] (see Exercise 12). Another method for handling part (c) of Theorem 13.1.1 is described in Exercise 14.

Exercises for Section 13.1

Exercise 1. Let $f \in F[x]$ be separable of degree n, and let $\alpha_1, \ldots, \alpha_n$ be the roots of f in a splitting field $F \subset L$ of f. In Section 6.3 we used the action of the Galois group on the roots to construct a one-to-one group homomorphism $\phi_1 : \mathrm{Gal}(L/F) \to S_n$. Now let $\beta_1, \ldots, \beta_n$ be the same roots, possibly written in a different order. This gives $\phi_2 : \mathrm{Gal}(L/F) \to S_n$. To relate ϕ_1 and ϕ_2, note that there is $\gamma \in S_n$ such that $\beta_i = \alpha_{\gamma(i)}$ for $1 \le i \le n$. Now define the conjugation map $\hat{\gamma} : S_n \to S_n$ by $\hat{\gamma}(\tau) = \gamma^{-1}\tau\gamma$.
(a) Prove that $\phi_2 = \hat{\gamma} \circ \phi_1$.
(b) Let $G \subset S_n$ be the image of ϕ_1. Explain why part (a) justifies the assertion made in the text that "if we change the labels, then G gets replaced with a conjugate subgroup."

Exercise 2. Prove that A_4 is the only subgroup of S_4 with 12 elements.

Exercise 3. Explain carefully why (13.6) follows from Exercise 9 of Section 2.4.

Exercise 4. Use Example 7.3.4 from Chapter 7 to show that (13.8) gives all subgroups of $\langle (1324), (12) \rangle$ of order 4 or 8.

Exercise 5. Let F be a field of characteristic $\neq 2$, and let $g \in F[x]$ be a monic cubic polynomial that has a root in F. Prove that g splits completely over F if and only if $\Delta(g) \in F^2$.

Exercise 6. This exercise is concerned with the proof of part (c) of Theorem 13.1.1. Let $f(x) = x^4 - c_1 x^3 + c_2 x^2 - c_3 x + c_4$ as in the theorem.
(a) Suppose that f has roots $\alpha_1, \alpha_2, \alpha_3, \alpha_4$ such that $\alpha_1 + \alpha_2 - \alpha_3 - \alpha_4 = \alpha_1\alpha_2 - \alpha_3\alpha_4 = 0$. Prove that f is not separable.
(b) Let β be a root of the resolvent $\theta_f(y)$. Use part (a) to prove that $4\beta + c_1^2 - 4c_2$ and $\beta^2 - 4c_4$ can't both vanish when f is separable.
(c) Suppose that $4\beta + c_1^2 - 4c_2 = 0$ in part (c) of Theorem 13.1.1. Prove carefully that G is conjugate to $\langle (1324), (12) \rangle$ if and only if $\Delta(f)(\beta^2 - 4c_4) \notin (F^*)^2$.

Exercise 7. In Exercise 18 of Section 12.1 you found the roots of $f = x^4 + 2x^2 - 4x + 2 \in \mathbb{Q}[x]$ using the formulas developed in that section. At the end of the exercise, we said that "this quartic is especially simple." Justify this assertion using Theorem 13.1.1.

Exercise 8. In Example 10.3.10, we showed that the roots of $f = 7m^4 - 16m^3 - 21m^2 + 8m + 4 \in \mathbb{Q}[m]$ can be constructed using origami. Show that the splitting field of f is an extension of $\mathbb{Q}$ of degree 24. By the results of Section 10.1, it follows that the roots of f are not constructible with straightedge and compass, since 24 is not a power of 2.

Exercise 9. As in Example 13.1.3, let $f = x^4 + ax^3 + bx^2 + ax + 1 \in F[x]$, and let α be a root of f in some splitting field of f over F. Show that α^{-1} is also a root of f, and then use (13.5) to conclude that 2 is a root of the resolvent $\theta_f(y)$.

Exercise 10. As in Example 13.1.4, let $f = x^4 + bx^2 + d \in F[x]$, where $d \notin F^2$. Compute $\Delta(f)$ and $\theta_f(y)$.

Exercise 11. In Example 13.1.6 we showed that if $f = x^4 + ax^3 + bx^2 + ax + 1 \in \mathbb{Z}[x]$ is irreducible over $\mathbb{Q}$, then its Galois group is $\mathbb{Z}/2\mathbb{Z} \times \mathbb{Z}/2\mathbb{Z}$ if and only if there is $c \in \mathbb{Q}$ such that $4a^2 + c^2 = (b + 2)^2$.
  (a) Show that $c \in \mathbb{Z}$, and use the irreducibility of f to prove that $c \neq 0$. Hence we may assume that $c > 0$, so that $(2a, c, b + 2)$ is a .
  (b) Show that $3^2 + 4^2 = 5^2$, $5^2 + 12^2 = 13^2$, $7^2 + 24^2 = 25^2$, and $8^2 + 15^2 = 17^2$ give two examples of polynomials with $\mathbb{Z}/2\mathbb{Z} \times \mathbb{Z}/2\mathbb{Z}$ as Galois group (two of the triples give reducible polynomials).

Exercise 12. As in part (c) of Theorem 13.1.1, assume that $\Delta(f) \notin F^2$ and that $\beta \in F$ is a root of $\theta_f(y)$. Then define the quartic polynomial

$$g(x) = (x^2 - \beta x + c_4)(x^2 + c_1 x + c_2 - \beta).$$

  (a) Use (13.6) to show that $F(\sqrt{\Delta(f)})$ is the splitting field E of $\theta_f(y)$ over F.
  (b) In [15], it is shown that $G \simeq \mathbb{Z}/4\mathbb{Z}$ if $g(x)$ splits completely over E, and $G \simeq D_8$ if it doesn't. Prove that this is equivalent to part (c) of Theorem 13.1.1.

Exercise 13. Suppose that $f \in F[x]$ satisfies the hypothesis of part (c) of Theorem 13.1.1, and let α be a root of f. Prove that $G \simeq \mathbb{Z}/4\mathbb{Z}$ if f splits completely over $F(\alpha)$, and $G \simeq D_8$ otherwise. This gives a version of part (c) that doesn't use resolvents. Since we can factor over extension fields by Section 4.2, this method is useful in practice.

Exercise 14. Use Theorem 13.1.1 to compute the Galois groups of the following polynomials in $\mathbb{Q}[x]$:
  (a) $x^4 + 4x + 2$.
  (b) $x^4 + 8x + 12$.
  (c) $x^4 + 1$.
  (d) $x^4 + x^3 + x^2 + x + 1$.
  (e) $x^4 - 2$.

Exercise 15. The goal of this exercise is to explain part (c) of Theorem 13.1.1 in terms of subfields of the splitting field $F \subset L = F(\alpha_1, \alpha_2, \alpha_3, \alpha_4)$ of $f = x^4 - c_1 x^3 + c_2 x^2 - c_3 x + c_4$. Assume that $\Delta(f) \notin F^2$ and that $\beta = \alpha_1\alpha_2 + \alpha_3\alpha_4 \in F$. In the text we showed that

$$4\beta + c_1^2 - 4c_2 = (\alpha_1 + \alpha_2 - \alpha_3 - \alpha_4)^2 \quad \text{and} \quad \beta^2 - 4c_4 = (\alpha_1\alpha_2 - \alpha_3\alpha_4)^2.$$

It follows that L contains the extensions

$$F \subset F\left(\sqrt{\Delta(f)}\right) \quad \text{and} \quad F \subset F\left(\sqrt{4\beta + c_1^2 - 4c_2}\right) \quad \text{and} \quad F \subset F\left(\sqrt{\beta^2 - 4c_4}\right).$$

The first has degree 2, and the second and third have degree at most 2.
  (a) Show that $F \subset F\left(\sqrt{4\beta + c_1^2 - 4c_2}\right)$ has degree 2 when $4\beta + c_1^2 - 4c_2 \neq 0$, and show that the same is true for $F \subset F\left(\sqrt{\beta^2 - 4c_4}\right)$. By part (a) of Exercise 6 it follows that at least one of these extensions has degree 2.

(b) Suppose that $u, v \in F$ give quadratic extensions $F \subset F(\sqrt{u})$ and $F \subset F(\sqrt{v})$. Prove that $F(\sqrt{u}) = F(\sqrt{v})$ if and only if $uv \in F^2$.

(c) If the Galois group of $F \subset L$ is $\mathbb{Z}/4\mathbb{Z}$, then L contains a unique subfield of degree 2 over F. Use this observation together with parts (a) and (b) to give a new proof of part (c) of Theorem 13.1.1.

Exercise 16. In the proof of Theorem 13.1.1, we used (13.5) to show that G is conjugate to a subgroup of D_8 when $\theta_f(y)$ has a root in F. If we don't know (13.5), then we can still prove that $|G| = 4$ or 8 using (12.17). Do this.

Exercise 17. Consider the subgroups $\langle (12), (34) \rangle$ and $\langle (12)(34), (13)(24) \rangle$ of S_4.

(a) Prove that these subgroups are isomorphic but not conjugate. This shows that when classifying subgroups of a given group, it can happen that nonconjugate subgroups can be isomorphic as abstract groups.

(b) Explain why the subgroup $\langle (12), (34) \rangle$ isn't mentioned in Theorems 13.1.1 and 13.1.5.

13.2   QUINTIC POLYNOMIALS

Polynomials of degree 5 have a richer Galois theory than those of degree 4. There are some obvious reasons for this: the computations are more complicated because the degree is higher, and the groups are more complicated because they need not be solvable. The surprise is that quintic equations have strong relations to many other areas of mathematics, including:

- **Geometry.** The rotational symmetry group of the icosahedron is A_5, and geometrically defined invariants of this group action have consequences for the quintic.
- **Iteration.** The "Galois theory" of Newton's method for solving polynomial equations due to Doyle and McMullen [6] uses A_5 in a crucial way.
- **Elliptic Functions.** These functions arise in complex analysis and number theory, yet can also be used to find roots of quintics that can't be solved by radicals.

Because of such connections, quintic equations are the subject of entire books, in particular those by King [16], Klein [17], and Shurman [26].

   The aims of this section are more modest. We will focus on computing the Galois group of a quintic and in particular on determining when a quintic is solvable by radicals. As we will see, this will involve some substantial mathematics.

A. Transitive Subgroups of S_5. If a quintic polynomial $f \in F[x]$ is separable and irreducible, then its Galois group $\mathrm{Gal}(L/F)$ (where $F \subset L$ is a splitting field) is isomorphic to a transitive subgroup of S_5. So we begin by determining these subgroups up to conjugacy. Our first result is elementary.

Lemma 13.2.1. *Let $G \subset S_5$ be a subgroup. Then the following are equivalent:*

(a) *G is transitive.*

(b) *$|G|$ is divisible by 5.*

(c) *G contains a 5-cycle.*

Proof. For (a) $\Rightarrow$ (b), recall that the order of an orbit divides the order of the group, by the Fundamental Theorem of Group Actions (Theorem A.4.9). Then we are done, since transitivity implies that $\{1, 2, 3, 4, 5\}$ is an orbit of the action of G.

The implication (b) $\Rightarrow$ (c) is proved using the argument given in the discussion following (6.8) in Section 6.4. Finally, for (c) $\Rightarrow$ (a), note that repeatedly applying the five cycle $(i_1\, i_2\, i_3\, i_4\, i_5)$ to i_1 gives

$$i_1 \mapsto i_2 \mapsto i_3 \mapsto i_4 \mapsto i_5 \mapsto i_1 \mapsto i_2 \mapsto \cdots .$$

Transitivity follows immediately, since $\{i_1, i_2, i_3, i_4, i_5\} = \{1, 2, 3, 4, 5\}$. $\qquad\square$

It turns out that we already know most of the transitive subgroups of S_5 up to conjugacy. More precisely, we have the following subgroups:

- The full symmetric group S_5 of order 120.
- The alternating group A_5 of order 60.
- The cyclic group $\langle(12345)\rangle$ of order 5.
- By Section 6.4, the one-dimensional affine linear group $\mathrm{AGL}(1, \mathbb{F}_p)$ is the group of order $p(p-1)$ consisting of maps $i \mapsto ai + b$ where $i, a, b \in \mathbb{F}_p$ and $a \neq 0$. If we set $p = 5$ and regard $\{1, 2, 3, 4, 5\}$ as congruence classes modulo 5, then

$$\mathrm{AGL}(1, \mathbb{F}_5) \subset S_5$$

is a subgroup of order 20. In particular, translation by 1 ($i \mapsto i + 1$) is the 5-cycle (12345) and multiplication by 2 ($i \mapsto 2i$) is the 4-cycle (1243). Be sure you understand this.

In Exercise 1 you will show that $\mathrm{AGL}(1, \mathbb{F}_5)$ is generated by (12345) and (1243). Furthermore, (12345) is an even permutation while (1243) is odd. Hence

(13.11) $$\mathrm{AGL}(1, \mathbb{F}_5) \cap A_5$$

is a proper subgroup of $\mathrm{AGL}(1, \mathbb{F}_5)$ containing $\langle(12345)\rangle$. The group (13.11) also contains $(1243)^2 = (14)(23)$ (multiplication by 4—do you see why?). In Exercise 1 you will show that

$$\mathrm{AGL}(1, \mathbb{F}_5) \cap A_5 = \langle(12345), (14)(23)\rangle \simeq D_{10},$$

where D_{10} is the dihedral group of order 10.

The subgroup (13.11) and the four subgroups described in the bullets give five subgroups of S_5, all transitive by Lemma 13.2.1 since their orders are divisible by 5. These groups fit together in the diagram

(13.12)

$$
\begin{array}{ccc}
 & & S_5 \\
 & \nearrow & \uparrow \\
\mathrm{AGL}(1, \mathbb{F}_5) & & A_5 \\
\uparrow & \nearrow & \\
\mathrm{AGL}(1, \mathbb{F}_5) \cap A_5 & & \\
\uparrow & & \\
\langle(12345)\rangle & &
\end{array}
$$

We now classify transitive subgroups of S_5 up to conjugacy.

Theorem 13.2.2. *Every transitive subgroup* $G \subset S_5$ *is conjugate to one of the groups in the diagram* (13.12).

Proof. By Lemma 13.2.1, G contains a 5-cycle. Hence, replacing G by a conjugate if necessary, we may assume that $(12345) \in G$. The key idea of the proof is to consider the number of cyclic subgroups of order 5 contained in G.

First suppose that $\langle (12345) \rangle$ is the only such subgroup of G. Then we have $g\langle (12345) \rangle g^{-1} = \langle (12345) \rangle$ for all $g \in G$. In the language of the Mathematical Notes to Section 7.2, this means that G is contained in the *normalizer*

$$N_{S_5}\big(\langle (12345) \rangle\big) = \{g \in S_5 \mid g\langle (12345) \rangle g^{-1} = \langle (12345) \rangle\}.$$

In Section 14.1 we will more generally consider the normalizer

$$N_{S_p}\big(\langle (12 \dots p) \rangle\big) = \{g \in S_p \mid g\langle (12 \dots p) \rangle g^{-1} = \langle (12 \dots p) \rangle\},$$

where p is now any prime, and we will prove in Lemma 14.1.2 that

$$N_{S_p}\big(\langle (12 \dots p) \rangle\big) = \mathrm{AGL}(1, \mathbb{F}_p).$$

This is part of Galois's brilliant analysis of which irreducible polynomials of prime degree are solvable by radicals. Rather than repeat the argument here, we will simply assume this result from Chapter 14.

It follows that if $\langle (12345) \rangle$ is the only subgroup of G of order 5, then G is a subgroup of $\mathrm{AGL}(1, \mathbb{F}_5)$. In Exercise 1 you will show that this implies that G is one of the groups

$$\langle (12345) \rangle, \quad \mathrm{AGL}(1, \mathbb{F}_5) \cap A_5, \quad \text{or} \quad \mathrm{AGL}(1, \mathbb{F}_5).$$

It remains to consider what happens when G contains more than one subgroup of order 5. In Exercise 2 you will prove that $\langle (12345) \rangle$ is a 5-Sylow subgroup of G. By the Third Sylow Theorem (see Theorem A.5.1 in Appendix A), the number of 5-Sylow subgroups of G is congruent to 1 modulo 5. Since we have more than one, we must have at least 6. Furthermore, each 5-Sylow subgroup has four 5-cycles, and any two such subgroups intersect in the identity. Since we have at least six such subgroups, G must contain at least twenty-four 5-cycles. Yet S_5 has exactly twenty-four 5-cycles by Exercise 2. It follows that G contains all 5-cycles.

We are almost done. The easily verified identity

$$(ijklm)(ijmlk) = (ikj), \quad \{i, j, k, l, m\} = \{1, 2, 3, 4, 5\},$$

shows that G contains all 3-cycles. We know from Section 8.4 that A_5 is generated by 3-cycles. It follows that G contains A_5, so that G is A_5 or S_5. This completes the proof. $\qquad\qquad\square$

In Exercise 3 we will give a more elementary version of the above argument that doesn't use the Third Sylow Theorem.

As a corollary of Theorem 13.2.2, we get the following criterion for an irreducible quintic to be solvable by radicals.

Corollary 13.2.3. *Assume that $f \in F[x]$ is irreducible of degree 5 and that F has characteristic 0. Then f is solvable by radicals over F if and only if its Galois group over F is isomorphic to a subgroup of $\mathrm{AGL}(1, \mathbb{F}_5)$.*

Proof. By Theorem 8.5.3, f is solvable by radicals if and only if its Galois group is solvable. The Galois group is isomorphic to a subgroup of S_5, but A_5 and S_5 aren't solvable, by Theorem 8.4.5. So one direction of the corollary follows immediately from Theorem 13.2.2. For the other direction, we note that $\mathrm{AGL}(1, \mathbb{F}_5)$ and hence all of its subgroups are solvable by Example 8.1.6 from Section 8.1. $\qquad\square$

Section 14.1 will discuss Galois's generalization of Corollary 13.2.3 in which 5 is replaced by an arbitrary prime p.

B.   Galois Groups of Quintics. Let $f \in F[x]$ be monic, separable, and irreducible of degree 5, where F is a field of characteristic $\neq 2$. We will determine the Galois group of f over F using a discriminant, a sextic resolvent, and a factorization.

The discriminant is the usual discriminant $\Delta(f)$, and the factorization will be described in the statement of Theorem 13.2.6. So let's turn our attention to the sextic resolvent. The idea is to find a polynomial

$$h \in F[x_1, \ldots, x_5]$$

with the property that

$$\mathrm{AGL}(1, \mathbb{F}_5) = \{\sigma \in S_5 \mid \sigma \cdot h = h\}.$$

Thus h should have $\mathrm{AGL}(1, \mathbb{F}_5)$ as its symmetry group. We will use the polynomial

$$h = u^2,$$

where

(13.13)
$$\begin{aligned}
u = {}& x_1 x_2 + x_2 x_3 + x_3 x_4 + x_4 x_5 + x_5 x_1 \\
& - x_1 x_3 - x_3 x_5 - x_5 x_2 - x_2 x_4 - x_4 x_1.
\end{aligned}$$

The signs are best explained using the diagram

(13.14)

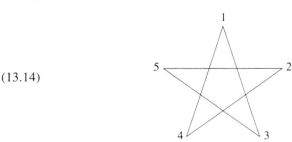

In the formula (13.13) for u, the coefficient of $x_i x_j$ (where $i \neq j$) is -1 if i and j are connected by a line segment in (13.14) and $+1$ otherwise. Since (12345) rotates the star by $2\pi/5$ radians, it follows that $(12345) \cdot u = u$.

On the other hand, (1243) takes (13.14) to the diagram

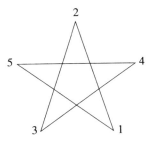

Here, i and j are connected by a line segment if and only if they are not connected in (13.14). Hence (1243) interchanges all signs, so that $(1243) \cdot u = -u$.

It follows that $h = u^2$ is fixed by (12345) and (1243). Since these generate $\mathrm{AGL}(1, \mathbb{F}_5)$, we see that h is invariant under $\mathrm{AGL}(1, \mathbb{F}_5)$. This is the full symmetry group of h, as we now prove.

Lemma 13.2.4. *Let $h = u^2$, where u is defined in* (13.13)*. Then* $\mathrm{AGL}(1, \mathbb{F}_5) = \{\sigma \in S_5 \mid \sigma \cdot h = h\}$.

Proof. Let $G = \{\sigma \in S_5 \mid \sigma \cdot h = h\}$. Then $\mathrm{AGL}(1, \mathbb{F}_5) \subset G$ by the above argument. If G were strictly bigger than $\mathrm{AGL}(1, \mathbb{F}_5)$, then (13.12) would show that $G = S_5$, that is, h would be symmetric. However, observe that

$$h = (x_1 x_2 + x_2 x_3 + \cdots - x_1 x_3 - \cdots)^2$$
$$= x_1^2 x_2^2 + x_2^2 x_3^2 + x_1^2 x_3^2 + 2x_1 x_2^2 x_3 - 2x_1^2 x_2 x_3 - 2x_1 x_2 x_3^2 + \cdots,$$

where terms involving x_4 or x_5 are not shown. This makes it easy to see that h is not symmetric. Thus $G \neq S_5$, which implies that $G = \mathrm{AGL}(1, \mathbb{F}_5)$. $\square$

By Exercise 4, left coset representatives of $\mathrm{AGL}(1, \mathbb{F}_5)$ in S_5 are

(13.15) $e, (123), (234), (345), (145), (125).$

Thus the orbit of S_5 acting on h consists of

$$h_1 = e \cdot h = h, \quad h_2 = (123) \cdot h, \quad h_3 = (234) \cdot h,$$
$$h_4 = (345) \cdot h, \quad h_5 = (145) \cdot h, \quad h_6 = (125) \cdot h$$

(be sure you can explain why). This enables us to form the *universal sextic resolvent*

(13.16) $$\theta(y) = \prod_{i=1}^{6} (y - h_i).$$

The methods of Section 12.1 imply that $\theta(y)$ has coefficients in $F[\sigma_1, \sigma_2, \sigma_3, \sigma_4, \sigma_5]$.

Our given monic separable irreducible quintic can be written as

$$f = x^5 - c_1 x^4 + c_2 x^3 - c_3 x^2 + c_4 x - c_5 \in F[x].$$

Let its splitting field be $F \subset L$, and let $\alpha_1, \alpha_2, \alpha_3, \alpha_4, \alpha_5$ be the roots of f in L. Under the evaluation map $x_i \mapsto \alpha_i$, we know that $\sigma_i \mapsto c_i \in F$. Thus (13.16) maps to the *sextic resolvent* of f

$$(13.17) \qquad \theta_f(y) = \prod_{i=1}^{6} (y - \beta_i) \in F[x],$$

where

$$\beta_i = h_i(\alpha_1, \alpha_2, \alpha_3, \alpha_4, \alpha_5) \in L.$$

The structure of $\theta_f(y)$ is described by the following proposition, whose proof we defer until later.

Proposition 13.2.5. *Given $f \in F[x]$ as above, its sextic resolvent can be written*

$$\theta_f(y) = (y^3 + b_2 y^2 + b_4 y + b_6)^2 - 2^{10} \Delta(f)\, y,$$

where $b_2, b_4, b_6 \in F$.

The Galois group of f is $\mathrm{Gal}(L/F)$ for a splitting field L of f over F. Also recall that $\mathrm{Gal}(L/F) \simeq G \subset S_5$, where G is transitive. Here is our main result.

Theorem 13.2.6. *Assume that $f \in F[x]$ is monic, separable, and irreducible of degree 5 and that F is a field of characteristic $\neq 2$. Then the subgroup $G \subset S_5$ defined above has the following properties:*
(a) *$G \subset A_5$ if and only if $\Delta(f) \in F^2$.*
(b) *G is conjugate to a subgroup of $\mathrm{AGL}(1, \mathbb{F}_5)$ if and only if the sextic resolvent $\theta_f(y)$ defined in (13.17) has a root in F.*
(c) *G is conjugate to $\langle (12345) \rangle$ if and only if f splits completely over $F(\alpha)$, where α is a root of f.*

Proof. Part (a) follows from Theorem 7.4.1 since F has characteristic $\neq 2$. Also, part (c) is relatively straightforward and is left as Exercise 5.

It remains to prove part (b). If G is conjugate to a subgroup of $\mathrm{AGL}(1, \mathbb{F}_5)$, then relabeling the roots if necessary, we may assume that $G \subset \mathrm{AGL}(1, \mathbb{F}_5)$. Let an arbitrary $\sigma \in \mathrm{Gal}(L/F)$ correspond to $\tau \in G$. Then

$$\begin{aligned}
\sigma(\beta_1) &= \sigma\big(h(\alpha_1, \alpha_2, \alpha_3, \alpha_4, \alpha_5)\big) \\
&= h(\sigma(\alpha_1), \sigma(\alpha_2), \sigma(\alpha_3), \sigma(\alpha_4), \sigma(\alpha_5)) \\
&= h(\alpha_{\tau(1)}, \alpha_{\tau(2)}, \alpha_{\tau(3)}, \alpha_{\tau(4)}, \alpha_{\tau(5)}) \\
&= (\tau \cdot h)(\alpha_1, \alpha_2, \alpha_3, \alpha_4, \alpha_5) = h(\alpha_1, \alpha_2, \alpha_3, \alpha_4, \alpha_5) = \beta_1,
\end{aligned}$$

where the last line follows from $\tau \in G \subset \mathrm{AGL}(1, \mathbb{F}_5)$ and Lemma 13.2.4. Since $F \subset L$ is a Galois extension, we must have $\beta_1 \in F$. Thus $\theta_f(y)$ has a root in F.

Conversely, suppose that $\theta_f(y)$ has a root in F. If G is conjugate to a subgroup of $AGL(1, \mathbb{F}_5)$, then we are done. So assume that this is not true. Since G is transitive, Theorem 13.2.2 implies that $A_5 \subset G$. Let τ_i be the 3-cycle in (13.15) with $\tau_i \cdot h = h_i$, and let $\sigma_i \in Gal(L/F)$ map to τ_i (the existence of σ_i follows from $A_5 \subset G$). Then arguing as above shows that

$$\sigma_i(\beta_1) = \beta_i, \quad i = 1, \ldots, 6$$

(be sure you can supply the details). By assumption, some $\beta_i \in F$, and then the above equations easily imply that $\beta_1 = \cdots = \beta_6$. Hence the sextic resolvent is

$$\theta_f(y) = (y - \beta_1)^6.$$

Comparing this with Proposition 13.2.5, we obtain the identity

(13.18) $$(y - \beta_1)^6 = (y^3 + b_2 y^2 + b_4 y + b_6)^2 - 2^{10} \Delta(f) \, y,$$

where $\beta_1, b_2, b_4, b_6 \in F$. Multiplying this out and comparing the coefficients of y^5, y^4, and y^3 gives the equations

$$-6\beta_1 = 2b_2,$$
$$15\beta_1^2 = b_2^2 + 2b_4,$$
$$-20\beta_1^3 = 2b_2 b_4 + 2b_6.$$

In Exercise 5 you will verify these equations and use them to show that

$$b_2 = -3\beta_1, \qquad b_4 = 3\beta_1^2, \qquad b_6 = -\beta_1^3,$$

since F has characteristic $\neq 2$. Then (13.18) becomes

$$\begin{aligned}(y - \beta_1)^6 &= (y^3 + b_2 y^2 + b_4 y + b_6)^2 - 2^{10} \Delta(f) \, y \\ &= (y^3 - 3\beta_1 y^2 + 3\beta_1^2 y - \beta_1^3)^2 - 2^{10} \Delta(f) \, y \\ &= (y - \beta_1)^6 - 2^{10} \Delta(f) \, y.\end{aligned}$$

Hence $2^{10} \Delta(f) = 0$. Yet F has characteristic $\neq 2$, and $\Delta(f) \neq 0$, since f is separable. This contradiction completes the proof of the theorem. ☐

The structure of the sextic resolvent plays a crucial role in the above proof. So it remains to prove Proposition 13.2.5.

Proof of Proposition 13.2.5. We will prove the proposition in the universal case and then specialize. As in the proof of Theorem 13.2.6, let τ_i be the 3-cycle in (13.15) with $\tau_i \cdot h = h_i$. Then set

$$u_i = \tau_i \cdot u,$$

where u is defined in (13.13). This gives the polynomial

$$\Theta(y) = \prod_{i=1}^{6} (y - u_i).$$

Using *Maple* or *Mathematica* and the methods of Section 2.3, one computes that

$$(13.19) \qquad \Theta(y) = y^6 + B_2 y^4 + B_4 y^2 + B_6 - 2^5 \sqrt{\Delta} \, y,$$

where $\sqrt{\Delta} = \prod_{1 \le i < j \le 5}(x_i - x_j)$ is the square root of the discriminant, and

$$
\begin{aligned}
B_2 ={}& 8\sigma_1\sigma_3 - 3\sigma_2^2 - 20\sigma_4, \\
B_4 ={}& 3\sigma_2^4 - 16\sigma_1\sigma_2^2\sigma_3 + 16\sigma_1^2\sigma_3^2 + 16\sigma_2\sigma_3^2 + 16\sigma_1^2\sigma_2\sigma_4 - 8\sigma_2^2\sigma_4 \\
& - 112\sigma_1\sigma_3\sigma_4 + 240\sigma_4^2 - 64\sigma_1^3\sigma_5 + 240\sigma_1\sigma_2\sigma_5 - 400\sigma_3\sigma_5 \\
B_6 ={}& 8\sigma_1\sigma_2^4\sigma_3 - \sigma_2^6 - 16\sigma_1^2\sigma_2^2\sigma_3^2 - 16\sigma_2^3\sigma_3^2 + 64\sigma_1\sigma_2\sigma_3^3 - 64\sigma_3^4 \\
& - 16\sigma_1^2\sigma_2^3\sigma_4 + 28\sigma_2^4\sigma_4 + 64\sigma_1^3\sigma_2\sigma_3\sigma_4 - 112\sigma_1\sigma_2^2\sigma_3\sigma_4 \\
& - 128\sigma_1^2\sigma_3^2\sigma_4 + 224\sigma_2\sigma_3^2\sigma_4 - 64\sigma_1^4\sigma_4^2 + 224\sigma_1^2\sigma_2\sigma_4^2 \\
& - 176\sigma_2^2\sigma_4^2 - 64\sigma_1\sigma_3\sigma_4^2 + 320\sigma_4^3 + 48\sigma_1\sigma_2^3\sigma_5 \\
& - 192\sigma_1^2\sigma_2\sigma_3\sigma_5 - 80\sigma_2^2\sigma_3\sigma_5 + 640\sigma_1\sigma_3^2\sigma_5 + 384\sigma_1^3\sigma_4\sigma_5 \\
& - 640\sigma_1\sigma_2\sigma_4\sigma_5 - 1600\sigma_3\sigma_4\sigma_5 - 1600\sigma_1^2\sigma_5^2 + 4000\sigma_2\sigma_5^2.
\end{aligned}
$$
(13.20)

You will do this computation in Exercise 6. The lovely ideas that underlie the formulas in (13.19) and (13.20) are explained in Exercises 7 and 8.

Since $h_i = u_i^2$, $\Theta(y)$ relates to the universal sextic resolvent $\theta(y)$ as follows:

$$
\begin{aligned}
\theta(y^2) &= \prod_{i=1}^{6}(y^2 - h_i) = \prod_{i=1}^{6}(y^2 - u_i^2) \\
&= \prod_{i=1}^{6}(y - u_i)(y + u_i) = (-1)^6 \prod_{i=1}^{6}(y - u_i)(-y - u_i) \\
&= \Theta(y)\,\Theta(-y).
\end{aligned}
$$

Combining this with (13.19), we see that $\theta(y^2)$ is the product

$$(y^6 + B_2 y^4 + B_4 y^2 + B_6 - 2^5 \sqrt{\Delta}\, y)(y^6 + B_2 y^4 + B_4 y^2 + B_6 + 2^5 \sqrt{\Delta}\, y),$$

which easily implies that

$$\theta(y^2) = (y^6 + B_2 y^4 + B_4 y^2 + B_6)^2 - 2^{10}\Delta\, y^2.$$

Replacing y^2 with y, we obtain the universal formula

$$\theta(y) = (y^3 + B_2 y^2 + B_4 y + B_6)^2 - 2^{10}\Delta\, y.$$

Then the evaluation $\sigma_i \mapsto c_i$ gives the desired formula for $\theta_f(y)$. $\qquad \square$

The proof just given shows that $\theta_f(y) = (y^3 + b_2 y^2 + b_4 + b_6)^2 - 2^{10}\Delta(f)\, y$, where b_2, b_4, b_6 are obtained from (13.20) by replacing σ_i with c_i. This will be useful in the examples computed below.

We can also describe the irreducible factorization of $\theta_f(y)$ as follows.

Proposition 13.2.7. *Let $f \in F[x]$ and $G \subset S_5$ be as in Theorem 13.2.6. Then:*

$$A_5 \subset G \iff \theta_f(y) \text{ is irreducible over } F,$$

$$\begin{array}{ccc} G \text{ is conjugate to a} & & \theta_f(y) = (y - \beta)g(y), \text{ where } \beta \in F \\ \text{subgroup of AGL}(1, \mathbb{F}_5) & \iff & \text{and } g(y) \in F[y] \text{ is irreducible over } F. \end{array}$$

Proof. You will prove this in Exercises 9 and 10.   □

C. Examples. We first note that Theorem 13.2.6 and diagram (13.12) lead to the following table for determining the Galois group $\text{Gal}(L/F) \simeq G \subset S_5$:

Is $\Delta(f)$ in F^2?	Does $\theta_f(y)$ have a root in F?	Does $f(x)$ split completely over $F(\alpha)$?	G up to conjugacy
No	No	—	S_5
Yes	No	—	A_5
No	Yes	—	$\text{AGL}(1, \mathbb{F}_5)$
Yes	Yes	No	$\text{AGL}(1, \mathbb{F}_5) \cap A_5$
Yes	Yes	Yes	$\langle (12345) \rangle$

In this table, α denotes a root of f, and the dashes in the third column indicate cases when the first two columns determine the Galois group. You will prove the correctness of the table in Exercise 11.

Here are three examples of how to use this table.

Example 13.2.8. In Section 6.4 we showed that S_5 is the Galois group of $f = x^5 - 6x + 3$ over $\mathbb{Q}$. We can verify this as follows. In Exercise 12 you will show that f is irreducible with discriminant

$$\Delta(f) = -1737531$$

and sextic resolvent

$$\theta_f(y) = (y^3 + 120y^2 + 8640y - 69120)^2 + 2^{10}1737531y.$$

You will also show that $\theta_f(y)$ is irreducible over $\mathbb{Q}$. Since $\Delta(f)$ is not in $\mathbb{Q}^2$, the table implies that the Galois group is S_5.   ◁▷

We will return to this example in Section 13.4. We next give an example that uses the third column of the table.

Example 13.2.9. Consider $f = x^5 - 2 \in \mathbb{Q}(\sqrt{5})[x]$. In Exercise 13 you will show that f is irreducible over $\mathbb{Q}(\sqrt{5})$ with discriminant

$$\Delta(f) = 50000 = 2^4 5^5$$

and sextic resolvent

$$\theta_f(y) = y^6 - 2^{10}50000y = y^6 - 2^{14}5^5y.$$

This obviously has a root in $\mathbb{Q}(\sqrt{5})$. Since $\Delta(f)$ is a square in $\mathbb{Q}(\sqrt{5})$, the table tells us that the Galois group is either cyclic of order 5 or dihedral of order 10. To distinguish these, we need to check if f splits completely when we adjoin a root to $\mathbb{Q}(\sqrt{5})$. But f can't split completely over $\mathbb{Q}(\sqrt{5}, \sqrt[5]{2})$, since its roots aren't all real. Hence the Galois group of f over $\mathbb{Q}(\sqrt{5})$ is $\mathrm{AGL}(1, \mathbb{F}_5) \cap A_5 \simeq D_{10}$.   ◁▷

In Exercise 13 you will redo this example using results from Chapters 6 and 7.

Example 13.2.10. A quintic polynomial studied by De Moivre and Euler is

$$f = x^5 + px^3 + \tfrac{1}{5}p^2 x + q \in \mathbb{Q}[x].$$

We will assume that f is irreducible over $\mathbb{Q}$. In Exercise 14 you will show that f has discriminant

$$\Delta(f) = \frac{(4p^5 + 3125q^2)^2}{3125} = \frac{(4p^5 + 3125q^2)^2}{2^5}$$

and sextic resolvent

$$\theta_f(y) = \left(y^3 - 7p^2 y^2 + 11p^4 y + \tfrac{3}{25}p^6 + 4000pq^2\right)^2 - 2^{10}\frac{(4p^5+3125q^2)^2}{3125}y.$$

You will also verify that $\theta_f(y)$ has a root $y = 5p^2 \in \mathbb{Q}$. Since $\Delta(f) \notin \mathbb{Q}^2$ (do you see why?), the table implies that the Galois group is $\mathrm{AGL}(1, \mathbb{F}_5)$.   ◁▷

In Exercise 13 you will give an elementary proof that the polynomial given in Example 13.2.10 is solvable by radicals.

D. Solvable Quintics. We first point out the following immediate consequence of Corollary 13.2.3 and Theorems 13.2.6.

Corollary 13.2.11. *Assume that $f \in F[x]$ is monic and irreducible of degree 5 and that F has characteristic 0. Then f is solvable by radicals over F if and only if its sextic resolvent $\theta_f(y)$ has a root in F.*   □

As an application of this corollary, we will determine when an irreducible quintic of the form

$$f = x^5 + ax + b \in F[x]$$

is solvable by radicals. We will assume that F has characteristic 0. Here is the somewhat surprising result.

Theorem 13.2.12. *Assume that $f = x^5 + ax + b \in F[x]$ is irreducible, where $a \neq 0$ and F has characteristic 0. Then f is solvable by radicals over F if and only if there are $\lambda, \mu \in F$ such that*

$$a = \frac{3125\lambda\mu^4}{(\lambda - 1)^4(\lambda^2 - 6\lambda + 25)}, \qquad b = \frac{3125\lambda\mu^5}{(\lambda - 1)^4(\lambda^2 - 6\lambda + 25)}.$$

Proof. In Exercise 15 you will show that f has discriminant

$$\Delta(f) = 256a^5 + 3125b^4$$

and sextic resolvent

$$\theta_f(y) = (y^3 - 20ay^2 + 240a^2y + 320a^3)^2 - 2^{10}(256a^5 + 3125b^4)y.$$

By Corollary 13.2.11, f is solvable by radicals if and only if $\theta_f(y)$ has a root in F.

Since $a \neq 0$, a root $\beta \in F$ of $\theta_f(y)$ can be written $\beta = a\lambda$ for $\lambda \in F$. We can also write $b = a\mu$ for some $\mu \in F$. With these substitutions, it follows that $\theta_f(y)$ has a root in F if and only if there is $\lambda \in F$ such that

$0 = \theta_f(a\lambda)$
$$= \big((a\lambda)^3 - 20a(a\lambda)^2 + 240a^2(a\lambda) + 320a^3\big)^2 - 2^{10}(256a^5 + 3125(a\mu)^4)(a\lambda),$$

which (after some algebra) simplifies to

$$0 = 2^{12}a^5\big((\lambda^6 - 10\lambda^5 + 55\lambda^4 - 140\lambda^3 + 175\lambda^2 - 106\lambda + 25)a - 3125\lambda\mu^4\big).$$

Since $a \neq 0$, this is equivalent to

$$a = \frac{3125\lambda\mu^4}{\lambda^6 - 10\lambda^5 + 55\lambda^4 - 140\lambda^3 + 175\lambda^2 - 106\lambda + 25}$$
$$= \frac{3125\lambda\mu^4}{(\lambda - 1)^4(\lambda^2 - 6\lambda + 25)},$$

where the factorization of the denominator is easily done in *Maple* or *Mathematica*. Using this and $b = a\mu$, we get the desired formulas for a and b.   □

We will say more about this result in the Historical Notes.

Mathematical Notes

There are three topics we need to discuss further.

▪ **Resolvents.** The sextic resolvent appearing in Theorem 13.2.6 used $h = u^2$, where u is given in (13.13). This resolvent appears in [Chebotarev] and [32], for example. The paper [1] contains an especially nice discussion of how h relates to the star diagram (13.14).

However, $h = u^2$ is not the only possibility. One can instead use

$$\theta_1 = x_1^2x_2x_5 + x_1^2x_3x_4 + x_2^2x_1x_3 + x_2^2x_4x_5 + x_3^2x_1x_5$$
$$+ x_3^2x_2x_4 + x_4^2x_1x_2 + x_4^2x_3x_5 + x_5^2x_1x_4 + x_5^2x_2x_3.$$

This leads to the sextic resolvent found in [3] and [7].

The table used in Examples 13.2.8–13.2.10 distinguishes between the groups $\langle(12345)\rangle$ and $AGL(1, \mathbb{F}_5) \cap A_5$ by factoring f over $F(\alpha)$, where α is a root of f. A natural question is whether this can be done with resolvents. The answer is yes, with a small complication. For example, if $G \subset \text{AGL}(1, \mathbb{F}_5) \cap A_5$, then the algorithm given in [3] computes

$$d = \big(\alpha_1\alpha_2(\alpha_2 - \alpha_1) + \alpha_2\alpha_3(\alpha_3 - \alpha_2) + \alpha_3\alpha_4(\alpha_4 - \alpha_3)$$
$$+ \alpha_4\alpha_5(\alpha_5 - \alpha_4) + \alpha_5\alpha_1(\alpha_1 - \alpha_5)\big)^2.$$

One can prove that $d \in F$ and that if $d \neq 0$, then $G = \langle (12345) \rangle$ if and only if $d \in F^2$. The latter condition is equivalent to the resolvent $y^2 - d$ having a root in F. Also note that $d \neq 0$ guarantees that the resolvent is separable.

The problem is that $d = 0$ can occur. When this happens, one performs a *Tschirnhaus transformation* to change f into a polynomial for which $d \neq 0$. We will say more about Tschirnhaus transformations in the Historical Notes. This complication is why we used the factorization of f over $F(\alpha)$ in part (c) of Theorem 13.2.6.

▪ **Radical Solutions.** When a quintic is solvable by radicals, it is natural to want explicit formulas for the roots. These can be complicated.

Example 13.2.13. Using our methods, it is straightforward to see that the Galois group over $\mathbb{Q}$ of $f = x^5 + 15x + 12 \in \mathbb{Q}[x]$ is isomorphic to $\text{AGL}(1, \mathbb{F}_5)$. In [1] and [29], it is shown that

$$\alpha = \left(\frac{-75 - 21\sqrt{10}}{125}\right)^{1/5} + \left(\frac{-75 + 21\sqrt{10}}{125}\right)^{1/5}$$
$$+ \left(\frac{225 - 72\sqrt{10}}{125}\right)^{1/5} + \left(\frac{225 + 72\sqrt{10}}{125}\right)^{1/5}$$

is a root of f.   ◁▷

For an arbitrary solvable quintic, an algorithm for writing down the roots explicitly is described in [7] and [18]. For the special case of solvable quintics of the form $x^5 + ax + b$, the solutions are described in [1] and [29].

▪ **Normal Forms.** A quintic of the form $x^5 + ax + b$ is said to be in *Bring–Jerrard* normal form. When F has characteristic 0, it can be shown that an arbitrary quintic in $F[x]$ can be transformed (using one of the Tschirnhaus transformations mentioned above) to one of Bring–Jerrard form, though in order to do so, one might need to replace F with a solvable extension. A procedure for doing this is described in [Dehn], [Postnikov], and [33]. In Exercise 16 you will show that in characteristic 5, not all quintics can be put into Bring–Jerrard form.

Two Bring–Jerrard quintics $x^5 + ax + b$ and $x^5 + a'x + b'$ in $F[x]$ are *equivalent* if there is $\lambda \in F^*$ such that

$$x^5 + a'x + b' = \lambda^{-5}\big((\lambda x)^5 + a(\lambda x) + b\big).$$

Hence the roots of $x^5 + a'x + b'$ are the roots of $x^5 + ax + b$ multiplied by λ^{-1}. Using a form of Theorem 13.2.12, [30] shows that when $F = \mathbb{Q}$, there are infinitely many inequivalent Bring–Jerrard quintics over $\mathbb{Q}$.

However, if we switch to quintics of the form $x^5 + ax^2 + b$, there is a similar notion of equivalence, but here, [30] shows that up to equivalence, there are only five such quintics. The argument reduces to finding $x, y \in \mathbb{Q}$ such that

$$y^2 = x^3 + \tfrac{89}{100}x^2 + \tfrac{8}{25}x + \tfrac{1}{25}.$$

This is an elliptic curve (see the Mathematical Notes to Section 13.1) with only finitely many solutions over $\mathbb{Q}$.

One can show that an arbitrary quintic in characteristic 0 can be transformed into a *Brioschi quintic*

$$x^5 - 10Wx^3 + 45W^2x - W^2.$$

The procedure for obtaining this normal form is described in [26, Ch. 5]. As for the Bring–Jerrard form, a solvable extension of F may be required to obtain the Brioschi form (see [26, Fig. 5.9.1]). The surprise is that the Brioschi quintic is deeply related to Section 7.5. This is because the rotational symmetries of the give an extension $K \subset \mathbb{C}(t)$ (t a variable) with Galois group A_5. The books [17] and [26] explain how a complete understanding of this extension enables one to find the roots of any quintic polynomial.

Historical Notes

The first serious attempt to find the roots of polynomials of degree $n \geq 5$ is due to Tschirnhaus in 1683. His idea was to simplify

$$x^n + a_{n-1}x^{n-1} + \cdots + a_0 = 0$$

by a substitution of the form

(13.21) $$y = b_0 + b_1x + \cdots + b_{n-1}x^{n-1},$$

now called a *Tschirnhaus transformation*. Eliminating x from these two equations gives an equation in y of degree n that can be significantly simpler if $b_0, \ldots, b_{n-1}$ are chosen carefully.

Example 13.2.14. Given $x^3 + 3x + 1 \in \mathbb{Q}[x]$, consider the Tschirnhaus transformation $y = a + bx + x^2$. In Exercise 17 you will show that eliminating x leads to the equation

(13.22) $$y^3 + (6 - 3a)y^2 + (9 + 3b + 3b^2 - 12a + 3a^2)y + P(a, b) = 0,$$

where $P(a, b)$ is a polynomial in a and b. You will also show that the coefficients of y^2 and y vanish if and only if a and b satisfy

$$a = 2 \quad \text{and} \quad b^2 + b - 1 = 0.$$

If we pick $b = (\sqrt{5} - 1)/2$, then the above cubic becomes

$$y^3 = \sqrt{5}^3 \frac{\sqrt{5} - 1}{2}.$$

In Exercise 17 you will use this to solve the original cubic. Note that the resulting Tschirnhaus transformation is defined over a degree 2 extension of $\mathbb{Q}$.   ◁▷

Tschirnhaus transformations can be used to solve all cubic and quartic equations. They fail in degree 5, though Bring in 1786 and Jerrard in 1834 showed that arbitrary quintic can be put into Bring–Jerrard form using Tschirnhaus transformations defined over suitable solvable extensions of the original field.

The quintic polynomial $x^5 + px^3 + \frac{1}{5}p^2 x + q$ from Example 13.2.10 was solved by De Moivre in 1706. The polynomial reappears in a 1764 paper of Euler devoted to algebraic equations. In this paper, Euler writes an equation of degree 5 as

$$x^5 = Ax^3 + Bx^2 + Cx + D.$$

(Can you explain why he omitted the x^4 term?) Let the roots be $\alpha, \beta, \gamma, \delta, \epsilon$ in some splitting field. Euler was seeking a formula of the form

(13.23) $$\alpha = \mathscr{A}\sqrt[5]{v} + \mathscr{B}\sqrt[5]{v^2} + \mathscr{C}\sqrt[5]{v^3} + \mathscr{D}\sqrt[5]{v^4},$$

where v is a root of an equation of degree < 4, and similar formulas for $\beta, \gamma, \delta, \epsilon$ with the radicals multiplied by suitable fifth roots of unity. Euler shows that this strategy works for polynomials of degree 2, 3, and 4, but for degree 5 he succeeds only in some special cases.

More precisely, he shows that if certain of the coefficients $\mathscr{A}, \mathscr{B}, \mathscr{C}, \mathscr{D}$ are zero in (13.23), then the original quintic reduces to one of the three special forms:

(13.24)
$$x^5 = D,$$
$$x^5 = 5Px^2 + 5Qx + Q^2/P + P^3/Q,$$
$$x^5 = 5Px^3 - 5P^2 x + D.$$

We can analyze the resulting Galois groups as follows. Assume that the polynomials of (13.24) have coefficients in $\mathbb{Q}$ and are irreducible. Then the Galois group of the first polynomial is isomorphic to $\mathrm{AGL}(1, \mathbb{F}_5)$ by Section 6.4, and the same is true for the third polynomial by Example 13.2.10. You will show in Exercise 18 that the Galois group is also $\mathrm{AGL}(1, \mathbb{F}_5)$ for the second polynomial in (13.24).

However, if we adjoin the fifth roots of unity (standard procedure in the eighteenth century), then you will show in Exercise 18 that the first two equations of (13.24) have cyclic Galois group over $\mathbb{Q}(\zeta_5)$ while the third has Galois group isomorphic to $\mathrm{AGL}(1, \mathbb{F}_5) \cap A_5$. Details of what Euler did can be found in [21].

The first sextic resolvents for the quintic are due to Lagrange, Malfatti, and Vandermonde around 1770. Lagrange used

$$\begin{aligned}
z = {} & 2\big(x_1^3(x_2 x_5 + x_3 x_5) + x_2^3(x_1 x_3 + x_4 x_5) + x_3^3(x_2 x_4 + x_1 x_5) \\
& + x_4^3(x_3 x_5 + x_1 x_2) + x_5^3(x_1 x_4 + x_2 x_3)\big) \\
& + 3\big(x_1(x_2^2 x_5^2 + x_3^2 x_5^2) + x_2(x_1^2 x_3^2 + x_4^2 x_5^2) + x_3(x_2^2 x_4^2 + x_1^2 x_5^2) \\
& + x_4(x_3^2 x_5^2 + x_1^2 x_2^2) + x_5(x_1^2 x_4^2 + x_2^2 x_3^2)\big).
\end{aligned}$$

Lagrange was led to this using the ideas discussed at the end of Section 12.1. The polynomial z is invariant under $\mathrm{AGL}(1, \mathbb{F}_5)$ and is the root of a sextic resolvent

$$z^6 - Az^5 + Bz^4 - Cz^3 + Dz^2 - Ez + F.$$

Lagrange computes A explicitly in terms of $\sigma_1, \sigma_2, \sigma_3, \sigma_4, \sigma_5$. The formulas are similar to (13.20), except that Lagrange computed them by hand (no computers back then!). Rather than continue with B, C, D, E, F, Lagrange comments that they can be computed by similar methods and goes to say:

> But we shall not insert here such details which, besides that they would require very long calculations, would moreover not cast any light on the resolution of equations of the fifth degree.

(See [Lagrange, p. 342].) For Lagrange, getting a resolvent of degree 6 was not helpful, since he wanted to reduce to equations of degree smaller than 5 (which we now know to be impossible). The irony is that one can use Lagrange's sextic to obtain a criterion similar to Corollary 13.2.11 for deciding which quintics are solvable by radicals. (This was done by Galois—see below.)

Malfatti's sextic resolvent is closely related to Lagrange's. He computed all of its coefficients in terms of the elementary symmetric polynomials and knew that the quintic was solvable by radicals when the sextic had a rational root. The converse was proved by Luther in 1847. The resolvent $h = u^2$ used in the text is due to Jacobi in 1835. He was the first to prove (13.19). In 1861, Cayley independently discovered this resolvent and related it to the star diagram (13.14) (see also [1]).

Galois was naturally the first to think about this in terms of the Galois group. He showed that Lagrange's sextic resolvent has a rational root if and only if the corresponding quintic is solvable by radicals. We will see in Section 14.1 that Galois also generalized this to irreducible polynomials of prime degree.

Theorem 13.2.12 about when a Bring–Jerrard quintic is solvable by radicals was first proved by Runge in 1885. In the same year, Glashan and Young published different versions of the same result. A modern proof appears in [29].

More on the history of the quintic equation can be found in [23], [34], and [35].

Exercises for Section 13.2

Exercise 1. As explained in the text, we can regard $\mathrm{AGL}(1, \mathbb{F}_5)$ as a subgroup of S_5.
(a) Prove that $\mathrm{AGL}(1, \mathbb{F}_5)$ is generated by (12345) and (1243).
(b) Prove that $\mathrm{AGL}(1, \mathbb{F}_5) \cap A_5$ is generated by (12345) and $(14)(23)$.
(c) Prove that the group of part (b) is isomorphic to the dihedral group D_{10}.
(d) Prove that $\langle(12345)\rangle$, $\mathrm{AGL}(1, \mathbb{F}_5) \cap A_5$, and $\mathrm{AGL}(1, \mathbb{F}_5)$ are the only subgroups of $\mathrm{AGL}(1, \mathbb{F}_5)$ containing $\langle(12345)\rangle$.

Exercise 2. This exercise will consider some simple properties of S_5.
(a) Prove that $\langle(12345)\rangle$ is a 5-Sylow subgroup of S_5 and more generally is a 5-Sylow subgroup of any subgroup $G \subset S_5$ containing $\langle(12345)\rangle$.
(b) Prove that S_5 has twenty-four 5-cycles.

Exercise 3. Let $G \subset S_5$ be transitive, and let N be the number of subgroups of G of order 5. In this exercise, you will use an argument from [Postnikov] to prove that $N = 1$ or 6 without using the Sylow Theorems. Let $C = \{\tau \in S_5 \setminus G \mid \tau \text{ is a 5-cycle}\}$.
(a) Prove that $\sigma \cdot \tau = \sigma \tau \sigma^{-1}$ defines an action of G on C.
(b) Let $\tau \in S_5$ be a 5-cycle. Prove that $\sigma \in S_5$ satisfies $\sigma \tau \sigma^{-1} = \tau$ if and only if $\sigma \in \langle \tau \rangle$.
(c) Use parts (a) and (b) to prove that $|G|$ divides $|C|$.

(d) Prove that $4N + |C| = 24$.

(e) Use parts (c) and (d) to prove that $N = 1$ or 6.

Exercise 4. Prove that (13.15) gives coset representatives of $AGL(1, \mathbb{F}_5)$ in S_5.

Exercise 5. Complete the proof of part (b) of Theorem 13.2.6. Then prove part (c).

Exercise 6. In this exercise, you will use *Maple* or *Mathematica* to prove (13.19) and (13.20).

(a) The first step is to enter (13.13) and call it, for example, u1. Then use substitution commands and (13.15) to create u2,...,u6. For example, u2 is obtained by applying (123) to u1. In *Maple*, this is done via the command

$$u2 := \text{subs}(\{x1 = x2, x2 = x3, x3 = x1\}, u1);$$

whereas in *Mathematica* one uses

$$u2 := u1 \ /. \ \{x1 {-}{>} x2, x2 {-}{>} x3, x3 {-}{>} x1\}$$

(b) Now multiply out $\Theta(y) = (y - u1) \cdots (y - u6)$ and use the methods of Section 2.3 to express the coefficients of $\Theta(y)$ in terms of the elementary symmetric polynomials.

(c) Show that your results imply (13.19) and (13.20).

Exercise 7. Consider $AGL(1, \mathbb{F}_5) \cap A_5 \subset S_5$, and let u be defined as in (13.13).

(a) Prove that the symmetry group of u is $AGL(1, \mathbb{F}_5) \cap A_5$.

(b) Prove that (13.15) gives coset representatives of $AGL(1, \mathbb{F}_5) \cap A_5$ in A_5.

Exercise 8. Let $u_1, \ldots, u_6$ be as in the proof of Proposition 13.2.5, and let $\tau \in S_5$ be a transposition.

(a) For each i, prove that $\tau \cdot u_i = -u_j$ for some j.

(b) Let $\Theta(y) = \prod_{i=1}^{6}(y - u_i)$ and write this polynomial as

$$\Theta(y) = y^6 + B_1 y^5 + B_2 y^4 + B_3 y^3 + B_4 y^2 + B_5 y + B_6.$$

Use part (a) to show that $\tau \cdot B_i = (-1)^i B_i$ for $i = 1, \ldots, 6$.

(c) Explain how part (b) and the results of Chapter 2 imply that the coefficients B_2, B_4, B_6 are polynomials in $\sigma_1, \sigma_2, \sigma_3, \sigma_4, \sigma_5$. This explains why the formulas (13.20) exist.

(d) Use Exercise 3 of Section 7.4 to show that the coefficients B_1, B_3, B_5 must be of the form $B \sqrt{\Delta}$, where B is a polynomial in $\sigma_1, \sigma_2, \sigma_3, \sigma_4, \sigma_5$.

(e) Note that $\sqrt{\Delta}$ has degree 10 as a polynomial in x_1, x_2, x_3, x_4, x_5. By considering the degrees of B_1, B_3, B_5 as polynomials in x_1, x_2, x_3, x_4, x_5, show that part (d) implies that $B_1 = B_3 = 0$ and that B_5 is a constant multiple of $\sqrt{\Delta}$. This explains (13.19).

Exercise 9. This exercise will prove the first equivalence of Proposition 13.2.7.

(a) First suppose that $\theta_f(y)$ is irreducible. Prove that $|G|$ is divisible by 6, and explain why this implies that $A_5 \subset G$.

(b) Now suppose that $A_5 \subset G$. Prove that $\text{Gal}(L/F)$ acts transitively on $\beta_1, \ldots, \beta_6$. However, we don't know that $\beta_1, \ldots, \beta_6$ are distinct.

(c) Let $p(y)$ be the minimal polynomial of β_1 over F. By part (b), it is also the minimal polynomial of $\beta_2, \ldots, \beta_6$. Prove that $\theta_f(y) = p(y)^m$, where $m = 1, 2, 3,$ or 6. The proof of Theorem 13.2.6 shows that $m = 6$ cannot occur, and $m = 1$ implies that $\theta_f(y)$ is irreducible over F. It remains to consider what happens when $m = 2$ or 3.

(d) Show that $(y^3 + ay^2 + by + c)^2 = \theta_f(y)$ implies that $\Delta(f) = 0$. Hence this case can't occur.

(e) Show that $(y^2 + ay + b)^3 = \theta_f(y)$ implies that $4b = a^2$, and then use this to show that $\Delta(f) = 0$.

Exercise 10. This exercise will prove the second equivalence of Proposition 13.2.7. Note that one direction follows trivially from Theorem 13.2.6. Hence we will assume that $G \subset$ AGL$(1, \mathbb{F}_5)$ and that $\theta_f(y) = (y - \beta_1)g(y)$ where $\beta_1 \in F$,

(a) Use $(12345) \in G$ to prove that Gal(L/F) acts transitively on $\beta_2, \ldots, \beta_6$. As in the previous exercise, we don't know if $\beta_2, \ldots, \beta_6$ are distinct.

(b) Let $p(y)$ be the minimal polynomial of β_2 over F. By part (a), it is also the minimal polynomial of $\beta_3, \ldots, \beta_6$. Prove that $\theta_f(y) = (y - \beta_1)p(y)^m$, where $m = 1$ or 5. If $m = 1$, then we are done. So we need to rule out $m = 5$.

(c) Show that $(y - \beta_1)(y - \beta_2)^5 = \theta_f(y)$ implies that $\beta_1 = \beta_2$, and then use this to show that $\Delta(f) = 0$.

Exercise 11. Show that the table preceding Example 13.2.8 follows from the diagram (13.12) and Theorem 13.2.6.

Exercise 12. Let $f = x^5 - 6x + 3 \in \mathbb{Q}[x]$ be as in Example 13.2.8. Compute $\Delta(f)$ and $\theta_f(y)$ and show that $\theta_f(y)$ is irreducible over $\mathbb{Q}$.

Exercise 13. Let $f = x^5 - 2 \in \mathbb{Q}(\sqrt{5})[x]$ be as in Example 13.2.9.

(a) Compute $\Delta(f)$ and $\theta_f(y)$.

(b) In Section 6.4 we showed that the Galois group of f over $\mathbb{Q}$ is isomorphic to AGL$(1, \mathbb{F}_5)$. Use this and the Galois correspondence to show that the Galois group over $\mathbb{Q}(\sqrt{5})$ is isomorphic to AGL$(1, \mathbb{F}_5) \cap A_5$.

Exercise 14. Let $f = x^5 + px^3 + \frac{1}{5}p^2x + q \in \mathbb{Q}[x]$ be as in Example 13.2.10, and assume that f is irreducible over $\mathbb{Q}$.

(a) Compute $\Delta(f)$ and $\theta_f(y)$.

(b) Factor $\theta_f(y) \in \mathbb{Q}[x]$, and conclude that $5p^2 \in \mathbb{Q}$ is a root of $\theta_f(y)$.

(c) Show that the substitution $x = z - \frac{p}{5z}$ transforms f into $z^5 - \frac{p^5}{5^5z^5} + q$.

(d) Use part (c) to give an elementary proof that f is solvable by radicals over $\mathbb{Q}$.

Exercise 15. As in Theorem 13.2.12, let $f = x^5 + ax + b$. Compute $\Delta(f)$ and $\theta_f(y)$.

Exercise 16. Let $f = x^5 + ax + b \in F[x]$, where f is separable and irreducible and F has characteristic 5. The goal of this exercise is to prove the observation of [24] that the Galois group of f over F is solvable.

(a) Prove that $a \neq 0$.

(b) Use Exercise 5 from Section 6.2 to show that Galois group of f over F is cyclic when $a = -1$.

(c) Show that there is a Galois extension $F \subset L$ with solvable Galois group such that f is equivalent (as defined in the Mathematical Notes) to a polynomial of the form $x^5 - x + b'$ for some $b' \in L$.

(d) Conclude that the Galois group of f over F is solvable.

(e) Show that there is a field F of characteristic 5 and a monic, separable, irreducible quintic $g \in F[x]$ that cannot be transformed to one in Bring–Jerrard form defined over any Galois extension $F \subset L$ with solvable Galois group.

In [24] Ruppert explores the geometric reasons why things go wrong in characteristic 5.

Exercise 17. Following Example 13.2.14, consider the equations $x^3 + 3x + 1 = 0$ and $y = a + bx + x^2$.

(a) Use *Maple* or *Mathematica* and Section 2.3 to eliminate x and obtain (13.22).

(b) Show that coefficients of y^2 and y in (13.22) both vanish if and only if $a = 2$ and $b^2 + b - 1 = 0$.

(c) The equation for y becomes trivial to solve when $a = 2$ and $b = (\sqrt{5} - 1)/2$. We could then solve for x using $y = a + bx + x^2$, but there is a better way to proceed. Note that

$$x^3 = -bx^2 - ax + yx$$

follows from $y = a + bx + x^2$. Furthermore, we can use $y = a + bx + x^2$ to eliminate the x^2 in the above equation. Then use $x^3 + 3x + 1 = 0$ to obtain an equation in which x appears only to the first power. Solving this gives a formula for x in terms of y. The general version of this argument can be found in [Lagrange, p. 223].

Exercise 18. This exercise is concerned with the polynomials (13.24). As in the Historical Notes, we will assume that they lie in $\mathbb{Q}[x]$ and are irreducible.

(a) Show that $\sqrt[5]{Q^2/P} + (P/Q)\sqrt[5]{Q^2/P}^2$ is a root of $x^5 - 5Px^2 - 5Qx - Q^2/P - P^3/Q$.

(b) Prove that the Galois group of $x^5 - 5Px^2 - 5Qx - Q^2/P - P^3/Q$ over $\mathbb{Q}$ is isomorphic to AGL$(1, \mathbb{F}_5)$.

(c) Prove that over $\mathbb{Q}(\sqrt{5})$, the first two polynomials of (13.24) have cyclic Galois group while the third has Galois group isomorphic to AGL$(1, \mathbb{F}_5) \cap A_5$.

Exercise 19. Use the methods of this section to compute the Galois group over $\mathbb{Q}$ of each of the following polynomials. Be sure to check that they are irreducible. Remember that in Section 4.2 we learned how to factor polynomials over a finite extension of $\mathbb{Q}$.

(a) $x^5 + x + 1$.

(b) $x^5 + 20x + 16$.

(c) $x^5 + 2$.

(d) $x^5 - 5x + 12$.

(e) $x^5 + x^4 - 4x^3 - 3x^2 + 3x + 1$.

Exercise 20. In the Mathematical Notes to Section 10.3, we noted that the roots of the polynomial $x^5 - 4x^4 + 2x^3 + 4x^2 + 2x - 6 \in \mathbb{Q}[x]$ can be constructed using a marked ruler and compass. Show that this polynomial is not solvable by radicals over $\mathbb{Q}$.

13.3 RESOLVENTS

So far, we have explained how to compute Galois groups of polynomials of degree 4 or 5. It is time to turn our attention to polynomials of higher degree. We will see that generalizations of the resolvents used in Sections 13.1 and 13.2 lead to a systematic strategy for computing Galois groups.

A. Jordan's Strategy. In Theorem 13.1.1 we used the Ferrari resolvent to help determine the Galois group of an irreducible polynomial of degree 4, and the sextic resolvent played a similar role in Theorem 13.2.6 for polynomials of degree 5. This strategy for computing Galois groups was first described by Jordan in 1870:

The path to follow to treat this question [computing the Galois group] will be the following: 1° one will form the various groups of the possible substitutions $G, G', \ldots$ among the roots of the equation; 2° let G be one of the these groups, chosen at will: one will affirm for oneself whether or not it contains the group of the equation by forming a function φ of the roots, invariant under the substitutions of G and variable for other substitutions, calculating by the method of symmetric functions the equation [the resolvent] that has for roots the various values of φ, and looking for a rational root. Among the groups of the series $G, G', \ldots$ that in this way contain the group of the equation, the smallest will be the group itself.

(See [Jordan1, p. 276].) The sextic resolvent $\theta_f(y)$ used Theorem 13.2.6 follows this model nicely: for $G = \mathrm{AGL}(1, \mathbb{F}_5)$, we have $\varphi = h = u^2$, whose symmetry group is precisely G, and the universal sextic $\theta(y)$ is the polynomial "that has for roots the various values of φ," which when evaluated at the coefficients of an irreducible quintic $f \in F[x]$ gives the resolvent $\theta_f(y)$. By Theorem 13.2.6, the question of whether the Galois group of f is conjugate to a subgroup of $\mathrm{AGL}(1, \mathbb{F}_5)$ is equivalent to "looking for a rational root," that is, a root of $\theta_f(y)$ in F.

As the discussion of the sextic $\theta_f(y)$ reveals, Jordan's description is not perfect. In fact, it omits three important things:

- First, one needs to distinguish between the resolvent in the universal case and its specialization to the given polynomial.
- Second, having a rational root only implies that G contains the Galois group up to conjugacy.
- Third, this can fail if the rational root is not simple.

We will say more about these items below. Nevertheless, Jordan's description is remarkably close to some of the modern methods used to compute the Galois group of an irreducible separable polynomial $f \in F[x]$ of degree n. Let $G_f \subset S_n$ correspond to the Galois group of f over F. Thus

$$\mathrm{Gal}(L/F) \simeq G_f \subset S_n,$$

where $F \subset L$ is a splitting field. In earlier sections, G_f was called G, but in the discussion that follows, G will instead denote an arbitrary transitive subgroup of S_n.

Here is the step-by-step process for determining G_f.

Step 1: Classify Groups. Transitive subgroups $G \subset S_n$ have been classified up to conjugacy for $n \leq 31$, though published tables [4] only go up to $n = 15$.

Step 2: Find Polynomials. For each $G \subset S_n$ from Step 1, we need to find a polynomial φ in $x_1, \ldots, x_n$ whose symmetry group is G. Stauduhar's 1973 paper [32], which pioneered the modern approach to Jordan's strategy, lists a polynomial φ for each transitive subgroup of S_4, S_5, and S_7 (and S_6 with some errors noted in [11]). We will follow the standard convention that φ has coefficients in $\mathbb{Z}$.

Step 3: Compute Resolvents. Take G and φ from Step 2. Following the method of Sections 13.1 and 13.2, we compute the resolvent in the universal case, write its coefficients in terms of the elementary symmetric polynomials, and then specialize to the coefficients of f. The problem is that the universal resolvent might be huge.

When $F = \mathbb{Q}$, we can avoid this difficulty as follows. Suppose that $f \in \mathbb{Z}[x]$ is irreducible of degree n. (Exercise 1 will explain why we can restrict to polynomials with integer coefficients.) Let $\varphi_1 = \varphi, \varphi_2, \ldots, \varphi_m$ be the orbit of φ under the action S_n (so m is the index of G in S_n). Then compute accurate numerical approximations $\alpha_1, \ldots, \alpha_n$ of the roots of f, and multiply out the approximate resolvent

(13.25) $$\big(y - \varphi_1(\alpha_1, \ldots, \alpha_n)\big) \cdots \big(y - \varphi_m(\alpha_1, \ldots, \alpha_n)\big).$$

However, since the true resolvent has integer coefficients in this case (you will prove this in Exercise 2), it follows that if we have computed the approximate roots α_i accurately enough, then the true resolvent is obtained from the approximate one by rounding its coefficients to the nearest integer. Doing this rigorously requires a careful understanding of the numerical issues involved. See [32] for the details.

Here is a simple example.

Example 13.3.1. Consider $f = x^5 - 6x + 3 \in \mathbb{Q}[x]$ from Example 13.2.8. To 16 decimal places, the roots of f are

$$\alpha_1 = -1.6709352644808655592,$$
$$\alpha_2 = -0.1181039225949867235 - 1.5874591621207593640i,$$
$$\alpha_3 = -0.1181039225949867235 + 1.5874591621207593640i,$$
$$\alpha_4 = 0.5055012304055246668,$$
$$\alpha_5 = 1.4016418792653143394.$$

Evaluating the polynomials $h_i = u_i^2$ from Section 13.2 at these numbers gives

$$\beta_1 = -43.4376362799772861 + 28.6930156587206645i,$$
$$\beta_2 = -71.5507381341784308 - 94.8067689529853707i,$$
$$\beta_3 = -71.5507381341784308 + 94.8067689529853707i,$$
$$\beta_4 = -5.0116255858442831 + 9.9920056672183422i,$$
$$\beta_5 = -5.0116255858442831 - 9.9920056672183422i,$$
$$\beta_6 = -43.4376362799772861 - 28.6930156587206645i$$

as approximate roots of the sextic resolvent $\theta_f(y)$. It follows that (13.25) becomes

$$\theta_f(y) \approx y^6 + 240.0000001y^5 + 31680.00001y^4 + 1935360.001y^3$$
$$+ 58060800.02y^2 + (584838144.3 - 0.07i)y$$
$$+ 4777574402 + 0.1109586026i.$$

However, multiplying out the formula for $\theta_f(y)$ given in Example 13.2.8 shows that

$$\theta_f(y) = y^6 + 240y^5 + 31680y^4 + 1935360y^3$$
$$+ 58060800y^2 + 584838144y + 4777574400.$$

Looking at the constant term, we see that our approximation is not good enough. Hence we need to increase the accuracy of the roots of f. ◁▷

The above calculation was done in *Maple*; *Mathematica* gives a similar result. The moral is that you need to know what you are doing when working numerically. Other methods for computing resolvents are discussed in [13].

Step 4: Use Resolvents. Suppose that $f \in F[x]$ is irreducible and separable of degree n and that $G \subset S_n$ and φ lead to the resolvent $R_f(y) \in F[y]$. We can use this to determine the location of the Galois group $G_f \subset S_n$ as follows.

Proposition 13.3.2. *Let $f \in F[x]$ be separable and irreducible of degree n.*
(a) *If G_f is conjugate to a subgroup of G, then $R_f(y)$ has a root in F.*
(b) *If $R_f(y)$ has a simple root in F, then G_f is conjugate to a subgroup of G.*

Proof. Recall from Section 5.3 that a root of $R_f(y)$ is *simple* if the corresponding linear factor appears exactly once in the factorization over a splitting field.

If G_f is conjugate to a subgroup of G, then $G_f \subset G$ follows by suitably relabeling the roots $\alpha_1, \dots, \alpha_n$ of f. Then $G_f \subset G$ easily implies that $\varphi(\alpha_1, \dots, \alpha_n)$ is invariant under $\mathrm{Gal}(L/F)$ and hence lies in F (be sure you can supply the details).

Conversely, let $\beta \in F$ be a simple root of $R_f(y)$. By relabeling the roots of f, we may assume that $\beta = \varphi(\alpha_1, \dots, \alpha_n)$. If $G_f \not\subset G$, then there is $\tau \in G_f$ such that $\tau \notin G$. Then $\tau \cdot \varphi \neq \varphi$, so that the resolvent may be written

$$R_f(y) = \big(y - \varphi(\alpha_1, \dots, \alpha_n)\big)\big(y - (\tau \cdot \varphi)(\alpha_1, \dots, \alpha_n)\big) \cdots$$
$$= \big(y - \varphi(\alpha_1, \dots, \alpha_n)\big)\big(y - \varphi(\alpha_{\tau(1)}, \dots, \alpha_{\tau(n)})\big) \cdots .$$

You will prove in Exercise 3 that $\beta = \varphi(\alpha_1, \dots, \alpha_n) \in F$ and $\tau \in G_f$ imply that

$$\varphi(\alpha_1, \dots, \alpha_n) = \varphi(\alpha_{\tau(1)}, \dots, \alpha_{\tau(n)}),$$

which is impossible because $\beta = \varphi(\alpha_1, \dots, \alpha_n)$ is a simple root of $R_f(y)$. This contradiction proves that $G_f \subset G$. $\qquad\square$

For irreducible quartics, the Ferrari resolvent used in Theorem 13.1.1 is separable when f is separable, and for irreducible quintics, the same is true for the sextic resolvent used in Theorem 13.2.6. Hence all roots of these resolvents are simple. But resolvents can have multiple roots. Here is an example.

Example 13.3.3. Let $n = 4$ and $\varphi = \sqrt{\Delta}\,(x_1 + x_2 - x_3 - x_4)$. In Exercise 4 you will verify that the symmetry group of φ is $G = \langle(1324)\rangle \subset S_4$. Thus the corresponding resolvent $R_f(y)$ has degree 6. For the polynomial $f = x^4 + bx^2 + d$, $d \notin F^2$, considered in Example 13.1.4, you will show in Exercise 4 that

$$R_f(y) = y^2\big((y^2 + 4b\Delta(f))^2 - 2^6 d\Delta(f)^2\big).$$

This has the rational root $0 \in F$, yet we showed in Example 13.1.4 that the Galois group is not contained in $\langle(1324)\rangle$ when $d(b - 4d^2) \notin F^2$. So $R_f(y)$ fails to give accurate information about the Galois group, because 0 is not a simple root.   ◁▷

This example shows the importance of the word "simple" in Proposition 13.3.2.

Step 5: Repair Resolvents. As we've seen, resolvents computed by the above process can fail if their rational roots are not simple. To fix this, the standard method is to use a Tschirnhaus transformation (see the Historical Notes to Section 13.2) to change f to a different polynomial g. In [12], it is shown that this can always be done in such a way that f and g have the same Galois group and the corresponding resolvent $R_g(y)$ is separable. Then redo Step 4 with g and $R_g(y)$.

Aside from some clever tricks, this method for computing Galois groups is the basis of the algorithm used in [3] for polynomials of degree 4, 5, 6, and 7. However, we will see below that the `galois` command in *Maple* computes Galois groups using a slightly different approach.

B. Relative Resolvents. The idea of a relative resolvent was introduced in Section 12.1 in the universal case. Relative resolvents also are implicit in Theorem 13.1.1, as we will now explain.

Example 13.3.4. According to Example 13.3.3, $\langle(1324)\rangle \subset S_4$ is the symmetry group of $\varphi = \sqrt{\Delta}\,(x_1 + x_2 - x_3 - x_4)$. In Exercise 4 you showed that in the universal case, φ leads to the resolvent

$$R(y) = \prod_{i=1}^{3} \left(y^2 - \Delta\,(4y_i + \sigma_1^2 - 4\sigma_2)\right),$$

where $y_1 = x_1x_2 + x_3x_4$, $y_2 = x_1x_3 + x_2x_4$, $y_3 = x_1x_4 + x_2x_3$ are the roots of the universal Ferrari resolvent $\theta(y)$. If $f = x^4 - c_1x^3 + c_2x^2 - c_3x + c_4 \in F[x]$ is irreducible and separable and has roots $\alpha_1, \alpha_2, \alpha_3, \alpha_4$, then as usual $x_i \mapsto \alpha_i$ gives the resolvent

$$(13.26) \qquad R_f(y) = \prod_{i=1}^{3} \left(y^2 - \Delta(f)\,(4\beta_i + c_1^2 - 4c_2)\right),$$

where $\beta_1, \beta_2, \beta_3$ are the roots of the Ferrari resolvent $\theta_f(y)$.

Suppose we've already computed $\theta_f(y)$ and found that it has a root in F. As usual, we may assume that $\beta_1 = \alpha_1\alpha_2 + \alpha_3\alpha_4 \in F$, so that $G_f \subset \langle(1324), (12)\rangle$. To decide if G_f lies in the subgroup $\langle(1324)\rangle$, we could use the above resolvent $R_f(y)$. But since we already know β_1, we could instead use the factor

$$(13.27) \qquad y^2 - \Delta(f)\,(4\beta_1 + c_1^2 - 4c_2) \in F[x],$$

which is an example of a *relative resolvent*. If (13.27) has a simple root in F, then $G_f = \langle(1324)\rangle$ follows by the relative version of Proposition 13.3.2.

Notice that (13.27) has a simple root in F when $4\beta_1 + c_1^2 - 4c_2 \neq 0$ and $\Delta(f)\,(4\beta_1 + c_1^2 - 4c_2) \in F^2$, exactly as described in part (c) Theorem 13.1.1. Do you see how part (c) explains what happens when the relative resolvent (13.27) has a multiple root? ◁▷

The general theory of relative resolvents is described in [13] and [32]. Their main advantage is that they have smaller degree and hence are easier to compute. You will prove a version of Proposition 13.3.2 for relative resolvents in Exercise 5.

C. Factoring Resolvents. So far, we have asked whether resolvents have a rational root. But there are situations where the irreducible factorization of a resolvent can be useful, even if none of the factors have degree 1. Here is an example.

Example 13.3.5. Let f be an irreducible separable quartic, and consider the sextic resolvent $R_f(y)$ given in (13.26). Assume that $R_f(y)$ is separable. In Exercise 6 you will prove that G_f is conjugate to $\langle (1324), (12) \rangle$ if and only if $R_f(y) = g(y)h(y)$, where $g(y), h(y) \in F[x]$ are irreducible of degrees 2 and 4 respectively.   ◁▷

A much more interesting example involves the group $GL(3, \mathbb{F}_2)$ of invertible 3×3 matrices with entries in $\mathbb{F}_2$. In Section 14.3 we will see that $GL(3, \mathbb{F}_2)$ is a simple group of order 168. The smallest non-Abelian simple group is A_5 of order 60, and one can show that $GL(3, \mathbb{F}_2)$ of order 168 is the next smallest.

Following [9] and [28] we will show that

$$g = x^7 - 154x + 99 \in \mathbb{Q}[x]$$

is an irreducible polynomial whose Galois group over $\mathbb{Q}$ is isomorphic to $GL(3, \mathbb{F}_2)$. Our tool will be the factorization of a resolvent of degree 35.

First observe that $GL(3, \mathbb{F}_2)$ acts on the eight-element vector space $\mathbb{F}_2^3$ by matrix multiplication. The origin $0 = (0, 0, 0)$ is fixed, but the seven nonzero vectors get permuted. In Exercise 7 you will show that labeling these vectors $v_1, \ldots, v_7$ induces a one-to-one group homomorphism

(13.28) $$GL(3, \mathbb{F}_2) \to S_7.$$

For simplicity we will identify $GL(3, \mathbb{F}_2)$ with its image under this map and hence regard $GL(3, \mathbb{F}_2)$ as a subgroup of S_7.

Now consider $\varphi = x_1 + x_2 + x_3 \in \mathbb{Q}[x_1, \ldots, x_7]$. If we are given a polynomial $f \in \mathbb{Q}[x]$ of degree 7 with roots $\alpha_1, \ldots, \alpha_7$, then we get the resolvent

$$R_f(y) = \prod_{1 \le i < j < k \le 7} \left(y - (\alpha_i + \alpha_j + \alpha_k) \right) \in \mathbb{Q}[y].$$

There is one factor for each three-element subset of $\{1, \ldots, 7\}$, so that $R_f(y)$ has degree $\binom{7}{3} = 35$. Then we have the following interesting result of [28].

Proposition 13.3.6. *Let $f \in \mathbb{Q}[x]$ be irreducible of degree 7, and let $R_f(y)$ be the above resolvent of degree 35, which we assume to be separable. Then the Galois group of f over $\mathbb{Q}$ is isomorphic to $GL(3, \mathbb{F}_2)$ if and only if $R_f(y) = g(y)h(y)$, where $g(y), h(y) \in \mathbb{Q}[x]$ are irreducible of degrees 7 and 28 respectively.*

Proof. First suppose that the Galois group of f over $\mathbb{Q}$ is isomorphic to $GL(3, \mathbb{F}_2)$. The transitive subgroups of S_7 are known (see [2, pp. 206–209]) and in particular, any subgroup of S_7 isomorphic to $\mathrm{GL}(3, \mathbb{F}_2)$ is conjugate to (13.28). By relabeling the roots, we may assume that

$$G_f = \mathrm{GL}(3, \mathbb{F}_2) \subset S_7.$$

Since $R_f(y)$ is separable, its irreducible factorization is governed by the action of the Galois group on its roots, which is equivalent to the action of $\mathrm{GL}(3, \mathbb{F}_2)$ on three-element subsets of $\{1, \ldots, 7\}$ (be sure you understand this). Hence we need to understand the action of $\mathrm{GL}(3, \mathbb{F}_2)$ on unordered triples of nonzero vectors of $\mathbb{F}_2^3$.

A one-dimensional subspace of $\mathbb{F}_2^3$ is a line through the origin, which consists of 0 and a nonzero vector, since we are over $\mathbb{F}_2$. Hence there are seven such subspaces. In Exercise 8 you will show that $\mathbb{F}_2^3$ also has seven two-dimensional subspaces, each of which consists of 0 together with three linearly dependent nonzero vectors.

It follows that of the 35 possible triples of nonzero vectors in $\mathbb{F}_2^3$, seven consist of linearly dependent vectors while the remaining 28 consist of linearly independent vectors. In Exercise 8 you will show that $\mathrm{GL}(3, \mathbb{F}_2)$ acts transitively on each of these sets of triples. As explained above, this describes the Galois action on the roots of $R_f(y)$, and the desired factorization follows.

The converse is proved in [28]. Let the Galois group of f be isomorphic to $G_f \subset S_7$. Since $R_f(y)$ is separable, the Galois action on its roots is equivalent to the action of G_f on three-element subsets of $\{1, \ldots, 7\}$. The conjugacy classes of subgroups S_7 are known, and for each conjugacy class, one can compute the orbits of its action on unordered triples. These are listed in Table I in [28]. Inspection of this list shows that $\mathrm{GL}(3, \mathbb{F}_2) \subset S_7$ is the only subgroup (up to conjugacy) such that the orbits have lengths 7 and 28. Thus G_f must be conjugate to $\mathrm{GL}(3, \mathbb{F}_2)$ when $R_f(y)$ has irreducible factors of degrees 7 and 28. $\qquad\square$

Here is the example mentioned earlier.

Example 13.3.7. For $f = x^7 - 154x + 99$, [9] computes that

$$
\begin{aligned}
R_f(y) = \; & y^{35} - 6160y^{29} + 29898y^{28} - 38277624y^{23} - 41255676y^{22} \\
& + 37518228y^{21} + 18524283008y^{17} + 6522421752y^{16} \\
& + 27295157736y^{15} + 35173338750y^{14} - 2894923232432y^{11} \\
& + 489571380144y^{10} - 4925879415072y^9 + 3933790086996y^8 \\
& - 702099623709y^7 + 149674336745472y^5 - 96219216479232y^4 \\
& - 25773004414080y^3 + 21354775085952y^2 + 946763427456y \\
& - 1217267263872.
\end{aligned}
$$

Using *Maple* or *Mathematica*, the irreducible factorization of $R_f(y)$ over $\mathbb{Q}$ is easily computed to be

$$R_f(y) = g(y)h(y),$$

where

$$g(y) = y^7 - 231y^3 - 462y^2 + 77y + 66$$

and $h(y)$ is the polynomial

$$
\begin{aligned}
&y^{28} + 231y^{24} + 462y^{23} - 6237y^{22} + 29832y^{21} + 53361y^{20} \\
&+ 213444y^{19} - 1245090y^{18} + 3958878y^{17} - 11719092y^{16} + 30817248y^{15} \\
&- 157564143y^{14} + 319312224y^{13} - 796323990y^{12} + 1481906118y^{11} \\
&- 2994381313y^{10} + 5889443406y^9 + 965064177y^8 - 4595839182y^7 \\
&+ 33180883659y^6 - 84492127566y^5 + 181691003340y^4 \\
&- 382065796728y^3 + 152613801648y^2 + 35862251040y - 18443443392.
\end{aligned}
$$

By Proposition 13.3.6, it follows that the Galois group of f over $\mathbb{Q}$ is isomorphic to $\mathrm{GL}(3, \mathbb{F}_2)$.   ◁▷

The galois command in *Maple* computes the Galois group over $\mathbb{Q}$ of an irreducible polynomial of degree ≤ 9 in $\mathbb{Q}[x]$. The algorithm used by *Maple* involves factoring resolvents of the above type. See [28] for more details. The computer algebra programs **GAP** [10] and Magma [20] can compute the Galois group over $\mathbb{Q}$ of an irreducible polynomial of degree ≤ 15 and ≤ 22 respectively.

Mathematical Notes

We will discuss two topics from this section.

■ **Trinomials of Degree 7.** In the text, we showed that the Galois group over $\mathbb{Q}$ of $x^7 - 154x + 99$ is isomorphic to $\mathrm{GL}(3, \mathbb{F}_2)$. In the late 1960s, $x^7 - 7x + 3$ was shown to have the same property. Recently, Elkies [8] studied the problem of finding *all* trinomials $f = ax^7 + bx + c \in \mathbb{Q}[x]$ with $\mathrm{GL}(3, \mathbb{F}_2)$ as Galois group. We say that another trinomial g is *equivalent* to f if

$$
g = \lambda\big(a(\mu x)^7 + b(\mu x) + c\big)
$$

for some $\lambda, \mu \in \mathbb{Q}^*$. Then Elkies shows that equivalence classes of trinomials over $\mathbb{Q}$ whose Galois group is contained in $\mathrm{GL}(3, \mathbb{F}_2)$ correspond to solutions $(x, y) \in \mathbb{Q}^2$ of the equation

(13.29) $y^2 = x(81x^5 + 396x^4 + 738x^3 + 660x^2 + 269x + 48).$

In 2001, Bruin found all solutions $(x, y) \in \mathbb{Q}^2$ (including points at infinity). This leads to the result that, up to equivalence, the only trinomials $ax^7 + bx + c \in \mathbb{Q}[x]$ with $\mathrm{GL}(3, \mathbb{F}_2)$ as Galois group are

$$
x^7 - 7x + 3, \quad x^7 - 154x + 99, \quad 37^2x^7 - 28x + 9, \quad 499^2x^7 - 23956x + 3^4113.
$$

Details and references can be found in [8]. Also, [9] and [28] include references to other papers on polynomials of degree 7 with $\mathrm{GL}(3, \mathbb{F}_2)$ as Galois group.

The equation (13.29) is another example of a Diophantine equation. In contrast to Example 13.1.7, this is not an elliptic curve. Instead, it has *genus* 2 (while elliptic curves have genus 1). By Faltings's proof of the Mordell Conjecture, it is known that equations of genus ≥ 2 have at most finitely many rational solutions, that is, solutions with $(x, y) \in \mathbb{Q}^2$. But the proof is not constructive, so that in practice it can be difficult to prove that one has found all rational solutions.

■ **Groups and Geometry.** The group $GL(3, \mathbb{F}_2)$ is important for reasons related to group theory and geometry. Let's begin with the group theory. In the Mathematical Notes to Section 11.1, we defined $GL(n, F)$ to be the group of invertible $n \times n$ matrices with entries in the field F. This group contains the subgroup $SL(n, F)$ of matrices of determinant 1. Furthermore, taking the quotient of each of these groups by the subgroup consisting of multiples of the identity matrix gives groups

$$PGL(n, F) \quad \text{and} \quad PSL(n, F).$$

We will say more about these groups in the Mathematical Notes to Section 14.3. There, we will see that aside from $PSL(2, \mathbb{F}_2) \simeq S_3$ and $PSL(2, \mathbb{F}_3) \simeq A_4$, the group $PSL(n, \mathbb{F}_q)$ is simple whenever $n \geq 2$.

In particular, $PSL(3, \mathbb{F}_2)$ is simple. However, by Exercise 9 we have

$$GL(3, \mathbb{F}_2) = SL(3, \mathbb{F}_2) \simeq PGL(3, \mathbb{F}_2) = PSL(3, \mathbb{F}_2).$$

This explains why $GL(3, \mathbb{F}_2)$ is simple. In Section 14.3 we will also see that

$$PSL(3, \mathbb{F}_2) \simeq PSL(2, \mathbb{F}_7).$$

Some papers, such as [9], write $PSL(2, \mathbb{F}_7)$ instead of $GL(3, \mathbb{F}_2)$.

For any field F, $GL(3, F)$ and $PGL(3, F)$ have interesting geometric properties. For $GL(3, F)$, the geometric object it acts on is the vector space F^3 of dimension 3 over F. For $PGL(3, F)$, the corresponding geometric object is the *projective plane* $\mathbb{P}^2$ over F. Although this is beyond the scope of the book, we will make one comment related to the proof of Proposition 13.3.6. There, we observed that $\mathbb{F}_2^3$ has seven one-dimensional subspaces and seven two-dimensional subspaces. Once you understand the geometry of $\mathbb{P}^2$ over $F = \mathbb{F}_2$, this follows immediately from *projective duality*. More on projective geometry can be found in [25].

Exercises for Section 13.3

Exercise 1. Let $f(x) = a_0 x^n + \cdots + a_n \in \mathbb{Q}[x]$.
(a) Prove that there are $\lambda, \mu \in \mathbb{Q}^*$ such that $g(x) = \lambda f(\mu x) \in \mathbb{Z}[x]$ is monic.
(b) Prove that f and g have isomorphic Galois groups over $\mathbb{Q}$.

Exercise 2. Let $f(x) = x^n + \cdots + a_n \in \mathbb{Z}[x]$, and let $R_f(y)$ be the resolvent built from $\varphi \in \mathbb{Z}[x_1, \ldots, x_n]$. Prove that $R_f(y) \in \mathbb{Z}[x]$.

Exercise 3. In the proof of Proposition 13.3.2, we asserted that

$$\varphi(\alpha_1, \ldots, \alpha_n) = \varphi(\alpha_{\tau(1)}, \ldots, \alpha_{\tau(n)})$$

follows from $\beta = \varphi(\alpha_1, \ldots, \alpha_n) \in F$ and $\tau \in G_f$. Prove this.

Exercise 4. As in Examples 13.3.3 and 13.3.4, let $\varphi = \sqrt{\Delta}\,(x_1 + x_2 - x_3 - x_4)$.
(a) Show that the symmetry group of φ is $G = \langle(1324)\rangle \subset S_4$.
(b) Show that in the universal case, φ leads to the resolvent

$$R(y) = \prod_{i=1}^{3}\left(y^2 - \Delta\,(4y_i + \sigma_1^2 - 4\sigma_2)\right),$$

where $y_1 = x_1x_2 + x_3x_4$, $y_2 = x_1x_3 + x_2x_4$, $y_3 = x_1x_4 + x_2x_3$ are the roots of the universal Ferrari resolvent $\theta(y)$.
(c) Let $R_f(y)$ be obtained by specializing the resolvent $R(y)$ of part (b) to $f = x^4 + bx^2 + d$. Show that

$$R_f(y) = y^2\left((y^2 + 4b\Delta(f))^2 - 2^6 d\Delta(f)^2\right).$$

Exercise 5. This problem will state and prove a relative version of Proposition 13.3.2. Fix a subgroup $H \subset S_n$ and suppose that $f \in F[x]$ is separable and that $G_f \subset H$. Now let $G \subset H$ be a subgroup. We want to know whether or not G_f lies in the smaller subgroup G. Let $\varphi \in F[x_1, \ldots, x_n]$ have G as its symmetry group and let $\varphi_1 = \varphi, \varphi_2, \ldots, \varphi_\ell$ be the orbit of H acting on φ. Then set

$$R^H(y) = \prod_{i=1}^{\ell}(y - \varphi_i) \in F[x_1, \ldots, x_n][y].$$

Finally, if $\alpha_1, \ldots, \alpha_n$ are the roots of f in a splitting field L, let

$$R_f^H(y) = \prod_{i=1}^{\ell}\left(y - \varphi_i(\alpha_1, \ldots, \alpha_n)\right) \in L[y]$$

be the polynomial obtained by $x_i \mapsto \alpha_i$.
(a) Explain why the degree of $R_f^H(y)$ is the index of G in H.
(b) Prove that $R_f^H(y) \in F[y]$.
(c) Assume that G_f is conjugate within H to a subgroup of G (this means that $\tau G_f \tau^{-1} \subset G$ for some $\tau \in H$). Prove that $R_f^H(y)$ has a root in F.
(d) Assume that $R_f^H(y)$ has a simple root in F. Prove that G_f is conjugate within H to a subgroup of G.
We call $R_f^H(y)$ a *relative resolvent*. You will verify in Exercise 10 that (13.27) from Example 13.3.4 is an example of a relative resolvent.

Exercise 6. Let $f \in F[x]$ be an irreducible separable quartic, where F has characteristic $\neq 2$. Also let $R_f(y)$ be the sextic resolvent defined in Example 13.3.4. The goal of this exercise is to show that $G_f \subset S_4$ determines the irreducible factorization of $R_f(y)$ over F. We will assume that $R_f(y)$ is separable.
(a) First suppose that $G_f = A_4$ or S_4. Prove that $R_f(y)$ is irreducible over F.
(b) Now suppose that $G_f = \langle(1324), (12)\rangle$. Prove that $R_f(y) = g(y)h(y)$, where $g(y)$, $h(y) \in F[x]$ are irreducible of degrees 2 and 4 respectively.
(c) Suppose that $G_f = \langle(12)(34), (13)(24)\rangle$. Prove that $R_f(y) = g_1(y)g_2(y)g_3(y)$, where $g_i(y) \in F[x]$ is irreducible of degree 2.
(d) Finally, suppose that $G_f = \langle(1324)\rangle$. Prove that $R_f(y) = g_1(y)g_2(y)g_3(y)$, where $g_1(y), g_2(y), g_3(y) \in F[x]$ are irreducible of degrees 1, 1, and 4 respectively.

(e) Explain why parts (a) through (d) enable one to determine G_f up to conjugacy using only $R_f(y)$ and $\Delta(f)$.

Notice that the claim made in Example 13.3.5 now follows immediately.

Exercise 7. The action of $GL(3, \mathbb{F}_2)$ on the nonzero vectors of $\mathbb{F}_2^3$ gives a group homomorphism $GL(3, \mathbb{F}_2) \to S_7$. Prove that this map is one to one.

Exercise 8. Consider the vector space $\mathbb{F}_2^3$.
(a) Prove that $\mathbb{F}_2^3$ has exactly seven two-dimensional subspaces.
(b) For a field F, let $B = \{\{v_1, v_2, v_3\} \subset F \mid v_1, v_2, v_3 \text{ are linearly independent over } F\}$. Prove that $GL(3, F)$ acts transitively on B.
(c) Let be F as in part (b). Prove that $GL(3, F)$ acts transitively on the set of two-dimensional subspaces of F^3.

Be sure you understand how parts (b) and (c) apply to the proof of Proposition 13.3.6.

Exercise 9. Prove that $GL(3, \mathbb{F}_2) = SL(3, \mathbb{F}_2) \simeq PGL(3, \mathbb{F}_2) = PSL(3, \mathbb{F}_2)$.

Exercise 10. Prove that (13.27) from Example 13.3.4 is an example of a relative resolvent in the sense of Exercise 5.

Exercise 11. In the proof of Proposition 13.3.6, we showed that when $GL(3, \mathbb{F}_2) \subset S_7$ acts on three-element subsets of $\{1, \ldots, 7\}$, the orbits have lengths 7 and 28. We also asserted that up to conjugacy, $GL(3, \mathbb{F}_2)$ is the only subgroup of S_7 with this property. In this exercise, you will study the action of some other subgroups of S_7.
(a) Prove that A_7 and S_7 act transitively on three-element subsets of $\{1, \ldots, 7\}$. Thus there is one orbit of length 35 for these groups.
(b) In Section 13.2, the group $AGL(1, \mathbb{F}_5) \subset S_5$ played an important role in understanding the Galois group of a quintic. In a similar way, we have $AGL(1, \mathbb{F}_7) \subset S_7$ provided we think of the indices as congruences classes modulo 7. Prove that the orbits of $AGL(1, \mathbb{F}_7)$ acting on the triples $\{0, 1, 2\}$ and $\{0, 1, 3\}$ have 21 and 14 elements respectively.

13.4   OTHER METHODS

This section will explore further tools for computing Galois groups. We begin with a result of Kronecker that works in complete generality but is not very efficient. However, his method also leads to a quick proof of a result of Dedekind that uses reduction modulo p to obtain useful information about Galois groups over $\mathbb{Q}$.

A. Kronecker's Analysis. In Section 12.3 we studied Kronecker's construction of the splitting field of a separable polynomial $f \in F[x]$ of degree n. Let's recall how this works.

Assume that F is infinite, and let $\alpha_1, \ldots, \alpha_n$ be the roots of f in a splitting field $F \subset L$. We saw in Section 12.2 that there are $t_1, \ldots, t_n \in F$ such that the $n!$ elements

$$t_1 \alpha_{\sigma(1)} + \cdots + t_n \alpha_{\sigma(n)}, \quad \sigma \in S_n,$$

are distinct. Thus

$$(13.30) \qquad s(y) = \prod_{\sigma \in S_n} \left(y - (t_1 \alpha_{\sigma(1)} + \cdots + t_n \alpha_{\sigma(n)}) \right) \in L[y]$$

is separable of degree $n!$. We showed in Section 12.3 that $s(y) \in F[y]$ and that if $h(y) \in F[y]$ is *any* irreducible factor of $s(y)$, then the quotient

$$F[y]/\langle h(y)\rangle$$

is a splitting field of f over F. It follows that the degree of $h(y)$ is the order of the Galois group of f over F.

This construction seems to require the roots $\alpha_1, \ldots, \alpha_n$. However, the universal version of $s(y)$ is given by

$$(13.31) \qquad S(y) = \prod_{\sigma \in S_n} \big(y - (t_1 x_{\sigma(1)} + \cdots + t_n x_{\sigma(n)})\big) \in F[x_1, \ldots, x_n, y].$$

The theory of symmetric polynomials tells us how to write $S(y)$ explicitly as a polynomial in $F[\sigma_1, \ldots, \sigma_n, y]$. Then specializing to the coefficients of f gives $s(y)$ as in (13.30). Furthermore, Exercises 4 and 5 of Section 12.3 show how to pick $t_1, \ldots, t_n \in F$ (without knowing the roots) so that $s(y)$ is separable.

Here is an example of this process.

Example 13.4.1. Consider $f = x^3 + x^2 - 2x - 1 \in \mathbb{Q}[x]$. In Exercise 1 you will show that if we set $t_1 = 1$, $t_2 = 1$, and $t_3 = 2$, then the universal polynomial (13.31) becomes

$$
\begin{aligned}
S(y) = {}& y^6 - 4\sigma_1 y^5 + (2\sigma_1^2 + 14\sigma_2)y^4 + (8\sigma_1^3 - 44\sigma_1\sigma_2 + 20\sigma_3)y^3 \\
& + (-7\sigma_1^4 + 18\sigma_1^2\sigma_2 + 49\sigma_2^2 - 40\sigma_1\sigma_3)y^2 + (-4\sigma_1^5 + 44\sigma_1^3\sigma_2 \\
& - 112\sigma_1\sigma_2^2 - 20\sigma_1^2\sigma_3 + 140\sigma_2\sigma_3)y + 4\sigma_1^6 - 32\sigma_1^4\sigma_2 + 55\sigma_1^2\sigma_2^2 \\
& + 36\sigma_2^3 + 76\sigma_1^3\sigma_3 - 322\sigma_1\sigma_2\sigma_3 + 343\sigma_3^2.
\end{aligned}
$$

Using $\sigma_1 \mapsto -1$, $\sigma_2 \mapsto -2$, $\sigma_3 \mapsto 1$, we obtain

$$
\begin{aligned}
s(y) &= y^6 + 4y^5 - 26y^4 - 76y^3 + 193y^2 + 240y - 377 \\
&= (y^3 + 2y^2 - 15y + 13)(y^3 + 2y^2 - 15y - 29),
\end{aligned}
$$

where the second line is the irreducible factorization in $\mathbb{Q}[y]$. This shows that $s(y)$ is separable (do you see why?). Thus the Galois group of f over $\mathbb{Q}$ has order 3. (This also follows from the theory of Chapter 9, since f is the minimal polynomial of $2\cos(2\pi/7) = \zeta_7 + \zeta_7^{-1}$.) ◁▷

Besides giving the order of the Galois group, Kronecker observed that by modifying the above construction, one can extract the entire Galois group from an irreducible factor of $s(y)$. The idea is that instead of letting $t_1, \ldots, t_n$ be elements of F, we let them be variables. To prevent confusion, we will label these variables $u_1, \ldots, u_n$ and write (13.30) as

$$(13.32) \qquad s_u(y) = \prod_{\sigma \in S_n} \big(y - (u_1 \alpha_{\sigma(1)} + \cdots + u_n \alpha_{\sigma(n)})\big) \in L[u_1, \ldots, u_n, y].$$

The subscript u is a reminder that $s_u(y)$ is a polynomial in the $n+1$ variables $u_1, \ldots, u_n, y$. In Exercise 2 you will show that the coefficients lie in F, so that

$$s_u(y) \in F[u_1, \ldots, u_n, y].$$

Furthermore, we can compute $s_u(y)$ by first working in the universal situation and then specializing to f. Thus we can find $s_u(y)$ without knowing the roots of f.

The polynomial ring $F[u_1, \ldots, u_n, y]$ has two key structures:

- $F[u_1, \ldots, u_n, y]$ has an S_n-action given by permutations of $u_1, \ldots, u_n$.
- $F[u_1, \ldots, u_n, y]$ is a UFD.

In addition, $L[u_1, \ldots, u_n, y]$ has the same two structures plus a third:

- $L[u_1, \ldots, u_n, y]$ has a $\mathrm{Gal}(L/F)$-action given by the Galois action on L.

As in the previous section, we write the Galois group of f over F as

$$\mathrm{Gal}(L/F) \simeq G_f \subset S_n.$$

The above structures, when applied to $s_u(y)$, give the following description of G_f.

Theorem 13.4.2. *Assume that $f \in F[x]$ is monic and separable of degree n, where F is an arbitrary field. Also let $h \in F[u_1, \ldots, u_n, y]$ be an irreducible factor of the polynomial $s_u(y) \in F[u_1, \ldots, u_n, y]$ constructed above. Then $G_f \subset S_n$ is conjugate to the subgroup*

$$G = \{\tau \in S_n \mid \tau \cdot h = h\} \subset S_n.$$

Proof. In Exercise 3 you will show that (13.32) is the irreducible factorization of $s_u(y)$ in $L[u_1, \ldots, u_n, y]$. Thus we can pick $\sigma \in S_n$ such that

(13.33) $$y - (u_1 \alpha_{\sigma(1)} + \cdots + u_n \alpha_{\sigma(n)})$$

is a factor of h in $L[u_1, \ldots, u_n, y]$. The permutation σ will be fixed for the remainder of the proof. Our goal will be to prove that $G = \sigma^{-1} G_f \sigma$.

Consider the polynomial

(13.34)
$$\begin{aligned}
\tilde{h} &= \prod_{\gamma \in \mathrm{Gal}(L/F)} \left(y - (u_1 \gamma(\alpha_{\sigma(1)}) + \cdots + u_n \gamma(\alpha_{\sigma(n)})) \right) \\
&= \prod_{\mu \in G_f} \left(y - (u_1 \alpha_{\mu\sigma(1)} + \cdots + u_n \alpha_{\mu\sigma(n)}) \right).
\end{aligned}$$

Standard arguments imply that $\tilde{h}$ is invariant under the action of $\mathrm{Gal}(L/F)$, so that $\tilde{h} \in F[u_1, \ldots, u_n, y]$ since $F \subset L$ is Galois. Now pick $\gamma \in \mathrm{Gal}(L/F)$. Since h has coefficients in F and (13.33) divides h in $L[u_1, \ldots, u_n, y]$, it follows that

$$y - (u_1 \gamma(\alpha_{\sigma(1)}) + \cdots + u_n \gamma(\alpha_{\sigma(n)}))$$

divides $\gamma \cdot h = h$ in $L[u_1, \ldots, u_n, y]$. Hence $\tilde{h}$ divides h in $L[u_1, \ldots, u_n, y]$, which by Exercise 3 implies that $\tilde{h}$ divides h in $F[u_1, \ldots, u_n, y]$. Since h is irreducible, it follows that $\tilde{h} = h$ after multiplying h by a suitable constant.

Now suppose that $\tau \in S_n$ satisfies $\tau \cdot h = h$. This implies that τ applied to (13.33) is a factor of h in $L[u_1, \ldots, u_n, y]$. Since (13.34) is the irreducible factorization of h in $L[u_1, \ldots, u_n, y]$, we must have

$$y - (u_{\tau(1)}\alpha_{\sigma(1)} + \cdots + u_{\tau(n)}\alpha_{\sigma(n)}) = y - (u_1\alpha_{\mu\sigma(1)} + \cdots + u_n\alpha_{\mu\sigma(n)})$$

for some $\mu \in G_f$. Since $u_1, \ldots, u_n$ are distinct variables, this implies that

$$\tau(i) = j \implies \alpha_{\sigma(i)} = \alpha_{\mu\sigma(j)}$$
$$\implies \alpha_{\sigma\tau^{-1}(j)} = \alpha_{\mu\sigma(j)}.$$

It follows that $\sigma\tau^{-1} = \mu\sigma$, since $\alpha_1, \ldots, \alpha_n$ are distinct. Thus

$$\tau = \sigma^{-1}\mu^{-1}\sigma \in \sigma^{-1}G_f\sigma.$$

This shows that $G \subset \sigma^{-1}G_f\sigma$. You will prove the opposite inclusion in Exercise 3, which implies $G = \sigma^{-1}G_f\sigma$. This completes the proof of the theorem. ☐

Theorem 13.4.2 gives an algorithm for computing the Galois group of f:

- Compute $s_u(y) \in F[u_1, \ldots, u_n, y]$.
- Factor $s_u(y)$ into irreducibles, and let h be an irreducible factor.
- For each $\tau \in S_n$, compute $\tau \cdot h$ and compare it with h.
- Then the Galois group of f over F is isomorphic to $\{\tau \in S_n \mid \tau \cdot h = h\}$.

For n large, this algorithm is extremely inefficient. For example, $s_u(y)$ has degree $10! = 3628800$ in y when $\deg(f) = 10$. Finding an irreducible factor h of $s_u(y)$ could take a long time. And even if we could find h, then we would need to compute $\tau \cdot h$ for all 3628800 permutations $\tau \in S_{10}$. Thus this algorithm is not useful in practice, although it is a completely general method for computing Galois groups.

Here is an example of how to use Theorem 13.4.2 when n is small.

Example 13.4.3. Consider $f = x^3 + x^2 - 2x - 1 \in \mathbb{Q}[x]$. In Exercise 4 you will show that $s_u(y)$ has the irreducible factorization

$$\begin{aligned}
s_u(y) = {}&(u_1^3 + 3u_1^2u_2 - 4u_1u_2^2 + u_2^3 - 4u_1^2u_3 - u_1u_2u_3 + 3u_2^2u_3 + 3u_1u_3^2 \\
&- 4u_2u_3^2 + u_3^3 + 2u_1^2y - 3u_1u_2y + 2u_2^2y - 3u_1u_3y - 3u_2u_3y + 2u_3^2y \\
&- u_1y^2 - u_2y^2 - u_3y^2 - y^3) \times (u_1^3 - 4u_1^2u_2 + 3u_1u_2^2 + u_2^3 + 3u_1^2u_3 \\
&- u_1u_2u_3 - 4u_2^2u_3 - 4u_1u_3^2 + 3u_2u_3^2 + u_3^3 + 2u_1^2y - 3u_1u_2y \\
&+ 2u_2^2y - 3u_1u_3y - 3u_2u_3y + 2u_3^2y - u_1y^2 - u_2y^2 - u_3y^2 - y^3)
\end{aligned}$$

in $\mathbb{Q}[u_1, u_2, u_3, y]$. (This calculation was done in *Mathematica*.) Let h be the first factor multiplied by -1. You will show in Exercise 4 that

$$\begin{aligned}
h = {}&y^3 + (u_1 + u_2 + u_3)y^2 + \big(7(u_1u_2 + u_1u_3 + u_2u_3) - 2(u_1 + u_2 + u_3)^2\big)y \\
&+ 7u_1u_2u_3 - (u_1 + u_2 + u_3)^3 + 7(u_1u_2^2 + u_1^2u_3 + u_2u_3^2).
\end{aligned}$$

In this formula for h, everything is symmetric in u_1, u_2, u_3 except for the last set of parentheses. Thus $\tau \in S_3$ fixes h if and only if τ fixes

$$u_1 u_2^2 + u_1^2 u_3 + u_2 u_3^2.$$

It follows easily that the group G of Theorem 13.4.2 is $\langle (123) \rangle \subset S_3$. ◁▷

Note that f is not required to be irreducible over F. Here is an example.

Example 13.4.4. For $f = x^3 - 1 \in \mathbb{Q}[x]$, you will show in Exercise 5 that

$$
\begin{aligned}
s_u(y) = {} & \left(y^2 + (u_1 + u_2 - 2u_3)y + u_1^2 + u_2^3 + u_3^2 - u_1 u_2 - u_1 u_3 - u_2 u_3 \right) \\
& \times \left(y^2 + (u_1 + u_3 - 2u_2)y + u_1^2 + u_2^3 + u_3^2 - u_1 u_2 - u_1 u_3 - u_2 u_3 \right) \\
& \times \left(y^2 + (u_2 + u_3 - 2u_1)y + u_1^2 + u_2^3 + u_3^2 - u_1 u_2 - u_1 u_3 - u_2 u_3 \right).
\end{aligned}
$$

In each factor, the terms of degree 0 in y are symmetric in u_1, u_2, u_3. So the coefficient of y is the crucial term. It follows that the first factor gives $G = \langle (12) \rangle$, the second gives $G = \langle (13) \rangle$, and the third gives $G = \langle (23) \rangle$. ◁▷

Although Example 13.4.4 is trivial from the point of view of Galois theory, it does show that Theorem 13.4.2 determines the Galois group only up to conjugacy.

In the earlier part of this section we assumed that F was infinite so that we could find $t_1, \ldots, t_n \in F$ such that the $n!$ elements $t_1 \alpha_{\sigma(1)} + \cdots + t_n \alpha_{\sigma(n)}$ were distinct. In contrast, Theorem 13.4.2 applies to all fields, even finite ones. This works because $u_1, \ldots, u_n$ are variables, so that the expressions $u_1 \alpha_{\sigma(1)} + \cdots + u_n \alpha_{\sigma(n)}$ are automatically distinct by the separability of f.

We will soon see that applying Theorem 13.4.2 over a finite field has some nice consequences.

B. Dedekind's Theorem. Given a polynomial $f \in \mathbb{Z}[x]$ and a prime p, we let $\bar{f} \in \mathbb{F}_p[x]$ be the polynomial obtained by reducing the coefficients of f modulo p. Then the following theorem of Dedekind shows how $\bar{f}$ can give information about the Galois group of f over $\mathbb{Q}$.

Theorem 13.4.5. *Let $f \in \mathbb{Z}[x]$ be monic and separable of degree n. Given a prime p such that $p \nmid \Delta(f)$, let*

$$\bar{f} = \bar{f}_1 \bar{f}_2 \cdots \bar{f}_r$$

be the irreducible factorization of $\bar{f}$ in $\mathbb{F}_p[x]$. Also set $d_i = \deg(\bar{f}_i)$. Then:

(a) *The Galois group of $\bar{f}$ over $\mathbb{F}_p$ is cyclic of order $\operatorname{lcm}(d_1, d_2, \ldots, d_r)$.*

(b) *The Galois group of f over $\mathbb{Q}$ contains an element that acts on the roots of f according to a product of disjoint cycles of the form*

$$\underbrace{(.\ .\ .)}_{d_1\text{-cycle}} \underbrace{(.\ .\ .)}_{d_2\text{-cycle}} \cdots \underbrace{(.\ .\ .)}_{d_r\text{-cycle}}.$$

Hence the Galois group of f contains an element of order $\operatorname{lcm}(d_1, d_2, \ldots, d_r)$.

Proof. First observe that $\bar{f}$ is separable, since $p \nmid \Delta(f)$ and $\Delta(\bar{f})$ is the reduction of $\Delta(f)$ modulo p by Exercise 4 of Section 5.3.

Part (a) is an easy application of Chapter 11. Since

$$x^{p^m} - x = \prod_{\alpha \in \mathbb{F}_{p^m}} (x - \alpha),$$

a separable polynomial in $\mathbb{F}_p[x]$ splits completely over $\mathbb{F}_{p^m}$ if and only if it divides $x^{p^m} - x$. Thus:

$$\bar{f} \text{ splits completely over } \mathbb{F}_{p^m} \iff \bar{f}_i \text{ splits completely over } \mathbb{F}_{p^m} \text{ for all } i$$
$$\iff \bar{f}_i \text{ divides } x^{p^m} - x \text{ for all } i$$
$$\iff d_i = \deg(\bar{f}_i) \text{ divides } m$$
$$\iff \mathrm{lcm}(d_1, d_2, \ldots, d_r) \text{ divides } m,$$

where the second equivalence uses our above observation and the third equivalence uses part (c) of Proposition 11.2.1. This easily implies that the splitting field of $\bar{f}$ over $\mathbb{F}_p$ is $\mathbb{F}_{p^d}$, $d = \mathrm{lcm}(d_1, d_2, \ldots, d_r)$. Since $\mathrm{Gal}(\mathbb{F}_{p^d}/\mathbb{F}_p)$ is cyclic of order d by Theorem 11.1.7, part (a) follows.

For later purposes, let's describe the action of $\mathrm{Gal}(\mathbb{F}_{p^d}/\mathbb{F}_p)$ on the roots of $\bar{f}$. By Theorem 11.1.7, the Galois group is generated by the Frobenius automorphism $\alpha \mapsto \alpha^p$. Since $\bar{f}_i$ is irreducible, Exercise 7 of Section 11.1 implies that if α is a root of $\bar{f}_i$, then all roots are given by

$$\alpha, \alpha^p, \alpha^{p^2}, \ldots, \alpha^{p^{d_i-1}}, \quad d_i = \deg(\bar{f}_i).$$

Hence the action of $\alpha \mapsto \alpha^p$ on the roots of $\bar{f}_i$ is given by a d_i-cycle. Since $\bar{f}$ is separable, it follows that $\alpha \mapsto \alpha^p$ acts on the roots of $\bar{f}$ according to a product of disjoint cycles of lengths $d_1, \ldots, d_r$, just as in part (b) of the theorem.

We turn to part (b). Consider the universal version of $s_u(y)$ defined by

$$S_u(y) = \prod_{\sigma \in S_n} \left(y - (u_1 x_{\sigma(1)} + \cdots + u_n x_{\sigma(n)}) \right) \in \mathbb{Z}[x_1, \ldots, x_n, u_1, \ldots, u_n, y].$$

This is symmetric in $x_1, \ldots, x_n$, so that

$$S_u(y) \in \mathbb{Z}[\sigma_1, \ldots, \sigma_n, u_1, \ldots, u_n, y]$$

by Fundamental Theorem of Symmetric Polynomials over $\mathbb{Z}$ (see Exercise 6 of Section 9.1).

Write the polynomials f and $\bar{f}$ as

$$f = x^n - c_1 x^{n-1} + \cdots + (-1)^n c_n \in \mathbb{Z}[x],$$
$$\bar{f} = x^n - \bar{c}_1 x^{n-1} + \cdots + (-1)^n \bar{c}_n \in \mathbb{F}_p[x].$$

This gives

$$s_u(y) \in \mathbb{Z}[u_1, \ldots, u_n, y] \quad \text{obtained from } S_u(y) \text{ via } \quad \sigma_i \mapsto c_i,$$
$$\bar{s}_u(y) \in \mathbb{F}_p[u_1, \ldots, u_n, y] \quad \text{obtained from } S_u(y) \text{ via } \quad \sigma_i \mapsto \bar{c}_i.$$

Thus $\bar{s}_u(y)$ is the reduction of $s_u(y)$ modulo p, since $\bar{c}_i$ is the reduction of c_i modulo p.

We relate $s_u(y)$ and $\bar{s}_u(y)$ to the Galois groups of f and $\bar{f}$ as follows. As usual, the Galois group of f over $\mathbb{Q}$ maps to a subgroup $G_f \subset S_n$ that records the Galois action on the roots. Given an irreducible factor h of $s_u(y)$ over $\mathbb{Q}$, Theorem 13.4.2 implies that G_f is conjugate to

$$G = \{\sigma \in S_n \mid \sigma \cdot h = h\}.$$

By Exercise 6 we may assume that h is an irreducible factor of $s_u(y)$ in the ring $\mathbb{Z}[u_1, \ldots, u_n, y]$. Reducing this modulo p gives $\bar{h} \in \mathbb{F}_p[u_1, \ldots, u_n, y]$. If $\bar{g}$ is an irreducible factor of $\bar{h}$, then it is also an irreducible factor of $\bar{s}_u(y)$, so that by Theorem 13.4.2, the Galois group of $\bar{f}$ over $\mathbb{F}_p$ gives a subgroup of S_n conjugate to

$$\overline{G} = \{\sigma \in S_n \mid \sigma \cdot \bar{g} = \bar{g}\}.$$

We claim that $\overline{G} \subset G$. To prove this, suppose that $\sigma \cdot \bar{g} = \bar{g}$ but $\sigma \cdot h = h_1 \neq h$. Then $\sigma \cdot s_u(y) = s_u(y)$ implies that h_1 is also an irreducible factor of $s_u(y)$. Since $\mathbb{Z}[u_1, \ldots, u_n, y]$ is a UFD, we must have

$$s_u(y) = h h_1 q$$

for some polynomial $q \in \mathbb{Z}[u_1, \ldots, u_n, y]$. Reducing this modulo p gives

$$\bar{s}_u(y) = \bar{h} \bar{h}_1 \bar{q} \in \mathbb{F}_p[u_1, \ldots, u_n, y].$$

Furthermore, the S_n-action is compatible with reduction modulo p, so that $h_1 = \sigma \cdot h$ implies $\bar{h}_1 = \sigma \cdot \bar{h}$. Since $\bar{g}$ divides $\bar{h}$, we see that $\bar{g} = \sigma \cdot \bar{g}$ divides $\sigma \cdot \bar{h} = \bar{h}_1$. By the above equation, this implies that $\bar{g}^2$ divides $\bar{s}_u(y)$. Yet over the splitting field L of $\bar{f}$, (13.32) implies that $\bar{s}_u(y)$ is a product of distinct irreducible factors. This easily implies that the same is true over $\mathbb{F}_p$. Hence we have a contradiction, which proves that $\sigma \cdot h = h$ whenever $\sigma \cdot \bar{g} = \bar{g}$. Thus $\overline{G} \subset G$.

By part (a) the Galois group of $\bar{f}$ over $\mathbb{F}_p$ contains an element whose action on the roots of $\bar{f}$ is given by a product of disjoint cycles of lengths $d_1, \ldots, d_r$. Since the conjugate of a product of disjoint cycles of lengths $d_1, \ldots, d_r$ is a permutation of the same form, we see that $\overline{G}$ and hence G contain a permutation of the desired form. Since G is conjugate to G_f, the Galois group of f must contain an automorphism whose action on the roots is as described in part (b) of the theorem. $\square$

Here are an example to illustrate part (b) of Theorem 13.4.5.

Example 13.4.6. Consider $f = x^5 + 20x + 16 \in \mathbb{Q}[x]$. In Exercise 7 you will verify that f is irreducible with discriminant $\Delta(f) = 2^{16}5^6$. This shows that the Galois group of f over $\mathbb{Q}$ is isomorphic to a subgroup of A_5.

Working modulo 7, we have the irreducible factorization

$$\bar{f} = (x + 2)(x + 3)(x^3 + 2x^2 + 5x + 5) \quad \text{in } \mathbb{F}_7[x].$$

Since $7 \nmid \Delta(f)$, part (b) of Theorem 13.4.5 implies that the order of the Galois group is divisible by 3. The classification of transitive subgroups of S_5 given in (13.12) shows that A_5 has no proper transitive subgroup of order divisible by 3. Hence the Galois group of f over $\mathbb{Q}$ is isomorphic to A_5.   ◁▷

Our next example uses the cycle decomposition of part (b) of Theorem 13.4.5.

Example 13.4.7. In Section 6.4 we showed that $f = x^5 - 6x + 3 \in \mathbb{Q}$ has S_5 as its Galois group over $\mathbb{Q}$. If you look carefully at the argument given in Section 6.4, you'll see that we showed first that the image of the Galois group in S_5 contains a 5-cycle and a 2-cycle, and second that any 5-cycle and 2-cycle generate S_5.

Using part (b) of Theorem 13.4.5, it is easy to get the required cycles. Consider the irreducible factorizations

$$\bar{f} = x^5 + 4x + 3 \quad \text{in } \mathbb{F}_5[x],$$
$$\bar{f} = (x + 2)(x + 7)(x + 13)(x^2 + 12x + 13) \quad \text{in } \mathbb{F}_{17}[x].$$

The first gives a 5-cycle and the second gives a 2-cycle. The theorem applies to these primes, since $\Delta(f) = -1737531 = -3^4 \cdot 19 \cdot 1129$.

In Exercise 8 you will give a different proof that the Galois group is S_5 by reducing modulo 11 and using the method of Example 13.4.6.   ◁▷

The paper [22] discusses an approach to computing Galois groups that uses Theorem 13.4.5 more systematically.

Mathematical Notes

We will discuss three topics related to Theorem 13.4.5.

■ **Reduction Modulo p.** Given a monic polynomial $f \in \mathbb{Z}[x]$, Theorem 13.4.5 shows that its factorization modulo p gives interesting information about the Galois group of f over $\mathbb{Q}$. The reduction is interesting for other reasons connected to what is known as *class field theory*. This is a large topic, so we will confine ourselves to two examples.

The first concerns the case when f is the quadratic polynomial $f = x^2 - a$, where $a \in \mathbb{Z}$. Since $\Delta(f) = 4a$, we know that $\bar{f} \in \mathbb{F}_p$ is separable when $p \nmid 4a$. For such a prime p, f splits completely modulo p if and only if the congruence

$$x^2 \equiv a \bmod p$$

has an integer solution. When the latter holds, we say that a is a *quadratic residue* modulo p. Quadratic residues play an important role in number theory and are related to the Legendre symbol defined in the Mathematical Notes to Section 9.2.

A deeper example is the following observation of Kronecker. The polynomial

$$f = (x^3 - 10x)^2 + 31(x^2 - 1)^2 \in \mathbb{Z}[x]$$

has discriminant $\Delta(f) = -2^6 \cdot 3^8 \cdot 31^3$. Now consider the following question: for which primes p does the factorization of f modulo p include a linear factor? In other words, when does f have a root modulo p? The amazing answer, due to Kronecker, is that if $p > 3$ is a prime different from 31, then

$$f(x) \equiv 0 \bmod 31 \text{ for some } x \in \mathbb{Z} \iff p = x^2 + 31y^2 \text{ for some } x, y \in \mathbb{Z}.$$

So our question characterizes primes of the form $x^2 + 31y^2$. Kronecker never published a proof of this result, which today is regarded as part of class field theory and complex multiplication. See [5] for an introduction to this rich subject.

■ **The Chebotarev Density Theorem.** Let $f \in \mathbb{Z}[x]$ be monic and separable of degree n with splitting field $\mathbb{Q} \subset L$. Given a prime p that doesn't divide $\Delta(f)$, Theorem 13.4.5 implies that $\mathrm{Gal}(L/\mathbb{Q})$ contains an element that corresponds to the Frobenius automorphism $\alpha \mapsto \alpha^p$ in the Galois group modulo p. This element of $\mathrm{Gal}(L/\mathbb{Q})$ is called the *Artin symbol* of p, denoted σ_p. Since the proof of Theorem 13.4.5 involves choices related to the ordering of the roots, the Artin symbol σ_p is well defined only up to conjugacy in $\mathrm{Gal}(L/\mathbb{Q})$. The *Chebotarev Density Theorem* describes the behavior of σ_p as we vary the prime p:

- Up to conjugacy, every element of $\mathrm{Gal}(L/\mathbb{Q})$ equals σ_p for some prime p.
- If we fix a conjugacy class C of $\mathrm{Gal}(L/\mathbb{Q})$, then the percentage of primes p whose Artin symbols σ_p lie in C is proportional to $|C|$.

In the second bullet, the "percentage of primes" needs to be defined carefully. This and the Chebotarev Density Theorem are discussed in §8.B of [5].

We can reformulate in terms of $\mathrm{Gal}(L/\mathbb{Q}) \simeq G \subset S_n$ as follows. A permutation $\tau \in S_n$ has *cycle type* $d_1, \ldots, d_r$, where $d_1 \le \cdots \le d_r$ and $d_1 + \cdots + d_r = n$, if τ is a product of disjoint cycles (including 1-cycles) of lengths $d_1, \ldots, d_r$. In Exercise 9 you will prove that two elements of S_n are conjugate if and only if they have the same cycle type. For a fixed cycle type $d_1, \ldots, d_r$, the set

(13.35) $\{\sigma \in G \mid \sigma \text{ has cycle type } d_1, \ldots, d_r\}$

is a union of conjugacy classes in G (see Exercise 10). Hence, if we fix the cycle type $d_1, \ldots, d_r$ of an element of G, then the Chebotarev density theorem implies the following:

- There is some prime p for which the irreducible factors of f modulo p have these degrees.
- The percentage of primes for which the irreducible factors of f modulo p have these degrees is proportional to number of elements of G with this cycle type.

Here is an example of what this looks like in practice.

Example 13.4.8. Consider $f = x^4 - 7x^3 + 19x^2 - 23x + 11 \in \mathbb{Z}[x]$, which has $\Delta(f) = 5^3$. For the 200 primes $p = 7, 11, \ldots, 1237$, it is straightforward to compute the degrees of the irreducible factors of f modulo p using *Mathematica* or *Maple*. When we tabulate the resulting degree patterns and the percentage of primes corresponding to the each pattern, we obtain:

(13.36)

irreducible factors	4 of degree 1	2 of degree 2	1 of degree 4
percentage of primes	25%	23%	52%

The last column shows that f remains irreducible for some primes, so that the Galois group of f contains a 4-cycle by Theorem 13.4.5. In S_4, a 4-cycle $(abcd)$ generates the subgroup

$$\langle (abcd) \rangle = \{e = (a)(b)(c)(d), \ (ac)(bd), (abcd), (dcba)\}.$$

For such a subgroup, the distribution of cycle types is:

(13.37)

cycle type	1,1,1,1	2,2	4
percentage of elements	25%	25%	50%

By the Chebotarev Density Theorem, the close match with (13.36) strongly suggests that the Galois group of f is cyclic of order 4. However, there could be a large prime whose degree pattern is not in (13.36). Hence this is not a rigorous computation of the Galois group.   ◁▷

The papers [13] and [22] discuss this method for computing Galois groups.

■ **Bad Reduction Modulo p.** When $f \in \mathbb{Z}[x]$ is monic and separable, Theorem 13.4.5 gives information about the Galois group of f by reducing f modulo primes p such that $p \nmid \Delta(f)$. If instead $p \mid \Delta(f)$, then $\bar{f}$ is not separable and our arguments fail.

When this happens, more advanced methods using the *decomposition group* and the *inertia group* can still provide useful information about the Galois group. For example, in Section 6.4 we mentioned that $f = x^n - x - 1$ has Galois group S_n over $\mathbb{Q}$ when $n \geq 2$. As explained on page 42 of reference [4] to Chapter 6, this is proved by reducing modulo the primes dividing the discriminant.

Exercises for Section 13.4

Exercise 1. Verify the computations given in Example 13.4.1.

Exercise 2. Prove that the polynomial $s_u(y)$ defined in (13.32) lies in $K[u_1, \ldots, u_n, y]$.

Exercise 3. This exercise is concerned with the proof of Theorem 13.4.2.
  (a) Let $\beta_1, \ldots, \beta_n \in L$. Prove that $y + \sum_{i=1}^r \beta_i u_i$ is irreducible in $L[u_1, \ldots, u_n, y]$. (This implies that (13.32) is the irreducible factorization of $s_u(y)$ in $L[u_1, \ldots, u_n, y]$.)
  (b) Let $g, h \in F[u_1, \ldots, u_n, y]$, and assume that in the larger ring $L[u_1, \ldots, u_n, y]$ we have $h = gq$ for some $q \in L[u_1, \ldots, u_n, y]$. Prove that $q \in F[u_1, \ldots, u_n, y]$.
  (c) In the final part of the proof of Theorem 13.4.2, we showed that $G \subset \sigma^{-1} G_f \sigma$. Prove the opposite inclusion.

Exercise 4. Consider the polynomial $s_u(y)$ when $f = x^3 + x^2 - 2x - 1$ from Example 13.4.3.
(a) Compute $s_u(y) \in \mathbb{Q}[u_1, u_2, u_3, y]$, and derive the factorization given in Example 13.4.3.
(b) Let h be the first factor of $s_u(y)$ given in Example 13.4.3, multiplied by -1 so that it is monic in y. Using SymmetricReduction in *Mathematica* or normalf in *Maple*, write h as a polynomial in y so that its coefficients are of the form

   a symmetric polynomial in u_1, u_2, u_3 + a remainder in u_1, u_2, u_3.

This should give the formula for h given in Example 13.4.3.

Exercise 5. Use the method of part (a) of Exercise 4 to derive the factorization of $s_u(y)$ given in Example 13.4.4.

Exercise 6. As in the proof of Theorem 13.4.5, suppose that we have $s_u(y) \in \mathbb{Z}[u_1, \ldots, u_n, y]$ and $h \in \mathbb{Q}[u_1, \ldots, u_n, y]$ is an irreducible factor of $s_u(y)$ when $s_u(y)$ is regarded as an element of $\mathbb{Q}[u_1, \ldots, u_n, y]$. In this exercise we will study how close h is to being an irreducible factor of $s_u(y)$ in $\mathbb{Z}[u_1, \ldots, u_n, y]$.
(a) We know that the rings $\mathbb{Z}[x_1, \ldots, x_n]$ and $\mathbb{Q}[x_1, \ldots, x_n]$ are both UFDs. Prove that if $f \in \mathbb{Z}[x_1, \ldots, x_n]$ is irreducible and nonconstant, then it is also irreducible when regarded as an element of $\mathbb{Q}[x_1, \ldots, x_n]$.
(b) Prove that if $s_u(y)$ and h are as above, then h is a $\mathbb{Q}$-multiple of an irreducible factor of $s_u(y)$ in $\mathbb{Z}[u_1, \ldots, u_n, y]$.

Exercise 7. Let $f = x^5 + 20x + 16 \in \mathbb{Q}[x]$ be the polynomial of Example 13.4.6. Show that f is irreducible over $\mathbb{Q}$, and compute its discriminant and irreducible factorization modulo 7.

Exercise 8. Compute the Galois group of $f = x^5 - 6x + 3$ over $\mathbb{Q}$ using reduction modulo 11 and the method of Example 13.4.6.

Exercise 9. Prove that two permutations in S_n are conjugate if and only if they have the same cycle type.

Exercise 10. Let G be a subgroup of S_n. For a fixed cycle type $d_1, \ldots, d_r$, consider the set (13.35) of all elements of G with this cycle type.
(a) Prove that this set is either empty or a union of conjugacy classes of G.
(b) Give an example where the set is empty, and give another example where it is a union of two conjugacy classes of G.

Exercise 11. This exercise will explore the ideas introduced in Example 13.4.8.
(a) For each transitive subgroup of S_4 make a table similar to (13.37) that lists the number of elements of each possible cycle type for that subgroup.
(b) For each polynomial in Exercise 15 of Section 13.1, compute its factorization modulo 200 primes, and record your results in table similar to (13.36). Use this to guess the Galois group of each polynomial.

REFERENCES

1. B. C. Berndt, B. K. Spearman, and K. S. Williams, *Commentary on an unpublished lecture by G. N. Watson on solving the quintic*, Math. Intelligencer **24**, no. 4 (2002), 15–33.

2. W. Burnside, *Theory of Groups of Finite Order*, Cambridge U. P., Cambridge, 1897.

3. H. Cohen, *A Course in Computational Algebraic Number Theory*, Springer-Verlag, New York, Berlin, Heidelberg, 1993.

4. J. Conway, A. Hulpke, and J. McKay, *On transitive permutation groups*, LMS J. Comput. Math. **1** (1998), 1–8.

5. D. Cox, *Primes of the Form $x^2 + ny^2$*, Wiley, New York, 1989.

6. P. Doyle and C. McMullen, *Solving the quintic by iteration*, Acta Math. **163** (1989), 151–180.

7. D. Dummit, *Solving solvable quintics*, Math. Comp. **57** (1991), 387–401.

8. N. D. Elkies, *Trinomials $ax^7 + bx + c$ with interesting Galois groups*. Available at http://www.math.harvard.edu/~elkies/trinomial.html.

9. D. W. Erbach, J. Fischer, and J. McKay, *Polynomials with* PSL$(2, 7)$ *as Galois group*, J. Number Theory **11** (1979), 69–75.

10. **GAP**: Groups, Algorithms and Programming, at http://www.gap-system.org/.

11. K. Girstmair, *On invariant polynomials and their applications in field theory*, Math. Comp. **48** (1987), 781–797.

12. K. Girstmair, *On the computation of resolvents of Galois groups*, Manuscripta Math. **43** (1983), 289–307.

13. A. Hulpke, *Techniques for computation of Galois groups*, in *Algorithmic Algebra and Number Theory (Heidelberg, 1997)*, edited by B. Matzat, G.-M. Greuel, and G. Hiss, Springer-Verlag, New York, Berlin, Heidelberg, 1999, 65–77.

14. Y. Ishibashi and T. Nakamizo, *A note on the Galois group of a quartic polynomial*, Bull. Fac. School Ed. Hiroshima Univ. **19** (1997), 39–41.

15. L.-C. Kappe and B. Warren, *An elementary test for the Galois group of a quartic polynomial*, Amer. Math. Monthly **96** (1989), 133–137.

16. R. B. King, *Beyond the Quartic Equation*, Birkhäuser, Boston, Basel, Berlin, 1996.

17. F. Klein, *Lectures on the ikosahedron and the solution of equations of the fifth degree*, English translation by George Gavin Morrice, Trübner & Co., London, 1888. Reprint by Chelsea, New York, 1956.

18. S. Kobayashi and H. Nakagawa, *Resolution of solvable quintic equation*, Math. Japon. **37** (1992), 883–886.

19. H. F. Kreimer, *Review of [14]*, Math. Reviews, 98a:12005.

20. Magma Computational Algebra System, at http://magma.maths.usyd.edu.au/magma/.

21. A. L. Maĭstrova, Решение адгебраических уравнений в работах Л. Эйлера [*Solution of algebraic equations in the works of L. Euler*], Istor.-Mat. Issled. No. 29 (1985), 189–199. Unpublished English translation by A. Shentizer.

22. J. McKay, *Some remarks on computing Galois groups*, SIAM J. Comput. **8** (1979), 344–347.

23. J. Pierpont, *Zur Geschichte der Gleichung des V. Grades (bis 1858)*, Monatsh. Math. Phys. **6** (1895), 15–68.

24. W. M. Ruppert, *On the Bring normal form of a quintic equation in characteristic 5*, Arch. Math. **58** (1992), 44–46.

25. P. Samuel, *Projective Geometry*, translated by S. Levy, Springer-Verlag, New York, Berlin, Heidelberg, 1988.

26. J. Shurman, *Geometry of the Quintic*, Wiley, New York, 1997.

27. J. H. Silverman and J. Tate, *Rational Points on Elliptic Curves*, Springer-Verlag, New York, Berlin, Heidelberg 1992.

28. L. Soicher and J. McKay, *Computing Galois groups over the rationals*, J. Number Theory **20** (1985), 273–281.

29. B. K. Spearman and K. S. Williams, *Characterization of solvable quintics $x^5 + ax + b$*, Amer. Math. Monthly **101** (1994), 986–992.

30. B. K. Spearman and K. S. Williams, *On solvable quintics $x^5 + ax + b$ and $x^5 + ax^2 + b$*, Rocky Mountain J. Math. **26** (1996), 753–772.

31. B. K. Spearman and K. S. Williams, *Quartic trinomials with Galois groups A_4 and V_4*, Far East J. Math. Sci. (FJMS) **2** (2000), 665–672.

32. R. P. Stauduhar, *The determination of Galois groups*, Math. Comp. **27** (1973), 981–996.

33. J. Stillwell, *Eisenstein's footnote*, Math. Intelligencer **17**, no. 2 (1995), 58–62.

34. E. Weisstein, *CRC Concise Encyclopeadia of Mathematics*, Second Edition, Chapman & Hall/CRC, Boca Raton, FL, 2003.

35. Wolfram Research, *Solving the Quintic*, Poster, Wolfram Research, Champaign, IL, 1995. Available at http://library.wolfram.com/examples/quintic/.

14

Solvable Permutation Groups

This chapter will study solvability by radicals for irreducible polynomials of degree p or p^2, where p is prime. These results go back to Galois and illustrate his amazing insight into group theory. We will also discover why Galois invented finite fields.

While Galois's result for degree p is relatively easy to prove, understanding the case of degree p^2 requires the theory of *permutation groups* (subgroups of S_n). We will see that in degree p^2, irreducible polynomials can be either *primitive* or *imprimitive*. The case of solvable imprimitive subgroups of S_{p^2} will be considered first, followed by the more complicated case of solvable primitive subgroups. The proofs will involve surprising amounts of group theory.

14.1 POLYNOMIALS OF PRIME DEGREE

The goal of this section is to prove the following wonderful theorem of Galois.

Theorem 14.1.1. *Let F be a field of characteristic 0, and let $f \in F[x]$ be irreducible of prime degree p. Then the following are equivalent:*

(a) *f is solvable by radicals over F.*

(b) *For every pair of roots $\alpha \neq \beta$ of f, $F(\alpha, \beta)$ is the splitting field of f over F.*

(c) *For some pair of roots $\alpha \neq \beta$ of f, $F(\alpha, \beta)$ is the splitting field of f over F.*

(d) *The Galois group of f over F is isomorphic to a subgroup of $\mathrm{AGL}(1, \mathbb{F}_p)$.*

The proof will be given later in the section. Recall from Section 6.4 that the group $\mathrm{AGL}(1, \mathbb{F}_p)$ consists of all functions

$$\gamma_{a,b}(u) = au + b, \qquad a \in \mathbb{F}_p^*, \ b \in \mathbb{F}_p.$$

407

If we identify the congruence classes $[1], [2], \ldots, [p] \in \mathbb{F}_p$ with the numbers $1, 2, \ldots, p$, then $\mathrm{AGL}(1, \mathbb{F}_p)$ becomes a subgroup of S_p. Thus

$$\mathrm{AGL}(1, \mathbb{F}_p) \subset S_p$$

is a subgroup of order $p(p-1)$, and an element $\gamma_{a,b} \in \mathrm{AGL}(1, \mathbb{F}_p)$ is the permutation

$$\gamma_{a,b} = \begin{pmatrix} 1 & 2 & \cdots & p \\ a+b & 2a+b & \cdots & pa+b \end{pmatrix} = \begin{pmatrix} i \\ ai+b \end{pmatrix},$$

where we interpret everything modulo p. In particular, let $\theta = \gamma_{1,1}$. Then we have the p-cycle

$$\theta = \begin{pmatrix} i \\ i+1 \end{pmatrix} = (12 \ldots p) \in \mathrm{AGL}(1, \mathbb{F}_p).$$

Here are two useful facts about θ and $\mathrm{AGL}(1, \mathbb{F}_p)$.

Lemma 14.1.2.
(a) $\mathrm{AGL}(1, \mathbb{F}_p)$ *is the normalizer of* $\langle \theta \rangle$ *in* S_p.
(b) *If* $\tau \in S_p$ *satisfies* $\tau \theta \tau^{-1} \in \mathrm{AGL}(1, \mathbb{F}_p)$, *then* $\tau \in \mathrm{AGL}(1, \mathbb{F}_p)$.

Proof. The normalizer of $\langle \theta \rangle$ in S_p consists of all $\tau \in S_p$ such that $\tau \langle \theta \rangle \tau^{-1} = \langle \theta \rangle$. In Exercise 1 you will show that τ lies in the normalizer if and only if

$$\tau \theta = \theta^\ell \tau \quad \text{for some } 1 \le \ell \le p - 1.$$

Since $\theta^\ell(i) = i + \ell$, the above equation is equivalent to the identity

(14.1) $$\tau(i+1) = \tau(i) + \ell, \quad i = 1, \ldots, p,$$

where as usual we interpret everything modulo p. This implies that

$$\tau(i+2) = \tau((i+1)+1) = \tau(i+1) + \ell = \tau(i) + \ell + \ell = \tau(i) + 2\ell,$$

and more generally, one easily proves that for any positive j,

$$\tau(i+j) = \tau(i) + j\ell, \quad i = 1, \ldots, p$$

(see Exercise 1). Then setting $i = p$ gives

$$\tau(j) = \tau(p+j) = \tau(p) + j\ell = \gamma_{\ell, \tau(p)}(j).$$

This shows that $\tau = \gamma_{\ell, \tau(p)} \in \mathrm{AGL}(1, \mathbb{F}_p)$. Conversely, it is easy to see that any $\gamma_{a,b} \in \mathrm{AGL}(1, \mathbb{F}_p)$ satisfies (14.1) with $\ell = a$. This proves part (a) of the lemma.

For part (b), first observe that $\langle \theta \rangle$ is a p-Sylow subgroup of $\mathrm{AGL}(1, \mathbb{F}_p)$, since $|\mathrm{AGL}(1, \mathbb{F}_p)| = p(p-1)$. Furthermore, it is unique by the Second Sylow Theorem (see Theorem A.5.1), since $\langle \theta \rangle$ is normal in $\mathrm{AGL}(1, \mathbb{F}_p)$ by part (a).

Now assume that $\tau \in S_p$ and $\tau \theta \tau^{-1} \in \mathrm{AGL}(1, \mathbb{F}_p)$. Then $\langle \tau \theta \tau^{-1} \rangle$ is a also p-Sylow subgroup of $\mathrm{AGL}(1, \mathbb{F}_p)$. By uniqueness, $\langle \theta \rangle = \langle \tau \theta \tau^{-1} \rangle = \tau \langle \theta \rangle \tau^{-1}$. Thus τ normalizes $\langle \theta \rangle$ and hence lies in $\mathrm{AGL}(1, \mathbb{F}_p)$ by part (a). $\square$

We will use the following lemma several times. See Exercise 2 for a proof.

Lemma 14.1.3. *Suppose that H is a normal subgroup of a finite group G and let $g \in G$. If the order of g is relatively prime to $[G : H]$, then $g \in H$.* □

We can now characterize solvable transitive subgroups of S_p.

Proposition 14.1.4. *Every solvable transitive subgroup $G \subset S_p$ is conjugate to a subgroup of $\mathrm{AGL}(1, \mathbb{F}_p)$ containing θ.*

Proof. Since G is transitive, the orbit of any $i \in \{1, \ldots, p\}$ is all of $\{1, \ldots, p\}$. Thus p divides the order of G by the Fundamental Theorem of Group Actions. Hence G contains an element of order p, by Cauchy's Theorem. This element must be a p-cycle, since we are in S_p (the order of a permutation is the least common multiple of its cycle lengths). By Exercise 9 of Section 13.4, any p-cycle in S_p is conjugate to θ. Replacing G with a suitable conjugate, we may assume that $\theta \in G$.

Since G is solvable, we can find subgroups

$$\{e\} = G_0 \subset G_1 \subset \cdots \subset G_{n-1} \subset G_n = G$$

such that $G_{\ell-1}$ is normal in G_ℓ and $[G_\ell : G_{\ell-1}]$ is prime for $\ell = 1, \ldots, n$.

Let i be the *smallest* index such that $\theta \in G_i$. Note that $i > 0$.

We first claim that $[G_i : G_{i-1}] = p$. To see why, suppose that $[G_i : G_{i-1}] = q$, where $q \neq p$ is prime. Since $\theta \in G_i$ has order p, Lemma 14.1.3 implies that $\theta \in G_{i-1}$, which contradicts the definition of i. Hence $[G_i : G_{i-1}] = p$.

We next claim that $i = 1$. If $i > 1$, then there is $\tau \in G_{i-1}$ such that $\tau(j) = k$ for some $j \not\equiv k \bmod p$. Then θ^{j-k} maps k to j, so that $\rho = \tau\theta^{j-k} \in G_i$ fixes k. Thus ρ is a product of disjoint cycles of lengths $< p$. Since p is prime, it follows that the order of ρ is relatively prime to $p = [G_i : G_{i-1}]$. Hence $\rho \in G_{i-1}$ by Lemma 14.1.3, and then $\theta^{j-k} \in G_{i-1}$ follows from $\tau \in G_{i-1}$. Since $j \not\equiv k \bmod p$, this implies that $\theta \in G_{i-1}$, which contradicts the definition of i. Hence $i = 1$.

Since $G_0 = \{e\}$, these claims imply that G_1 has order p and contains θ. Since θ has order p, we conclude that $G_1 = \langle \theta \rangle$. It follows that $G_1 \subset \mathrm{AGL}(1, \mathbb{F}_p)$.

Now let $1 \leq j \leq n$ be the *largest* index such that $G_j \subset \mathrm{AGL}(1, \mathbb{F}_p)$. Suppose that $j < n$, and take $\tau \in G_{j+1}$. Then $\theta \in G_1 \subset G_j$ implies that $\tau\theta\tau^{-1} \in G_j$, since G_j is normal in G_{j+1}. This gives $\tau\theta\tau^{-1} \in \mathrm{AGL}(1, \mathbb{F}_p)$, so that $\tau \in \mathrm{AGL}(1, \mathbb{F}_p)$ by part (b) of Lemma 14.1.2. Since $\tau \in G_{j+1}$ was arbitrary, we conclude that $G_{j+1} \subset \mathrm{AGL}(1, \mathbb{F}_p)$. This contradicts the definition of j. Hence we must have $j = n$, which gives the desired inclusion $G = G_n \subset \mathrm{AGL}(1, \mathbb{F}_p)$. □

We now have the tools needed to prove Galois's theorem.

Proof of Theorem 14.1.1. Let f have roots $\alpha_1, \ldots, \alpha_p$ in a splitting field L. Then $\mathrm{Gal}(L/F)$ is isomorphic to $G \subset S_p$, where $\sigma \in \mathrm{Gal}(L/F)$ maps to $\tau \in G$ such that $\sigma(\alpha_i) = \alpha_{\tau(i)}$. By Proposition 6.3.7, G is transitive, since F is irreducible.

First consider (a) $\Leftrightarrow$ (d). If f is solvable by radicals over F, then G is transitive (by the above) and solvable (by Theorem 8.5.3). Using Proposition 14.1.4, we conclude that G is conjugate to a subgroup of $\mathrm{AGL}(1, \mathbb{F}_p)$. This proves (a) $\Rightarrow$ (d). For the converse, note that $\mathrm{AGL}(1, \mathbb{F}_p)$ is solvable by Example 8.1.6, and then any subgroup is also solvable by Proposition 8.1.3. Thus the Galois group of f over F is solvable, so that f is solvable by radicals over F by Theorem 8.5.3.

We next prove (b) $\Rightarrow$ (c) $\Rightarrow$ (d) $\Rightarrow$ (b). The first implication is obvious. For the second, observe that in the proof of Proposition 14.1.4, the first paragraph applies to any transitive subgroup of S_p. Thus we may assume that $\langle \theta \rangle \subset G$. Now suppose that $L = F(\alpha_i, \alpha_j)$ is the splitting field of f over F for some $i \neq j$. This gives the extensions

$$F \subset F(\alpha_i) \subset F(\alpha_i, \alpha_j) = L,$$

where the first has degree p, since f is irreducible over F, and the second has degree at most $p - 1$, since α_j is a root of $f/(x - \alpha_i) \in F(\alpha_i)[x]$. By the Tower Theorem,

$$(14.2) \qquad |G| = |\mathrm{Gal}(L/F)| = [L : F] = pm, \quad 1 \leq m \leq p - 1.$$

Since p is prime, it follows that $\langle \theta \rangle$ is a p-Sylow subgroup of G. According to the Third Sylow Theorem (see Theorem A.5.1), the number of p-Sylow subgroups of G divides $|G|$ and is congruent to 1 modulo p. In Exercise 3 you will use this and (14.2) to show that $\langle \theta \rangle$ is the unique p-Sylow subgroup of G and hence is normal in G. It follows that G is contained in the normalizer of $\langle \theta \rangle$ in S_p. By Lemma 14.1.2, the normalizer is $\mathrm{AGL}(1, \mathbb{F}_p)$, so that $G \subset \mathrm{AGL}(1, \mathbb{F}_p)$. Thus $\mathrm{Gal}(L/F)$ is isomorphic to a subgroup of $\mathrm{AGL}(1, \mathbb{F}_p)$.

For (d) $\Rightarrow$ (b), relabel the roots of f so that $G \subset \mathrm{AGL}(1, \mathbb{F}_p)$. We need to show that $F(\alpha_i, \alpha_j)$ is the splitting field of f over F for any $i \neq j$. By the Galois correspondence, it suffices to show that the only element $\sigma \in \mathrm{Gal}(L/F)$ fixing $F(\alpha_i, \alpha_j)$ is the identity. Since σ corresponds to $\tau \in G$ and $G \subset \mathrm{AGL}(1, \mathbb{F}_p)$, we see that $\tau = \gamma_{a,b}$ for some $a \in \mathbb{F}_p^*$ and $b \in \mathbb{F}_p$. Then

$$\sigma(\alpha_i) = \alpha_i \;\Rightarrow\; \alpha_{\tau(i)} = \alpha_i \;\Rightarrow\; \alpha_{ai+b} = \alpha_i,$$
$$\sigma(\alpha_j) = \alpha_j \;\Rightarrow\; \alpha_{\tau(j)} = \alpha_j \;\Rightarrow\; \alpha_{aj+b} = \alpha_j.$$

This gives the equations $ai + b = i$ and $aj + b = j$, which modulo p have the unique solution $a = 1, b = 0$, since $i \neq j$ modulo p. Thus $\tau = \gamma_{1,0}$ is the identity, so that σ is the identity in $\mathrm{Gal}(L/F)$. Hence $F(\alpha_i, \alpha_j)$ is the splitting field. $\square$

We first encountered the affine linear group $\mathrm{AGL}(1, \mathbb{F}_p)$ as the Galois group of $\mathbb{Q} \subset \mathbb{Q}(\zeta_p, \sqrt[p]{2})$ in Section 6.4. This extension is the splitting field of $x^p - 2 \in \mathbb{Q}[x]$, which is obviously solvable by radicals. Hence what we did in Section 6.4 is a perfect illustration of Theorem 14.1.1.

Mathematical Notes

The proof of Theorem 14.1.1 uses the following concept from group theory.

■ **Frobenius Groups.** We showed above that $F(\alpha_i, \alpha_j)$ is the splitting field by arguing that the identity is the only element of $\mathrm{AGL}(1, \mathbb{F}_p)$ that fixes i and j. This generalizes as follows. If a finite group G acts transitively on a set X such that $1 < |X| < |G|$, and for every $x \neq y$ in X the identity is the only element of G fixing x and y, then we say that G is a *Frobenius group*. When this happens, the isotropy subgroup G_x of any $x \in X$ is called a *Frobenius complement*. A discussion of Frobenius groups can be found in [2, Sec. 3.4] and [12, p. 90]. See also Exercise 4.

Historical Notes

Galois considered Theorem 14.1.1 to be one the best applications of his theory. His version of the theorem is as follows [Galois, p. 69]:

PROPOSITION VIII

THEOREM. In order that an irreducible equation of prime degree be solvable by radicals, it is necessary and sufficient that when any two of the roots are known, the others can be deduced from them rationally.

If we are working over a field F, then "deduced from them rationally" means that the other roots are rational functions with coefficients in F in the known roots α, β. This implies that $F(\alpha, \beta)$ is the splitting field. Thus Galois's theorem is (a) $\Leftrightarrow$ (b) of Theorem 14.1.1. Galois especially liked this result because its statement doesn't mention Galois theory, yet the Galois group is crucial to the proof.

As for part (d) of Theorem 14.1.1, Galois says the following [Galois, p. 67]:

Therefore, "if an irreducible equation of prime degree is solvable by radicals, then the group of the equation contains only substitutions of the form

$$x_k \quad x_{ak+b}$$

a and b being constants."

Reciprocally, if this holds then I say that the equation will be solvable by radicals.

This is (a) $\Leftrightarrow$ (d) of Theorem 14.1.1. Galois denotes the roots of the polynomial as $x_0, \ldots, x_{n-1}$, where n is prime (his n is our p). Furthermore, on page 65, Galois says "We set in general $x_n = x_0, x_{n+1} = x_1, \ldots$" Thus Galois treats the indices modulo n just as we treat them modulo p.

Galois published his Proposition VIII separately in 1830, before he had worked out his general theory of solvability. It is possible that thinking about this case (and the group $\text{AGL}(1, \mathbb{F}_p)$ in particular) led Galois to the general idea of normal subgroups and solvability. The irony is that by focusing attention on the special case of Theorem 14.1.1, Galois distracted his contemporaries from the real depth of his innovations.

Galois also formulated solvability by radicals in terms of the resolvents discussed in Chapters 12 and 13. Let f be irreducible of degree p. Since $\text{AGL}(1, \mathbb{F}_p)$ has index $(p-2)!$ in S_p, the theory of Section 13.3 constructs a resolvent $R_f(y) \in F[y]$ such that if the Galois group of f is isomorphic to a subgroup of $\text{AGL}(1, \mathbb{F}_p)$, then $R_f(y)$ has a root in F, and the converse is true provided that the root is simple. Because of this, Galois asserts that to check solvability by radicals, it suffices to know whether or not this auxiliary equation [our $R_f(y)$] of degree $1.2.3\ldots(n-2)$ has a rational root.

(See [Galois, p. 69].) Here, Galois's n is our p, and "rational root" means root in F. Although Galois (like Jordan) missed the importance of simple roots, this result can be regarded as the generalization of Corollary 13.2.11 for an arbitrary prime. Galois also knew how to build the resolvent $R_f(y)$ using the methods discussed in Section 12.1. For example, the polynomial (12.20) appears in Galois's memoir.

Exercises for Section 14.1

Exercise 1. This exercise is concerned with the proof of part (a) of Lemma 14.1.2. Let $\theta = (12\ldots p) \in S_p$.
(a) Prove that $\tau \in S_p$ lies in the normalizer of $\langle\theta\rangle$ if and only if $\tau\theta = \theta^\ell\tau$ for some $1 \le \ell \le p - 1$.
(b) Prove that (14.1) implies that $\tau(i + j) = \tau(i) + j\ell$ for all positive integers j.

Exercise 2. Let H be a normal subgroup of a finite group G and let $g \in G$. The goal of this exercise is to prove Lemma 14.1.3.
(a) Explain why $(gH)^{o(g)} = (gH)^{[G:H]} = gH$ in the quotient group G/H.
(b) Now assume that $\gcd(o(g), [G:H]) = 1$. Prove that $g \in H$.

Exercise 3. Use (14.2) and the Third Sylow Theorem to prove that $\langle\theta\rangle$ is the unique p-Sylow subgroup of G. Also explain why this implies that $\langle\theta\rangle$ is normal in G.

Exercise 4. The definition of Frobenius group given in the Mathematical Notes involves a group G acting transitively on a set X. Prove that a group G is a Frobenius group if and only if G has a subgroup H such that $1 < |H| < |G|$ and $H \cap gHg^{-1} = \{e\}$ for all $g \notin H$.

Exercise 5. Let F be a subfield of the real numbers, and let $f \in F[x]$ be irreducible of degree $p > 2$, where p is prime. Also assume that f is solvable by radicals. Prove that f has either a single real root or p real roots.

Exercise 6. By Example 8.5.5, $f = x^5 - 6x + 3$ is not solvable by radicals over $\mathbb{Q}$. Give a new proof of this fact using the previous exercise together with the irreducibility of f and part (b) of Exercise 6 from Section 6.4,

Exercise 7. Use Lemma 14.1.3 and part (a) of Lemma 14.1.2 to give a proof of part (b) of Lemma 14.1.2 that doesn't use the Sylow Theorems.

14.2   IMPRIMITIVE POLYNOMIALS OF PRIME-SQUARED DEGREE

Having studied polynomials of prime degree p, we turn our attention to polynomials of degree p^2. In this section, we will see that such polynomials can be either *primitive* or *imprimitive*. Our main result (Theorem 14.2.15) will describe the Galois group of an irreducible imprimitive polynomial of degree p^2 that is solvable by radicals. The primitive case will be considered in Sections 14.3 and 14.4.

The proof of Theorem 14.2.15 will require that we study *permutation groups*, that is, subgroups of S_n. After defining *primitive* and *imprimitive* permutation groups, we will concentrate on the imprimitive case and use *wreath products* to classify all solvable transitive imprimitive subgroups of S_{p^2}. Primitive permutation groups will be considered in Section 14.3.

A. Primitive and Imprimitive Groups. By Section 6.3, the Galois group of a separable polynomial f of degree n gives a permutation group $G \subset S_n$ that records the Galois action on the roots of f. An important idea in Galois theory is that properties of f should be reflected in the properties of G. For example, Proposition 6.3.7 says that f is irreducible over F if and only if G is transitive. Thus "transitive" is the permutation group analog of "irreducible" for polynomials.

We next consider the concepts of *imprimitive* and *primitive*, which apply to both polynomials and permutation groups. We will begin with the former, where the idea is that separable polynomials come in two flavors, according to whether or not the roots break up into "blocks" under the action of the Galois group.

Before giving the general definition, let's consider an example.

Example 14.2.1. The polynomial $f = x^4 - 2 \in \mathbb{Q}[x]$ is separable with roots

$$\sqrt[4]{2}, -\sqrt[4]{2} \quad \text{and} \quad i\sqrt[4]{2}, -i\sqrt[4]{2}.$$

We have written the roots in two blocks that have the following nice property: if we apply $\sigma \in \mathrm{Gal}(\mathbb{Q}(i, \sqrt[4]{2})/\mathbb{Q})$ to the first block of roots, then the result is either the first block or the second block, and the same is true if we apply σ to the second block. This follows because $\sigma(-\alpha) = -\sigma(\alpha)$. Hence the action of the Galois group respects the block structure of the roots.   ◁▷

This leads to the following general definition.

Definition 14.2.2. *Let* $f \in F[x]$ *be a separable polynomial with splitting field* L.
(a) f *is **imprimitive** if the set of roots of* f *can be written as a disjoint union*

$$R_1 \cup \cdots \cup R_k$$

such that for every $\sigma \in \mathrm{Gal}(L/F)$ *and* $1 \le i \le k$, *we have* $\sigma(R_i) = R_j$ *for some* $1 \le j \le k$. *We also require that* $k > 1$ *and* $|R_i| > 1$ *for some* i.
(b) f *is **primitive** if it is not imprimitive.*

In the definition of imprimitive, the R_i are the blocks, and $\sigma(R_i) = R_j$ means that the Galois group preserves the block structure of the roots. The requirements that $k > 1$ and some $|R_i| > 1$ exclude the trivial block structures where there is only one block or where every block consists of a single root.

When the polynomial is also irreducible, we get some useful information about the size of the blocks in the imprimitive case.

Lemma 14.2.3. *Let* $f \in F[x]$ *be irreducible and separable of degree* n. *Assume that* f *is imprimitive with roots* $R_1 \cup \cdots \cup R_k$ *as in Definition 14.2.2. Then every* R_i *has the same number of elements, say* l. *Thus* $n = kl$, *where* $k > 1$ *and* $l > 1$.

Proof. Given blocks R_i and R_j with $i \ne j$, pick $\alpha \in R_i$ and $\beta \in R_j$. Since f is irreducible and L is its splitting field over F, $\mathrm{Gal}(L/F)$ acts transitively on the roots.

Thus there is $\sigma \in \mathrm{Gal}(L/F)$ such that $\sigma(\alpha) = \beta$. Since f is imprimitive, we have $\sigma(R_i) = R_j$, so that $|R_i| = |R_j|$, since σ is one to one. If $l = |R_i|$, then

$$n = |R_1| + \cdots + |R_k| = kl,$$

since f is separable. Then $k > 1$ and $l > 1$ follow from Definition 14.2.2.   □

Here are some easy examples.

Example 14.2.4. If f is irreducible and separable of prime degree p, then f cannot be imprimitive, since it is impossible to write $p = kl$ with $k > 1$ and $l > 1$. Hence irreducible separable polynomials of prime degree are automatically primitive.

However, if f is irreducible and separable of degree p^2, then f can be either primitive or imprimitive. In the latter case, we must have p blocks, each consisting of p roots. When $p = 2$, we saw an instance of this in Example 14.2.1.   ◁▷

We translate these concepts into group theory as follows.

Definition 14.2.5. *Let G be a subgroup of S_n. Then:*
(a) *G is **imprimitive** if there is a disjoint union*

$$\{1, \ldots, n\} = R_1 \cup \cdots \cup R_k$$

such that for every $\tau \in G$ and every $1 \le i \le k$, we have $\tau(R_i) = R_j$ for some $1 \le j \le k$. We also require that $k > 1$ and $|R_i| > 1$ for some i.
(b) *G is **primitive** if it is not imprimitive.*

Here is an example of an imprimitive permutation group.

Example 14.2.6. The subgroup $G = \langle (1324), (34) \rangle \subset S_4$ is imprimitive via the blocks $R_1 = \{1, 2\}$ and $R_2 = \{3, 4\}$. This follows because (1324) maps each block to the other while (34) takes each block to itself.

If we label the roots of $x^4 - 2$ as $\alpha_1 = \sqrt[4]{2}, \alpha_2 = -\sqrt[4]{2}, \alpha_3 = i\sqrt[4]{2}, \alpha_4 = -i\sqrt[4]{2}$, then $\mathrm{Gal}(\mathbb{Q}(i, \sqrt[4]{2})/\mathbb{Q}) \simeq G \subset S_4$. Do you see how the above blocks relate to those used in Example 14.2.1?   ◁▷

Lemma 14.2.3 has the following group-theoretic analog.

Lemma 14.2.7. *Let G be a transitive subgroup of S_n. Assume that G is imprimitive with blocks $R_1, \ldots, R_k$ as in Definition 14.2.5. Then every R_i has the same number of elements, say l. Thus $n = kl$, where $k > 1$ and $l > 1$.*   □

We omit the proof because it is identical to the proof of Lemma 14.2.3. Be sure you understand this.

B. Wreath Products. Let $f \in F[x]$ be separable, irreducible, and imprimitive of degree n. How does being imprimitive restrict the Galois group of f? As we will see, the answer involves the concept of *wreath product*.

By what we've done so far, our question reduces to the study of transitive imprimitive subgroups $G \subset S_n$. By Lemma 14.2.3, we have $k > 1$ blocks, each consisting of $l > 1$ elements, where $n = kl$. To begin our analysis, we will regard $S_n = S_{kl}$ as permutations of the product $\{1, \ldots, k\} \times \{1, \ldots, l\}$ and use the blocks

$$(14.3) \qquad \{1, \ldots, k\} \times \{1, \ldots, l\} = R_1 \cup \cdots \cup R_k, \quad R_i = \{i\} \times \{1, \ldots, l\}.$$

Then consider

$$(14.4) \quad S_k \wr S_l = \{\sigma \in S_{kl} \mid \text{there is } \tau \in S_k \text{ with } \sigma(R_i) = R_{\tau(i)} \text{ for } 1 \leq i \leq k\}.$$

It is easy to see that $S_k \wr S_l$ is an imprimitive subgroup of $S_{kl} = S_n$ with respect to the blocks $R_1, \ldots, R_k$. We call $S_k \wr S_l$ the *wreath product* of S_k with S_l.

We can describe an element $\sigma \in S_k \wr S_l$ as follows. Since $R_i = \{i\} \times \{1, \ldots, l\}$ and $\sigma(R_i) = R_{\tau(i)}$, there is a unique $\mu_i \in S_l$ such that for all $(i, j) \in R_i$, we have

$$\sigma(i, j) = (\tau(i), \mu_i(j)) \in R_{\tau(i)}.$$

Thus μ_i describes how σ maps R_i to $R_{\tau(i)}$. If we write $\sigma = (\tau; \mu_1, \ldots, \mu_k)$, then

$$(14.5) \qquad S_k \wr S_l = \{(\tau; \mu_1, \ldots, \mu_k) \mid \tau \in S_k, \, \mu_1, \ldots, \mu_k \in S_l\}.$$

In more concrete terms, think of a dresser with k drawers and l items in each drawer. Then elements of the wreath product (14.5) permute the items in the dresser by permuting the drawers via τ and permuting the items in each drawer via the μ_j.

Using (14.5), we can describe some interesting subgroups of $S_k \wr S_l$. Given subgroups $A \subset S_k$ and $B \subset S_l$, define the set

$$A \wr B = \{(\tau; \mu_1, \ldots, \mu_k) \mid \tau \in A, \, \mu_1, \ldots, \mu_k \in B\}.$$

This is a subgroup with the following properties.

Lemma 14.2.8. *Let $A \subset S_k$ and $B \subset S_l$ be subgroups. Then:*
(a) *$A \wr B$ is a subgroup of $S_k \wr S_l$.*
(b) *The map $(\tau; \mu_1, \ldots, \mu_k) \mapsto \tau$ defines a group homomorphism $A \wr B \to A$ that is onto and whose kernel is isomorphic to $B^k = B \times \cdots \times B$ (k times).*

Proof. Given $\sigma = (\tau; \mu_1, \ldots, \mu_k) \in A \wr B$, we first show that $\sigma^{-1} \in A \wr B$. Since σ maps R_i to $R_{\tau(i)}$ via μ_i, it is clear that σ^{-1} maps $R_{\tau(i)}$ to R_i via μ_i^{-1}. If we set $j = \tau(i)$, then $i = \tau^{-1}(j)$, so that σ^{-1} maps R_j to $R_{\tau^{-1}(j)}$ via $\mu_{\tau^{-1}(j)}^{-1}$. It follows that

$$(14.6) \qquad \sigma^{-1} = \left(\tau^{-1}; \mu_{\tau^{-1}(1)}^{-1}, \ldots, \mu_{\tau^{-1}(k)}^{-1}\right).$$

This obviously lies in $A \wr B$. In Exercise 1 you will show that

$$(14.7) \qquad (\tau; \mu_1, \ldots, \mu_k)(\tau'; \mu_1', \ldots, \mu_k') = (\tau\tau'; \mu_{\tau'(1)}\mu_1', \ldots, \mu_{\tau'(k)}\mu_k').$$

Hence $A \wr B$ is closed under multiplication, and part (a) follows.

It remains to prove part (b) of the lemma. The multiplication formula (14.7) shows that $(\tau; \mu_1, \ldots, \mu_k) \mapsto \tau$ is a homomorphism, which is clearly onto by the definition of $A \wr B$. Furthermore, its kernel is clearly the set

$$\{(e; \mu_1, \ldots, \mu_k) \mid \mu_1, \ldots, \mu_k \in B\}.$$

Then (14.7) shows that the obvious map to B^k is a group isomorphism. $\qquad\square$

The subgroups $S_k \wr S_l \subset S_{kl} = S_n$ have the following important property.

Proposition 14.2.9. *Every transitive imprimitive subgroup of S_n is conjugate to a subgroup of $S_k \wr S_l$ for some nontrivial factorization $n = kl$.*

Proof. Let $G \subset S_n$ be transitive and imprimitive. By Lemma 14.2.7, we have blocks

$$R'_1 \cup \cdots \cup R'_k,$$

where each R'_i has l elements and every $\sigma \in G$ maps R'_i to some R'_j. To compare this to (14.3), pick $\tau \in S_{kl}$ with the property that $\tau(R_i) = R'_i$ for $1 \le i \le k$. Such a τ exists because $|R_i| = |R'_i|$. One easily checks that if $\sigma \in G$ maps R'_i to R'_j, then $\tau^{-1}\sigma\tau$ maps R_i to R_j. It follows that $\tau^{-1}G\tau \subset S_k \wr S_l$. $\qquad\square$

Proposition 14.2.9 implies the following result about the Galois group of an imprimitive polynomial.

Corollary 14.2.10. *Let $f \in F[x]$ be separable, irreducible, and imprimitive of degree n. Then n has a nontrivial factorization $n = kl$ such that the Galois group of f over F is isomorphic to a subgroup of $S_k \wr S_l$.* $\qquad\square$

Here is an example of this result in degree 6.

Example 14.2.11. Suppose that $f = x^6 + bx^4 + cx^2 + d \in F[x]$ is irreducible and separable over a field of characteristic $\neq 2$. It is easy to see that this polynomial is imprimitive, for if α is a root, then so is $-\alpha$. Hence the roots can be partitioned into the three blocks

$$R_1 = \{\alpha, -\alpha\}, \qquad R_2 = \{\beta, -\beta\}, \qquad R_3 = \{\gamma, -\gamma\}$$

that are obviously permuted by the Galois group. Thus the Galois group of f over F is isomorphic to a subgroup of $S_3 \wr S_2 \subset S_6$. This group has order $6 \cdot 2^3 = 48$. For $x^6 - 4x^2 + 1 \in \mathbb{Q}[x]$, the `galois` command of *Maple* shows that its Galois group over $\mathbb{Q}$ has order 48. Hence the Galois group is isomorphic to $S_3 \wr S_2$.

For $x^6 - 4x^2 - 1 \in \mathbb{Q}[x]$, the discriminant is $2^6 229^2$. Thus its Galois group over $\mathbb{Q}$ is isomorphic to a subgroup of $(S_3 \wr S_2) \cap A_6$, which has order 24 (you will verify this in Exercise 2). The `galois` command shows that the Galois group has order 24, so that the Galois group is isomorphic to $(S_3 \wr S_2) \cap A_6$ in this case. See Exercise 3 for more on the structure of these groups. $\qquad\triangleleft\triangleright$

When $n = p^2$ and p is prime, the only nontrivial factorization of n is $n = p \cdot p$. This gives the following corollary of Proposition 14.2.9 that will be useful later in the section.

Corollary 14.2.12. *Every transitive imprimitive subgroup of S_{p^2} is conjugate to a subgroup of $S_p \wr S_p$.* □

For an irreducible imprimitive polynomial of degree p^2, it follows that the Galois group is isomorphic to a subgroup of $S_p \wr S_p$. By (14.5), the order of this group is

$$|S_p \wr S_p| = (p!)^{p+1}.$$

This may seem like a large number, but it is actually quite small in comparison with $|S_{p^2}| = (p^2)!$. Here is an example.

Example 14.2.13. When $p = 17$, we have

$$|S_{17^2}| = 289! \approx 2.1 \times 10^{587}, \quad \text{while} \quad |S_{17} \wr S_{17}| = (17!)^{18} \approx 8.3 \times 10^{261}.$$

So $|S_{17^2}|$ is much bigger than $|S_{17} \wr S_{17}|$. ◁▷

We conclude by determining the structure of $S_2 \wr S_2$.

Example 14.2.14. The order of $S_2 \wr S_2$ is $(2!)^3 = 8$. To figure out which group of order 8 this is, recall from Example 14.2.6 that $\langle (1324), (34) \rangle \subset S_4$ is imprimitive. This has order 8 and by Corollary 14.2.12 is conjugate to a subgroup of $S_2 \wr S_2$. It follows that

$$S_2 \wr S_2 \simeq \langle (1324), (34) \rangle.$$

In particular, $S_2 \wr S_2$ is a dihedral group of order 8. ◁▷

C. The Solvable Case. We now have all of the tools needed to classify solvable imprimitive subgroups of S_{p^2}. The key player is the solvable permutation group $\mathrm{AGL}(1, \mathbb{F}_p) \subset S_p$. Using Lemma 14.2.8, we obtain the wreath product

$$\mathrm{AGL}(1, \mathbb{F}_p) \wr \mathrm{AGL}(1, \mathbb{F}_p) \subset S_p \wr S_p \subset S_{p^2}.$$

This allows us to describe *all* transitive imprimitive solvable groups of S_{p^2}.

Theorem 14.2.15. *Let G be a transitive subgroup of S_{p^2}. Then the following are equivalent:*

(a) *G is solvable and imprimitive.*
(b) *G is conjugate to a subgroup of $\mathrm{AGL}(1, \mathbb{F}_p) \wr \mathrm{AGL}(1, \mathbb{F}_p)$.*

Proof. Since $\mathrm{AGL}(1, \mathbb{F}_p)$ is solvable, $\mathrm{AGL}(1, \mathbb{F}_p) \wr \mathrm{AGL}(1, \mathbb{F}_p)$ is also solvable by Exercise 4. Then (b) $\Rightarrow$ (a) follows easily, since every subgroup of $S_p \wr S_p$ is imprimitive and every subgroup of $\mathrm{AGL}(1, \mathbb{F}_p) \wr \mathrm{AGL}(1, \mathbb{F}_p)$ is solvable.

We now consider (a) $\Rightarrow$ (b). Let $G \subset S_{p^2}$ be transitive, solvable, and imprimitive. By Corollary 14.2.12, we may assume that $G \subset S_p \wr S_p$. Let G' be the image of

G under the homomorphism $S_p \wr S_p \to S_p$ of part (b) of Lemma 14.2.8. We claim that G' is transitive. To prove this, take any i and j, and pick $u \in R_i$ and $v \in R_j$. Since G is transitive, there is $\sigma = (\tau; \mu_1, \ldots, \mu_p) \in G$ such that $\sigma(u) = v$. Then $\sigma(R_i) = R_j$, which implies that $\tau(i) = j$. Hence $G' \subset S_p$ is transitive. Since G' is solvable by Theorem 8.1.4, Proposition 14.1.4 implies that $\delta G' \delta^{-1} \subset \mathrm{AGL}(1, \mathbb{F}_p)$ for some $\delta \in S_p$. It follows that after conjugating G by $(\delta; e, \ldots, e) \in S_p \wr S_p$, we have $G' \subset \mathrm{AGL}(1, \mathbb{F}_p)$. Thus

$$(14.8) \qquad G \subset \mathrm{AGL}(1, \mathbb{F}_p) \wr S_p \subset S_p \wr S_p,$$

and we are halfway done with the proof.

Now fix i between 1 and p, and consider the group

$$G_i = \{\sigma \in G \mid \sigma(R_i) = R_i\}.$$

In Exercise 5 you will show that the map $G_i \to S_p$ defined by

$$(14.9) \qquad (\tau; \mu_1, \ldots, \mu_p) \mapsto \mu_i$$

is a group homomorphism. By Exercise 5 the image $G'_i \subset S_p$ of this map is transitive and solvable. Then Proposition 14.1.4 implies that there is $\delta_i \in S_p$ such that $\theta = (12 \ldots p) \in \delta_i G'_i \delta_i^{-1} \subset \mathrm{AGL}(1, \mathbb{F}_p)$. Hence, after we conjugate G by $(e; \delta_1, \ldots, \delta_p)$, we may assume that

$$(14.10) \qquad G'_i \subset \mathrm{AGL}(1, \mathbb{F}_p) \quad \text{and} \quad \theta \in G'_i$$

for all i. Notice that (14.8) continues to hold.

Now let $\sigma = (\tau; \mu_1, \ldots, \mu_p) \in G$ be arbitrary, and fix j between 1 and p. We will prove that $\mu_j \in \mathrm{AGL}(1, \mathbb{F}_p)$ as follows. By (14.10) with $i = \tau(j)$, we can find $(\rho; \nu_1, \ldots, \nu_p) \in G$ such that $\rho(i) = i$ and $\nu_i = \theta$. Using (14.6) and (14.7), we obtain the element

$$(14.11) \qquad \gamma = \sigma^{-1}(\rho; \nu_1, \ldots, \nu_p)\sigma = (\tau^{-1}\rho\tau; \lambda_1, \ldots, \lambda_p) \in G,$$

where $\lambda_j = \mu_j^{-1}\theta\mu_j$ (see Exercise 5). Since

$$\tau^{-1}\rho\tau(j) = \tau^{-1}\rho(i) = \tau^{-1}(i) = j,$$

we see that $\gamma \in G_j$, so that by (14.9),

$$\lambda_j = \mu_j^{-1}\theta\mu_j \in G'_j \subset \mathrm{AGL}(1, \mathbb{F}_p).$$

It follows that $\mu_j \in \mathrm{AGL}(1, \mathbb{F}_p)$ by part (b) of Lemma 14.1.2. Since j was arbitrary and $\tau \in \mathrm{AGL}(1, \mathbb{F}_p)$ by (14.8), it follows that $G \subset \mathrm{AGL}(1, \mathbb{F}_p) \wr \mathrm{AGL}(1, \mathbb{F}_p)$. $\square$

By Theorem 8.5.3, a polynomial is solvable by radicals if and only if its Galois group is solvable. Hence we have the following corollary of Theorem 14.2.15.

Corollary 14.2.16. *Let* $f \in F[x]$ *be irreducible and imprimitive of degree* p^2, *where* F *has characteristic* 0. *Then* f *is solvable by radicals over* F *if and only if the Galois group of* f *over* F *is isomorphic to a subgroup of the wreath product* $\mathrm{AGL}(1, \mathbb{F}_p) \wr \mathrm{AGL}(1, \mathbb{F}_p)$. $\square$

This corollary shows that the size of the Galois group of an irreducible solvable imprimitive polynomial of degree p^2 is bounded by

$$|\mathrm{AGL}(1, \mathbb{F}_p) \wr \mathrm{AGL}(1, \mathbb{F}_p)| = p^{p+1}(p-1)^{p+1}.$$

As p gets larger, this becomes very small in comparison with the size of S_{p^2}.

Example 14.2.17. When $p = 17$, we have $|S_{17^2}| \approx 2.1 \times 10^{587}$ and $|S_{17} \wr S_{17}| \approx 8.3 \times 10^{261}$ by Example 14.2.13. In contrast,

$$|\mathrm{AGL}(1, \mathbb{F}_{17}) \wr \mathrm{AGL}(1, \mathbb{F}_{17})| = 17^{18} 16^{18} \approx 6.6 \times 10^{43}.$$

Hence, while a random polynomial of degree 17^2 can have a Galois group as large as S_{17^2}, an irreducible solvable imprimitive polynomial of this degree has a much smaller Galois group. ◁▷

Mathematical Notes

Here are some further remarks on wreath products.

■ **Wreath Products.** The wreath product defined in the text can be generalized as follows. Let G be any group and let $A \subset S_n$ be a permutation group. Then set

$$A \wr G = \{(\tau; g_1, \ldots, g_n) \mid \tau \in A, \ g_1, \ldots, g_n \in G\}.$$

Following (14.7), we define a group operation on this set via

$$(\tau; g_1, \ldots, g_n)(\tau'; g_1', \ldots, g_n') = (\tau\tau'; g_{\tau'(1)}g_1', \ldots, g_{\tau'(n)}g_n').$$

In Exercise 6 you will show that this makes $A \wr G$ into a group that satisfies part (b) of Lemma 14.2.8. You will also show that if G is finite, then

$$|A \wr G| = |A||G|^n.$$

One surprise is that we can represent a wreath product as a semidirect product. See Exercise 7 for the details. Further information about wreath products can be found in [5, p. 81] and [12, pp. 219–228].

Historical Notes

The term "primitive" is due to Galois, though he said "not primitive" instead of "imprimitive." Like us, he began with polynomials. Here is his definition of "not primitive" [Galois, p. 163]:

One calls equations not primitive the equations that are, for example, of degree mn, which decompose into m factors of degree n by means of a single equation of degree m. Such are the equations of Gauss. Primitive equations are those that do not possess such a simplification.

To understand this, suppose that $f \in F[x]$ is separable of degree mn with splitting field $F \subset L$. Having "a single equation of degree m" means that we adjoin *all* roots of such a polynomial of degree m. This gives a subfield

$$F \subset K \subset L$$

such that $F \subset K$ is Galois, and having f decompose "into m factors of degree n by means of" this subfield means that there is a factorization

(14.12)
$$f = \prod_{i=1}^{m} f_i, \quad f_i \in K[x].$$

We can relate this to our definition of imprimitive as follows. Let $f \in F[x]$ be monic, separable, and imprimitive, and assume also that f is irreducible (as is implicit in Galois's definition). By Lemma 14.2.3, we can assume that the roots of f fall into m blocks $R_1, \ldots, R_m$, each consisting of n roots (so that f has degree mn). Then let f_i be the monic polynomial whose roots are the elements of R_i. In Exercise 8 you will show that $f_i \in K[x]$, where $F \subset K$ is a Galois extension determined by the block structure of the roots. Hence we recover (14.12).

In the above quotation, Galois claims that the cyclotomic equations considered by Gauss are primitive. You will prove this in Exercises 9 and 10.

We also note that Galois applied the terms "primitive" and "not primitive" to both polynomials and groups. See, for example, [Galois, p. 79].

Exercises for Section 14.2

Exercise 1. Prove (14.7).

Exercise 2. The wreath product $S_3 \wr S_2 \subset S_6$ can be thought of as the subgroup of all permutations that preserve the blocks $R_1 = \{1, 2\}$, $R_2 = \{3, 4\}$, $R_3 = \{5, 6\}$. As noted in Example 14.2.11, $S_3 \wr S_2$ has order $6 \cdot 2^3 = 48$.
(a) Show that $(S_3 \wr S_2) \cap A_6$ has order 24.
(b) Show that $S_3 \wr S_2$ is the centralizer of $(12)(34)(56)$ in S_6 (meaning that $S_3 \wr S_2$ consists of all permutations in S_6 that commute with $(12)(34)(56)$).
(c) Use part (b) to show that $S_3 \wr S_2$ is isomorphic to $((S_3 \wr S_2) \cap A_6) \times S_2$.
See the next exercise for more on $S_3 \wr S_2$ and $(S_3 \wr S_2) \cap A_6$.

Exercise 3. One of the challenges of group theory is that the same group can have radically different descriptions. For instance, S_4 and the group $G = (S_3 \wr S_2) \cap A_6$ appearing in Example 14.2.11 both have order 24. In this exercise, you will prove that they are isomorphic. We will use the notation of Exercise 2.
(a) There is a natural homomorphism $G \to S_3$ given by how elements of G permute the blocks R_1, R_2, R_3. Show that this map is onto, and express the elements of the kernel as products of disjoint cycles.

(b) Use the Sylow Theorems to show that G has one or four 3-Sylow subgroups.

(c) Show that A_6 has no element of order 6.

(d) Use part (c) and the kernel of the map $G \to S_3$ from part (a) to show that G has four 3-Sylow subgroups.

(e) G acts by conjugation on its four 3-Sylow subgroups. Use this to prove that $G \simeq S_4$.

(f) Using Exercise 2, conclude that $S_3 \wr S_2 \simeq S_4 \times S_2$.

We note without proof that $S_3 \wr S_2 \simeq S_4 \times S_2$ is also isomorphic to the full symmetry group (rotations and reflections) of the octahedron.

Exercise 4. Let A and B be solvable permutation groups. Prove that their wreath product $A \wr B$ is also solvable.

Exercise 5. This exercise will complete the proof of Theorem 14.2.15.

(a) Let $G_i \to S_p$ be the map defined in (14.9). Prove that it is a group homomorphism and that its image $G'_i \subset S_p$ is transitive and solvable.

(b) Let $\sigma = (\tau; \mu_1, \dots, \mu_p)$ and $(\rho; \nu_1, \dots, \nu_p)$ be as in the proof of Theorem 14.2.15. Thus we have a fixed j such that $i = \tau(j)$, $\nu_i = \theta$, and $\rho(i) = i$. Now let $\gamma = (\tau^{-1}\rho\tau; \lambda_1, \dots, \lambda_p)$ be as in (14.11). Prove carefully that $\lambda_j = \mu_j^{-1}\theta\mu_j$.

Exercise 6. Let A be a subgroup of S_n, and let G be any group. Then define $A \wr G$ as in the Mathematical Notes.

(a) Prove that $A \wr G$ is a group under the multiplication defined in the Mathematical Notes.

(b) State and prove a version of part (b) of Lemma 14.2.8 for $A \wr G$.

(c) Prove that $|A \wr G| = |A||G|^n$ when G is finite.

Exercise 7. Let $A \wr G$ be as in Exercise 6, and let H be the set of all functions

$$\phi : \{1, \dots, n\} \to G.$$

(a) Given $\phi, \chi \in H$, define $\phi\chi \in H$ by $(\phi\chi)(i) = \phi(i)\chi(i)$. Prove that this makes H into a group isomorphic to the product group G^n.

(b) Elements of $A \wr G$ can be written (τ, ϕ), where $\phi \in H$. Prove that in this notation, (14.7) becomes

$$(\tau, \phi)(\tau', \phi') = \big(\tau\tau', ((\tau')^{-1}) \cdot \phi)\phi'\big).$$

(c) $A \subset S_n$ acts on $\{1, \dots, n\}$. Show that this induces an action of A on H via $(\tau \cdot \phi)(i) = \phi(\tau^{-1}(i))$. Be sure you understand why the inverse is necessary.

(d) The action of part (c) enable us to define the semidirect product $H \rtimes A$. Using the description of $A \wr G$ given in part (b), prove that the map

$$(\tau, \phi) \mapsto (\tau \cdot \phi, \tau)$$

defines a group isomorphism $A \wr G \simeq H \rtimes A$. This shows that wreath products can be represented as semidirect products.

Exercise 8. The goal of this exercise is to relate Definition 14.2.2 to Galois's definition of not primitive. Let $f \in F[x]$ be monic, separable, and irreducible with splitting field $F \subset L$. Also assume that f is imprimitive with blocks of roots given by $R_1, \dots, R_m$, where each block has n elements (thus $\deg(f) = mn$). Let f_i be the monic polynomial whose roots are the elements of R_i, and let $K \subset L$ be the fixed field of $\{\sigma \in \mathrm{Gal}(L/F) \mid \sigma(R_i) = R_i \text{ for all } i\}$.

(a) Show that $f = \prod_{i=1}^m f_i$ and that $f_i \in K[x]$ for all i.

(b) In Galois's definition, K is obtained by adjoining the roots of a separable polynomial of degree m. In modern terms, Galois wants $F \subset K$ to be a Galois extension such that $\mathrm{Gal}(K/L)$ is isomorphic to a subgroup of S_m. Prove that the field K defined in part (a) has these properties.

See Exercise 14 for some examples.

Exercise 9. Assume that $G \subset S_n$ is transitive and Abelian.
(a) Prove that $|G| = n$ by considering the isotropy subgroups of G.
(b) Prove that G is primitive if and only if $|G|$ is prime.

Thus a transitive Abelian permutation group is imprimitive unless it is cyclic of prime order.

Exercise 10. Let $\Phi_p(x)$ be the cyclotomic polynomial whose roots are the primitive pth roots of unity, where p is prime. We know that $\Phi_p(x)$ is irreducible of degree $p - 1$. In the quotation given in the Historical Notes, Galois asserts that $\Phi_p(x)$ is imprimitive.
(a) Prove Galois's claim for $p > 3$ using Exercise 9.
(b) Explain why we need to assume that $p > 3$ in part (a).

Exercise 11. Given a prime p, let $C_p \subset S_p$ be the cyclic subgroup generated by the p-cycle $(12 \ldots p)$. As explained in the text, this gives the wreath product $C_p \wr C_p \subset S_{p^2}$. Prove that $C_p \wr C_p$ is a p-Sylow subgroup of S_{p^2}.

Exercise 12. Let f be an irreducible imprimitive polynomial of degree 6, 8, or 9 over a field of characteristic 0. Prove that f is solvable by radicals over F.

Exercise 13. Let $f = x^6 + bx^3 + c \in F[x]$ be irreducible, where F has characteristic different from 2 or 3. We will study the size of the Galois group of f over F.
(a) Show that f is separable. Thus we can think of the Galois group as a subgroup of S_6.
(b) Show that $x^6 + bx^3 + c$ is imprimitive and that its Galois group lies in $S_2 \wr S_3$. Also show that $|S_2 \wr S_3| = 72$. Thus the Galois group has order ≤ 72.
(c) Let $F \subset L$ be the splitting field of f over F. Use the Tower Theorem to show that $[L : F] \leq 36$. Hence the Galois group has order at most 36.

Using *Maple*, one can show that the Galois group of $x^6 + 2x^3 - 2$ over $\mathbb{Q}$ has order 36 and hence is as large as possible.

Exercise 14. Here are some examples to illustrate Galois's definition of imprimitive. We will use the notation of Exercise 8. Let F be a field of characteristic different from 2 or 3.
(a) Let $f = x^6 + bx^4 + cx^2 + d \in F[x]$ be irreducible with splitting field $F \subset L$. Show that the splitting field of $x^3 + bx^2 + cx + d$ gives an intermediate field $F \subset K \subset L$ such that $F \subset K$ is Galois and $f = f_1 f_2 f_3$, where $f_i \in K[x]$ has degree 2 for $i = 1, 2, 3$. Also explain how K relates to the field K constructed in Exercise 8.
(b) Work out the analogous theory when $f = x^6 + bx^3 + c \in F[x]$ is irreducible.

Exercise 15. Let $G \subset S_n$ be transitive. Prove that G is primitive if and only if the isotropy subgroups of G are maximal with respect to inclusion.

14.3   PRIMITIVE PERMUTATION GROUPS

We now consider primitive permutation groups. Our main result is a powerful theorem of Galois on the structure of solvable primitive permutation groups. In order to prove this, we will define *doubly transitive groups* and use finite fields to construct some interesting permutation groups. We will also study the *minimal normal subgroups* of a solvable group. As an added bonus, we will learn why Galois was interested in finite fields.

The theory developed in this section will also be used in Section 14.4 when we classify solvable primitive subgroups of S_{p^2}.

A. Doubly Transitive Permutation Groups.
For permutation groups, double transitivity is defined as follows.

Definition 14.3.1. *A subgroup $G \subset S_n$ is **doubly transitive** if whenever we have $i, i', j, j' \in \{1, \ldots, n\}$ such that*

$$i \neq i' \quad and \quad j \neq j',$$

there is $\sigma \in G$ such that

$$\sigma(i) = j \quad and \quad \sigma(i') = j'.$$

We already know an example of a doubly transitive group.

Example 14.3.2. In Section 14.1, we considered $AGL(1, \mathbb{F}_p)$ as a subgroup of S_p. To prove that this is doubly transitive, consider $i \neq i'$ and $j \neq j'$, where we now regard these as elements of $\mathbb{F}_p$. Since $i \neq i'$, there are $a, b \in \mathbb{F}_p$ such that

$$ai + b = j \quad and \quad ai' + b = j',$$

and $j \neq j'$ implies that $a \neq 0$. Thus the condition of Definition 14.3.1 is satisfied by $\gamma_{a,b} \in AGL(1, \mathbb{F}_p)$. We will generalize this example later in the section. ◁▷

The concepts of doubly transitive, primitive, and transitive are related as follows.

Proposition 14.3.3. *Let $G \subset S_n$ be a subgroup. Then:*

$$G \text{ doubly transitive} \Rightarrow G \text{ primitive} \Rightarrow G \text{ transitive}.$$

Proof. First suppose that G is doubly transitive and imprimitive. Then we have blocks $R_1, \ldots, R_k$, where $k > 1$ and $|R_i| > 1$ for some i. For this i, pick $i_1 \neq i_2$ in R_i and also pick $i_3 \in R_j$ for some $j \neq i$. Then we have pairs $i_1 \neq i_2$ and $i_1 \neq i_3$, so that by double transitivity we can find $\sigma \in G$ such that

$$\sigma(i_1) = i_1 \quad and \quad \sigma(i_2) = i_3.$$

Now consider $\sigma(R_i)$, which by assumption is one of the blocks $R_1, \ldots, R_k$. Then $\sigma(i_1) = i_1 \in R_i$ implies that $\sigma(R_i) = R_i$, while $\sigma(i_2) = i_3 \in R_j$ implies that $\sigma(R_i) = R_j$. This contradiction proves the first implication.

The second implication will be proved in Exercise 1. □

Doubly transitive permutation groups also have the following property.

Proposition 14.3.4. *If $G \subset S_n$ is doubly transitive, then $|G|$ is divisible by $n(n-1)$.*

Proof. Let $P = \{(i, j) \mid 1 \le i, j \le n, i \ne j\}$ be the set of pairs of distinct elements of $\{1, \ldots, n\}$. This set has $n(n-1)$ elements, and G acts on P via $\sigma \cdot (i, j) = (\sigma(i), \sigma(j))$. The crucial observation is that G acts transitively on P because G is doubly transitive on $\{1, \ldots, n\}$. Thus the G-orbit of any $(i, j) \in P$ has $n(n-1)$ elements. By the Fundamental Theorem of Group Actions, this implies that $n(n-1)$ divides the order of G. $\qquad\square$

B. Affine Linear and Semilinear Groups. The finite fields introduced in Chapter 11 lead to some important permutation groups. Let $\mathbb{F}_q$ be a finite field with $q = p^m$ elements, p prime, and let $\mathbb{F}_q^n$ be the standard n-dimensional vector space over $\mathbb{F}_q$. As in the Mathematical Notes to Section 11.1, $\mathrm{GL}(n, \mathbb{F}_q)$ is the group of invertible $n \times n$ matrices with entries in $\mathbb{F}_q$. This acts on $\mathbb{F}_q^n$ by matrix multiplication when elements of $\mathbb{F}_q^n$ are regarded as column vectors.

Using $\mathrm{GL}(n, \mathbb{F}_q)$, we construct the larger group $\mathrm{AGL}(n, \mathbb{F}_q)$ of *affine linear transformations*, which are maps $\gamma_{A,v} : \mathbb{F}_q^n \to \mathbb{F}_q^n$ defined by

$$\gamma_{A,v}(u) = Au + v, \quad A \in \mathrm{GL}(n, \mathbb{F}_q), \ v \in \mathbb{F}_q^n.$$

Thus $\mathrm{AGL}(n, \mathbb{F}_q)$ combines linear maps with translations. Note that $\mathrm{GL}(1, \mathbb{F}_q) = \mathbb{F}_q^*$, so that when $q = p$, $\mathrm{AGL}(1, \mathbb{F}_p)$ is the one-dimensional affine linear group studied in Sections 6.4 and 14.1.

The group $\mathrm{AGL}(n, \mathbb{F}_q)$ contains the subgroups

$$(14.13) \qquad \begin{aligned} \mathbb{F}_q^n &\simeq \{\gamma_{I_n, v} \mid v \in \mathbb{F}_q^n\} \subset \mathrm{AGL}(n, \mathbb{F}_q), \\ \mathrm{GL}(n, \mathbb{F}_q) &\simeq \{\gamma_{A,0} \mid A \in \mathrm{GL}(n, \mathbb{F}_q)\} \subset \mathrm{AGL}(n, \mathbb{F}_q), \end{aligned}$$

where $I_n \in \mathrm{GL}(n, \mathbb{F}_q)$ is the identity matrix and $0 \in \mathbb{F}_q^n$ is the zero vector. For simplicity, we will write (14.13) as

$$\mathbb{F}_q^n \subset \mathrm{AGL}(n, \mathbb{F}_q) \quad \text{and} \quad \mathrm{GL}(n, \mathbb{F}_q) \subset \mathrm{AGL}(n, \mathbb{F}_q).$$

In Exercise 2 you will show that $\mathbb{F}_q^n$ is a normal subgroup of $\mathrm{AGL}(n, \mathbb{F}_q)$ with quotient isomorphic to $\mathrm{GL}(n, \mathbb{F}_q)$. You will also express $\mathrm{AGL}(n, \mathbb{F}_q)$ as a semidirect product via the action of $\mathrm{GL}(n, \mathbb{F}_q)$ on $\mathbb{F}_q^n$.

By using the Galois group $\mathrm{Gal}(\mathbb{F}_q/\mathbb{F}_p)$, we can enlarge $\mathrm{AGL}(n, \mathbb{F}_q)$ as follows. An *affine semilinear transformation* is a map $\gamma_{A,\sigma,v} : \mathbb{F}_q^n \to \mathbb{F}_q^n$ defined by

$$\gamma_{A,\sigma,v}(u) = A\sigma(u) + v, \quad A \in \mathrm{GL}(n, \mathbb{F}_q), \ \sigma \in \mathrm{Gal}(\mathbb{F}_q/\mathbb{F}_p), \ v \in \mathbb{F}_q^n.$$

These maps form the *affine semilinear group* $\mathrm{A\Gamma L}(n, \mathbb{F}_q)$.

When $q = p$, it is easy to see that $\mathrm{A\Gamma L}(n, \mathbb{F}_q) = \mathrm{AGL}(n, \mathbb{F}_q)$, and when $q = p^m, m > 1$, you will prove in Exercise 3 that $\mathrm{AGL}(n, \mathbb{F}_q)$ is normal of index m in $\mathrm{A\Gamma L}(n, \mathbb{F}_q)$. Furthermore, we have inclusions

$$\mathbb{F}_q^n \subset \mathrm{AGL}(n, \mathbb{F}_q) \subset \mathrm{A\Gamma L}(n, \mathbb{F}_q),$$

and $\mathbb{F}_q^n$ is a normal subgroup of $A\Gamma L(n, \mathbb{F}_q)$ (see Exercise 3).

These groups act on $\mathbb{F}_q^n$, which means that they can be regarded as subgroups of S_{q^n}. As permutation groups, they have the following important properties.

Proposition 14.3.5. *The groups* $AGL(n, \mathbb{F}_q)$ *and* $A\Gamma L(n, \mathbb{F}_q)$ *are doubly transitive subgroups of* S_{q^n}. *They are also primitive.*

Proof. In Exercise 4 you will prove that $AGL(n, \mathbb{F}_q)$ is doubly transitive when acting on $\mathbb{F}_q^n$. Hence the same is true for the larger group $A\Gamma L(n, \mathbb{F}_q)$. Both groups are then primitive by Proposition 14.3.3. $\qquad\square$

In Section 14.4, we will study solvable subgroups of S_{p^2}. Applying the above theory, we get subgroups

$$
\text{(14.14)} \qquad
\begin{aligned}
q = p^2 \text{ and } n = 1 &\Rightarrow \mathbb{F}_{p^2} \subset AGL(1, \mathbb{F}_{p^2}) \subset A\Gamma L(1, \mathbb{F}_{p^2}), \\
q = p \text{ and } n = 2 &\Rightarrow \mathbb{F}_p^2 \subset AGL(2, \mathbb{F}_p).
\end{aligned}
$$

However, $\mathbb{F}_{p^2}$ is a vector space over $\mathbb{F}_p$ of dimension 2. In Exercise 3 you will show that elements of $A\Gamma L(1, \mathbb{F}_{p^2})$ are *linear* when considered as maps between vector spaces over $\mathbb{F}_p$. It follows that if we use a basis to identify $\mathbb{F}_{p^2}$ with $\mathbb{F}_p^2$, then (14.14) gives the inclusions

$$
\text{(14.15)} \qquad \mathbb{F}_p^2 = \mathbb{F}_{p^2} \subset AGL(1, \mathbb{F}_{p^2}) \subset A\Gamma L(1, \mathbb{F}_{p^2}) \subset AGL(2, \mathbb{F}_p) \subset S_{p^2}.
$$

C. Minimal Normal Subgroups. Before proving Galois's theorem on solvable primitive permutation groups, we need to take a detour into pure group theory.

Definition 14.3.6. *A normal subgroup N of a group $G \neq \{e\}$ is **minimal** if $N \neq \{e\}$ and all nontrivial subgroups of N (i.e., subgroups of N different from $\{e\}$ and N) are not normal in G.*

Here are some examples of minimal normal subgroups.

Example 14.3.7. Let $n \geq 5$. Then A_n is clearly a minimal normal subgroup of S_n, since A_n is simple. $\qquad\triangleleft\triangleright$

Example 14.3.8. The translation subgroup $\mathbb{F}_p^n$ is a normal subgroup of the affine linear group $AGL(n, \mathbb{F}_p)$, where as above we identify $v \in \mathbb{F}_p^n$ with the translation $\gamma_{I_n, v} \in AGL(n, \mathbb{F}_p)$. Since $\mathbb{F}_p^n$ is Abelian, any subgroup of $\mathbb{F}_p^n$ is normal in $\mathbb{F}_p^n$. But when is such a subgroup normal in $AGL(n, \mathbb{F}_p)$? To answer this, note that

$$
\gamma_{A, w} \circ \gamma_{I_n, v} \circ \gamma_{A, w}^{-1} = \gamma_{I_n, Av}
$$

by part (b) of Exercise 2. Since $GL(n, \mathbb{F}_p)$ acts transitively on $\mathbb{F}_p^n \setminus \{0\}$ by part (c) of Exercise 4, it follows easily that if $\{0\} \neq H \subset \mathbb{F}_p^n$ is normal in $AGL(n, \mathbb{F}_p)$, then $H = \mathbb{F}_p^n$. Thus $\mathbb{F}_p^n \subset AGL(n, \mathbb{F}_p)$ is a minimal normal subgroup. $\qquad\triangleleft\triangleright$

Example 14.3.9. Consider the wreath product $G = S_2 \wr A_l \subset S_{2l}$, where $l \geq 5$. Part (b) of Lemma 14.2.8 shows that the subgroup

$$A_l \times A_l \simeq \{(e; \mu_1, \mu_2) \mid \mu_i \in A_l\} \subset S_2 \wr A_l = G$$

is normal in G. We will regard $N = A_l \times A_l$ as a subgroup of G. In Exercise 5 you will prove that N has the following properties:

- The nontrivial normal subgroups of N are $\{e\} \times A_l$ and $A_l \times \{e\}$.
- The factors of $A_l \times A_l$ get permuted under conjugation by elements of

$$S_2 \simeq \{(\tau; e, e) \mid \tau \in S_2\} \subset S_2 \wr A_l = G.$$

Now suppose that a nontrivial subgroup $H \subset N = A_l \times A_l$ is normal in G. Then H is normal in N, so that $H = \{e\} \times A_l$ or $A_l \times \{e\}$ by the first bullet. But these subgroups can't be normal in G, by the second bullet. We conclude that $N = A_l \times A_l$ is a minimal normal subgroup of G.   ◁▷

The minimal normal subgroups in these examples are simple (Example 14.3.7) or products of simple groups (Examples 14.3.8 and 14.3.9). The following result shows that this is no accident.

Proposition 14.3.10. *Let N be a minimal normal subgroup of a finite group G. Then there is a simple group A such that we have an isomorphism*

$$N \simeq A^n = \underbrace{A \times \cdots \times A}_{n \text{ times}}$$

for some $n \geq 1$.

Proof. Let A be a minimal normal subgroup of N. Given $g \in G$, set $A_g = gAg^{-1}$. Exercise 6 shows that A_g is a minimal normal subgroup of N isomorphic to A.

We will first prove that $N \simeq A^n$ for some $n \geq 1$. If $A = N$, then we are done. So suppose that $A \neq N$. By the minimality of N, we know that $A_{g_1} \neq A$ for some $g_1 \in G$. Since the intersection of normal subgroups of N is normal in N and since A is minimal in N, we must have $A_{g_1} \cap A = \{e\}$. Then Exercise 7 implies that

$$A A_{g_1} = \{a a_1 \mid a \in A, \ a_1 \in A_{g_1}\} \subset N$$

is a normal subgroup of N isomorphic to the product group $A \times A_{g_1}$. If $A A_{g_1} = N$, then we are done, since $A_{g_1} \simeq A$.

Suppose that $A A_{g_1} \neq N$. If $A_g \subset A A_{g_1}$ for all $g \in G$, then it is easy to show that $A A_{g_1}$ is normal in G (see Exercise 6). This is impossible by the minimality of N. Hence there is $g_2 \in G$ such that $A_{g_2} \not\subset A A_{g_1}$. Then $(A A_{g_1}) \cap A_{g_2} = \{e\}$, since the left-hand side is normal in N and lies in the minimal normal subgroup A_{g_2}. Arguing as in the previous paragraph, N contains the subgroup

$$A A_{g_1} A_{g_2} \simeq A \times A_{g_1} \times A_{g_2} \simeq A^3.$$

If $A A_{g_1} A_{g_2} = N$, then we are done, and if not, we continue as above. In Exercise 6 you will show that this eventually leads to the desired isomorphism $N \simeq A^n$.

It remains to prove that A is simple. The isomorphism $N \simeq A^n$ takes $A \subset N$ to $A \times \{e\} \times \cdots \times \{e\} \subset A^n$. If $B \subset A$ is a nontrivial normal subgroup, then $N \simeq A^n$ takes B to the subgroup

$$B \times \{e\} \times \cdots \times \{e\} \subset A \times A \times \cdots \times A = A^n,$$

which is easily seen to be normal, since B is normal in A. It follows that B is normal in N. This is impossible, since A is a minimal normal subgroup of N. Hence A must be simple, and the proposition is proved. $\qquad\square$

When N is a minimal normal subgroup of a solvable group G, the simple group A appearing in $N \simeq A^n$ must also be solvable. The only solvable simple groups are cyclic of prime order, so that $A \simeq \mathbb{F}_p$ as groups. Thus we have proved the following corollary of Proposition 14.3.10.

Corollary 14.3.11. *Let N be a minimal normal subgroup of a finite solvable group. Then there is a prime p such that $N \simeq \mathbb{F}_p^n$ for some $n \geq 1$.* $\qquad\square$

D. The Solvable Case. Before proving Galois's structure theorem for solvable primitive permutation groups, we need some preliminary definitions and results.

We first want to say more about inclusions such as $\mathrm{AGL}(n, \mathbb{F}_q) \subset S_{q^n}$, which we obtain by identifying $\mathbb{F}_q^n$ with $\{1, \ldots, q^n\}$. This is done carefully as follows. Given a set T, let $S(T) = \{\varphi : T \to T \mid \varphi \text{ is one to one and onto}\}$. This is a group under composition, called the *symmetry group* of T. Here are some examples.

Example 14.3.12. Since every affine linear or semilinear transformation of $\mathbb{F}_q^n$ is one to one and onto, we have natural inclusions

$$(14.16) \qquad \mathrm{AGL}(n, \mathbb{F}_q) \subset \mathrm{A\Gamma L}(n, \mathbb{F}_q) \subset S(\mathbb{F}_q^n).$$

For a more basic example, note that $S(\{1, \ldots, \ell\})$ is the symmetric group S_ℓ. ◁▷

If T has ℓ elements, then there is a one to one onto map $\gamma : T \to \{1, \ldots, \ell\}$. It is easy to check that $\hat{\gamma}(\varphi) = \gamma \circ \varphi \circ \gamma^{-1}$ defines a group isomorphism

$$\hat{\gamma} : S(T) \simeq S_\ell.$$

Under $\hat{\gamma}$, a subgroup of $G \subset S(T)$ maps to a subgroup of S_ℓ. In Exercise 8 you will show that if we use a different map $\gamma' : T \to \{1, \ldots, \ell\}$, then G maps to a second subgroup of S_ℓ conjugate to the first.

In particular, a one-to-one onto map $\gamma : \mathbb{F}_q^n \to \{1, \ldots, q^n\}$ gives a group isomorphism $\hat{\gamma} : S(\mathbb{F}_q^n) \simeq S_{q^n}$. Applying $\hat{\gamma}$ to (14.16) gives subgroups of S_{q^n} also called $\mathrm{AGL}(n, \mathbb{F}_q)$ and $\mathrm{A\Gamma L}(n, \mathbb{F}_q)$. Since γ is not unique, these subgroups are only defined up to conjugacy in S_{q^n}.

We next define what it means for a permutation group to be *regular*. Given a group G and $g \in G$, define $\varphi_g : G \to G$ by $\varphi_g(h) = gh$. One easily shows that $\varphi_g \in S(G)$. Since $\varphi_g \circ \varphi_h = \varphi_{gh}$, the mapping $g \mapsto \varphi_g$ gives an isomorphism

$$G \simeq \{\varphi_g \mid g \in G\} \subset S(G).$$

In general, if T is any set, then a subgroup $G \subset S(T)$ is *regular* if there is a one-to-one onto map $\gamma : G \to T$ such that the isomorphism $\hat{\gamma} : S(G) \simeq S(T)$ takes $\{\varphi_g \mid g \in G\} \subset S(G)$ to $G \subset S(T)$. When T is finite, it follows that every regular subgroup of $S(T)$ has $|T|$ elements.

Make sure you understand how this definition captures the idea that $G \subset S(T)$ is regular when the action of G on T looks like the action of G on itself given by the group operation of G. Here are some examples.

Example 14.3.13. Let G be a group with n elements. In Section 7.4 we used the Cayley table of G to show that G is isomorphic to a subgroup of S_n. In Exercise 9 you will show that this subgroup is regular.   ◁▷

Example 14.3.14. Consider $\mathrm{AGL}(n, \mathbb{F}_q) \subset S(\mathbb{F}_q^n)$. If $v \in \mathbb{F}_q^n$, then φ_v is translation by v, so that when we identify v with translation by v, we see that the translation subgroup

$$\mathbb{F}_q^n \subset \mathrm{AGL}(n, \mathbb{F}_q)$$

is a regular subgroup of $S(\mathbb{F}_q^n)$. Furthermore, if we use $\gamma : \mathbb{F}_q^n \simeq \{1, \ldots, q^n\}$ to regard $\mathbb{F}_q^n \subset \mathrm{AGL}(n, \mathbb{F}_q)$ as subgroups of S_{q^n}, then $\mathbb{F}_q^n$ is regular in S_{q^n}.   ◁▷

The following lemma will be useful in our proof of Galois's structure theorem.

Lemma 14.3.15. *Suppose that $G \subset S_\ell$ is a subgroup. Then:*
(a) *If G is primitive and $N \neq \{e\}$ is normal is G, then N is transitive.*
(b) *If G is transitive and Abelian, then G is regular.*

Proof. For part (a), consider the orbits of N acting on $\{1, \ldots, \ell\}$. Fix an orbit $N \cdot j$, $j \in \{1, \ldots, \ell\}$, and take $\sigma \in G$. Since N is normal in G, we have

$$\sigma(N \cdot j) = \sigma N \sigma^{-1} \cdot \sigma(j) = N \cdot \sigma(j).$$

This shows that G preserves the block structure given by the orbits of N. Since G is primitive, the block structure is trivial, so that either there is only one orbit or every orbit has only one element. The latter is impossible (since $N \neq \{e\}$), and the former implies that N is transitive. This proves part (a).

Turning to part (b), consider the isotropy subgroup G_j of $j \in \{1, \ldots, \ell\}$. We claim that $G_j = \{e\}$. To prove this, let $\tau \in G$ and observe that

$$G_{\tau(j)} = \tau G_j \tau^{-1} = G_j,$$

where we use (A.19) and the fact that G is Abelian. Since G is transitive, we conclude that the isotropy subgroups of G are equal. Thus $\tau \in G_j$ fixes not only j but also every element of $\{1, \ldots, \ell\}$. Hence $\tau = e$, so that $G_j = \{e\}$.

In Exercise 10 you will show that a subgroup of S_ℓ is regular if and only if it is transitive with trivial isotropy subgroups. It follows that G is regular.   □

We can now prove the following great theorem of Galois.

Theorem 14.3.16. *Let $G \subset S_\ell$ be a solvable primitive permutation group. Then $\ell = p^n$, p prime, and (up to conjugacy)*

$$\mathbb{F}_p^n \subset G \subset \mathrm{AGL}(n, \mathbb{F}_p) \subset S_{p^n}.$$

Proof. Let N be a minimal normal subgroup of G. Since G is primitive, part (a) of Lemma 14.3.15 implies that N is transitive, and since G is solvable, Corollary 14.3.11 implies that $N \simeq \mathbb{F}_p^n$.

In particular, N is transitive and Abelian, so that N is regular by part (b) of Lemma 14.3.15. It follows immediately that $\ell = |N| = p^n$, as claimed in the theorem. Furthermore, since $N \simeq \mathbb{F}_p^n$, being regular means that $N \subset S_{p^n}$ is the image of

$$\mathbb{F}_p^n \subset S(\mathbb{F}_p^n)$$

under the isomorphism $\hat{\gamma} : S(\mathbb{F}_p^n) \simeq S_{p^n}$ coming from some one-to-one onto map $\gamma : \mathbb{F}_p^n \to \{1, \dots, p^n\}$. Hence, to study $N \subset G \subset S_{p^n}$, we will consider

$$\mathbb{F}_p^n \subset G' \subset S(\mathbb{F}_p^n),$$

where G' maps to G under $\hat{\gamma}$. It suffices to prove that $G' \subset \mathrm{AGL}(n, \mathbb{F}_p) \subset S(\mathbb{F}_p^n)$. Be sure you understand this.

To describe the action of G' on $\mathbb{F}_p^n$, let $G'_0 \subset G'$ be the isotropy subgroup of $0 \in \mathbb{F}_p^n$. Let $g \in G'_0$ and consider the map from $\mathbb{F}_p^n$ to itself given by

$$(14.17) \qquad\qquad v \mapsto g \cdot v.$$

To understand (14.17), write translation by v as $\gamma_{I_n, v}$. Since $\mathbb{F}_p^n \subset G'$ is normal, we have $g \gamma_{I_n, v} g^{-1} = \gamma_{I_n, w}$ for some $w \in \mathbb{F}_p^n$. Using $g \cdot 0 = 0$, we compute the action of g on v as follows:

$$\begin{aligned} g \cdot v = g \cdot (\gamma_{I_n, v} \cdot 0) &= (g\,\gamma_{I_n, v}) \cdot 0 \\ &= (g\,\gamma_{I_n, v}\,g^{-1}) \cdot (g \cdot 0) = \gamma_{I_n, w} \cdot 0 = w. \end{aligned}$$

This shows that the map $v \mapsto g \cdot v$ corresponds to conjugation by g on the normal subgroup $\mathbb{F}_p^n \subset G'$. Since conjugation is a group homomorphism, $v \mapsto g \cdot v$ must also be a group homomorphism. Such a map is automatically linear over $\mathbb{F}_p$ by Exercise 11. Thus (14.17) gives a element of $\mathrm{GL}(n, \mathbb{F}_p)$ when the latter is regarded as consisting of permutations of $\mathbb{F}_p^n$. In other words, any element of G'_0 is of the form $\gamma_{A,0}$ for some $A \in \mathrm{GL}(n, \mathbb{F}_p)$.

We now prove that $G' \subset \mathrm{AGL}(n, \mathbb{F}_p)$. As above, translation by $v \in \mathbb{F}_p^n$ is $\gamma_{I_n, v}$. Given $g \in G'$, let $v = g \cdot 0$. Then $\gamma_{I_n, -v}\, g \in G'$ maps 0 to 0 and hence lies in G'_0. Thus $\gamma_{I_n, -v}\, g = \gamma_{A,0}$, which implies that $g = \gamma_{I_n, v} \gamma_{A,0} = \gamma_{A,v} \in \mathrm{AGL}(n, \mathbb{F}_p)$. This shows that $G' \subset \mathrm{AGL}(n, \mathbb{F}_p)$ and completes the proof of the theorem. $\qquad\square$

When applied to polynomials, Theorems 8.5.3 and 14.3.16 imply the following structure theorem for the Galois group of a primitive solvable polynomial.

Corollary 14.3.17. *Let F be a field of characteristic 0, and let $f \in F[x]$ be primitive. If f is solvable by radicals over F, then f has degree p^n for some prime p and integer $n \geq 1$, and the Galois group of f over F is isomorphic to a subgroup of $\mathrm{AGL}(n, \mathbb{F}_p)$ containing the translation subgroup $\mathbb{F}_p^n$.* $\qquad\square$

Theorem 14.3.16 shows that a solvable primitive permutation group G satisfies

$$\mathbb{F}_p^n \subset G \subset \mathrm{AGL}(n, \mathbb{F}_p) \subset S_{p^n}.$$

Furthermore, the final part of the proof shows that the isotropy subgroup G_0 of $0 \in \mathbb{F}_p^n$ can be regarded as a subgroup of $\mathrm{GL}(n, \mathbb{F}_p)$ such that

$$G = \{\gamma_{A,v} \mid v \in \mathbb{F}_p^n, A \in G_0\}.$$

Thus G is uniquely determined by G_0. So it makes sense to ask if there is anything special that we can say about G_0. As we will see, the answer involves the following definition.

Definition 14.3.18. $G_0 \subset \mathrm{GL}(n, \mathbb{F}_p)$ *is **irreducible** if there is no nontrivial subspace $V \subset \mathbb{F}_p^n$ (i.e., no subspace $V \neq \{0\}$ and $\neq \mathbb{F}_p^n$) such that $g(V) \subset V$ for all $g \in G_0$.*

Using this, we get the following useful result.

Proposition 14.3.19. *Assume that G is a permutation group satisfying*

$$\mathbb{F}_p^n \subset G \subset \mathrm{AGL}(n, \mathbb{F}_p) \subset S_{p^n},$$

and let $G_0 \subset \mathrm{GL}(n, \mathbb{F}_p)$ be the isotropy subgroup of 0. Then:
(a) *G is primitive if and only if G_0 is irreducible.*
(b) *G is solvable if and only if G_0 is solvable.*

Proof. For part (a), we will prove that G is imprimitive if and only if G_0 is reducible. First assume that G is imprimitive with blocks $R_1, \ldots, R_k$. Since $\mathbb{F}_p^n \subset G$ and $\mathbb{F}_p^n$ acts transitively on itself, we know that G is transitive. By Lemma 14.2.7, it follows that $1 < |R_1| = \cdots = |R_k| < p^n$.

Suppose for simplicity that $0 \in R_1$. We claim that R_1 is a subspace of $\mathbb{F}_p^n$. To prove this, take $v \in R_1$ and observe that $v \cdot 0 = v$, since $\mathbb{F}_p^n$ acts by translation. Since G preserves the blocks, we must have $v \cdot R_1 = R_1$, which means $v + w \in R_1$ for all $w \in R_1$. Since $v \in R_1$ was arbitrary, R_1 is closed under addition and hence is a subgroup, because R_1 is finite. Exercise 11 then implies that R_1 is a subspace.

However, every $g \in G_0$ maps 0 to 0 and hence R_1 to R_1, since G preserves the blocks. This shows that R_1 is a nontrivial subspace of $\mathbb{F}_p^n$ such that $g(R_1) = R_1$ for all $g \in G_0$. Hence G_0 is reducible.

Conversely, suppose that there is a nontrivial subspace V such that $g(V) \subset V$ for all $g \in G_0$. Then $1 < |V| < p^n$, and $g(V) = V$ for all g. Now let $R_1, \ldots, R_k$ be the cosets of V in $\mathbb{F}_p^n$. In Exercise 12 you will show that G is imprimitive with respect to the blocks $R_1, \ldots, R_k$. This completes the proof of part (a).

The proof of part (b) is a straightforward application of the results of Section 8.1. See Exercise 12 for the details. $\qquad\square$

Theorem 14.3.16 and Proposition 14.3.19 imply that classifying solvable primitive subgroups of S_{p^n} reduces to the study of solvable irreducible subgroups of $GL(n, \mathbb{F}_p)$. We will use the $n = 2$ case of this strategy in Section 14.4 when we consider solvable primitive subgroups of S_{p^2}.

Mathematical Notes

Some important ideas from group theory appear in this section.

- **Multiply Transitive Groups.** Besides transitive and doubly transitive groups, one can define k-transitive subgroups of S_n for $1 \le k \le n$ as follows. A subgroup $G \subset S_n$ acts on the set P_k of ordered k-tuples of distinct elements of $\{1, \ldots, n\}$ by

$$\sigma \cdot (i_1, \ldots, i_k) = (\sigma(i_1), \ldots, \sigma(i_k)), \quad \sigma \in G, \ (i_1, \ldots, i_k) \in P_k.$$

Then G is k-*transitive* if G acts transitively on P_k. In Exercise 13 you will show that S_n is n-transitive and A_n is $(n-2)$-transitive, and in Proposition 14.3.5 we showed that $AGL(n, \mathbb{F}_{p^m})$ is 2-transitive (i.e., doubly transitive).

An example of a 4-transitive group is the Mathieu group

$$M_{11} = \langle (2\ 10)(4\ 11)(5\ 7)(8\ 9), (1\ 4\ 3\ 8)(2\ 5\ 6\ 9) \rangle \subset S_{11}.$$

This is a simple group of order 7920 and is the smallest *sporadic* group in the classification of finite simple groups. Some of the many interesting aspects of multiply transitive groups are discussed in [2, Ch. 7] and [5, Secs. 5.7, 5.8].

- **Finite Simple Groups.** The group $GL(n, \mathbb{F}_q)$ is finite whenever $\mathbb{F}_q$ is a finite field. This leads to an interesting finite simple group as follows. First observe that $GL(n, \mathbb{F}_q)$ contains the normal subgroups

$$SL(n, \mathbb{F}_q) = \{A \in GL(n, \mathbb{F}_q) \mid \det(A) = 1\},$$
$$\mathbb{F}_q^* I_n = \{\lambda I_n \mid \lambda \in \mathbb{F}_q^*\},$$

where I_n is the $n \times n$ identity matrix. The group $SL(n, \mathbb{F}_q)$ is normal because it is the kernel of the homomorphism $\det : GL(n, \mathbb{F}_q) \to \mathbb{F}_q^*$, and $\mathbb{F}_q^* I_n$ is normal because its elements commute with all $n \times n$ matrices.

The *projective linear group* is the quotient group

$$PGL(n, \mathbb{F}_q) = GL(n, \mathbb{F}_q)/\mathbb{F}_q^* I_n,$$

which is also finite. Furthermore, inside this group we have the subgroup

$$PSL(n, \mathbb{F}_q) \subset PGL(n, \mathbb{F}_q)$$

consisting of all elements of $PGL(n, \mathbb{F}_q)$ represented by an element of $SL(n, \mathbb{F}_q)$. The remarkable fact is that $PSL(n, \mathbb{F}_q)$ is almost always simple.

Theorem 14.3.20. *Let $\mathbb{F}_q$ be a finite field and $n > 1$ be an integer. Then $PSL(n, \mathbb{F}_q)$ is a simple group except when $n = 2$ and $q = 2$ or 3.* $\qquad\square$

A proof can be found in [1] or [6]. In Exercises 14 and 15 you will show that $\mathrm{PSL}(2, \mathbb{F}_2) \simeq S_3$ and $\mathrm{PSL}(2, \mathbb{F}_3) \simeq A_4$, which are not simple groups. You will also show that

$$|\mathrm{PSL}(2, \mathbb{F}_4)| = |\mathrm{PSL}(2, \mathbb{F}_5)| = 60 \quad \text{and} \quad |\mathrm{PSL}(2, \mathbb{F}_7)| = |\mathrm{PSL}(3, \mathbb{F}_2)| = 168.$$

One can prove that $\mathrm{PSL}(2, \mathbb{F}_4) \simeq \mathrm{PSL}(2, \mathbb{F}_5) \simeq A_5$ and that every non-Abelian simple group of order < 200 is isomorphic to either A_5 or $\mathrm{PSL}(2, \mathbb{F}_7) \simeq \mathrm{PSL}(3, \mathbb{F}_2)$ (see [6, Satz 6.15] and [12, pp. 106–107]). In Example 13.3.7 and Exercise 9 of Section 13.3 we showed that the Galois group of $x^7 - 154x + 99$ over $\mathbb{Q}$ is

$$\mathrm{GL}(3, \mathbb{F}_2) \simeq \mathrm{PSL}(3, \mathbb{F}_2).$$

Finally, we should mention that other finite simple groups can be constructed using matrices over finite fields. These groups play an important role in the classification of finite simple groups. See [4] for an introduction.

■ **The O'Nan–Scott Theorem.** Theorem 14.3.16 describes the structure of solvable primitive permutation groups. This is a special case of the *O'Nan–Scott Theorem*, which describes the structure of arbitrary primitive permutation groups. The O'Nan–Scott Theorem is a basic tool in the study of primitive permutation groups. The full statement of the theorem (see [2, Ch. 4]) is beyond the scope of this book.

However, it is possible to give a brief glimpse into what this theorem says. We need the following concept.

Definition 14.3.21. *The **socle** of a finite group G is the subgroup H generated by the minimal normal subgroups of G.*

In Exercise 16 you will show that the socle is a product of finite simple groups. It is also obviously normal in G. For a primitive permutation group $G \subset S_\ell$, one can prove the stronger result that the socle $H \subset G$ is a transitive subgroup such that $H \simeq A^m$ for some finite simple group A.

The O'Nan–Scott Theorem classifies a primitive permutation group $G \subset S_\ell$ according to its socle $H \simeq A^m$. There are two cases, each with several subcases:

Regular Socle. If H is regular, then G falls into one of two classes:
- *Abelian Socle.* $A = \mathbb{F}_p$ and $H = \mathbb{F}_p^m \subset G \subset \mathrm{AGL}(m, \mathbb{F}_p)$.
- *Non-Abelian Socle.* $H \simeq A^m$, where $m \geq 6$ and A is non-Abelian, and G is a "twisted wreath product" with restricted isotropy subgroups (see [2, Sec. 4.7]).

Nonregular Socle. Here, H is non-Abelian and G falls into one of three classes:
- *Almost Simple.* $H = A$, where A is non-Abelian, and $A \subset G \subset \mathrm{Aut}(A)$, where $\mathrm{Aut}(A)$ is the group of all automorphisms of A, and G/A is solvable.
- *Diagonal.* $H \simeq A^m$, where $m \geq 2$, and G is a "subgroup of a wreath product with diagonal action" (see [2, Sec. 4.7]).
- *Product.* $H \simeq A^m$, where $m > 2$ and G is a "subgroup of a wreath product with product action" (see [2, Sec. 4.7]).

One way to think of Theorem 14.3.16 is that it explains how solvable primitive permutation groups relate to the larger class of all primitive permutation groups: they fit into the "regular Abelian socle" class of the O'Nan–Scott theorem.

The O'Nan–Scott theorem has many applications in group theory. For example, we know from Section 14.2 that doubly transitive groups are primitive. One can show that doubly transitive groups belong to the *Abelian Socle* or *Almost Simple* cases of the O'Nan–Scott theorem. This and the classification of finite simple groups lead to a classification of all doubly transitive permutation groups. See [2, Sec. 7.7] for a discussion of this result.

Historical Notes

Why did Galois invent finite fields? After all, the main focus of his research was on the roots of polynomials. This question is now easy to answer using Galois's own words. Before giving the quotation, we recall from the Historical Notes to Section 11.1 that Galois considered elements of finite fields as "imaginary solutions" of congruences. In this language, here is what Galois had to say about the importance of finite fields [Galois, p. 125]:

> It is mainly in the theory of permutations ... that the consideration of imaginary roots of congruences appears to be indispensable. They give a simple and easy method to recognize in which case a primitive equation is solvable by radicals, as I will now try to give the idea in a few words.
>
> Given an algebraic equation $fx = 0$ of degree p^ν, suppose that the p^ν roots are denoted by x_k, where the index k has the p^ν values determined by the congruence $b^{p^\nu} = b \pmod{p}$.
>
> Take any arbitrary rational function of V of the p^ν roots x_k. One transforms this function by substituting everywhere the index k with the index $(ak + b)^{p^r}$, a, b, r being arbitrary constants satisfying $a^{p^\nu - 1} = 1$ $b^{p^\nu} \equiv b \pmod{p}$ and r an integer.

This is taken from Galois's article on finite fields. In the second paragraph of the quotation, Galois explains how elements of S_{p^ν} can be regarded a permutations of the finite field $\mathbb{F}_{p^\nu}$. The function V in the third paragraph is an element of the splitting field of f, and the substitutions described by Galois form the affine semilinear group $A\Gamma L(1, \mathbb{F}_{p^\nu})$. The formula $(ak + b)^{p^r}$ differs from the definition of semilinear given in the text, but later in the article Galois explains that when using this group,

> ... the value substituted for k in every index can be put in the three forms

$$(ak + b)^{p^r} = (a\{k + b'\})^{p^r} = a'k^{p^r} + b'' = a'(k + b')^{p^r}.$$

(See [Galois, p. 125].) The formula $a'k^{p^r} + b''$ is the one we used to define $A\Gamma L(1, \mathbb{F}_{p^\nu})$. This group uses both the field $\mathbb{F}_{p^\nu}$ and the Galois group $\mathrm{Gal}(\mathbb{F}_{p^\nu}/\mathbb{F}_p)$.

These quotes show that Galois's reason for introducing $A\Gamma L(1, \mathbb{F}_{p^\nu})$ is that he wants to "recognize in which case a primitive equation is solvable by radicals." Galois knew that $A\Gamma L(1, \mathbb{F}_{p^\nu})$ is solvable and plays an important role in determining when a primitive polynomial is solvable by radicals. We will have more to say about this in the next section.

Theorem 14.3.16 is the major result of this section and is due to Galois, though he stated only the polynomial version given in Corollary 14.3.17. In his letter to Chevalier written the night before his fatal duel, Galois describes his theorem as follows [Galois, p. 177]:

> $1°$ In order that a primitive equation be solvable by radicals, it must be of degree p^v, p being prime.
>
> $2°$ All of the permutations of such an equation are of the form
>
> $$x_{k,l,m,\ldots} / x_{ak+bl+cm+\ldots+f, a_1 k+b_1 l+c_1 m+\ldots+g, \ldots}$$
>
> k, l, m being v indices that take the p values indicating all of the roots. The indices are taken modulo p, that is to say, the roots are the same when one adds a multiple of p to one of the indices.
>
> The group obtained by using all substitutions of this linear form contains all together $p^n (p^n - 1)(p^n - p) \ldots (p^n - p^{n-1})$ permutations.

Notice how item $2°$ describes $\mathrm{AGL}(v, \mathbb{F}_p)$. Also observe that the final sentence replaces v with n. In Exercise 17 you will prove Galois's assertion that

$$(14.18) \qquad |\mathrm{AGL}(n, \mathbb{F}_p)| = p^n (p^n - 1)(p^n - p) \cdots (p^n - p^{n-1}).$$

We should also mention the observation of [10, p. 133] that Abel knew Galois's assertion $1°$ about the degree of a primitive polynomial solvable by radicals. Here is how Abel stated the result [Abel, Vol. II, p. 222]:

> If an irreducible equation of degree μ, divisible by prime numbers distinct from each other, is solvable algebraically, then one can always decompose μ into two factors μ_1 and μ_2, such that the given equation is decomposable into μ_1 equations, each of degree μ_2, and whose coefficients depend on equations of degree μ_1.

When we compare this with Galois's definition of primitive given in the Historical Notes to Section 14.2, we see that Abel is saying that if an irreducible polynomial f is solvable by radicals, then f is imprimitive whenever its degree is not a prime power. The above passage appears in an unfinished manuscript that Abel wrote shortly before his death. It shows how Abel was also struggling to understand what it means for a polynomial to be solvable by radicals.

Finally, the simple groups coming from finite fields were first studied by Jordan. In 1870, Jordan gave an incomplete proof that $\mathrm{PSL}(n, \mathbb{F}_p)$ is simple except for $n = 2$ and $p = 2$ or 3. In his proof, Jordan used what we now call *Jordan canonical form* to study matrices in $\mathrm{GL}(n, \mathbb{F}_p)$. This canonical form uses the eigenvalues of the matrix, which are roots of the characteristic polynomial. Hence the eigenvalues lie in finite extensions of $\mathbb{F}_p$. This shows that more general finite fields arise naturally when analyzing $\mathrm{GL}(n, \mathbb{F}_p)$. Jordan went on to consider $\mathrm{GL}(n, \mathbb{F}_{p^m})$, though the first complete proof of Theorem 14.3.20 is due to Dickson in 1897.

Exercises for Section 14.3

Exercise 1. The goal of this exercise is to prove that primitive permutation groups are transitive. Assume that $G \subset S_n$ is primitive but not transitive, and derive a contradiction as follows.

(a) Explain why $n > 1$.

(b) Let the orbits of G acting on $\{1, \ldots, n\}$ be $R_1, \ldots, R_k$ (see Section A.4 if you have forgotten about orbits). Explain why $k > 1$ and why elements of G map every orbit to itself.

(c) Conclude that G is imprimitive. Be sure to take into account the case when every orbit consists of a single element.

Exercise 2. Let $\gamma_{I_n, v} \in \mathrm{AGL}(n, \mathbb{F}_q)$ be translation by $v \in \mathbb{F}_q^n$, and let $\gamma_{A, w} \in \mathrm{AGL}(n, \mathbb{F}_q)$ be arbitrary.

(a) Prove that $\gamma_{A, w}^{-1} = \gamma_{A^{-1}, -A^{-1}w}$.

(b) Prove that $\gamma_{A, w} \circ \gamma_{I_n, v} \circ \gamma_{A, w}^{-1} = \gamma_{I_n, Av}$.

(c) Part (b) shows that the translation subgroup $\mathbb{F}_q^n \subset \mathrm{AGL}(n, \mathbb{F}_q)$ is normal. Prove that the quotient group $\mathrm{AGL}(n, \mathbb{F}_q)/\mathbb{F}_q^n$ is isomorphic to $\mathrm{GL}(n, \mathbb{F}_q)$.

(d) Prove that $\mathrm{AGL}(n, \mathbb{F}_q)$ is isomorphic to the semidirect product $\mathbb{F}_q^n \rtimes \mathrm{GL}(n, \mathbb{F}_q)$, where $\mathrm{GL}(n, \mathbb{F}_q)$ acts on $\mathbb{F}_q^n$ by matrix multiplication.

Exercise 3. Consider the affine semilinear group $\mathrm{A\Gamma L}(n, \mathbb{F}_q)$ for $q = p^m$.

(a) Prove that $\mathrm{AGL}(n, \mathbb{F}_q)$ is a normal subgroup of $\mathrm{A\Gamma L}(n, \mathbb{F}_q)$ of index m.

(b) Prove that $\mathbb{F}_q^n$ is a normal subgroup of $\mathrm{A\Gamma L}(n, \mathbb{F}_q)$.

(c) Prove that elements of $\mathrm{A\Gamma L}(n, \mathbb{F}_q)$ give maps $\mathbb{F}_q^n \to \mathbb{F}_q^n$ that are affine linear over $\mathbb{F}_p$.

Exercise 4. Let F be any field. The definition of $\mathrm{AGL}(n, \mathbb{F}_q)$ given in the text extends to $\mathrm{AGL}(n, F)$. The goal of this exercise is to prove that $\mathrm{AGL}(n, F)$ is doubly transitive when we regard elements of $\mathrm{AGL}(n, F)$ as permutations of the vector space F^n.

(a) Use $F^n \subset \mathrm{AGL}(n, F)$ to show that $\mathrm{AGL}(n, F)$ acts transitively on F^n.

(b) Inside $\mathrm{AGL}(n, F)$, we have the isotropy subgroup of $0 \in F^n$. Prove that this isotropy subgroup is $\mathrm{GL}(n, F)$.

(c) Prove that $\mathrm{GL}(n, F)$ acts transitively on $F^n \setminus \{0\}$.

(d) Use Exercise 19 below to conclude that $\mathrm{AGL}(n, F)$ is doubly transitive.

Exercise 5. Let A and B be non-Abelian simple groups. You will show that $A \times \{e_B\}$ and $\{e_A\} \times B$ are the only nontrivial normal subgroups of $A \times B$. Let $N \subset A \times B$ be a normal subgroup different from $\{(e_A, e_B)\}$, $A \times \{e_B\}$, and $\{e_A\} \times B$.

(a) Prove that $A \times \{e_B\}$ and $\{e_A\} \times B$ are normal in $A \times B$. Hence, if we can show that $N = A \times B$, then we will be done.

(b) Prove that we can find $(a, b) \in N$ such that $e_A \neq a \in A$ and $e_B \neq b \in B$.

(c) Let $(a, b) \in N$ be as in part (b). Show that $(aa_1a^{-1}a_1^{-1}, e_B) \in N$ for any $a_1 \in A$.

(d) Given $e_A \neq a \in A$, prove that there is $a_1 \in A$ such that $aa_1 \neq a_1a$. Then combine this with parts (b) and (c) to show that $N \cap (A \times \{e_B\}) = A \times \{e_B\}$.

(e) Part (d) implies that $A \times \{e_B\} \subset N$, and the inclusion $\{e_A\} \times B \subset N$ is proved similarly. Use this to prove that $N = A \times B$.

Exercise 18 will explore various aspects of this argument.

Exercise 6. Let $A \subset N$ be a minimal normal subgroup, where N is normal in a larger group G. Given $g \in G$, we set $A_g = gAg^{-1}$.

(a) Prove that A_g is isomorphic to A and is a minimal normal subgroup of N.

(b) Fix $g_1 \in G$ and consider AA_{g_1}. By Exercise 7, we know that AA_{g_1} is a subgroup of N. Assume that $A_g \subset AA_{g_1}$ for all $g \in G$. Prove that AA_{g_1} is normal in G.

(c) Use the following idea to complete the proof of Proposition 14.3.10. Let $\mathscr{A}$ be the set of all subgroups of N of the form $A_{g_1} \cdots A_{g_n}$ such that the map $(a_1, \ldots, a_n) \mapsto a_1 \cdots a_n$ defines an isomorphism

$$A_{g_1} \times \cdots \times A_{g_n} \simeq A_{g_1} \cdots A_{g_n}.$$

Note that $A = A_e \in \mathscr{A}$. Then pick an element of $\mathscr{A}$ of maximal order.

Exercise 7. Let H and K be normal subgroups of a group G. Let $HK = \{hk \mid h \in H, k \in K\}$.
(a) Prove that HK is a normal subgroup of G.
(b) Assume that $H \cap K = \{e\}$. Prove that $hk = kh$ for all $h \in H, k \in K$.
(c) As in part (b), assume that $H \cap K = \{e\}$. Prove that the map $H \times K \to HK$ defined by $(h, k) \mapsto hk$ is a group isomorphism.

Exercise 8. Suppose that $\gamma, \gamma' : T \to \{1, \ldots, \ell\}$ are one to one and onto. As explained in the text, these give isomorphisms $\hat{\gamma}, \hat{\gamma}' : S(T) \simeq S_\ell$.
(a) Explain why $\sigma = \gamma \circ (\gamma')^{-1}$ is an element of S_ℓ.
(b) Let $\sigma \in S_\ell$ be as in part (a), and let $\hat{\sigma} : S_\ell \to S_\ell$ be conjugation by σ. Thus $\hat{\sigma}(\tau) = \sigma \tau \sigma^{-1}$ for $\tau \in S_\ell$. Prove that $\hat{\gamma} = \hat{\sigma} \circ \hat{\gamma}'$.
This proves that $\hat{\gamma}$ and $\hat{\gamma}'$ differ by conjugation by an element of S_ℓ.

Exercise 9. Let G be a group of order n. In Section 7.4 we constructed a subgroup $H \subset S_n$ isomorphic to G. Prove that H is regular in S_n.

Exercise 10. A permutation group $G \subset S_\ell$ is *regular* if there is a one-to-one onto map $\gamma : G \to \{1, \ldots, \ell\}$ such that $\hat{\gamma} : S(G) \simeq S_\ell$ maps $\{\varphi_g \mid g \in G\} \subset S(G)$ to $G \subset S_\ell$. Recall that $\varphi_g \in S(G)$ is defined by $\varphi_g(h) = gh$ for $h \in G$. The goal of this exercise is to show that G is regular if and only if it is transitive with trivial isotropy subgroups.
(a) Let $G \subset S_\ell$ be regular. Prove that G is transitive and that the isotropy subgroups of G are trivial.
(b) For the rest of the exercise, assume that G is transitive with trivial isotropy subgroups. Define $\gamma : G \to \{1, \ldots, \ell\}$ by $\gamma(\tau) = \tau(1)$ for $\tau \in G$. Prove that this map is one-to-one and onto.
(c) The map γ of part (b) gives $\hat{\gamma} : S(G) \simeq S_\ell$. Show that $\hat{\gamma}(\varphi_g) = g$, and conclude that G is regular.

Exercise 11. We can regard $\mathbb{F}_p^n$ as both a group (under addition) and a vector space over $\mathbb{F}_p$ (under addition and scalar multiplication). However, since we are over $\mathbb{F}_p$, scalar multiplication can be built out of addition. Use this observation to prove the following:
(a) Any subgroup of $\mathbb{F}_p^n$ is a subspace.
(b) Any group homomorphism $\gamma : \mathbb{F}_p^n \to \mathbb{F}_p^n$ is linear.

Exercise 12. This exercise will use the notation of the proof of Proposition 14.3.19.
(a) Suppose that $V \subset \mathbb{F}_p^n$ is a nontrivial subspace such that $g(V) \subset V$ for all $g \in G_0$. Use the cosets of V in $\mathbb{F}_p^n$ to prove that G is imprimitive.
(b) Explain why $\mathbb{F}_p^n$ is normal in G, and prove that $G/\mathbb{F}_p^n \simeq G_0$. Use this to prove part (b) of Proposition 14.3.19.

Exercise 13. Consider the definition of k-transitive given in the Mathematical Notes.
(a) Prove that S_n is n-transitive.
(b) Prove that A_n is $(n-2)$-transitive when $n \geq 3$.

Exercise 14. Consider the groups $\mathrm{GL}(2, \mathbb{F}_q)$, $\mathrm{SL}(2, \mathbb{F}_q)$, $\mathrm{PGL}(2, \mathbb{F}_q)$, and $\mathrm{PSL}(2, \mathbb{F}_q)$ defined in the Mathematical Notes.
(a) Prove that $|\mathrm{GL}(2, \mathbb{F}_q)| = q(q-1)(q^2-1)$.
(b) Prove that $|\mathrm{SL}(2, \mathbb{F}_q)| = |\mathrm{PGL}(2, \mathbb{F}_q)| = q(q^2-1)$
(c) Prove that $\mathrm{PSL}(2, \mathbb{F}_q) = \mathrm{SL}(2, \mathbb{F}_q)/\{\pm I_2\}$, and conclude that

$$|\mathrm{PSL}(2, \mathbb{F}_q)| = \begin{cases} \frac{1}{2}q(q^2-1), & q \neq 2^n, \\ q(q^2-1), & q = 2^n. \end{cases}$$

(d) Compute $|\mathrm{PSL}(2, \mathbb{F}_q)|$ for $q = 2, 3, 4, 5, 7$.
(e) Show that $|\mathrm{GL}(3, \mathbb{F}_2)| = |\mathrm{PSL}(3, \mathbb{F}_2)| = 168$.

Exercise 15. Prove that $\mathrm{GL}(2, \mathbb{F}_2) = \mathrm{SL}(2, \mathbb{F}_2) \simeq \mathrm{PSL}(2, \mathbb{F}_2) \simeq S_3$ and $\mathrm{PSL}(2, \mathbb{F}_3) \simeq A_4$.

Exercise 16. Let G be a finite group with socle H. Prove that H is isomorphic to a product of finite simple groups.

Exercise 17. Prove Galois's formula (14.18) for $|\mathrm{AGL}(n, \mathbb{F}_p)|$.

Exercise 18. Here are some observations related to Exercise 5.
(a) Give an example to show that Exercise 5 is false if we drop the assumption that A and B are non-Abelian.
(b) Let $A_1, \ldots, A_r$ be non-Abelian simple groups. Determine all nontrivial normal subgroups of $A_1 \times \cdots \times A_r$.

Exercise 19. Let $G \subset S_n$ be transitive, and let G_i be the isotropy subgroup of $i \in \{1, \ldots, n\}$. Thus $G_i = \{\sigma \in G \mid \sigma(i) = i\}$.
(a) Prove that G is doubly transitive if and only if G_i acts transitively on $\{1, \ldots, n\} \setminus \{i\}$.
(b) More generally, let $k \geq 2$. Prove that G is k-transitive if and only if G_i acts $(k-1)$-transitively on $\{1, \ldots, n\} \setminus \{i\}$.

Exercise 20. Let $G \subset S_n$ be doubly transitive. Proposition 14.3.3 implies that G is transitive. Prove that G is transitive directly from the definition of doubly transitive.

Exercise 21. Generalize (14.15) by showing that we have inclusions

$$\mathbb{F}_p^{nm} = \mathbb{F}_{p^m}^n \subset \mathrm{AGL}(n, \mathbb{F}_{p^m}) \subset \mathrm{A\Gamma L}(n, \mathbb{F}_{p^m}) \subset \mathrm{AGL}(nm, \mathbb{F}_p) \subset S_{p^{nm}}.$$

Exercise 22. Show that $\mathrm{AGL}(n, \mathbb{F}_q)$ is isomorphic to the subgroup

$$\left\{ \begin{pmatrix} A & v \\ 0 & 1 \end{pmatrix} \mid A \in \mathrm{GL}(n, \mathbb{F}_q),\ v \in \mathbb{F}_q^n \right\} \subset \mathrm{GL}(n+1, \mathbb{F}_q),$$

where $\begin{pmatrix} A & v \\ 0 & 1 \end{pmatrix}$ is the $(n+1) \times (n+1)$ matrix such that the upper left $n \times n$ corner is A, the first n entries of the last column are v, and the first n entries of the last row are all zero.

Exercise 23. Use Theorem 14.3.20 to show that $\mathrm{AGL}(2, \mathbb{F}_p)$ is not solvable for $p > 3$.

Exercise 24. The action of $\mathrm{PGL}(2, F)$ on $\widehat{F} = F \cup \{\infty\}$ was introduced in Section 7.5. In particular, Exercise 11 of that section implies that the isotropy subgroup of $\mathrm{PGL}(2, F)$ at the point ∞ can be identified with $\mathrm{AGL}(1, F)$. Use part (c) of Exercise 4 and Exercise 19 to prove that the action of $\mathrm{PGL}(2, F)$ on $\widehat{F}$ is 3-transitive (also called *triply transitive*).

Exercise 25. Prove that $\mathrm{AGL}(1, \mathbb{F}_4) \simeq A_4$ and $\mathrm{A\Gamma L}(1, \mathbb{F}_4) \simeq S_4$.

Exercise 26. Compute the orders of the groups in (14.15).

14.4 PRIMITIVE POLYNOMIALS OF PRIME-SQUARED DEGREE

Let $f \in F[x]$ be a primitive polynomial of degree p^2, where p is prime and F has characteristic 0. The main goal of this section is to understand which Galois groups can occur when f is solvable by radicals over F.

By Chapter 8 and Section 14.2, this is equivalent to classifying the solvable primitive subgroups of S_{p^2} up to conjugacy. The answer is more complicated than in the imprimitive case. Instead of the single subgroup $\mathrm{AGL}(1, \mathbb{F}_p) \wr \mathrm{AGL}(1, \mathbb{F}_p) \subset S_{p^2}$ used in Theorem 14.2.15, the primitive case will require three subgroups, denoted M_1, M_2, and M_3.

Our strategy will be to first describe the M_i and then show that every primitive solvable subgroup of S_{p^2} is conjugate to a subgroup of one of them. The results of Section 14.3 imply that most of the proofs will take place in $\mathrm{AGL}(2, \mathbb{F}_p)$ and $\mathrm{GL}(2, \mathbb{F}_p)$. You will see a lot of 2×2 matrices in this section.

A. The First Two Subgroups. The subgroups M_1 and M_2 are relatively easy to describe. The first subgroup is the affine semilinear group

$$(14.19) \qquad M_1 = \mathrm{A\Gamma L}(1, \mathbb{F}_{p^2}) \subset S_{p^2}$$

from (14.15). This subgroup has the following properties.

Proposition 14.4.1. *The subgroup $M_1 = \mathrm{A\Gamma L}(1, \mathbb{F}_{p^2}) \subset S_{p^2}$ is solvable, doubly transitive, and primitive. Furthermore, $|M_1| = 2p^2(p^2 - 1)$.*

Proof. In Exercise 1 you will prove that M_1 is solvable and compute its order. Then we are done, since M_1 is doubly transitive and primitive by Proposition 14.3.5. $\square$

The second subgroup is constructed as follows. A pair of affine linear transformations $\gamma, \gamma' \in \mathrm{AGL}(1, \mathbb{F}_p)$ give

$$\delta = (\gamma, \gamma') : \mathbb{F}_p^2 \to \mathbb{F}_p^2$$

defined by

$$(14.20) \qquad \delta(\alpha, \beta) = (\gamma(\alpha), \gamma'(\beta)).$$

In Exercise 2 you will show that δ is an affine linear transformation of $\mathbb{F}_p^2$. Thus we have an inclusion

$$\mathrm{AGL}(1, \mathbb{F}_p) \times \mathrm{AGL}(1, \mathbb{F}_p) \subset \mathrm{AGL}(2, \mathbb{F}_p).$$

The first $\mathrm{AGL}(1, \mathbb{F}_p)$ acts on the first coordinate of a point in $\mathbb{F}_p^2$, and the second $\mathrm{AGL}(1, \mathbb{F}_p)$ acts on the second coordinate. To get a more interesting group, we add the matrix $\left(\begin{smallmatrix} 0 & 1 \\ 1 & 0 \end{smallmatrix}\right)$ that switches the coordinates. This gives the group

$$(14.21) \qquad M_2 = \left\langle \mathrm{AGL}(1, \mathbb{F}_p) \times \mathrm{AGL}(1, \mathbb{F}_p), \left(\begin{smallmatrix} 0 & 1 \\ 1 & 0 \end{smallmatrix}\right) \right\rangle \subset \mathrm{AGL}(2, \mathbb{F}_p) \subset S_{p^2},$$

where the last inclusion is from (14.15). This subgroup has the following properties.

Proposition 14.4.2. *The subgroup $M_2 \subset S_{p^2}$ described in* (14.21) *is solvable and, when $p > 2$, primitive. Furthermore, $|M_2| = 2p^2(p-1)^2$.*

Proof. In Exercise 2 you will verify that $\left(\begin{smallmatrix} 0 & 1 \\ 1 & 0 \end{smallmatrix}\right)$ has order 2 and satisfies

$$\left(\begin{smallmatrix} 0 & 1 \\ 1 & 0 \end{smallmatrix}\right)\Big(\mathrm{AGL}(1, \mathbb{F}_p) \times \mathrm{AGL}(1, \mathbb{F}_p)\Big)\left(\begin{smallmatrix} 0 & 1 \\ 1 & 0 \end{smallmatrix}\right)^{-1} = \mathrm{AGL}(1, \mathbb{F}_p) \times \mathrm{AGL}(1, \mathbb{F}_p).$$

It follows that $\mathrm{AGL}(1, \mathbb{F}_p) \times \mathrm{AGL}(1, \mathbb{F}_p) \subset M_2$ is a subgroup of index 2. From here, it is easy to compute $|M_2|$ and show that M_2 is solvable (see Exercise 2).

It remains to prove that M_2 is primitive. First note that M_2 contains the translation subgroup $\mathbb{F}_p^2$, since $\mathbb{F}_p \subset \mathrm{AGL}(1, \mathbb{F}_p)$. Hence, by Proposition 14.3.19, M_2 is primitive if and only if the isotropy subgroup $(M_2)_0 \subset \mathrm{GL}(2, \mathbb{F}_p)$ is irreducible. In Exercise 2 you will verify that $(M_2)_0$ is generated by the matrices

$$(14.22) \qquad \left(\begin{smallmatrix} 0 & 1 \\ 1 & 0 \end{smallmatrix}\right), \left(\begin{smallmatrix} \lambda & 0 \\ 0 & \mu \end{smallmatrix}\right), \qquad \lambda, \mu \in \mathbb{F}_p^*.$$

Let $\{0\} \neq V \subset \mathbb{F}_p^2$ be a subspace such that $\gamma(V) \subset V$ for all matrices γ in (14.22). If we can show that $V = \mathbb{F}_p^2$, then $(M_2)_0$ will be irreducible and we will be done.

Take $(a, b) \neq (0, 0)$ in V. Using $\left(\begin{smallmatrix} \lambda & 0 \\ 0 & \mu \end{smallmatrix}\right)$ from (14.22), we see that $(\lambda a, \mu b) \in V$ for all $\lambda, \mu \in \mathbb{F}_p^*$. When a and b are both nonzero, this gives $(p-1)^2$ elements of V. Since $p > 2$ implies that $(p-1)^2 > p$, we conclude that $V = \mathbb{F}_p^2$ in this case. On the other hand, if $a = 0$, then $b \neq 0$, and using $\left(\begin{smallmatrix} 0 & 1 \\ 1 & 0 \end{smallmatrix}\right)$ from (14.22) shows that $(b, 0) \in V$. Adding this to $(0, b) \in V$, we obtain $(b, b) \in V$ with both coordinates nonzero. Hence we are reduced to the previous case, so that $V = \mathbb{F}_p^2$. The case when $b = 0$ is handled similarly. $\qquad\square$

Notice how the proof of primitivity uses $\left(\begin{smallmatrix} 0 & 1 \\ 1 & 0 \end{smallmatrix}\right) \in M_2$. In fact, it is easy to see that the smaller group $\mathrm{AGL}(1, \mathbb{F}_p) \times \mathrm{AGL}(1, \mathbb{F}_p)$ is imprimitive (see Exercise 2).

It is also interesting to compare the subgroups M_1 and M_2. By Propositions 14.4.1 and 14.4.2, we have

$$\frac{|M_1|}{|M_2|} = \frac{2p^2(p^2 - 1)}{2p^2(p-1)^2} = \frac{p+1}{p-1}.$$

Thus $|M_1| > |M_2|$. In Exercise 3 you will show that when $p > 3$, M_2 is not doubly transitive and is not isomorphic to a subgroup of M_1. So M_1 and M_2 are quite distinct as subgroups of S_{p^2}.

B. The Third Subgroup. The third subgroup M_3 is harder to describe than the first two. We begin with a lemma about 2×2 matrices that will prove to be surprisingly useful. Recall that in any group G, the *centralizer* $C_G(g)$ of $g \in G$ is the subgroup consisting of all elements of G that commute with g. Also let $I_2 \in \mathrm{GL}(2, \mathbb{F}_p)$ denote the identity matrix.

Lemma 14.4.3. *If $g \in \mathrm{GL}(2, \mathbb{F}_p) \setminus \mathbb{F}_p^* I_2$, then*

$$C_{\mathrm{GL}(2, \mathbb{F}_p)}(g) = \{m \in \mathrm{GL}(2, \mathbb{F}_p) \mid m = a I_2 + b g \text{ for some } a, b \in \mathbb{F}_p\}.$$

Proof. Every $aI_2 + bg \in \mathrm{GL}(2, \mathbb{F}_p)$ obviously commutes with g. Now take m in $C_{\mathrm{GL}(2,\mathbb{F}_p)}(g)$. Since $g \notin \mathbb{F}_p^* I_2$, you will prove in Exercise 4 that there is $v \in \mathbb{F}_p^2$ such that v and gv form a basis of $\mathbb{F}_p^2$. Hence there exist $a, b \in \mathbb{F}_p$ such that

$$mv = av + bgv = (aI_2 + bg)(v).$$

Using $mg = gm$, we obtain

$$mgv = g(mv) = g(av + bgv) = agv + bg^2v = (aI_2 + bg)(gv).$$

This implies that $m = aI_2 + bg$, since their corresponding linear maps agree on a basis of $\mathbb{F}_p^2$. $\qquad\square$

The subgroup M_3 is constructed using the projective linear group $\mathrm{PGL}(2, \mathbb{F}_p)$ discussed in the Mathematical Notes to Section 14.3. The normal subgroup

$$\mathbb{F}_p^* I_2 = \{\lambda I_2 \mid \lambda \in \mathbb{F}_p^*\} \subset \mathrm{GL}(2, \mathbb{F}_p)$$

gives the quotient group

$$\mathrm{PGL}(2, \mathbb{F}_p) = \mathrm{GL}(2, \mathbb{F}_p)/\mathbb{F}_p^* I_2,$$

where the image of $m = \begin{pmatrix} a & b \\ c & d \end{pmatrix} \in \mathrm{GL}(2, \mathbb{F}_p)$ in the quotient will be denoted

$$[m] = \begin{bmatrix} a & b \\ c & d \end{bmatrix} \in \mathrm{PGL}(2, \mathbb{F}_p).$$

To define M_3, we will construct a subgroup of $\mathrm{PGL}(2, \mathbb{F}_p)$ isomorphic to S_4 when $p > 2$. By Exercise 1 of Section 8.1, S_4 has the normal subgroup

$$\langle (12)(34), (13)(24) \rangle \simeq (\mathbb{Z}/2\mathbb{Z})^2.$$

Our strategy for finding a subgroup of $\mathrm{PGL}(2, \mathbb{F}_p)$ isomorphic to S_4 uses a carefully chosen subgroup isomorphic to $(\mathbb{Z}/2\mathbb{Z})^2$. Here is the precise result.

Proposition 14.4.4. *Assume that $p > 2$. Then:*
(a) *There exist $g, h \in \mathrm{GL}(2, \mathbb{F}_p)$ such that $gh = -hg$ and $\det(g) = \det(h) = 1$.*
(b) *Let g, h be as in part (a). Then $g^2 = h^2 = -I_2$ and $[g], [h] \in \mathrm{PGL}(2, \mathbb{F}_p)$ generate a subgroup H such that*

$$H = \langle [g], [h] \rangle \simeq (\mathbb{Z}/2\mathbb{Z})^2.$$

Furthermore, the centralizer $C(H) = C_{\mathrm{PGL}(2,\mathbb{F}_p)}(H)$ (consisting of all elements of $\mathrm{PGL}(2, \mathbb{F}_p)$ that commute with every element of H) satisfies

$$C(H) = H,$$

and the normalizer $N(H) = N_{\mathrm{PGL}(2,\mathbb{F}_p)}(H)$ satisfies

$$N(H) \simeq S_4.$$

(c) *The subgroups H and N(H) defined in part (b) are unique up to conjugacy by elements of* $\mathrm{PGL}(2, \mathbb{F}_p)$.

Proof. In Exercise 5 you will prove that there are $s, t \in \mathbb{F}_p$ such that $s^2 + t^2 = -1$. Then let

$$g = \begin{pmatrix} 0 & -1 \\ 1 & 0 \end{pmatrix} \quad \text{and} \quad h = \begin{pmatrix} s & t \\ t & -s \end{pmatrix}.$$

One easily computes that g, h have the desired properties. This proves part (a).

For part (b), we first show that $g^2 = -I_2$. Since $\det(g) = 1$, the characteristic polynomial $P(x) = \det(g - xI_2)$ can be written $P(x) = x^2 + ax + 1$. Then the Cayley–Hamilton Theorem implies that

$$g^2 + ag + I_2 = 0.$$

(Do Exercise 6 if you didn't study this in your linear algebra course.) Conjugating by h and using $hgh^{-1} = -g$ easily implies that

$$g^2 - ag + I_2 = 0.$$

Adding these equations and dividing by 2 $(p > 2)$ implies that

$$g^2 = -I_2,$$

and reversing the roles of g and h gives $h^2 = -I_2$. Note also that neither g nor h lies in $\mathbb{F}_p^* I_2$ since $gh = -hg \neq hg$ $(p > 2)$. It follows easily that the subgroup $H = \langle [g], [h] \rangle \subset \mathrm{PGL}(2, \mathbb{F}_p)$ is isomorphic to $(\mathbb{Z}/2\mathbb{Z})^2$.

To study $C(H)$, first note that if $m_1, m_2 \in \mathrm{GL}(2, \mathbb{F}_p)$, then

(14.23) $\qquad m_1 m_2 = \pm m_2 m_1 \iff [m_1][m_2] = [m_2][m_1].$

One direction is obvious, since $[\pm m_2] = [m_2]$. For the other direction, observe that $[m_1][m_2] = [m_2][m_1]$ implies that $m_1 m_2 = \lambda m_2 m_1$ for some $\lambda \in \mathbb{F}_p^*$. Taking the determinant of each side shows that $\lambda^2 = 1$, so that $\lambda = \pm 1$.

Now let $[m] \in C(H)$. Then $mg = \pm gm$ and $mh = \pm hm$ by (14.23). If both signs are +, then $m \in C_{\mathrm{GL}(2,\mathbb{F}_p)}(g) \cap C_{\mathrm{GL}(2,\mathbb{F}_p)}(h)$. Lemma 14.4.3 implies that

$$m = aI_2 + bg = cI_2 + dh, \quad a, b, c, d \in \mathbb{F}_p.$$

If $b \neq 0$, then g would be a linear combination of I_2 and h and hence would commute with h. This is impossible, since $gh = -hg$ and $p > 2$. Thus $b = 0$, which shows that m is a multiple of I_2. It follows that $[m] = [I_2] \in H$.

On the other hand, if $mg = -gm$ and $mh = hm$, then

$$(mh)g = m(hg) = m(-gh) = (-mg)h = (gm)h = g(mh),$$
$$(mh)h = (hm)h = h(mh),$$

so that $mh \in C_{\mathrm{GL}(2,\mathbb{F}_p)}(g) \cap C_{\mathrm{GL}(2,\mathbb{F}_p)}(h)$. By the above paragraph, we conclude that $[m] = [h] \in H$. The remaining possibilities for the signs are handled similarly and imply that $[m] = [g]$ or $[gh]$. See Exercise 7 for the details. This shows that $C(H) \subset H$. The other inclusion is trivial, since H is Abelian. Thus $C(H) = H$.

To describe $N(H)$, first observe that $N(H)$ acts on H by conjugation. The identity element is fixed, so that conjugation permutes the three nonidentity elements $[g], [h], [gh]$. It follows that we have a group homomorphism

$$\varphi : N(H) \to S_3,$$

where an element $m \in N(H)$ maps to the permutation of $[g], [h], [gh]$ given by conjugation by m.

The kernel of φ consists of those $m \in N(H)$ that conjugate every element of H to itself. In other words, $\mathrm{Ker}(\varphi) = C(H)$, which is H by the above. Since $|H| = 4$, it follows immediately that $|N(H)| \leq 24$, with equality if and only if φ is onto.

To prove that φ is onto, note that $(I_2 + g)(I_2 - g) = 2I_2$ since $g^2 = -I_2$. Thus

$$(I_2 + g)^{-1} = \tfrac{1}{2}(I_2 - g)$$

since $p > 2$. Then conjugating h by $I_2 + g$ gives

$$(I_2 + g)h(I_2 + g)^{-1} = \tfrac{1}{2}(I_2 + g)h(I_2 - g) = \tfrac{1}{2}(h - hg + gh - ghg)$$
$$= \tfrac{1}{2}(h + gh + gh + hg^2) = gh,$$

where we use $gh = -hg$ and $g^2 = -I_2$. It follows easily that $[I_2 + g]$ conjugates $[g]$ to itself and interchanges $[h]$ and $[gh]$. Thus $[I_2 + g]$ is an element of $N(H)$ that maps to a 2-cycle in S_3. Similarly, $[I_2 + h]$ conjugates $[h]$ to itself and interchanges $[g]$ and $[gh]$, so that $[I_2 + h] \in N(H)$ maps to a different 2-cycle. Since S_3 is generated by any two distinct 2-cycles, we see that φ is onto and $|N(H)| = 24$.

The final step is to show that $N(H) \simeq S_4$. In Exercise 8 you will show that $N(H)$ has four 3-Sylow subgroups. Then the action of $N(H)$ on its 3-Sylow subgroups gives a group homomorphism $N(H) \to S_4$. In Exercise 8 you will prove that this map is an isomorphism. The proof of part (b) is now complete.

For part (c), first suppose that $g \notin \mathbb{F}_p^* I_2$ satisfies $g^2 = -I_2$. By Exercise 4, there is $v \in \mathbb{F}_p^2$ such that v, gv form a basis of $\mathbb{F}_p^2$. Since g takes v to gv and gv to $g^2 v = -v$, it follows that

$$Q^{-1}gQ = \begin{pmatrix} 0 & -1 \\ 1 & 0 \end{pmatrix},$$

where Q is the matrix whose columns are v and gv. This easily implies that all elements $g \notin \mathbb{F}_p^* I_2$ satisfying $g^2 = -I_2$ are conjugate.

Now suppose that we have g, h and g', h' as in the statement of the proposition. Then the above paragraph shows that g and g' are conjugate, say $g = Qg'Q^{-1}$. Replacing g', h' with their conjugates by Q, we may assume that $g = g'$. We need to conjugate h to h' in a way that preserves g.

Since $gh = -hg$ and $gh' = -h'g$, it is easy to see that $h^{-1}h'$ commutes with g and hence lies in $C(g) = C_{\mathrm{GL}(2,\mathbb{F}_p)}(g)$. Note also that $\det(h^{-1}h') = 1$, since $\det(h) = \det(h') = 1$. In Exercise 9, you will show that this implies that

(14.24) $$h^{-1}h' = \det(m)\, m^{-2}, \quad m \in C(g).$$

Since $g \notin \mathbb{F}_p^* I_2$, Lemma 14.4.3 implies that m is a linear combination of I_2 and g. Using this together with $gh = -hg$ and $g^2 = -I_2$, one easily computes that

$$mhm = c h, \quad c \in \mathbb{F}_p.$$

Taking determinants gives $c = \pm \det(m)$. Combining this with (14.24), we obtain

$$mhm^{-1} = (mhm)m^{-2} = (\pm \det(m) \, h)m^{-2}$$
$$= \pm h(\det(m) \, m^{-2}) = \pm h(h^{-1}h') = \pm h'.$$

Since $m \in C(g)$, it follows immediately that $[m]$ conjugates $H = \langle [g], [h] \rangle$ to $H' = \langle [g], [h'] \rangle$. This easily implies the corresponding statement for $N(H)$ and $N(H')$. The proposition is now proved. □

We can give explicit generators for the subgroup $N(H)$ described in Proposition 14.4.4. When $p \equiv 1 \bmod 4$, we can find an element $i \in \mathbb{F}_p^*$ of order 4. In Exercise 10 you will show that $N(H)$ is generated by the images of the matrices

$$(14.25) \qquad \begin{pmatrix} 0 & 1 \\ 1 & 0 \end{pmatrix}, \ \begin{pmatrix} i & 0 \\ 0 & 1 \end{pmatrix}, \ \begin{pmatrix} 1 & -1 \\ 1 & 1 \end{pmatrix}.$$

Replacing i with $\sqrt{-1} \in \mathbb{C}$ in (14.25) gives the matrices from Example 7.5.10. There, we showed that the images of these matrices in $\mathrm{PGL}(2, \mathbb{C})$ generate the rotational symmetry group of the octahedron. Furthermore, this symmetry group is isomorphic to S_4 by Exercise 10 of Section 7.5. So it is nice to see that the same matrices work in $\mathbb{F}_p$ when $p \equiv 1 \bmod 4$. Explicit generators for $N(H)$ when $p \equiv 3 \bmod 4$ are described in Exercise 10.

We can finally construct M_3. Assume that $p > 2$ and consider the homomorphism

$$\pi : \mathrm{AGL}(2, \mathbb{F}_p) \longrightarrow \mathrm{GL}(2, \mathbb{F}_p) \longrightarrow \mathrm{PGL}(2, \mathbb{F}_p),$$

where the first map takes $\gamma_{A,v}$ to A and the second is the quotient map that takes A to $[A]$. By Proposition 14.4.4, we have $S_4 \simeq N(H) \subset \mathrm{PGL}(2, \mathbb{F}_p)$. Then M_3 is defined to be the inverse image of this subgroup under π. Thus

$$(14.26) \qquad M_3 = \pi^{-1}(N(H)) \subset \mathrm{AGL}(2, \mathbb{F}_p) \subset S_{p^2}.$$

The subgroup M_3 has the following properties.

Proposition 14.4.5. *Let $p > 2$. The subgroup $M_3 \subset S_{p^2}$ described in (14.26) is solvable and primitive. Furthermore, $|M_3| = 24p^2(p - 1)$.*

Proof. It is straightforward to show that M_3 is solvable because S_4 is, and the order of M_3 is also easy to compute. We leave this as Exercise 11.

It remains to prove that M_3 is primitive. By Proposition 14.3.19, it suffices to show that $(M_3)_0$ is irreducible. First observe that $(M_3)_0$ is the inverse image of $N(H) \subset \mathrm{PGL}(2, \mathbb{F}_p)$ under the quotient map $\mathrm{GL}(2, \mathbb{F}_p) \to \mathrm{PGL}(2, \mathbb{F}_p)$. Let $V \subset$

$\mathbb{F}_p^2$ be a one-dimensional subspace, and suppose that $g(V) \subset V$ for all $g \in (M_3)_0$. In Exercise 11 you will show that by extending a basis of V to a basis of $\mathbb{F}_p^2$ and replacing $(M_3)_0$ by a suitable conjugate, we have

$$(14.27) \qquad (M_3)_0 \subset \left\{ \begin{pmatrix} a & b \\ 0 & c \end{pmatrix} \mid a, c \in \mathbb{F}_p^*, \ b \in \mathbb{F}_p \right\}.$$

Notice that both groups contain $\mathbb{F}_p^* I_2$. When we take the quotient of $(M_3)_0$ by $\mathbb{F}_p^* I_2$, we get $N(H) \subset \mathrm{PGL}(2, \mathbb{F}_p)$, and by Exercise 11 the quotient of the larger group in (14.27) is isomorphic to $\mathrm{AGL}(1, \mathbb{F}_p)$. Since $S_4 \simeq N(H)$, (14.27) implies that

$$S_4 \text{ is isomorphic to a subgroup of } \mathrm{AGL}(1, \mathbb{F}_p).$$

The idea is that these groups are sufficiently different that this is impossible. See Exercise 11 for the details. $\qquad\qquad\square$

It is possible to describe M_3 more explicitly. First, one can show that M_3 is isomorphic to the semidirect product $\mathbb{F}_p^2 \rtimes (M_3)_0$. Furthermore, a careful description of the structure of $(M_3)_0$ can be found in [13, Ch. 5], where $(M_3)_0$ is denoted by M_4 when $p \equiv 1 \bmod 4$ and by M_3 when $p \equiv 3 \bmod 4$.

One important observation is that except for certain small primes p, the subgroups M_1, M_2, and M_3 of $\mathrm{AGL}(2, \mathbb{F}_p)$ satisfy

$$(14.28) \qquad\qquad M_i \not\subset M_j \quad \text{when } i \neq j.$$

Hence we really need three subgroups. We showed above that (14.28) holds for M_1 and M_2 when $p > 3$. Then comparing $|M_3| = 24p^2(p-1)$ with $|M_1| = 2p^2(p^2-1)$ and $|M_2| = 2p^2(p-1)^2$ shows that $M_1 \not\subset M_3$ and $M_2 \not\subset M_3$ when $p > 13$. Furthermore, $(M_1)_0$ and $(M_2)_0$ have Abelian subgroups of index 2, which easily implies that $M_3 \not\subset M_1$ and $M_3 \not\subset M_2$. See Exercise 12 for more details, including a precise list of the exceptions to (14.28).

C. The Solvable Case. We can now state our main result concerning solvable primitive subgroups of S_{p^2}. Since every subgroup of $S_{2^2} = S_4$ is solvable, we will assume that $p > 2$.

Theorem 14.4.6. *Let $G \subset S_{p^2}$ be primitive, where $p > 2$ is prime. Then the following are equivalent:*
(a) *G is solvable.*
(b) *G is conjugate to a subgroup of one of the groups M_1, M_2, M_3 defined in (14.19), (14.21), (14.26) respectively.*

Proof. The proof of (b) $\Rightarrow$ (a) is easy, since we know that M_1, M_2, and M_3 are solvable by Propositions 14.4.1, 14.4.2, and 14.4.5.

To prove (a) $\Rightarrow$ (b), first note that G is conjugate to a subgroup of $\mathrm{AGL}(2, \mathbb{F}_p)$ containing $\mathbb{F}_p^2$, by Theorem 14.3.16. Furthermore, Proposition 14.3.19 implies that the isotropy subgroup G_0 of $0 \in \mathbb{F}_p^2$ is irreducible and solvable. It is also straightforward to show that G is uniquely determined by $G_0 \subset \mathrm{GL}(2, \mathbb{F}_p)$. Thus it suffices to prove that in $\mathrm{GL}(2, \mathbb{F}_p)$, G_0 is conjugate to a subgroup of $(M_i)_0$, $i \in \{1, 2, 3\}$.

We also note that we can assume that $\mathbb{F}_p^* I_2 \subset G_0$. To see why, note that matrices in $\mathbb{F}_p^* I_2$ commute with all elements of $\mathrm{GL}(2, \mathbb{F}_p)$. This makes it easy to see that the subgroup of $\mathrm{GL}(2, \mathbb{F}_p)$ generated by G_0 and $\mathbb{F}_p^* I_2$ is solvable (you will prove this carefully in Exercise 13). If this larger group lies in some $(M_i)_0$ up to conjugacy, then so does G_0. Hence we may assume that $\mathbb{F}_p^* I_2 \subset G_0$. In particular, $\mathbb{F}_p^* I_2$ is an Abelian normal subgroup of G_0.

Let $A \subset G_0$ be an Abelian normal subgroup containing $\mathbb{F}_p^* I_2$ of maximal order. The proof now breaks up into two cases, depending on A.

Case 1: First suppose that $A \neq \mathbb{F}_p^* I_2$ and pick $g \in A \setminus \mathbb{F}_p^* I_2$. Then consider the centralizer

$$C(g) = C_{\mathrm{GL}(2,\mathbb{F}_p)}(g)$$

and its normalizer

$$N(C(g)) = N_{\mathrm{GL}(2,\mathbb{F}_p)}(C(g)).$$

Our strategy will be to first prove that

(14.29) $$G_0 \subset N(C(g))$$

and then show that $N(C(g))$ is conjugate to either $(M_1)_0$ or $(M_2)_0$.

To prove (14.29), take $m \in G_0$. Then $mgm^{-1} \in A$, since $g \in A$ and A is normal in G_0. We also know that $A \subset C(g)$, since A is Abelian. Hence $mgm^{-1} \in C(g)$. By Lemma 14.4.3, this implies that

$$mgm^{-1} = aI_2 + bg, \quad a, b \in \mathbb{F}_p.$$

Now take an arbitrary element $h \in C(g)$. Using Lemma 14.4.3 again, we can write $h = cI_2 + dg$, where $c, d \in \mathbb{F}_p$. Then

$$mhm^{-1} = m(cI_2 + dg)m^{-1} = cI_2 + dmgm^{-1}$$
$$= cI_2 + d(aI_2 + bg) = (c + da)I_2 + dbg.$$

This lies in $C(g)$ by Lemma 14.4.3. Thus m normalizes $C(g)$, so that $m \in N(C(g))$. This completes the proof of (14.29).

The next step is to study $N(C(g))$. Here, our main tool will be the characteristic polynomial $P(x) = \det(g - xI_2)$ of g. This is a quadratic polynomial with coefficients in $\mathbb{F}_p$. There are three possible behaviors for $P(x)$:

- $P(x)$ is reducible and separable.
- $P(x)$ is reducible and nonseparable.
- $P(x)$ is irreducible.

We will consider each possibility separately.

Reducible and Separable. We will show that $N(C(g))$ is conjugate to $(M_2)_0$. By hypothesis, the eigenvalues of g are $\alpha \neq \beta$ in $\mathbb{F}_p$, which means that g is diagonalizable (be sure you can explain why). Hence there is $Q \in \mathrm{GL}(2, \mathbb{F}_p)$ such that

$$QgQ^{-1} = \begin{pmatrix} \alpha & 0 \\ 0 & \beta \end{pmatrix}.$$

If we replace G_0 with its conjugate QG_0Q^{-1}, then we may assume that

$$g = \begin{pmatrix} \alpha & 0 \\ 0 & \beta \end{pmatrix}.$$

We will show that $N(C(g)) = (M_2)_0$ in this situation.

Using Lemma 14.4.3, it is easy to see that

(14.30) $$C(g) = \left\{ \begin{pmatrix} \mu & 0 \\ 0 & \nu \end{pmatrix} \,\middle|\, \mu, \nu \in \mathbb{F}_p^* \right\}$$

(see Exercise 14). Now let $m = \begin{pmatrix} a & b \\ c & d \end{pmatrix} \in N(C(g))$. Then $mgm^{-1} \in C(g)$, which by the above description of $C(g)$ implies that

$$m \begin{pmatrix} \alpha & 0 \\ 0 & \beta \end{pmatrix} m^{-1} = \begin{pmatrix} \mu & 0 \\ 0 & \nu \end{pmatrix},$$

where $\mu \neq \nu$ because $\alpha \neq \beta$. If we multiply on the right by m and compare entries, then it is straightforward to show that $b = c = 0$ or $a = d = 0$. Hence

$$m = \begin{pmatrix} a & 0 \\ 0 & d \end{pmatrix} \quad \text{or} \quad m = \begin{pmatrix} 0 & b \\ c & 0 \end{pmatrix} = \begin{pmatrix} 0 & 1 \\ 1 & 0 \end{pmatrix} \begin{pmatrix} c & 0 \\ 0 & d \end{pmatrix}.$$

Since $(M_2)_0$ is generated by the matrices (14.22), it follows that $m \in (M_2)_0$. Thus

$$N(C(g)) \subset (M_2)_0.$$

The opposite inclusion is straightforward to prove (see Exercise 14). We conclude that $N(C(g)) = (M_2)_0$.

Reducible and Nonseparable. We will show that this case can't occur, since G_0 is irreducible. By hypothesis, the only eigenvalue of g is $\alpha \in \mathbb{F}_p$. Hence there is $Q \in \mathrm{GL}(2, \mathbb{F}_p)$ such that

$$QgQ^{-1} = \begin{pmatrix} \alpha & \beta \\ 0 & \alpha \end{pmatrix},$$

where $\beta \neq 0$ because $g \notin \mathbb{F}_p^* I_2$. Now replace G_0 with QG_0Q^{-1} and note that G_0 remains irreducible. Hence we may assume that

$$g = \begin{pmatrix} \alpha & \beta \\ 0 & \alpha \end{pmatrix}.$$

In Exercise 15, you will use Lemma 14.4.3 to show that

(14.31) $$C(g) = \left\{ \begin{pmatrix} \mu & \nu \\ 0 & \mu \end{pmatrix} \,\middle|\, \mu, \nu \in \mathbb{F}_p, \mu \neq 0 \right\},$$

and you will prove that the normalizer of $C(g)$ is

(14.32) $$N(C(g)) = \left\{ \begin{pmatrix} \mu & \nu \\ 0 & \lambda \end{pmatrix} \,\middle|\, \mu, \nu, \lambda \in \mathbb{F}_p, \mu\lambda \neq 0 \right\}.$$

Also recall from (14.29) that $G_0 \subset N(C(g))$.

We obtain a contradiction as follows. Let $V \subset \mathbb{F}_p^2$ be the subspace spanned by the vector $\binom{1}{0} \in \mathbb{F}_p^2$. Since every element of (14.32) takes V to itself, we see that $G_0 \subset N(C(g))$ cannot be irreducible. This gives the desired contradiction.

Irreducible. We will show that $N(C(g))$ is conjugate to $(M_1)_0$. Since $M_1 = A\Gamma L(1, \mathbb{F}_{p^2})$, it is easy to see that $(M_1)_0$ is the group $\Gamma L(1, \mathbb{F}_{p^2})$ consisting of semilinear maps $\gamma_{\alpha,\sigma} : \mathbb{F}_{p^2} \to \mathbb{F}_{p^2}$, $\alpha \in \mathbb{F}_{p^2}^*$, $\sigma \in \mathrm{Gal}(\mathbb{F}_{p^2}/\mathbb{F}_p)$, defined by

$$(14.33) \qquad \gamma_{\alpha,\sigma}(u) = \alpha\sigma(u), \quad u \in \mathbb{F}_{p^2}.$$

While $\gamma_{\alpha,\sigma}$ need not be linear over $\mathbb{F}_{p^2}$, it is always linear over $\mathbb{F}_p$ (do you see why?). To represent $\gamma_{\alpha,\sigma}$ as an element of $\mathrm{GL}(2, \mathbb{F}_p)$, we will use an isomorphism

$$(14.34) \qquad T : \mathbb{F}_{p^2} \simeq \mathbb{F}_p^2$$

of vector spaces over $\mathbb{F}_p$. Let $\mathrm{Aut}_{\mathbb{F}_p}(\mathbb{F}_{p^2})$ be the group of vector space isomorphisms $\mathbb{F}_{p^2} \to \mathbb{F}_{p^2}$ that are linear over $\mathbb{F}_p$. Then we have a group isomorphism

$$(14.35) \qquad \mathrm{Aut}_{\mathbb{F}_p}(\mathbb{F}_{p^2}) \simeq \mathrm{GL}(2, \mathbb{F}_p)$$

where an $\mathbb{F}_p$-linear isomorphism $\phi : \mathbb{F}_{p^2} \to \mathbb{F}_{p^2}$ maps to the matrix representing

$$T \circ \phi \circ T^{-1} : \mathbb{F}_p^2 \to \mathbb{F}_p^2.$$

(you will verify this in Exercise 16). Under (14.35), the subgroup

$$\Gamma L(1, \mathbb{F}_{p^2}) \subset \mathrm{Aut}_{\mathbb{F}_p}(\mathbb{F}_{p^2}).$$

maps to

$$(M_1)_0 \subset \mathrm{GL}(2, \mathbb{F}_p).$$

Different isomorphisms T in (14.34) give different isomorphisms (14.35) that are related by conjugation in $\mathrm{GL}(2, \mathbb{F}_p)$ (see Exercise 16).

By assumption, $g \in \mathrm{GL}(2, \mathbb{F}_p)$ has irreducible characteristic polynomial $P(x)$. To analyze $N(C(g))$, we will make a special choice of T in (14.34). Consider the following bases of $\mathbb{F}_{p^2}$ and $\mathbb{F}_p^2$:

- Since $P(x)$ has degree 2, it splits completely in $\mathbb{F}_{p^2}$. Let $\alpha \in \mathbb{F}_{p^2}$ be a root, and note that $\alpha \notin \mathbb{F}_p$, since $P(x)$ is irreducible over $\mathbb{F}_p$. Then $1, \alpha$ form a basis of $\mathbb{F}_{p^2}$ as a vector space over $\mathbb{F}_p$.
- Write $g = \left(\begin{smallmatrix} a & b \\ c & d \end{smallmatrix}\right)$, and observe that $c \neq 0$, since otherwise $P(x)$ would have roots $a, d \in \mathbb{F}_p$. Then $\binom{1}{0}, \binom{a}{c}$ form a basis of $\mathbb{F}_p^2$ as a vector space over $\mathbb{F}_p$.

Using these bases, define $T : \mathbb{F}_{p^2} \simeq \mathbb{F}_p^2$ by

$$T(1) = \binom{1}{0} \quad \text{and} \quad T(\alpha) = \binom{a}{c}.$$

We claim that for this choice of T, the element of $\mathrm{Aut}_{\mathbb{F}_p}(\mathbb{F}_{p^2})$ corresponding to $g \in \mathrm{GL}(2, \mathbb{F}_p)$ via (14.35) is multiplication by α.

More precisely, define $\gamma_\alpha : \mathbb{F}_{p^2} \to \mathbb{F}_{p^2}$ by $\gamma_\alpha(\beta) = \alpha\beta$ for $\beta \in \mathbb{F}_{p^2}$. We must show that

$$T \circ \gamma_\alpha \circ T^{-1} = g, \quad \text{that is,} \quad T \circ \gamma_\alpha = g \circ T,$$

where we now think of g as the linear map given by matrix multiplication. To prove this, first note that

(14.36) $$T \circ \gamma_\alpha(1) = T(\alpha) = \begin{pmatrix} a \\ c \end{pmatrix} = \begin{pmatrix} a & b \\ c & d \end{pmatrix}\begin{pmatrix} 1 \\ 0 \end{pmatrix} = g\begin{pmatrix} 1 \\ 0 \end{pmatrix} = g \circ T(1).$$

If the characteristic polynomial of g is $P(x) = x^2 + ax + b$, then

$$g^2 + ag + bI_2 = 0$$

by the Cayley–Hamilton Theorem. Using this and $T(\alpha) = g \circ T(1)$ from (14.36), we obtain

$$g \circ T(\alpha) = g^2 \circ T(1) = (-ag - bI_2) \circ T(1) = -ag \circ T(1) - bT(1)$$
$$= -aT(\alpha) - bT(1).$$

Since $\alpha^2 + a\alpha + b = 0$, we also have

$$T \circ \gamma_\alpha(\alpha) = T(\alpha^2) = T(-a\alpha - b) = -aT(\alpha) - bT(1).$$

Thus $T \circ \gamma_\alpha(\alpha) = g \circ T(\alpha)$. This and (14.36) imply that $T \circ \gamma_\alpha = g \circ T$. We conclude that g corresponds to γ_α under (14.35), as claimed.

It follows that $N(C(g)) \subset \mathrm{GL}(2, \mathbb{F}_p)$ corresponds to $N(C(\gamma_\alpha)) \subset \mathrm{Aut}_{\mathbb{F}_p}(\mathbb{F}_{p^2})$ under (14.35), where in the latter inclusion, the centralizer and normalizer are now computed relative to $\mathrm{Aut}_{\mathbb{F}_p}(\mathbb{F}_{p^2})$. Thus, if we can prove that

(14.37) $$N(C(\gamma_\alpha)) = \Gamma\mathrm{L}(1, \mathbb{F}_{p^2})$$

when $\alpha \in \mathbb{F}_{p^2} \setminus \mathbb{F}_p$, then it will follow that $N(C(g)) = (M_1)_0$. Be sure you understand this.

We now prove (14.37). In Exercise 17 you will show that if $\alpha \in \mathbb{F}_{p^2} \setminus \mathbb{F}_p$, then

(14.38) $$C(\gamma_\alpha) = \{aI_2 + b\gamma_\alpha \in \mathrm{Aut}_{\mathbb{F}_p}(\mathbb{F}_{p^2}) \mid a, b \in \mathbb{F}_p\} = \{\gamma_\beta \mid \beta \in \mathbb{F}_{p^2}^*\}.$$

Fix $m \in N(C(\gamma_\alpha))$. Then $m \circ \gamma_\alpha \circ m^{-1} \in C(\gamma_\alpha)$, so that $m \circ \gamma_\alpha \circ m^{-1} = \gamma_\beta$ for some $\beta \in \mathbb{F}_{p^2}^*$. This implies that

$$m(\alpha) = m \circ \gamma_\alpha(1) = \gamma_\beta \circ m(1) = \beta m(1),$$
$$m(\alpha^2) = m \circ \gamma_\alpha(\alpha) = \gamma_\beta \circ m(\alpha) = \beta m(\alpha) = \beta^2 m(1),$$

where the last equality of the second line uses the first line. Thus

$$0 = m(0) = m(\alpha^2 + a\alpha + b) = m(\alpha^2) + am(\alpha) + bm(1)$$
$$= \beta^2 m(1) + a\beta m(1) + bm(1) = (\beta^2 + a\beta + b)m(1).$$

Since $m(1) \neq 0$ and $\mathbb{F}_{p^2}$ is a field, we must have

(14.39) $$\beta^2 + a\beta + b = 0.$$

To relate this to $\Gamma L(1, \mathbb{F}_{p^2})$, write the Galois group of $\mathbb{F}_{p^2}$ over $\mathbb{F}_p$ as

$$\text{Gal}(\mathbb{F}_{p^2}/\mathbb{F}_p) = \{e, \sigma\} \simeq \mathbb{Z}/2\mathbb{Z},$$

where e is the identity and σ has order 2. Then the roots of $P(x) = x^2 + ax + b$ are α and $\sigma(\alpha)$. Hence (14.39) implies that

$$\beta = \alpha \quad \text{or} \quad \beta = \sigma(\alpha).$$

In Exercise 17 you will show that if we set $\delta = m(1)$, then

(14.40)
$$\beta = \alpha \implies m = \gamma_\delta = \gamma_{\delta,e} \in \Gamma L(1, \mathbb{F}_{p^2}),$$
$$\beta = \sigma(\alpha) \implies m = \gamma_\delta \circ \sigma = \gamma_{\delta,\sigma} \in \Gamma L(1, \mathbb{F}_{p^2})$$

in the notation of (14.33). This proves that $N(C(\gamma_\alpha)) \subset \Gamma L(1, \mathbb{F}_{p^2})$. The opposite inclusion is straightforward (see Exercise 17), and (14.37) follows.

Case 2: Finally, suppose that $A = \mathbb{F}_p^* I_2$. We will show that G_0 is conjugate to a subgroup of $(M_3)_0$. First note that $(M_3)_0 \subset GL(2, \mathbb{F}_p)$ is the inverse image of the subgroup $N(H) \subset PGL(2, \mathbb{F}_p)$ from Proposition 14.4.4. Let $G_0' \subset PGL(2, \mathbb{F}_p)$ be the image of $G_0 \subset GL(2, \mathbb{F}_p)$. Since $\mathbb{F}_p^* I_2 \subset G_0$, it suffices to prove that $G_0' \subset N(H)$ after a suitable conjugation.

Fix a minimal normal subgroup $B' \subset G_0'$ as defined in Section 14.3, and let B be the inverse image of B' in $GL(2, \mathbb{F}_p)$. Since $\mathbb{F}_p^* I_2 \subset G_0$, we have

$$\mathbb{F}_p^* I_2 \subset B \subset G_0.$$

Note also that B is normal in G_0. We now prove some basic facts about B' and B.

For B', first note that G_0' is nontrivial, since G_0 is irreducible (be sure you can fill in the details). Then B' is a minimal normal subgroup of the nontrivial group G_0'. This has two useful consequences:

- B' is generated by the conjugates (with respect to G_0') of any of its nonidentity elements (Exercise 18).
- B' is Abelian, since G_0' is solvable (Corollary 14.3.11).

[The solvability of G is used twice in the proof of Theorem 14.4.6: at the beginning of the proof to reduce to $G_0 \subset GL(2, \mathbb{F}_p)$, and here to imply that B' is Abelian.]

For B, recall that its *center* $Z(B)$ consists of all elements of B commuting with every element of B. We claim that $Z(B)$ is as small as possible, that is,

(14.41) $$Z(B) = \mathbb{F}_p^* I_2.$$

To see why, observe that $Z(B)$ is normal in G_0 because B is (you will prove this in Exercise 18). Note also that $Z(B)$ is Abelian and contains $\mathbb{F}_p^* I_2$. But the hypothesis of Case 2 is that $A = \mathbb{F}_p^* I_2$, which means that $\mathbb{F}_p^* I_2$ is the maximal Abelian normal subgroup of G_0 containing $\mathbb{F}_p^* I_2$. The equality (14.41) follows immediately.

The next step is to find some interesting elements of B. More precisely, we claim that there are $g, h \in B$ such that

$$(14.42) \qquad\qquad gh = -hg, \quad \det(g) = \det(h) = 1.$$

To prove this, take $[m_1] \in B'$, $[m_1] \neq [I_2]$. The conjugates of $[m_1]$ generate B', so that B is generated by $\mathbb{F}_p^* I_2$ and the conjugates of m_1. Since $m_1 \notin \mathbb{F}_p^* I_2$, (14.41) implies that m_1 doesn't commute with at least one if its conjugates, say m_2. Then $g = m_1 m_2^{-1} \in B$ has $\det(g) = 1$, since $\det(m_1) = \det(m_2)$. It is also easy to see that g doesn't commute with m_1, so that $g \notin \mathbb{F}_p^* I_2$. Hence the conjugates of $[g]$ generate B', which means that B is generated by $\mathbb{F}_p^* I_2$ and the conjugates of g. Arguing as above, g has a conjugate h such that $gh \neq hg$. Also note that $\det(h) = 1$. Since $[g][h] = [h][g]$ (B' is Abelian), (14.23) implies the desired equation $gh = -hg$.

Let $g, h \in B$ satisfy (14.42). Then $[g]$ and $[h]$ generate the subgroup H defined in Proposition 14.4.4. Thus

$$H \subset B' \subset G_0'.$$

Since B' is Abelian, we have $B' \subset C(H) = H$, where the last equality is by Proposition 14.4.4. Thus $H = B'$. Since B' is normal in G_0', we also have $G_0' \subset N(B') = N(H)$. As noted earlier, this completes the proof of the theorem. $\qquad\square$

A much more sophisticated proof of Theorem 14.4.6 can be found in [14, §21]. This reference studies solvable subgroups of $\mathrm{GL}(n, \mathbb{F}_{p^\ell})$ for arbitrary n and ℓ.

When p is large, Theorem 14.4.6 implies that solvable primitive subgroups of S_{p^2} are relatively small in size. Here is an example.

Example 14.4.7. When $p = 17$, Propositions 14.4.1, 14.4.2, and 14.4.5 imply that the orders of M_1, M_2, M_3 are

$$|M_1| = 2 \cdot 17^2(17^2 - 1) = 166464 \approx 1.7 \times 10^5,$$
$$|M_2| = 2 \cdot 17^2(17 - 1)^2 = 147968 \approx 1.5 \times 10^5,$$
$$|M_3| = 24 \cdot 17^2(17 - 1) = 110976 \approx 1.1 \times 10^5.$$

By Theorem 14.4.6, solvable primitive subgroups of S_{17^2} are extremely small when compared to $|S_{17^2}| \approx 2.1 \times 10^{587}$. In contrast, recall from Example 14.2.17 that the largest solvable imprimitive subgroup of S_{17^2} has order

$$|\mathrm{AGL}(1, \mathbb{F}_{17}) \wr \mathrm{AGL}(1, \mathbb{F}_{17})| = 17^{18} 16^{18} \simeq 6.6 \times 10^{43}.$$

Thus being solvable and primitive is much more restrictive than being solvable and imprimitive. $\qquad\triangleleft\triangleright$

Combining Corollary 14.2.16 and Theorem 14.4.6, we get the following criterion for when an irreducible polynomial of degree p^2 is solvable by radicals.

Corollary 14.4.8. *Let $f \in F[x]$ be irreducible of degree p^2, where F is a field of characteristic 0. Then f is solvable by radicals over F if and only if either*

(a) *f is imprimitive and the Galois group of f over F is isomorphic to a subgroup of the wreath product $\mathrm{AGL}(1, \mathbb{F}_p) \wr \mathrm{AGL}(1, \mathbb{F}_p)$, or*

(b) *f is primitive and the Galois group of f over F is isomorphic to a subgroup of the groups M_1, M_2, and M_3 defined in (14.19), (14.21), and (14.26).*

Mathematical Notes

This section includes some interesting ideas from group theory.

▪ **Solvable Linear Groups.** For most of the proof of Theorem 14.4.6, we worked with the group $G_0 \subset \mathrm{GL}(2, \mathbb{F}_p)$. From this point of view, the argument showed that every solvable irreducible subgroup of $\mathrm{GL}(2, \mathbb{F}_p)$ is conjugate to a subgroup of $(M_1)_0$, $(M_2)_0$, or $(M_3)_0$. A systematic approach to the study of solvable linear groups can be found in [7] and [14].

▪ **Doubly Transitive Solvable Permutation Groups.** In Proposition 14.4.1, we showed that $M_1 = \mathrm{A\Gamma L}(1, \mathbb{F}_{p^2})$ is solvable and doubly transitive. What is more surprising is that, with some exceptions for small primes, this group contains *all* solvable doubly transitive subgroups of S_{p^2}.

Theorem 14.4.9. *Let $p > 23$ be prime. Then every solvable doubly transitive subgroup $G \subset S_{p^2}$ is conjugate to a subgroup of $M_1 = \mathrm{A\Gamma L}(1, \mathbb{F}_{p^2})$.*

Proof. Since G is solvable, Theorem 14.4.6 implies that G is conjugate to a subgroup of M_1, M_2, or M_3. Furthermore, since G is doubly transitive, Proposition 14.3.4 implies that $|G|$ is divisible by $p^2(p^2 - 1)$. However,

$$|M_2| = 2p^2(p - 1)^2 \quad \text{and} \quad |M_3| = 24p^2(p - 1)$$

are not divisible by $p^2(p^2 - 1)$ when $p > 23$. This proves the theorem. $\square$

The following much stronger result was proved by Huppert in 1957.

Theorem 14.4.10. *Let $G \subset S_\ell$ be solvable and doubly transitive. Then $\ell = p^m$ for some prime p. Furthermore, if $p^m \notin \{3^2, 5^2, 7^2, 11^2, 23^2, 3^4\}$, then G is conjugate to a subgroup of $\mathrm{A\Gamma L}(1, \mathbb{F}_{p^m})$.*

Proof. Our hypothesis implies that G is solvable and primitive, and then $\ell = p^m$ by Theorem 14.3.16. This is the easy part of the proof. For the rest of the argument, see [9, §7 of Ch. XII]. $\square$

In fact, one can prove that up to conjugacy, all solvable doubly transitive permutation groups lie in $\mathrm{A\Gamma L}(1, \mathbb{F}_{p^m})$, except for the 13 groups described in [8].

▪ **Classifying Permutation Groups.** Besides the two classes of groups just discussed (solvable linear groups and solvable doubly transitive groups), there has been a lot of work on classifying other sorts of interesting groups. Here is a brief sample of what has been done:

- Solvable primitive subgroups of S_n for $n \le 256$. See [13].
- Primitive subgroups of S_n for $n \le 1000$. See [11].
- Transitive subgroups of S_{p^2} for all primes p. See [3].
- All subgroups of $\mathrm{PSL}(2, \mathbb{F}_{p^m})$. See [1, Ch. XII] or [6, §8 of Ch. II].

The last bullet has some unexpected relations with Section 7.5 and Lemma 14.4.4. See Exercise 19 for an interesting subgroup of $\mathrm{PSL}(2, \mathbb{F}_p)$ when $p \equiv 1 \bmod 8$.

Historical Notes

Galois worked very hard to understand solvable primitive subgroups, though his research was incomplete at the time of his death. In the Historical Notes to Section 14.3, we gave quotations from Galois's paper on finite fields describing $A\Gamma L(1, \mathbb{F}_{p^n})$ and his version of Theorem 14.3.16, which asserts that up to conjugacy, a solvable primitive group G satisfies

$$\mathbb{F}_p^n \subset G \subset \mathrm{AGL}(n, \mathbb{F}_p) \subset S_{p^n}.$$

In this paper, Galois notes that $A\Gamma L(1, \mathbb{F}_{p^n})$ is solvable and that any polynomial whose Galois group is a subgroup of this group is solvable by radicals. He also makes the following intriguing statement [Galois, p. 125]:

> This remark would be of little importance if I had not already demonstrated that reciprocally, a primitive equation would not be known to be solvable by radicals, without satisfying the conditions that I have just stated. (I exclude equations of the 9th and 25th degree.)

Galois seems to be saying that, with a few exceptions, a solvable primitive permutation group satisfies

(14.43) $G \subset A\Gamma L(1, \mathbb{F}_{p^n})$.

As we know from Theorem 14.4.6, this is not correct, for the groups M_2 and M_3 are counterexamples when $n = 2$. On other other hand, if we replace "primitive" with "doubly transitive," then we get a statement close to Huppert's theorem about doubly transitive solvable groups (Theorem 14.4.10). Furthermore, in his letter to Chevalier, Galois indicates that the above assertion is "too restricted. There are few exceptions, but there are some" [Galois, p. 177]. So it is hard to know exactly what Galois was thinking. Nevertheless, the results of this chapter make it abundantly clear that Galois's insight into permutation groups was nothing short of astonishing. The reader may wish to consult [10] for further discussion of these issues.

The proof of Theorem 14.4.6 given in the text is based on suggestions of Walt Parry and Jordan's 1868 paper *Sur la résolution algébrique des équations primitives des degré p^2 (p étant premier impair)* [Jordan2, pp. 171–195]. Jordan was aware that his results provide counterexamples to some of Galois's assertions.

Readers interested in learning more about the history of transitive permutation groups should consult the introduction to [11] and Appendix A of [13].

Exercises for Section 14.4

Exercise 1. Prove that $M_1 = A\Gamma L(1, \mathbb{F}_{p^2})$ is solvable, and compute its order.

Exercise 2. This exercise will study the subgroup $M_2 \subset AGL(2, \mathbb{F}_p)$ defined in (14.21).
(a) Prove that the map δ defined in (14.20) gives an element of $AGL(2, \mathbb{F}_p)$.
(b) Prove that $\begin{pmatrix} 0 & 1 \\ 1 & 0 \end{pmatrix}$ has order 2 and normalizes $AGL(1, \mathbb{F}_p) \times AGL(1, \mathbb{F}_p) \subset AGL(2, \mathbb{F}_p)$.
(c) Prove that M_2 is solvable, and compute its order.
(d) Prove that $(M_2)_0$ is generated by the matrices in (14.22).
(e) Prove that $AGL(1, \mathbb{F}_p) \times AGL(1, \mathbb{F}_p) \subset AGL(2, \mathbb{F}_p)$ is imprimitive in S_{p^2}.

Exercise 3. Let M_1 and M_2 be the groups defined in the text, and assume that $p > 3$. Prove that M_2 is not doubly transitive and not isomorphic to a subgroup of M_1.

Exercise 4. Let V be a vector space of dimension 2 over a field F, and let $T : V \to V$ be a linear map that is not a multiple of the identity. Also assume that T is an isomorphism. Prove that there is $v \in V$ such that v and $T(v)$ form a basis of V over F.

Exercise 5. Fix $a \in \mathbb{F}_p$, $p > 2$. The goal of this exercise is to find $s, t \in \mathbb{F}_p$ with $s^2 + t^2 = a$.
(a) Let $S = \{s^2 \mid s \in \mathbb{F}_p\}$. Prove that $|S| = (p + 1)/2$.
(b) Let $S' = \{a - s^2 \mid s \in \mathbb{F}_p\}$. Show that $S \cap S' \neq \emptyset$, and use this to prove the existence of $s, t \in \mathbb{F}_p$ such that $s^2 + t^2 = a$.

Exercise 6. Let $A = \begin{pmatrix} a & b \\ c & d \end{pmatrix}$ be a 2×2 matrix with entries in a field F.
(a) Prove that the characteristic polynomial of A is $P(x) = x^2 - \text{tr}(A)x + \det(A)$, where $\text{tr}(A) = a + d$ and $\det(A) = ad - bc$ are the trace and determinant of A.
(b) Prove that $P(A) = A^2 - \text{tr}(A)A + \det(A)I_2$ is the zero matrix.
The Cayley–Hamilton Theorem generalizes part (b) by showing that $P(A)$ is the zero matrix when $P(x)$ is the characteristic polynomial of an $n \times n$ matrix A.

Exercise 7. Complete the proof of $C(H) = H$ from Proposition 14.4.4 begun in the text.

Exercise 8. Let G be a group with a normal subgroup $H \simeq (\mathbb{Z}/2\mathbb{Z})^2$ such that $C_G(H) = H$ and the map $G \to \text{Aut}(H)$ given by conjugation is onto. The goal of this exercise is to prove that $G \simeq S_4$. Note that $|G| = 24$ by the proof of Proposition 14.4.4.
(a) Use the Sylow Theorems to show that G has one or four 3-Sylow subgroups. Then use $C_G(H) = H$ to show that the number is four.
(b) Let H_1 be a 3-Sylow subgroup of G. Use the Sylow Theorems to show that its normalizer has order 6.
(c) Now consider the homomorphism $\phi : G \to S_4$ given by the action of G by conjugation on the 3-Sylow subgroups. Use part (b) to prove that $\text{Ker}(\phi)$ cannot contain an element of order 3.
(d) Conclude that the image of ϕ contains A_4. It follows that if ϕ is not an isomorphism, then G contains a normal subgroup of order 2.
(e) Prove that G cannot contain a normal subgroup of order 2. Thus $\phi : G \simeq S_4$.
This exercise is closely related to Exercise 3 of Section 14.2.

Exercise 9. Let g and $C(g)$ be as in the proof of part (c) of Proposition 14.4.4.
(a) Show that $C(g)$ is Abelian and contains $\mathbb{F}_p^* I_2$.
(b) If $m \in C(g)$, then it is easy to see that $\det(m) m^{-2} \in C(g)$. By part (a), it follows that $\phi(m) = \det(m) m^{-2}$ defines a group homomorphism $\phi : C(g) \to C(g)$. Prove that $\text{Ker}(\phi) = \mathbb{F}_p^* I_2$ and $|\text{Im}(\phi)| = |C(g)|/(p - 1)$.

(c) Prove that $\text{Im}(\phi) \subset \{w \in C(g) \mid \det(w) = 1\}$.
(d) Explain why we may assume that $g = \left(\begin{smallmatrix} 0 & -1 \\ 1 & 0 \end{smallmatrix}\right)$. Then use Lemma 14.4.3 and Exercise 5 to show that $\det : C(g) \to \mathbb{F}_p^*$ is onto. Conclude that

$$\text{Im}(\phi) = \{w \in C(g) \mid \det(w) = 1\}.$$

The equality proved in part (d) shows that every element of $C(g)$ of determinant 1 is of the form $\det(m) \, m^{-2}$ for some $m \in C(g)$. This will be used in the proof of part (c) of Proposition 14.4.4.

Exercise 10. Consider the subgroup $N(H) \subset \text{PGL}(2, \mathbb{F}_p)$ defined in Proposition 14.4.4.
(a) Prove that the images of the matrices (14.25) generate $N(H)$ when $p \equiv 1 \bmod 4$.
(b) Prove that generators of H and the images of the matrices

$$\begin{pmatrix} 1 & -1 \\ 1 & 1 \end{pmatrix} \quad \text{and} \quad \begin{pmatrix} s & t-1 \\ t+1 & -s \end{pmatrix}$$

from [14, p. 163] generate $N(H)$ when $p \equiv 3 \bmod 4$.

Exercise 11. Let M_3 be as in Proposition 14.4.5.
(a) Show that M_3 is solvable and compute its order.
(b) Suppose that $\mathbb{F}_p^2$ has a one-dimensional subspace V such that $g(V) \subset V$ for all g in $(M_3)_0$. Prove carefully that $(M_3)_0$ can be conjugated so that (14.27) holds.
(c) Consider the quotient of the larger group in (14.27) by $\mathbb{F}_p^* I_2$. Prove that this group is isomorphic to $\text{AGL}(1, \mathbb{F}_p)$.
(d) Prove that S_4 is not isomorphic to a subgroup of $\text{AGL}(1, \mathbb{F}_p)$.

Exercise 12. Consider the subgroups M_1, M_2, and M_3 defined in the text.
(a) Show that $(M_1)_0$ and $(M_2)_0$ have Abelian subgroups of index 2, and use this to prove that neither can contain $(M_3)_0$. This proves that $M_3 \not\subset M_1$ and $M_3 \not\subset M_2$.
(b) Explain why $M_3 = \text{AGL}(2, \mathbb{F}_3)$ when $p = 3$.
(c) Show that $(M_1)_0/\mathbb{F}_p^* I_2$ has an element of order $p+1$, and use this to prove that $M_1 \not\subset M_3$ when $p > 3$.
(d) Show that $M_2 \not\subset M_3$ when $p > 5$.
(e) Show that $M_2 \subset M_3$ when $p = 5$.
It follows that the only exceptions to (14.28) are $M_1 \subset M_3$ and $M_2 \subset M_3$ when $p = 3$ and $M_2 \subset M_3$ when $p = 5$. This result is due to Jordan.

Exercise 13. Let $G_0 \subset \text{GL}(2, \mathbb{F}_p)$ be solvable. Prove that the subgroup generated by G_0 and $\mathbb{F}_p^* I_2$ is also solvable.

Exercise 14. Let $g = \left(\begin{smallmatrix} \alpha & 0 \\ 0 & \beta \end{smallmatrix}\right)$, where $\alpha, \beta \in \mathbb{F}_p^*$ and $\alpha \neq \beta$.
(a) Prove (14.30).
(b) Let $m = \left(\begin{smallmatrix} a & b \\ c & d \end{smallmatrix}\right) \in N(C(g))$. In the argument following (14.30), we claimed that $b = c = 0$ or $a = d = 0$. Supply the missing details.
(c) Prove that $(M_2)_0 \subset N(C(g))$.

Exercise 15. Prove (14.31) and (14.32).

Exercise 16. Let V, W be vector spaces over a field F, and let $\text{Aut}_F(V)$ be the group of vector space isomorphisms $V \simeq V$. Also let $T : V \to W$ be a vector space isomorphism.
(a) Prove that $\phi \mapsto T \circ \phi \circ T^{-1}$ induces a group isomorphism $\gamma_T : \text{Aut}_F(V) \simeq \text{Aut}_F(W)$.

(b) Suppose that $T' : V \to W$ is another isomorphism. Prove that there is $\Phi \in \text{Aut}_F(W)$ such that $T' = \Phi \circ T$. In the notation of part (a), $\gamma_\Phi : \text{Aut}_F(W) \simeq \text{Aut}_F(W)$ is conjugation by Φ.

(c) In the situation of part (b), prove that $\gamma_{T'} = \gamma_\Phi \circ \gamma_T$.

Exercise 17. Fix $\alpha \in \mathbb{F}_{p^2} \setminus \mathbb{F}_p$, and let γ_α be as defined just before (14.36).

(a) Prove (14.38) and (14.40). For (14.38), you should use the argument from the proof of Lemma 14.4.3.

(b) Prove that $\Gamma L(1, \mathbb{F}_{p^2}) \subset N(C(\gamma_\alpha))$.

Exercise 18. Let M be a finite group.

(a) Let $A \subset M$ be a minimal normal subgroup, and let $g \neq e$ be in A. Prove that A is generated by the elements hgh^{-1} as h varies over all elements of M.

(b) Let $A \subset M$ be a normal subgroup. Prove that the center $Z(A)$ of A is normal in M.

Exercise 19. In the Mathematical Notes, we mentioned that all subgroups of $\text{PSL}(2, \mathbb{F}_q)$ are known up to conjugacy. We will do a small part of this classification by proving that $\text{PSL}(2, \mathbb{F}_p)$ contains a subgroup isomorphic to S_4 when $p \equiv 1 \mod 8$. By Exercise 10 the images of the matrices (14.25) generate a subgroup of $\text{PGL}(2, \mathbb{F}_p)$ isomorphic to S_4.

(a) Explain why $\mathbb{F}_p^*$ has an element ζ of order 8. Then $i = \zeta^2$ has order 4.

(b) Compute $(1 + i)^2$ and use this to prove that there is $\alpha \in \mathbb{F}_p$ such that $\alpha^2 = 2$.

(c) Show that the matrices (14.25) lie in $\text{SL}(2, \mathbb{F}_p)$ after multiplication by suitable elements of $\mathbb{F}_p^*$. Hence their images generate a subgroup of $\text{PSL}(2, \mathbb{F}_p)$ isomorphic to S_4.

(d) Over $\mathbb{C}$, $\zeta_8 = \cos(2\pi/8) + i \sin(2\pi/8) = (1 + i)/\sqrt{2}$. How does this relate to part (b)? More generally, one can prove that if $q = p^m$ and $p > 2$, then $\text{PSL}(2, \mathbb{F}_q)$ always contains a copy of A_4, and it contains a copy of S_4 if and only if $q \equiv \pm 1 \mod 8$ (see [1, Ch. XII] or [6, §8 of Ch. II]). You should compare the list of groups given in these references with (7.29), which asserts that the finite subgroups of $\text{PSL}(2, \mathbb{C}) = \text{PGL}(2, \mathbb{C})$ are cyclic, dihedral, or isomorphic to A_4, S_4, or A_5.

Exercise 20. Assume that $g, h \in \text{GL}(2, \mathbb{F}_p)$ satisfy $gh = -hg$ and $\det(g) = \det(h) = 1$, as in part (a) of Proposition 14.4.4. Also assume that $p > 2$.

(a) Prove that the subgroup $\langle g, h \rangle \subset \text{GL}(2, \mathbb{F}_p)$ is isomorphic to the quaternion group $Q = \{\pm 1, \pm i, \pm j, \pm k\}$, where $i^2 = j^2 = k^2 = -1$, $ij = -ji = k$, and $-1 \in Z(Q)$.

(b) Prove that $(M_3)_0$ is the normalizer of $\langle g, h \rangle$ in $\text{GL}(2, \mathbb{F}_p)$.

The quaternion group is an example of an *extraspecial* 2-*group*. The normalizer of an extraspecial 2-group in $\text{GL}(2, \mathbb{F}_q)$ is part of Aschbacher's classification of subgroups of $\text{GL}(n, \mathbb{F}_q)$. This is explained (briefly) in [11].

REFERENCES

1. L. E. Dickson, *Linear Groups with an Exposition of the Galois Field Theory*, B. G. Teubner, Leipzig, 1901. Reprint by Dover, New York, 1958.

2. J. D. Dixon and B. Mortimer, *Permutation Groups*, Springer-Verlag, New York, Berlin, Heidelberg, 1996.

3. E. Dobson and D. Witte, *Transitive permutation groups of prime-squared degree*, J. Algebraic Combin. **16** (2002), 43–69.

4. D. Gorenstein, *Classifying the finite simple groups*, Bull. Amer. Math. Soc. **14** (1986), 1–98.

5. M. Hall, *The Theory of Groups*, Macmillan, New York, 1959. Reprint by Chelsea, New York, 1972.

6. B. Huppert, *Endliche Gruppen I*, Springer-Verlag, New York, Berlin, Heidelberg, 1967.

7. B. Huppert, *Lineare auflösbare Gruppen*, Math. Z. **67** (1957), 479–518.

8. B. Huppert, *Zweifach transitive, auflösbare Permutationsgruppen*, Math. Z. **68** (1957), 126–150.

9. B. Huppert and N. Blackburn, *Finite Groups III*, Springer-Verlag, New York, Berlin, Heidelberg, 1982.

10. I. Radloff, *Évariste Galois: Principles and Applications*, Historia Math. **29** (2002), 114–137.

11. C. M. Roney-Dougal and W. R. Unger, *The affine primitive permutation groups of degree less than* 1000, J. Symbolic Comput. **35** (2003), 421–439.

12. J. S. Rose, *A Course on Group Theory*, Cambridge U. P., Cambridge, 1978.

13. M. W. Short, *The Primitive Soluble Permutation Groups of Degree less than 256*, Springer-Verlag, New York, Berlin, Heidelberg, 1992.

14. D. A. Suprunenko, *Matrix Groups*, Translations of Mathematical Monographs, Volume 45, AMS, Providence, RI, 1976.

15

The Lemniscate

The lemniscate is the plane curve defined by the equation $(x^2 + y^2)^2 = x^2 - y^2$. Here is a picture:

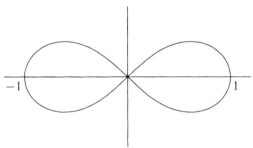

We will consider the Galois groups of polynomials arising from division of the lemniscate into arcs of equal length. This will allow us to prove the following wonderful theorem of Abel [Abel, Vol. I, p. 314]:

> One can divide the entire circumference of the lemniscate into m equal parts *by ruler and compass alone*, if m is of the form 2^n or $2^n + 1$, the last number being at the same time prime; or as well if m is a product of several numbers of these two forms.

Abel goes on to say that this theorem is "precisely the same as that of M. Gauss, relative to the circle." You will verify this in Exercise 1.

To prove Abel's theorem, we will study doubly periodic functions of a complex variable and the theory of complex multiplication. We will also learn why Eisenstein proved his irreducibility criterion.

15.1   DIVISION POINTS AND ARC LENGTH

To formulate Abel's theorem on the lemniscate carefully, we need to define the *n-division points* of the lemniscate and study the arc length of this curve.

A. Division Points of the Lemniscate. In Section 10.2 we used the nth roots of unity to determine when a regular n-gon can be constructed by straightedge and compass. In terms of the unit circle centered at the origin, the nth roots of unity divide the circle into n segments of equal length, starting from $(1, 0)$. For $n = 5$, the fifth roots of unity $1, \zeta_5, \zeta_5^2, \zeta_5^3, \zeta_5^4$ give the picture

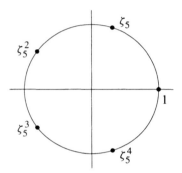

In general, the nth roots of unity $\zeta_n^i, i = 0, \ldots, n-1$, are the *n-division points* of the unit circle. Then Gauss's theorem of Section 10.2 can be restated as the assertion that the n-division points of the unit circle are constructible by straightedge and compass if and only if n is a power of 2 times a product of distinct Fermat primes.

Abel, following hints of Gauss, asked the same question for the lemniscate. Here, the *n-division points* of the lemniscate are obtained as follows. Begin at the origin and follow the curve into the first quadrant, down into the fourth quadrant, back through the origin into the second quadrant, down into the third quadrant, and finally back to the origin. As we do this, we mark those points that give one-nth of the total arc length, two-$n$ths of the arc length, and so on. For $n = 5$, the 5-division points give the picture

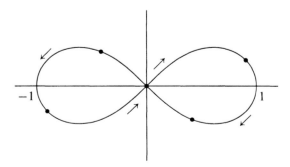

The n-division points divide the lemniscate into n segments of equal length. When n is odd, as in the above picture, the middle segment straddles the origin. When n is even, the n-division points are symmetric about the x- and y-axes, with the middle division point at the origin. For $n = 6$, the 6-division points give the picture

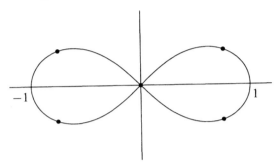

The n-division points on the lemniscate will lead to some remarkable polynomials analogous to the cyclotomic polynomials. The Galois theory of these polynomials will enable us to understand when the n-division points can be constructed by straightedge and compass.

At the beginning of the chapter, we defined the lemniscate using the Cartesian equation $(x^2+y^2)^2 = x^2 - y^2$. In Exercise 2 you will show that in polar coordinates, the lemniscate is given by the equation

(15.1) $$r^2 = \cos(2\theta).$$

The polar coordinate r will play a central role in this chapter. One reason is that in order to construct a point on the lemniscate, we only need r. This might seem obvious in that we get the desired point (and its mirror images about the x- and y-axes) by intersecting the lemniscate with the circle of radius r. But the lemniscate is actually unnecessary. In other words, if $0 < r < 1$ is constructible in the sense of Section 10.1, then so are the x and y-coordinates of the four points on the lemniscate of distance r from the origin. To see this, we use $(x^2 + y^2)^2 = x^2 - y^2$ and $r^2 = x^2 + y^2$. This gives the equations

$$r^4 = x^2 - y^2 \quad \text{and} \quad r^2 = x^2 + y^2.$$

Solving for x and y in terms of r, we obtain

$$x = \pm\sqrt{\tfrac{1}{2}(r^2 + r^4)} \quad \text{and} \quad y = \pm\sqrt{\tfrac{1}{2}(r^2 - r^4)}.$$

Since the constructible numbers form a subfield of $\mathbb{C}$ closed under square roots, we see that x and y are constructible when r is. Thus, to prove that a given point on the lemniscate is constructible by straightedge and compass, it suffices to show that the corresponding polar coordinate r is constructible. Also note that the converse holds: if x and y are constructible, then so is $r = \sqrt{x^2 + y^2}$. We have thus proved the following result.

Proposition 15.1.1. *Let P be a point on the lemniscate, and let r be the distance from P to the origin. Then P can be constructed by straightedge and compass if and only if r is a constructible number.*   □

B. Arc Length of the Lemniscate. The *n*-division points of the lemniscate are defined in terms of arc length. Hence we need to study the arc length of the lemniscate. By (15.1), the polar equation of the lemniscate is

$$r^2 = \cos(2\theta).$$

If we focus on the first quadrant, then we get the picture

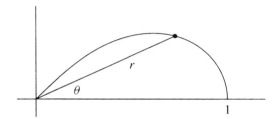

Solving the above equation for θ gives $\theta = \frac{1}{2}\cos^{-1}\left(r^2\right)$. This makes θ into a function of r. Note that θ decreases from $\frac{\pi}{4}$ to 0 as r increases from 0 to 1.

Recall that arc length in Cartesian and polar coordinates is given by

$$ds = \sqrt{dx^2 + dy^2} = \sqrt{dr^2 + r^2 d\theta^2}.$$

It follows that the arc length of the lemniscate from the origin to the point in the first quadrant with polar coordinates (r_0, θ_0) is given by

$$\text{arc length} = \int_0^{r_0} \sqrt{1 + r^2 \left(\frac{d\theta}{dr}\right)^2}\, dr.$$

Differentiating $r^2 = \cos(2\theta)$ with respect to r gives $2r = -\sin(2\theta) \cdot 2\frac{d\theta}{dr}$, so that

$$1 + r^2 \left(\frac{d\theta}{dr}\right)^2 = 1 + r^2 \left(-\frac{r}{\sin(2\theta)}\right)^2 = 1 + \frac{r^4}{\sin^2(2\theta)}.$$

Since $\sin^2(2\theta) = 1 - \cos^2(2\theta) = 1 - r^4$, we obtain

$$1 + r^2 \left(\frac{d\theta}{dr}\right)^2 = 1 + \frac{r^4}{1 - r^4} = \frac{1}{1 - r^4}.$$

Hence our arc length formula becomes

(15.2) $$\text{arc length} = \int_0^{r_0} \frac{1}{\sqrt{1 - r^4}}\, dr.$$

The integral (15.2) is improper when $r = 1$. Since it converges (see Exercise 3), $\int_0^1 (1 - r^4)^{-1/2}\, dr$ is the arc length of the first-quadrant portion of the lemniscate. In the eighteenth century, this number was denoted $\frac{\varpi}{2}$, where ϖ is a variant of the Greek letter π. Thus

$$\varpi = 2 \int_0^1 \frac{1}{\sqrt{1 - r^4}}\, dr \approx 2.62206.$$

It follows that the arc length of the lemniscate is 2ϖ and the arc length between successive n-division points is $\frac{2\varpi}{n}$.

We will write (15.2) as

(15.3)
$$s = \int_0^r \frac{1}{\sqrt{1 - t^4}}\, dt,$$

where s represents the arc length along the lemniscate from the origin to the point in the first quadrant with polar coordinates (r, θ). Then (15.3) expresses s as a function of r. Following Abel the inverse function will be written $r = \varphi(s)$, so that

(15.4)
$$r = \varphi(s) \iff s = \int_0^r \frac{1}{\sqrt{1 - t^4}}\, dt.$$

Since

$$0 \le r \le 1 \quad \text{corresponds to} \quad 0 \le s \le \tfrac{\varpi}{2},$$

we see that φ is defined on the interval $[0, \frac{\varpi}{2}]$. In Section 15.2 we will extend φ to a periodic function on $\mathbb{R}$, and in Section 15.3 we will further extend φ to a doubly periodic meromorphic function on $\mathbb{C}$.

In particular, when $n \ge 4$, the first n-division point of the lemniscate lies in the first quadrant. Since its arc length from the origin is $\frac{2\varpi}{n}$, Proposition 15.1.1 implies that the first n-division point is constructible by straightedge and compass if and only if

$$r_0 = \varphi\left(\tfrac{2\varpi}{n}\right)$$

is a constructible number. In Section 15.2 we will develop multiplication formulas for $\varphi(ns)$, $n \in \mathbb{Z}$ and use them to show that:

- $\varphi\left(\frac{2\varpi}{n}\right)$ is the root of a polynomial with coefficients in $\mathbb{Z}$.
- $\varphi\left(\frac{2\varpi}{n}\right)$ is constructible if and only if *all* n-division points are constructible by straightedge and compass.

In Section 15.5 we will consider the Galois group of the extension

$$\mathbb{Q}(i) \subset \mathbb{Q}\left(i, \varphi\left(\tfrac{2\varpi}{n}\right)\right).$$

The appearance of $i = \sqrt{-1}$ is unexpected but will make perfect sense once we study the complex multiplication formulas for $\varphi\left((n + im)s\right)$, $n + im \in \mathbb{Z}[i]$, in Section 15.4. Using this and some clever ideas of Eisenstein, we will then be able to prove Abel's theorem on the lemniscate.

Mathematical Notes

Here are comments about two topics from this section.

▪ **Integrals and Inverse Functions.** The definition of Abel's function $r = \varphi(s)$ involves an integral defining s in terms of r and then an inverse function to get r in terms of s.

The idea of an inverse function of an integral is more common than you might expect. For example, one standard definition of e^x is to first define the natural logarithm via the integral

$$\ln(x) = \int_1^x \frac{1}{t} \, dt, \quad x > 0,$$

and then define e^x to be the inverse function of $\ln(x)$. So e^x is the inverse function of an integral. Another example from calculus is the indefinite integral

$$\int \frac{1}{\sqrt{1 - x^2}} \, dx = \sin^{-1}(x) + C.$$

In terms of definite integrals, this can be written

$$\sin^{-1}(x) = \int_0^x \frac{1}{\sqrt{1 - t^2}} \, dt, \quad -1 \le x \le 1.$$

The inverse function of $\sin^{-1}(x)$ is of course $\sin(x)$. So $\sin(x)$ is the inverse function of an integral.

Now comes an intriguing idea. Suppose that we knew neither $\sin$ nor $\sin^{-1}$. How can we understand the integral $\int (1 - x^2)^{-1/2} \, dx$? One way would be to *define* $\sin(x)$ to be the inverse function of

$$x \mapsto \int_0^x \frac{1}{\sqrt{1 - t^2}} \, dt.$$

Furthermore, if we *define* $\cos(x) = \frac{d}{dx} \sin(x)$ and $\pi = 2 \int_0^1 (1 - t^2)^{-1/2} \, dt$, then all standard properties of $\sin(x)$ and $\cos(x)$ can be derived from these definitions.

One way to regard (15.4) is that Abel's function $\varphi(s)$ is obtained by applying the same idea to the integral $\int (1 - x^4)^{-1/2} \, dx$. There are many analogies between $\sin(x)$ and $\varphi(s)$, and it is possible to develop the properties of these functions in parallel. This is done nicely in [10, pp. 240–243]. See also [14, pp. 1–9].

▪ **Elliptic Integrals and Elliptic Functions.** The integral $\int (1 - x^4)^{-1/2} \, dx$ is an example of an *elliptic integral*, and *elliptic functions* are inverse functions of elliptic integrals. In general, an elliptic integral is an indefinite integral of the form

$$\text{(15.5)} \qquad \int \frac{A(x) + B(x)\sqrt{P(x)}}{C(x) + D(x)\sqrt{P(x)}} \, dx,$$

where $A(x), B(x), C(x), D(x)$ are polynomials in x, and $P(x)$ is a polynomial in x of degree 3 or 4. If $P(x)$ has degree 1 or 2, then the integral (15.5) can be evaluated using standard techniques of integration. So elliptic integrals, where $P(x)$ has degree 3 or 4, are the next integrals to consider. It follows that $\int (1 - x^4)^{-1/2} \, dx$ is an especially simple example of an elliptic integral.

We will say more about elliptic integrals and elliptic functions in the next section. For now, we conclude with another example of an elliptic integral. A standard ellipse with center at the origin is given by an equation of the form $\frac{x^2}{a^2} + \frac{y^2}{b^2} = 1$. Assume that $a = 1$ and $0 < b < 1$. If we set $k = \sqrt{1 - b^2}$, then in Exercise 4 you will show that the ellipse

$$x^2 + \frac{y^2}{b^2} = 1$$

has arc length given by

$$(15.6) \quad \int \sqrt{1 + \left(\frac{dy}{dx}\right)^2} \, dx = \int \sqrt{\frac{1 - k^2 x^2}{1 - x^2}} \, dx = \int \frac{\sqrt{(1 - x^2)(1 - k^2 x^2)}}{1 - x^2} \, dx.$$

This special case of (15.5) is where the "elliptic" in "elliptic integral" comes from.

Historical Notes

The lemniscate first appeared in the mathematical literature as part of the *ovals of Cassini*, described by the French astronomer Cassini in 1680. In Cartesian coordinates, the ovals are the family of curves defined by the equation

$$(15.7) \quad \left((x - a)^2 + y^2\right)\left((x + a)^2 + y^2\right) = b^4.$$

The lemniscate we've been studying corresponds to $a = b = 1/\sqrt{2}$ (Exercise 5). In general, $a < b$ gives a dumbbell-shaped curve and $a > b$ gives two ovals, as in the following picture:

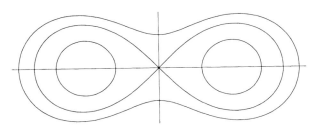

Unaware of Cassini's work, in 1694 Jacob (or James) Bernoulli gave the equation of the lemniscate as

$$xx + yy = a\sqrt{xx - yy}.$$

He described the curve as having "the form of a figure 8 on its side, as of a band folded into a knot, or of a lemniscus, or of a knot of a French ribbon." Here, "lemniscus" is a Latin word (taken from the Greek) meaning a hanging ribbon attached to the garland worn by the winner of an athletic contest.

Bernoulli was led to this curve by an indirect route. In 1691 he encountered the integral $2 \int_0^1 (1 - t^4)^{-1/2} \, dt$ (this should look familiar) in his study of the *elastic curve*. To represent this geometrically, he looked for a curve defined by an algebraic equation whose arc length equals $2 \int_0^1 (1 - t^4)^{-1/2} \, dt$. In 1694 he showed that the lemniscate has the desired arc length, using the polar description of the lemniscate, as we did earlier in the section. The reader should consult [2] for a discussion of the elastic curve and Bernoulli's priority dispute with his brother Johann, who independently discovered the lemniscate in 1694 in a different context.

Bernoulli's use of polar coordinates to compute the arc length of the lemniscate represents the first use of arc length in polar coordinates. It is ironic that calculus students study the lemniscate in one part of the course and arc length in polar coordinates in another, but they never put the two together, since the resulting integral can't be evaluated by the usual methods of calculus.

Our discussion shows that the lemniscate and its arc length were well known by the beginning of the eighteenth century. Thus Abel's theorem on dividing the lemniscate into arcs of equal length deals with a topic familiar to the mathematical community of the time.

We will say more about elliptic integrals and elliptic functions in the Historical Notes to the next section.

Exercises for Section 15.1

Exercise 1. Prove that the numbers described in Abel's theorem at the beginning of the chapter are precisely those in Theorem 10.2.1, provided we replace "product of several numbers" with "product of distinct numbers" in Abel's statement of the theorem.

Exercise 2. Show that in polar coordinates, the equation of the lemniscate is $r^2 = \cos(2\theta)$.

Exercise 3. Prove that the integrals $\int_0^{\pm 1} (1 - t^4)^{-1/2} \, dt$ converge.

Exercise 4. Prove the arc length formula stated in (15.6).

Exercise 5. Show that (15.7) reduces to $(x^2 + y^2)^2 = x^2 - y^2$ when $a = b = 1/\sqrt{2}$.

Exercise 6. Let $n > 0$ be an odd integer, and assume that the n-division points of the lemniscate can be constructed with straightedge and compass. Prove that the same is true for the $2n$-division points. Your proof should include a picture.

Exercise 7. Recall that in Greek geometry, the ellipse is defined to be the locus of all points whose *sum* of distances to two given points is constant. Suppose instead we consider the locus of all points whose *product* of distances to two given points is constant. Show that this leads to (15.7) when the given points are $(a, 0)$, $(-a, 0)$ and the constant is b^4.

15.2   THE LEMNISCATIC FUNCTION

In (15.4) we defined Abel's function $\varphi(s)$ by

(15.8) $$r = \varphi(s) \iff s = \int_0^r \frac{1}{\sqrt{1 - t^4}}\, dt.$$

Since s represents arc length from the origin along the first quadrant portion of the lemniscate, we see that $\varphi(s)$ is defined on $[0, \frac{\varpi}{2}]$, where

$$\varpi = 2 \int_0^1 \frac{1}{\sqrt{1 - t^4}}\, dt.$$

In this section, we will extend $\varphi(s)$ to a function on $\mathbb{R}$ of period 2ϖ and show that it satisfies some remarkable addition and multiplication formulas. We will also apply these formulas to straightedge-and-compass constructions on the lemniscate.

A. A Periodic Function. Our first task is to define $\varphi(s)$ as a function of period 2ϖ on $\mathbb{R}$. We will do this by extending the arc length interpretation of $\varphi(s)$ given in Section 15.1.

The arc length parametrization of the lemniscate is defined by sending a real number s to the point P on the lemniscate such that:

- If $s = 0$, then P is the origin.
- If $s > 0$, then move from the origin into the first quadrant portion of the lemniscate and continue along the curve until we reach the point P whose cumulative arc length from the origin is s.
- If $s < 0$, then move from the origin into the third quadrant portion and continue until we reach the point P whose cumulative arc length from the origin is $-s$.

We call s the *signed arc length variable* of the lemniscate. When $|s|$ is large, we may need to loop around several times before reaching the point P. Since the total arc length of the lemniscate is 2ϖ, we see that s and $s \pm 2\varpi$ give the same point on the lemniscate for any $s \in \mathbb{R}$. This is similar to measuring angles on the unit circle, where s and $s \pm 2\pi$ give the same point on the circle.

The lemniscate is $r^2 = \cos(2\theta)$ in polar coordinates. Recall that r is allowed to be negative as well as positive or zero. We will restrict θ to lie in $[-\frac{\pi}{4}, \frac{\pi}{4}]$, so that $0 \le r \le 1$ gives the right half of the lemniscate and $-1 \le r \le 0$ gives the left half. We call r the *polar distance* of the corresponding point on the lemniscate. Strictly speaking, r is really the *signed polar distance*, since r is negative on the left half of the lemniscate. We will use the shorter term "polar distance" for simplicity.

Now consider (15.8). It is easy to see that the signed arc length s satisfies

$$s = \int_0^r \frac{1}{\sqrt{1 - t^4}}\, dt$$

for $-\frac{\varpi}{2} \le s \le \frac{\varpi}{2}$ and $-1 \le r \le 1$. This implies that (15.8) can be used to define $\varphi(s)$ for $-\frac{\varpi}{2} \le s \le \frac{\varpi}{2}$. In other words, for s in this range, Abel's function $\varphi(s)$

is simply the polar distance (with the above convention on r) of the point on the lemniscate with signed arc length s.

It is now easy to extend φ to all of $\mathbb{R}$: given $s \in \mathbb{R}$, $\varphi(s)$ is the polar distance of the point on the lemniscate whose signed arc length from the origin is s. Thus

$$\varphi(s) = r,$$

where s and r are related according to the diagram

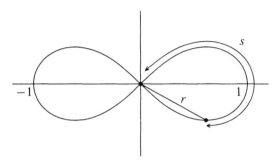

Note that $\varphi(s)$ has period 2ϖ, since s and $s + 2\varpi$ give the same point on the lemniscate. Furthermore, $\varphi(s)$ also satisfies the following identities:

$$(15.9) \qquad \begin{aligned} \varphi(-s) &= -\varphi(s), \\ \varphi(\varpi - s) &= \varphi(s). \end{aligned}$$

The first follows because s and $-s$ correspond to points on the lemniscate symmetric about the origin, and the second follows because s and $\varpi - s$ correspond to points symmetric about the x-axis (recall that each half of the lemniscate has length ϖ). See Exercise 1 for the details. Using the arc length interpretation of $\varphi(s)$, one can show that $\varphi(s)$ is infinitely differentiable for all $s \in \mathbb{R}$, though we omit the proof.

The function $\sin\left(\frac{\pi}{\varpi}s\right)$ has the same period and amplitude as $\varphi(s)$ (check this). If we plot $\varphi(s)$ and $\sin\left(\frac{\pi}{\varpi}s\right)$ for $0 \le s \le 2\varpi$, then we get the following graphs:

Abel's Function   The Sine Function

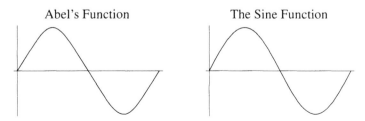

The function $\sin(x)$ satisfies identities similar to (15.9), but the full theory of $\sin(x)$ requires $\sin'(x) = \cos(x)$ as well. The same is true for $\varphi(s)$, where we will use $\varphi'(s)$. We will also need the following crucial identity.

Proposition 15.2.1. *Let $\varphi(s)$ be defined as above. Then*

$$\varphi'^2(s) = 1 - \varphi^4(s).$$

Proof. Using (15.9) and the periodicity of $\varphi(s)$, you will show in Exercise 2 that it suffices to prove that

$$\varphi'(s) = \sqrt{1 - \varphi^4(s)}, \quad 0 \le s \le \tfrac{\varpi}{2}.$$

To derive this equation, first observe that (15.8) gives the identity

$$s = \int_0^{\varphi(s)} \frac{1}{\sqrt{1 - t^4}} \, dt, \quad 0 \le s \le \tfrac{\varpi}{2}.$$

Differentiating each side with respect to s and using the Chain Rule and the Fundamental Theorem of Calculus, we obtain

$$1 = \frac{1}{\sqrt{1 - \varphi^4(s)}} \, \varphi'(s), \quad 0 \le s < \tfrac{\varpi}{2}$$

(be sure you understand why $s = \tfrac{\varpi}{2}$ is excluded). It follows that

$$\varphi'(s) = \sqrt{1 - \varphi^4(s)}, \quad 0 \le s < \tfrac{\varpi}{2}.$$

For $s = \tfrac{\varpi}{2}$, note that $\sqrt{1 - \varphi^4(s)}$ vanishes at $\tfrac{\varpi}{2}$ since $\varphi(\tfrac{\varpi}{2}) = 1$. Since 1 is the maximum value of $\varphi(s)$ (can you explain why?), we see that $\varphi'(s)$ also vanishes at $\tfrac{\varpi}{2}$. This completes the proof. $\square$

Other properties of $\varphi'(s)$ will be developed in Exercise 3, and in Exercise 4 you will adapt the method used in Proposition 15.2.1 to derive the standard trigonometric identity $\cos^2(x) = 1 - \sin^2(x)$.

B. Addition Laws. The addition law for $\sin(x)$ states that

$$\sin(x + y) = \sin(x)\cos(y) + \cos(x)\sin(y).$$

For $\varphi(x)$, the addition law goes back to Euler, who in 1753 proved the identity

(15.10)
$$\int_0^\alpha \frac{1}{\sqrt{1 - t^4}} \, dt + \int_0^\beta \frac{1}{\sqrt{1 - t^4}} \, dt = \int_0^\gamma \frac{1}{\sqrt{1 - t^4}} \, dt$$
$$\text{when } \alpha, \beta \in [0, 1] \text{ and } \gamma = \frac{\alpha\sqrt{1 - \beta^4} + \beta\sqrt{1 - \alpha^4}}{1 + \alpha^2\beta^2}.$$

To state this in terms of φ, let x, y, and z represent the three integrals in (15.10), so that $\varphi(x) = \alpha$, $\varphi(y) = \beta$, and $\varphi(z) = \gamma$. Then $x + y = z$ implies that

$$\varphi(x + y) = \varphi(z) = \gamma = \frac{\alpha\sqrt{1 - \beta^4} + \beta\sqrt{1 - \alpha^4}}{1 + \alpha^2\beta^2},$$

which when combined with $\varphi(x) = \alpha$ and $\varphi(y) = \beta$ gives

$$(15.11) \qquad \varphi(x+y) = \frac{\varphi(x)\sqrt{1-\varphi^4(y)} + \varphi(y)\sqrt{1-\varphi^4(x)}}{1 + \varphi^2(x)\varphi^2(y)}.$$

Furthermore, $\sqrt{1-\varphi^4(x)} = \varphi'(x)$ for $0 \le x \le \frac{\varpi}{2}$ by Proposition 15.2.1. Thus

$$(15.12) \qquad \varphi(x+y) = \frac{\varphi(x)\varphi'(y) + \varphi(y)\varphi'(x)}{1 + \varphi^2(x)\varphi^2(y)}.$$

Rather than use Euler's result, Abel gave a different proof that (15.12) holds for all $x, y \in \mathbb{R}$. You will explore Abel's argument in Exercise 5.

In Exercise 6 you will use $\varphi(-x) = -\varphi(x)$ to show that (15.12) implies the subtraction law

$$\varphi(x-y) = \frac{\varphi(x)\varphi'(y) - \varphi(y)\varphi'(x)}{1 + \varphi^2(x)\varphi^2(y)}.$$

When this is combined with (15.12), we easily obtain the identity

$$(15.13) \qquad \varphi(x+y) + \varphi(x-y) = \frac{2\varphi(x)\varphi'(y)}{1 + \varphi^2(x)\varphi^2(y)}.$$

This will be useful later in the section.

The addition laws give some nice straightedge-and-compass constructions.

Example 15.2.2. Let's divide the lemniscate into eight pieces of length $\frac{2\varpi}{8} = \frac{\varpi}{4}$. Here is a picture of $r_0 = \varphi\left(\frac{\varpi}{4}\right)$ and the 8-division points:

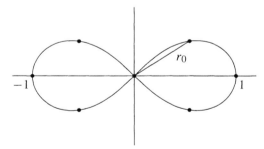

This picture and Proposition 15.1.1 show that to construct the 8-division points, we need only construct r_0. Since $\varphi\left(\frac{\varpi}{2}\right) = 1$, the addition law (15.11) implies that

$$1 = \varphi\left(\tfrac{\varpi}{2}\right) = \varphi\left(\tfrac{\varpi}{4} + \tfrac{\varpi}{4}\right) = \frac{2\varphi\left(\frac{\varpi}{4}\right)\sqrt{1 - \varphi^4\left(\frac{\varpi}{4}\right)}}{1 + \varphi^4\left(\frac{\varpi}{4}\right)} = \frac{2r_0\sqrt{1 - r_0^4}}{1 + r_0^4}.$$

Solving this equation in *Maple* or *Mathematica* shows that the unique real positive solution is given by

$$r_0 = \sqrt{\sqrt{2} - 1} \approx .643594.$$

This is obviously constructible and hence gives the desired construction.   ◁▷

The reasoning behind Example 15.2.2 can be generalized as follows.

Proposition 15.2.3. *If $\varphi(x_0)$ is constructible, then so is $\varphi(\frac{x_0}{2})$.*

Proof. Setting $x = y$ in (15.12) gives the duplication formula

$$(15.14) \qquad \varphi(2x) = \frac{2\varphi(x)\varphi'(x)}{1 + \varphi^4(x)}.$$

Let $r_0 = \varphi(\frac{x_0}{2})$ and $a = \varphi(x_0)$. Then (15.14) and $\varphi'^2(\frac{x_0}{2}) = 1 - \varphi^4(\frac{x_0}{2})$ imply that

$$a^2 = \Big(\frac{2\,r_0\,\varphi'(\frac{x_0}{2})}{1 + r_0^4}\Big)^2 = \frac{4r_0^2(1 - r_0^4)}{(1 + r_0^4)^2}.$$

To solve this equation for r_0, let $t \in \mathbb{C}$ satisfy

$$(15.15) \qquad t^2 = \frac{2ir_0^2}{1 - r_0^4}, \quad i = \sqrt{-1}$$

and observe that

$$(15.16) \qquad \frac{-2it^2}{1 - t^4} = \frac{-2i\,\frac{2ir_0^2}{1-r_0^4}}{1 - \big(\frac{2ir_0^2}{1-r_0^4}\big)^2} = \frac{4r_0^2(1 - r_0^4)}{(1 + r_0^4)^2} = a^2.$$

Solving (15.16) for t^2 by the quadratic formula shows that t^2 is constructible because a is, and then solving (15.15) for r_0 completes the proof. $\square$

The formulas (15.15) and (15.16) in the above proof seem to come out of nowhere. In Section 15.4 we will use *complex multiplication* and $2 = (1+i)(1-i)$ to "factor" the duplication formula for $\varphi(2x)$ into (15.15) and (15.16). So these formulas will eventually make perfect sense.

Here is a slightly more complicated example.

Example 15.2.4. Dividing the lemniscate into six pieces of equal length gives the following picture:

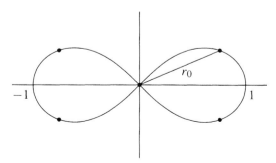

To compute $r_0 = \varphi\left(\frac{2\varpi}{6}\right) = \varphi\left(\frac{\varpi}{3}\right)$, first observe that (15.13) with $2x$ and x in place of x and y gives

$$\varphi(3x) + \varphi(x) = \varphi(2x + x) + \varphi(2x - x) = \frac{2\varphi(2x)\varphi'(x)}{1 + \varphi^2(2x)\varphi^2(x)}$$

$$= \frac{2\frac{2\varphi(x)\varphi'(x)}{1+\varphi^4(x)}\varphi'(x)}{1 + \left(\frac{2\varphi(x)\varphi'(x)}{1+\varphi^4(x)}\right)^2\varphi^2(x)},$$

where the last line uses (15.14). Using $\varphi'^2(x) = 1 - \varphi^4(x)$ and a bit of algebra, we obtain the tripling formula

(15.17) $$\varphi(3x) = -\varphi(x)\frac{\varphi^8(x) + 6\varphi^4(x) - 3}{1 + 6\varphi^4(x) - 3\varphi^8(x)}.$$

Since $\varphi(\varpi) = 0$, substituting $x = \frac{\varpi}{3}$ into (15.17) shows that $r_0 = \varphi\left(\frac{\varpi}{3}\right)$ satisfies

$$r_0^8 + 6r_0^4 - 3 = 0,$$

which is easily seen to have the unique real positive solution

$$r_0 = \sqrt[4]{2\sqrt{3} - 3} \approx .825379.$$

This is clearly constructible and hence gives the desired straightedge-and-compass construction.   ◁▷

C. Multiplication by Integers.

The doubling and tripling formulas

$$\varphi(2x) = \frac{2\varphi(x)\varphi'(x)}{1 + \varphi^4(x)},$$

$$\varphi(3x) = -\varphi(x)\frac{\varphi^8(x) + 6\varphi^4(x) - 3}{1 + 6\varphi^4(x) - 3\varphi^8(x)}$$

from (15.14) and (15.17) can be generalized to formulas that express $\varphi(nx)$ in terms of $\varphi(x)$ and $\varphi'(x)$ for any positive integer n.

Theorem 15.2.5. *Given an integer $n > 0$, there are relatively prime polynomials $P_n(u), Q_n(u) \in \mathbb{Z}[u]$ such that if n is odd, then*

$$\varphi(nx) = \varphi(x)\frac{P_n(\varphi^4(x))}{Q_n(\varphi^4(x))},$$

and if n is even, then

$$\varphi(nx) = \varphi(x)\frac{P_n(\varphi^4(x))}{Q_n(\varphi^4(x))}\varphi'(x).$$

Furthermore, $Q_n(0) = 1$.

Proof. We will prove the theorem by induction on n. Setting $P_1(u) = Q_1(u) = 1$ gives the desired formula for $n = 1$. For $n = 2$, note that (15.14) can be written

$$\varphi(2x) = \varphi(x) \frac{2}{1 + \varphi^4(x)} \varphi'(x).$$

Thus the theorem holds for $n = 2$ with $P_2(u) = 1$, $Q_2(u) = 1 + u$. Now assume that it holds for $n - 1$ and n. Using (15.13) with nx and x in place of x and y, we obtain

$$\varphi\big((n + 1)x\big) = -\varphi\big((n - 1)x\big) + \frac{2\varphi(nx)\varphi'(x)}{1 + \varphi^2(nx)\varphi^2(x)}.$$

If n is even, then $n - 1$ is odd, so that our inductive hypothesis implies that

$$\varphi\big((n + 1)x\big) = -\left(\varphi(x) \frac{P_{n-1}\big(\varphi^4(x)\big)}{Q_{n-1}\big(\varphi^4(x)\big)}\right) + \frac{2\left(\varphi(x) \dfrac{P_n\big(\varphi^4(x)\big)}{Q_n\big(\varphi^4(x)\big)} \varphi'(x)\right)\varphi'(x)}{1 + \left(\varphi(x) \dfrac{P_n\big(\varphi^4(x)\big)}{Q_n\big(\varphi^4(x)\big)} \varphi'(x)\right)^2 \varphi^2(x)}.$$

Using $\varphi'^2(x) = 1 - \varphi^4(x)$ and clearing denominators, this simplifies to

$$\varphi\big((n + 1)x\big) = \varphi(x) \frac{P_{n+1}\big(\varphi^4(x)\big)}{Q_{n+1}\big(\varphi^4(x)\big)},$$

where

(15.18) $Q_{n+1}(u) = Q_{n-1}(u)\big(Q_n^2(u) + u P_n^2(u)(1 - u)\big)$

and $P_{n+1}(u)$ is given by a similar recursive formula (see Exercise 7). It follows that $P_{n+1}(u), Q_{n+1}(u) \in \mathbb{Z}[u]$ by our inductive hypothesis. Note also that $Q_{n+1}(0) = 1$ follows from $Q_n(0) = Q_{n-1}(0) = 1$. Finally, dividing $P_{n+1}(u)$ and $Q_{n+1}(u)$ by their greatest common divisor shows that we may assume that they are relatively prime in $\mathbb{Z}[u]$. In Exercise 7 you will show $Q_{n+1}(0) = 1$ continues to hold, after multiplying $P_{n+1}(u), Q_{n+1}(u)$ by -1 if necessary.

The case when n is odd is similar and will be covered in Exercise 7.   □

Theorem 15.2.5 has some nice consequences concerning the division points on the lemniscate. The polar distances of the n-division points are

$$\varphi\big(m \tfrac{2\varpi}{n}\big), \quad m = 0, 1, \ldots, n - 1.$$

When n is odd, the periodicity of φ and Theorem 15.2.5 imply that

$$0 = \varphi(m \cdot 2\varpi) = \varphi\big(n \cdot m \tfrac{2\varpi}{n}\big) = \varphi\big(m \tfrac{2\varpi}{n}\big) \frac{P_n\big(\varphi^4\big(m \tfrac{2\varpi}{n}\big)\big)}{Q_n\big(\varphi^4\big(m \tfrac{2\varpi}{n}\big)\big)},$$

so that the polar distance $\varphi\big(m \tfrac{2\varpi}{n}\big)$ is a root of $u P_n(u^4)$ when n is odd. In Exercise 8 you will show that when n is even, the polar distances are roots of $u P_n(u^4)(1 - u^2)$. We call these polynomials the *n-division polynomials*. We have thus proved the following corollary of Theorem 15.2.5.

Corollary 15.2.6. *Let $n \in \mathbb{Z}$ be positive. Then the polar distances of the n-division points of the lemniscate are roots of the n-division polynomials defined above.*   □

We also have the following result about straightedge-and-compass constructions.

Corollary 15.2.7. *Let n be a positive integer such that $\varphi\left(\frac{2\varpi}{n}\right)$ is constructible.*

(a) *$\varphi\left(m\frac{2\varpi}{n}\right)$ is constructible for every $m \in \mathbb{Z}$.*

(b) *The n-division points of the lemniscate are constructible by straightedge and compass.*

(c) *If in addition $\varphi\left(\frac{2\varpi}{m}\right)$ is constructible for a positive integer m, then so is $\varphi\left(\frac{2\varpi}{N}\right)$, where $N = \operatorname{lcm}(n, m)$.*

Proof. If $\varphi\left(\frac{2\varpi}{n}\right)$ is constructible, then so is $\varphi'\left(\frac{2\varpi}{n}\right)$ since $\varphi'(x) = \pm\sqrt{1 - \varphi^4(x)}$. Part (a) is obvious for $n = 1$ and 2, so we may assume that $n > 2$.

When $m > 0$, Theorem 15.2.5 implies that $\varphi\left(m\frac{2\varpi}{n}\right)$ is a rational function of $\varphi\left(\frac{2\varpi}{n}\right)$ and $\varphi'\left(\frac{2\varpi}{n}\right)$ with coefficients in $\mathbb{Z}$. In Exercise 9 you will show that the denominator is nonvanishing, since $n > 2$ and the polynomials $P_m(u)$, $Q_m(u)$ in Theorem 15.2.5 are relatively prime. Hence $\varphi\left(m\frac{2\varpi}{n}\right)$ is constructible, since the constructible numbers form a subfield of $\mathbb{C}$.

The case $m = 0$ is obvious, and $m < 0$ follows from $m > 0$ because φ is an odd function. This completes the proof of part (a).

Part (b) follows immediately from part (a) and Proposition 15.1.1.

For part (c), let $d = \gcd(n, m)$. Then $N = \operatorname{lcm}(n, m) = \frac{nm}{d}$. It follows that if integers μ, ν satisfy $\mu n + \nu m = d$, then

$$\mu\frac{2\varpi}{n} + \nu\frac{2\varpi}{m} = (\mu m + \nu n)\frac{2\varpi}{nm} = d\frac{2\varpi}{nm} = \frac{2\varpi}{N}.$$

By part (a), $\varphi\left(\mu\frac{2\varpi}{n}\right)$ and $\varphi\left(\nu\frac{2\varpi}{m}\right)$ are constructible, and—as above—the same is true for $\varphi'\left(\mu\frac{2\varpi}{n}\right)$ and $\varphi'\left(\nu\frac{2\varpi}{m}\right)$. Then the addition law (15.12) expresses $\varphi\left(\frac{2\varpi}{N}\right)$ as a rational expression with coefficients in $\mathbb{Z}$ in the constructible numbers given by the values of φ and φ' at $\mu\frac{2\varpi}{n}$ and $\nu\frac{2\varpi}{m}$. Since the denominator of this rational expression is

$$1 + \varphi^2\left(\mu\frac{2\varpi}{n}\right)\varphi^2\left(\nu\frac{2\varpi}{m}\right) \neq 0,$$

it follows that $\varphi\left(\frac{2\varpi}{N}\right)$ is constructible.   □

Parts (b) and (c) of Corollary 15.2.7 imply that if the n-division points and m-division points of the lemniscate are constructible by straightedge and compass, then the same is true of the N-division points for $N = \operatorname{lcm}(n, m)$. This fact will be useful in Section 15.5.

Here are some applications of Corollary 15.2.7.

Example 15.2.8. Since $\varphi(2\varpi) = 0$, Proposition 15.2.3 implies that $\varphi\left(\frac{2\varpi}{2^n}\right)$ is constructible for $n \geq 0$. Then part (b) of Corollary 15.2.7 shows that the 2^n-division points of the lemniscate can be constructed by straightedge and compass.   ◁▷

Example 15.2.9. When $n = 5$, one can show that

$$\varphi(5x) = \varphi(x) \frac{P_5(\varphi^4(x))}{Q_5(\varphi^4(x))}, \quad \text{where}$$

(15.19)

$$P_5(u) = u^6 + 50u^5 - 125u^4 + 300u^3 - 105u^2 - 62u + 5,$$

$$Q_5(u) = 1 + 50u - 125u^2 + 300u^3 - 105u^4 - 62u^5 + 5u^6$$

(see [12, p. 82]). Note the "reverse symmetry" of the coefficients of $P_5(u)$ and $Q_5(u)$. For the 5-division points of the lemniscate, the discussion preceding Corollary 15.2.6 implies that $r_0 = \varphi\left(\frac{2\varpi}{5}\right)$ is a root of the 5-division polynomial $u P_5(u^4)$. Thus

$$0 = r_0 P_5(r_0^4) = r_0(r_0^{24} + 50r_0^{20} - 125r_0^{16} + 300r_0^{12} - 105r_0^8 - 62r_0^4 + 5).$$

You will show in Exercise 10 that the real positive solutions are constructible, though this is not obvious from the above equation. By Corollary 15.2.7, it follows that the 5-division points are constructible by straightedge and compass.   ◁▷

This discussion makes it clear that understanding the n-division points of the lemniscate is intimately related to the multiplication formula for $\varphi(nx)$. But to unleash the full power of these formulas, we will need to extend φ to a function of a complex variable. This is will be done in the next section.

Historical Notes

Although the link between arc length and the lemniscate goes back to Bernoulli, the first person to make substantial progress in this area was Fagnano. In 1718 he proved the case $\alpha = \beta$ of Euler's addition law (15.10), namely

$$2\int_0^\alpha \frac{1}{\sqrt{1 - t^4}} \, dt = \int_0^\gamma \frac{1}{\sqrt{1 - t^4}} \, dt \quad \text{when} \quad \gamma = \frac{2\sqrt{1 - \alpha^4}}{1 + \alpha^4}.$$

Using this and other results, Fagnano was able to divide one arch of the lemniscate into two, three, and five segments of equal length by straightedge and compass. Fagnano's results and methods are discussed in [1].

Things got really interesting when Fagnano's papers were submitted to the Berlin Academy as part of his application for membership. Euler was asked to read these papers in December 1751, and by 1753 he was able to show that Fagnano's duplication formula was a special case of (15.10). More importantly, he also realized that $\sqrt{1 - t^4}$ could be replaced by $\sqrt{P(t)}$, where $P(t)$ is any separable polynomial of degree 4 with real coefficients. This led to the theory of *elliptic integrals*, which was developed at great length by Lagrange and Legendre. Eventually the integrals were put in the standard form

(15.20)

$$\int \frac{1}{\sqrt{1 - k^2 \sin^2 \theta}} \, d\theta,$$

which after the substitution $t = \sin\theta$ gives

(15.21)
$$\int \frac{1}{\sqrt{(1-t^2)(1-k^2t^2)}} \, dt.$$

We call k the *modulus*, so that $\int (1-t^4)^{-1/2} \, dt$ corresponds to the modulus $k = i$.

The first person to consider the inverse function of $\int (1-t^4)^{-1/2} \, dt$ was Gauss in 1797, though this work was not published until after his death in 1855. Abel and Jacobi introduced the inverse functions of elliptic integrals in 1827. In Abel's great paper *Recherches sur les fonctions elliptiques* [Abel, Vol. I, pp. 263–388], he considered the inverse function of an elliptic integral of the form

$$\int \frac{1}{\sqrt{(1-c^2t^2)(1+e^2t^2)}} \, dt,$$

so that $c = e = 1$ gives the lemniscatic function we've been studying. Jacobi, on the other hand, used the integral (15.20) and wrote its inverse function as $\theta = \operatorname{am} u$. Thus $\sin\theta = \sin\operatorname{am} u$ is the inverse function of (15.21). These days, we write $\sin\operatorname{am} u$ as $\operatorname{sn}(u,k)$, or simply $\operatorname{sn}(u)$ if the modulus is understood, though *Mathematica* writes $\operatorname{sn}(u,k)$ as `JacobiSN[u, k²]`. In the text, we used `JacobiSN[u, −1]` to draw the graph of the lemniscatic function $\varphi(u) = \operatorname{sn}(u,i)$.

One of the critical discoveries of Gauss, Abel, and Jacobi is that inverse functions of elliptic integrals are doubly periodic functions of a complex variable. We will consider a special case of this in the next section. More on the history of elliptic integrals can be found in [1], [5, pp. 3–16], and [10, pp. 267–268]. A nice introduction to the duplication formula (15.14) appears in [14].

Exercises for Section 15.2

Exercise 1. Give a careful proof of (15.9) using the hints given in the text.

Exercise 2. Supply the details needed to complete the proof of Proposition 15.2.1.

Exercise 3. Here are some useful properties of $\varphi'(s)$.
(a) $\varphi(s)$ has period 2ϖ. Explain why this implies that the same is true for $\varphi'(s)$.
(b) $\varphi(s)$ is an odd function by (15.9). Explain why this implies that $\varphi'(s)$ is even.
(c) Use (15.9) to prove that $\varphi'(\varpi - s) = -\varphi'(s)$.
(d) Use Proposition 15.2.1 to prove that $\varphi''(s) = -2\varphi^3(s)$.

Exercise 4. Suppose that we define $\sin(x)$ by $y = \sin(x) \iff x = \int_0^y (1-t^2)^{-1/2} \, dt$. Then define $\cos(x)$ to be $\sin'(x)$. Use the method of Proposition 15.2.1 to prove the standard trigonometric identity $\cos^2(x) = 1 - \sin^2(x)$.

Exercise 5. Here is Abel's proof of the addition law for φ.
(a) Let $g(x, y)$ be differentiable on $\mathbb{R}^2$, and set $h(u, v) = g\left(\frac{1}{2}(u+v), \frac{1}{2}(u-v)\right)$. Use the Chain Rule to prove that

$$\frac{\partial h}{\partial v}(u, v) = \frac{1}{2}\frac{\partial g}{\partial x}\left(\tfrac{1}{2}(u+v), \tfrac{1}{2}(u-v)\right) - \frac{1}{2}\frac{\partial g}{\partial y}\left(\tfrac{1}{2}(u+v), \tfrac{1}{2}(u-v)\right).$$

(b) Use part (a) to show that $g(x, y) = g(x + y, 0)$ on $\mathbb{R}^2$ if and only if $\frac{\partial g}{\partial x} = \frac{\partial g}{\partial y}$ on $\mathbb{R}^2$.

(c) Prove the addition law for φ by applying part (b) to

$$g(x, y) = \frac{\varphi(x)\varphi'(y) + \varphi(y)\varphi'(x)}{1 + \varphi^2(x)\varphi^2(y)}.$$

Part (d) of Exercise 3 will be useful.

Exercise 6. Show that the subtraction law

$$\varphi(x - y) = \frac{\varphi(x)\varphi'(y) - \varphi(y)\varphi'(x)}{1 + \varphi^2(x)\varphi^2(y)}$$

follows from the addition law together with (15.9) and Exercise 3.

Exercise 7. The proof of Theorem 15.2.5 uses induction on n.

(a) Assume that n is even. In (15.18), we gave a formula for $Q_{n+1}(u)$ in terms of $Q_n(u)$ and $Q_{n-1}(u)$. Derive the corresponding formula for $P_{n+1}(u)$.

(b) Suppose that polynomials $P_n(u)$, $Q_n(u)$ satisfy all of the conditions of the theorem except for the requirement that they be relatively prime. Since $\mathbb{Z}[u]$ is a UFD, we can write $P_n(u) = C_n(u)\widetilde{P}_n(u)$, $Q_n(u) = C_n(u)\widetilde{Q}_n(u)$, where $C_n(u)$, $\widetilde{P}_n(u)$, $\widetilde{Q}_n(u) \in \mathbb{Z}[u]$ and $\widetilde{P}_n(u)$, $\widetilde{Q}_n(u)$ are relatively prime. Prove that we can assume that $\widetilde{Q}_n(0) = 1$ and that $\widetilde{P}_n(u)$, $\widetilde{Q}_n(u)$ satisfy all conditions of Theorem 15.2.5.

(c) Complete the inductive step of the proof when n is odd.

Exercise 8. Let n be even, and let $P_n(u)$ be the polynomial from Theorem 15.2.5. Complete the proof of Corollary 15.2.6 by showing that the polar distances of the n-division points of the lemniscate are roots of $u P_n(u^4)(1 - u^2)$.

Exercise 9. This exercise is concerned with the proof of Corollary 15.2.7.

(a) Suppose that $P(u)$, $Q(u) \in \mathbb{Z}[u]$ are relatively prime and $Q(0) = 1$. Prove that $u P(u^4)$ and $Q(u^4)$ have no common roots in any extension of $\mathbb{Q}$.

(b) Fix x in $\mathbb{R}$ and $m > 0$ in $\mathbb{Z}$, and let $P_m(u)$, $Q_m(u) \in \mathbb{Z}[u]$ be as in Theorem 15.2.5. Thus $\varphi(mx)Q_m(\varphi^4(x)) = \varphi(x)P_m(\varphi^4(x))$. Prove that $Q_m(\varphi^4(x)) \neq 0$ when $\varphi(x) \neq 0$.

(c) Show that $\varphi(\frac{2\varpi}{n}) \neq 0$ when $n > 2$ is in $\mathbb{Z}$ and conclude that $Q_m(\varphi^4(\frac{2\varpi}{n})) \neq 0$.

Exercise 10. The polar distances of the 5-division points of the lemniscate satisfy the equation

$$0 = r_0(r_0^{24} + 50r_0^{20} - 125r_0^{16} + 300r_0^{12} - 105r_0^8 - 62r_0^4 + 5).$$

This equation was first derived by Fagnano in 1718.

(a) Show that the r_0 corresponding to the 10-division points also satisfy this equation.

(b) Use *Maple* or *Mathematica* to show that this equation factors as

$$0 = r_0(r_0^8 - 2r_0^4 + 5)(r_0^{16} + 52r_0^{12} - 26r_0^8 - 12r_0^4 + 1)$$

and that the only positive real solutions are

$$\sqrt[4]{-13 + 6\sqrt{5} \pm 2\sqrt{85 - 38\sqrt{5}}}.$$

Explain (with a picture) how these solutions relate to the 5- and 10-division points.

Exercise 11. Use $\sin(x + y) = \sin x \cos y + \sin y \cos x$ to show that if $\alpha, \beta \in [0, 1]$, then

$$\int_0^\alpha \frac{1}{\sqrt{1 - t^2}}\, dt + \int_0^\beta \frac{1}{\sqrt{1 - t^2}}\, dt = \int_0^\gamma \frac{1}{\sqrt{1 - t^2}}\, dt,$$

where γ is the real number defined by

$$\gamma = \alpha\sqrt{1 - \beta^2} + \beta\sqrt{1 - \alpha^2}.$$

Note the similarity to (15.10).

Exercise 12. Show that the substitution $t = \sin\theta$ transforms (15.20) into (15.21), and use this to prove carefully that $\varphi(u) = \sin\operatorname{am}(u)$ when the modulus is $k = i$.

15.3   THE COMPLEX LEMNISCATIC FUNCTION

Corollary 15.2.6 implies that the polar distances $r = \varphi\left(m\frac{2\varpi}{n}\right)$ of the n-division points of the lemniscate are roots of the n-division polynomials. To prove Abel's theorem on the lemniscate, we need to represent *all* roots of these polynomials using φ. Since many of the roots are complex, this requires that we follow Gauss and Abel and extend φ to a function defined on $\mathbb{C}$.

Abel began by considering $\varphi(iy)$ for $y \in \mathbb{R}$. We know that $r = \varphi(y)$ is the inverse function of $y = \int_0^r (1 - t^4)^{-1/2}\, dt$. The change of variables $t = iu$ shows that

$$\int_0^{ir} \frac{1}{\sqrt{1 - t^4}}\, dt = i\int_0^r \frac{1}{\sqrt{1 - u^4}}\, du = iy.$$

This suggests that $\varphi(iy)$ can be defined to be $\varphi(iy) = ir = i\varphi(y)$. Then Abel used the addition law to define $\varphi(x + iy)$ as

$$\varphi(x + iy) = \frac{\varphi(x)\varphi'(iy) + \varphi(iy)\varphi'(x)}{1 + \varphi^2(x)\varphi^2(iy)}.$$

Since $\varphi(iy) = i\varphi(y)$ easily implies that $\varphi'(iy) = \varphi(y)$ (see Exercise 1), the formula for $\varphi(x + iy)$ simplifies to

$$(15.22) \qquad \varphi(z) = \varphi(x + iy) = \frac{\varphi(x)\varphi'(y) + i\varphi(y)\varphi'(x)}{1 - \varphi^2(x)\varphi^2(y)}.$$

To make Abel's approach rigorous, we will *define* $\varphi(z)$ using (15.22). Over $\mathbb{R}$, $\varphi(x)$ is periodic and defined everywhere; over $\mathbb{C}$, we will see that $\varphi(z)$ is doubly periodic and has poles. The properties of $\varphi(z)$ will play a crucial role in Sections 15.4 and 15.5.

This section will assume familiarity with standard topics from complex analysis, including the Cauchy–Riemann equations and Laurent series. We will refer to [11], though the results we need are in most introductory texts on the subject.

A.   A Doubly Periodic Function. As above, we define $\varphi(z) = \varphi(x + iy)$ using equation (15.22). Here are some basic properties of this function.

Proposition 15.3.1. *The function $\varphi(z)$ satisfies the following:*

(a) $\varphi(z)$ *is analytic for all* $z \neq (m + in)\frac{\varpi}{2}$, *m, n odd.*

(b) *The addition law*

$$\varphi(z + w) = \frac{\varphi(z)\varphi'(w) + \varphi(w)\varphi'(z)}{1 + \varphi^2(z)\varphi^2(w)}$$

holds for all $z, w \in \mathbb{C}$ *such that both sides are defined.*

(c) *For* $z \in \mathbb{C}$ *and* $m, n \in \mathbb{Z}$, *we have*

$$\varphi(z + m\varpi + n\varpi i) = (-1)^{m+n}\varphi(z).$$

Proof. First observe that $\varphi(z)$ is defined whenever the denominator $1 - \varphi^2(x)\varphi^2(y)$ in (15.22) is nonzero. The polar distance interpretation of $\varphi(x)$ shows that $\varphi^2(x) \leq 1$ for all $x \in \mathbb{R}$, with equality if and only if x is an odd multiple of $\frac{\varpi}{2}$. Hence $\varphi(z)$ is defined on the open set $\Omega = \{z \in \mathbb{C} \mid z \neq (m + in)\frac{\varpi}{2}, m, n \in \mathbb{Z} \text{ odd}\}$.

Write $\varphi(z) = \varphi(x + iy) = u(x, y) + iv(x, y)$, where $u(x, y)$ and $v(x, y)$ are the real and imaginary parts of the right-hand side of (15.22). It is easy to see that $u(x, y)$ and $v(x, y)$ are differentiable on Ω as functions of x, y, since $\varphi(x)$ is infinitely differentiable on $\mathbb{R}$. Furthermore, using the identity $\varphi'^2(x) = 1 - \varphi^4(x)$ for $x \in \mathbb{R}$, it is straightforward to verify that $u(x, y)$ and $v(x, y)$ satisfy the Cauchy–Riemann equations

$$\frac{\partial u}{\partial x} = \frac{\partial v}{\partial y}, \quad \frac{\partial u}{\partial y} = -\frac{\partial v}{\partial x}$$

(see Exercise 2). By [11, 1.5.8], it follows that $\varphi(z)$ is analytic on Ω.

For part (b), let z and w be complex variables, and define

$$g(z, w) = \frac{\varphi(z)\varphi'(w) + \varphi(w)\varphi'(z)}{1 + \varphi^2(z)\varphi^2(w)}.$$

When $x_0 \in \mathbb{R}$ is fixed, $\varphi(x_0 + w)$ and $g(x_0, w)$ are analytic in w and coincide when $w \in \mathbb{R}$ by the addition law (15.12). By the Identity Theorem [11, 6.1.1], $\varphi(x_0 + w) = g(x_0, w)$ for all w such that both are defined. It follows that when $w_0 \in \mathbb{C}$ is fixed, $\varphi(z + w_0)$ and $g(z, w_0)$ are analytic in z and coincide when $z \in \mathbb{R}$. Using the Identity Theorem again, we see that $\varphi(z + w_0) = g(z, w_0)$ for all z such that both are defined. Since $w_0 \in \mathbb{C}$ is arbitrary, this proves the addition law.

The proof of part (c) requires a series of facts about $\varphi(z)$ and $\varphi'(z)$ that you will verify in Exercise 2. We begin with the following table of values:

(15.23)

x	$\varphi(x)$	$\varphi'(x)$
0	0	1
$\frac{\varpi}{2}$	1	0
ϖ	0	-1
$\frac{3\varpi}{2}$	-1	0

We also need the identities for $z \in \mathbb{C}$:

(15.24)
$$\varphi(iz) = i\varphi(z),$$
$$\varphi'(iz) = \varphi'(z).$$

Earlier, these identities were "proved" in order to motivate (15.22). Here, they are instead rigorous consequences of (15.22).

Since φ and φ' have period 2ϖ on $\mathbb{R}$, (15.23) and (15.24) easily imply that

(15.25)
$$\varphi(m\varpi) = \varphi(m\varpi i) = 0,$$
$$\varphi'(m\varpi) = \varphi'(m\varpi i) = (-1)^m.$$

Using the addition law, it is now straightforward to show that

(15.26) $\varphi(z + m\varpi) = (-1)^m \varphi(z)$ and $\varphi(z + n\varpi i) = (-1)^n \varphi(z)$

for $m, n \in \mathbb{Z}$. The desired identity for $\varphi(z + m\varpi + n\varpi i)$ follows immediately.   □

Part (c) of Proposition 15.3.1 implies that φ is *doubly periodic*:

(15.27) $$\varphi(z) = \varphi\big(z + (1 + i)\varpi\big) = \varphi\big(z + (1 - i)\varpi\big).$$

Note that the periods $(1 + i)\varpi$ and $(1 - i)\varpi$ are linearly independent over $\mathbb{R}$. The picture is the following:

(15.28)

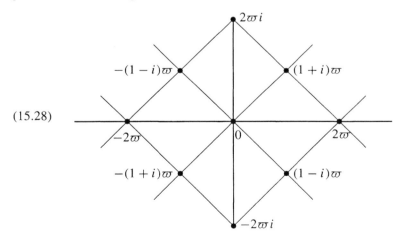

The dots in this picture are the complex numbers in the set

$$\mathscr{L} = \{(m + ni)\varpi \mid m + n \equiv 0 \bmod 2\} = \{m(1 + i)\varpi + n(1 - i)\varpi \mid m, n \in \mathbb{Z}\}.$$

This is the *period lattice* of φ. Double periodicity means that once we know the values of $\varphi(z)$ for all z in one of the tilted squares, we know its values for all $z \in \mathbb{C}$.

B. Zeros and Poles. Our next task is to study the zeros and poles of $\varphi(z)$. Recall that $z_0 \in \mathbb{C}$ is a *simple zero* of an analytic function $g(z)$ if $g(z_0) = 0$ and $g'(z_0) \neq 0$. This is equivalent to saying that the power series expansion of $g(z)$ at z_0 is

$$g(z) = a_1(z - z_0) + \sum_{n=2}^{\infty} a_n(z - z_0)^n, \quad a_1 \neq 0.$$

As defined in [11, 3.3.2], z_0 is a *simple pole* of a meromorphic function $g(z)$ if the Laurent expansion of $g(z)$ at z_0 is

$$g(z) = \frac{a_{-1}}{z - z_0} + \sum_{n=0}^{\infty} a_n(z - z_0)^n, \quad a_{-1} \neq 0.$$

Theorem 15.3.2. $\varphi(z)$ *is meromorphic on* $\mathbb{C}$ *with the following zeros and poles:*
(a) *The zeros are all simple and occur at* $z = (m + in)\varpi$ *for* $m, n \in \mathbb{Z}$.
(b) *The poles are all simple and occur at* $z = (m + in)\frac{\varpi}{2}$ *for* m, n *odd.*

Proof. Since $\varphi(0) = 0$ and $\varphi'(0) = 1$, part (c) of Proposition 15.3.1 easily implies that φ has a simple zero at $(m + in)\varpi$ for all $m, n \in \mathbb{Z}$.

Using the addition law together with (15.23) and (15.24), we see that

$$\varphi(z + \tfrac{\varpi}{2}) = \frac{\varphi(z)\varphi'(\frac{\varpi}{2}) + \varphi(\frac{\varpi}{2})\varphi'(z)}{1 + \varphi^2(z)\varphi^2(\frac{\varpi}{2})} = \frac{\varphi'(z)}{1 + \varphi^2(z)}.$$

Similarly,

$$\varphi(z \pm \tfrac{\varpi}{2}i) = \pm i \frac{\varphi'(z)}{1 - \varphi^2(z)}$$

(see Exercise 3). Multiplying these two equations gives the remarkable identity

$$\varphi(z + \tfrac{\varpi}{2}) \varphi(z \pm \tfrac{\varpi}{2}i) = \left(\frac{\varphi'(z)}{1 + \varphi^2(z)} \right) \left(\pm i \frac{\varphi'(z)}{1 - \varphi^2(z)} \right) = \pm i,$$

since $\varphi'^2(z) = 1 - \varphi^4(z)$.

Replacing z with $z + \frac{\varpi}{2}$ and using $\varphi(z + \varpi) = -\varphi(z)$ (prove this), we obtain

(15.29) $$\varphi(z) \varphi\big(z + (1 \pm i)\tfrac{\varpi}{2}\big) = \mp i.$$

If $\varphi(z_0) = 0$, then $\varphi\big(z_0 + (1 + i)\frac{\varpi}{2}\big)$ is undefined by (15.29). Hence

$$z_0 + (1 + i)\tfrac{\varpi}{2} = (m + in)\tfrac{\varpi}{2}, \quad m, n \text{ odd}$$

by Proposition 15.3.1. It follows easily that z_0 is one of our known simple zeros.

To analyze the poles of φ, we write (15.29) as

$$\varphi\big(z + (1 \pm i)\tfrac{\varpi}{2}\big) = \frac{\mp i}{\varphi(z)}.$$

Since $\varphi(z)$ has a simple zero at $z = 0$, we see that φ has simple poles at $z = (1 \pm i)\frac{\varpi}{2}$. Using the double periodicity of φ, we conclude that φ has simple poles at $(m + in)\frac{\varpi}{2}$ for m, n odd. Then we are done, since these are the only possible singularities of φ by Proposition 15.3.1. $\square$

Our next result will play an important role in the next section.

Theorem 15.3.3. *Fix a complex number w_0. Then the equation $\varphi(z) = w_0$ has a solution $z_0 \in \mathbb{C}$. Furthermore, if z_0 is one solution, then all solutions are given by*

$$z = (-1)^{m+n}z_0 + (m + in)\varpi, \quad m, n \in \mathbb{Z}.$$

Proof. Let $g(z)$ be analytic in a region $\Omega \subset \mathbb{C}$, and let $C \subset \Omega$ be a simple closed curve, oriented counterclockwise. The Zero–Pole Theorem [11, 6.2.1] says that if $g(z)$ has no zeros or poles on C, then

$$\frac{1}{2\pi i} \int_C \frac{g'(z)}{g(z)}\,dz = Z - P,$$

where Z is the numbers of zeros of $g(z)$ inside C, counted with multiplicity, and P is the numbers of poles of $g(z)$ inside C, also counted with multiplicity.

The function $g(z) = \varphi(z) - w_0$ has the same poles as φ, which are $(m + in)\frac{\varpi}{2}$, m, n odd, by Theorem 15.3.2. This means that we can't use the tilted squares from (15.28). However, since the zeros of $g(z)$ are isolated (see [11, 6.1.2]), we can shift one of the squares to the left as pictured below to obtain a curve C such that $g(z)$ has neither zeros nor poles on C:

(15.30)

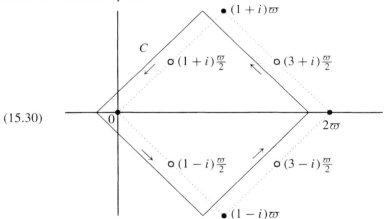

The open circles are poles of $g(z)$ and are simple by Theorem 15.3.2. Exactly two lie in the interior of C, so that $P = 2$.

Since $g(z) = \varphi(z) - w_0$ has periods $(1 \pm i)\varpi$, the same is true for $g'(z)$ and $g'(z)/g(z)$. Opposite edges of C differ by $(1 \pm i)\varpi$, so that $g'(z)/g(z)$ takes the same values on opposite edges. Hence the integrals along opposite edges cancel, since they have opposite orientations. This gives

$$Z - 2 = Z - P = \frac{1}{2\pi i} \int_C \frac{g'(z)}{g(z)}\,dz = 0.$$

We conclude that inside C, $g(z) - w_0$ has either two simple zeros or one double zero. In particular, $g(z) = w_0$ must have a solution z_0 inside C.

From z_0, Proposition 15.3.1 gives the additional solutions

$$\varphi\big((-1)^{m+n}z_0 + (m+in)\varpi\big) = (-1)^{m+n}\varphi\big((-1)^{m+n}z_0\big) = \varphi(z_0) = w_0,$$

where the second equality follows since φ is odd. We must show that there are no other solutions. Let D be the region enclosed by C (including the boundary). Translating D by elements of the period lattice $\mathscr{L} = \{(m+ni)\varpi \mid m+n \text{ is even}\}$ covers the entire complex plane. In particular, $-z_0 + \varpi$ has a translate by $\mathscr{L}$ that lies in the interior of D, that is, there are $m, n \in \mathbb{Z}$ with $m+n$ even such that

$$(15.31) \qquad -z_0 + \varpi + (m+in)\varpi = (-1)^{n+m+1}z_0 + \big((m+1)+in\big)\varpi$$

lies inside the curve C. If (15.31) differs from z_0, then we have found all zeros of $g(z)$ inside C. Since every other zero has a translate by $\mathscr{L}$ that lies inside C, it follows that all solutions of $\varphi(z) = w_0$ have the desired form. Finally, if (15.31) coincides with z_0, then it is easy to see that

$$z_0 = (a+ib)\tfrac{\varpi}{2}, \qquad a, b \in \mathbb{Z}, \ a+b \text{ odd}.$$

In Exercise 4 you will show that this implies that $\varphi'(z_0) = 0$. By what we proved above, it follows that z_0 is the only zero of $g(z)$ inside C. As before, we conclude that the solutions have the desired form. $\qquad \square$

Mathematical Notes

Two ideas implicit in this section require further comment.

▪ **Elliptic Functions.** By Proposition 15.3.1, φ is a meromorphic function on $\mathbb{C}$ with periods $(1+i)\varpi$, $(1-i)\varpi$ that are linearly independent over $\mathbb{R}$. In general, an *elliptic function* is a meromorphic function on $\mathbb{C}$ with periods ω_1, ω_2 that are linearly independent over $\mathbb{R}$. While the basic ideas of elliptic functions go back to Abel and Jacobi, these days most texts follow the approach of Weierstrass, who defined the *Weierstrass $\wp$-function* to be

$$\wp(z; \omega_1, \omega_2) = \frac{1}{z^2} + \sum_{\substack{m,n\in\mathbb{Z} \\ (m,n)\neq(0,0)}} \left(\frac{1}{\big(z-(n\omega_1 + m\omega_2)\big)^2} - \frac{1}{(n\omega_1 + m\omega_2)^2} \right).$$

For example, if we let $\wp_1(z)$ denote the $\wp$-function with periods $(1+i)\varpi$, $(1-i)\varpi$ (this is the notation of [13]), then one can show that

$$(15.32) \qquad \varphi(z) = -2\frac{\wp_1(z)}{\wp_1'(z)} \quad \text{and} \quad \varphi'(z) = \frac{4\wp_1^2(z) - 1}{4\wp_1^2(z) + 1}.$$

Furthermore, the relation

$$\varphi'^2(z) = 1 - \varphi^4(z)$$

translates into the relation

(15.33) $$\wp_1'^2(z) = 4\wp_1^3(z) + \wp_1(z).$$

In general, the $\wp$-function $\wp(z) = \wp(z; \omega_1, \omega_2)$ satisfies

(15.34) $$\wp'^2(z) = 4\wp^3(z) - g_2 \wp(z) - g_3,$$

where g_2 and g_3 are constants determined by the periods ω_1, ω_2. There is also an addition law for $\wp(z + w)$. Introductions to elliptic functions can be found in [3, §10], [7, Ch. 3], [10, Sec. 8.3], [12, Ch. 2], [14, Ch. 1], and [18, Ch. 9].

▪ **Elliptic Curves.** The primary geometric object of this chapter is the lemniscate, which is the curve defined by the equation $(x^2 + y^2)^2 = x^2 - y^2$. However, the elliptic functions we've been studying lead to other curves of interest. For example, the relation

$$\varphi'^2(z) = 1 - \varphi^4(z)$$

shows that the map $z \mapsto (\varphi(z), \varphi'(z))$ parametrizes the curve $y^2 = 1 - x^4$. Similarly, the relation (15.33) for the Weierstrass $\wp$-function $\wp_1(z)$ shows that $z \mapsto (\wp_1(z), \wp_1'(z))$ parametrizes the curve

$$y^2 = 4x^3 + x,$$

and for a general $\wp$-function, (15.34) shows that $z \mapsto (\wp(z), \wp'(z))$ parametrizes

$$y^2 = 4x^3 - g_2 x - g_3.$$

These are *elliptic curves*. They have an intrinsic addition law compatible with the addition law of the $\wp$-function. Some of the most important theorems and conjectures of modern number theory involve elliptic curves. Introductions to this wonderful area of mathematics can be found in [6], [8], [12], [16], and [18].

Historical Notes

In the Historical Notes to Section 15.3, we saw that Abel's theory of elliptic functions began with the integral

$$\int \frac{1}{\sqrt{(1 - c^2 t^2)(1 + e^2 t^2)}} \, dt.$$

He denoted the inverse function by $\varphi(x)$ and then defined $f(x) = \sqrt{1 - c^2 \varphi^2(x)}$ and $F(x) = \sqrt{1 + e^2 \varphi^2(x)}$. These functions are related via $\varphi'(x) = f(x)F(x)$. Abel gave addition laws for $\varphi(x)$, $f(x)$, $F(x)$ and multiplication formulas for $\varphi(nx)$, $f(nx)$, $F(nx)$ similar in spirit to Theorem 15.2.5. He also extended these functions to functions of $z \in \mathbb{C}$ and determined their periods, zeros, and poles. Abel's paper [Abel, Vol. I, pp. 263–388] contains many wonderful formulas and is fun to read.

Jacobi developed a similar theory based on the integral (15.21). He defined functions sin am x, cos am x, and Δ am x, later simplified to sn(x), cn(x), dn(x). His version of the theory became very influential, though it was eventually superseded by the $\wp$-function introduced by Weierstrass in 1882. One nice result of Weierstrass is that every elliptic function with the same periods as $\wp(z)$ is a rational function in $\wp(z)$ and $\wp'(z)$. So once the period lattice is fixed, only two elliptic functions are needed in order to get all others.

Gauss anticipated most of the work of Abel and Jacobi on elliptic functions but never published his results. As he wrote in 1828 [Gauss, Vol. X.1, p. 248],

> I shall most likely not soon prepare my investigations on transcendental functions that I have had for many years—since 1798—because I have many other matters that must be cleared up. Herr Abel has now, I see, anticipated me and relieved me of the burden in regard to one third of these matters, particularly since he carried out all these developments with great concision and elegance.

What led Gauss and Abel to work over the complex numbers? It appears that they were inspired to define $\varphi(z)$ for $z \in \mathbb{C}$ in order to represent *all* roots of the n-division polynomials of the lemniscate. The high degree of these polynomials suggests that the roots can't be all real. In the next section, we will use the theory of complex multiplication to describe the roots of the n-division polynomials.

More on the history of elliptic functions can be found in [3] and [5]. A classic treatment of the Jacobian elliptic functions appears in [19].

Exercises for Section 15.3

Exercise 1. Suppose that $g(z)$ is an analytic function satisfying $g(iz) = ig(z)$. Prove that $g'(iz) = g'(z)$.

Exercise 2. This exercise is concerned with the proof of Proposition 15.3.1.
(a) Prove that $\varphi(x + iy)$, as defined by (15.22), satisfies the Cauchy–Riemann equations.
(b) Prove (15.23), (15.24), (15.25), and (15.26).

Exercise 3. Prove the formula for $\varphi\left(z \pm \frac{\varpi}{2}i\right)$ stated in the proof of Theorem 15.3.2.

Exercise 4. Prove that $\varphi'(z)$ vanishes at all points of form $(m + in)\frac{\varpi}{2}$, $m + n$ odd.

Exercise 5. A useful observation is that an identity for φ proved over $\mathbb{R}$ automatically becomes an identity over $\mathbb{C}$.
(a) Prove this carefully, using results from complex analysis such as [11, 6.1.1].
(b) Explain why $\varphi'^2(z) = 1 - \varphi^4(z)$ holds for all $z \in \mathbb{C}$.

Exercise 6. By Theorem 15.3.3, $\varphi(z) = \varphi(z_0)$ if and only if $z = (-1)^{m+n}z_0 + (m + in)\varpi$. Following Abel, prove this using (15.13).

15.4 COMPLEX MULTIPLICATION

By Exercise 5 of Section 15.3, the multiplication formulas for $\varphi(nx)$, $x \in \mathbb{R}$, extend to give formulas for $\varphi(nz)$, $z \in \mathbb{C}$. Over $\mathbb{C}$ we also have the formula (15.24) given by

$$(15.35) \qquad \varphi(iz) = i\varphi(z), \quad i = \sqrt{-1}.$$

So besides multiplying by $n \in \mathbb{Z}$, we can also multiply by i. Combining these with the addition law gives formulas for $\varphi\big((n + im)z\big)$, where $n + im \in \mathbb{Z}[i]$ is any Gaussian integer. In other words, $\varphi(z)$ has *complex multiplication* by $\mathbb{Z}[i]$.

Before developing the general theory, let's give an example to illustrate the power of complex multiplication.

Example 15.4.1. In Exercise 1 you will use the addition law together with (15.35) and $\varphi(-z) = -\varphi(z)$ to prove that

(15.36)
$$\varphi\big((1 + i)z\big) = \frac{(1 + i)\varphi(z)\varphi'(z)}{1 - \varphi^4(z)},$$
$$\varphi\big((1 - i)z\big) = \frac{(1 - i)\varphi(z)\varphi'(z)}{1 - \varphi^4(z)}.$$

These are simple examples of complex multiplication.

To see the relevance of (15.36), square each side and apply $\varphi'^2(z) = 1 - \varphi^4(z)$. This gives

(15.37)
$$\varphi^2\big((1 + i)z\big) = \frac{2i\varphi^2(z)}{1 - \varphi^4(z)},$$
$$\varphi^2\big((1 - i)z\big) = \frac{-2i\varphi^2(z)}{1 - \varphi^4(z)}.$$

The surprise is that we've seen disguised versions of these formulas in the proof of Proposition 15.2.3. To explain why, let $r_0 = \varphi(\frac{x_0}{2})$ and $a = \varphi(x_0)$ as in the proof. Then set $t = \varphi\big((1 + i)\frac{x_0}{2}\big)$ and apply the first formula of (15.37) to obtain

$$t^2 = \frac{2ir_0^2}{1 - r_0^4}.$$

Since $2 = (1 - i)(1 + i)$, the second formula of (15.37) implies that

$$a^2 = \varphi^2(x_0) = \varphi^2\big((1 - i)(1 + i)\tfrac{x_0}{2}\big) = \frac{-2i\varphi^2\big((1 + i)\frac{x_0}{2}\big)}{1 - \varphi^4\big((1 + i)\frac{x_0}{2}\big)} = \frac{-2it^2}{1 - t^4}.$$

The above two equations are (15.15) and (15.16) from the proof of Proposition 15.2.3. Earlier, they seemed to appear out of nowhere, but now that we know complex multiplication, they are no longer so mysterious.   ◁▷

The proof of Proposition 15.2.3 used the duplication formula for $\varphi(2x)$. Example 15.37 shows that factoring 2 in $\mathbb{Z}[i]$ enables us to factor the duplication formula into equations that are simpler to understand. We will use similar factorizations in Section 15.5 when we prove Abel's theorem on the lemniscate.

The theory of *complex multiplication* gives formulas for $\varphi(\beta z)$, where $z \in \mathbb{C}$ and $\beta = n + im \in \mathbb{Z}[i]$ is a Gaussian integer. In this section we will first review some basic facts about $\mathbb{Z}[i]$ and then derive formulas for $\varphi(\beta z)$, paying special attention to the case when β is prime in $\mathbb{Z}[i]$.

A. The Gaussian Integers. The ring of Gaussian integers is defined by

$$\mathbb{Z}[i] = \{a + ib \mid a, b \in \mathbb{Z}\}.$$

The *units* of $\mathbb{Z}[i]$ form the group $\mathbb{Z}[i]^* = \{\pm 1, \pm i\} = \{i^\varepsilon \mid \varepsilon = 0, 1, 2, 3\}$, and nonzero Gaussian integers α, β are *associate* if $\alpha = i^\varepsilon \beta$ for some $i^\varepsilon \in \mathbb{Z}[i]^*$. Furthermore, $\mathbb{Z}[i]$ is a UFD with the following primes (up to associates):

- $2 = (1 + i)(1 - i)$, where $1 + i$ and $1 - i$ are associate primes in $\mathbb{Z}[i]$.
- When $p \equiv 3 \bmod 4$ is prime in $\mathbb{Z}$, p is also prime in $\mathbb{Z}[i]$.
- When $p \equiv 1 \bmod 4$ is prime in $\mathbb{Z}$, there are $a, b \in \mathbb{Z}$ such that $p = a^2 + b^2 = (a + bi)(a - bi)$, where $a + bi$ and $a - bi$ are nonassociate primes in $\mathbb{Z}[i]$.

Also, $\mathbb{Z}[i]$ is a PID, so that every ideal is of the form $\beta \mathbb{Z}[i]$ for some $\beta \in \mathbb{Z}[i]$. All of these facts are proved in most books on abstract algebra. See, for example, [Herstein, Sec. 3.8].

Given $\alpha \in \mathbb{Z}[i]$, we say that $\beta \equiv \gamma \bmod \alpha$ if α divides $\beta - \gamma$ in $\mathbb{Z}[i]$. To understand the quotient ring $\mathbb{Z}[i]/\alpha \mathbb{Z}[i]$, recall that $\alpha = a + ib \in \mathbb{Z}[i]$ has *norm*

$$N(\alpha) = \alpha \overline{\alpha} = |\alpha|^2 = a^2 + b^2 \in \mathbb{Z}$$

such that $N(\alpha \beta) = N(\alpha)N(\beta)$. Then we have the following result.

Lemma 15.4.2. *Let α be a nonzero element of $\mathbb{Z}[i]$. Then:*
(a) $\mathbb{Z}[i]/\alpha \mathbb{Z}[i]$ *is a finite ring with $N(\alpha)$ elements.*
(b) *If α is prime, then $\mathbb{Z}[i]/\alpha \mathbb{Z}[i]$ is the finite field*

$$\mathbb{Z}[i]/\alpha \mathbb{Z}[i] \simeq \mathbb{F}_{N(\alpha)}.$$

Proof. You will prove this in Exercises 2 and 3. $\qquad\qquad\qquad\qquad\square$

We say that a Gaussian integer $a + bi \in \mathbb{Z}[i]$ is *odd* if $a + b$ is odd and *even* if $a + b$ is even. If $\alpha, \beta \in \mathbb{Z}[i]$, then

$$\alpha\beta \text{ is odd } \Leftrightarrow \alpha \text{ and } \beta \text{ are odd,}$$
(15.38) $\qquad\quad \alpha + \beta \text{ is even } \Leftrightarrow \alpha, \beta \text{ are both even or both odd,}$
$$\alpha \text{ is even } \Leftrightarrow 1 + i \text{ divides } \alpha$$

(see Exercise 4). Since $1 + i$ is prime in $\mathbb{Z}[i]$, the last line of (15.38) can be stated as

$$\alpha \text{ is odd } \Leftrightarrow 1 + i \text{ and } \alpha \text{ are relatively prime.}$$

B. Multiplication by Gaussian Integers. When $n \in \mathbb{Z}$, Theorem 15.2.5 expresses $\varphi(nz)$ in terms of $\varphi(z)$ when n is odd and in terms of $\varphi(z)$ and $\varphi'(z)$ when n is even. Here, we will generalize on the former case by giving formulas for $\varphi(\beta z)$ in terms of $\varphi(z)$ when $\beta \in \mathbb{Z}[i]$ is odd.

In one sense, the formulas are easy—the proof of Theorem 15.4.4 given below shows that they are simple consequences of the addition law, the multiplication formulas for $\varphi(nz)$ from Theorem 15.2.5, and the identity $\varphi(iz) = i\varphi(z)$. However, in order to prove Abel's theorem on the lemniscate, we need to understand the fine structure of these formulas.

Here is an example to illustrate the issues involved.

Example 15.4.3. In Exercise 5 you will use the addition formula to derive the formula

$$\varphi\big((2+i)z\big) = \varphi(z) \frac{(-2+i)\varphi^8(z) - 6i\varphi^4(z) + 2 + i}{5\varphi^8(z) - 2\varphi^4(z) + 1}.$$

The numerator and denominator have a common factor that can be canceled. In Exercise 5 you will show that this leads to the simpler formula

(15.39) $$\varphi\big((2+i)z\big) = -i\varphi(z) \frac{\varphi^4(z) + (-1+2i)}{(-1+2i)\varphi^4(z) + 1}.$$

We pulled out a factor of $-i$ to ensure that the numerator is monic and the denominator has constant term 1. Note also the "reverse symmetry" of the coefficients of numerator and denominator. This will be important below.   ◁▷

The following theorem generalizes the formula (15.39) for $\varphi\big((2+i)z\big)$.

Theorem 15.4.4. *Let $\beta \in \mathbb{Z}[i]$ be odd. Then there exist relatively prime polynomials $P_\beta(u)$, $Q_\beta(u)$ in the polynomial ring $\mathbb{Z}[i][u]$ and $\varepsilon \in \{0, 1, 2, 3\}$ such that*
(a) *For all $z \in \mathbb{C}$, we have*

$$\varphi(\beta z) = i^\varepsilon \varphi(z) \frac{P_\beta\big(\varphi^4(z)\big)}{Q_\beta\big(\varphi^4(z)\big)}.$$

(b) *$\beta \equiv i^\varepsilon \mod 2(1 + i)$.*
(c) *$P_\beta(u)$ and $Q_\beta(u)$ have degree $d = (N(\beta) - 1)/4$, where $N(\beta)$ is the norm of β.*
(d) *The roots of the β-division polynomial $uP_\beta(u^4)$ are the complex numbers $\varphi\big(\alpha \frac{\varpi}{\beta}\big)$ for $\alpha \in \mathbb{Z}[i]$ odd.*
(e) *$P_\beta(u)$ is monic, $Q_\beta(0) = 1$, and $Q_\beta(u) = u^d P_\beta(1/u)$, where d is from part (c).*

Before beginning the proof, let's explain what the theorem says about $\varphi(\beta z)$ when $\beta \in \mathbb{Z}[i]$ is odd. Let $P_\beta(u)$, $Q_\beta(u) \in \mathbb{Z}[i][u]$ be the polynomials given in the theorem. Parts (c) and (e) imply that $P_\beta(u)$, $Q_\beta(u)$ can be written in the form

$$P_\beta(u) = u^d + a_1 u^{d-1} + \cdots + a_d,$$
$$Q_\beta(u) = u^d P_\beta(1/u)$$
$$= u^d\big((1/u)^d + a_1(1/u)^{d-1} + \cdots + a_d\big)$$
$$= 1 + a_1 u + \cdots + a_d u^d,$$

where $d = (N(\beta) - 1)/4$ and $a_1, \ldots, a_d \in \mathbb{Z}[i]$. This is the "reverse symmetry" mentioned above. Then the complex multiplication formula for $\varphi(\beta z)$ can be written

$$\varphi(\beta z) = i^\varepsilon \varphi(z) \frac{\varphi^{4d}(z) + a_1 \varphi^{4d-4}(z) + \cdots + a_d}{1 + a_1 \varphi^4(z) + \cdots + a_d \varphi^{4d}(z)},$$

where $\beta \equiv i^{\varepsilon} \mod 2(1 + i)$ by part (b). Here is an example.

Example 15.4.5. Suppose that $\beta = 2 + i$. Since $d = (N(\beta) - 1)/4 = (5 - 1)/4 = 1$ and $\beta \equiv -i \mod 2(1 + i)$, the above formula reduces to

$$\varphi\big((2 + i)z\big) = -i\varphi(z)\,\frac{\varphi^4(z) + a_1}{a_1\varphi^4(z) + 1},$$

where $a_1 \in \mathbb{Z}[i]$. Comparing this to (15.39), we see that $a_1 = -1 + 2i$.   ◁▷

The following lemma will be useful in the proof of Theorem 15.4.4.

Lemma 15.4.6. *Let $\beta \in \mathbb{Z}[i]$ be odd. Then the set*

$$R_\beta = \{\varphi(z) \mid z \in \mathbb{C},\, \varphi(\beta z) = 0\}$$

has precisely $N(\beta)$ elements and consists of all complex numbers of the form

$$\varphi\big(\alpha \tfrac{\varpi}{\beta}\big), \quad \alpha \in \mathbb{Z}[i] \text{ odd.}$$

Proof. First observe that if $\alpha \in \mathbb{Z}[i]$ is odd, then $\varphi\big(\alpha \tfrac{\varpi}{\beta}\big) \in R_\beta$, since $\varphi\big(\beta \cdot \alpha \tfrac{\varpi}{\beta}\big) = \varphi(\alpha\varpi) = 0$, where the last equality is by Theorem 15.3.2. Going the other way, suppose that $\varphi(\beta z) = 0$. Then Theorem 15.3.2 implies that

$$\beta z = (a + ib)\varpi, \quad a, b \in \mathbb{Z}.$$

Let $\alpha = a + ib \in \mathbb{Z}[i]$. Then $z = \alpha \tfrac{\varpi}{\beta}$, so that $\varphi(z) = \varphi\big(\alpha \tfrac{\varpi}{\beta}\big)$. If α is odd, then we are done. On the other hand, if α is even, then $\beta - \alpha$ is odd. Using the identity $\varphi(\varpi - z) = \varphi(z)$ from (15.9), we obtain

$$\varphi\big((\beta - \alpha)\tfrac{\varpi}{\beta}\big) = \varphi\big(\varpi - \alpha\tfrac{\varpi}{\beta}\big) = \varphi\big(\alpha\tfrac{\varpi}{\beta}\big).$$

This shows that the elements of R_β have the desired form.

To determine the size of R_β, fix $\varphi\big(\alpha\tfrac{\varpi}{\beta}\big) \in R_\beta$, where $\alpha \in \mathbb{Z}[i]$ is odd. We claim that α is unique modulo $\beta\mathbb{Z}[i]$. To see why, suppose that

$$\varphi\big(\alpha\tfrac{\varpi}{\beta}\big) = \varphi\big(\tilde{\alpha}\tfrac{\varpi}{\beta}\big), \quad \alpha, \tilde{\alpha} \in \mathbb{Z}[i] \text{ odd.}$$

By Theorem 15.3.3, there is $a + ib \in \mathbb{Z}[i]$ such that

$$\tilde{\alpha}\tfrac{\varpi}{\beta} = (-1)^{a+b}\alpha\tfrac{\varpi}{\beta} + (a + ib)\varpi.$$

This implies that

$$\tilde{\alpha} = (-1)^{a+b}\alpha + (a + ib)\beta.$$

Since α, $\tilde{\alpha}$, and β are odd, $a + ib$ is even by (15.38). Thus $(-1)^{a+b} = 1$ and hence

$$\tilde{\alpha} = \alpha + (a + ib)\beta,$$

so that α and $\tilde{\alpha}$ give the same element of $\mathbb{Z}[i]/\beta\mathbb{Z}[i]$. Since every coset of $\mathbb{Z}[i]/\beta\mathbb{Z}[i]$ can be represented by an odd Gaussian integer (given any α, either α or $\alpha + \beta$ is odd), it follows that

$$|R_\beta| = |\mathbb{Z}[i]/\beta\mathbb{Z}[i]| = N(\beta),$$

where the last equality follows from Lemma 15.4.2. $\square$

We now turn to the proof of the theorem.

Proof of Theorem 15.4.4. We will prove the theorem in five steps.

Step 1: Existence of $P_\beta(u)$ and $Q_\beta(u)$ for all β. Given $\beta \in \mathbb{Z}[i]$, we claim that there are $P_\beta(u), Q_\beta(u) \in \mathbb{Z}[i][u]$ such that $Q_\beta(0) = 1$ and

(15.40)
$$\varphi(\beta z) = \varphi(z)\,\frac{P_\beta(\varphi^4(z))}{Q_\beta(\varphi^4(z))}$$

when β is odd, and

(15.41)
$$\varphi(\beta z) = \varphi(z)\,\frac{P_\beta(\varphi^4(z))}{Q_\beta(\varphi^4(z))}\,\varphi'(z)$$

when β is even. We will prove (15.40) and (15.41) using the multiplication formulas from Theorem 15.2.5 together with the identities

(15.42)
$$\varphi(iz) = i\varphi(z),$$
$$\varphi\big((1+i)z\big) = \frac{(1+i)\varphi(z)\varphi'(z)}{1 - \varphi^4(z)},$$
$$\varphi\big((\beta+1)z\big) = -\varphi\big((\beta-1)z\big) + \frac{2\varphi(\beta z)\varphi'(z)}{1 + \varphi^2(\beta z)\varphi^2(z)},$$
$$\varphi\big((\beta+i)z\big) = -\varphi\big((\beta-i)z\big) + \frac{2i\varphi(\beta z)\varphi'(z)}{1 - \varphi^2(\beta z)\varphi^2(z)}.$$

We already know the first and second lines, and the third and fourth lines follow from the first and (15.13) (see Exercise 6).

The formulas for $\varphi(iz)$ and $\varphi\big((1+i)z\big)$ from (15.42) satisfy (15.40) and (15.41). From here, repeated use of the third line of (15.42) shows that for all integers $n \geq 0$, there are polynomials $P_{n+i}(u), Q_{n+i}(u) \in \mathbb{Z}[i][u]$ that give the desired formula for $\varphi\big((n+i)z\big)$. The argument is similar to what we did in the proof of Theorem 15.2.5. In particular, when n is even, we get the recursion

$$Q_{n+1+i}(u) = Q_{n-1+i}(u)\big(Q_{n+i}^2(u) + u P_{n+i}^2(u)(1 - u)\big)$$

similar to (15.18). This makes it easy to show that $Q_{n+i}(0) = 1$ for all $n \geq 0$ even, and the argument that $Q_{n+i}(0) = 1$ for $n \geq 0$ odd is similar.

Now fix an integer $n \geq 0$. We have formulas for $\varphi\big((n+i)z\big)$ (just proved) and $\varphi(nz)$ (by Theorem 15.2.5). Then repeated use of the fourth line of (15.42) shows that for all $m \geq 0$, there are polynomials $P_{n+im}(u)$, $Q_{n+im}(u) \in \mathbb{Z}[i][u]$ that give the desired formula for $\varphi\big((n+im)z\big)$ and satisfy $Q_{n+im}(0) = 1$. See Exercise 7.

Hence we have formulas for $\varphi\big((n+im)z\big)$ for all integers $n, m \geq 0$. Then

(15.43)
$$
\begin{aligned}
\varphi\big((-m+in)z\big) &= \varphi\big(i\,(n+im)z\big) &&= i\varphi\big((n+im)z\big),\\
\varphi\big((-n-im)z\big) &= \varphi\big(-(n+im)z\big) &&= -\varphi\big((n+im)z\big),\\
\varphi\big((m-in)z\big) &= \varphi\big(-i(n+im)z\big) &&= -i\varphi\big((n+im)z\big)
\end{aligned}
$$

make it easy to construct the desired $P_\beta(u)$, $Q_\beta(u) \in \mathbb{Z}[i][u]$ for all $\beta \in \mathbb{Z}[i]$.

Step 2: Remove Common Factors. For the rest of the proof, we will assume that β is odd. The polynomials $P_\beta(u)$, $Q_\beta(u)$ constructed in Step 1 might have a common factor. Since $\mathbb{Z}[i]$ is a UFD, the same is true for $\mathbb{Z}[i][u]$ by Theorem A.5.6. Thus

$$
P_\beta(u) = C_\beta(u)\widetilde{P}_\beta(u) \quad \text{and} \quad Q_\beta(u) = C_\beta(u)\widetilde{Q}_\beta(u),
$$

where $C_\beta(u)$, $\widetilde{P}_\beta(u)$, $\widetilde{Q}_\beta(u) \in \mathbb{Z}[i][u]$ and $\widetilde{P}_\beta(u)$, $\widetilde{Q}_\beta(u)$ are relatively prime. Since $Q_\beta(0) = 1$, we can multiply $C_\beta(u)$, $\widetilde{P}_\beta(u)$, $\widetilde{Q}_\beta(u)$ by suitable units in $\mathbb{Z}[i]^* = \{\pm 1, \pm i\}$ so that $\widetilde{Q}_\beta(0) = 1$. Since β is odd, we have

$$
\varphi(\beta z) = \varphi(z)\,\frac{P_\beta\big(\varphi^4(z)\big)}{Q_\beta\big(\varphi^4(z)\big)} = \varphi(z)\,\frac{C_\beta\big(\varphi^4(z)\big)\widetilde{P}_\beta\big(\varphi^4(z)\big)}{C_\beta\big(\varphi^4(z)\big)\widetilde{Q}_\beta\big(\varphi^4(z)\big)} = \varphi(z)\,\frac{\widetilde{P}_\beta\big(\varphi^4(z)\big)}{\widetilde{Q}_\beta\big(\varphi^4(z)\big)}.
$$

Hence we may assume that $P_\beta(u)$, $Q_\beta(u)$ are relatively prime in $\mathbb{Z}[i][u]$.

Step 3: The Constant i^ε. In Exercise 8 you will show that $\big(\mathbb{Z}[i]/2(1+i)\mathbb{Z}[i]\big)^* = \{\pm[1], \pm[i]\}$, so that $\beta \equiv i^\varepsilon \bmod 2(1+i)$ for some $\varepsilon \in \{0, 1, 2, 3\}$. Multiplying $P_\beta(u)$ by a suitable unit of $\mathbb{Z}[i]^*$, we obtain the equation

(15.44)
$$
\varphi(\beta z) = i^\varepsilon \varphi(z)\frac{P_\beta\big(\varphi^4(z)\big)}{Q_\beta\big(\varphi^4(z)\big)}.
$$

In Exercise 8 you will also show that

(15.45)
$$
\varphi\big(\beta\tfrac{\varpi}{2}\big) = i^\varepsilon.
$$

This will be useful later in the proof. It follows that the relatively prime polynomials $P_\beta(u)$, $Q_\beta(u) \in \mathbb{Z}[i][u]$ satisfy parts (a) and (b) of the theorem together with the condition $Q_\beta(0) = 1$ from part (e). Steps 4 and 5 will show that $P_\beta(u)$, $Q_\beta(u)$ satisfy the remaining conditions of the theorem.

Step 4: The Roots of $uP_\beta(u^4)$. We will use Lemma 15.4.6 to determine the roots of the β-division polynomial $A_\beta(u) = uP_\beta(u^4)$. Also let $B_\beta(u) = Q_\beta(u^4)$. Since β is odd, (15.44) implies that

(15.46)
$$
\varphi(\beta z) = i^\varepsilon \frac{A_\beta\big(\varphi(z)\big)}{B_\beta\big(\varphi(z)\big)}.
$$

In Exercise 9 you will show that $A_\beta(u)$ and $B_\beta(u)$ have no common roots in $\mathbb{C}$, since $Q_\beta(0) = 1$ and $P_\beta(u)$, $Q_\beta(u)$ are relatively prime in $\mathbb{Z}[i][u]$. Using this and (15.46), it follows that

$$A_\beta(\varphi(z)) = 0 \iff \varphi(\beta z) = 0.$$

Since any root of $A_\beta(u)$ is of the form $\varphi(z)$ for some $z \in \mathbb{C}$ by Theorem 15.3.3, we conclude that the roots of $A_\beta(u)$ form the set

$$R_\beta = \{\varphi(z) \mid z \in \mathbb{C}, \varphi(\beta z) = 0\}$$

from Lemma 15.4.6. Then the lemma implies that the roots can be written in the form described in part (d) of the theorem.

We next show that all roots have multiplicity 1. So assume that $u_0 = \varphi(z_0)$ is a multiple root. Then $A_\beta(u_0) = A'_\beta(u_0) = 0$, and hence $B_\beta(u_0) \neq 0$ by the previous paragraph. Differentiating (15.46) with respect to z and substituting $z = z_0$ gives

$$\varphi'(\beta z_0)\,\beta = \frac{B_\beta(u_0)A'_\beta(u_0)\varphi'(z_0) - B'_\beta(u_0)A_\beta(u_0)\varphi'(z_0)}{B_\beta(u_0)^2} = 0$$

(note that $\varphi'(z_0)$ is defined because $\varphi(z_0)$ is). Since $\varphi(\beta z_0) = 0$, we see that φ has a multiple zero at βz_0. This is impossible by Theorem 15.3.2.

We conclude that the degree of $A_\beta(u)$ is the number of elements in R_β. By Lemma 15.4.6, it follows that $A_\beta(u) = u P_\beta(u^4)$ has degree $N(\beta)$, so that $P_\beta(u)$ has degree $d = (N(\beta) - 1)/4$. This proves part (c) for $P_\beta(u)$.

Step 5: Relate P_β and Q_β. Once we show that

(15.47) $$Q_\beta(u) = u^d P_\beta(1/u), \quad d = (N(\beta) - 1)/4,$$

it will follow immediately that $Q_\beta(u)$ has degree d and $P_\beta(u)$ is monic (since $Q_\beta(u)$ has constant term 1). Thus we need only prove (15.47) to complete the proof.

The identity (15.29) implies that

$$\varphi(z)\varphi\left(z + (1+i)\tfrac{\varpi}{2}\right) = -i = i^3.$$

Setting $w = z + (1+i)\tfrac{\varpi}{2}$, we obtain

(15.48) $$\varphi(z)\varphi(w) = i^3.$$

In Exercise 10 you will use (15.48) and $\beta \equiv i^\varepsilon \bmod 2(1+i)$ to show that

(15.49) $$\varphi(\beta z)\varphi(\beta w) = i^{3+2\varepsilon}.$$

Then

(15.50) $$\frac{\varphi(\beta z)}{i^\varepsilon \varphi(z)} = \frac{i^\varepsilon \varphi(w)}{\varphi(\beta w)} = \frac{Q_\beta\left(\varphi^4(w)\right)}{P_\beta\left(\varphi^4(w)\right)} = \frac{Q_\beta\left(1/\varphi^4(z)\right)}{P_\beta\left(1/\varphi^4(z)\right)},$$

where the first equality uses (15.48) and (15.49), the second uses (15.44) with w in place of z, and the third follows by raising (15.48) to the fourth power to obtain $\varphi^4(w) = 1/\varphi^4(z)$. Comparing (15.50) with (15.44), we conclude that

$$\frac{Q_\beta(1/u^4)}{P_\beta(1/u^4)} = \frac{P_\beta(u^4)}{Q_\beta(u^4)}$$

as rational functions in u with coefficients in $\mathbb{Q}(i)$. Thus

$$\frac{Q_\beta(1/u)}{P_\beta(1/u)} = \frac{P_\beta(u)}{Q_\beta(u)}.$$

Recall from Step 4 that $\deg(P_\beta(u)) = d$, where $d = (N(\beta) - 1)/4$. In Exercise 11 you will show that the above equation implies that

(15.51) $$u^d P_\beta(1/u) = \lambda Q_\beta(u)$$

for some nonzero constant $\lambda \in \mathbb{Q}(i)$. However, if we evaluate (15.44) at $z = \frac{\varpi}{2}$ and use (15.45) and $\varphi\left(\frac{\varpi}{2}\right) = 1$, then we obtain

$$i^\varepsilon = i^\varepsilon \frac{P_\beta(1)}{Q_\beta(1)}.$$

Thus $P_\beta(1) = Q_\beta(1) \neq 0$. Then substituting $u = 1$ into (15.51) implies that $\lambda = 1$, so that $Q_\beta(u) = u^d P_\beta(1/u)$. This completes the proof. $\qquad\square$

Here are two examples of Theorem 15.4.4 from earlier in the chapter.

Example 15.4.7. When $\beta = 3$, equation (15.17) gives

$$\varphi(3z) = -\varphi(z) \frac{\varphi^8(z) + 6\varphi^4(z) - 3}{1 + 6\varphi^4(z) - 3\varphi^8(z)}.$$

In the notation of Theorem 15.4.4, this means

$$P_3(u) = u^2 + 6u - 3 \quad \text{and} \quad Q_3(u) = u^2 P_3(1/u) = 1 + 6u - 3u^2.$$

These polynomials have degree $(N(3) - 1)/4 = 2$. Note also that $i^\varepsilon = -1$, since $3 \equiv -1 \bmod 2(1 + i)$.

When $\beta = 5$, equation (15.19) gives

$$\varphi(5z) = \varphi(z) \frac{P_5(\varphi^4(z))}{Q_5(\varphi^4(z))}, \quad \text{where}$$

$$P_5(u) = u^6 + 50u^5 - 125u^4 + 300u^3 - 105u^2 - 62u + 5,$$

$$Q_5(u) = 1 + 50u - 125u^2 + 300u^3 - 105u^4 - 62u^5 + 5u^6.$$

These polynomials have degree $(N(5) - 1)/4 = 6$ and satisfy $Q_5(u) = u^6 P_6(1/u)$. Furthermore, we have $i^\varepsilon = 1$, since $5 \equiv 1 \bmod 2(1 + i)$. $\qquad\triangleleft\triangleright$

In general, one can show that when $n > 0$ is in $\mathbb{Z}$, the polynomials $P_n(u)$ and $Q_n(u)$ from Theorem 15.4.4 lie in $\mathbb{Z}[u]$.

C. Multiplication by Gaussian Primes. When β is an odd prime in $\mathbb{Z}[i]$, Theorem 15.4.4 has the following important refinement due to Eisenstein. This result will play a crucial role in the proof of Abel's theorem.

Theorem 15.4.8. *Let $\beta \in \mathbb{Z}[i]$ be an odd prime, and let*

$$P_\beta(u) = u^d + a_1 u^{d-1} + \cdots + a_d \in \mathbb{Z}[i][u], \quad d = (N(\beta) - 1)/4,$$

be the corresponding polynomial from Theorem 15.4.4. Then:
(a) $a_1, \ldots, a_d$ *are divisible by β and $a_d = i^{-\varepsilon}\beta$, where $\beta \equiv i^\varepsilon$ mod $2(1+i)$.*
(b) $P_\beta(u^4)$ *is irreducible over $\mathbb{Q}(i)$.*

Proof. Our proof will follow [12] and is based on Eisenstein's original proof from 1850 [Eisenstein, pp. 556–619]. We first observe that the Schönemann–Eisenstein criterion, stated in Theorem 4.2.3 for polynomials in $\mathbb{Z}[u]$ and primes in $\mathbb{Z}$, also applies to polynomials in $\mathbb{Z}[i][u]$ and primes in $\mathbb{Z}[i]$. You will prove this in Exercise 12. Then part (a) implies that

$$P_\beta(u^4) = u^{4d} + a_1 u^{4(d-1)} + \cdots + a_d \in \mathbb{Z}[i][u]$$

satisfies the criterion for the Gaussian prime β and hence is irreducible over $\mathbb{Q}(i)$. Thus part (b) of the theorem follows from part (a).

Proving part (a) will be harder. Since β is odd, Theorem 15.4.4 implies that

$$(15.52) \qquad \varphi(\beta z) = i^\varepsilon \varphi(z) \frac{\varphi^{4d}(z) + a_1 \varphi^{4(d-1)}(z) + \cdots + a_d}{1 + a_1 \varphi(z) + \cdots + a_d \varphi^{4d}(z)},$$

where the coefficients $a_1, \ldots, a_d \in \mathbb{Z}[i]$ depend on β. To prove part (a), we will analyze the relation between $a_1, \ldots, a_d$ and β by expanding each side of (15.52) as a power series in z.

Several power series will appear in the proof. The first comes from

$$i^\varepsilon \frac{u^d + a_1 u^{d-1} + \cdots + a_d}{1 + a_1 u + \cdots + a_d u^d},$$

which we write as

$$i^\varepsilon \frac{u^d + a_1(\beta)u^{d-1} + \cdots + a_d(\beta)}{1 + a_1(\beta)u + \cdots + a_d(\beta)u^d}$$

to emphasize the dependence on β. This rational function is analytic at $u = 0$ (the denominator doesn't vanish at 0) and hence has a power series expansion

$$(15.53) \qquad i^\varepsilon \frac{u^d + a_1(\beta)u^{d-1} + \cdots + a_d(\beta)}{1 + a_1(\beta)u + \cdots + a_d(\beta)u^d} = \sum_{k=0}^{\infty} b_k(\beta) u^k$$

$$= b_0(\beta) + b_1(\beta)u + b_2(\beta)u^2 + \cdots.$$

In Exercise 13 you will prove that $b_k(\beta) \in \mathbb{Z}[i]$ for all k. This follows because the constant term in the denominator is 1 and the other coefficients lie in $\mathbb{Z}[i]$. Using the power series (15.53), the multiplication formula (15.52) can be written

$$
\begin{aligned}
(15.54) \qquad \varphi(\beta z) &= \varphi(z)\big(b_0(\beta) + b_1(\beta)\varphi^4(z) + b_2(\beta)\varphi^8(z) + \cdots\big) \\
&= b_0(\beta)\varphi(z) + b_1(\beta)\varphi^5(z) + b_2(\beta)\varphi^9(z) + \cdots .
\end{aligned}
$$

The second power series comes from $\varphi(z)$. Since $\varphi(z)$ is analytic at $z = 0$, it can be expanded in a power series in z. In Exercise 14 you will use $\varphi(iz) = i\varphi(z)$, $\varphi'(0) = 1$, and $\varphi'^2(z) = 1 - \varphi^4(z)$ to prove that the power series has the form

$$
(15.55) \qquad \varphi(z) = z + \sum_{j=1}^{\infty} c_j z^{4j+1} = z + c_1 z^5 + c_2 z^9 + \cdots , \qquad c_j \in \mathbb{Q}.
$$

You will also show that $c_1 = -\frac{1}{10}$ and $c_2 = \frac{1}{120}$. Then replacing z with βz in (15.55) gives the third power series

$$
(15.56) \qquad \varphi(\beta z) = \sum_{j=0}^{\infty} c_j \beta^{4j+1} z^{4j+1} = \beta z + c_1 \beta^5 z^5 + c_2 \beta^9 z^9 + \cdots .
$$

From here, the proof proceeds in three steps. Here is an overview of what we will do in each step:

- **Step 1.** Derive a formula for $b_k(\beta)$ in terms of β that holds for all odd $\beta \in \mathbb{Z}[i]$. This will follow by substituting the series for $\varphi(z)$ and $\varphi(\beta z)$ into (15.54).
- **Step 2.** Prove that β divides $b_0(\beta), \ldots, b_{d-1}(\beta)$ when β is an odd prime. This will be done by analyzing the formula of Step 1 using a clever idea of Eisenstein.
- **Step 3.** Relate $a_1(\beta), \ldots, a_d(\beta)$ to $b_0(\beta), \ldots, b_{d-1}(\beta)$ and conclude that β divides $a_1(\beta), \ldots, a_d(\beta)$. This will follow easily from (15.53).

We now turn to the first step.

Step 1: Express $b_k(\beta)$ in terms of β. If we substitute (15.55) and (15.56) into the identity (15.54), then we obtain

$$
(15.57) \qquad
\begin{aligned}
\beta z + c_1 \beta^5 z^5 + c_2 \beta^9 z^9 + \cdots = \; &b_0(\beta)(z + c_1 z^5 + c_2 z^9 + \cdots) + \\
&b_1(\beta)(z + c_1 z^5 + c_2 z^9 + \cdots)^5 + \\
&b_2(\beta)(z + c_1 z^5 + c_2 z^9 + \cdots)^9 + \cdots .
\end{aligned}
$$

When we expand the right-hand side of (15.57), a given power of z appears only finitely often, since all terms of

$$
(z + c_1 z^5 + c_2 z^9 + \cdots)^{4j+1} = z^{4j+1}(1 + c_1 z^4 + c_2 z^8 + \cdots)^{4j+1}
$$

have degree $\geq 4j + 1$ in z. In Exercise 15 you will show that up to degree 9 in z, the right-hand side of (15.57) begins with

$$
(15.58) \quad b_0(\beta)z + \big(b_0(\beta)c_1 + b_1(\beta)\big)z^5 + \big(b_0(\beta)c_2 + 5b_1(\beta)c_1 + b_2(\beta)\big)z^9 + \cdots .
$$

Since this equals $\beta z + c_1\beta^5 z^5 + c_2\beta^9 z^9 + \cdots$, comparing coefficients gives

$$\beta = b_0(\beta),$$
$$c_1\beta^5 = b_0(\beta)c_1 + b_1(\beta),$$
$$c_2\beta^9 = b_0(\beta)c_2 + 5b_1(\beta)c_1 + b_2(\beta),$$

and then solving for $b_0(\beta), b_1(\beta), b_2(\beta)$ yields

$$b_0(\beta) = \beta,$$
$$b_1(\beta) = \beta(c_1\beta^4 - c_1),$$
$$b_2(\beta) = \beta(c_2\beta^8 - 5c_1^2\beta^4 + 5c_1^2 - c_2).$$

These equations hold for all odd $\beta \in \mathbb{Z}[i]$. We will see below that $b_0(\beta) = \beta$ is very important.

In general, one can prove (see Exericse 16) that for any k, there is a polynomial $S_k(u) \in \mathbb{Q}[u]$ of degree $4k$ such that

$$(15.59) \qquad b_k(\beta) = \beta S_k(\beta), \quad \beta \in \mathbb{Z}[i]\text{ odd}.$$

This follows because the c_j all lie in $\mathbb{Q}$. The crucial thing here is that the same polynomial $S_k(u)$ works for *all* odd β. For example, since $c_1 = -\frac{1}{10}$, the above equations imply that

$$b_1(\beta) = \beta S_1(\beta), \quad S_1(u) = -\tfrac{1}{10}u^4 + \tfrac{1}{10}.$$

Step 2: Prove that β divides $b_0(\beta), \ldots, b_{d-1}(\beta)$ when β is an odd prime. The equation (15.59) seems to imply that $b_k(\beta)$ is a multiple of β for all $k \geq 0$. The problem is that $S_k(u) \in \mathbb{Q}[u]$ need not have integer coefficients, as shown by $S_1(u)$. Hence we need to study the denominators of the coefficients of $S_k(u)$.

Let s_k be the least common multiple of these denominators. Then

$$S_k(u) = \tfrac{1}{s_k} T_k(u),$$

where $s_k \in \mathbb{Z} \setminus \{0\}$, $T_k(u) \in \mathbb{Z}[u]$, and ± 1 are the only integers dividing s_k and all coefficients of $T_k(u)$. Eisenstein observed that if $\alpha \in \mathbb{Z}[i]$ is an odd prime, then

$$(15.60) \qquad \alpha|s_k \implies N(\alpha) \leq 4k + 1.$$

To prove this, first observe that (15.59) implies that

$$(15.61) \qquad s_k b_k(\beta) = \beta T_k(\beta), \quad \beta \in \mathbb{Z}[i]\text{ odd}.$$

We noted above that $b_k(\beta)$ always lies in $\mathbb{Z}[i]$. This means that if an odd Gaussian prime α divides s_k, then α also divides $\beta T_k(\beta)$. It follows that

$$(15.62) \qquad \beta T_k(\beta) \equiv 0 \bmod \alpha, \quad \beta \in \mathbb{Z}[i]\text{ odd}.$$

Then consider the following:

- Since α is odd, the proof of Lemma 15.4.6 shows that elements of $\mathbb{Z}[i]/\alpha\mathbb{Z}[i]$ are of the form $[\beta]$, β odd. Thus (15.62) implies that the reduction of $uT_k(u)$ modulo α is a polynomial with at least $|\mathbb{Z}[i]/\alpha\mathbb{Z}[i]|$ roots.
- Since α divides s_k, the definition of s_k shows that the reduction of $uT_k(u)$ modulo α is a nonzero polynomial of degree at most $4k+1$. Hence the reduction has at most $4k+1$ roots since $\mathbb{Z}[i]/\alpha\mathbb{Z}[i]$ is field by Lemma 15.4.2.

These bullets imply that $|\mathbb{Z}[i]/\alpha\mathbb{Z}[i]| \leq 4k+1$. However, $|\mathbb{Z}[i]/\alpha\mathbb{Z}[i]| = N(\alpha)$ by Lemma 15.4.2. Thus $N(\alpha) \leq 4k+1$, and (15.60) follows.

Now fix an odd Gaussian prime β. Then (15.60), applied to β, tells us that

$$N(\beta) > 4k+1 \implies \beta \nmid s_k.$$

Note that $N(\beta) > 4k+1$ if and only if $k < d = (N(\beta)-1)/4$. It follows that $\beta \nmid s_k$ for $k = 0, \ldots, d-1$. Since β is prime, (15.61) implies that β divides $b_k(\beta)$ for $k = 0, \ldots, d-1$. This is what we needed to prove.

Step 3: Relate $a_1(\beta), \ldots, a_d(\beta)$ **to** $b_0(\beta), \ldots, b_{d-1}(\beta)$**.** This is easy, for if we write (15.54) in the form

$$i^\varepsilon \left(u^d + a_1(\beta)u^{d-1} + \cdots + a_d(\beta) \right)$$
$$= \left(1 + a_1(\beta)u + \cdots + a_d(\beta)u^d \right)\left(\textstyle\sum_{k=0}^{\infty} b_k(\beta)u^k \right)$$

and multiply out the right-hand side, then comparing coefficients of the powers of u gives the equations

$$i^\varepsilon a_d(\beta) = b_0(\beta),$$
$$i^\varepsilon a_{d-1}(\beta) = a_1(\beta)b_0(\beta) + b_1(\beta),$$
$$i^\varepsilon a_{d-2}(\beta) = a_2(\beta)b_0(\beta) + a_1(\beta)b_1(\beta) + b_2(\beta),$$
$$\vdots$$
$$i^\varepsilon a_1(\beta) = a_{d-1}(\beta)b_0(\beta) + a_{d-2}(\beta)b_1(\beta) + \cdots + b_{d-1}(\beta).$$

The $a_j(\beta)$ lie in $\mathbb{Z}[i]$, and $b_0(\beta), \ldots, b_{d-1}(\beta)$ are divisible by β by Step 2. It follows that in the above equations, the right-hand side is always divisible by β. This shows that β divides $a_1(\beta), \ldots, a_d(\beta)$, since i^ε is a unit. Furthermore, we proved earlier that $b_0(\beta) = \beta$, so that the first equation implies that $a_d(\beta) = i^{-\varepsilon}\beta$. This completes the proof of part (a). □

Mathematical Notes

Here are some further comments about complex multiplication.

■ **Complex Multiplication.** In our discussion of elliptic functions in Section 15.3, we mentioned that the Weierstrass $\wp$-function $\wp(z; \omega_1, \omega_2)$ for periods ω_1, ω_2 has an addition law. It follows easily that it also satisfies multiplication formulas for

$n \in \mathbb{Z}$ that generalize Theorem 15.2.5. However, the $\wp$-function rarely has complex multiplication. More precisely, $\wp(z; \omega_1, \omega_2)$ has complex multiplication by some $\beta \in \mathbb{C} \setminus \mathbb{Z}$ if and only if ω_2/ω_1 is a root of a quadratic polynomial with integer coefficients. This means that ω_2/ω_1 lies in an *imaginary quadratic field*, which is a field of the form $\mathbb{Q}(\sqrt{-m})$ for some $m > 0$ in $\mathbb{Z}$. For example, the periods $\omega_1 = (1 - i)\varpi$, $\omega_2 = (1 + i)\varpi$ of Abel's function $\varphi(z)$ have ratio

$$\frac{\omega_2}{\omega_1} = \frac{(1+i)\varpi}{(1-i)\varpi} = i,$$

which is a root of $x^2 + 1 = 0$. So the associated imaginary quadratic field is $\mathbb{Q}(i)$. In general, elliptic functions with complex multiplication have a deep relation to imaginary quadratic fields. This is discussed in books such as [3], [9], [15], and [18]. This is also related to *class field theory*, which will be discussed in the Mathematical Notes to Section 15.5.

Historical Notes

In addition to the general theory of elliptic functions, Abel also considered the lemniscatic function $\varphi(z)$ we've been studying. Let $m + \mu i \in \mathbb{Z}[i]$ be odd, and set $x = \varphi(\delta)$. Then Abel states complex multiplication by $m + \mu i$ as "one has

$$\varphi(m + \mu i)\delta = x.T,$$

where T is a rational function of x^4" [Abel, Vol. I, p. 354]. As an example, he writes the formula for complex multiplication by $2 + i$ as

$$\varphi(2 + i)\delta = x \, \frac{2 - 2x^8 + i(1 - 6x^4 + x^8)}{1 - 2x^4 + 5x^8} = xi \, \frac{1 - 2i - x^4}{1 - (1 - 2i)x^4}.$$

This is remarkably close to what we did in Example 15.4.5.

Eisenstein also has an important role to play in this story since he was the first to prove Theorems 15.4.4 and 15.4.8. Here is an extract of a letter that he wrote to Gauss in 1847 [Eisenstein, p. 845]:

> If $m = a + bi$ is an odd complex integer, p is its norm and $y = \frac{U}{V} = \frac{A_0 x + A_1 x^5 + \cdots + A_{(p-1)/4} x^p}{1 + B_1 x^4 + \cdots + B_{(p-1)/4} x^p}$ is the algebraic integral of the equation
>
> $$\int dy / \sqrt{1 - y^4} = m \int dx / \sqrt{1 - x^4},$$
>
> then I have further shown that for an *two-term* complex prime number m the coefficients of the numerator, except for the last which is a complex unit, and the coefficients of the denominator, except for the first which $= 1$, are all divisible by m. I conjecture that the theorem is also correct, when m is a *one-term* prime number …

Here, a "two-term" odd complex prime is $m = a + bi$ such that $p = a^2 + b^2$ is prime in $\mathbb{Z}$ with $p \equiv 1 \bmod 4$, and a "one-term" complex prime is a prime in $\mathbb{Z}$ such that

$p \equiv 3 \bmod 4$. In this letter, Eisenstein could prove part (a) of Theorem 15.4.8 only in the "two-term" case, though later he obtained a general proof. Also, if we think of $\varphi(z)$ as the inverse function of the elliptic integral $\int (1 - t^4)^{-1/2} \, dt$, then it should be clear that the displayed equation in Eisenstein's letter refers to the multiplication formula for $\varphi(mz)$ in terms of $\varphi(z)$.

The clearest statement of Eisenstein's irreducibility criterion appears in a paper he wrote in 1850 [Eisenstein, p. 542], where we find the following theorem:

> If in a polynomial $F(x)$ of x of arbitrary degree whose coefficient of the highest term is $= 1$, and all following coefficients are (real, complex) integers, in which a certain (real resp. complex) prime number m appears, if in addition the last coefficient is $= \varepsilon m$, where ε represents a number not divisible by m; then it is impossible to bring $F(x)$ into the form
>
> $$(x^\mu + a_1 x^{\mu-1} + \cdots + a_\mu)(x^\nu + b_1 x^{\nu-1} + \cdots + b_\nu),$$
>
> where μ and $\nu \geq 1$, $\mu + \nu = $ the degree of $F(x)$, and all a and b are (real resp. complex) integers; and the equation $F(x) = 0$ is accordingly irreducible.

The reason Eisenstein states the theorem for both $\mathbb{Z}$ and $\mathbb{Z}[i]$ is that he probably discovered it first over $\mathbb{Z}[i]$ in his study of complex multiplication on the lemniscate and then realized that it also applies over $\mathbb{Z}$.

When we discussed the Schönemann–Eisenstein criterion in the Historical Notes to Section 4.2, it was easy to explain what led Schönemann to the criterion, namely, does reducibility modulo p imply reducibility modulo p^2. But as we've seen in this chapter, it was a much richer mathematical context that led Eisenstein to his discovery.

Exercises for Section 15.4

Exercise 1. Prove (15.36).

Exercise 2. Let $\alpha \in \mathbb{Z}[i]$ be nonzero. The goal of this exercise is to prove part (a) of Lemma 15.4.2, which asserts that $|\mathbb{Z}[i]/\alpha\mathbb{Z}[i]| = N(\alpha)$. The idea is to forget multiplication and think of $\mathbb{Z}[i]$ and $\mathbb{Z}[i]/\alpha\mathbb{Z}[i]$ as groups under addition. Let m be the greatest common divisor of the real and imaginary parts of α, so that $\alpha = m(a + bi)$, where $\gcd(a, b) = 1$. Then pick $c, d \in \mathbb{Z}$ such that $ad - bc = 1$.
(a) Show that the map $\mathbb{Z}[i] \to \mathbb{Z} \oplus \mathbb{Z}$ defined by

$$\mu + \nu i \longmapsto \mu(d, -b) + \nu(-c, a) = (\mu d - \nu c, -\mu b + \nu a)$$

is a group isomorphism under addition.
(b) Show that the map of part a takes α and $i\alpha$ to $(m, 0)$ and $(-m(ac + bd), m(a^2 + b^2))$ respectively. Then use this to show that the map takes $\alpha\mathbb{Z}[i] \subset \mathbb{Z}[i]$ to the subgroup

$$m\mathbb{Z} \oplus m(a^2 + b^2)\mathbb{Z} \subset \mathbb{Z} \oplus \mathbb{Z}.$$

(c) Use part b to conclude that $|\mathbb{Z}[i]/\alpha\mathbb{Z}[i]| = N(\alpha)$.

Exercise 3. Prove part (b) of Lemma 15.4.2.

Exercise 4. Prove (15.38).

Exercise 5. Derive the two formulas for $\varphi\big((2+i)z\big)$ stated in Example 15.4.3.

Exercise 6. Prove the third and fourth lines of (15.42).

Exercise 7. Supply the details omitted in the proof of Step 1 of Theorem 15.4.4.

Exercise 8. Consider the finite ring $\mathbb{Z}[i]/2(1+i)\mathbb{Z}[i]$, and let $\beta \in \mathbb{Z}[i]$ be odd.
 (a) Prove that $\big(\mathbb{Z}[i]/2(1+i)\mathbb{Z}[i]\big)^* = \{\pm[1], \pm[i]\}$, and explain why this implies that $\beta \equiv i^\varepsilon \bmod 2(1+i)$ for some $\varepsilon \in \{0, 1, 2, 3\}$.
 (b) Prove that $\varphi\big(\beta\frac{\varpi}{2}\big) = i^\varepsilon$.

Exercise 9. Suppose that we have relatively prime polynomials $P_\beta(u)$, $Q_\beta(u) \in \mathbb{Z}[i][u]$ such that $Q_\beta(0) = 1$. Prove that $u P_\beta(u^4)$ and $Q_\beta(u^4)$ no common roots in $\mathbb{C}$.

Exercise 10. Let $w = z + (1+i)\frac{\varpi}{2}$. Use (15.48) and $\beta \equiv i^\varepsilon \bmod 2(1+i)$ to show that

$$\varphi(\beta z)\varphi(\beta w) = i^{3+2\varepsilon}.$$

Exercise 11. Let F be a field, and let $A(u), B(u) \in F[u]$ be nonzero relatively prime polynomials such that

$$\frac{B(1/u)}{A(1/u)} = \frac{A(u)}{B(u)}$$

in $F(u)$. Let $d = \deg(A)$. Prove that $d = \deg(B)$ and that there is a constant $\lambda \in F^*$ such that $u^d A(1/u) = \lambda B(u)$.

Exercise 12. Let $\beta \in \mathbb{Z}[i]$ be prime, and let $f = a_0 u^d + a_1 u^{d-1} + \cdots + a_d \in \mathbb{Z}[i][u]$. Prove the Schönemann–Eisenstein criterion over $\mathbb{Z}[i]$, which states that if $\beta \nmid a_0$, $\beta | a_1, \ldots, \beta | a_d$, and $\beta^2 \nmid a_d$, then f is irreducible over $\mathbb{Q}(i)$.

Exercise 13. Prove that the coefficients $b_k(\beta)$ defined in (15.54) lie in $\mathbb{Z}[i]$.

Exercise 14. The function $\varphi(z)$ is analytic at $z = 0$ and hence has a power series expansion.
 (a) In Exercise 3 of Section 15.2, you used $\varphi'^2(z) = 1 - \varphi^4(z)$ to show that $\varphi''(z) = -2\varphi^3(z)$. Use these two identities to prove by induction that for every $n \geq 1$, there is a polynomial $G_n(u) \in \mathbb{Z}[u]$ such that $\varphi^{(n)}(z)$ equals $G_n\big(\varphi(z)\big)$ if n is even and $G_n\big(\varphi(z)\big)\varphi'(z)$ if n is odd.
 (b) Use part (a) to prove that the coefficients of the power series expansion of $\varphi(z)$ at $z = 0$ lie in $\mathbb{Q}$.
 (c) Use part (b) and $\varphi(iz) = i\varphi(z)$ to show that $\varphi(z) = \sum_{j=0}^\infty c_j z^{4j+1}$, $c_j \in \mathbb{Q}$.
 (d) Show that $c_0 = 1$, $c_1 = -\frac{1}{10}$, and $c_2 = \frac{1}{120}$.

Exercise 15. Show carefully that (15.58) follows from (15.57).

Exercise 16. Prove that for each integer $k \geq 0$ there exists a polynomial $S_k(u) \in \mathbb{Q}[u]$ of degree $4k$ such that (15.59) holds for all odd $\beta \in \mathbb{Z}[i]$.

Exercise 17. Let $n \in \mathbb{Z}$ be an odd integer. Prove that $n \equiv (-1)^{(n-1)/2} \bmod 2(1+i)$. This shows that when n is an odd integer, we have $i^\varepsilon = (-1)^{(n-1)/2}$ in the formula for $\varphi(nz)$ given in Theorem 15.4.4.

15.5   ABEL'S THEOREM

In this final section of the book, we will prove Abel's theorem about straightedge and compass constructions on the lemniscate. The tools used will include Galois theory and the theory of complex multiplication developed in Section 15.4.

A. The Lemniscatic Galois Group. Let n be an odd positive integer and consider

$$L = \mathbb{Q}\left(i, \varphi\left(\tfrac{\varpi}{n}\right)\right).$$

We will see that the Galois group of $\mathbb{Q}(i) \subset L$ involves the group

$$\left(\mathbb{Z}[i]/n\mathbb{Z}[i]\right)^*$$

of units in $\mathbb{Z}[i]/n\mathbb{Z}[i]$. Since $\mathbb{Z}[i]$ is a PID, a coset $[\alpha]$ lies in $\left(\mathbb{Z}[i]/n\mathbb{Z}[i]\right)^*$ if and only if α is relatively prime to n in $\mathbb{Z}[i]$ (see Exercise 1).

Theorem 15.5.1. $\mathbb{Q}(i) \subset L$ *is a Galois extension and there is a one-to-one group homomorphism*

$$\mathrm{Gal}\left(L/\mathbb{Q}(i)\right) \to \left(\mathbb{Z}[i]/n\mathbb{Z}[i]\right)^*.$$

In particular, $\mathrm{Gal}\left(L/\mathbb{Q}(i)\right)$ *is Abelian.*

Proof. Let $A_n(u) = u\,P_n(u^4)$ be the n-division polynomial defined in part (d) of Theorem 15.4.4. The theorem tells us that the roots of $A_n(u)$ are given by

$$(15.63) \qquad\qquad \varphi\left(\alpha\tfrac{\varpi}{n}\right), \quad \alpha \in \mathbb{Z}[i] \text{ odd}$$

and the proof of Lemma 15.4.6 shows that for each root, the associated $\alpha \in \mathbb{Z}[i]$ is unique modulo $n\mathbb{Z}[i]$.

   Since each α in (15.63) is odd, the complex multiplication formula for $\varphi(\alpha z)$ given by Theorem 15.4.4 shows that $\varphi\left(\alpha\tfrac{\varpi}{n}\right)$ is a rational function in $\varphi\left(\tfrac{\varpi}{n}\right)$ with coefficients in $\mathbb{Q}(i)$. It follows that $A_n(u)$ splits completely in $L = \mathbb{Q}\left(i, \varphi\left(\tfrac{\varpi}{n}\right)\right)$. Since one of the roots is $\varphi\left(\tfrac{\varpi}{n}\right)$, it follows immediately that L is the splitting field of $A_n(u)$ over $\mathbb{Q}(i)$. Thus $\mathbb{Q}(i) \subset L$ is a Galois extension.

   Now take $\sigma \in \mathrm{Gal}\left(L/\mathbb{Q}(i)\right)$. Then $\sigma\left(\varphi\left(\tfrac{\varpi}{n}\right)\right)$ is a root of $A_n(u)$ and hence is one of the numbers (15.63). Thus there is $\alpha \in \mathbb{Z}[i]$ odd such that

$$(15.64) \qquad\qquad \sigma\left(\varphi\left(\tfrac{\varpi}{n}\right)\right) = \varphi\left(\alpha\tfrac{\varpi}{n}\right).$$

As noted above, α is unique modulo $n\mathbb{Z}[i]$.

   In Exercise 2 you will use Theorem 15.4.4 to show that if $\beta \in \mathbb{Z}[i]$ is odd, then

$$(15.65) \qquad\qquad \sigma\left(\varphi\left(\beta\tfrac{\varpi}{n}\right)\right) = \varphi\left(\alpha\beta\tfrac{\varpi}{n}\right).$$

We now prove that α is relatively prime to n. Let m be the order of σ in $\mathrm{Gal}\left(L/\mathbb{Q}(i)\right)$, so that σ^m is the identity. Then repeatedly applying (15.65) yields

$$\varphi\left(\tfrac{\varpi}{n}\right) = \sigma^m\left(\varphi\left(\tfrac{\varpi}{n}\right)\right) = \varphi\left(\alpha^m\tfrac{\varpi}{n}\right).$$

By uniqueness, we conclude that

$$1 \equiv \alpha^m \bmod n.$$

Hence α is relatively prime to n in $\mathbb{Z}[i]$, so that $\sigma \mapsto [\alpha]$ gives a well-defined map

(15.66) $\mathrm{Gal}\big(L/\mathbb{Q}(i)\big) \to \big(\mathbb{Z}[i]/n\mathbb{Z}[i]\big)^*.$

If σ and τ map to α and β respectively, then (15.65) easily implies that $\sigma\tau\big(\varphi(\frac{\varpi}{n})\big) = \varphi\big(\alpha\beta\frac{\varpi}{n}\big)$. Thus $\sigma\tau$ maps to $\alpha\beta$, which shows that the map is a group homomorphism. Furthermore, if $[\alpha] = [\beta]$ in $\big(\mathbb{Z}[i]/n\mathbb{Z}[i]\big)^*$, then

$$\alpha = \beta + (a + ib)n$$

where $a + ib$ is even because α, β, and n are odd. Then Proposition 15.3.1 implies that

$$\sigma\big(\varphi(\tfrac{\varpi}{n})\big) = \varphi\big(\alpha\tfrac{\varpi}{n}\big) = \varphi\big(\beta\tfrac{\varpi}{n}\big) = \tau\big(\varphi(\tfrac{\varpi}{n})\big),$$

from which we conclude that the map is one-to-one since $\varphi(\frac{\varpi}{n})$ generates L over $\mathbb{Q}(i)$. This completes the proof. $\square$

Since Abelian groups are solvable, one corollary of Theorem 15.5.1 and Chapter 8 is that the coordinates of the n-division points of the lemniscate are expressible by radicals over $\mathbb{Q}$. (You will prove this assertion carefully in Exercise 3.)

The homomorphism (15.66) constructed in Theorem 15.5.1 is the lemniscatic analog of the homomorphism

$$\mathrm{Gal}\big(\mathbb{Q}(\zeta_n)/\mathbb{Q}\big) \to (\mathbb{Z}/n\mathbb{Z})^*$$

studied in Chapter 9. We will say more about this analogy in the Mathematical Notes at the end of the section.

B. Straightedge-and-Compass Constructions. We now have the tools needed to prove Abel's theorem on the lemniscate.

Theorem 15.5.2. *Let n be a positive integer. Then the following are equivalent:*
(a) *The n-division points of the lemniscate can be constructed using straightedge and compass.*
(b) $\varphi\big(\frac{2\varpi}{n}\big)$ *is constructible.*
(c) *n is an integer of the form*

$$n = 2^s p_1 \cdots p_r,$$

where $s \geq 0$ is an integer and $p_1, \ldots, p_r$ are $r \geq 0$ distinct Fermat primes.

Proof. The implication (a) $\Rightarrow$ (b) is easy, since $\varphi\big(\frac{2\varpi}{n}\big)$ is the polar distance of an n-division point. The converse (b) $\Rightarrow$ (a) follows from part (b) of Corollary 15.2.7.

The proof of (c) ⇒ (b) will be a nice application of Theorem 15.5.1 together with some results of Section 15.2. We first observe that by part (c) of Corollary 15.2.7, $\varphi\left(\frac{2\varpi}{n}\right)$ is constructible provided that

$$\varphi\left(\tfrac{2\varpi}{2^s}\right),\ \varphi\left(\tfrac{2\varpi}{p_1}\right),\ \ldots,\ \varphi\left(\tfrac{2\varpi}{p_r}\right) \text{ are constructible.}$$

Since $\varphi\left(\frac{2\varpi}{2^s}\right)$ is constructible by Proposition 15.2.3, we need only show that $\varphi\left(\frac{2\varpi}{p}\right)$ is constructible when p is a Fermat prime.

By part (a) of Corollary 15.2.7, $\varphi\left(\frac{2\varpi}{p}\right)$ is constructible whenever $\varphi\left(\frac{\varpi}{p}\right)$ is. For the latter, Theorem 15.5.1 gives the Galois extension $\mathbb{Q}(i) \subset L = \mathbb{Q}\left(i, \varphi\left(\frac{\varpi}{p}\right)\right)$ with

$$\mathrm{Gal}\left(L/\mathbb{Q}(i)\right) \simeq \text{ a subgroup of } \left(\mathbb{Z}[i]/p\mathbb{Z}[i]\right)^*.$$

In Exercise 4 you will use the methods of Chapter 10 to prove that if

(15.67) $$\left|\left(\mathbb{Z}[i]/p\mathbb{Z}[i]\right)^*\right| = \text{ a power of 2,}$$

then $\varphi\left(\frac{\varpi}{p}\right)$ is constructible.

We will show that (15.67) holds whenever $p = 2^{2^m} + 1$ is a Fermat prime. The case $p = 3$ is easy (see Exercise 5). If $p > 3$, then $m \geq 1$, so that

$$p = 2^{2^m} + 1 = (2^{2^{m-1}} + i)(2^{2^{m-1}} - i) = \beta\overline{\beta},$$

where $\beta, \overline{\beta}$ are nonassociate primes in $\mathbb{Z}[i]$ of norm p. In this case, Exercise 6 and Lemma 15.4.2 give isomorphisms

$$\mathbb{Z}[i]/p\mathbb{Z}[i] = \mathbb{Z}[i]/\beta\overline{\beta}\mathbb{Z}[i] \simeq \mathbb{Z}[i]/\beta\mathbb{Z}[i] \times \mathbb{Z}[i]/\overline{\beta}\mathbb{Z}[i] \simeq \mathbb{F}_p \times \mathbb{F}_p.$$

Thus

$$\left|\left(\mathbb{Z}[i]/p\mathbb{Z}[i]\right)^*\right| = |\mathbb{F}_p^* \times \mathbb{F}_p^*| = (p-1)^2 = 2^{2^{m+1}}.$$

This proves (15.67) for all Fermat primes and completes the proof of (c) ⇒ (b).

It remains to prove (b) ⇒ (c). This is where we will use the irreducibility result proved in Theorem 15.4.8. Let n be an integer such that $\varphi\left(\frac{2\varpi}{n}\right)$ is constructible. We may assume that $n > 1$ since the theorem is trivially true when $n = 1$. Furthermore, the doubling formula (15.14) implies that we may assume that n is odd (be sure you can explain why), and Proposition 15.2.3 shows that $\varphi\left(\frac{\varpi}{n}\right)$ is constructible.

Let p be a prime dividing n. Then p is odd because n is. Let β be a complex prime such that $p = \beta$ if $p \equiv 3 \bmod 4$ and $p = \beta\overline{\beta}$ if $p \equiv 1 \bmod 4$. Then $n/\beta \in \mathbb{Z}[i]$ is odd (since n and β are), so that $\frac{\varpi}{\beta}$ is an odd multiple of $\frac{\varpi}{n}$. This makes it easy to show that

$$\varphi\left(\tfrac{\varpi}{\beta}\right) \in \mathbb{Q}\left(i, \varphi\left(\tfrac{\varpi}{n}\right)\right)$$

(see Exercise 7). It follows that $\varphi\left(\frac{\varpi}{\beta}\right)$ is constructible, since i and $\varphi\left(\frac{\varpi}{n}\right)$ are. By Corollary 10.1.8 from Chapter 10, the minimal polynomial of $\varphi\left(\frac{\varpi}{\beta}\right)$ over $\mathbb{Q}$ has degree equal to a power of 2. Then the Tower Theorem shows that the minimal polynomial of $\varphi\left(\frac{\varpi}{\beta}\right)$ over $\mathbb{Q}(i)$ also has degree equal to a power of 2.

Theorem 15.4.4 implies that $\varphi\left(\frac{\varpi}{\beta}\right)$ is a root of $u\,P_\beta(u^4)$. It is easy to see that $\varphi\left(\frac{\varpi}{\beta}\right) \neq 0$ (see Exercise 8), so that $\varphi\left(\frac{\varpi}{\beta}\right)$ is a root of $P_\beta(u^4)$. Since β is an odd prime, $P_\beta(u^4)$ has degree $N(\beta) - 1$ by Theorem 15.4.4 and is irreducible over $\mathbb{Q}(i)$ by Theorem 15.4.8. This proves that the minimal polynomial of $\varphi\left(\frac{\varpi}{\beta}\right)$ over $\mathbb{Q}(i)$ has degree $N(\beta) - 1$.

When $p = \beta$ for $p \equiv 3 \bmod 4$, we have $N(\beta) - 1 = p^2 - 1 = (p+1)(p-1)$. One easily sees that this is a power of 2 if and only if $p = 3$ (see Exercise 9). On the other hand, when $p = \beta\overline{\beta}$ for $p \equiv 1 \bmod 4$, we have $N(\beta) - 1 = p - 1$, which is a power of 2 if and only if p is a Fermat prime.

Thus the only primes dividing n are Fermat primes. To complete the proof of the theorem, we need to show that $p^2 | n$ cannot occur. So assume that $p^2 | n$, where p is prime. Then there is an odd complex prime β such that $\beta^2 | n$. By Exercise 7,

$$u_0 = \varphi\left(\tfrac{\varpi}{\beta^2}\right) \in L = \mathbb{Q}\left(i, \varphi\left(\tfrac{\varpi}{n}\right)\right),$$

which implies as above that u_0 is constructible. Hence the degree of its minimal polynomial over $\mathbb{Q}(i)$ is a power of 2. We will prove that the minimal polynomial has degree $N(\beta)(N(\beta) - 1)$. This is not be a power of 2 since $N(\beta) = p$ or p^2.

Since β is odd, Theorem 15.4.4 implies that

$$\varphi\left(\tfrac{\varpi}{\beta}\right) = \varphi\left(\beta\,\tfrac{\varpi}{\beta^2}\right) = i^\varepsilon \varphi\left(\tfrac{\varpi}{\beta^2}\right) \frac{P_\beta\left(\varphi^4\left(\tfrac{\varpi}{\beta^2}\right)\right)}{Q_\beta\left(\varphi^4\left(\tfrac{\varpi}{\beta^2}\right)\right)} = i^\varepsilon u_0 \, \frac{P_\beta(u_0^4)}{Q_\beta(u_0^4)}.$$

Since $\varphi\left(\frac{\varpi}{\beta}\right)$ is a root of $P_\beta(u^4)$, this formula for $\varphi\left(\frac{\varpi}{\beta}\right)$ gives the equation

$$0 = P_\beta\left(\left(\varphi\left(\tfrac{\varpi}{\beta}\right)\right)^4\right) = P_\beta\left(u_0^4 \, \frac{P_\beta(u_0^4)^4}{Q_\beta(u_0^4)^4}\right) = 0.$$

If we write $P_\beta(u) = u^d + a_1 u^{d-1} + \cdots + a_d$, $d = (N(\beta) - 1)/4$, then clearing denominators in the above equation shows that u_0 is a root of

$$P(u) = u^{4d}\,P_\beta(u^4)^{4d} + a_1 u^{4d-4} P_\beta(u^4)^{4d-4} Q_\beta(u^4)^4 + \cdots + a_d Q_\beta(u^4)^{4d}.$$

This has coefficients in $\mathbb{Z}[i]$ and degree $4d(4d+1) = N(\beta)(N(\beta) - 1)$, since $P_\beta(u)$, $Q_\beta(u) \in \mathbb{Z}[i][u]$ have degree d. Furthermore, Theorem 15.4.8 implies that

(15.68) β divides $a_1, \dots, a_d$.

Thus $P_\beta(u) \equiv u^d \bmod \beta$. Using this and (15.68), we see that

$$P(u) \equiv u^{N(\beta)(N(\beta)-1)} \bmod \beta,$$

since $4d(4d+1) = N(\beta)(N(\beta) - 1)$. Furthermore, $Q_\beta(0) = 1$ by Theorem 15.4.4, so that the constant term of $P(u)$ is

$$P(0) = 0 + \cdots + 0 + a_d Q_\beta(0)^{4d} = a_d.$$

Theorem 15.4.8 shows that a_d is not divisible by β^2, so that by the Schönemann–Eisenstein criterion over $\mathbb{Q}(i)$ (proved in Exercise 12 of Section 15.4), $P(u)$ is irreducible over $\mathbb{Q}(i)$. Thus the minimal polynomial of u_0 over $\mathbb{Q}(i)$ has degree $N(\beta)(N(\beta) - 1)$. The proof is now complete.   □

Mathematical Notes

Here are comments about some ideas related to this section.

▪ **The Lemniscatic Galois Extension.** Let $n \in \mathbb{Z}$ be odd and positive. The field

$$L = \mathbb{Q}\left(i, \varphi\left(\tfrac{\varpi}{n}\right)\right)$$

played an important role in our treatment of the lemniscate. This field has a nice relation to the elliptic curve $y^2 = 4x^3 + x$ discussed in the Mathematical Notes to Section 15.3. To explain this, first note the surprising fact that $\varphi'\left(\tfrac{\varpi}{n}\right) \in L$. You will prove this in Exercise 10. This means that

(15.69) $$L = \mathbb{Q}\left(i, \varphi\left(\tfrac{\varpi}{n}\right), \varphi'\left(\tfrac{\varpi}{n}\right)\right).$$

Then, using the formulas (15.32) and (15.33), one can show that L is the extension of $\mathbb{Q}(i)$ generated by the x- and y-coordinates of the $(1 + i)n$-torsion points on the elliptic curve $y^2 = 4x^3 + x$.

In general, extensions generated by torsion points of elliptic curves are an important topic in number theory. See [16] or [18] for a nice introduction.

▪ **Abelian Extensions of $\mathbb{Q}(i)$.** In Theorem 15.5.1, we constructed a one-to-one group homomorphism

$$\mathrm{Gal}\left(L/\mathbb{Q}(i)\right) \longrightarrow \left(\mathbb{Z}[i]/n\mathbb{Z}[i]\right)^*$$

when n is odd and positive. Thus $\mathbb{Q}(i) \subset L$ is an Abelian extension. A remarkable fact is that as n ranges over all positive integers, the fields L defined in (15.69) contain *all* Abelian extensions of $\mathbb{Q}(i)$, in the sense that if $\mathbb{Q}(i) \subset K$ is a Galois extension with Abelian Galois group, then there is an integer $n > 0$ such that

$$\mathbb{Q}(i) \subset K \subset L = \mathbb{Q}\left(i, \varphi\left(\tfrac{\varpi}{n}\right), \varphi'\left(\tfrac{\varpi}{n}\right)\right).$$

The proof of this result uses class field theory and complex multiplication. See, for example, [15, Ch. II, Example 5.8].

▪ **Class Field Theory.** A number field K is a finite extension of $\mathbb{Q}$. The main goal of *class field theory* is to describe all Abelian extensions of K. For example, when $K = \mathbb{Q}$, the Kronecker–Weber Theorem from the Historical Notes to Section 6.5 states that every Abelian extension of $\mathbb{Q}$ is a subfield of the cyclotomic extension $\mathbb{Q}(\zeta_n)$ for some n. Similarly, we noted above that every Abelian extension of $\mathbb{Q}(i)$ is a subfield of the lemniscatic extension (15.69) for some n.

The general version of class field theory describes Abelian extensions of a number field K, though the description uses the language of algebraic number theory and is not as explicit as for $K = \mathbb{Q}$ or $\mathbb{Q}(i)$. See [3, §8], [9, Sec. 8.4], or [15, §II.3] for a brief review of class field theory. In the special case of an imaginary quadratic field K, the theory of *complex multiplication* uses certain elliptic curves to give an explicit description of the Abelian extensions of K and their Galois groups. This is described in [9, Ch. 10] and [15, Ch. II].

For example, the theory of *ray class fields* implies that if n is odd, then

$$\mathbb{Q}(i) \subset L' = \mathbb{Q}\big(i, \varphi'\big(\tfrac{\varpi}{n}\big)\big)$$

is a Galois extension with Galois group

(15.70) $\mathrm{Gal}\big(L'/\mathbb{Q}(i)\big) \simeq \big(\mathbb{Z}[i]/(1+i)n\mathbb{Z}[i]\big)^{*}/\{\pm[1], \pm[i]\}.$

Using (15.70), one can get a shorter proof of Theorem 15.5.2 that doesn't require the hard work of Theorems 15.4.4 and 15.4.8. This is closely related to the elegant treatment of Abel's theorem given in [13].

The theory of elliptic curves is an important and beautiful area of number theory. There are also many unsolved problems of great interest. But only certain elliptic curves—those with complex multiplication—have the special link to Abelian extensions of quadratic number fields.

Historical Notes

The story of this section begins with Article 335 of *Disquisitiones* [4], where Gauss introduces his theory of geometric constructions and cyclotomic fields. He then goes on to say

> The principles of the theory that we are going to explain actually extend much farther than we will indicate. For they can be applied not only to circular functions but just as well to other transcendental functions, e.g. to those that depend on the integral $\int [1/\sqrt{(1 - x^4)}]\,dx$ and also to various types of congruences. Since, however, we are preparing a large work on those transcendental functions ..., we have decided to consider only circular functions here.

Gauss's "large work" never appeared, though the reference to the lemniscate would have been unmistakable to any nineteenth-century reader.

Abel was clearly intrigued by Gauss's remark. He read *Disquisitiones* carefully and understood Gauss's method for solving cyclotomic equations by radicals. He also defined a version of the function $\varphi(z)$ for the integral

$$\int \frac{1}{\sqrt{(1 - c^2 t^2)(1 + e^2 t^2)}}\,dt$$

and gave formulas for multiplication by n. The resulting n-division polynomials lead to certain algebraic equations, and one of Abel's goals in *Recherches sur les fonctions elliptiques* [Abel, Vol. I, pp. 263–388] is to determine whether these equations are solvable by radicals. Abel notes on page 352 that the n-division polynomial, taken in its full generality, is probably not solvable algebraically for arbitrary values of e and c, but nevertheless, there are particular cases when one can solve it completely ...

For Abel, the case of greatest interest was the lemniscate given by $e = c = 1$, though he also knew that $e = c\sqrt{3}$ and $e = c(2 \pm \sqrt{3})$ give polynomials that are solvable by radicals. From the modern point of view, these "particular cases" correspond to elliptic curves with complex multiplication.

As an example of Abel's methods, let $\varphi(z)$ be the lemniscatic function and fix a prime $p \equiv 1 \bmod 4$. Write $p = 4\nu + 1 = \alpha^2 + \beta^2$, $\alpha, \beta \in \mathbb{Z}$. Then, on pages 357 and 358, Abel asserts that

one has an equation

$$R = 0$$

of degree $\frac{\alpha^2 + \beta^2 - 1}{2} = 2\nu$, whose roots are

$$\varphi^2(\delta), \varphi^2(2\delta), \varphi^2(3\delta) \ldots \varphi^2(2\nu\delta),$$

where for brevity one supposes $\delta = \frac{\omega}{\alpha + \beta i}$.

Given this, one can easily solve the equation $R = 0$, by aid of the method of M. Gauss.

Letting ε be a primitive root of $\alpha^2 + \beta^2$, I say that one can express the roots as follows:

$$\varphi^2(\delta), \varphi^2(\varepsilon\delta), \varphi^2(\varepsilon^2\delta), \varphi^2(\varepsilon^3\delta) \ldots \varphi^2(\varepsilon^{2\nu-1}\delta).$$

Here, ω is what we call ϖ, and ε is an integer whose congruence class modulo $p = \alpha^2 + \beta^2$ generates the multiplicative group $\mathbb{F}_p^* \simeq \big(\mathbb{Z}[i]/(\alpha + \beta i)\mathbb{Z}[i]\big)^*$.

This quotation shows Abel using $\alpha + \beta i$ to study the p-division points on the lemniscate. Furthermore, the roots listed above have an important structure. We may assume that ε is odd, so that the multiplication formula for $\varphi(\varepsilon z)$ easily implies that

$$\varphi^2(\varepsilon z) = \theta\big(\varphi^2(z)\big)$$

for some rational function $\theta(u)$ with coefficients in $\mathbb{Z}$. It follows that if we let $x_0 = \varphi^2(\delta)$, then the roots of $R = 0$ can be written

$$x_0, \theta(x_0), \theta^2(x_0), \theta^3(x_0), \ldots, \theta^{2\nu}(x_0),$$

where the exponents refer to composition, that is, $\theta^2(x_0) = \theta(\theta(x_0))$. Compare this with Abel's 1829 paper *Mémoire sur une classe particulière d'équations résolubles algébriquement*, where he says that radical solutions exist

if all of the roots of an equation can be expressed by

$$x, \theta x, \theta^2 x, \theta^3 x, \ldots \theta^{n-1} x, \text{ where } \theta^n x = x,$$

θx being a rational function of x, and $\theta x, \theta^2 x, \ldots$ the functions of the same form as θx, taken two times, three times, etc.

(See [Abel, Vol. I, pp. 478–479].) Abel proves that any equation whose roots satisfy this condition is solvable by radicals. These quotations show that Abel's condition arises naturally from his work on the lemniscate.

We saw in Section 6.5 that Abel considered a more general class of equations in his "classe particulière" paper. Rather than assume as above that all of the roots are generated by iterating a single rational function, suppose instead that $f(x) = 0$, $f \in F[x]$, has a root x_0 with that property that any other root is of the form $\theta_i(x_0)$ for some rational function $\theta_i \in F(u)$. If we further assume that

$$\theta_i(\theta_j(x_0)) = \theta_j(\theta_i(x_0))$$

for all i and j, then Theorem 6.5.3 implies that the Galois group of the splitting field of f over F is Abelian and hence solvable by radicals over F by Chapter 8.

The Historical Notes to Section 6.5 describe how Abel's equations led to the modern *Abelian group* via the nineteenth century *Abelian equation*. But in Chapter 6, we didn't know what led Abel to these particular equations. Now we do—it was his work on the lemniscate! Thus the term "Abelian group," known to every beginning algebra student, has an unexpectedly rich history.

Kronecker was the first to realize the full power of the equations described by Abel. In the 1853 paper where he introduced the term "Abelian equation," Kronecker conjectured that all Abelian extensions of $\mathbb{Q}$ are contained in cyclotomic extensions (this is the Kronecker–Weber Theorem from Section 6.5), and he also asserts the following [Kronecker, Vol. I, p. 11]:

> There also exists a close relation between the roots of Abelian equations whose coefficients are complex integers of the form $a + b\sqrt{-1}$ and the roots of equations arising from the division of the lemniscate ...

Kronecker speculated that similar results might hold over imaginary quadratic fields. He called this "mein liebsten Jugendtraum" ("the dearest dream of my youth") in a letter written to Dedekind in 1880 [Kronecker, Vol. V, p. 455]. The first complete proofs of the theorems of class field theory and complex multiplication were given by Tagaki and Fueter in the 1920s. A nice discussion of Gauss, Abel, Eisenstein, and Kronecker appears in [17, Ch. 3 and 4]. See also [10, Sec. 8.6].

Exercises for Section 15.5

Exercise 1. Let $\beta \in \mathbb{Z}[i]$ be nonzero. Then $\alpha \in \mathbb{Z}[i]$ gives $[\alpha] \in \mathbb{Z}[i]/\beta\mathbb{Z}[i]$. Prove that $[\alpha] \in (\mathbb{Z}[i]/\beta\mathbb{Z}[i])^*$ if and only if α is relatively prime to β.

Exercise 2. As in the proof of Theorem 15.5.1, let $u_0 = \varphi(\frac{\varpi}{n})$, and assume that $\sigma \in \mathrm{Gal}(L/\mathbb{Q}(i))$ satisfies $\sigma(u_0) = \varphi(\alpha \frac{\varpi}{n})$, where $\alpha \in \mathbb{Z}[i]$ is odd. Use the multiplication formula for $\beta \in \mathbb{Z}[i]$ odd to prove (15.65).

Exercise 3. Use Theorem 15.5.1 and Chapter 8 to prove that the x- and y-coordinates of the n-division points of the lemniscate are expressible by radicals over $\mathbb{Q}$.

Exercise 4. Give a careful proof that (15.67) implies that $\varphi(\frac{\varpi}{n})$ is constructible.

Exercise 5. Prove that $|(\mathbb{Z}[i]/3\mathbb{Z}[i])^*| = 8$.

Exercise 6. Let $\alpha, \beta \in \mathbb{Z}[i]$ be nonzero and relatively prime. Prove the Chinese Remainder Theorem for $\mathbb{Z}[i]$, which asserts that there is a ring isomorphism

$$\mathbb{Z}[i]/\alpha\beta\mathbb{Z}[i] \simeq \mathbb{Z}[i]/\alpha\mathbb{Z}[i] \times \mathbb{Z}[i]/\beta\mathbb{Z}[i].$$

Exercise 7. When evaluating the multiplication formula for $\varphi(\alpha z)$ at a complex number z_0, one needs to worry about poles and vanishing denominators.

(a) Let $\alpha \in \mathbb{Z}[i]$ be odd, and assume that z_0 is a pole of neither $\varphi(z)$ nor $\varphi(\alpha z)$. Prove carefully that $Q_\alpha(\varphi^4(z_0)) \neq 0$ and that

$$\varphi(\alpha z_0) = i^\varepsilon \varphi(z_0) \frac{P_\alpha(\varphi^4(z_0))}{Q_\alpha(\varphi^4(z_0))}.$$

Exercise 9 of Section 15.4 will be useful.

(b) Let n be odd, and let p be a prime dividing n. Then let β be a Gaussian prime such that $p = \beta$ if $p \equiv 3 \bmod 4$ and $p = \beta\overline{\beta}$ if $p \equiv 1 \bmod 4$. Use part (a) to prove carefully that

$$\varphi\left(\tfrac{\varpi}{\beta}\right) \in \mathbb{Q}\left(i, \varphi\left(\tfrac{\varpi}{n}\right)\right).$$

Theorem 15.3.2 will be helpful.

(c) Let n be odd, and let p be a prime such that p^2 divides n. Also define β as in part (b). Prove that

$$\varphi\left(\tfrac{\varpi}{\beta^2}\right) \in \mathbb{Q}\left(i, \varphi\left(\tfrac{\varpi}{n}\right)\right).$$

Exercise 8. Let $\beta \in \mathbb{Z}[i]$ be an odd prime. Prove that $\varphi\left(\tfrac{\varpi}{\beta}\right) \neq 0$.

Exercise 9. Let $p \in \mathbb{Z}$ be prime. Prove that $p^2 - 1$ is a power of 2 if and only if $p = 3$.

Exercise 10. Let $n \in \mathbb{Z}$ be odd and positive, and let $L = \mathbb{Q}\left(i, \varphi\left(\tfrac{\varpi}{n}\right)\right)$. Use (15.9) and the multiplication law for $\varphi((n-1)z)$ to prove that $\varphi'\left(\tfrac{\varpi}{n}\right) \in L$.

REFERENCES

1. R. Ayoub, *The lemniscate and Fagnano's contributions to elliptic integrals*, Arch. Hist. Exact Sci. **29** (1984), 131–149.

2. D. Cox, *The arithmetic–geometric mean of Gauss*, L'Ens. Math. **30** (1984), 275–330.

3. D. Cox, *Primes of the Form $x^2 + ny^2$*, Wiley, New York, 1989.

4. C. F. Gauss, *Disquisitiones Arithmeticae*, Leipzig, 1801. Republished in 1863 as Volume I of [Gauss]. French translation, *Recherches Arithmétiques*, Paris, 1807. Reprint by Hermann, Paris, 1910. German translation, *Untersuchungen über Höhere Arithmetik*, Berlin, 1889. Reprint by Chelsea, New York, 1965. English translation, Yale U. P., New Haven, 1966. Reprint by Springer-Verlag, New York, Berlin, Heidelberg, 1986.

5. C. Houzel, *Fonctions elliptiques et intégrales abéliennes*, in *Abrégé d'Histoire des Mathématiques*, edited by J. Dieudonné, Hermann, Paris, 1978, 1–113.

6. K. Ireland and M. Rosen, *A Classical Introduction to Modern Number Theory*, Springer-Verlag, New York, Berlin, Heidelberg, 1982.

7. G. A. Jones and D. Singerman, *Complex Functions: An Algebraic and Geometric Viewpoint*, Cambridge U. P., Cambridge, 1987.

8. N. Koblitz, *Introduction to Elliptic Curves and Modular Forms*, Springer-Verlag, New York, Berlin, Heidelberg, 1984.

9. S. Lang, *Elliptic Functions*, Addison-Wesley, Reading, MA, 1973.

10. F. Lemmermeyer, *Reciprocity Laws*, Springer-Verlag, New York, Berlin, Heidelberg, 2000.

11. J. E. Marsden and M. J. Hoffman, *Basic Complex Analysis*, Third Edition, W. H. Freeman, New York, 1999.

12. V. Prasolov and Y. Solovyev, *Elliptic Functions and Elliptic Integrals*, AMS, Providence, RI, 1997.

13. M. Rosen, *Abel's theorem on the lemniscate*, Amer. Math. Monthly **88** (1981), 387–395.

14. C. L. Siegel, *Topics in Complex Function Theory, Vol. I*, Wiley, New York, 1969.

15. J. H. Silverman, *Advanced Topics in the Arithmetic of Elliptic Curves*, Springer-Verlag, New York, Berlin, Heidelberg, 1994.

16. J. H. Silverman and J. Tate, *Rational Points on Elliptic Curves*, Springer-Verlag, New York, Berlin, Heidelberg, 1992.

17. S. G. Vlăduţ, *Kronecker's Jugentraum and Modular Functions*, Gordon and Breach, New York, 1991.

18. L. C. Washington, *Elliptic Curves: Number Theory and Cryptography*, Chapman and Hall/CRC, Boca Raton, FL, 2003.

19. E. T. Whittaker and G. N. Watson, *A Course of Modern Analysis*, Fourth Edition, Cambridge U. P., Cambridge, 1963.

Appendix A
Abstract Algebra

This appendix summarizes most of the abstract algebra needed for Galois theory. Section A.1 reviews basic material on groups, rings, fields, and polynomials. Most of this material should be familiar, though it might be a good idea to review the notation. Before reading Chapter 1, students should also read about complex numbers and the nth roots of unity from Section A.2.

The other sections cover a variety of topics. Section A.3 discusses polynomials with coefficients in $\mathbb{Q}$. Section A.4 deals with group actions, which are used in several places in the text. Section A.5 includes the Sylow theorems, the Chinese Remainder Theorem, the multiplicative group of a field, and unique factorization domains.

A.1 BASIC ALGEBRA

We recall some basic material from abstract algebra.

A. Groups. We assume that the student is familiar with groups and subgroups. We usually write the group operation in a group G as gh for $g, h \in G$, and the identity element is denoted e. If G is finite, then $|G|$ is called the *order* of G.

If $g \in G$, then the *order* of g, denoted $o(g)$, is the smallest positive integer n such that $g^n = e$, if it exists. If $g^n \neq e$ for all positive integers n, then $o(g) = \infty$.

Given a subgroup H of a group G, the *left coset* determined by $g \in G$ is

$$gH = \{gh \mid h \in H\}$$

and the *right coset* determined by g is

$$Hg = \{hg \mid h \in H\}.$$

Two left cosets $g_1 H$ and $g_2 H$ are equal if and only if $g_1^{-1} g_2 \in H$. Similarly, $Hg_1 = Hg_2$ if and only if $g_2 g_1^{-1} \in H$.

The left cosets gH of $H \subset G$ partition G into disjoint subsets. Furthermore, if H is finite, then each left coset has the same number of elements as H, that is, $|gH| = |H|$ for all $g \in H$. If G is also finite, then the number of cosets is finite. This leads to *Lagrange's Theorem*, which is stated as follows.

Theorem A.1.1. *If H is a subgroup of a finite group G, then $|H|$ divides $|G|$.*   □

The quotient $|G|/|H|$ equals the number of left cosets. This number is the *index* of H in G and is denoted $[G : H]$. We discuss Lagrange's version of Theorem A.1.1 in Chapter 12

The above statements also apply to right cosets. In general, the partition of G into right cosets can differ from its partition into left cosets. Galois was the first to recognize the importance of when these partitions agree. This happens when the subgroup H is normal. As is well known,

$$H \text{ is normal in } G \iff gH = Hg \text{ for all } g \in G \iff gHg^{-1} = H \text{ for all } g \in G.$$

When $H \subset G$ is normal, the left (= right) cosets form a group under the operation $g_1 H \cdot g_2 H = g_1 g_2 H$. This is called the *quotient group* and is denoted G/H. The identity element of G/H is the coset $eH = H$.

Example A.1.2. The integers modulo n under addition form the quotient group $\mathbb{Z}/n\mathbb{Z}$. Elements of $\mathbb{Z}/n\mathbb{Z}$ are sometimes called *congruence classes*. The congruence class of $i \in \mathbb{Z}$ is denoted $[i] \in \mathbb{Z}/n\mathbb{Z}$.   ◁▷

We also assume that the student knows the definition of group homomorphism $\varphi : G_1 \to G_2$. Given such a φ, its *kernel* is

$$\mathrm{Ker}(\varphi) = \{g \in G_1 \mid \varphi(g) = e_2\},$$

where e_2 is the identity of G_2, and its *image* is

$$\mathrm{Im}(\varphi) = \{\varphi(g) \mid g \in G_1\}.$$

Then $\mathrm{Ker}(\varphi)$ is a normal subgroup of G_1 and $\mathrm{Im}(\varphi)$ is a subgroup of G_2.

If a group homomorphism $\varphi : G_1 \to G_2$ is one to one and onto, then the inverse function $\varphi^{-1} : G_2 \to G_1$ is also a group homomorphism. Thus φ is a *group isomorphism*. In this situation, we often write $\varphi : G_1 \simeq G_2$.

Given a group homomorphism $\varphi : G_1 \rightarrow G_2$, the *Fundamental Theorem of Group Homomorphisms* relates $\mathrm{Ker}(\varphi)$ and $\mathrm{Im}(\varphi)$ as follows.

Theorem A.1.3. *Let $\varphi : G_1 \rightarrow G_2$ be a group homomorphism. Then there is a unique group isomorphism $\overline{\varphi} : G_1/\mathrm{Ker}(\varphi) \simeq \mathrm{Im}(\varphi)$ such that $\overline{\varphi}(g\,\mathrm{Ker}(\varphi)) = \varphi(g)$ for all $g \in G_1$.* $\square$

A group G is *cyclic* if there is $g \in G$ such that $G = \{g^l \mid l \in \mathbb{Z}\}$. When G is cyclic, recall that

$$G \simeq \begin{cases} \mathbb{Z}, & \text{if } G \text{ is infinite,} \\ \mathbb{Z}/n\mathbb{Z}, & \text{if } |G| = n < \infty. \end{cases}$$

We have the following result about the subgroups of a cyclic group.

Theorem A.1.4. *Let G be a cyclic group. Then:*

(a) *Every subgroup of G is cyclic.*

(b) *If $|G| = n < \infty$, then for every positive divisor d of n, G has a unique subgroup of order d.* $\square$

One way to create cyclic groups is to pick $g \in G$ and consider the subgroup generated by g, namely

$$\langle g \rangle = \{g^l \mid l \in \mathbb{Z}\}.$$

If g has finite order $o(g) < \infty$, then $\langle g \rangle$ is a cyclic group of order $o(g)$. It follows that if G is a finite group, then applying Lagrange's Theorem to $\langle g \rangle \subset G$ shows that $o(g)$ divides $|G|$.

A partial converse is the following classic theorem of Cauchy.

Theorem A.1.5. *If a prime p divides the order $|G|$ of a finite group G, then G has an element of order p.* $\square$

For us, one of the most important groups is the *symmetric group* S_n. This is the group of permutations of n objects, usually thought of as elements of the set $\{1, \ldots, n\}$. Thus S_n is the set of functions

$$S_n = \{\sigma : \{1, \ldots, n\} \rightarrow \{1, \ldots, n\} \mid \sigma \text{ is one to one and onto}\},$$

where the group operation is given by composition of functions, and the identity element is the identity function $e(i) = i$ for $1 \leq i \leq n$.

If $\sigma \in S_n$ is given by $\sigma(j) = i_j$ for $j = 1, \ldots, n$, then following Cauchy, we write σ in the form

$$\sigma = \begin{pmatrix} 1 & 2 & \cdots & n \\ i_1 & i_2 & \cdots & i_n \end{pmatrix}.$$

Also recall cycle notation. Given distinct numbers $i_1, \ldots, i_l \in \{1, \ldots, n\}$ with $l \geq 2$, the *l-cycle* $\sigma = (i_1 \, i_2 \cdots i_l) \in S_n$ is the permutation defined by

(A.1)
$$\begin{aligned}
\sigma(i_1) &= i_2, \\
\sigma(i_2) &= i_3, \\
&\vdots \\
\sigma(i_{l-1}) &= i_l, \\
\sigma(i_l) &= i_1, \\
\sigma(i) &= i, \quad i \notin \{i_1, \ldots, i_l\}.
\end{aligned}$$

Note also that

$$(i_1 \, i_2 \cdots i_l) = (i_2 \cdots i_l \, i_1) = (i_3 \cdots i_l \, i_1 \, i_2) = \cdots = (i_l \, i_1 \, i_2 \cdots i_{l-1}).$$

As usual, a 2-cycle is called a *transposition*. Every element of S_n can be written uniquely as a product of disjoint cycles.

When multiplying cycles, it is important to remember that the operation is composition of functions. For example, consider

$$(345)(123)(12) = (1453).$$

When we apply the left-hand side, we first operate using (12), then using (123), and finally using (345). So we move right to left through the cycles, while inside an individual cycle, we move in the opposite direction (e.g., (345) takes 4 to 5). Note that some books use different conventions for multiplying cycles.

Also recall the identity

(A.2) $$(i_1 \, i_2 \cdots i_l) = (i_1 \, i_l)(i_1 \, i_{l-1}) \cdots (i_1 \, i_3)(i_1 \, i_2),$$

which expresses an *l*-cycle as a product of $l - 1$ transpositions.

A permutation $\sigma \in S_n$ is *even* if it is a product of an even number of transpositions, and *odd* otherwise. It follows from (A.2) that an *l*-cycle is even when *l* is odd, and is odd when *l* is even. The *sign* of σ is defined by

(A.3) $$\mathrm{sgn}(\sigma) = \begin{cases} +1, & \text{if } \sigma \text{ is even,} \\ -1, & \text{if } \sigma \text{ is odd.} \end{cases}$$

Note that $\mathrm{sgn} : S_n \to \{\pm 1\}$ is a group homomorphism.

The most important subgroup of S_n is the *alternating group* A_n, which is the subgroup consisting of all even permutations. It is a normal subgroup of S_n of index 2. This follows from $A_n = \mathrm{Ker}(\mathrm{sgn})$.

Example A.1.6. Note that

$$S_3 = \{e, (12), (13), (23), (123), (132)\},$$
$$A_3 = \{e, (123), (132)\} = \langle (123) \rangle.$$

Furthermore, one can show that

$$S_3, \ A_3, \ \langle (12) \rangle, \ \langle (13) \rangle, \ \langle (23) \rangle, \ \{e\}$$

are all subgroups of S_3.

A group G is *Abelian* if the group operation is commutative, that is, if $gh = hg$ for all $g, h \in G$. Recall that every subgroup of an Abelian group is normal. The reason for the name "Abelian" involves the Galois theory of an interesting class of equations studied by Abel. This is explained in Sections 6.5 and 15.5.

Given groups G and H, their *direct product*, or *product*, is the set $G \times H = \{(g, h) \mid g \in G, h \in H\}$ with group operation given by $(g, h)(g', h') = (gg', hh')$. Products enable us to create new groups from old ones and are used in structure theorems, such as Theorem A.1.7 below. The Mathematical Notes to Section 6.4 introduce a generalization of the direct product called the *semidirect product*.

Most courses in abstract algebra prove the following structure theorem for finite Abelian groups.

Theorem A.1.7. *Every finite Abelian group is isomorphic to a product of cyclic groups of prime power order.* $\square$

B. Rings. The student should also be familiar with rings and ideals from abstract algebra. For us, *all rings are commutative and have a multiplicative identity*. We write the additive identity of a ring R as 0 and the multiplicative identity as 1.

Since R is commutative, a subset $I \subset R$ is an ideal if and only if I is a subgroup under addition and $ra \in I$ whenever $r \in R$ and $a \in I$.

An ideal I is *principal* if there is $r \in R$ such that $I = \{rs \mid s \in R\}$. We say that r *generates* I. Principal ideals of R are denoted either rR or $\langle r \rangle$.

The cosets of an ideal I in R are sets of the form $r + I = \{r + s \mid s \in I\}$ for $r \in R$. Two cosets $r + I$ and $s + I$ are equal if and only if $r - s \in I$. The set of all cosets is denoted R/I and is a ring under the operations

$$(r + I) + (s + I) = (r + s) + I,$$
$$(r + I) \cdot (s + I) = rs + I.$$

We call R/I a *quotient ring*. Since R is commutative with a multiplicative identity, the same is true for R/I. The additive and multiplicative identities of R/I are $0 + I = I$ and $1 + I$ respectively.

Example A.1.8. Every ideal of $\mathbb{Z}$ is principal, so that the ideals of $\mathbb{Z}$ are $n\mathbb{Z}$ for integers $n \geq 0$. Integers modulo n under addition and multiplication form the quotient ring $\mathbb{Z}/n\mathbb{Z}$, where the congruence class $[i]$ is the coset $i + n\mathbb{Z}$.

For us, a ring homomorphism $\varphi : R \to S$ is a function satisfying the usual conditions $\varphi(r + s) = \varphi(r) + \varphi(s)$ and $\varphi(rs) = \varphi(r)\varphi(s)$ for all $r, s \in R$. In this book, *all ring homomorphisms preserve the multiplicative identity*, unless explicitly stated otherwise. This means that $\varphi(1_R) = 1_S$, where 1_R and 1_S are the multiplicative identities of R and S respectively.

Given such a φ, its *kernel* is

$$\mathrm{Ker}(\varphi) = \{r \in R \mid \varphi(r) = 0\},$$

and its *image* is

$$\mathrm{Im}(\varphi) = \{\varphi(r) \mid r \in R\}.$$

Then $\mathrm{Ker}(\varphi)$ is an ideal of R and $\mathrm{Im}(\varphi)$ is a subring of S.

If a ring homomorphism $\varphi : R \to S$ is one to one and onto, then the inverse function $\varphi^{-1} : S \to R$ is also a ring homomorphism. Thus φ is a *ring isomorphism*. In this situation, we often write $\varphi : R \simeq S$.

Given a ring homomorphism $\varphi : R \to S$, the *Fundamental Theorem of Ring Homomorphisms* is as follows.

Theorem A.1.9. *Let $\varphi : R \to S$ be a ring homomorphism. Then there is a unique ring homomorphism $\overline{\varphi} : R/\mathrm{Ker}(\varphi) \simeq \mathrm{Im}(\varphi)$ such that $\overline{\varphi}(r + \mathrm{Ker}(\varphi)) = \varphi(r)$ for all $r \in R$.* □

An *integral domain* is a ring R such that $rs = 0$, $r, s \in R$, implies that $r = 0$ or $s = 0$. Section A.5 will discuss a special class of integral domains called *unique factorization domains*.

Example A.1.10. The ring of integers $\mathbb{Z}$ is an integral domain, but $\mathbb{Z}/6\mathbb{Z}$ is not, since $[2] \cdot [3] = [6] = [0]$, yet $[2]$ and $[3]$ are nonzero in $\mathbb{Z}/6\mathbb{Z}$. ◁▷

C. Fields. A *field* F is a ring such that every nonzero element has a multiplicative inverse. To avoid trivial examples, we assume that $0 \neq 1$ in F. Commonly used fields include:

$$\mathbb{Q} = \text{the field of rational numbers,}$$
$$\mathbb{R} = \text{the field of real numbers,}$$
$$\mathbb{C} = \text{the field of complex numbers.}$$

Note that a field is always an integral domain. Also recall that the only ideals of a field F are $\{0\}$ and F itself.

One way to create fields is via the *field of fractions* of an integral domain R. This is defined to be the set

$$K = \left\{ \frac{r}{s} \mid r, s \in R, \ s \neq 0 \right\},$$

where we regard r/s and t/u as equal if and only if $ru = st$. This becomes a ring under the operations

$$\frac{r}{s} + \frac{t}{u} = \frac{ru + st}{su},$$
$$\frac{r}{s} \cdot \frac{t}{u} = \frac{rt}{su}.$$

If $r/s \neq 0$, then the multiplicative inverse of r/s is s/r. Thus K is a field. We call K the "field of fractions" of R, though the term "quotient field" is also used.

Note that the function

$$\varphi : R \longrightarrow K$$

defined by $\varphi(r) = r/1$ is a one-to-one ring homomorphism, so that $R \simeq \varphi(R)$. In this situation, we usually identify R with $\varphi(R)$. This allows us to regard an integral domain R as a subset of its field of fractions K.

Example A.1.11. The field of fractions of $\mathbb{Z}$ is the field of rational numbers $\mathbb{Q}$.◁▷

A second important method for creating fields is by means of maximal ideals. An ideal $M \subset R$ is *maximal* if $M \neq R$ and for all ideals J of R, $M \subset J \subset R$ implies $J = M$ or $J = R$. Most abstract algebra courses prove the following theorem that characterizes maximal ideals in terms of their quotient rings.

Theorem A.1.12. *Let M be an ideal of a ring R. Then R/M is a field if and only if M is a maximal ideal.*
□

For $\mathbb{Z}$, we can determine the maximal ideals as follows.

Example A.1.13. One easily checks that $n\mathbb{Z} \subset m\mathbb{Z}$ if and only if m divides n. It follows that $p\mathbb{Z}$ is a maximal ideal of $\mathbb{Z}$ if and only if p is prime (be sure you see why). By Theorem A.1.12, $\mathbb{Z}/p\mathbb{Z}$ is a field. It is customary to denote this field by $\mathbb{F}_p$. This field has p elements. In Chapter 11, we describe *all* finite fields.   ◁▷

Theorem A.1.12 is used in Chapter 3 to prove that any polynomial with coefficients in a field has roots in some possibly larger field.

We next discuss the *characteristic* of a field F. Given a positive integer n, define

$$n \cdot 1 = \underbrace{1 + \cdots + 1}_{n \text{ times}} \in F,$$

where 1 is the multiplicative identity of F.

The distributive law implies that $(n \cdot 1)(m \cdot 1) = (nm) \cdot 1$. If $n \cdot 1 = 0$ for some positive n, then let p be the least such number. We claim that p is prime. This is easy to see, for if we had $p = ab$ with $0 < a, b < p$, then

$$0 = p \cdot 1 = (ab) \cdot 1 = (a \cdot 1)(b \cdot 1).$$

Since F is an integral domain, we would have $a \cdot 1 = 0$ or $b \cdot 1 = 0$, which would contradict the minimality of p. Thus p is prime.

Because of this, we say that F has *characteristic* 0 if $n \cdot 1 \neq 0$ for all positive integers n and has *characteristic* p if $p \cdot 1 = 0$ and p is prime.

Thus $\mathbb{Q}$, $\mathbb{R}$, and $\mathbb{C}$ all have characteristic 0, while $\mathbb{F}_p$ has characteristic p. In general, Galois theory is easier in characteristic 0 than in characteristic p.

D. Polynomials. A polynomial in x with coefficients in a field F is an expression

$$f = a_n x^n + a_{n-1} x^{n-1} + \cdots + a_1 x + a_0$$

where $a_n, a_{n-1}, \ldots, a_1, a_0 \in F$. If $a_n \neq 0$, then we say that f has *degree n*, written $\deg(f) = n$. If $a_n = 1$, then we say that f is *monic*.

If f and g are nonzero polynomials, then fg is also nonzero, since F is an integral domain. It follows easily that

(A.4) $$\deg(fg) = \deg(f) + \deg(g).$$

Notice also that we have not defined the degree of the zero polynomial. One might be tempted to set $\deg(0) = 0$, but this would not be consistent with (A.4) (do you see why?). For this reason, we prefer to leave $\deg(0)$ undefined.

The set of all polynomials in x with coefficients in F forms a ring $F[x]$ under addition and multiplication of polynomials. Note that $F[x]$ is an integral domain.

The following *division algorithm* is proved in most abstract algebra texts.

Theorem A.1.14. *Let $f, g \in F[x]$, and assume that g is nonzero. Then there are polynomials $q, r \in F[x]$ such that*

$$f = qg + r, \quad \text{where } r = 0 \text{ or } \deg(r) < \deg(g).$$

Furthermore, q and r are unique. □

As an application of this theorem, consider the case when $g = x - a$ for some $a \in F$. The division algorithm implies that $f = q \cdot (x - a) + r$ where $r \in F$ (be sure you see why). Evaluating this equation at $x = a$ yields

$$f(a) = q(a) \cdot 0 + r,$$

so that $r = f(a)$. Thus $f = q \cdot (x - a) + f(a)$. This leads to the following result.

Corollary A.1.15. *Given $f \in F[x]$ and $a \in F$, the linear polynomial $x - a$ is a factor of f if and only if $f(a) = 0$, that is, if a is a root of f.* □

Using this corollary and induction, one easily obtains the following bound on the number of roots of a polynomial.

Corollary A.1.16. *Let $f \in F[x]$ be nonconstant. Then f has at most $\deg(f)$ roots in the field F.* □

In Chapter 3, we show that by going to a larger field, a polynomial $f \in F[x]$ has exactly $\deg(f)$ roots, provided that we take the multiplicities of the roots into account.

Another application of Theorem A.1.14 is the *Euclidean algorithm* for computing the greatest common divisor (or gcd) of two polynomials $f, g \in F[x]$, at least one of which is nonzero. Recall that $\gcd(f, g)$ is the monic polynomial of maximum degree in $F[x]$ which divides both f and g. If $g \neq 0$, we compute $\gcd(f, g)$ by repeatedly applying the division algorithm until we get a zero remainder:

$$f = q_0 g + r_0, \quad \deg(r_0) < \deg(g),$$
$$g = q_1 r_0 + r_1, \quad \deg(r_1) < \deg(r_0),$$
$$r_0 = q_2 r_1 + r_2, \quad \deg(r_2) < \deg(r_1),$$
$$\vdots$$
$$r_n = q_{n+2} r_{n+1} + r_{n+2}, \quad \deg(r_{n+2}) < \deg(r_{n+1}),$$
$$r_{n+1} = q_{n+3} r_{n+2} + 0.$$

Then one can prove that $\gcd(f, g)$ is the monic polynomial obtained by multiplying r_{n+2} by a suitable constant. On the other hand, if $g = 0$, then one easily sees that $\gcd(f, 0) = f$. In general, the greatest common divisor has the following three properties:

- For any $h \in F[x]$, h divides $\gcd(f, g)$ $\iff$ h divides both f and g.
- $\gcd(f, g) = 1$ $\iff$ f and g are relatively prime in $F[x]$.
- $\gcd(f, g) = Af + Bg$ for some $A, B \in F[x]$.

One can also use Theorem A.1.14 to determine the ideals of $F[x]$.

Theorem A.1.17. *Every ideal of $F[x]$ is of the form $\langle f \rangle = \{fg \mid g \in F[x]\}$ for some $f \in F[x]$.* □

This is proved in most abstract algebra courses. Recall that the basic idea of the proof is that if $I \subset F[x]$ is a nonzero ideal, then pick $f \in I \setminus \{0\}$ of minimal degree. Then one proves $I = \langle f \rangle$ using the division algorithm.

In general, an integral domain in which every ideal is principal is called a *principal ideal domain*, or PID. It follows that $\mathbb{Z}$ and $F[x]$ are both PIDs.

One can also find unique generators for ideals in $F[x]$. For the zero ideal, the unique generator is of course 0. For nonzero ideals, we can use monic polynomials to give unique generators as follows.

Proposition A.1.18. *Every nonzero ideal of $F[x]$ can be written uniquely as $\langle f \rangle$ where f is monic.* □

Be sure you can prove this proposition.

In the ring of integers $\mathbb{Z}$, prime numbers play a central role. For $F[x]$, the corresponding objects are *irreducible polynomials*. Recall that a nonconstant polynomial in $F[x]$ is *irreducible over F* if it is not a product of polynomials in $F[x]$ of strictly smaller degree.

An important result proved in most abstract algebra texts is that every nonconstant polynomial in $F[x]$ can be factored into a product of irreducibles, where the factorization is unique up to order and multiplication by constants. In the terminology of Section A.5, $F[x]$ is a *unique factorization domain*, or UFD.

Another important result is that the ideal $\langle f \rangle \subset F[x]$ is maximal if and only if the polynomial $f \in F[x]$ is irreducible over F. This is proved in Chapter 3 when we study the existence of roots.

In general, it is not easy to test whether a given polynomial $f \in F[x]$ is irreducible over F. When $\deg(f) = 2$ or 3, any nontrivial factorization of f must have a factor of degree 1. By Corollary A.1.15, having a factor of degree 1 in $F[x]$ is equivalent to having a root in F. Thus we have proved the following.

Lemma A.1.19. *If $f \in F[x]$ has $\deg(f) = 2$ or 3, then f is irreducible over F if and only if f has no roots in F.* □

See Sections A.3, A.5, and 4.2 for more about factorization.

A.2   COMPLEX NUMBERS

In this appendix, we take a naive point of view and regard $\mathbb{C}$ as the set of numbers $a + bi$, where $i = \sqrt{-1}$ and $a, b \in \mathbb{R}$. A rigorous algebraic construction of $\mathbb{C}$ is presented in Chapter 3. Given $z = a + bi$, we define

$$\begin{aligned}
\operatorname{Re}(z) &= a \quad \text{the } \textit{real part} \text{ of } z, \\
\operatorname{Im}(z) &= b \quad \text{the } \textit{imaginary part} \text{ of } z, \\
\overline{z} &= a - bi \quad \text{the } \textit{complex conjugate} \text{ of } z.
\end{aligned}$$

Furthermore, the *absolute value* of $z = a + bi$ is

$$|z| = \sqrt{z\overline{z}} = \sqrt{a^2 + b^2}.$$

A.   Addition, Multiplication, and Division.   Addition and multiplication of complex numbers are defined by

$$\begin{aligned}
(a + bi) + (c + di) &= (a + c) + (b + d)i, \\
(a + bi)(c + di) &= (ac - bd) + (bc + ad)i
\end{aligned}$$

and satisfy

(A.5)
$$\begin{aligned}
\overline{z + w} &= \overline{z} + \overline{w}, \\
\overline{zw} &= \overline{z}\,\overline{w}.
\end{aligned}$$

Under these operations, $\mathbb{C}$ is a ring with additive identity $0 = 0 + 0i$ and multiplicative identity $1 = 1 + 0i$. To see that $\mathbb{C}$ is a field, note that if $z = a + bi \neq 0$ (which means that a and b are not both 0), then

$$\frac{1}{z} = \frac{1}{a + bi} = \frac{1}{a + bi} \cdot \frac{a - bi}{a - bi} = \frac{a - bi}{a^2 + b^2} = \frac{a}{a^2 + b^2} - \frac{b}{a^2 + b^2} i.$$

If we think of $z = a + bi$ as the point (a, b) in the plane, we can also represent z using polar coordinates (r, θ). Since $r = \sqrt{a^2 + b^2} = |z|$, we get the picture

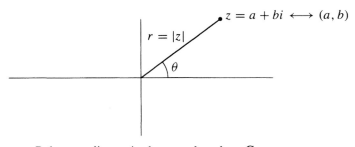

Polar coordinates in the complex plane $\mathbb{C}$

In this situation, we follow Euler and define

(A.6) $$e^{i\theta} = \cos\theta + i\sin\theta.$$

The relation between polar and Cartesian coordinates implies that $a = |z|\cos\theta$ and $b = |z|\sin\theta$. Hence

$$z = a + bi = |z|\cos\theta + |z|\sin\theta\, i = |z|\, e^{i\theta}.$$

This is the *polar representation* of z. In Exercise 1, you will prove that

(A.7) $$|zw| = |z||w|,$$
$$e^{i\theta}e^{i\phi} = e^{i(\theta+\phi)}.$$

It follows that if $z = |z|\, e^{i\theta}$ and $w = |w|\, e^{i\phi}$, then the polar representation of zw is

(A.8) $$zw = |z||w|\, e^{i(\theta+\phi)}.$$

Thus we multiply lengths and add angles when we multiply two complex numbers.

B. Roots of Complex Numbers. We next consider the roots of the polynomial $x^n - a$, where $a \in \mathbb{C}$ and $n \in \mathbb{Z}$ is positive. The solutions of $x^n - a = 0$ are called the *nth roots of a*. To describe the nth roots, write our given complex number a as $a = |a|\, e^{i\theta}$. We will assume that $a \neq 0$, so that $|a|$ is positive. We seek a complex number w such that $w^n = a$. If we write $w = |w|\, e^{i\phi}$, then Exercise 2 implies that

(A.9) $$w^n = |w|^n e^{in\phi},$$

so that the equation $w^n = a$ becomes

$$|w|^n e^{in\phi} = |a|\, e^{i\theta}.$$

This equation is clearly satisfied if $|w|^n = |a|$ and $n\phi = \theta$, that is, if $|w| = \sqrt[n]{|a|}$ and $\phi = \theta/n$. In other words, the complex number

(A.10) $$w = \sqrt[n]{|a|}\, e^{i\theta/n}$$

is an nth root of a.

In polar coordinates, we can change the angle by an integer multiple of 2π without changing the point. For the polar representation $a = |a|\, e^{i\theta}$, this means that we can write $a = |a|\, e^{i(\theta+2\pi m)}$ for any $m \in \mathbb{Z}$. Then, if we apply (A.10) to this representation of a, we get the nth root

(A.11) $$w = \sqrt[n]{|a|}\, e^{i(\theta+2\pi m)/n}.$$

As we vary $m \in \mathbb{Z}$, we claim that this gives precisely n distinct nth roots of a.

To prove this, note that $(\theta + 2\pi m_1)/n$ and $(\theta + 2\pi m_2)/n$ differ by an integer multiple of 2π if and only if $m_1 \equiv m_2 \bmod n$. Hence, in (A.11), we can assume that $m = 0, 1, \ldots, n - 1$, which gives the nth roots

(A.12) $$\sqrt[n]{|a|}\, e^{i\theta/n}, \ \sqrt[n]{|a|}\, e^{i(\theta+2\pi)/n}, \ldots, \sqrt[n]{|a|}\, e^{i(\theta+2\pi(n-1))/n}.$$

Note that θ/n, $(\theta + 2\pi)/n$, ..., $(\theta + 2\pi(n-1))/n$ are n distinct angles in the plane since no two differ by an integer multiple of 2π. Thus we have proved the following.

Proposition A.2.1. *Every $a \neq 0$ in $\mathbb{C}$ has n distinct nth roots (A.12). These are the roots of the polynomial $x^n - a \in \mathbb{C}[x]$.*   □

By Corollary A.1.15, each root gives a linear factor of $x^n - a$. This implies that

(A.13) $$x^n - a = (x - \sqrt[n]{|a|}\, e^{i\theta/n}) \cdots (x - \sqrt[n]{|a|}\, e^{i(\theta + 2\pi(n-1))/n}).$$

We can simplify the above formulas using the nth roots of unity. If we set

$$\zeta_n = e^{2\pi i/n},$$

then (A.9) implies that $\zeta_n^m = e^{2\pi i m/n}$. It follows that when $a = 1$, (A.12) shows that the roots of $x^n - 1$ are given by

$$1, \zeta_n, \zeta_n^2, \ldots, \zeta_n^{n-1}.$$

These are the nth *roots of unity*. In this case, the factorization (A.13) becomes

(A.14) $$x^n - 1 = (x - 1)(x - \zeta_n) \cdots (x - \zeta_n^{n-1}).$$

Returning to the nth roots of $a \in \mathbb{C}$, we can now simplify (A.12). By (A.7),

$$e^{i(\theta + 2\pi m)/n} = e^{i\theta/n} e^{2\pi i m/n} = e^{i\theta/n} \zeta_n^m.$$

Then the nth roots of a given by (A.12) can be written as

$$w_1, \ \zeta_n w_1, \ \zeta_n^2 w_1, \ldots, \ \zeta_n^{n-1} w_1, \quad \text{where } w_1 = \sqrt[n]{|a|}\, e^{i\theta/n},$$

and the factorization (A.13) simplifies to

(A.15)
$$x^n - a = x^n - w_1^n$$
$$= (x - w_1)(x - \zeta_n w_1)(x - \zeta_n^2 w_1) \cdots (x - \zeta_n^{n-1} w_1).$$

For small n, the root of unity $\zeta_n = e^{2\pi i/n} = \cos(\frac{2\pi}{n}) + i \sin(\frac{2\pi}{n})$ is easy to work out. For example, standard facts from trigonometry imply that

$$\zeta_2 = \cos \pi + i \sin \pi = -1,$$
$$\zeta_3 = \cos(\tfrac{2\pi}{3}) + i \sin(\tfrac{2\pi}{3}) = \tfrac{1}{2}(-1 + i\sqrt{3}),$$
$$\zeta_4 = \cos(\tfrac{\pi}{2}) + i \sin(\tfrac{\pi}{2}) = i.$$

Thus the square roots of unity are $1, -1$; the cube roots are $1, \zeta_3, \zeta_3^2$; and the fourth roots are $\pm 1, \pm i$. We often denote the cube root of unity ζ_3 by ω. Exercise 6 below will show how the formula for $\omega = \zeta_3$ follows from the quadratic formula.

Roots of unity appear in several places in the text: in Chapter 8, where we study solvability by radicals; in Chapter 9, where we compute the minimal polynomial of ζ_n; and in Chapter 10, where we explore Gauss's work on the constructibility of regular polygons.

Exercises for Section A.2

Exercise 1. Prove (A.7).

Exercise 2. Let $z = |z| e^{i\theta}$ be the polar representation of $z \in \mathbb{C}$. Prove that

$$z^n = |z|^n e^{in\theta}, \quad n > 0 \text{ in } \mathbb{Z},$$

using induction on n and (A.8).

Exercise 3. This exercise will discuss *De Moivre's formula*, which states that

$$(\cos\theta + i\sin\theta)^n = \cos n\theta + i\sin n\theta, \quad n > 0 \text{ in } \mathbb{Z}.$$

(a) Show that De Moivre's formula follows from (A.6) and Exercise 2.
(b) Use De Moivre's formula and the binomial theorem for $n = 4$ and 5 to express $\cos 4\theta$, $\sin 4\theta$, $\cos 5\theta$, and $\sin 5\theta$ in terms of $\cos\theta$ and $\sin\theta$.
(c) Use De Moivre's formula and the binomial theorem to prove that $\cos n\theta$ can be written as a polynomial in $\cos\theta$ with integer coefficients.

Exercise 4. Use a calculator to find a seventh root of $3 + 2i$. Note that $\theta = \tan^{-1}(\frac{2}{3})$.

Exercise 5. For $n = 4$, 5, and 6, draw a picture to show how the nth roots of unity form the vertices of a regular n-gon inscribed in the unit circle in the complex plane.

Exercise 6. The cube root of unity $\omega = \zeta_3$ is a root of $x^3 - 1 = (x - 1)(x^2 + x + 1)$. Use the quadratic formula to show that ω and ω^2 are given by $\frac{1}{2}(-1 \pm i\sqrt{3})$.

Exercise 7. Use $\zeta_n^n = 1$ and $|\zeta_n| = 1$ to show that $\overline{\zeta_n} = \zeta_n^{n-1} = 1/\zeta_n$.

Exercise 8. This exercise will derive an explicit formula for the fifth root of unity ζ_5 using the factorization $x^5 - 1 = (x - 1)(x^4 + x^3 + x^2 + x + 1)$.
(a) Use Exercise 7 to show that if $x = \zeta_5$, then $x + 1/x = 2\cos(\frac{2\pi}{5})$
(b) Explain why $x = \zeta_5$ satisfies $x^2 + x + 1 + 1/x + 1/x^2 = 0$. Then show that $y = x + 1/x$ is a root of $y^2 + y - 1$.
(c) Use part (b) to conclude that $\cos(\frac{2\pi}{5}) = \frac{1}{4}(-1 + \sqrt{5})$. Then show that

$$\zeta_5 = \frac{-1 + \sqrt{5}}{4} + \frac{i}{2}\sqrt{\frac{5 + \sqrt{5}}{2}}.$$

Chapter 10 explains how this relates to the straightedge-and-compass construction of the regular pentagon.

Exercise 9. In this exercise, you will give two proofs of the identity

$$1 + \zeta_n + \zeta_n^2 + \cdots + \zeta_n^{n-1} = 0, \quad n > 0 \text{ in } \mathbb{Z}.$$

(a) Show that this identity follows from (A.14) by comparing the coefficients of x^{n-1}.
(b) Give a second proof using the factorization $x^n - 1 = (x - 1)(x^{n-1} + \cdots + 1)$.

(c) More generally, use part (b) to show that $m \not\equiv 0$ mod n implies that

$$1 + \zeta_n^m + \zeta_n^{2m} + \cdots + \zeta_n^{(n-1)m} = 0.$$

Also determine the sum on the left-hand side when $m \equiv 0$ mod n.

Exercise 10. The eighth root of unity ζ_8 is given by $\cos(\frac{\pi}{4}) + i \sin(\frac{\pi}{4}) = \frac{1}{\sqrt{2}}(1 + i)$.

(a) Show that the eighth roots of unity are given by $\pm 1, \pm i, \frac{1}{\sqrt{2}}(\pm 1 \pm i)$.

(b) Use the factorization of $x^4 + 1$ given at the end of Section A.3 to show that

$$x^8 - 1 = (x - 1)(x + 1)(x^2 + 1)(x^2 + \sqrt{2}x + 1)(x^2 - \sqrt{2}x + 1),$$

and explain how this factorization relates to part (a).

A.3 POLYNOMIALS WITH RATIONAL COEFFICIENTS

We next discuss the polynomial ring $\mathbb{Q}[x]$. In this case, we often take a polynomial with rational coefficients and multiply it by a constant to clear denominators, giving a polynomial with integer coefficients. In general, we let $\mathbb{Z}[x]$ denote the ring of polynomials in x with coefficients in $\mathbb{Z}$.

As is well known, we can describe the rational roots of $f \in \mathbb{Z}[x]$ as follows.

Proposition A.3.1. *Let $f = a_n x^n + \cdots + a_0 \in \mathbb{Z}[x]$ be nonconstant. If $p/q \in \mathbb{Q}$ is a root of f, where $p, q \in \mathbb{Z}$ are relatively prime, then $p|a_0$ and $q|a_n$.*   □

Note that combining Lemma A.1.19 and Proposition A.3.1 gives an algorithm for deciding whether a polynomial in $\mathbb{Q}[x]$ of degree 2 or 3 is irreducible over $\mathbb{Q}$. In Section 4.2, we show that a similar algorithm exists when the degree is greater than 3. The crucial result, due to Gauss, is that we can reduce factorization in $\mathbb{Q}[x]$ to factorization in $\mathbb{Z}[x]$. This is *Gauss's Lemma*, which is stated as follows.

Theorem A.3.2. *Suppose that $f \in \mathbb{Z}[x]$ is nonconstant and that $f = gh$ where $g, h \in \mathbb{Q}[x]$. Then there is a nonzero $\delta \in \mathbb{Q}$ such that $\tilde{g} = \delta g$ and $\tilde{h} = \delta^{-1}h$ have integer coefficients. Thus $f = \tilde{g}\tilde{h}$ in $\mathbb{Z}[x]$.*

Proof. Let $s \in \mathbb{Z}$ be a common denominator for the coefficients of g, so that $g = \frac{1}{s}g_1$ for $g_1 \in \mathbb{Z}[x]$. Then let $r \in \mathbb{Z}$ be the greatest common divisor of the coefficients of g_1. Factoring out r enables us to write

$$g = \frac{1}{s}g_1 = \frac{r}{s}g_2,$$

where $g_2 \in \mathbb{Z}[x]$ has relatively prime coefficients. Similarly, we can write

$$h = \frac{1}{u}h_1 = \frac{t}{u}h_2,$$

where $h_1, h_2 \in \mathbb{Z}[x]$ and h_2 has relatively prime coefficients.

Let $\delta = \frac{s}{r}$, and observe that

$$\delta g = \frac{s}{r} \cdot \frac{r}{s} g_2 = g_2 \in \mathbb{Z}[x],$$

$$\delta^{-1} h = \frac{r}{s} \cdot \frac{t}{u} h_2 = \frac{rt}{su} h_2.$$

If we can show that $su|rt$, then $\frac{rt}{su} h_2 \in \mathbb{Z}[x]$ will follow, since $h_2 \in \mathbb{Z}[x]$. This will prove the theorem.

Hence it remains to show that $su|rt$. For this purpose, we will prove that if p is prime, then $p^a|su$ implies that $p^a|rt$ (do you see why this implies $su|rt$?). So pick a prime p and suppose that $p^a|su$. Then write

(A.16)
$$g_2 = b_l x^l + \cdots + b_0,$$
$$h_2 = c_m x^m + \cdots + c_0.$$

Since $b_l, \ldots, b_0$ have no nontrivial common factors, we can find an index $i \geq 0$ such that $p|b_0, \ldots, p|b_{i-1}$ and $p \nmid b_i$. Similarly, there is an index $j \geq 0$ such that $p|c_0, \ldots, p|c_{j-1}$ and $p \nmid c_j$.

Multiplying the expressions for g_2 and h_2 given in (A.16), we see that the coefficient of x^{i+j} in $g_2 h_2$ is $d_{i+j} = b_0 c_{i+j} + \cdots + b_{i+j} c_0$. We can write this in the following form:

(A.17) $d_{i+j} = \underbrace{b_0 c_{i+j} + \cdots + b_{i-1} c_{j+1}}_{p \text{ divides } b_0, \ldots, b_{i-1}} + b_i c_j + \underbrace{b_{i+1} c_{j-1} + \cdots + b_{i+j} c_0}_{p \text{ divides } c_{j-1}, \ldots, c_0}.$

Since $p \nmid b_i c_j$, this shows that d_{i+j} is relatively prime to p. Thus $\gcd(p^a, d_{i+j}) = 1$.

Next observe that $f = gh = \frac{r}{s} g_2 \cdot \frac{t}{u} h_2$ implies that

(A.18) $suf = rt g_2 h_2.$

Since $p^a|su$ and $f \in \mathbb{Z}[x]$, we see that p^a divides the coefficient of x^{i+j} on the left-hand side of (A.18). However, the coefficient of x^{i+j} on the right-hand side of (A.18) is $rt d_{i+j}$, and it follows that p^a divides $rt d_{i+j}$. Since p^a is relatively prime to d_{i+j}, we conclude that p^a must divide rt, which is what we needed to show. This completes the proof. $\square$

We will generalize in Section A.5.

A.4   GROUP ACTIONS

Some of the most interesting groups arise as the symmetries of particular mathematical objects. This leads to the notion of a *group action*. Here is the precise definition.

Definition A.4.1. *Let G be a group and X be a set. Then an **action** of G on X is a function $G \times X \to X$, written $(g, x) \mapsto g \cdot x$, such that*

(a) $e \cdot x = x$ for all $x \in X$.

(b) $g \cdot (h \cdot x) = (gh) \cdot x$ for all $g, h \in G$ and $x \in X$.

Here are some simple examples of group actions.

Example A.4.2. The symmetric group S_n acts on the set $X = \{1, 2, \ldots, n\}$. If $\sigma \in S_n$ and $i \in X$, then $\sigma \cdot i$ is just $\sigma(i)$. ◁▷

Example A.4.3. Let $GL(n, \mathbb{R})$ be the set of invertible $n \times n$ matrices with real entries. This is a group under matrix multiplication and acts on $\mathbb{R}^n$ as follows: if $A \in GL(n, \mathbb{R})$ and $v \in \mathbb{R}^n$, then $A \cdot v$ is the matrix product Av, where we think of v as a column vector. ◁▷

Example A.4.4. Let $S^1 = \{e^{i\theta} \mid \theta \in \mathbb{R}\}$ be the set of complex numbers of absolute value 1. This is a group under multiplication of complex numbers, and S^1 acts on $\mathbb{C}$ by multiplication. ◁▷

We next define some important concepts related to a group action.

Definition A.4.5. *Let a group G act on a set X, and let $x \in X$.*

(a) *The **orbit** of x is the set $G \cdot x = \{g \cdot x \mid g \in G\}$.*

(b) *The **isotropy subgroup** of x is the subgroup $G_x = \{g \in G \mid g \cdot x = x\}$.*

Here are some examples of orbits and isotropy subgroups.

Example A.4.6. In the action of S_n on $X = \{1, 2, \ldots, n\}$, the orbit of any $i \in X$ is all of X, and the isotropy subgroup of i consists of all permutations which fix i. Do you see why the isotropy subgroup is isomorphic to S_{n-1}? ◁▷

Example A.4.7. Let $H = \langle \sigma \rangle$ be the cyclic subgroup generated by $\sigma \in S_n$. Then H acts on $X = \{1, 2, \ldots, n\}$. In Exercise 1, you will show that the orbits of this action correspond to the decomposition of σ into a product of disjoint cycles. ◁▷

Example A.4.8. In the action of S^1 on $\mathbb{C}$, consider a point $z \neq 0$. Then the orbit $S^1 \cdot z$ of z is the circle of radius $|z|$ centered at the origin, and the isotropy subgroup of z is trivial. ◁▷

If G acts on X, then one can easily show that

$$x \sim y \iff x = g \cdot y \text{ for some } g \in G$$

is an equivalence relation on X whose equivalence classes are the orbits of G. It follows that X is a disjoint union of orbits. Furthermore, the isotropy subgroups of points on the same orbit are related as follows

(A.19) $$G_{g \cdot x} = g G_x g^{-1}$$

for $g \in G$ and $x \in X$. Proofs of these assertions can be found in Section 1.12 of Volume I of [Jacobson].

The *Fundamental Theorem of Group Actions* relates orbits to cosets of the isotropy subgroup as follows.

Theorem A.4.9. *Let G act on X, and let $G_x \subset G$ be the isotropy subgroup of $x \in X$. Then:*

(a) *There is a one-to-one correspondence*

$$\{\text{left cosets of } G_x \text{ in } G\} \simeq G \cdot x.$$

(b) *If G is finite, then*

$$[G : G_x] = |G \cdot x|.$$

Thus $|G| = |G_x||G \cdot x|$, so that $|G|$ is divisible by both $|G_x|$ and $|G \cdot x|$.

Proof. Let $G/G_x = \{gG_x \mid g \in G\}$ be the set of left cosets of the isotropy subgroup. Then define $\varphi : G/G_x \to G \cdot x$ by

$$\varphi(gG_x) = g \cdot x.$$

We first need to show that this map is well defined. If $g_1 G_x = g_2 G_x$, then $g_1 = g_2 h$ for some $h \in G_x$ (be sure you know why). Then

$$g_1 \cdot x = (g_2 h) \cdot x = g_2 \cdot (h \cdot x) = g_2 \cdot x,$$

where the second equality follows from Definition A.4.1, and the third follows from $h \in G_x$. This proves that φ is well defined.

Since every $y \in G \cdot x$ is of the form $y = g \cdot x$ for some $g \in G$, we see that

$$y = g \cdot x = \varphi(gG_x).$$

Thus φ is onto. To show that φ is also one to one, suppose that $\varphi(g_1 G_x) = \varphi(g_2 G_x)$. By the definition of φ, this implies that

$$g_1 \cdot x = g_2 \cdot x.$$

Using the properties of group actions, we obtain

$$x = e \cdot x = (g_1^{-1} g_1) \cdot x = g_1^{-1} \cdot (g_1 \cdot x) = g_1^{-1} \cdot (g_2 \cdot x) = (g_1^{-1} g_2) \cdot x.$$

Thus $g_1^{-1} g_2 \in G_x$, so that $g_1 G_x = g_2 G_x$. Hence φ is one to one. From here, the rest of the theorem follows easily. □

Chapter 12 discusses the special case of this theorem discovered by Lagrange in his study of the roots of polynomials.

The following definition is used in Section 6.3 when we study how the Galois group acts on the roots of a polynomial.

Definition A.4.10. *The action of G on X is **transitive** if for every $x, y \in X$, there is $g \in G$ such that $g \cdot x = y$.*

More on group actions may be found in Section 1.12 of Volume I of [Jacobson].

Exercises for Section A.4

Exercise 1. As in Example A.4.7, let $H = \langle \sigma \rangle$ be the cyclic subgroup generated by $\sigma \in S_n$. Assume that $\tau \neq e$ and that

$$\sigma = \tau_1 \tau_2 \cdots \tau_r$$

is the decomposition of σ into a product of disjoint cycles. Suppose that $\tau_1 = (i_1 \cdots i_l)$.
(a) Use (A.1) to prove that $\{i_1, \ldots, i_l\}$ is the orbit of i_1 under the action of H.
(b) Explain why Theorem A.4.9 implies that l divides the order of σ.

Exercise 2. In the group action considered in Example A.4.8, find the orbit and isotropy subgroup of $0 \in \mathbb{C}$.

Exercise 3. The symmetric group S_3 is sometimes introduced as the symmetry group of an equilateral triangle $\triangle$.
(a) In the language of this section, explain how S_3 acts on $\triangle$. You may assume that the vertices of $\triangle$ are labeled 1, 2, 3.
(b) For each subgroup of S_3 given in Example A.1.6, determine all points $p \in \triangle$ whose isotropy subgroup is the given subgroup of S_3. Also describe the orbit of p.

Exercise 4. Given a group G, define $g \cdot h = ghg^{-1}$ for $g, h \in G$. Prove that this is a group action of G on itself. We say that G acts on itself by conjugation. Then:
(a) Prove that the orbit $G \cdot g$ is the conjugacy class C_g of g and that the isotropy subgroup is the subgroup $C(g)$ consisting of all elements of G that commute with g.
(b) Let G be finite. Prove that $[G : C(g)] = |C_g|$.

Exercise 5. Prove that a group G acts transitively on a set X if and only if $G \cdot x = X$ for all $x \in X$ if and only if $G \cdot x = X$ for some $x \in X$.

A.5   MORE ALGEBRA

Here are some further results about groups, rings, fields, and polynomials that will be used in the text.

A. The Sylow Theorems. Let G be a finite group, and let p be a prime dividing the order of G. Then a subgroup $H \subset G$ is called *p-Sylow subgroup* if $|H| = p^n$, where p^n is the highest power of p dividing $|G|$. Here is the basic result concerning Sylow subgroups.

Theorem A.5.1. *Let p be a prime dividing the order of a finite group G. Then:*
(a) *(First Sylow Theorem) G has a p-Sylow subgroup.*
(b) *(Second Sylow Theorem) Any two p-Sylow subgroups of G are conjugate in G.*
(c) *(Third Sylow Theorem) Let N be the number of p-Sylow subgroups of G. Then $N \equiv 1 \bmod p$, and N divides $|G|$.* □

A proof of the First Sylow Theorem can be found in [Herstein, Thm. 2.12.1], and the Second and Third Sylow Theorems are proved in [Herstein, Thms. 2.12.2 and 2.12.3 and Lem. 2.12.6].

The Sylow Theorems have some nice applications in Chapters 8 and 14.

B. The Chinese Remainder Theorem. The following result will be useful in several places in the text. Given a positive integer n, let $[a]_n \in \mathbb{Z}/n\mathbb{Z}$ denote the congruence class of a modulo n.

Lemma A.5.2. *Let n and m be relatively prime positive integers. Then the map $[a]_{nm} \mapsto ([a]_n, [a]_m)$ gives a well-defined ring isomorphism*

$$\mathbb{Z}/nm\mathbb{Z} \simeq \mathbb{Z}/n\mathbb{Z} \times \mathbb{Z}/m\mathbb{Z}.$$

Proof. If $[a]_{nm} = [b]_{nm}$, then $nm \mid a - b$, from which we conclude that $([a]_n, [a]_m) = ([b]_n, [b]_m)$. Hence the map is well defined, and it is easy to see that it is a ring homomorphism. Furthermore, if $[a]_{nm}$ is in the kernel, then $n \mid a$ and $m \mid a$, which implies that $nm \mid a$, since n and m are relatively prime. Thus $[a]_{nm} = [0]_{nm}$, so that the map is one to one. It is then onto, since both rings have order nm. $\qquad\square$

C. The Multiplicative Group of a Field. Given a field F, its *multiplicative group* is $F^* = F \setminus \{0\}$, which is a group under multiplication by the definition of field. The fact that a polynomial of degree m has at most m roots in a field implies the following interesting fact about F^*.

Proposition A.5.3. *Let $G \subset F^*$ be a finite subgroup of the multiplicative group of a field F. Then G is cyclic.*

Proof. First observe that G is Abelian because F is a field. Then Theorem A.1.7 implies that G is isomorphic to a product of cyclic groups, say

$$G \simeq \mathbb{Z}/m_1\mathbb{Z} \times \cdots \times \mathbb{Z}/m_r\mathbb{Z},$$

where $m_1, \ldots, m_r$ are integers > 1. Thus $|G| = m_1 \cdots m_r$. If $r = 1$, then we are done. So assume that $r \geq 2$.

Let $m = \mathrm{lcm}(m_1, \ldots, m_r)$ be the least common multiple of the m_i. It is then easy to verify that $g^m = 1$ for every $g \in G$. Since G is a subgroup of F^*, it follows that every $g \in G$ is a root of $x^m - 1 \in F[x]$. Hence this polynomial has at least $|G| = m_1 \cdots m_r$ roots in F. But, as noted above, $x^m - 1$ has at most m roots in F, since F is a field. Thus

$$m = \mathrm{lcm}(m_1, \ldots, m_r) \geq m_1 \cdots m_r,$$

which clearly implies that $\mathrm{lcm}(m_1, \ldots, m_r) = m_1 \cdots m_r$. This in turn implies that $m_1, \ldots, m_r$ are pairwise relatively prime (be sure you understand why). However, if n and m are relatively prime, then by Lemma A.5.2, there is a ring isomorphism

$$\mathbb{Z}/n\mathbb{Z} \times \mathbb{Z}/m\mathbb{Z} \simeq \mathbb{Z}/nm\mathbb{Z},$$

which is also a group isomorphism if we forget multiplication. Using this repeatedly, we obtain a group isomorphism

$$G \simeq \mathbb{Z}/m_1\mathbb{Z} \times \cdots \times \mathbb{Z}/m_r\mathbb{Z} \simeq \mathbb{Z}/(m_1 \cdots m_r)\mathbb{Z}.$$

This completes the proof of the proposition. $\qquad\square$

D. Unique Factorization Domains. Given a ring R, a *unit* of R is an element of R that has a multiplicative inverse in R. The set of all units of R is denoted R^*. Note that R^* is a group under multiplication.

Now let R be an integral domain. We say that $r \in R$ is *irreducible* if it is not a unit and $r = ab, a, b \in R$, implies that a or b is in R^*.

Example A.5.4. Given a field F, the units of the polynomial ring $F[x]$ are the nonzero elements of F, i.e., $F[x]^* = F^*$. Furthermore, the irreducible elements of $F[x]$, as defined above, are precisely the irreducible polynomials of $F[x]$.   ◁▷

Here is the precise definition of unique factorization domain.

Definition A.5.5. *An integral domain R is a **unique factorization domain**, or **UFD**, if the following two conditions hold:*
(a) *Every nonzero element of R is either a unit or a product of irreducibles.*
(b) *If $r_1 \cdots r_k = s_1 \cdots s_l$, where $r_1, \ldots, r_k, s_1, \ldots, s_l \in R$ are irreducible, then $k = l$, and there is a permutation $\sigma \in S_k$ such that for each $1 \le i \le k$ there is a unit $a_i \in R^*$ such that $r_i = a_i s_{\sigma(i)}$.*

The basic example of a UFD is the ring of integers $\mathbb{Z}$. Another important class of examples come from polynomial rings. Here is the basic result.

Theorem A.5.6. *Let R be a UFD, and let $R[x]$ be the ring of polynomials in a variable x with coefficients in R. Then $R[x]$ is a UFD.*   □

A proof can be found in [Herstein, Thm. 3.11.1] or [Jacobson, Vol. I, Thm. 2.25]. This result implies, for example, that $\mathbb{Z}[x]$ is a UFD. In Chapter 2 we will discuss the ring $F[x_1, \ldots, x_n]$ of polynomials in $x_1, \ldots, x_n$ with coefficients in F. Using Theorem A.5.6 and induction on the number of variables, it is straightforward to prove the following.

Corollary A.5.7. *If F is a field, then $F[x_1, \ldots, x_n]$ is a UFD.*   □

In the course of proving Theorem A.5.6, one needs the following generalization of Gauss's Lemma (Theorem A.3.2).

Theorem A.5.8. *Let R be a UFD with field of fractions K. Suppose that $f \in R[x]$ is nonconstant and that $f = gh$ where $g, h \in K[x]$. There is a nonzero $\delta \in K$ such that $\tilde{g} = \delta g$ and $\tilde{h} = \delta^{-1}h$ have coefficients in R. Thus $f = \tilde{g}\tilde{h}$ in $R[x]$.*   □

The proof is identical to the proof of Theorem A.3.2 given in Section A.3. An immediate corollary of Theorem A.5.8 is that if $f \in R[x]$ is irreducible and nonconstant, then it is also irreducible in $K[x]$. Furthermore, if $f, g \in R[x]$ are relatively prime and nonconstant, then Theorem A.5.8 implies that they are also relatively prime in $K[x]$.

Appendix B
Hints to Selected
Exercises

This appendix contains hints to selected exercises in the text.

Section 1.1 (pages 9–10)

Exercise 2. Hint: Explain why $\overline{\omega} = \omega^2$.

Exercise 3. Hint: By choosing the correct square root of q^2, show that Cardan's formulas reduce to $y_1 = \sqrt[3]{-q}$, $y_2 = \omega^2 \sqrt[3]{-q}$, and $y_3 = \omega \sqrt[3]{-q}$ when $p = 0$.

Exercise 7. Hint: First show that all three polynomials give the same z_1 but a different z_2.

Exercise 8. Hint: Use Example 1.1.1 and Exercise 7.

Section 1.3 (pages 20–22)

Exercise 1. (c) Hint: $f'(y_1) = 3(y_1 - \alpha)(y_1 - \beta)$ and $f(\alpha) = (\alpha - y_1)(\alpha - y_2)(\alpha - y_3)$.

Exercise 2. (a) Hint: Use (1.22).

Exercise 3. (a) Hint: Remember that y_1, y_2, y_3 are distinct.

Exercise 4. (b) Hint: Part (c) of Exercise 1 and $\Delta \neq 0$ imply that $f(\alpha)$ and $f(\beta)$ are nonzero.

Exercise 6. Hint: By part (a) of Exercise 3, $\Delta > 0$ implies that $p \neq 0$.

Exercise 7. Hint: Dividing $y^3 - 15y - 4$ by $y - 4$ leads to a quadratic equation.

Exercise 11. (b) Hint: Use Exercise 3 of Section A.2. (c) Hint: $\omega = e^{2\pi i/3}$.

Section 2.1 (pages 29–30)

Exercise 1. Hint: Give a proof by contradiction. Unique factorization will be useful. You may assume that x and y are irreducible in $F[x, y]$.

Exercise 3. (c) Hint: Let $\alpha_1 = \cdots = \alpha_n = -\alpha$ in the corollary.

Section 2.2 (pages 38–41)

Exercise 1. Hint: Let $i_1 < \cdots < i_r$. If $i_r > 1$, then the exponent of x_1 in $x_{i_1} \cdots x_{i_r}$ is 0.

Exercise 2. (b) Hint: Use part (a).

Exercise 4. Hint: Express the coefficients of (2.17) in terms of $\sigma_1, \sigma_2, \sigma_3$ evaluated at the α_i.

Exercise 7. Hint: If $\tau \in S_n$ is fixed, you will need to explain why $\tau\sigma$ ranges over all elements of S_n as σ does.

Exercise 9. (c) Hint: Use the *well-ordering* property of the nonnegative integers, which states that any strictly decreasing sequence of nonnegative integers is finite.

Exercise 11. Hint: Use the method of Example 2.2.6 and Exercise 4.

Exercise 16. (c) Hint: You can't use the formulas of Chapter 1. So you need to compute $(1 - \omega)^2(1 - \omega^2)^2(\omega - \omega^2)^2$.

Exercise 18. Hint: Use the Newton identities and explain why every s_k is a polynomial in the σ_i with coefficients in $\mathbb{Z}$.

Exercise 20. Hint: Suppose that $\sigma_2 = P(s_1, \ldots, s_n)$. Then evaluate this at $x_1 = \cdots = x_n = 0$ and at $x_1 = x_2 = 1, x_3 = \cdots = x_n = 0$.

Section 2.3 (pages 45–46)

Exercise 1. Hint: To find the roots of $y^3 + 2y^2 - 3y + 5$, use the *Mathematica* command

$$\text{N[Solve[y\textasciicircum3} + 2\text{y\textasciicircum2} - 3\text{y} + 5 == 0, \text{y}]]$$

or the *Maple* command

$$\text{fsolve(y\textasciicircum3} + 2 * \text{y\textasciicircum2} - 3 * \text{y} + 5 = 0, \text{y}, \text{complex});$$

Note that fsolve normally only finds real roots, but by specifying the complex option, it will find all roots, real and complex.

Exercise 4. Hint: If the roots are x_1, x_2, x_3, then one way for this to happen is $x_1 = (x_2+x_3)/2$, which gives the equation $2x_1 - x_2 - x_3 = 0$. There are two other ways this can happen. Then take the product of all three ways.

Section 2.4 (pages 51–52)

Exercise 2. Hint: Use Theorem 2.4.4.

Exercise 5. Hint: Use Proposition 2.4.1.

Exercise 8. (e) Hint: F has characteristic $\neq 2$. (f) Hint: Use Exercise 5.

Exercise 10. (a) Hint: $F[\sigma_1, \ldots, \sigma_n] \simeq F[u_1, \ldots, u_n]$ is a UFD.

Section 3.1 (page 62)

Exercise 2. Hint: The definition of ring homomorphism given in Section A.1 requires that φ preserve the multiplicative identity. Also remember that a homomorphism is one to one if and only if its kernel is $\{0\}$.

Section 3.2 (page 69)

Exercise 2. (c) Hint: Solve the second equation of part (b) for y, and substitute the result into the first. Also remember that $b \neq 0$.

Exercise 3. Hint: Apply the IVT to $x^2 - a$ on a suitably chosen interval.

Exercise 5. (b) Hint: This follows from Lemma 3.2.3 and Exercise 4 with $\mathbb{R}$ replaced by F.

Exercise 6. Hint: Use part (b) of Exercise 1.

Section 4.1 (pages 80–81)

Exercise 1. Hint: When $f(\alpha) = 0$ for $f \in F[x]$ nonzero, what equation is satisfied by $1/\alpha$?

Exercise 4. Hint: First use Lemma 4.1.9 to show that $F(\alpha_1, \ldots, \alpha_r) \subset F(\alpha_1, \ldots, \alpha_n)$. Then use the lemma a second time.

Exercise 6. Hint: Show that the ring homomorphism $F[x_1, \ldots, x_n] \to F[\alpha_1, \ldots, \alpha_n]$ given by $x_i \mapsto \alpha_i$ is an isomorphism. Then explain why this extends to an isomorphism of the fields of fractions.

Exercise 7. (a) Hint: Remember that $g(\alpha) \neq 0$.

Exercise 8. (a) Hint: First explain why it suffices to show that $\sqrt{3} \notin \mathbb{Q}(\sqrt{2})$. You may assume that $\sqrt{2}$, $\sqrt{3}$, and $\sqrt{6}$ are irrational. (b) Hint: What is $\alpha - \sqrt{3}$?

Section 4.2 (page 88)

Exercise 1. (c) Hint: How many roots does $h - g$ have? What is its degree? Corollary A.1.16 will be useful.

Exercise 5. (b) Hint: Note that ζ_2, ζ_3, ζ_4, ζ_6, ζ_8, and ζ_{12} are also roots of $x^{24} - 1$. What are the minimal polynomials of these numbers?

Exercise 9. Hint: Suppose that $(g/h)^p = t$, where $g, h \in k[t]$ are relatively prime. Show that $g^p = th^p$ would imply that first g and then h are divisible by t.

Section 4.3 (page 94)

Exercise 6. Hint: Compute $[F(\alpha, \beta) : F]$ in two ways.

Section 4.4 (page 98)

Exercise 2. (a) Hint: Consider the field extensions

$$\mathbb{Q} \subset \mathbb{Q}(\sqrt{2}) \subset \mathbb{Q}(\sqrt{2}, \sqrt{5}) \subset \mathbb{Q}(\sqrt{2}, \sqrt{5}, \sqrt[4]{12}) \subset \cdots \subset L.$$

Exercise 3. (a) Hint: Use Gauss's Lemma. (b) Hint: The minimal polynomial of ω over $\mathbb{Q}$ is $x^2 + x + 1$.

Exercise 6. Hint: If an element of $F(x)$ is algebraic over F, write it as p/q where $p, q \in F[x]$ are relatively prime. If $a_n (p/q)^n + \cdots + a_0 = 0$, where $a_i \in F$, then clear denominators and use unique factorization to conclude that $p, q \in F$.

Exercise 7. Hint: Take $\alpha \in L$ and consider the minimal polynomial of α over F.

Exercise 9. Hint: Use Exercise 3.

Section 5.1 (page 106)

Exercise 3. Hint: Use Exercise 4 of Section 4.3 to show that $L = F(\alpha)$ for some $\alpha \in L$. Then let f be the minimal polynomial of α over F.

Exercise 4. Hint: Consider $-\omega$, where $\omega = e^{2\pi i/3}$.

Exercise 5. Hint: In Section 4.2, we used *Maple* and *Mathematica* to factor f in $L[x]$.

Exercise 6. Hint: Compute $\sqrt{2 + \sqrt{2}} \cdot \sqrt{2 - \sqrt{2}}$.

Exercise 7. (b) Hint: If $\alpha \in L$ is a root of f, then compute $f(\alpha + 1)$.

Exercise 8. (b) Hint: Use Proposition 4.2.5 and the method of Exercise 5 of Section 4.3.

Exercise 9. (a) Hint: Combine $[L : F] = n!$ with the proof of Theorem 5.1.5.

Exercise 11. (a) Hint: Consider $F \subset F(\alpha) \subset L$, where $\alpha \in L$ is a root of F.

Exercise 13. Hint: Apply the proposition to $\mathbb{Q}(\sqrt{2}) \subset L$. Part (a) of Exercise 7 from Section 4.1 will be useful.

Section 5.2 (page 109)

Exercise 3. (c) Hint: Compute $(x - \alpha)^3$.

Exercise 4. Hint: Use Theorem 4.4.10 and Exercise 1 from Section 4.4.

Section 5.3 (pages 117–118)

Exercise 2. (a) Hint: Treat the cases $p = 2$ and $p > 2$ separately.

Exercise 3. (b) Hint: Use Lemma 5.3.10.

Exercise 4. Hint: Recall how $\Delta(f)$ and $\Delta(f_p)$ are obtained from Δ. You may assume that Δ is a polynomial in $\mathbb{Z}[\sigma_1, \ldots, \sigma_n]$.

Exercise 6. Hint: Remember that the given polynomial need not be irreducible.

Exercise 7. (b) Hint: Look at the exponents of the nonzero terms of f.

Exercise 8. (a) Hint: Write $F = k(u)(t)$ and use Exercise 9 of Section 4.2. (b) Hint: Remember that k has characteristic 3.

Exercise 9. (b) Hint: Express β as a polynomial in α.

Exercise 12. Hint: Use Lemma 5.3.5.

Section 5.4 (page 123)

Exercise 4. (c) Hint: Use part (b) and Example 5.4.4.

Exercise 5. (a) Hint: First explain why $\alpha + \lambda\beta$ and $\alpha + \mu\beta$ lie in $F(\alpha + \lambda\beta)$.

Exercise 7. Hint: Use Theorem 5.4.1 and the previous exercise.

Section 6.1 (pages 129–130)

Exercise 4. (b) Hint: Think about kernels of ring homomorphisms and ideals of fields.

Exercise 6. Hint: What is $\sqrt{6} \cdot \sqrt{15}$?

Exercise 7. (a) Hint: See part (b) of Exercise 4. (b) Hint: Regard L as a vector space over F, and show that σ is a linear map. Now use standard results of linear algebra.

Section 6.2 (page 132)

Exercise 3. (a) Hint: Use the method of Exercise 5 of Section 4.3.

Exercise 6. Hint: Use Exercise 11 of Section 5.1.

Section 6.3 (pages 135–6)

Exercise 4. Hint: Write down the roots of f explicitly, and determine how the elements of $\text{Gal}(L/\mathbb{Q})$, as described in the proof of Theorem 6.2.1, act on the roots. Then look at the corresponding permutations in S_4.

Exercise 6. Hint: Use Theorem A.4.9 from Section A.4.

Section 6.4 (page 142)

Exercise 4. Hint: Can you find an inverse function?

Section 6.5 (page 145)

Exercise 3. Hint: Lemma 6.1.3 will be useful.

Exercise 4. Hint: Use the nth roots of unity, and show that $\theta_i(x)$ can be chosen to be x^i provided the roots of $x^n - 1 = 0$ are labeled appropropriately.

Exercise 7. Hint: Use Exercise 6.

Section 7.1 (page 153)

Exercise 3. Hint: Use Proposition 7.1.6.

Exercise 5. Hint: Use part (b) of Proposition 7.1.7 twice.

Exercise 7. Hint: This is similar to Exercise 5.

Exercise 8. Hint: Let $\alpha_1 = \alpha, \alpha_2, \ldots, \alpha_r$ be as in the definition of h, and consider the subgroup $H = \{\sigma \in \mathrm{Gal}(L/F) \mid \sigma(\alpha) = \alpha\}$. Then study how the left cosets of H act on α.

Exercise 12. (a) Hint: Do not use the Theorem of the Primitive Element—give a direct proof. (b) Hint: Use Lemma 5.3.5.

Section 7.2 (pages 160–161)

Exercise 2. Hint: Show that $\sigma^{-1}\mathrm{Gal}(L/\sigma K)\sigma \subset \mathrm{Gal}(L/K)$ follows from $\sigma^{-1}(\sigma K) = K$.

Exercise 7. Hint: This follows from the argument used to prove (a) $\Leftrightarrow$ (b) in Theorem 7.2.5.

Exercise 9. (b) Hint: Use Exercise 7.

Section 7.3 (pages 166–167)

Exercise 5. (b) Hint: Use the automorphism $L \simeq L$ which sends t to it.

Exercise 7. (a) Hint: Proposition 4.2.5 will be useful.

Exercise 9. Hint for (a) $\Rightarrow$ (b): If $\mathrm{Gal}(L/F) = \{e, \sigma, \tau, \sigma\tau\}$, then consider the fixed fields of $\langle\sigma\rangle$ and $\langle\tau\rangle$. See also Exercise 12 of Section 7.1.

Exercise 10. Hint for (c) $\Rightarrow$ (a): If $\mathbb{Q}(\alpha) \neq \mathbb{Q}(\beta)$, then let $L = \mathbb{Q}(\alpha, \beta)$, and show that there are $\sigma, \tau \in \mathrm{Gal}(L/\mathbb{Q})$ such that $\mathbb{Q}(\alpha)$ is the fixed field of $\langle\sigma\rangle$, and $\mathbb{Q}(\beta)$ is the fixed field of $\langle\tau\rangle$. Then see where $\mathbb{Q}(\alpha + \beta)$ fits in the Galois correspondence, and explore how $\sigma, \tau, \sigma\tau$ act on $\alpha + \beta$.

Exercise 11. Hint: Use the Galois correspondence and Exercise 8 of Section 7.2.

Exercise 13. Hint: Show that the Galois closure constructed in Proposition 7.1.7 can be realized as a subfield of L. Then use the Galois correspondence and Exercise 12.

Exercise 14. Hint: First use part (d) of Exercise 4 of Section 6.2 to show that $\mathrm{Gal}(\mathbb{Q}(\zeta_n)/\mathbb{Q})$ is Abelian.

Section 7.4 (page 173)

Exercise 7. Hint: Use the Galois correspondence and Proposition 6.3.7.

Section 7.5 (pages 185–187)

Exercise 2. Hint: $R = F[y]$ is a UFD with field of fractions $K = F(y)$.

Exercise 3. (a) Hint: In A and B, the coefficient of each power of x is a rational function in y.

Exercise 12. (c) Hint: In part (b), we "broke" one of the symmetries of the polyhedron by moving some of the vertices. To obtain the groups in part (c), you need to "break" some of the symmetries in a similar way.

Section 8.1 (page 196)

Exercise 4. Hint: The proof is similar to part (a) of Exercise 2.

Section 8.2 (page 200)

Exercise 1. (b) Hint: See Exercise 8 of Section 7.3.

Exercise 2. Hint: Use Definition 8.2.1, the Tower Theorem, and $[L:\mathbb{Q}] = 3$.

Exercise 3. (a) Hint: Consider the intersection of all subfields of L containing K_1 and K_2.

Exercise 4. (a) Hint: Adapt the proof of Theorem 7.1.7 to show that $F(\alpha_1, \ldots, \alpha_r)$ is a Galois closure. (b) Hint: Use Proposition 5.1.8.

Section 8.3 (page 210)

Exercise 2. Hint: Splitting fields.

Exercise 3. Hint: See Exercise 9 in Section A.2.

Exercise 4. (b) Hint: Remember that $\zeta_i \in F_{i-1}$.

Exercise 7. (a) Hint: Use Exercise 6.

Exercise 8. (a) Hint: As in Exercise 3 of Section 7.4, $\alpha_1^3 + \beta_1^3 = 2A$. Also, when using the computer algebra system, you should write ω as $\frac{1}{2}(-1 + i\sqrt{3})$. (b) Hint: Recall that $\sigma_1 = x_1 + x_2 + x_3$. Also, what is $1 + \omega + \omega^2$?

Section 8.4 (page 215)

Exercise 1. Hint: Use Cauchy's Theorem (Theorem A.1.5).

Exercise 2. Hint: When i, j, k, l are distinct, verify that $(i\ j)(k\ l) = (i\ j\ k)(j\ k\ l)$. You will also need to consider the case when i, j, k, l are not distinct.

Exercise 5. Hint: If σ, τ are elements of H different from e, then what can you say about $\sigma^2, \sigma\tau, \tau^2$?

Exercise 6. Hint: If $H_1 \subset G/H$ is normal, then what can you say about $\pi^{-1}(H_1)$, where $\pi : G \to G/H$ takes g to the coset gH? See Exercises 3 and 4 from Section 8.1.

Section 8.5 (pages 219–220)

Exercise 1. Hint: Suppose that $F \subset L_1 \subset M_1$, where $F \subset M_1$ is radical. Explain why we can assume that M_1 is the splitting field of some polynomial $g \in F[x]$. Then let $L_2 \subset M_2$ be a splitting field of g regarded as a polynomial in $L_2[x]$. Prove that $F \subset M_2$ is radical. Theorem 5.1.6 will be useful.

Exercise 3. Hint: Use Proposition 5.3.8.

Section 8.6 (page 226)

Exercise 3. Hint: If you use a computer to draw the graph of f, it seems clear that there are four real roots. To make this rigorous, you should use the Intermediate Value Theorem.

Exercise 5. (a) Hint: What is $(\alpha + i)^p$ in characteristic p?

Exercise 6. (a) Hint: Unique factorization.

Exercise 7. Hint: Follow the proof of (b) $\Rightarrow$ (a) of Theorem 8.3.3. You will also need to explain why primitive mth roots of unity exist for all m not divisible by p.

Section 9.1 (pages 236–237)

Exercise 1. When $n = 1$, note that $[0] = [1]$ in $\mathbb{Z}/1\mathbb{Z}$, so that $[0]$ is in $(\mathbb{Z}/1\mathbb{Z})^*$.

Exercise 5. Hint: *Maple* and *Mathematica* can factor polynomials over $\mathbb{Q}$. Also, what is the degree of $\Phi_{105}(x)$? (See Exercise 13 for another approach to computing $\Phi_{105}(x)$.)

Exercise 6. (a) Hint: Analyze the proof of Theorem 2.2.2. (b) Hint: Use Lemma 5.3.10 and Lemma 9.1.2.

Exercise 8. (f) Hint: $p \nmid n$.

Exercise 11. Hint: Use (9.4).

Exercise 15. Hint: Use Exercise 14 and (9.4).

Exercise 16. (b) Hint: Use part (a) and Lemma 9.1.1.

Section 9.2 (pages 252–253)

Exercise 6. Hint: Use Lemma 9.2.4 and Exercise 9 of Section A.2.

Exercise 9. (b) Hint: To five decimal places, $(-1 + \sqrt{17})/2 = 1.56155$ and $(-1 - \sqrt{17})/2 = -2.56155$, yet the quadratic formula says that $(8, 1) = (-1 \pm \sqrt{17})/2$ for some choice of sign. (c) Hint: When computing $\sqrt{(4, 1)^2 - 4(4, 3)}$, do not compute $(4, 1)^2$ using part (b). Rather, use Proposition 9.2.9 to express $(4, 1)^2$ in terms of $(4, 2)$, $(4, 3)$, and $(4, 17) = 4$.

Exercise 10. Hint: See Exercise 3.

Exercise 11. (a) Hint: $[g^e]$ generates H_f, and $[g^{e/q}]$ generates H_{fq}. (b) Hint: First prove that $[L_f(\omega) : L_{fq}(\omega)] = [L_f : L_{fq}]$ using the method of Exercise 5 of Section 4.3. Then look at the argument used in the proof of the General Case of Theorem 8.3.3. (c) Hint: First use Propostion 9.2.6 to study how σ' acts on f-periods.

Exercise 12. (a) Hint: Use Exercise 16 of Section 9.1.

Exercise 13. Hint: First show that $H_8 \subset (\mathbb{Z}/17\mathbb{Z})^*$ is the subgroup of squares. Then, for $p \nmid a$, explain why $x^2 \equiv a \bmod 17$ has a solution if and only if $[a] \in H_8$.

Exercise 14. (a) Hint: Label the roots as $\alpha_i = r^{g^{i-1}}$ for $i = 1, \ldots, n - 1$.

Section 10.1 (pages 268–269)

Exercise 4. (a) Hint: You can think of ℓ_1 as the line through the points (u_1, v_1) and (u_2, v_2) in the plane $\mathbb{R}^2$. Consider the cases $u_1 = u_2$ and $u_1 \neq u_2$ separately.

Exercise 5. (a) Hint: Do you remember how to construct an equilateral triangle?

Exercise 6. Hint: Argue as in Example 10.1.9 that such a trisection implies that $\cos 20°$ is constructible. Then use $\cos(\frac{\pi}{3}) = \frac{1}{2}$ and the identity $\cos(3\theta) = 4\cos^3 \theta - 3\cos\theta$ from Section 1.3.

Exercise 7. Hint: Combine the construction used in Proposition 7.1.7 with the fact that $\mathbb{C}$ is algebraically closed.

Exercise 8. (a) Hint: Mimic the proof of Theorem 10.1.6.

Exercise 10. (d) Hint: Use part (c) and $x = a/3$.

Exercise 12. (b) Hint: $r/3$.

Section 10.2 (pages 272–273)

Exercise 1. Hint: Note that if m is odd, then $x^m + 1 = x^m - (-1)^m$.

Exercise 5. (a) Hint: Use $\zeta_{2^s}^2 = \zeta_{2^{s-1}}$ and Theorem 10.1.6. (b) Hint: Use Exercise 16 of Section 9.1.

Section 10.3 (pages 283–286)

Exercise 1. (a) Hint: Q_1 and Q_2 are the reflections of P_1 and P_2 about l.

Exercise 3. (c) Hint: How does part (b) relate to (10.8)?

Exercise 4. Hint: Use implicit differentiation.

Exercise 5. (b) Hint: What are the roots of $x^3 + 2x + 1$? (c) Hint: Use (10.11).

Exercise 6. Hint: The distance between α_1 and α_2 equals the distance between the reflections of these points about ℓ.

Exercise 13. Hint: Let M be the midpoint of QR. Use the circle with center M and radius $1/2$ to show that $\angle QMP = 2\angle QRP$.

Exercise 14. (a) Hint: A perpendicular from R to the x-axis will meet the x-axis at a point S. This gives $\triangle ROS$. Then let T bisect the segment $\overline{RP}$, and prove that $\triangle ROT$ is congruent to $\triangle ROS$.

Exercise 15. (b) Hint: This is very challenging. A solution can be found on page 128 of [12] in the references to Chapter 10.

Section 11.1 (page 298)

Exercise 2. Hint: See Section A.1.

Exercise 3. Hint: Show that $x^{p^m} - x$ is a factor of $x^{p^n} - x$ whenever m divides n.

Exercise 9. (a) Hint: Use Exercise 4 of Section 9.1. (b) Hint: $\langle p, f \rangle = p\mathbb{Z}[\alpha] + f\mathbb{Z}[\alpha]$.

Exercise 11. Hint: If α is a root of f in some splitting field, then $[\mathbb{F}_p(\alpha) : \mathbb{F}_p] = n$.

Section 11.2 (pages 306–307)

Exercise 4. (b) Hint: Divide by the smallest power of p appearing in the formula and then work modulo p.

Exercise 5. Hint: If $\alpha^m = 1$, then write $m = p^s d$ where $s \geq 0$ and $\gcd(d, p) = 1$.

Exercise 10. (b) Hint: Explain why f_i and f_j are relatively prime for $i \neq j$.

Exercise 11. (b) Hint: Show that each R_i is a field.

Exercise 12. (c) Hint: What is the factorization of $x^p - x$?

Section 12.1 (pages 330–332)

Exercise 1. Hint: Look at the proof of (7.1).

Exercise 2. (b) Hint: Another root of the Ferrari resolvent is $y_2 = (23) \cdot y_1 = x_1 x_3 + x_2 x_4$. How are $H(y_1)$ and $H(y_2)$ related?

Exercise 5. (a) Hint: For the choice of sign with x_i, x_j as roots, the constant term is $x_i x_j$. But what is the constant term of the corresponding equation in (12.11)? Then do the same for the other equation.

Exercise 9. (b) Hint: First show that $\sigma \cdot f = f$ for $\sigma \in H$, using the fact that multiplication by σ permutes the elements of H. Then show that $\sigma \cdot f \neq f$ for all $\sigma \notin H$, using the fact that the exponents are distinct. Remember that two polynomials are different if one has a term which doesn't appear in the other.

Exercise 11. Hint: If you have done Exercise 19, then use part (f) of that exercise. Otherwise, follow the proof of Theorem 12.1.10, and pick $f \in L$ with H as isotropy subgroup. Then let

$$f_1 = f, f_2, \ldots, f_s$$

be the distinct rational functions $\sigma \cdot f$ for $\sigma \in A_n$. Show that $s = [A_n : H]$ and that

$$G = \{\sigma \in A_n \mid \sigma \cdot f_i = f_i \quad \text{for all } i = 1, \ldots, s\}$$

is a normal subgroup of A_n. Then use Theorem 8.4.3 and the argument of Theorem 12.1.10.

Exercise 12. (b) Hint: Use the Galois correspondence.

Exercise 13. Hint: First explain why $\sigma \cdot \alpha^n = \alpha^n$ implies that $\sigma \cdot \alpha = \zeta^j \alpha$ for some j.

Exercise 14. (b) Hint: Look at the proof of Theorem 12.1.4.

Exercise 20. Hint: Use part (f) of the previous exercise.

Exercise 21. (a) Hint: What obvious subgroup of S_n has $(n - 1)!$ elements? (b) Hint: Exercise 3.

Section 12.2 (pages 343–346)

Exercise 1. Hint: If $W_1 \subset W_2 \cup \cdots \cup W_m$, then intersect both sides with W_1 and use the inductive assumption.

Exercise 3. Hint: In the second proof, replace f_i and g_i, $i = 1, \ldots, s$, with V_σ and $\alpha_{\sigma(j)}$, $\sigma \in S_n$. Thus the polynomial $\psi(x)$ from (12.5) will be a sum of $n!$ terms. Show that $\psi(x) \in F[x]$ by arguments similar to those used in the proof of Proposition 5.2.1 in Section 5.2. Be sure that your argument explains where the separability of $s(y)$ is used.

Exercise 4. Hint: Let g be the minimal polynomial of β, and let M be a splitting field of fg over F.

Exercise 6. (d) Hint: Use (8.3).

Exercise 9. Hint: Use Exercise 14 and $F \subset K \cap L \subset K$.

Exercise 12. (b) Hint: Use what you did in Exercise 10.

Exercise 15. (a) Hint: See the proof of Theorem 7.2.7.

Section 12.3 (pages 352–354)

Exercise 1. Hint: We are in characteristic 0.

Exercise 5. (a) Hint: A nonzero polynomial in one variable of degree N has at most N roots in a field. Also, when $n > 1$, note that $g = \sum_{j=0}^{N_n} g_j(x_1, \ldots, x_{n-1}) x_n^j$, where at least one $g_j \in F[x_1, \ldots, x_{n-1}]$ is nonzero.

Exercise 6. (b) Hint: Use (12.7). (c) Hint: Use Exercise 4 of Section 9.1.

Section 13.1 (pages 365–367)

Exercise 1. (a) Hint: Take $\sigma \in \text{Gal}(L/F)$ and let $\tau = \phi_1(\sigma)$. Also, explain why $\alpha_i = \beta_{\gamma^{-1}(i)}$.

Exercise 5. Hint: If g doesn't split completely over F, then show that the roots of g in some splitting field are $b, u \pm \sqrt{v}$ where $b, u, v \in F$ and $v \notin F^2$.

Exercise 6. (a) Hint: Consider $(x - \alpha_1)(x - \alpha_2)$ and $(x - \alpha_3)(x - \alpha_4)$.

Exercise 16. Hint: First show that $|G|$ is a power of 2.

Section 13.2 (pages 381–383)

Exercise 3. (b) Hint: Let $\tau = (i_1 \ldots i_5)$. Show that $\sigma(i_1) = i_k$ implies that $\sigma(i_2) = i_{k+1}$ and so on. Then show that $\sigma = \tau^{k-1}$. (d) Hint: How many 5-cycles are there in S_5? (e) Hint: Remember that 5 divides $|G|$.

Exercise 8. (a) Hint: Explain why $\tau \cdot u_i = \pm u_j$ for some j, and prove that some even permutation takes u_i to u_j. Then use Exercise 7 to determine the sign.

Exercise 9. (a) Hint: See the proof of part (a) of Theorem 13.1.1. (e) Hint: Comparing the coefficients of y^5, y^4, and y^3 gives equations which can be solved using *Maple* or *Mathematica* to express b_2, b_4, b_6 in terms of a, b. Then compare the coefficients of y^2 and substitute the formulas for b_2, b_4, b_6 to get an equation involving only a and b. Now factor. A different argument is needed in characteristic 3.

Exercise 10. (c) Hint: Use the method described in the hint to part (e) of Exercise 9. Here, you will need a different argument in characteristic 5.

Exercise 16. (e) Hint: Use Section 7.4.

Section 13.3 (pages 392–394)

Exercise 1. (a) Hint: First multiply f by a suitable integer so that $a_0, \ldots, a_n \in \mathbb{Z}$. Then multiply by a_0^{n-1}.

Exercise 2. Hint: Use Exercise 6 of Section 9.1.

Exercise 5. (b) Hint: Use Galois theory.

Exercise 6. (b) Hint: Let $\varphi_2 = (234) \cdot \varphi$ and $\varphi_3 = (34) \cdot \varphi_2$. Show that $G_f \cdot \varphi = \{\pm\varphi\}$ and $G_f \cdot \varphi_2 = \{\pm\varphi_2, \pm\varphi_3\}$. (c) Hint: Study the action of G_f on φ, φ_2, and φ_3, where φ_2 and φ_3 are defined in the hint to part (b). (d) Hint: Use the hints to parts (b) and (c).

Exercise 7. Hint: By linear algebra, the map sending a matrix to the corresponding linear map is one to one.

Exercise 8. (a) Hint: If F is a field, then every subspace of F^n of dimension $n - 1$ is defined by an equation $a_1 x_1 + \cdots + a_n x_n = 0$ where $a_1, \ldots, a_n \in F$ are not all zero. Furthermore, this equation is unique up to multiplication by a nonzero element of F. (c) Hint: If $V \subset F^3$ has dimension 2, then it has a basis which can be completed to a basis of F^3. Now use part (b).

Exercise 9. Hint: For a field F, let $F^* I_n$ be the subgroup of $GL(n, F)$ consisting of multiples of the identity matrix. Then $PGL(n, F) = GL(n, F)/F^* I_n$ and $PSL(n, F) = SL(n, F)/(F^* I_n \cap SL(n, F))$.

Exercise 11. (b) Hint: Compute isotropy subgroups.

Section 13.4 (page 404)

Exercise 4. (a) Hint: Use Section 2.3 to express the universal version of s_u in terms of $\sigma_1, \sigma_2, \sigma_3, y$. Then specialize to the coefficients of f. You will be surprised at the size of the polynomials involved.

Exercise 6. (a) Hint: If $f = gh$ in $\mathbb{Q}[x_1, \ldots, x_n]$, then pick positive integers r, s as small as possible such that $rg, sh \in \mathbb{Z}[x_1, \ldots, x_n]$. Now apply unique factorization to $rsf = (rg)(sh)$, and remember that a prime $p \in \mathbb{Z}$ is irreducible (in the UFD sense) in $\mathbb{Z}[x_1, \ldots, x_n]$. (b) Hint: Take an irreducible factorization of s_u in $\mathbb{Z}[u_1, \ldots, u_n, y]$, and apply part (a) together with the fact that $\mathbb{Q}[u_1, \ldots, u_n, y]$ is a UFD.

Exercise 10. (b) Hint: Look at subgroups of S_4 with four elements.

Section 14.1 (page 412)

Exercise 2. (b) Hint: $m \, o(g) + n \, [G : H] = 1$.

Exercise 4. Hint: Remember that $g \cdot hH = (gh)H$ gives an action of G on the set of left cosets of H in G.

Section 14.2 (pages 420–422)

Exercise 2. (a) Hint: Show that $S_3 \wr S_2$ contains a transposition.

Exercise 3. (d) Hint: Use part (b) of Exercise 7 of Section 14.3.

Exercise 8. (b) Hint: For $\sigma \in Gal(L/F)$, we have $\tau \in S_m$ such that $\sigma(R_i) = R_{\tau(i)}$. What is the kernel of the map sending σ to τ? The Galois correspondence from Chapter 7 will also be useful.

Exercise 9. (a) Hint: Use Proposition 9.2.8 with $f = 1$ and $f' = f$.

Exercise 10. Hint: What power of p divides $(p^2)!$?

Exercise 11. Hint: Use Exercise 4.

Exercise 15. Hint: Show that $|G_i| n = |G|$, where G_i is the isotropy subgroup of $i \in \{1, \ldots, n\}$. Given a subgroup $G_i \subset H \subset G$, consider the subsets $(\tau H) \cdot i, \tau \in G$. Show that these partition $\{1, \ldots, n\}$ into blocks that are stable under the action of G.

Section 14.3 (pages 435–437)

Exercise 2. (c) Hint: Consider the map $\gamma_{A,v} \mapsto A$. (d) Hint: Semidirect products are discussed in the Mathematical Notes to Section 6.4, and (6.10) does the special case when $n = 1$ and $q = p$.

Exercise 3. (a) Hint: Consider the map which sends $\gamma_{A,\sigma,v}$ to σ.

Exercise 4. (c) Hint: Use linear algebra.

Exercise 5. (b) Hint: You need to study what happens when $N \subset A \times \{e_B\}$ or $N \subset \{e_A\} \times B$. (c) Hint: Conjugate $(a^{-1}, b^{-1}) \in N$ by $(a_1, b^{-1}) \in A \times B$.

Exercise 6. (a) Hint: Explain why conjugation by g gives an automorphism of N.

Exercise 7. (b) Hint: Consider $hkh^{-1}k^{-1}$.

Exercise 12. (b) Hint: Use part (c) of Exercise 2.

Exercise 13. (b) Hint: Given two $(n-2)$-tuples of elements of $\{1, \ldots, n\}$ consisting of distinct points, show that there are exactly two elements $\tau, \tau' \in S_n$ which map one $(n-2)$-tuple to the other. Then show that τ and τ' differ by a transposition.

Exercise 14. (a) Hint: A 2×2 matrix has determinant 0 if and only if either its first column is zero or its first column is nonzero and the second column is a multiple of the first. (e) Hint: Apply the Fundamental Theorem of Group Actions to $GL(3, \mathbb{F}_2)$ acting on $\mathbb{F}_2^3 \setminus \{0\}$.

Exercise 15. Hint: First show that $\mathbb{F}_p^2$ has $p + 1$ lines through the origin, and explain why $PSL(2, \mathbb{F}_p)$ permutes these lines. Then study the cases $p = 2$ and $p = 3$ in detail.

Exercise 16. Hint: In the proof of Proposition 14.3.10, replace A and its conjugates with the minimal normal subgroups of G.

Exercise 17. Hint: Use induction and the action of $GL(n, \mathbb{F}_p)$ on $\mathbb{F}_p^n \setminus \{0\}$.

Section 14.4 (pages 453–454)

Exercise 1. Hint: Use Exercise 2 of Section 14.3.

Exercise 3. Hint: Use Proposition 14.3.4.

Exercise 9. (a) Hint: Use Lemma 14.4.3. (b) Hint: To analyze the kernel, write $m \in \text{Ker}(\phi)$ as a linear combination of I_2 and g. For the image, use the Fundamental Theorem of Group Homomorphisms. (d) Hint: Use $g = \begin{pmatrix} 0 & -1 \\ 1 & 0 \end{pmatrix}$ to show that $C(g) = \{\begin{pmatrix} r & -s \\ s & r \end{pmatrix} \mid r, s \in \mathbb{F}_p, r^2 + s^2 \neq 0\}$. Then use the Fundamental Theorem of Group Homomorphisms to determine the size of the kernel.

Exercise 11. (c) Hint: The can be proved using Exercise 20 of Section 14.3. (d) Hint: Suppose that such a subgroup exists. Then consider its image under the map $AGL(1, \mathbb{F}_p) \to \mathbb{F}_p^*$. Remember that you know all normal subgroups of S_4.

Exercise 12. (c) Hint: The multiplicative group of a finite field is cyclic. (d) Hint: The only hard case is when $p = 7$. Show that $M_2 \subset M_3$ implies that M_2 maps to $A_4 \subset S_4$ when you map to $PGL(2, \mathbb{F}_7)$. Then use part (a). (e) Show that the image of $(M_2)_0$ in $PGL(2, \mathbb{F}_5)$ has order 8 and contains the subgroup H of Lemma 14.4.4. What do you know about subgroups of index 2?

Exercise 13. Hint: Consider the map $\mathbb{F}_p^* \times G_0 \to GL(2, \mathbb{F}_p)$ defined by $(\lambda, g) \mapsto \lambda g$.

Exercise 13. (d) Hint: Use (14.42).

Section 15.2 (pages 475–476)

Exercise 9. (a) Hint: Gauss's Lemma implies that $P(u)$, $Q(u)$ are relatively prime in $\mathbb{Q}[u]$. Thus $R(u)P(u) + S(u)Q(u) = 1$ for some $R(u)$, $S(u) \in \mathbb{Q}[u]$. Now replace u with u^4.

Exercise 11. Hint: Let $\alpha = \sin x$ and $\beta = \sin y$.

Section 15.3 (page 483)

Exercise 2. (b) Hint: Use Proposition 15.2.1 and part (d) of Exercise 3 of Section 15.2.

Exercise 6. Hint: Use the identity (15.13) with $x = \frac{1}{2}(z + z_0)$ and $y = \frac{1}{2}(z - z_0)$. See also [Abel, Vol. I, pp. 277–278].

Section 15.4 (page 497–498)

Exercise 3. Hint: See the proof of (a) $\Leftrightarrow$ (b) of Proposition 3.1.1. Also recall if $\alpha \in \mathbb{Z}[i]$ is prime, then $\alpha = \beta\gamma$, β, $\gamma \in \mathbb{Z}[i]$, implies that β or γ is a unit, that is, lies in $\mathbb{Z}[i]^*$.

Exercise 9. Hint: See the hint for Exercise 9 of Section 15.2.

Exercise 12. Hint: Follow the proof of Theorem 4.2.3. You will also need Gauss's Lemma over $\mathbb{Z}[i]$, which holds by Theorem A.5.8 because $\mathbb{Z}[i]$ is a UFD.

Exercise 13. Hint: Use the geometric series $1/(1 + x) = \sum_{k=0}^{\infty}(-1)^k x^k$ to write

$$\frac{1}{1 + a_1 u + \cdots + a_d u^d} = \sum_{k=0}^{\infty}(-1)^k \left(a_1 u + \cdots + a_d u^d\right)^k.$$

The series on the right-hand side makes sense because

$$(a_1 u + \cdots + a_d u^d)^k = u^k (a_1 + \cdots + a_d u^{d-1})^k.$$

So a given power of u appears in only finitely many terms.

Exercise 14. (d) Hint: For $c_0 = 1$, use $\varphi'(0) = 1$, and for $c_1 = -\frac{1}{10}$, use $\varphi'^2(z) = 1 - \varphi^4(z)$ with $\varphi(z) = z + \cdots$ and $\varphi'(z) = 1 + 5c_1 z^4 + \cdots$. What is the coefficient of z^4?

Section 15.5 (page 506)

Exercise 1. Hint: Use the fact that $\mathbb{Z}[i]$ is a PID to prove that if α, β are relatively prime, then $\gamma\alpha + \delta\beta = 1$ for some γ, $\delta \in \mathbb{Z}[i]$.

Exercise 5. Hint: Use part (b) of Lemma 15.4.2.

Exercise 6. Hint: Show that the obvious map $\mathbb{Z}[i]/\alpha\beta\mathbb{Z}[i] \to \mathbb{Z}[i]/\alpha\mathbb{Z}[i] \times \mathbb{Z}[i]/\beta\mathbb{Z}[i]$ is one to one. Then use part (a) of Lemma 15.4.2.

References

The references are divided into three categories and are cited in the text using the last name of the author. Each chapter has its own references cited numerically.

A. Books and Monographs on Galois Theory

[Artin] E. Artin, *Galois Theory*, Univ. of Notre Dame Press, Notre Dame, 1942. Reprint by Dover, New York, 1998.

[Chebotarev] N. G. Chebotarev, *Grundzüge der Galois'schen Theorie von N. Tschebotaröw*, translated and edited by H. Schwerdtfeger, Noordhoff, Groningen, 1950. Note: The correct English transliteration is "Chebotaryov." We use "Chebotarev" because it is the version most often used in the literature.

[Dehn] E. Dehn, *Algebraic Equations*, Columbia U. P., New York, 1930.

[Dickson] L. E. Dickson, *Introduction to the Theory of Algebraic Equations*, Wiley, New York, 1903. Reprinted in *Congruence of Sets and Other Monographs*, Chelsea, New York, 1967.

[Edwards] H. M. Edwards, *Galois Theory*, Springer-Verlag, New York, Berlin, Heidelberg, 1984.

[Escofier] J.-P. Escofier, *Galois Theory*, Springer-Verlag, New York, Berlin. Heidelberg, 2001.

[Garling] D. J. H. Garling, *A Course in Galois Theory*, Cambridge U. P., Cambridge, 1986.

[Hadlock] C. R. Hadlock, *Field Theory and its Classical Problems*, Carus Monographs, Volume 19, MAA, Washington, DC, 1978.

[Jordan1] C. Jordan, *Traité des substitutions et des équations algébriques*, Gauthier-Villars, Paris, 1870.

[Postnikov] M. M. Postnikov, Основы теории Галуа [*Foundations of Galois Theory*], Gosudarstv. Izdat. Fiz.-Mat. Lit., Moscow, 1960. Enlarged edition, 1963. English translations of the 1960 edition by Hindustan Pub. Corp., New Delhi and Gordon & Breach, New York, 1961; Pergamon Press, Oxford, 1962; and Noordhoff, Groningen, 1962. Unpublished English translation of the additions to the 1963 edition by A. Shenitzer.

[Stewart] I. Stewart, *Galois Theory*, Third Edition, Chapman & Hall/CRC, Boca Raton, FL, 2003.

[Tignol] J.-P. Tignol, *Galois' Theory of Algebraic Equations*, Longman Scientific and Technical, Harlow, England, and Wiley, New York, 1988. Corrected reprint by World Scientific, Singapore, 2001.

B. Books on Abstract Algebra

[Grillet] P. Grillet, *Algebra*, Wiley, New York, 1999.

[Herstein] I. N. Herstein, *Topics in Algebra*, Second Edition, Wiley, New York, 1975.

[Jacobson] N. Jacobson, *Basic Algebra*, Volumes I and II, W. H. Freeman, San Francisco, 1980.

[van der Waerden] B. L. van der Waerden, *Algebra*, Volumes I and II, Springer-Verlag, New York, Berlin, Heidelberg, 1991.

C. Collected Works

[Abel] N. H. Abel, *Œuvres complètes de Niels Henrik Abel*, edited by L. Sylow and S. Lie, Grøndahl & Søn, Christiana, 1881.

[Cauchy] A. L. Cauchy, *Œuvres complètes*, Series 1, Volume X, Gauthier-Villars, Paris, 1897.

[Eisenstein] F. G. Eisenstein, *Mathematische Werke*, Volume II, Chelsea, New York, 1975. Reprint by Amer. Math. Soc., Providence, RI, 1989.

[Galois] É. Galois, *Écrits et mémoires mathématiques d'Évariste Galois*, edited by R. Bourgne and J.-P. Azra, Gauthier-Villars, Paris, 1962.

[Gauss] C. F. Gauss, *Werke*, K. Ges. Wiss., Göttingen, 1863–1927.

[Jordan2] C. Jordan, *Œuvres de Camille Jordan*, Volume I, edited by J. Dieudonné, Gauthier-Villars, Paris, 1961,

[Kronecker] L. Kronecker, *Werke*, Leipzig, 1895–1931. Reprint by Chelsea, New York, 1968.

[Lagrange] J. L. Lagrange, *Œuvres de Lagrange*, Volume 3, Gauthier-Villars, Paris, 1869.

Index

This index uses the following conventions:
- Page numbers of definitions are <u>underlined</u>.
- Page numbers of theorems are in **bold**.

PURE AND APPLIED MATHEMATICS

A Wiley-Interscience Series of Texts, Monographs, and Tracts

Founded by RICHARD COURANT
Editors: MYRON B. ALLEN III, DAVID A. COX, PETER LAX
Editors Emeriti: PETER HILTON, HARRY HOCHSTADT, JOHN TOLAND

*Now available in a lower priced paperback edition in the Wiley Classics Library.
†Now available in paperback.

*Now available in a lower priced paperback edition in the Wiley Classics Library.
†Now available in paperback.

CPSIA information can be obtained at www.ICGtesting.com
Printed in the USA
BVOW030202281211

278877BV00002B/1/P